FUNDAMENTALS OF LINEAR SYSTEMS FOR PHYSICAL SCIENTISTS AND ENGINEERS

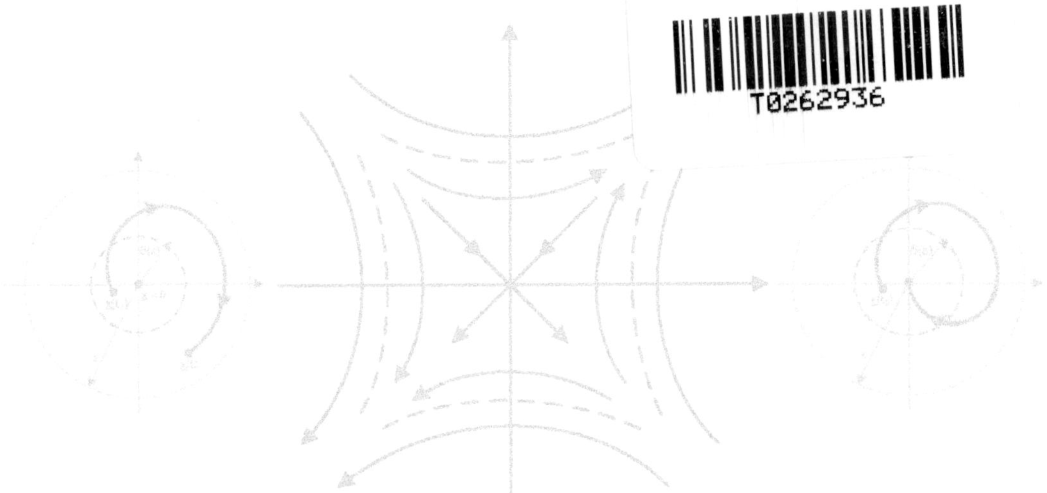

N. N. Puri

CRC Press
Taylor & Francis Group
Boca Raton London New York

CRC Press is an imprint of the
Taylor & Francis Group, an **informa** business

CRC Press
Taylor & Francis Group
6000 Broken Sound Parkway NW, Suite 300
Boca Raton, FL 33487-2742

First issued in paperback 2018

ISBN-13: 978-1-4398-1157-3 (hbk)
ISBN-13: 978-1-138-37418-8 (pbk)

Library of Congress Cataloging-in-Publication Data

Puri, N. N.
Fundamentals of linear systems for physical scientists and engineers / N.N. Puri.
p. cm.
Includes bibliographical references and index.
ISBN 978-1-4398-1157-3 (hardcover : alk. paper)
1. Vector spaces. 2. Discrete-time systems. 3. System analysis. I. Title.

QA186.P87 2010
512'.52--dc22 2009021572

Visit the Taylor & Francis Web site at
http://www.taylorandfrancis.com

and the CRC Press Web site at
http://www.crcpress.com

Preface

I present this work in humility and service to the academic community.

> Ich bin jetzt matt, mein Weg war weit,
> vergieb mir, ich vergass,
> was Er, der gross in Goldgeschmeid
> wie in der Sonne sass,
> dir kuenden liess, du Sinnende,
> (verwirrt hat mich der Raum).
> Sieh: ich bin das Beginnende,
> du aber bist der Baum.
> —Rainer Maria Rilke

Dedicated to

My mother Parkash Vati, who nurtured the qualities of my heart.

My father Amar Nath, to whom I am grateful for nurturing my academic ambitions.

To my wife Kamal, who stands by me through all the trials and tribulations.

To my children Monica and Lalit, Tony, Ashish, Serena, Sanjiv, and Poonam who made me realize life can be very beautiful and full of fun.

To my grandchildren, Saarika, Cimrin, Nishi baba, Selene "bean," Shaya, Reyna, and Annika, who made me realize that the interest is more joyful than even the capital.

To my friends and relatives in India, the United States, and Switzerland who make me feel wanted, particularly friends like "Shonyka."

To my students who made me realize that listening is a virtue.

To Professors Voelker and Weygandt who guided me during my career and without whose encouragement I would be still struggling in the university— a very thin edge separates success from failure.

To my country of birth, India, which gave me goodly heritage, and my adopted country, the United States, which made me realize even the sky is not the limit for free people.

This book is my striving, my "Nekrolog."

Acknowledgments

This work could not have been finished without the unselfish, dedicated effort of my student, Manish Mahajan, who painstakingly did all the typing and diagrams.

"Der Herr Gott, boese ist er boesehaft ist er aber nicht." —Albert Einstein

"I yield freely to the sacred frenzy. I dare to confess that I have stolen vessels of the Egyptians to build a tabernacle for my *Gods* far from the bounds of Egypt. If you pardon me, I shall rejoice, if you reproach me, I shall endure. The die is cast and I am writing this book — to be either read now or by posterity, it matter not. It can wait a century for a reader, as God had waited thousands of years for a witness."

— Kepler (quoted by V.M. Tikhomirov).

We meditate on the glory of the Creator;
Who has created the Universe;
Who is worthy of Worship;
Who is the embodiment of Knowledge and Light;
Who is the remover of all Sin and Ignorance;
May He enlighten our Intellect.

Contents

1 System Concept Fundamentals and Linear Vector Spaces **1**

1.1 Introduction 1

1.2 System Classifications and Signal Definition 3

1.2.1 Linear Systems 5

1.2.2 Linear Time Invariant (LTI) Systems 6

1.2.3 Causal Systems 7

1.2.4 Dynamical–Nondynamical Systems 7

1.2.5 Continuous and Discrete Systems 8

1.2.6 Lumped Parameter vs. Distributed Parameter Systems . . . 8

1.2.7 Deterministic vs. Stochastic Signals and Systems 8

1.2.8 Input–Output Description of a Dynamic System 9

1.3 Time Signals and Their Representation 10

1.3.1 Impulse or Delta Function Signals 11

1.3.2 Discrete Delta Function or Delta Function Sequence 21

1.3.3 General Signal Classifications 22

1.4 Input–Output Relations (System Response) 31

1.4.1 Superposition or Convolution Integral 31

1.5 Signal Representation via Linear Vector Spaces 47

1.5.1 Definition of a Vector Space 47

2 Linear Operators and Matrices **93**

2.1 Introduction 93

2.2 Matrix Algebra - Euclidian Vector Space 96

2.2.1 Fundamental Concepts 101

2.3 Systems of Linear Algebraic Equations 122

2.3.1 Geometric Interpretation of Simultaneous Equations 122

2.3.2 Eigenvalues and Eigenvectors of Matrices 128

2.3.3 Generalized Eigenvectors—Matrix of Multiplicity k 139

2.4 Diagonalization—Eigenvalue Decomposition 141

2.4.1 Solution to the Equation $\boldsymbol{Ax} = \boldsymbol{b}$, Revisited 143

2.4.2 Singular Value Decomposition of a Matrix (SVD) 144

2.5 Multiple Eigenvalues—Jordan Canonical Form 149

2.5.1 Cayley-Hamilton Theorem 153

2.6 Co-efficients of Characteristic Polynomial 154

2.7 Computation of Matrix Polynomial Function 155

2.8 S-N Decomposition of a Non-singular Matrix 158

2.9 Computation of $\boldsymbol{A}^n$ without Eigenvectors 160

2.10 Companion Matrices 162

2.11 Choletsky Decomposition (***LU*** Decomposition) 163

2.12 Jacobi and Gauss-Seidel Methods 164

2.13 Least Squares (Pseudo Inverse Problem) 165

2.14 Hermitian Matrices and Definite Functions 165

2.15 Summary of Useful Facts and Identities 167

2.16 Finite and Infinite Dimensional Operators 182

2.16.1 Operators and Matrices in Infinite Dimensional Space . . . 187

3 Ordinary Differential and Difference Equations **211**

3.1 Introduction 211

3.2 System of Differential and Difference Equations . . . 212
3.2.1 First Order Differential Equation Systems . . . 212
3.2.2 First Order Difference Equation . . . 215
3.2.3 n-th Order Constant Coefficient Differential Equation . . . 218
3.2.4 k-th Order Difference Equations . . . 225
3.3 Matrix Formulation of the Differential Equation . . . 230
3.3.1 Solution of Equation $\dot{\boldsymbol{x}} = \boldsymbol{A}_E\boldsymbol{x} + \boldsymbol{b}f(t)$ . . . 231
3.4 Matrix Formulation of the Difference Equation . . . 235
3.5 Time Varying Linear Differential Equations . . . 238
3.6 Computing $e^{\boldsymbol{A}t}, \boldsymbol{A}^N, f(\boldsymbol{A})$ without Determination of Eigenvectors . . 245
3.7 Stability of Autonomous Differential Equations . . . 252
3.7.1 Liapunov's Second Method (Direct Method) . . . 264
3.7.2 Stability Studies for Nonlinear Difference Equations Using Liapunov's Second Method . . . 268

4 Complex Variables for Transform Methods 297

4.1 Introduction . . . 297
4.2 Complex Variables and Contour Integration . . . 298
4.2.1 Definition of a Complex Variable . . . 298
4.2.2 Analytic Function . . . 299
4.2.3 Derivative of Complex Variable at a Point . . . 300
4.2.4 Path of Integration . . . 302
4.2.5 Useful Facts about Complex Variable Differentiation . . . 306
4.2.6 Cauchy's Integration Theorem . . . 307
4.2.7 Modified Cauchy's Integral Theorem . . . 308
4.2.8 Taylor Series Expansion and Cauchy's Integral Formula . . 310
4.2.9 Classification of Singular Points . . . 312
4.2.10 Calculation of Residue of $f(z)$ at $z = z_0$ . . . 313

4.2.11 Contour Integration . . . 314
4.2.12 Contour Integral Computation . . . 315
4.2.13 Summary on Evaluation of Residues—Special Results . . . 316
4.2.14 Laurent Series Expansion of a Function $f(z)$ . . . 321
4.2.15 Evaluation of Real Integrals by Residues . . . 327
4.2.16 Branch Points—Essential Singularities . . . 350
4.3 Poisson's Integral on Unit Circle (or Disk) . . . 359
4.4 Positive Real Functions . . . 374
4.4.1 Bilinear Transformation . . . 375

5 Integral Transform Methods 381

5.1 Introduction . . . 381
5.2 Fourier Transform Pair Derivation . . . 382
5.3 Another Derivation of Fourier Transform . . . 384
5.4 Derivation of Bilateral Laplace Transform L_b . . . 385
5.5 Another Bilateral Laplace Transform Derivation . . . 387
5.6 Single-Sided Laplace Transform . . . 388
5.7 Summary of Transform Definitions . . . 390
5.8 Laplace Transform Properties . . . 391
5.9 Recovery of the Original Time Function . . . 400
5.9.1 Partial Fraction Expansion Method . . . 402
5.9.2 Laplace Inverse via Contour Integration . . . 406
5.10 Constant Coefficient Differential Equations . . . 410
5.11 Computation of $x(t)$ for Causal Processes . . . 411
5.12 Inverse Bilateral Laplace Transform $F_b(s)$ . . . 412
5.13 Transfer Function . . . 416
5.14 Impulse Response . . . 419
5.15 Convolution for Linear Time-Invariant System . . . 420

5.16 Frequency Convolution in Laplace Domain 423
5.17 Parseval's Theorem 426
5.18 Generation of Orthogonal Signals 428
5.19 The Fourier Transform 431
5.20 Fourier Transform Properties 439
5.21 Fourier Transform Inverse 451
5.22 Hilbert Transform 455
5.22.1 Hilbert Transform—Inversion of Singular Integrals 458
5.22.2 Physical Realization of Hilbert Tranform of a Function . . . 459
5.23 The Variable Parameter Differential Equations 463
5.24 Generalized Error Function 469

6 Digital Systems, Z-Transforms, and Applications **477**

6.1 Introduction 477
6.2 Discrete Systems and Difference Equations 478
6.2.1 k-th Order Difference of a Discrete Function 480
6.2.2 Building Blocks of the Discrete Systems 482
6.3 Realization of a General Discrete System 483
6.4 Z-Transform for the Discrete Systems 484
6.5 Fundamental Properties of Z-Transforms 486
6.6 Evaluation of $f(n)$, Given Its Z-Transform 504
6.7 Difference Equations via Z-Transforms 510
6.7.1 Causal Systems Response $y(n)$ ($Y(z) = \hat{N}(z)/\hat{D}(z)$) 511
6.7.2 Digital Transfer Function 513
6.7.3 Representation of Digital Transfer Function 514
6.8 Computation for the Sum of the Squares 516
6.8.1 Sum of Squared Sampled Sequence 516
6.9 Bilateral Z-Transform $f(n) \leftrightarrow F_b(z)$ 526

6.10 Summation of the Series via Z-Transforms 529
6.11 Sampled Signal Reconstruction 531
6.11.1 Introduction 531
6.11.2 Sampling of a Band-Limited Time-Continuous Signal and Its Exact Reconstruction from Sampled Values 533
6.11.3 Fourier Series Revisited 537
6.11.4 Discrete Fourier Transforms or Discrete Fourier Series and Fast Fourier Transform Computation Algorithm 547
6.11.5 Computation of $F_N(n)$ from $f_N(k)$ and Vice Versa 554
6.11.6 Aliasing Error of Numerical Computation of DFT Due to the Time-Limited Signal Restriction 555
6.11.7 The Fast Fourier Transform (FFT) 561
6.11.8 Numerical Computation of DFT via Fast Fourier Transform—FFT 563
6.11.9 FFT in Two Dimensions 571
6.11.10 Appendix: Accelerating Power Series Convergence 572

7 State Space Description of Dynamic Systems 577
7.1 Introduction 577
7.2 State Space Formulation 578
7.2.1 Definition of the State of a System 578
7.2.2 State Variable Formulation—n-th Order System 580
7.2.3 State Variable Formulation of a General System 581
7.3 State Variables Selection 583
7.4 Methods of Deriving State Variable Equations 584
7.4.1 Lagrangian Set of Equations of Motion 584
7.4.2 Formulation of the State Variable Equations of an Electric Network Using Linear Graph Theory 588

7.5 State Space Concepts 595
7.5.1 State Space Similarity Transformations 595
7.5.2 Transfer Function Matrix from State Space Equations 596
7.5.3 Canonical Realizations of a Given Transfer Function 598
7.5.4 Controllability and Observability Concepts 611
7.5.5 Controllability Definition and Criterion 612
7.5.6 Observability Definition and Criterion 613
7.5.7 Controllability–Observability Geometric Interpretation . . . 614
7.5.8 Geometric Controllability–Observability Criterion 618
7.5.9 MIMO Systems Observability–Controllability Criterion . . 621
7.5.10 Canonical Controllable–Observable Decomposition 626
7.5.11 Kalman Decomposition for SISO–Geometric Viewpoint . . 632
7.5.12 Controllability and Observability Grammian 636
7.5.13 State Variable Feedback Control via State Observers 651
7.5.14 Controllability–Observability Time-Varying Systems 654
7.5.15 SISO Controller Design–Closed-Loop Poles Placement . . . 659
7.5.16 Minimal Realization of Time-Invariant Linear Systems . . . 662

8 Calculus of Variations **677**

8.1 Introduction 677
8.2 Maxima, Minima, and Stationary Points 677
8.2.1 Extremal of a Function Subject to Single Constraint 682
8.2.2 Extremal of a Function Subject to Multiple Constraints . . . 686
8.3 Definite Integral Extremal (Functional)—Euler-Lagrange, Variable Endpoints 687
8.4 Integral Extremal with Multiple Constraints 698
8.5 Mayer's Form 699
8.6 Bolza's Form 701

8.7 Variational Principles and Optimal Control 702
8.8 Hamilton-Jacobi—Euler-Lagrange Equations 704
8.9 Pontryagin's Extremum Principle 705
8.10 Dynamic Programming 717

9 Stochastic Processes and Their Linear Systems Response 729

9.1 Preliminaries 730
9.1.1 Probability Concepts and Definitions 730
9.1.2 Random Variables 735
9.2 Continous RV and Probability Density Function 736
9.2.1 Expected Value, Variance, and Standard Deviation of RV . . 737
9.2.2 Discrete Random Variable 738
9.2.3 n-Dimensional Random Variables Distribution 738
9.2.4 Two-Dimensional Random Variables (Bivariate) 739
9.2.5 Bivariate Expectation, Covariance 740
9.2.6 Lindeberg–Feller Central Limit Theorem 744
9.3 Random Walk, Brownian, and Wiener Process 749
9.3.1 Stochastic Differential and Integral Equations (SDE) 759
9.3.2 Simplified Ito's Theorem and Ito's Differential Rules 763
9.3.3 Optimal Control of the Stochastic Process 769
9.3.4 General Random Walk 773
9.3.5 Martingale's Stochastic Process 774
9.4 Markov Chains and the Law of Large Numbers 776
9.4.1 Markov Chains 776
9.4.2 Markov's Inequality 780
9.4.3 Tchebychev's Inequality 780
9.4.4 Law of Large Numbers 782
9.4.5 Sterling's Formula (Approximation) 782

9.4.6 Some Important Probability Density Functions 785
9.5 Stochastic Hilbert Space 787
9.5.1 Vector Space of Random Variables 787
9.5.2 Moment Generating Function or Characteristic Function . . 790
9.6 Random or Stochastic Processes 793
9.6.1 Stochastic Process PDF and pdf 795
9.6.2 Mean, Correlation Functions, and Spectra 796
9.6.3 Types of Random Processes 797
9.6.4 Autocorrelation Properties of an Ergodic Process 801
9.6.5 Cross-correlation Functions of Stationary Ergodic Process . 802
9.6.6 Wiener-Kinchin Theorem on Correlation Functions 805
9.6.7 Spectral Power Density 808
9.6.8 Karhunen-Loeve (K-L) Expansion of a Random Process . . 810
9.6.9 Determination of Eigenvalues and Eigenvectors of $S_{xx}(s^2)$. 813
9.6.10 LTI System Response to Stochastic Processes 821
9.7 Wiener Filters 825
9.7.1 Optimal Estimation with Noise (Memoryless System) . . . 825
9.7.2 Wiener Filter Stochastic Signal Estimation without Noise . . 832
9.7.3 Wiener Filter Estimation of the Signal with Additive Noise . 834
9.8 Estimation, Control, Filtering and Prediction 841
9.8.1 Estimation and Control 841
9.8.2 Filtering-Prediction Problem (Kalman-Bucy Filter) 847
9.8.3 Prediction Problem 849
9.8.4 Discrete Time Kalman Filter 850
9.8.5 Wiener Filter in State Variable—Kalman Form 854

Index **861**

List of Figures

1.1 System as an "Operator" 3

1.2 *R*-*L* Linear Circuit 4

1.3 Dynamic Systems Input–Output Description 9

1.4 Pulse Function 11

1.5 Delta Function 12

1.6 Sifting Property 14

1.7 Geometrical Interpretation of Fourier Series 25

1.8 Step Function 27

1.9 Ramp Function 27

1.10 Rectangular Pulse 28

1.11 Signum Function 28

1.12 Sinc Pulse 28

1.13 Gaussian Pulse 29

1.14 Impulse Train 29

1.15 Sine Integral Function 30

1.16 Triangular Pulse 30

1.17 Linear System 32

1.18 Dynamic System Response to $\delta(t - n\Delta\tau$ Input 34

1.19 Dynamic Electrical Response 34

1.20 Input–Output Dynamic System Response 34

1.21 Time Varying Linear System 37
1.22 Linear Time Varying Cascaded Systems 37
1.23 Cascaded System—Convolved 38
1.24 Time Convolution of Two Pulses Shifted by Time, T 39
1.25 LTI Step Response 42
1.26 Circuit with Step Input 43
1.27 Cascaded Systems 43
1.29 Metric Space, Vectors Represented as "Points" 50
1.30 Normed Vectors Represented as "Geometric Objects" 51
1.31 Geometric Interpretation of the Inner Product of Two Vectors 52
1.32 Euclidean Metric Triangular Inequality 55
1.33 Coordinate Representation of a Vector 56
1.34 Hierarchy of Linear Spaces 71
1.35 The First 8 Haar Functions on a 0 to 1 Interval 75
1.36 The First 8 Walsh Functions on a 0 to 1 Interval. 76
1.37 Projection of $\boldsymbol{f}$ on Subspaces $\boldsymbol{E}_N$ and $\boldsymbol{W}_N$ 78
1.38 Error Vector $\boldsymbol{e}$ and Estimated Vector $a\boldsymbol{x}$ Are Orthogonal 80
1.39 Orthogonal Decomposition of $\boldsymbol{y}$ in n-Dimensions 81
1.40 Transformation between the Unit Disk and the s-Plane Known as Bilinear Transformation 82
1.41 Function $f(t)$ in Example 1.21 87
1.42 Best Approximation from a Function in $\boldsymbol{L}^2$ to $\boldsymbol{H}^2$ 89

2.1 Domain and Range Subspaces 94
2.2 Operator Transformation of Vectors from One Space to Another . . 97
2.3 The Vectors Spanning the Matrix $\boldsymbol{A}$ Are Orthogonal to the Vector $\boldsymbol{x}$ 123
2.4 Projection Operator. 182

3.1 Stable and Asymptotically Stable Equilibrium Points 254

3.2 Phase Portrait: Both λ_1, λ_2 Real and of the Same Sign 259
3.3 Phase Portrait: Diagonal Matrix 260
3.4 Saddle Point λ_1, λ_2 Real and Different Signs 261
3.5 Focus: λ_1, λ_2 Complex Conjugates 261
3.6 Center Point Singularity 262

4.1 Complex Plane 298
4.2 Mapping from Complex Plane z to Complex Plane $f(z)$ 299
4.3 Complex Function Integration Path 303
4.4 Integration Independent of the Path 304
4.5 Cauchy's Integral Theorem 308
4.6 Modified Cauchy's Integral Theorem 309
4.7 Taylor Series Expansion about Analytic Point 310
4.8 Isolated Singularity of $f(z)$ at z_0 314
4.9 Several Isolated Singularities 315
4.10 Laurent Series 322
4.11 Proof of Laurent Series 323
4.12 Computation of $I_2(z)$ 324
4.13 Computation of $I_1(z)$ 325
4.14 Explanation for Jordan Lemma 1 330
4.15 Explanation for Jordan Lemma 2 332
4.16 Singularities of Function $f(z)$ 334
4.17 Complex Integration Contour 335
4.18 Computation of $I = \int_0^{+\infty} \frac{\mathrm{d}x}{1+x^2}$ 337
4.19 Computation of $\int_{-\infty}^{+\infty} \frac{dx}{1+x^4}$ 338
4.20 Integration $I = \int_0^{\infty} \frac{dx}{1+x^3}$ 340

4.21 Integration $I = \int_0^{2\pi} \frac{d\theta}{a + b\cos\theta}, a > b > 0$ . . . 342
4.22 Singularities of the Function $f(z/j)e^{zt}$ . . . 347
4.23 $I = \int_{-j\infty}^{+j\infty} \frac{e^z}{z} dz$ . . . 348
4.24 $\int_{-\infty}^{+\infty} \frac{e^{jax}}{x^2 + b^2} dx, \quad b > 0$ . . . 349
4.25 Branch Point Singularity . . . 350
4.26 Branch Cut . . . 352
4.27 $I = \int_{-\infty - jr}^{\infty - jr} e^{-\alpha z^2} dz$ . . . 354
4.28 $\int_0^{+\infty} e^{-x^{2k}} dx, 2k \geq 1$ . . . 355
4.29 $I_1 = \oint \frac{e^{jz^2}}{\sin \sqrt{\pi} z}$ . . . 357
4.30 Poisson's Integral . . . 360
4.31 Poisson-Jensen's Integral . . . 366
4.32 Proof of Fundamental Theorem of Algebra . . . 369
4.33 Mean Value Theorem . . . 371
4.34 Preliminaries . . . 372
4.35 Representation of s and $F(s)$ . . . 374
4.36 Mapping of Outer Region of Unit Disk in z-Plane to RHS in s-Plane 375
4.37 Transformation of Bounded PR Function to a Bounded PR Function 376

5.1 Double-Sided Laplace Transform Region of Convergence . . . 385
5.2 Single Laplace Transform Region of Convergence . . . 389
5.3 Step Function . . . 393
5.4 Ramp Function . . . 394
5.5 $f(t) = t^n, \ n = 0, 1, 2, \ldots$ . . . 395
5.6 Time Function $e^{-\alpha t}u(t)$ and ROC . . . 395
5.7 Region of Integration . . . 398
5.8 Jordan's Lemma for Double-Sided Laplace-Sided Inverse Application 406

5.9 Jordan's Lemma for Single-Sided Laplace Inverse Application . . . 408
5.10 Jordan's Lemma . . . 413
5.11 Strip of Convergence . . . 415
5.12 Strip of Convergence . . . 416
5.13 Transfer Function Concept . . . 417
5.14 Input–Output Relation via Transfer Function . . . 418
5.15 Region of Convergence of $f_1(t)f_2(t)$ Transform. . . . 424
5.16 Contour Integration . . . 426
5.17 Causal Time Exponential . . . 432
5.18 Noncausal Time Exponential . . . 433
5.19 Pulse Function . . . 434
5.20 Unit Function for All Times . . . 434
5.21 Decaying Exponential for Positive and Negative Times . . . 434
5.22 Signum Function . . . 435
5.23 Gaussian Function . . . 435
5.24 Infinite Train of Impulses $S_T(t)$ . . . 438
5.25 Triangular Pulse . . . 442
5.26 Time-Frequency Signals Spectrum . . . 449
5.27 Time-Frequency Signals Spectrum Continued . . . 450
5.28 Computation of Fourier Transform Inverse for $t > 0$ . . . 451
5.29 Computation of Fourier Transform Inverse for $t < 0$ . . . 452
5.30 Computation of R–L Circuit Response via Fourier Transforms . . . 453
5.31 Hilbert Transform Realization . . . 459
5.32 Hilbert Transform Derivation via Contour Integration . . . 461
5.33 Computation of $F(\omega)$ Given $R(\omega)$ . . . 463
5.34 The Path of Integration . . . 464
5.35 Integration Path, C (Fractional Negative λ) for Solution $y_1(t, \lambda)$ . . . 472

5.36 Integration Path, C^* for Solution $y_2(t, \lambda)$ of Eq. 5.104 474

6.1 Continuous System with Sampler at the Input and the Output 478
6.2 Equivalent Discrete System 479
6.3 k-th order Discrete System Description 481
6.4 Delay Element 482
6.5 Adder Element 482
6.6 Gain Element 482
6.7 Accumulator 482
6.8 Realization of a General Discrete System 483
6.9 Z-Transform Convergence Domain of Exponential Growth Functions 485
6.10 Z-Transform of a Delayed Step Function 490
6.11 Region of Analyticity of $F(\lambda)$ Outside the Circle $c, |\lambda| > R_1$ 502
6.12 Analytic Region for $F(z)$ and $F(1/\lambda)$ 505
6.13 Location of Poles of $F(z)$ 509
6.14 Transfer Function 513
6.15 Transfer Function of Cascaded Systems 514
6.16 State Variable Representation of a Discrete System 515
6.17 Block Diagram for Z-Transfer Function 515
6.18 Optimal Minimum Sum of Squared Error Design 516
6.19 Region of Convergence $F_b^+(z)$ and $F_b^-(z)$ 527
6.20 Region of Convergence of $F_b(z)$ 527
6.21 Location of the Poles of $F_b(z)$ and Region of Convergence 528
6.22 $\sin \omega t$ Sampled Every $2\pi/\omega$ Seconds 532
6.23 Sampling of the Frequency Band-Limited Signals 534
6.24 Transfer Function $H(j\omega)$ of a Low-Pass Recovery Filter (Reconstruction Filter). 535
6.25 Distortionless Transmission 537

6.26 Reconstruction of a Band-Limited Signal 537
6.27 Filter with Comb Function $S_T(t)$ Input. 543
6.28 N-Point Staircase Representation of $f(t)$ and Its Fourier Spectra $F(f)$548
6.29 Time-Limited Aperiodic Signal and Its Periodic Representation . . . 551
6.30 Sampled Cosinusoid 558
6.31 Fourier Coefficients for a 4-Point FFT 558
6.32 Fourier Coefficients for a 16-Point FFT 560
6.33 Computational Complexity of Multiplications 562
6.34 N-th Roots of Unity 564
6.35 $(N/2)$ Point Transforms for the k-th Time Function—Remove N! . . 567
6.36 FFT for the 8-Point Data 570
6.37 Binary Tree for a Sequence of an 8-Point FFT 570

7.1 R, L, C Network Illustrating State Variables 578
7.2 Schematic n-State Variable Description 581
7.3 Block Diagram for State Space Representation 583
7.4 State Variable Equations for a Mechanical System 586
7.5 State Variable Equations for the Electric Circuit 587
7.6 A Simple Network, Capacitors, and Inductors Replaced by Their Respective Voltage and Current Sources 589
7.7 An Electric Network and Its Corresponding Linear Graph 591
7.8 Normal State Tree (Heavy Lines) for the Circuit 7.7 593
7.9 State Variables from Cutsets and Loops. 593
7.10 Transfer Function to be Realized 600
7.11 Controller Realization of a Transfer Function 601
7.12 Controllability Canonical Form—Realization #2 602
7.13 Observer Canonical Form—Realization #3 604
7.14 Observability Canonical Form—Realization #4 606

7.15 Vectors $\boldsymbol{x}(t)$, $\boldsymbol{b}$, and $\boldsymbol{c}$ in State, Dual Space-Geometric Representation 614
7.16 Decomposition—Observable and Controllable Subsystem 633
7.17 Dynamic System Kalman Decomposition 634
7.18 State Variable Feedback Controller. 651
7.19 Observer Design. 653

8.1 Maximum Area Rectangle inside a Circle—Kepler's Problem 684
8.2 Extremal Curve for a Definite Integral 688
8.3 Brachistochrone Problem 692
8.4 Poisson's Equation 695
8.5 Corner Point at $t = t_c$ 701
8.6 Switching Curves 708
8.7 Simple Modeling of the Rocket Launch 709
8.8 Rocket Launch Trajectory 711
8.9 Optimal Trajectory 718
8.10 Riccati Computer 721

9.1 Union and Intersection 730
9.2 $A \cup A^c \equiv F$ 730
9.3 $A - B \equiv A - (A \cap B)$ 731
9.4 Cumulative Distribution Function, CDF 736
9.5 Random Walk—Wiener Process 750
9.6 Wiener Process as an Integration of White Noise 759
9.7 Transformation of t^n via Kernel e^{-t} 783
9.8 Random Function of Time from Ensemble of Random Process . . . 793
9.9 Generation of Wiener Process by Integrating White Noise 819
9.10 Response to Stochastic Input 821
9.11 White Noise through a Lag Network 824
9.12 Signal with Given Autocorrelation Function through a Lag Network 825

9.13 Optimal Estimator—Memoryless System 826

9.14 Feedback Implementation of Minimum Error Variance Estimation . 831

9.15 Noncausal Wiener Filter 832

9.16 Wiener Filter with Uncorrelated Noise Signal and Noise 834

9.17 Causal Wiener Filter 836

9.18 Dynamical System—Kalman Filtering Problem 841

9.19 Optimal Estimate Equations via Calculus of Variations 845

9.20 Discrete Kalman Filter 853

9.21 WSS Process with Noise and Wiener Filter 855

9.22 Equivalent Innovation Representation of the Random Process 855

List of Tables

3.1 Method Used: Routh Canonical Transformation 279

3.2 Method Used: Integration by Parts 282

3.3 Method Used: Variable Multiplier 284

3.4 Method Used: Variable Gradient 288

5.1 Transform Definitions 391

5.2 Laplace Transform Properties 392

5.3 Table of Laplace Transform Pairs 401

5.4 Fourier Transform Properties 446

6.1 Z-Transform Properties 487

6.2 Table of the Single-Sided Z-Transform Pairs 503

6.3 Some Important Series 529

6.4 Properties of Fourier Series 539

6.5 Short Collection of the Fourier Series 543

6.6 Properties of **DFT** 556

6.7 Bit Reversal 571

Chapter 1

System Concept Fundamentals and Linear Vector Spaces

1.1 Introduction

In this chapter, we introduce some fundamental concepts of the systems, signals, and their interaction in a precise mathematical form. System inputs and outputs are signals that can be considered as time functions or frequency functions or geometrical objects in vector spaces. The system is an "**operator**," which acts on the input signals and modifies them to result in the output signals. Linear systems represent a very large and most important class of systems that obey certain laws represented by a very rich theory. As a first approach, if we want to examine input–output time behavior, we need to represent the operation of a system operator via "convolution" of the system input with the "**impulse response of the system**." Secondly, if we want to learn about input-ouput frequency behavior, we use the so-called "**Transform Methods**." The third representation involves the differential equation approach. In addition, the fourth approach considers vector space theory in which the input–output signals are considered as "vectors" in the linear vector spaces and the system is considered as an "operator" mapping the input vector functions into the output vector functions.

We shall also introduce various "test" input signals, and definitions of such terms as "linear" and "convolution." From a mathematical point of view, there is a one-to-one parallel between continous and discrete signals. Therefore, the same methodology will be used to study both the continous and the discrete systems. Furthermore, the linear vector spaces will be used as a tool for systems analysis and synthesis. The objective is to establish a mathematical framework in which seemingly different engineering, financial, economic, and biological systems can be modeled by similar analytical and simulation methods to ensure optimum system performance before and after their realization. In essence, the book is developed with a view to study the generation, stability, and the accuracy of both continous and discrete signals with systems acting as "operators," thereby modifying the signal behavior. The main theme can be summarized as:

Basic Problem: Given a physical system, how do you compute the set of output responses from a set of inputs and initial conditions (system with initial energy)?

There is no clear-cut answer to this question for a completely general system, namely, nonlinear and time varying, etc. We will have more to say about it later on. At the moment we are dealing with **Linear Systems** only. They adhere to a well-defined property called "superposition." For these **linear** systems the answer to this basic problem is provided by the following complementary approaches:

1. Superposition principle or the so-called **Convolution Integral**.

2. Solution to the differential equations representing the system outputs in time domain and the associated state variable approach.

3. The frequency domain transform methods involving Fourier Transforms,

Laplace Transforms, Z (known as zee) Transforms of the input and the output signals.

In this chapter, we shall only deal with the convolution integral and its properties. Each of the other two methods merit a detailed discussion in their own right and will be dealt with accordingly.

1.2 System Classifications and Signal Definition

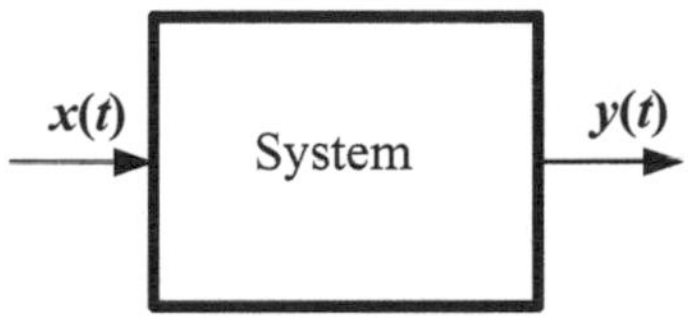

Figure 1.1: System as an "Operator"

Figure 1.1 shows a simple system that mathematically speaking takes an input signal (or signals) and results in an output signal (or signals). The output $y(t)$ and the input $x(t)$ are interrelated via some transformation or "mapping" process, called an **operator**, which represents the inherent characteristic of a given system. Thus, a locomotive can be viewed as a system transforming the fuel input into the motion as an output. This is a very general definition. If t is a continuous time variable and n, a discrete independent variable, then a system defines a mathematical relationship:

$$\boldsymbol{T}[x(t)(\text{or } x(n))] = y(t)(\text{or } y(n)) \qquad \text{Definition of an operator} \qquad (1.1)$$

"$\boldsymbol{T}$" represents the operator symbol.

Example 1.1:

Figure 1.2 shows an *R–L* circuit and its operator representation.

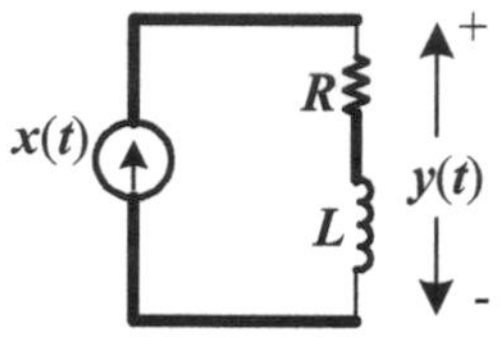

Figure 1.2: *R-L* Linear Circuit

$x(t)$ = Input Current

$y(t)$ = Output voltage across *R* and *L* in series

From Kirchoff's laws:

$$y(t) = \left(R + L\frac{d}{dt}\right)x(t)$$

Thus, for the circuit in Figure 1.2

$$\boldsymbol{T} = \left[R + L\frac{\mathrm{d}}{\mathrm{d}t}\right]$$

In general, the set of inputs $x(t)$ are referred to as the "Domain" of the operator $\boldsymbol{T}$, which maps them into the set of outputs $y(t)$ called the "Range" of the operator $\boldsymbol{T}$.

1.2.1 Linear Systems

A linear system is characterized by an operator $\boldsymbol{T}$ with the following two properties:

(i) $\boldsymbol{T}[kx(t)] = k\boldsymbol{T}[x(t)]$ Amplification property (Homogeneity)

(ii) $\boldsymbol{T}[k_1x_1(t) + k_2x_2(t)] = k_1\boldsymbol{T}[x_1(t)] + k_2\boldsymbol{T}[x_2(t)]$ Additive property

Both these properties are necessary and sufficient to avoid pathological situations such as k_1 or k_2 being irrational numbers.

If either of these relations do not hold good, the system is defined as nonlinear.

Note: For a system to be linear, it is necessary that $x(t) = 0$ implies $y(t) = 0$.

A simple equation like $y = ax + b$ where a and b are nonzero constants, is not a linear system. The system can be made linear by making $b = 0$.

Example 1.2:

$$\boldsymbol{T} = \left[R + L\frac{\mathrm{d}}{\mathrm{d}t}\right] \quad \text{is a linear operator}$$

because

$$\left[R + L\frac{\mathrm{d}}{\mathrm{d}t}\right][x_1(t) + x_2(t)] = \left[Rx_1(t) + L\frac{\mathrm{d}}{\mathrm{d}t}x_1(t)\right] + \left[Rx_2(t) + L\frac{\mathrm{d}}{\mathrm{d}t}x_2(t)\right] \quad (1.2)$$

$$\left[R + L\frac{\mathrm{d}}{\mathrm{d}t}\right][kx(t)] = k\left[Rx(t) + L\frac{\mathrm{d}}{\mathrm{d}t}x(t)\right] \quad (1.3)$$

Note that the superposition property is valid for the linear operators only.

Example 1.3:

$$\boldsymbol{T} = [\cdot]^2 + \frac{\mathrm{d}^2}{\mathrm{d}t^2} \quad \text{is a nonlinear operator}$$

because

$$\boldsymbol{T}[x_1(t) + x_2(t)] = [x_1 + x_2]^2(t) + \frac{\mathrm{d}^2}{\mathrm{d}t^2}[x_1(t) + x_2(t)]$$

but,

$$\boldsymbol{T}[x_1(t)]+\boldsymbol{T}[x_2(t)] = \left[x_1^2(t) + \frac{\mathrm{d}^2}{\mathrm{d}t^2}x_1(t)\right]+\left[x_2^2(t) + \frac{\mathrm{d}^2}{\mathrm{d}t^2}x_2(t)\right] \neq \boldsymbol{T}[x_1(t)+x_2(t)] \quad (1.4)$$

Thus, the superposition property is invalid for this operator which is **nonlinear**.

1.2.2 Linear Time Invariant (LTI) Systems

The operator representing this type of system, in addition to amplification and superposition, has the following **additional property**:

$$\boldsymbol{T}[x(t)] = y(t) \quad \text{implies} \quad \boldsymbol{T}[x(t-\tau)] = y(t-\tau) \quad (\tau \text{ is a constant})$$

Thus, a time shift in the input results in a corresponding time shift in the output.

A linear system that does not meet this requirement is called a **linear time varying (LTV) system.**

Example 1.4:

$$\boldsymbol{T} = \sum_{i=0}^{n} a_i \frac{\mathrm{d}^i}{\mathrm{d}t^i} \qquad (a_i \text{ are constants})$$

represents a linear time invariant system **LTI**

$$\boldsymbol{T} = \sum_{i=0}^{n} a_i(t) \frac{\mathrm{d}^i}{\mathrm{d}t^i} \qquad (a_i(t) \text{ are time functions})$$

represents a linear time varying system called **LTV**

time shift property does not hold here.

1.2.3 Causal Systems

If

$$\boldsymbol{T}[x(t)] = y(t)$$

and

$$x(t) = 0, \quad \text{for} \quad t < 0,$$

implies

$$y(t) = 0, \quad t < 0$$

Then such a system is called **Causal.**

Because of this property, when an input is applied to such a system at $t = 0$, its response is identically zero for $t < 0$. Thus, the response of a causal system at any time t is dependent only on the input function up to that instant t and not upon the past values of the input signal.

All real-time physical systems are causal. Such systems are also known as physically **realizable** or nonanticipative. On the other hand, noncausal temporal systems are physically unrealizable in the real-time but can be approximated by causal systems with a delay. However, there are other noncausal systems where the independent variable is other than time. Such is the case with "stochastic" or "deterministic" systems where the domain of the independent variable varies between $-\infty$ and $+\infty$. These are noncausal but may be realizable systems (explained later).

1.2.4 Dynamical–Nondynamical Systems

If the present output depends upon the present input only and not upon its past history, then such a system is referred to as "Inertialess" or "Memoryless" or a Nondynamical system. On the other hand, if the system response at the present time depends upon the present as well as the past history, then the system is called a **Dynamical** or **With Memory** system.

Example 1.5:

$$y(t) = kx(t) \quad (k \text{ constant}) \text{ is a } \textbf{Memoryless System} \tag{1.5}$$

$$y(t) = Rx(t) + L\frac{\mathrm{d}}{\mathrm{d}t}x(t) \quad \text{is a } \textbf{Dynamical Systems with Memory} \tag{1.6}$$

1.2.5 Continuous and Discrete Systems

If the independent variable t is continuous, then the system is referred to as **Continuous** and is represented by a continous signal $x(t)$. But if the variable t takes only discrete values n, then the system is called **discrete** or **digital** and is represented by a discrete sequence $x[n]$.

1.2.6 Lumped Parameter vs. Distributed Parameter Systems

Systems involving masses, springs, dashpots, resistors, capacitors and inductors, etc, are known as lumped parameter systems and are described by ordinary differential equations known as **ODEs**.

Other systems such as transmission lines, cardiovascular system, nuclear reactors, Maxwell's electromagnetic equations applications, etc, which deal with partial differential equations or **PDEs** are called **Distributed Parameter Systems**. These can still be linear or nonlinear.

1.2.7 Deterministic vs. Stochastic Signals and Systems

In many physical systems, both the input and the output signals may be contaminated with extraneous signals or errors, which are characterized as unwanted "disturbances." If these measured signals are repeatable, we call such signals

Deterministic, but if they are not repeatable and have a probabilistic character we call them **Random** or **Stochastic** signals. The extraneous signals in these systems may be referred to as **Noise**. In such cases, the data is **Processed** and **Useful Information** is extracted in some "optimal fashion" using Stochastic Methods.

1.2.8 Input–Output Description of a Dynamic System

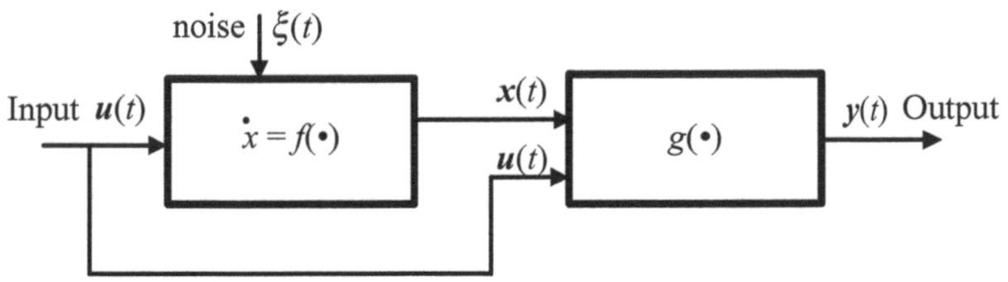

Figure 1.3: Dynamic Systems Input–Output Description

It is customary to describe a system (Figure 1.3) by a block diagram with the input vector $\boldsymbol{u}(t)$, state vector $\boldsymbol{x}(t)$ (to be defined later), and the output vector $\boldsymbol{y}(t)$. Vectors $\boldsymbol{\xi}(t)$ and $\omega(t)$ represent the input noise (disturbances) and the measurement noise respectively. The bold variables represent a multivariable quantity (explained later as Euclidean vectors with components x_1, x_2 etc.).

The system equations are:

$$\dot{\boldsymbol{x}} = \boldsymbol{f}(\boldsymbol{x}, \boldsymbol{u}, \boldsymbol{\xi}, t) \qquad \textbf{State equations} \tag{1.7}$$

$$y = \boldsymbol{g}(\boldsymbol{x}, \boldsymbol{u}, \omega, t) \qquad \textbf{Output equations} \tag{1.8}$$

The domain and range of the "vector" signals are properly defined.

Determination of the functions $\boldsymbol{f}(\cdot)$ and $\boldsymbol{g}(\cdot)$ is dependent on the physics of a particular system and in general requires a lots of ingenuity, experience, and understanding of the physical phenomena governing the system. Our treatment of the subject only involves systems where the functions of $\boldsymbol{f}$ and $\boldsymbol{g}$ are linear functions of the dependent variables $\boldsymbol{x}, \boldsymbol{u}$, and possibly $\boldsymbol{t}$.

There are four different but complimentary approaches to study the input–output behavior of these systems:

1. Input–output behavior as an **integral operator**. This is known as the "convolution" or convolution integral approach.

2. Input–output behavior as a **differential operator**. This involves the study of linear inhomogenous differential or difference equations. We shall devote a whole chapter to this topic because of its importance.

3. Input–output behavior as a sum of **frequency exponential signals** (or sinusoids). This aspect of system theory is studied via **Laplace** and **Fourier** transforms and deserves a full chapter for each topic in its own right.

4. The input–output signals are represented as "vectors" in **linear vector spaces** and system as a mathematical **operator** or a "mapping" of input signal from one vector space to another vector space representing the output. This is a very rich and important approach for studing the communication and control problems.

This chapter concentrates on convolution operators and linear vector spaces. The other approaches will be studied in later chapters.

1.3 Time Signals and Their Representation

System modeling requires a mathematical description of its system components and various input signals. A few signals and their most important properties are discussed below.

1.3.1 Impulse or Delta Function Signals

There is a rigorous mathematical treatment of a family of these functions developed by [Schwartz, L.]. Impulse or delta function is a building block for most of the time signals and it is denoted by $\delta(t)$. It is also referred to as **Dirac Delta Function** or **Generalized Function**.

The unit impulse defines an "assignment rule" that **assigns** to a function $x(t)$, a number $x(0)$ according to the following rule:

$$\int_{-\infty}^{+\infty} x(t)\delta(t)\,\mathrm{d}t = x(0) \tag{1.9}$$

It should be understood that Eq. 1.9 has meaning only as an integral.
Its working properties can be better understood as following:
Consider the pulse function $P_T(t - t_0)$ shown below:

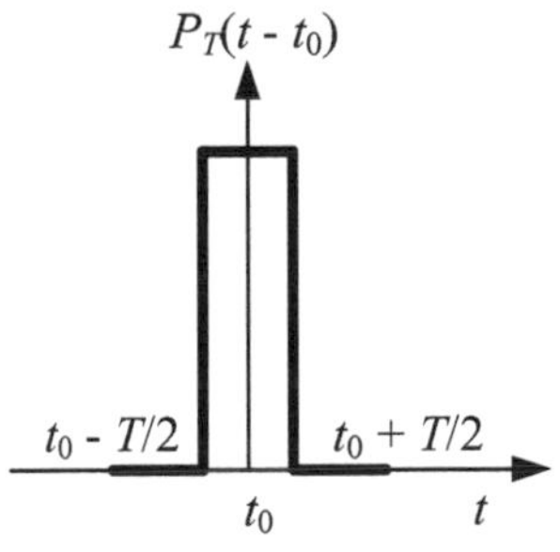

Figure 1.4: Pulse Function

This function is 0 for all values of t except between the interval $t_0 - T/2$ to $t_0 + T/2$ where it takes on the value 1.

Define

$$\delta(t - t_0) = \lim_{T \to 0} \left[\frac{1}{T} P_T(t - t_0) \right]$$

This definition, though not mathematically rigorous, is in engineering, quite useful. For rigorous treatment, we need the generalized function theory of [Schwartz, L.].

Figure 1.5 shows a graphical representation of $\delta(t - t_0)$.

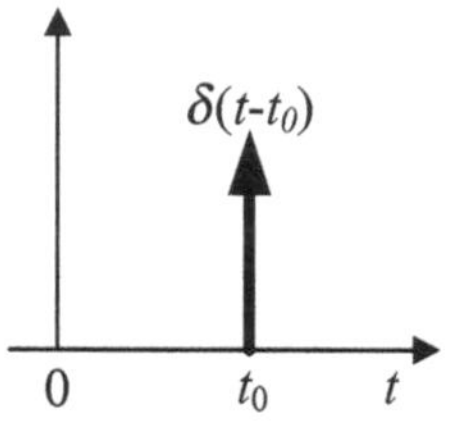

Figure 1.5: Delta Function

Thus, $\delta(t - t_0)$ is an impulse, which originates at $t = t_0$. Note that $\frac{1}{T}P_T(t - t_0)$ is a function of unit area, which in the limit $T \to 0$ has infinite height and zero width.

In fact, there is a family of functions $f_T(t - t_0)$ which in the limit yields:

$$\delta(t - t_0) = \lim_{T \to 0} f_T(t - t_0)$$

such that

$$\int_{t_0 - T/2}^{t_0 + T/2} f_T(t - t_0)\ \mathrm{d}t = 1$$

$$\begin{aligned} \lim_{T \to \infty} [T f_T(t - t_0)] &= 1 \qquad \text{for} \quad t_0 - T/2 \le t \le t_0 + T/2 \\ &= 0 \qquad \text{otherwise} \end{aligned}$$

Various Examples of $f_T(t - t_0)$, $T > 0$

1.

$$\delta(t) = \lim_{T\to 0}\left[\frac{1}{\pi}\frac{T}{T^2 + t^2}\right]$$

2.

$$\delta(t) = \lim_{T\to 0}\left[\frac{1}{\sqrt{2\pi}T}e^{-t^2/2T^2}\right]$$

3.

$$\delta(t) = \lim_{T\to 0}\left[\frac{1}{\pi}\frac{\sin(t/T)}{t}\right]$$

4.

$$\delta(t) = \lim_{T\to 0}\left[\frac{1}{T}e^{-|t|/T}\right]$$

5.

$$\delta(t) = \lim_{\epsilon\to 0}\frac{1}{\pi t}\sin(t/\epsilon)$$

6.

$$\delta(t) = \lim_{n\to\infty}\frac{1}{2\pi}\left[\frac{\sin\left[(n + 1/2)t\right]}{\sin(1/2t)}\right] = \lim_{\epsilon\to 0}\frac{1}{2\pi}\left[\frac{\sin\left[(1/\epsilon + 1/2)t\right]}{\sin(1/2t)}\right]$$

7.

$$\delta(t) = \lim_{\epsilon\to 0}|t|^{\epsilon - 1}$$

8.

$$\delta(t) = -\frac{1}{\pi}\int_{-\infty}^{+\infty}\frac{1}{\tau(t-\tau)}\,\mathrm{d}\tau$$

Sifting Property of the Impulse Function

We shall show that

$$\int_{-\infty}^{\infty} f(t)\delta(t - t_0)\,\mathrm{d}t = f(t_0)$$

This is known as the **Sifting Property** of $\delta(t)$ and plays an important role as discussed below. Consider the integral

$$\int_{-\infty}^{\infty} f(t)\left(\lim_{T\to 0}\frac{1}{T}P_T(t-t_0)\right) dt$$

the function f(t) being continuous at t_0.

Figure 1.6 illustrates the above integral as "sifting" the function $f(t)$.

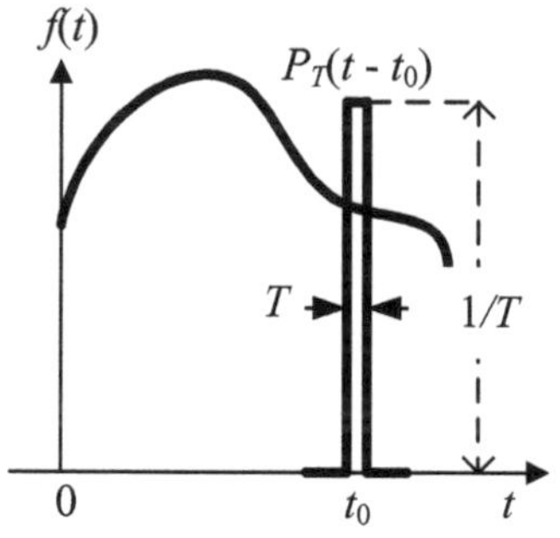

Figure 1.6: Sifting Property

If T is very small, $f(t)$ can be approximated by $f(t_0)$ for $(t_0 - T/2) \le t \le (t_0 + T/2)$. Therefore,

$$\int_{-\infty}^{\infty} f(t)\delta(t-t_0)\, dt = \lim_{T\to 0}\int_{-\infty}^{\infty} f(t)\left(\frac{1}{T}P_T(t_0)\right) dt \approx \lim_{T\to 0} f(t_0)\int_{-\infty}^{\infty}\frac{1}{T}P_T(t_0)\, dt = f(t_0)$$

In fact any function $f(t)$ can be manufactured as

$$f(t) = \int_{-\infty}^{\infty} f(\tau)\delta(t-\tau)\, dt = \int_{-\infty}^{\infty} f(t-\tau)\delta(\tau)\, d\tau \qquad (1.10)$$

Equation 1.10 is a superposition integral of the delayed impulses $\delta(t-\tau)$ of the strength of $f(\tau)$, as well as an undelayed impulse $\delta(\tau)$ of strength $f(t-\tau)$,

$-\infty < \tau < \infty$. Both the interpretations are useful.

Note: The impulse function is meaningful only when used under the integral sign and is associated with the convolution of the signals.

Thus,

$$f(t_0)\delta(t - t_0) \equiv f(t)\delta(t - t_0)$$

implies

$$\int_{-\infty}^{\infty} f(t_0)\delta(t - t_0)\,\mathrm{d}t = \int_{-\infty}^{\infty} f(t)\delta(t - t_0)\,\mathrm{d}t_0 = f(t_0) \tag{1.11}$$

The equation 1.11 is also referred to as a convolution of $f(t)$ with $\delta(t)$, the convolution process symbolized by "$*$" (Not multiplication). Thus,

$$f(t) = f(t)^*\delta(t) = \delta(t)^*f(t) = \int_{-\infty}^{\infty} f(\tau)\delta(t - \tau)\,\mathrm{d}t = \int_{-\infty}^{\infty} f(t - \tau)\delta(\tau)\,\mathrm{d}\tau \tag{1.12}$$

The convolution process is discussed at great length in the later pages.

Delta Function and Its Connection with the Dirichlet Kernel

Consider a set of functions

$$f_{2n+1}(t) = \left\{\frac{\Delta w}{2\pi} e^{jn\Delta\omega t} : n = 0,\ \pm 1,\ \ldots \pm N\right\} \tag{1.13}$$

The Dirichlet kernel can be written as

$$\delta_N(t - \tau) = \sum_{n=-N}^{N} f_{2n+1}(t - \tau) = \frac{\Delta\omega}{2\pi} \sum_{n=-N}^{N} e^{jn\Delta\omega(t-\tau)} \tag{1.14}$$

In the limit as $N \to \infty$, $\Delta\omega \to \mathrm{d}\omega$, and $n\Delta\omega \to \omega$ (real frequency), the Dirichlet kernel takes the form:

$$\lim_{N\to\infty} \delta_N(t-\tau) = \delta(t-\tau) = \frac{1}{2\pi}\int_{-\infty}^{\infty} e^{j\omega(t-\tau)}\,\mathrm{d}\omega \qquad \text{(important relation)}$$

If we let

$$s = c + j\omega \qquad \text{(complex frequency)}$$

Then

$$\delta(t-\tau) = \frac{1}{2\pi j}\int_{c-j\infty}^{c+j\infty} e^{s(t-\tau)}\,\mathrm{d}s$$

If we let

$$z = e^{j\Delta\omega(t-\tau)} \quad t = n, \quad \tau = m, \quad n\Delta\omega = \omega$$

Then the Eq. 1.14 can describe a discrete delta function:

$$\delta(n-m) = \frac{1}{2\pi j}\oint \frac{(z^{n-m})}{z}\,\mathrm{d}z \qquad \text{Important Relation} \qquad (1.15)$$

Some Useful Properties of Impulse Function (Delta Function)

1. **Delta function is an even function of time**

   From Figure 1.4

$$\delta(t) = \delta(-t)$$

2. **Derivative property of delta function**

$$f(t)\left[\frac{\mathrm{d}^n\delta(t-t_0)}{\mathrm{d}t^n}\right] = f(t)\left[\delta^{(n)}(t-t_0)\right] = (-1)^n\frac{\mathrm{d}^n f(t)}{\mathrm{d}t^n}\left[\delta(t-t_0)\right] \qquad (1.16)$$

   This result is easily proven using integration by parts. The n-th derivative $\delta^{(n)}(t)$ is an even function of time for n even and an odd function of time for odd n. The function $\delta^{(1)}$ is an odd function and is known as a **Doublet**.

3. **Delta function whose argument is a general function of time** t

Given a function $g(t)$, $-\infty < t < \infty$, such that:

$$g(t_i) = 0, \ \frac{\mathrm{d}}{\mathrm{d}t} g(t_i) \neq 0 \quad -\infty < t_i < \infty, \quad i = 1, 2, \cdots, n \text{ finite}$$

We shall prove that:

$$\delta(g(t)) \equiv \sum_{i=1}^{n} \left\| \left[\frac{\mathrm{d}}{\mathrm{d}t} g(t_i) \right] \right\|^{-1} \delta(t - t_i)$$

and

$$I = \int_{-\infty}^{\infty} f(t)\delta(g(t))\,\mathrm{d}t = \sum_{i=1}^{n} f(t_i) \left\| \left[\frac{\mathrm{d}}{\mathrm{d}t} g(t_i) \right] \right\|^{-1} \tag{1.17}$$

Proof:

Examine the integral

$$I = \int_{-\infty}^{\infty} f(t)\delta(g(t))\,\mathrm{d}t$$

Let

$$x = g(t), \quad t = g^{-1}(x) = t(x)$$

$$\mathrm{d}x = \mathrm{d}g(t) = \left[\frac{\mathrm{d}}{\mathrm{d}t} g(t) \right] \mathrm{d}t$$

Therefore,

$$\mathrm{d}t = \left[\frac{\mathrm{d}}{\mathrm{d}t} g(t) \right]^{-1} \mathrm{d}x$$

The integral I can be rewritten as

$$I = \int_{-\infty}^{\infty} f(t)\delta(g(t))\,\mathrm{d}t = \sum_{i=1}^{n} \int_{t_i-\epsilon}^{t_i+\epsilon} f(t)\delta(g(t))\,\mathrm{d}t = \sum_{i=1}^{n} I_i \quad \text{for} \quad \epsilon > 0\,, \quad \epsilon \to 0$$

where

$$I_i = \int_{t_i-\epsilon}^{t_i+\epsilon} f(t)\delta(g(t))\,\mathrm{d}t$$

Case 1: $\frac{\mathrm{d}}{\mathrm{d}t}g(t_i) = \left.\frac{\mathrm{d}x}{\mathrm{d}t}\right|_{t=t_i} > 0$ In this case both $\mathrm{d}t$ and $\mathrm{d}x$ are of the same sign at t_i and hence as t changes from $t_i - \epsilon$ to $t_i + \epsilon$, x changes from $-\epsilon(t_i)$ to $+\epsilon(t_i)$

Therefore,

$$\begin{aligned} I_i &= \int_{-\epsilon(t_i)}^{\epsilon(t_i)} f(t(x))\left[\frac{\mathrm{d}}{\mathrm{d}t(x)}g(t(x))\right]^{-1}\delta(x)\,\mathrm{d}x \\ &= f(t_i)\left[\frac{\mathrm{d}}{\mathrm{d}t}g(t_i)\right]^{-1} \end{aligned}$$

Case 2: $\frac{\mathrm{d}}{\mathrm{d}t}g(t_i) = \left.\frac{\mathrm{d}x}{\mathrm{d}t}\right|_{t=t_i} < 0$ In this case both $\mathrm{d}t$ and $\mathrm{d}x$ are of the opposite sign at t_i and hence as t changes from $t_i - \epsilon$ to $t_i + \epsilon$, x changes from $+\epsilon(t_i)$ to $-\epsilon(t_i)$

Therefore,

$$\begin{aligned} I_i &= \int_{+\epsilon(t_i)}^{-\epsilon(t_i)} f(t(x))\left[\frac{\mathrm{d}}{\mathrm{d}t(x)}g(t(x))\right]^{-1}\delta(x)\,\mathrm{d}x \\ &= -f(t_i)\left[\frac{\mathrm{d}}{\mathrm{d}t}g(t_i)\right]^{-1} \end{aligned}$$

Combining the two cases

$$I = \int_{-\infty}^{\infty} f(t)\delta(g(t))dt = \sum_{i=1}^{n} f(t_i)\left|\left[\frac{d}{dt}g(t_i)\right]^{-1}\right|$$

or

$$\delta(g(t)) = \sum_{i=1}^{n}\left|\left[\frac{\mathrm{d}}{\mathrm{d}t}g(t_i)\right]\right|^{-1}\delta(t - t_i)$$

Example 1.6:

Let

$$g(t) = a(t - t_0), \quad a \text{ is a constant}$$

then

$$\delta(a(t - t_0)) \equiv |a|^{-1}\delta(t - t_0)$$

Example 1.7:

Let

$$g(t) = t^2 - a^2 = (t - a)(t + a)$$

then

$$\delta(t^2 - a^2) \equiv 2|a|^{-1}\,[\delta(t - |a|) + \delta(t + |a|)]$$

4. **Convolution with $\delta^{(n)}(t)$**

$$\delta(t)^{*}\delta^{(n)}(t) \equiv \delta^{(n)}(t)$$

5. **Multiplication with function t**

$$t\delta(t) \equiv 0$$

Obviously, $t\delta(t)$ is an odd function of time and when integrated from $-\infty$ to ∞ yields 0.

6. **Relationship between step function $u(t)$ and impulse function $\delta(t)$.**

Let

$$u(t) = \begin{cases} 1 & t > 0 \\ \frac{1}{2} & t = 0 \\ 0 & t < 0 \end{cases}$$

Differentiating this function, yields:

$$\frac{\mathrm{d}u(t)}{\mathrm{d}t} = \delta(t)$$

7. **Transform definitions of impulse function** $\delta(t)$

The following two representations of the delta function are very useful as explained earlier and their true meaning will be understood only after transform calculus is presented. For the sake of completeness additional expressions for delta functions are:

$$\delta(t-\tau) = \frac{1}{2\pi}\int_{-\infty}^{\infty} e^{j\omega(t-\tau)}\,\mathrm{d}\omega \qquad \textbf{(Fourier Transform Definition of } \delta(t)\textbf{)}$$

$$\delta(t-\tau) = \frac{1}{2\pi j}\int_{c-j\infty}^{c+j\infty} e^{s(t-\tau)}\,\mathrm{d}s \qquad \textbf{(Laplace Transform Definition of } \delta(t)\textbf{)}$$

$$\left.\begin{aligned}\delta(t-\tau) &= \lim_{\Omega\to\infty}\left[\frac{1}{2\pi}\int_{-\Omega}^{\Omega} e^{j\omega(t-\tau)}\,\mathrm{d}\omega\right]\\ &= \lim_{\Omega\to\infty}\left[\frac{1}{2\pi}\int_{-\Omega}^{\Omega} \cos(\omega(t-\tau))\,\mathrm{d}\omega\right]\\ &= \lim_{\Omega\to\infty}\left[\frac{1}{\pi}\frac{\sin(\Omega(t-\tau))}{(t-\tau)}\right]\end{aligned}\right\} \begin{array}{l}\textbf{(Dirichlet kernel definition}\\ \textbf{of the Delta function)}\end{array}$$

The imaginary part under the integral sign is an odd function and therefore it vanishes while the real part is an even function and therefore doubles up.

The concept of impulse function is also useful in such diversified areas as electrostatics or heat or fluid mechanics. For example, a point charge in space can be represented as an impulse function in three dimensions.

1.3.2 Discrete Delta Function or Delta Function Sequence

- **Discrete impulse function:**

$$\delta(n-k) = \begin{cases} 1 & n = k \\ 0 & n \neq k \end{cases}$$

Function $\delta(n-k)$ plays the same role for discrete systems as impulse function $\Delta(t-\tau)$ in the continuous (or analog) systems.

Thus,

$$x(n) = \sum_{k=-\infty}^{\infty} x(k)\delta(n-k) = \sum_{k=-\infty}^{\infty} x(n-k)\delta(k)$$

Furthermore,

$$\delta(n-k) \equiv z^k\delta(n) \qquad z^k \text{ is called the } k \text{ right-shift operator}$$

$$\delta(n+k) \equiv z^{-k}\delta(n) \qquad z^{-k} \text{ is called the } k \text{ left-shift operator}$$

- **Unit step sequence**:

$$u(n) = \begin{cases} 1 & n \geq 0 \\ 0 & n < 0 \end{cases}$$

$$u(n) = \sum_{i=0}^{\infty} \delta(i)$$

$$\delta(n) = u(n) - u(n-1)$$

- **Another useful definition of $\delta(n)$:**

$$\delta(n) = \left[\frac{1}{2\pi}\int_{-\pi}^{\pi} e^{j\omega n}\,\mathrm{d}\omega\right] = \lim_{\omega_c \to \pi} \int_{-\omega_c}^{\omega_c} e^{j\omega n}\,\mathrm{d}\omega = \lim_{\omega_c \to \pi}\left[\frac{\sin(\omega_c(n))}{\pi n}\right]$$

This definition is useful in the study of discrete Fourier transforms.

- **Two-dimensional impulse function**

  Two-dimensional impulse function can be visualized as a limit of sequence of functions $\delta_k(r)$ such that:

$$\delta(x-\xi, y-\eta) = \lim_{r\to 0} \delta_k(r)$$

where

$$r = \sqrt{(x-\xi)^2 + (y-\eta)^2}$$

Just as in one dimension, a two-dimensional function $f(x, y)$ that is continous over a region Ω containing the points ξ and η, yield

$$\iint_{\Omega} \delta(x-\xi, y-\eta) f(x, y)\, \mathrm{d}x\, \mathrm{d}y = f(\xi, \eta)$$

Examples of $\delta_k(r)$ are:

1)

$$\delta_k(r) = \begin{cases} \dfrac{k^2}{\pi^2} & r < \frac{1}{k} \\ 0 & r \geq \frac{1}{k} \end{cases}$$

2)

$$\delta_k(r) = \frac{ke^{-kr^2}}{\pi}$$

1.3.3 General Signal Classifications

In the following, there are some signals that play an important role in the communication and the control theory. Below, we describe the frequently used signals.

1. **Periodic signals and their Fourier series expansion**

A periodic signal is defined as:

$$f(t) = f(t + nT), \quad n = 1, 2, 3, \ldots, \qquad T = \text{period of the signal.}$$

For example

$$f(t) = A\sin(\omega_o t + \phi) \quad \text{is a periodic signal with } T = \frac{2\pi}{\omega_0}$$

Let us study the Fourier expansion of a bounded general periodic signal satisfying the following two conditions:

(a)

$$\int_{nT-T/2}^{nT+T/2} |f(t)|\, \mathrm{d}t < \infty \quad n = 1, 2, 3, \cdots, \qquad T = \text{period}$$

(b) The signal $f(t)$ has a finite number of maxima and minima and a finite number of bounded discontinuities for $-\infty < t < \infty$.

Such a function can be expanded in the complex Fourier series as:

$$f(t) = \sum_{k=-\infty}^{\infty} c_k e^{j\omega_o kt} \tag{1.18}$$

$$\left(nT - \frac{T}{2}\right) < t \leq \left(nT + \frac{T}{2}\right) \qquad \omega_o = \frac{2\pi}{T}, \quad n = 1, 2, \ldots$$

Notice:

$$\int_{nT-T/2}^{nT+T/2} e^{j\omega_o(k-m)}\, \mathrm{d}t = T\delta_{km} = \begin{cases} 0 & k \neq m \\ T & k = m \end{cases} \tag{1.19}$$

Multiplying both sides of the Eq. 1.18 with $e^{-j\omega_o mt}$ and using the results of Eq. 1.19.

$$c_k = \frac{1}{T} \int_{nT-T/2}^{nT+T/2} f(t)e^{-j\omega_o kt}\, dt = |c_k|e^{j\phi_k} = \text{complex coefficient} \qquad (1.20)$$

furthermore,

$$\overline{c}_k = c_{-k} = |c_k|e^{-j\phi_k} \quad k = 0, \pm 1, \pm 2, \cdots \infty$$

The coefficients $\overline{c}_k$ and c_{-k} are complex conjugates of each other.

This series can also be written in sine, cosine form:

$$f(t) = \sum_{k=0}^{\infty} [a_k \cos(k\omega_o t) + b_k \sin(k\omega_o t)] = \sum_{k=0}^{\infty} |c_k| \cos(k\omega_o t + \phi_k)$$

$$a_k = c_k + c_{-k}, \qquad b_k = j(c_k - c_{-k})$$

$$|c_k| = \sqrt{a_k^2 + b_k^2}, \qquad \phi_k = \tan^{-1} \frac{b_k}{a_k}$$

(c) **Signal Power**

If $f(t)$ is thought of as a current in a 1-ohm resistor, the average power dissipated is called the average signal power p_a, where:

$$p_a = \int_{nT-T/2}^{nT+T/2} |f(t)|^2\, dt = \int_{nT-T/2}^{nT+T/2} \left[\bar{f}(t)f(t)\right] dt = \sum_{k=-\infty}^{\infty} c_k c_{-k} = \sum_{k=-\infty}^{\infty} |c_k|^2$$

(d) **Orthogonality**

Let

$$\phi_k(t) = \frac{1}{\sqrt{T}} e^{j\omega_o kt}$$

be represented by a vector $\boldsymbol{\phi}_k$ in infinite dimensional space $\boldsymbol{V}$, Then

$$(\boldsymbol{\phi}_k, \boldsymbol{\phi}_m) = (\phi_k(t), \phi_m(t)) = \int_{nT-T/2}^{nT+T/2} \frac{1}{T} e^{j\omega_o(k-m)t}\, \mathrm{d}t \tag{1.21}$$

$$= \frac{1}{T} \int_{-T/2}^{T/2} e^{j\omega_o(k-m)\tau}\, \mathrm{d}\tau = \delta_{km}$$

The set $\{\phi_k(t)\}_{-\infty}^{\infty}$ represents a set of orthonormal functions playing an important role in the communications and the control systems. In fact, we are in an infinite dimensional vector space whose coordinates (or basis) are a set of vectors $\boldsymbol{\phi}_k$ represented by a set of time functions $\phi_k(t)$, c_k denoting the magnitude of the components of the vector $\boldsymbol{f}$ representing $f(t)$ with basis $\phi_k(t)$ (see vector spaces Chapter 2).

Figure 1.7 shows the geometrical interpretation of the Fourier Series in vector space.

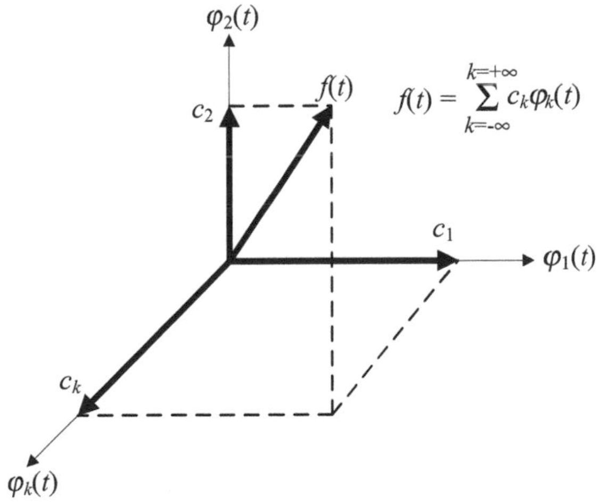

Figure 1.7: Geometrical Interpretation of Fourier Series

(e) A very interesting connection can be found between the difference equations, Fourier series, and Z-transforms as following:

Let

$$e^{j\omega_o t} = z^{-1}, \qquad |z| = 1, \ z \text{ being a complex variable.}$$

$$c_k = x_k$$

$$X(z) = f(t), \quad t = (j\omega_0)^{-1} \ln z^{-1}$$

Then Eq. 1.18 can be written as

$$X(z) = \sum_{k=-\infty}^{\infty} x(k) z^{-k} \qquad \text{(Z-Transform of } x(k)\text{)}$$

and

$$x(k) = \frac{1}{2\pi j} \oint_{|z|=1} X(z) z^{k-1}\, \mathrm{d}z \quad \text{(counterclockwise integration)}$$

Complex variables and complex integration is explained in Chapter 4.

(f) **Energy relation from** $X(z)$

$$X(z) = \sum_{k=-\infty}^{\infty} x(k) z^{-k} = \sum_{m=-\infty}^{\infty} x(m) z^{-m}$$

From the above equation, we can easily see that

$$\oint_{|z|=1} \frac{X(z)X(z^{-1})}{z}\, \mathrm{d}z = \oint_{|z|=1} \sum_{k=-\infty}^{\infty} x(k) z^{-k} \left(\sum_{m=-\infty}^{\infty} x(m) z^{-k+m-1}\, \mathrm{d}z \right)$$

$$= \sum_{k=-\infty}^{\infty} x^2(k) = \sum_{k=-\infty}^{\infty} c_k^2 \quad (c_k \text{ is real})$$

This is also known as Parseval's Theorem for discrete functions.

2. **Step Function** $u(t)$

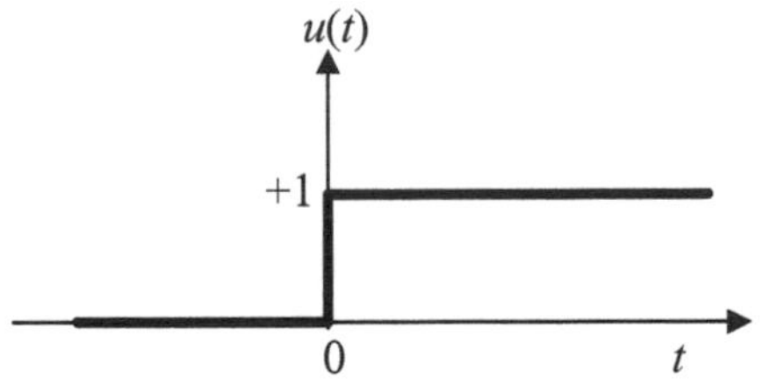

Figure 1.8: Step Function

$$u(t) = \begin{cases} 1 & t \geq 0 \\ 0 & t < 0 \end{cases}$$

Therefore,

$$u(-t) = 0 \quad t > 0$$

3. **Ramp Function** $r(t)$

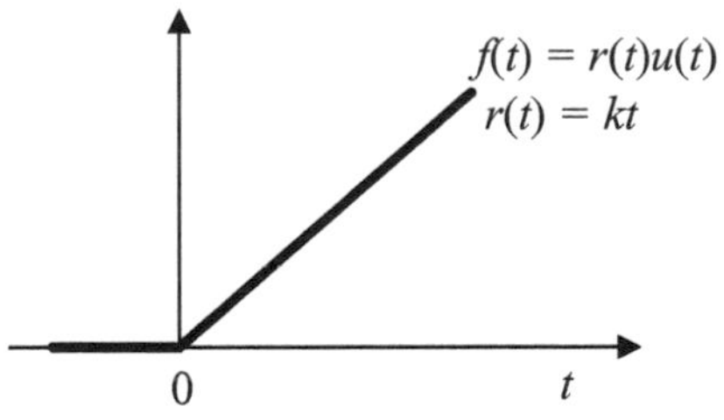

Figure 1.9: Ramp Function

$$r(t) = ku(t) \qquad \text{where } k \text{ is a constant}$$

4. **Rectangular Pulse Function** $p_T(t - t_0)$

$$p_T(t - t_o) = u(t - t_o + \frac{T}{2}) - u(t - t_o - \frac{T}{2}) = \begin{cases} 1 & |t - t_o| \leq \frac{T}{2} \\ 0 & \text{otherwise} \end{cases}$$

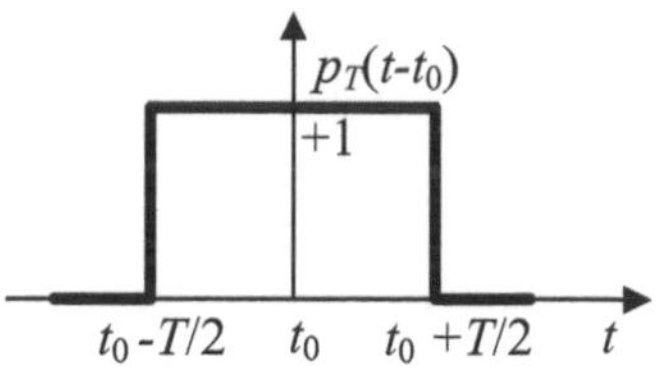

Figure 1.10: Rectangular Pulse

5. **Signum Function** sgn (t)

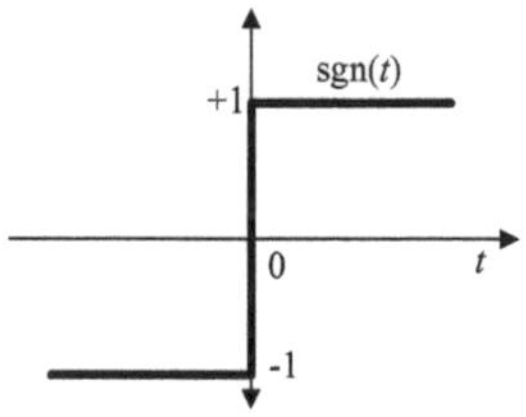

Figure 1.11: Signum Function

$$\text{sgn}\,(t) = \frac{|t|}{t} = \begin{cases} 1 & t > 0 \\ 0 & t = 0 \\ -1 & t < 0 \end{cases}$$

$$u(t) = \frac{1}{2} + \frac{1}{2}\text{sgn}\,(t)$$

6. **Sinc Function** Sinc (t)

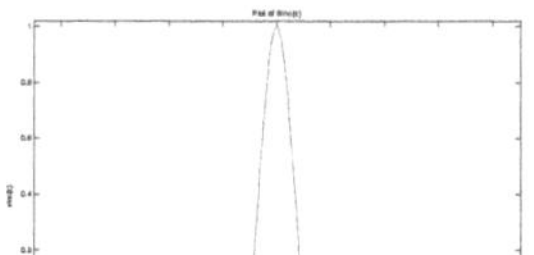

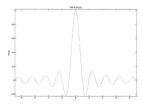

Figure 1.12: Sinc Pulse

$$\text{Sinc}\,(\Omega t) = \frac{\sin(\Omega t)}{t} \qquad -\infty < t < \infty$$

7. **Gaussian Function** $g(t, \sigma)$

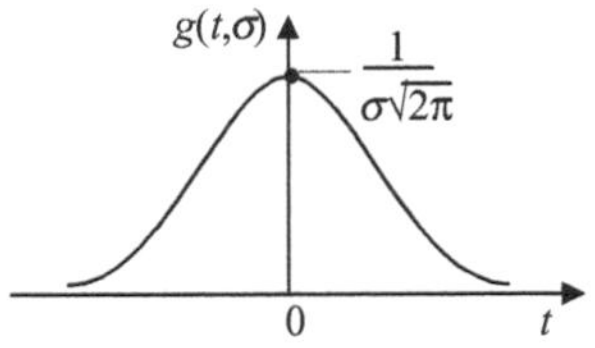

Figure 1.13: Gaussian Pulse

$$g(\sigma, t) = \frac{1}{\sqrt{2\pi}\sigma} \exp\left[-\frac{t^2}{2\sigma^2}\right] \qquad -\infty < t < \infty$$

Note: Gaussian function plays an important role in Communication theory.

8. **Impulse Train Function (also called Comb Function)** $s_T(t)$

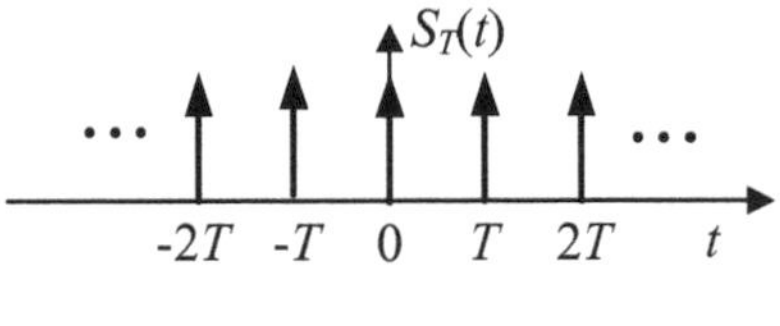

Figure 1.14: Impulse Train

Being a periodic function, its Fourier series expansion is

$$s_T(t) = \sum_{n=-\infty}^{\infty} \delta(t - nT) = \frac{1}{T} \sum_{n=-\infty}^{\infty} e^{j\omega_o nT}, \quad \omega_o = \frac{2}{T}$$

9. **Sine Integral Function** $s_i(t)$

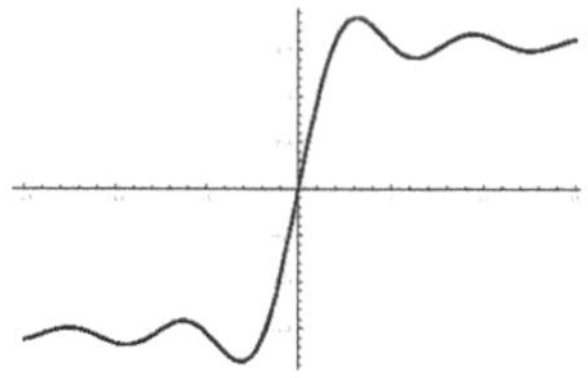

Figure 1.15: Sine Integral Function

$$s_i(t) = \int_0^t \frac{\sin(\tau)}{\tau} \, d\tau$$

10. **Triangular Pulse Function** $q_T(t)$

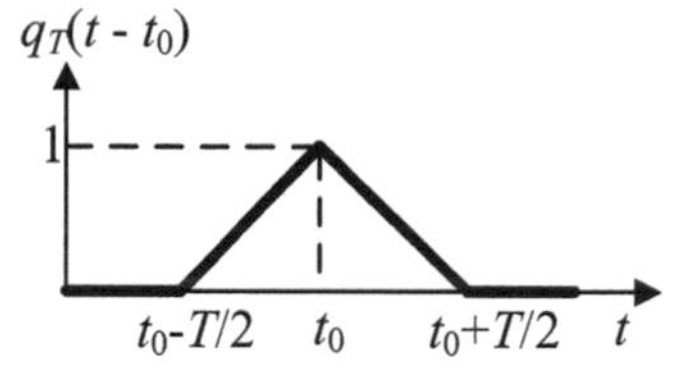

Figure 1.16: Triangular Pulse

$$q_T(t) = \begin{cases} 1 - \dfrac{|t - t_0|}{T} & |t - t_0| < T \\ 0 & \text{otherwise} \end{cases}$$

1.4 Input–Output Relations (System Response)

1.4.1 Superposition or Convolution Integral

Consider the following L, R, C series circuit described by the following differential equation

$$L\frac{d^2q(t)}{dt^2} + R\frac{dq(t)}{dt} + \frac{1}{C}q(t) = v(t)$$

The differential equation describing the above circuit represents a fairly general linear dynamical system. $v(t)$ is an **input function** (such as voltage) and q(t) is the **system output** (such as charge). In what follows, we shall assume that the system is at rest having no initial energy, namely, no initial charge on the capacitor and no initial current through the inductor.

We ask the following question:

If the response of the system to some test input can be determined, then is it possible to find the system response to any general input utilizing the results of this test input?

The answer is "Yes" provided the system is linear.

As an example, if the test function is a delayed impulse $\delta(t - \tau), 0 \leq \tau \leq t$ and the system is linear, then the response of the system can be computed for any general input via the convolution integral discussed below.

We could have used some other test input such as sinusoid or step.

The response of the system to the delayed impulse, $\delta(t - \tau)$ is called the **impulse response** $h(t, \tau)$.

Figure 1.17 represents the above system with parameters L, R, C as the physical characteristics of the system and are reflected in the delayed impulse response $h(t, \tau)$.

When the system parameters such as $R, L,$ and C are constants rather than time dependent, we call it a linear time invariant system.

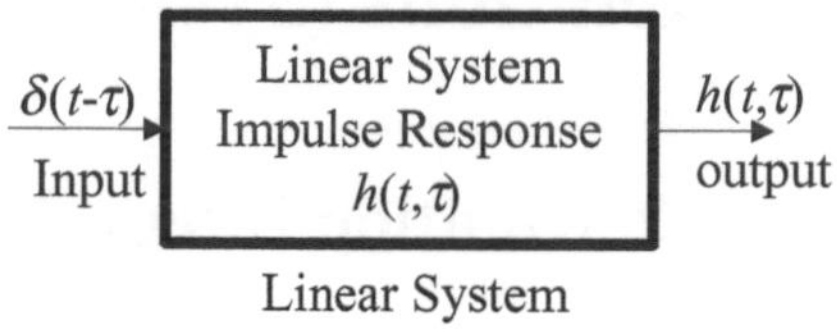

Figure 1.17: Linear System

For a linear time invariant (LTI) system

$$h(t, \tau) = h(t - \tau)$$

Important: When the parameters are time dependent, $h(t, \tau) \neq h(t - \tau)$.

The delayed impulse response $h(t, \tau)$, representing the convolution relationship between the input and the output of the sytem can be elaborated as following:

- **Convolution as an integral operator in the time domain** $(-\infty \leq t \leq \infty)$

  Define $\boldsymbol{A}$, on $-\infty \leq t \leq \infty$ as a convolution operator

$$\boldsymbol{A}[\boldsymbol{x}] = \boldsymbol{y}$$

  such that

$$\boldsymbol{A}[\boldsymbol{x}(t)] = y(t) = \int_{-\infty}^{+\infty} h(t, \tau)x(\tau)\, \mathrm{d}\tau$$

  where

$$\boldsymbol{A}[\boldsymbol{\delta}(t)] = \int_{-\infty}^{+\infty} h(t, \tau)\delta(\tau)\, \mathrm{d}\tau = h(t)$$

$$\boldsymbol{A}[\boldsymbol{\delta}(t - \tau)] = \int_{-\infty}^{+\infty} h(t, \sigma)\delta(\sigma - \tau)\, \mathrm{d}\sigma = h(t, \tau)$$

The output $y(t)$ is defined as the convolution operation of the input $x(t)$ and the impulse response $h(t)$.

$$A[\boldsymbol{x}] \equiv \boldsymbol{x} * \boldsymbol{h} = \boldsymbol{h} * \boldsymbol{x} = \boldsymbol{y}$$

where

$$y(t) = \int_{-\infty}^{+\infty} h(t, \tau)x(\tau)\,\mathrm{d}\tau$$

$\boldsymbol{x}$ and $\boldsymbol{y}$ are vector symbols for $\boldsymbol{x}(t)$ and $\boldsymbol{y}(t)$ respectively.

" * " represents the operation of convolution between two operands $\boldsymbol{x}$ and $\boldsymbol{h}$ (or $\boldsymbol{h}$ and $\boldsymbol{x}$).

When the system is invariant,

$$y(t) = \int_{-\infty}^{+\infty} h(t-\tau)x(\tau)\,\mathrm{d}\tau$$

- **Derivation of the relationship between the input $v(t)$, system impulse response $h(t)$ and the system output $q(t)$**

  Let the input $\delta(t-\tau)$ result in system response $h(t, \tau)$.

  The input function $v(t)$ can be expanded as

$$v(t) = \int_{\infty}^{\infty} v(\tau)\delta(t-\tau)d\tau \approx \sum_{n=-\infty}^{\infty} v(n\Delta\tau)\delta(t-n\Delta\tau)\Delta\tau$$

If the input $v(n\Delta\tau)\delta(t-n\Delta\tau)\Delta\tau$ is applied to the system then the resultant output of this linear system is $v(n\Delta\tau)h(t, n\Delta\tau)\Delta\tau$.

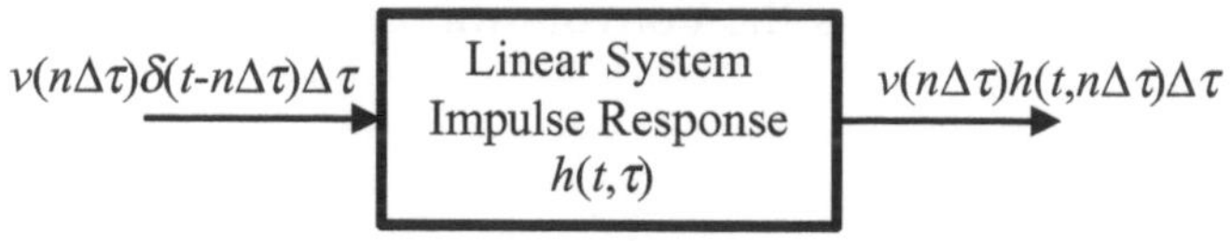

Figure 1.18: Dynamic System Response to $\delta(t - n\Delta\tau$ Input

Invoking the linearity property, the resulting input–output relation is:

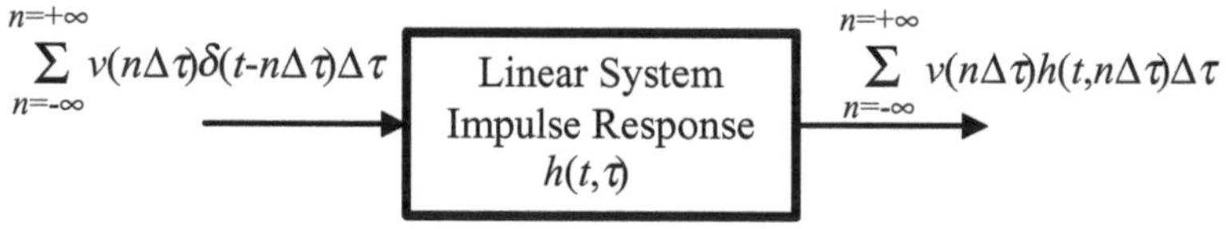

Figure 1.19: Dynamic Electrical Response

In the limit as $\Delta\tau \to \mathrm{d}\tau$ and the summation changes to integration, the input–output relationship takes the form:

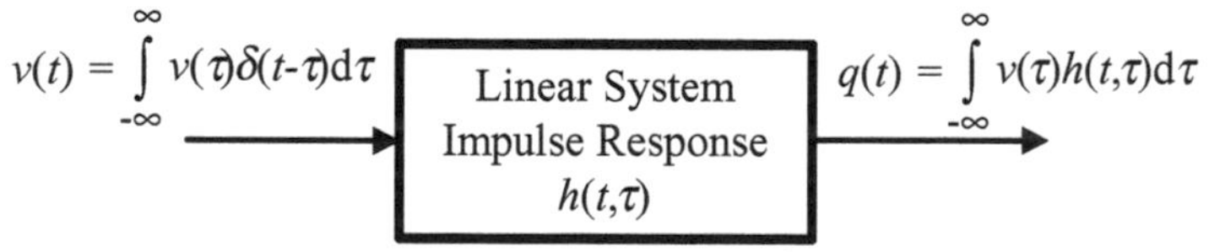

Figure 1.20: Input–Output Dynamic System Response

Summary:

$$\text{System Input} \qquad v(t) = \int_{-\infty}^{\infty} v(\tau)\delta(t-\tau)\,\mathrm{d}\tau$$

$$\text{System Output} \qquad q(t) = \int_{-\infty}^{\infty} v(\tau)h(t,\tau)\,\mathrm{d}\tau$$

$$h(t,\tau)|_{\tau=0} = h(t) \qquad \text{the impulse response}$$

The above equation is known as **Superposition Integral** (or **Convolution integral**).

The delayed impulse response $h(t, \tau)$ uniquely represents the system characteristics for a linear system whose parameters such as resistance, inductance, capacitance, mass etc., may be functions of time or constants.

Summarizing: For LTI Systems only

$$h(t,\tau) = h(t-\tau)$$

System Output

$$q(t) = \int_{\infty}^{\infty} h(t-\tau)v(\tau)\,\mathrm{d}\tau = \int_{\infty}^{\infty} h(\tau)v(t-\tau)\,\mathrm{d}\tau \quad \text{for the system input } v(t)$$

$$q(t) = h(t) * v(t) \quad \text{for system input } v(t)$$

" * " represents the convolution operation.

Another notation for convolution of two signals is

$$f_1 \otimes f_2 = f_1 * f_2$$

"⊗" is used when convolution is performed in more than one-dimension.

Time Invariant Convolution Operator

- **Noncausal System—Noncausal Input**

$$q(t) = h(t) {}^{*}v(t) = \int_{-\infty}^{+\infty} h(t-\tau)v(\tau)\,\mathrm{d}\tau \qquad \text{Noncausal output}$$

- **Causal System—Noncausal Input**

$$h(t-\tau) = 0 \qquad (t-\tau) < 0$$

$$q(t) = h(t) {}^{*}v(t) = \int_{-\infty}^{t} h(t-\tau)v(\tau)\,\mathrm{d}\tau \qquad \text{Noncausal output}$$

- **Causal System—Causal Input**

$$h(t-\tau) \equiv 0 \qquad (t-\tau) < 0$$

$$q(\tau) \equiv 0 \qquad \tau < 0$$

$$q(t) = \int_{0}^{t} h(t-\tau)v(\tau)\,\mathrm{d}\tau \qquad \text{Causal output}$$

- **Convolution operation properties**

The following convolution properties can be easily proven:

1. **Commutative Property**

   Convolution operation is commutative.

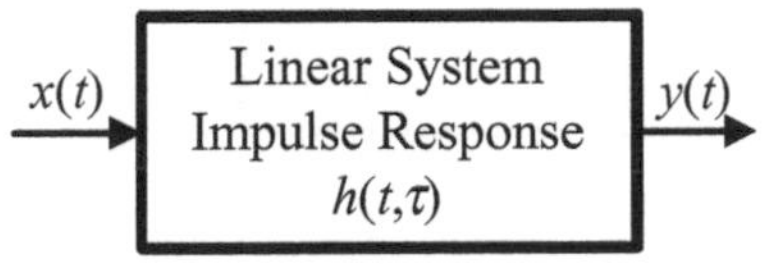

Figure 1.21: Time Varying Linear System

$$y(t) = h(t) * x(t) = x(t) * h(t) = \int_{-\infty}^{\infty} h(\tau)x(t,\tau)d\tau = \int_{-\infty}^{\infty} x(\tau)h(t,\tau)d\tau$$

2. **Associative Property, Cascaded Systems**

   Convolution operations are associative.

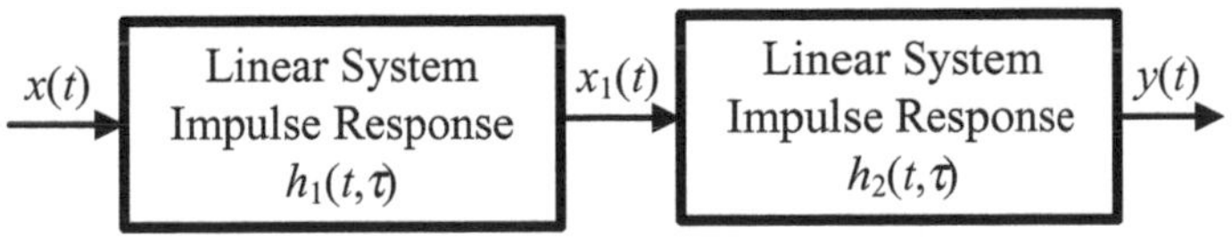

Figure 1.22: Linear Time Varying Cascaded Systems

$$\begin{aligned} y(t) &= x_1 * h_2(t) = (x(t) * h_1(t)) * h_2(t) \\ &= x(t) * (h_1(t) * h_2(t)) = (h_1(t) * h_2(t)) * x(t) \end{aligned}$$

where

$$h_1(t) * h_2(t) = \int_{-\infty}^{\infty} h_1(\tau_1) h_2(t, \tau_1)\, d\tau_1 = h(t)$$

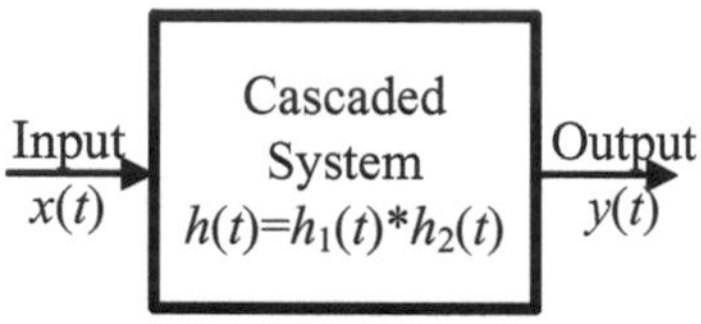

Figure 1.23: Cascaded System—Convolved

$$y(t) = (h_1(t) * h_2(t)) * x(t) = \int_{-\infty}^{\infty}\int_{-\infty}^{\infty}\int_{-\infty}^{\infty} h_1(\tau_1) h_2(\tau_2, \tau_1) x(t, \tau_2)\, d\tau_1\, d\tau_2$$

3. **Distributive Property**

$$(x * y) + (z * y) = (x + z) * y$$

4. **Derivative Property**

$$\frac{d}{dt}(h(t) * x(t)) = \frac{d}{dt} h(t) * x(t) + h(t) * \frac{d}{dt} x(t)$$

5. **Area under Convolution**

   The area under convolution is the product of the areas under each function.

$$\int_{-\infty}^{\infty} h(t) * x(t)\, dt = \left[\int_{-\infty}^{\infty} h(t)\, dt\right]\left[\int_{-\infty}^{\infty} x(t)\, dt\right]$$

6. A very interesting characterization of **impulse function** is **convolution**

$$\delta(t) = \left(\frac{-1}{\sqrt{(\pi)}t}\right) * \left(\frac{1}{\sqrt{(\pi)}t}\right) = -\frac{1}{\pi}\int_{-\infty}^{+\infty} \frac{1}{\tau(t-\tau)}\,\mathrm{d}\tau$$

This integral can be evaluated via contour integration involving Cauchy's residues theorem. We shall evaluate this integral in Chapter 4, on complex variables. The integrand is a complex function that involves two poles on the real axis at $\tau = 0$ and $\tau = t$. When $t \neq 0$, the residues of the poles cancel each other. On the other hand, for $t = 0$, there is a double pole at the origin resulting in infinite value.

Example 1.8:

Given

$$h_1(t) = p_{T_1}\left(t - \frac{T_1}{2}\right)$$
$$h_2(t) = p_{T_2}\left(t - \frac{T_2}{2}\right) \qquad T_2 > T_1$$

Required:

$$f(t) = h_1(t) * h_2(t) = \int_{-\infty}^{\infty} h_1(t-\tau)h_2(\tau)\,\mathrm{d}\tau$$

Solution:

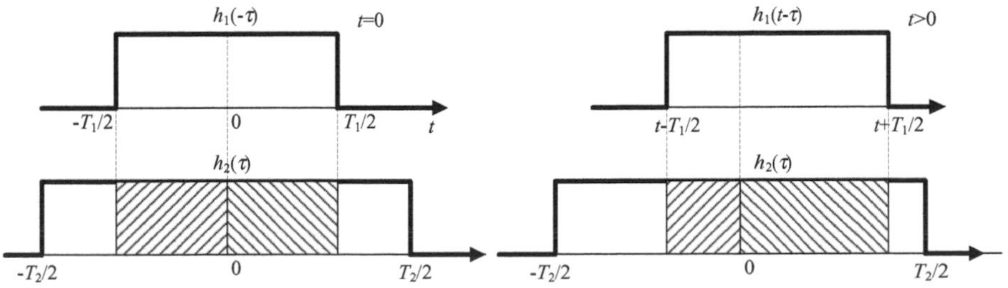

Figure 1.24: Time Convolution of Two Pulses Shifted by Time, T

The reader can verify that:

$$\begin{aligned}
f(t) &= 0, && t \leq 0 \\
f(t) &= t, && 0 \leq t < T_1 \\
f(t) &= 1, && T_1 \leq t \leq T_2 \\
f(t) &= 1 - \frac{T_2 - t}{T_1}, && T_1 < t \leq T_1 + T_2 \\
f(t) &= 0, && T_1 + T_2 < t
\end{aligned}$$

Note: If

$$\text{duration of the pulse } f_1(t) = T_1$$
$$\text{duration of the pulse } f_2(t) = T_2$$

Then

$$\text{duration of the convolution } (f_1(t) * f_2(t)) = (f_2(t) * f_1(t)) = T_1 + T_2$$

Example 1.9:

$$\left.\begin{aligned}
h_1(t) &= \frac{1}{\sqrt{2\pi\sigma_1^2}} e^{-(t-m_1)^2/2\sigma_1^2} \\
h_2(t) &= \frac{1}{\sqrt{2\pi\sigma_2^2}} e^{-(t-m_2)^2/2\sigma_2^2}
\end{aligned}\right\} \text{(Gaussian Signals)}$$

After some tedious algebra, it can be seen that:

$$h_1(t) * h_2(t) = \frac{1}{\sqrt{2\pi\sigma^2}} e^{-(t-m)^2/2\sigma^2}$$

where

$$m = m_1 + m_2 \quad \text{(mean of the convolved signal – sum)}$$

$$\sigma^2 = \sigma_1^2 + \sigma_2^2 \quad \text{(variance of the convolved signal – sum)}$$

Hint:

$$\int_{-\infty}^{+\infty} e^{-(ax^2+bx+c)}\,\mathrm{d}x = \frac{\sqrt{2\pi}}{a} e^{-(c-b^2/4a)}$$

- **2-D Space Convolution**

$$f(x,y) = h_1(x,y) \otimes h_2(x,y) = \int_{-\infty}^{+\infty}\int_{-\infty}^{+\infty} h_1(\xi,\eta)h_2(x-\xi,y-\eta)\,\mathrm{d}\xi\,\mathrm{d}\eta$$

- **Discrete System Convolution**

For one-dimensional systems

$$f(n) = h(n) * x(n) = \sum_{m=-\infty}^{+\infty} h(m)x(n-m) = \sum_{m=-\infty}^{+\infty} h(n-m)x(m)$$

For two-dimensional systems

$$f(n,m) = h_1(n,m) \otimes h_2(n,m) = \sum_{j=-\infty}^{\infty}\sum_{k=-\infty}^{\infty} h_1(j,k)h_2(n-j,m-k)$$

Example 1.10

Let $u(t)$ be a step input to a system with impulse response $h(t)$.

The step response $y(t)$ of the system is

$$y(t) = u(t) * h(t) = \int_{-\infty}^{+\infty} u(\tau)h(t-\tau)\,\mathrm{d}\tau = \int_{0}^{\infty} h(t-\tau)\,\mathrm{d}\tau$$

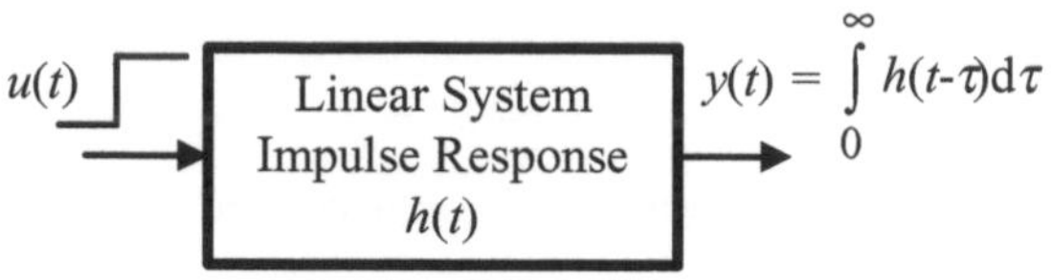

Figure 1.25: LTI Step Response

If the system is causal, then the step response $y(t)$ is

$$y(t) = \int_{-\infty}^{\infty} h(t-\tau)\,\mathrm{d}\tau = \int_{-\infty}^{t} h(\tau)\,\mathrm{d}\tau = \int_{0}^{t} h(\tau)\,\mathrm{d}\tau$$

Example 1.11:

Let the input to the system with impulse response $h(t)$ be

$$f(t) = u(t) - u(t-\tau) \quad \text{a pulse centered at } t = T/2.$$

The output $y(t)$ is

$$y(t) = \int_{0}^{\infty} h(t-\tau)\,[u(\tau) - u(\tau - T)]\,\mathrm{d}\tau$$

or

$$y(t) = \int_{0}^{t} h(\tau)\,\mathrm{d}\tau - \int_{T}^{t} h(\tau)\,\mathrm{d}\tau = \int_{0}^{t-T} h(\tau)\,\mathrm{d}\tau$$

Example 1.12:

Consider the following circuit with step input.

The differential equation governing the circuit is

$$\frac{\mathrm{d}y(t)}{\mathrm{d}t} + \frac{R}{L}y(t) = u(t), \qquad y(0) = 0$$

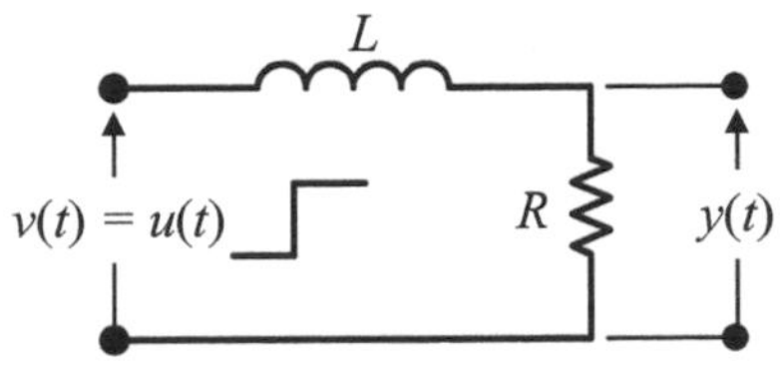

Figure 1.26: Circuit with Step Input

To compute the impulse response, we solve

$$\frac{\mathrm{d}}{\mathrm{d}t}h(t) + \frac{R}{L}h(t) = \delta(t), \qquad h(0) = 0$$

It is easy to see that (in Chapter 3)

$$h(t) = e^{-(R/L)t}$$

The response to the step input is

$$y(t) = u(t) * h(t) = \int_0^t e^{-(R/L)\tau}\,\mathrm{d}\tau = \frac{L}{R}\left[1 - e^{-(L/R)t}\right]$$

Example 1.13:

Consider the following discrete system in which two subsystems with impulse response $h_1(n)$ and $h_2(n)$ are cascaded.

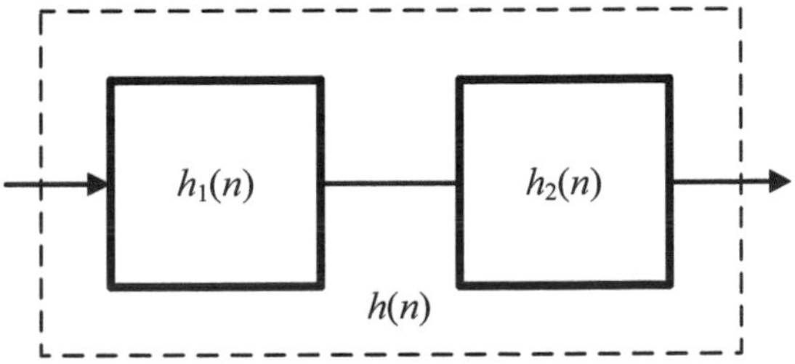

Figure 1.27: Cascaded Systems

$$h_1(n) = \alpha_1\delta(n) + (\beta_1)^n\, u(n) \qquad \|\beta_1\| < 1, \quad h_1(n) = 0, \quad n < 0$$

$$h_2(n) = (\beta_2)^n\, u(n) \qquad \|\beta_2\| < 1, \quad h_2(n) = 0, \quad n < 0$$

$$h(n) = h_1(n) * h_2(n) = \sum_{m=0}^{\infty} \alpha_1\beta_2^{n-m}\delta(m)u(m-n) + \sum_{m=0}^{\infty} \beta_1^m\beta_2^{n-m}u(m)u(m-n)$$

or

$$h(n) = \alpha_1\beta_2^n + \beta_2^n \sum_{m=0}^{n} \left(\frac{\beta_1}{\beta_2}\right)^m = \beta_2^m \left[\alpha_1 + \frac{1-\left(\frac{\beta_1}{\beta_2}\right)^n}{1-(\beta_1/\beta_2)}\right] u(n)$$

- **Impulse Response, Transfer Function, Frequency Response, and Their Relationships**

Let $\boldsymbol{A}$ be an operator that acts upon a function $x(t)$, $-\infty < t < \infty$, resulting in an output $y(t)$ such that

$$\boldsymbol{A}[x(t)] = y(t)$$

From a set of functions $x(t)$, let us choose a **particular function** $x(t)$ such that

$$\boldsymbol{A}[x(t)] = \lambda x(t), \qquad \lambda \text{ being a scalar}$$

Such a function $x(t)$ has a very special "connection" with the operator $\boldsymbol{A}$ and is called the **eigenfunction** of the operator $\boldsymbol{A}$, belonging to the eigenvalue λ. Essentially, the operator $\boldsymbol{A}$ only changes the "length" of the function $x(t)$ without changing its "orientation." We are interested in the eigenfunctions and the corresponding eigenvalues of the Impulse Response operator. To achieve this objective, consider an input function $x(t) = e^{st}$, to a system having an Impulse Response $h(t)$.

The resultant output $y(t)$ of this system is :

$$y(t) = \int_0^{\infty} h(\tau)e^{s(t-\tau)}\, \mathrm{d}\tau = \left[\int_0^{\infty} h(\tau)e^{-s\tau}\, \mathrm{d}\tau\right] e^{st}$$

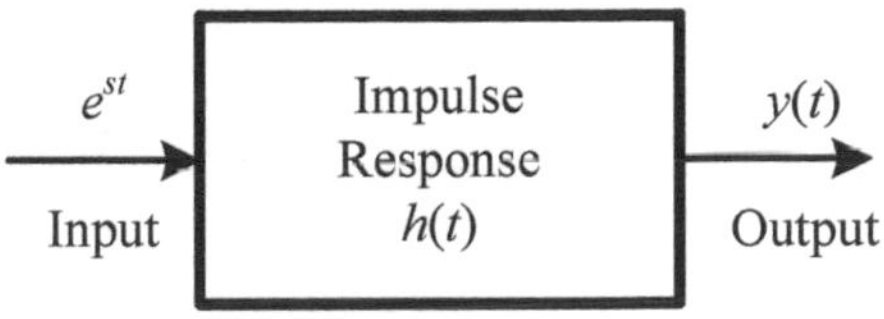

Figure 1.28: Eigenfunctions of the Impulse Response as a Transfer Function

Let

$$\int_0^\infty h(\tau)e^{-s\tau}\,\mathrm{d}\tau = \int_0^\infty h(t)e^{-st}\,\mathrm{d}t = H(s)$$

Thus,

$$y(t) = \mathbf{A}[e^{st}] = H(s)e^{st}$$

The function $H(s)$ is called the **Transfer Function** of the system with impulse response $h(t)$. Thus, Transfer Function $H(s)$ is the eigenvalue of the system impulse response operator $h(t)$ with the eigenfunction e^{st}. This is an extremely interesting result.

In fact,

The function e^{st} is the only eigenfunction of a linear time invariant system with impulse response $h(t)$ and the transfer function $H(s)$ is the corresponding eigenvalue. Variable s can take real or complex values.

Continuing the same reasoning, $e^{j\omega t}$ is the only eigenfunction of the operator $h(t)$ with the eigenvalue $H(j\omega)$, where

$$H(j\omega) = \int_0^\infty h(t)e^{-j\omega t}\,\mathrm{d}t \quad \Bigg\} \begin{array}{l}\text{Frequency Response of the System}\\ \text{with impulse response } h(t).\end{array}$$

We shall elaborate on these concepts in the Chapter 2.

- **Stable LTI systems**

For stable LTI systems, bounded input results in a bounded output.

Thus, if

$$h(t) = \text{Impulse response of a system}$$

$$x(t) = \text{Input to the system such that } |x(t)| \leq M < \infty$$

$$y(t) = \text{Output of the system} |y(t)| < \infty$$

Such an LTI system is called **stable**, implying:

$$\int_{-\infty}^{+\infty} |h(t)| \, \mathrm{d}t < \infty$$

This can be seen as:

$$y(t) = \int_{-\infty}^{+\infty} h(\tau)x(t-\tau) \, \mathrm{d}\tau$$

or

$$\begin{aligned} |y(t)| &\leq \int_{-\infty}^{+\infty} |h(\tau)x(t-\tau)| \, \mathrm{d}\tau \\ &\leq \int_{-\infty}^{+\infty} |h(\tau)||x(t-\tau)| \, \mathrm{d}\tau \\ &\leq M \int_{-\infty}^{+\infty} |h(\tau)| \, \mathrm{d}\tau \end{aligned}$$

Hence, LTI stable systems have the following property:

$$\int_{-\infty}^{+\infty} |h(t)| \, \mathrm{d}t < \infty$$

1.5 Signal Representation via Linear Vector Spaces

Linear vector spaces represent a rich structure for studying linear systems. The signals $x(t)$ and $y(t)$ are represented as vectors $\boldsymbol{x}$ and $\boldsymbol{y}$ in the vector spaces and the operation of the system is a mathematical operator or a "mapping" from one space to another space or onto itself. System analysis and synthesis requires optimal approximation with respect to (w.r.t) some desired output performance index. For this purpose, we need to study the signal strength, its energy, and the frequency contents along with its amplitude content and phase information. A natural setting for such problems arising in communication and control systems is the mathematical structure of the linear vector spaces that provide a geometrical interpretation. The time signals are viewed as "vectors" or "geometric objects" moving with time, in this vector space. When the additional structure such as "metric" or distance with "completeness" (Banach) and "inner product with completeness" (Hilbert) is added, the vector space becomes useful.

As the chapter develops, we shall see that the **Fourier Series is the key to the understanding of the linear vector spaces**.

1.5.1 Definition of a Vector Space

A vector space $\boldsymbol{V}$ over a field $\mathcal{F}$, is defined as a set of objects (known as vectors) that is "closed" under the operations of **vector addition** and **scalar multiplication**. Essentially, it implies that the resulting objects after being operated upon have the same properties as the original set. Scalars are members of the field $\mathcal{F}$. A typical example of a vector for our purpose is a time function that can be viewed as a moving point in multi-dimensional or infinite-dimensional spaces. Another simple example of a vector space is a set of real or complex numbers.

We define bold letters, $\boldsymbol{u}, \boldsymbol{v}, \boldsymbol{w}$ as the elements of a vector space $\boldsymbol{V}$ and α, β as the scalar elements of a field $\mathcal{F}$. With each vector, we associate a "magnitude" and a "direction." Let us start with the definition of a field.

Fields

A field $\mathcal{F}$ is a basic algebraic structure with the following properties:

(i) Operation of addition (+)

(ii) Operation of multiplication $(\cdot)$ and division $(\cdot)^{-1}$

(iii) Identity 1 and zero

In addition to these properties the elements a, b, c, etc., of the field $\mathcal{F}$, satisfy the following axioms:

1. **Additive Associativity:** $a + (b + c) = (a + b) + c$
2. **Additive Commutativity:** $a + b = b + a$
3. **Left Distributivity:** $a \cdot (b + c) = a \cdot b + a \cdot c$
4. **Additive Identity:** $a + 0 = 0 + a = a$
5. **Additive Inverse:** $a + (-a) = 0$
6. **Multiplication with Zero:** $a \cdot 0 = 0$
7. **Multiplicative Associativity:** $a \cdot (b \cdot c) = (a \cdot b) \cdot c$
8. **Multiplicative Commutativity:** $a \cdot b = b \cdot a$
9. **Multiplicative Identity:** $a \cdot 1 = 1 \cdot a = a$
10. **Multiplicative Inverse:** $a\,(a)^{-1} = (a)^{-1}a = 1, \quad a \neq 0$

A subset of the field that does not satisfy the last five multiplicative axioms is called a "**ring**." Real numbers $\mathcal{R}$, rational numbers Q and complex numbers C are well-known fields.

Vectors

The following axioms govern the vector operations:

1. **Addition Commutativity:** $\boldsymbol{u} + \boldsymbol{v} = \boldsymbol{v} + \boldsymbol{u}$

2. **Addition Associativity:** $(\boldsymbol{u} + \boldsymbol{v}) + \boldsymbol{w} = \boldsymbol{u} + (\boldsymbol{v} + \boldsymbol{w})$

3. **Scalar Multiplication Associativity:** $\alpha(\beta \boldsymbol{v}) = (\alpha\beta)\boldsymbol{v}$

4. **Null Vector:** $\mathbf{0} + \boldsymbol{v} = \boldsymbol{v} + \mathbf{0} = \boldsymbol{v}$

5. **Addition Inverse:** $\boldsymbol{v} + (-\boldsymbol{v}) = 0$

   These five axioms generate a general vector space, which has limited applications. Most of the vector spaces that are useful in the signal analysis have the following additional structural properties.

6. **Multiplication:** Multiplication of two vectors $\boldsymbol{u}$ and $\boldsymbol{v}$ defined as $\boldsymbol{uv}$ does not have any meaning but requires additional structure such as "dot" (inner product) and "cross" (vector product) discussed below.

7. **Metric** or distance function $\mathrm{d}(\cdot,\cdot)$ with the following properties:

   a. $\mathrm{d}(\boldsymbol{v}, \boldsymbol{u}) = \mathrm{d}(\boldsymbol{u}, \boldsymbol{v}) \geq 0$ for $\boldsymbol{v}, \boldsymbol{u}$ in $\boldsymbol{V}$, also

      $\mathrm{d}(\boldsymbol{v}, \boldsymbol{u}) = 0$ if and only if $\boldsymbol{v} = \boldsymbol{u}$;

   b. $\mathrm{d}(\boldsymbol{v}, \boldsymbol{w}) \leq \mathrm{d}(\boldsymbol{v}, \boldsymbol{u}) + \mathrm{d}(\boldsymbol{u}, \boldsymbol{w})$ {triangular inequality} for all $\boldsymbol{v}, \boldsymbol{w}$, and $\boldsymbol{u}$ in $\boldsymbol{V}$.

      The distance function $d(\cdot,\cdot)$ is called the **metric of the space. It is a measure of "distance" between two vector space elements, a nonnegative number. In general, different spaces may have many different metrics.**

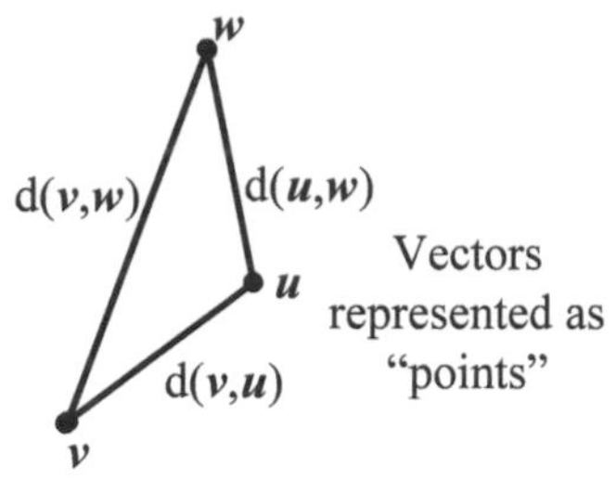

Figure 1.29: Metric Space, Vectors Represented as "Points"

Exercise 1.1:

$V = \{\boldsymbol{u}, \boldsymbol{v}, \boldsymbol{w}\}$ be a space of three elements:

1) $d(\boldsymbol{u}, \boldsymbol{v}) = 1,\ d(\boldsymbol{v}, \boldsymbol{w}) = 10,\ d(\boldsymbol{w}, \boldsymbol{u}) = 5$

   Verify that $\boldsymbol{v}$ is not a metric space (hint: fails triangular inequality).

2) $d(\boldsymbol{u}, \boldsymbol{v}) = 3,\ d(\boldsymbol{v}, \boldsymbol{w}) = 4,\ d(\boldsymbol{w}, \boldsymbol{u}) = 5$

   Verify that $\boldsymbol{v}$ is a metric space.

Example 1.14:

Consider $x - y$ plane with points (x_1, y_1) and (x_2, y_2).

Let $V = \{(x_1, y_1), (x_2, y_2)\}$

Define a metric $d((x_1, y_1), (x_2, y_2)) = \max\{|x_1 - x_2|, |y_1 - y_2|\}$

It is easy to see that this is a metric space.

8. **Norm of a vector**

   A norm on a vector space $\boldsymbol{V}$ is a linear functional of its element $\boldsymbol{v}$ and is denoted by $||\boldsymbol{v}||$. It has the following three properties:

   a. $||\boldsymbol{v}|| \geq 0 \Rightarrow$ Positive definiteness;

   b. $||\alpha\boldsymbol{v}|| = |\alpha|||\boldsymbol{v}|| \Rightarrow$ Homogeniety;

   c. $||\boldsymbol{v} + \boldsymbol{u}|| \leq ||\boldsymbol{v}|| + ||\boldsymbol{u}||\{$ triangular inequality$\}$ for all $\boldsymbol{v}$ and $\boldsymbol{u}$ in $\boldsymbol{V}$.

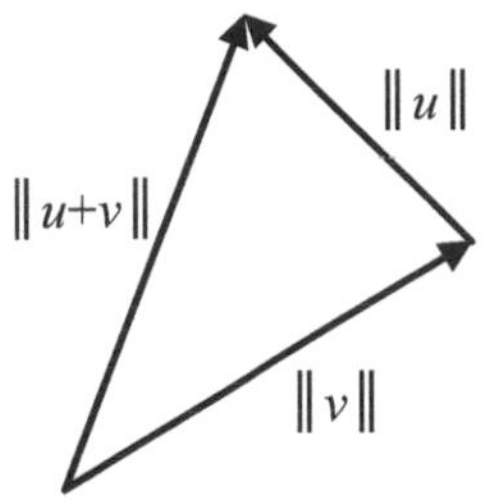

Figure 1.30: Normed Vectors Represented as "Geometric Objects"

The metric was introduced by a famous Polish mathematician, Stephen Banach and is known as the **Banach Space**. Distance and norm look alike but are different in general. There are many different types of norms. In the finite dimensional space, all norms are equivalent for a given space. But in an infinite dimensional space, there can be many different vector norms. Introduction of a norm to a vector space increases its usefulness regarding energy and power calculations depending upon the structure of the norm.

9. **Inner product metric** $(\cdot, \cdot)$**, (inner product on** $V \times V$ **for** $u, v \in V$**)**

   A normed vector space may have an additional structure known as the "inner product," which is a uniquely defined complex number $(\boldsymbol{v}, \boldsymbol{u})$ satisfying the following axioms:

   - $(\boldsymbol{v}, \boldsymbol{u}) = (\bar{\boldsymbol{u}}, \boldsymbol{v})$, $\bar{\boldsymbol{u}}$ is complex conjugate of $\boldsymbol{u}$
   - $(\boldsymbol{v}, \boldsymbol{v}) = ||\boldsymbol{v}|| > 0$
   - $(\alpha\boldsymbol{v}, \boldsymbol{u}) = \bar{\alpha}(\boldsymbol{u}, \boldsymbol{v})$
   - $(\boldsymbol{v}, \alpha\boldsymbol{u}) = \alpha(\boldsymbol{v}, \boldsymbol{u})$
   - $(\boldsymbol{v}, \alpha_1\boldsymbol{u}_1 + \alpha_2\boldsymbol{u}_2) = \alpha_1(\boldsymbol{v}, \boldsymbol{u}_1) + \alpha_2(\boldsymbol{v}, \boldsymbol{u}_2)$
   - $||\boldsymbol{v} + \boldsymbol{u}|| + ||\boldsymbol{v} - \boldsymbol{u}|| = 2||\boldsymbol{v}||^2 + 2||\boldsymbol{u}||^2$ (Parallelogram Law)
   - $|(\boldsymbol{v}, \boldsymbol{u})| \leq ||\boldsymbol{v}||||\boldsymbol{u}||$ (Cauchy–Schwartz Inequality)

- The "inner product" space can be metricized if in addition to the inner product norm the distance between vectors is defined as:
  $d(\boldsymbol{u}, \boldsymbol{v}) = ||\boldsymbol{u} - \boldsymbol{v}||$
- Inner product is very often referred to as a **dot product** or **scalar product**,
  $(\boldsymbol{v}, \boldsymbol{u}) = \boldsymbol{v} \cdot \boldsymbol{u} = |\boldsymbol{v}||\boldsymbol{u}| \cos \angle(\boldsymbol{v}, \boldsymbol{u})$

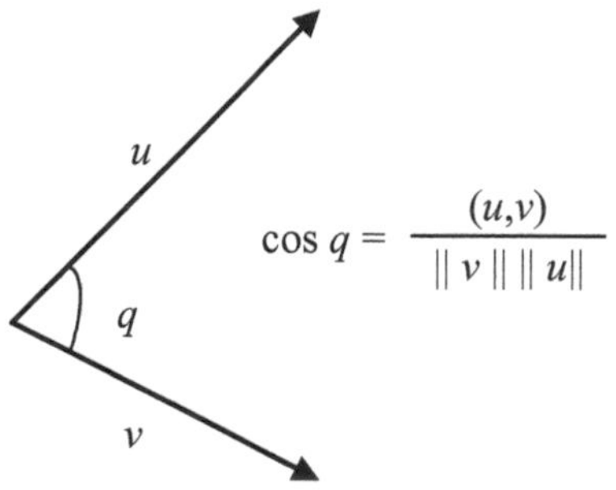

Figure 1.31: Geometric Interpretation of the Inner Product of Two Vectors

Example 1.15

Consider $\boldsymbol{f}$ and $\boldsymbol{g}$ as two vectors in the space $C[a, b]$ of complex valued functions on the interval (a, b). The inner product of these two vectors is defined as:

$$(\boldsymbol{f}, \boldsymbol{g}) = \int_a^b \bar{f}(t) g(t)\, \mathrm{d}t$$

In order for these spaces to be useful in signal study they need another additional property known as "**Cauchy Completeness**" (defined later). We shall show later on that this space $C[a, b]$ is not Cauchy Complete.

Importance of Inner Product Spaces

The properties of a set of time signals can be studied and analyzed in a systematic manner in linear vector spaces known as **signal space** by utilizing geometric arguments. The inner product space is a precursor to the signal space. We introduce the idea of signals as vectors with **magnitude** and **direction** along with the concepts of **inner product** and **orthogonality**. Inner or scalar product of two signals $f_1(t), f_2(t)$ in the following three different spaces is defined as:

- Signals Energy Relation

$$(\boldsymbol{f}_1, \boldsymbol{f}_2) = (f_1(t), f_2(t)) = \lim_{T\to\infty} \int_{-\infty}^{+\infty} \bar{f}_1(t) f_2(t)\, \mathrm{d}t$$

- Signals Power Relation

$$(\boldsymbol{f}_1, \boldsymbol{f}_2) = \lim_{T\to\infty} \frac{1}{2T}(f_1(t), f_2(t)) = \lim_{T\to\infty} \frac{1}{2T} \int_{-\infty}^{+\infty} \bar{f}_1(t) f_2(t)\, \mathrm{d}t$$

- Periodic Signals Power Relation

$$(\boldsymbol{f}_1, \boldsymbol{f}_2) = (f_1(t), f_2(t)) = \frac{1}{T_0} \int_{0}^{T_0} \bar{f}_1(t) f_2(t)\, \mathrm{d}t$$

$$\boldsymbol{f}_1 \to f_1(t), \boldsymbol{f}_2 \to f_2(t)$$

To fully exploit the effectiveness of these vector spaces, we need some additional concepts.

10. **Linear independence of vectors:**

A set of vectors $\boldsymbol{v}_1, \boldsymbol{v}_2, \cdots, \boldsymbol{v}_n$ in a space $\boldsymbol{V}$ are linearly independent if

$$c_1\boldsymbol{v}_1 + c_2\boldsymbol{v}_2 + \cdots + c_n\boldsymbol{v}_n = \boldsymbol{0} \quad \text{is true if and only if}$$

$$c_1 = c_2 = \cdots = c_n = 0$$

Otherwise, the set formed by $\boldsymbol{v}_1, \boldsymbol{v}_2, \cdots, \boldsymbol{v}_n$ is called a **Linearly Dependent Set**. Each vector of the linearly independent set has some "information" not available in the other vectors.

11. **Dimension of a vector space:**

    A vector space has an innumerable number of vectors. If the vector space has a maximum of only n linearly independent vectors, then the dimension of the space is called n. In other words, any vector in the n-dimensional space can be expressed as a linear combination of any n linearly independent vectors in this space.

    When n is finite, we have a finite dimensional space, otherwise the space is infinite dimensional.

Example 1.16: Euclidean n-dimensional space

Consider $\boldsymbol{v}, \boldsymbol{u} \in \boldsymbol{V}$

$$\left.\begin{aligned} \boldsymbol{v} &= (\alpha_1, \alpha_2, \cdots, \alpha_n) \\ \boldsymbol{u} &= (\beta_1, \beta_2, \cdots, \beta_n) \end{aligned}\right\} \leftarrow n\text{-tuples of scalar complex variables}$$

Then

$$\boldsymbol{v} + \boldsymbol{u} = (\alpha_1 + \beta_1, \alpha_2 + \beta_2, \cdots, \alpha_n + \beta_n)$$

$$\gamma \boldsymbol{v} = (\gamma\alpha_1, \gamma\alpha_2, \cdots, \gamma\alpha_n),\ \gamma \text{ is a scalar.}$$

$$(\boldsymbol{v}, \boldsymbol{u}) = \sum_{i=1}^{n} \bar{\alpha}_i \beta_i$$

The vectors in Euclidean space satisfy the following properties associated with a metric space:

$$\mathrm{d}^2(\boldsymbol{v}, \boldsymbol{u}) = (\boldsymbol{v} - \boldsymbol{u}, \boldsymbol{v} - \boldsymbol{u}) = \|\boldsymbol{v} - \boldsymbol{u}\|^2$$

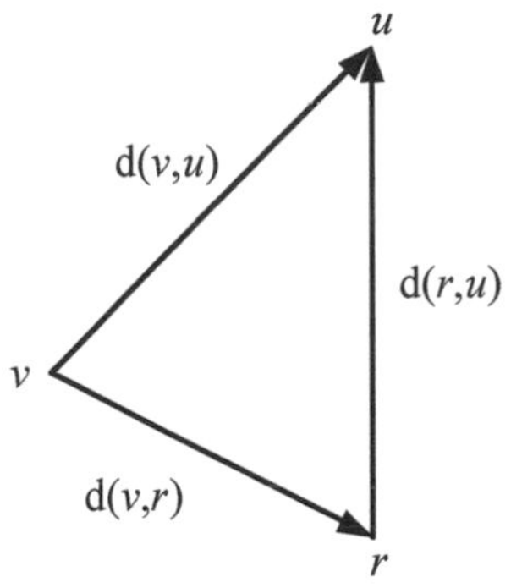

Figure 1.32: Euclidean Metric Triangular Inequality

$$(\boldsymbol{v}, \boldsymbol{u}) = 0 \Rightarrow \boldsymbol{v} \perp \boldsymbol{u}, \quad (\boldsymbol{v}, \boldsymbol{u} \text{ orthogonal})$$

$$\cos(\boldsymbol{v}, \boldsymbol{u}) = \frac{(\boldsymbol{v}, \boldsymbol{u})}{(\boldsymbol{v}, \boldsymbol{v})(\boldsymbol{u}, \boldsymbol{u})}$$

$$(\boldsymbol{v}, \boldsymbol{u}) = ||\boldsymbol{v}||\, ||\boldsymbol{u}|| \cos(\boldsymbol{v}, \boldsymbol{u})$$

$$\mathrm{d}(\boldsymbol{v}, \boldsymbol{u}) \leq \mathrm{d}(\boldsymbol{v}, \boldsymbol{r}) + \mathrm{d}(\boldsymbol{r}, \boldsymbol{u})$$

12. **Basis vectors of a space and its coordinate representation** (Fig. 1.33)
    Consider an n-dimensional vector space $\boldsymbol{V}$. Any collection of the n linearly independent vectors $\boldsymbol{e}_1, \boldsymbol{e}_2, \cdots, \boldsymbol{e}_n$, form a set of basis vectors (just like unit orthogonal vectors *i, j, k* in three- dimensional space). Such a set is known as the span of the vector space $\boldsymbol{V}$. Basis vectors in a space play the role of general coordinates. **No basis vector can be written as a linear combination of other basis vectors**. Any general vector $\boldsymbol{x}$ in the space $\boldsymbol{V}$ can be represented as a linear combination of these basis vectors. Thus,

$$\boldsymbol{x} = \sum_{i=1}^{n} x_i \boldsymbol{e}_i \tag{1.22}$$

As discussed earlier, the symbol $\boldsymbol{x}$ can represent a "vector" or a "point" in n-dimensional space (both visualizations are useful). The vectors, $\boldsymbol{e}_i$, $i = 1, 2, \cdots, n$ can be thought of as coordinates and x_i as the components (projections) of $\boldsymbol{x}$ along the coordinate axis $\boldsymbol{e}_i$.

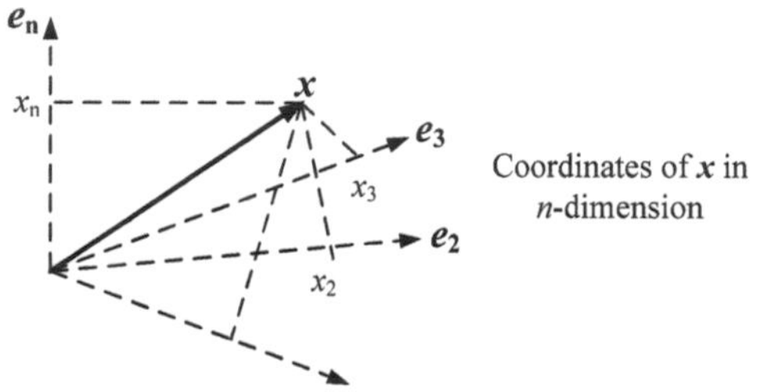

Figure 1.33: Coordinate Representation of a Vector

In general, $\boldsymbol{e}_i$ need not be orthogonal but still must be **independent**. Later on we shall show how the basis vectors can be orthonormalized.

Example 1.17:

Show that a set of complex numbers form a vector space.

Let $\boldsymbol{x}$ be a set of complex numbers in infinite dimensional space. Basis vectors $\boldsymbol{e}_i$ are:

$$\boldsymbol{e}_i = \begin{bmatrix} 0 \cdots 0 \; 1 \; 0 \cdots 0 \end{bmatrix}$$

$$\uparrow$$

i-th component

Then

$$\boldsymbol{x} = \sum_{i=1}^{\infty} x_i \boldsymbol{e}_i, \text{ where } x_i \text{ are complex numbers.}$$

13. **Orthonormal Basis**

Two vectors $\boldsymbol{x}$ and $\boldsymbol{y}$ are called orthogonal if

$$(\boldsymbol{x}, \boldsymbol{y}) = 0 \quad \text{and represented as } \boldsymbol{x} \perp \boldsymbol{y}. \tag{1.23}$$

This represents the single most important concept in signal theory emphasizing thc fact that the two signals represented by these vectors are absolutely uncorrelated.

Consider a basis vector set $\boldsymbol{e}_1, \boldsymbol{e}_2, \cdots, \boldsymbol{e}_n$ is such that

$$(\boldsymbol{e}_i, \boldsymbol{e}_j) = \delta_{ij} = \begin{cases} 0 & i \neq j \\ 1 & i = j \end{cases} \quad \text{Orthonormal Set} \tag{1.24}$$

The set is called an orthonormal basis.

For the orthonormal basis

$$\boldsymbol{x} = \sum_{i=1}^{n} x_i \boldsymbol{e}_i$$
$$x_i = (\boldsymbol{x}, \boldsymbol{e}_i)$$

Important: Basis vectors need not be orthogonal.

14. **Dual Space**

Corresponding to a n-dimensional vector space $\boldsymbol{V}$ with a general basis $\{\boldsymbol{h}_i\}_{i=1}^n$, there exists a dual space $\boldsymbol{V}^*$ with a basis $\{\boldsymbol{\gamma}_i\}_{i=1}^n$ such that

$$(\boldsymbol{h}_i, \boldsymbol{\gamma}_j) = \delta_{ij} \tag{1.25}$$

Thus, the basis vector of $\boldsymbol{V}$ are orthonormal to the basis vector of $\boldsymbol{V}^*$.

Thus, if $\boldsymbol{x} \in \boldsymbol{V}$ and $\boldsymbol{y} \in \boldsymbol{V}^*$, we can write

$$\boldsymbol{x} = \sum_{i=1}^{n} x_i \boldsymbol{h}_i$$

$$\boldsymbol{y} = \sum_{i=1}^{n} y_i \boldsymbol{\gamma}_i = \sum_{j=1}^{n} y_j \boldsymbol{\gamma}_j$$

$$(\boldsymbol{x}, \boldsymbol{y}) = \sum_{i=1}^{n} \sum_{j=1}^{n} \bar{x}_i y_j (\boldsymbol{h}_i, \boldsymbol{\gamma}_j) = \sum_{i=1}^{n} \bar{x}_i y_i \tag{1.26}$$

$$(\boldsymbol{x}, \boldsymbol{x}) = \sum_{i=1}^{n} |\boldsymbol{x}|^2 = \sum_{i=1}^{n} |(\boldsymbol{x}, \boldsymbol{h}_i)|^2 = ||\boldsymbol{x}||^2 \quad \textbf{Parseval Equality}$$

Furthermore, if $\{\boldsymbol{h}_i\}_{i=1}^{n}$ is a orthonormal set, then

$$\boldsymbol{h}_i = \boldsymbol{\gamma}_i = \boldsymbol{e}_i$$

Example 1.18:

Consider an n-dimensional space **V** in which $f_1(t), f_2(t), ..., f_n(t)$ are linearly independent. Since any (n+1) functions in this space are **linearly dependent**, a general vector $\boldsymbol{f} \in \mathbf{V}$ takes the form:

$$\mathbf{f} = \sum_{i=1}^{n} \lambda_i \mathbf{f_i}$$

$$f(t) = \sum_{i=1}^{n} \lambda_i f_i(t)$$

$$\begin{bmatrix} \lambda_1 \\ \lambda_2 \\ \vdots \\ \lambda_n \end{bmatrix} = \begin{bmatrix} (\mathbf{f}_1, \mathbf{f}_1) & (\mathbf{f}_2, \mathbf{f}_1) & \cdots & (\mathbf{f}_n, \mathbf{f}_1) \\ (\mathbf{f}_1, \mathbf{f}_2) & (\mathbf{f}_2, \mathbf{f}_2) & \cdots & (\mathbf{f}_n, \mathbf{f}_2) \\ \vdots & & & \\ (\mathbf{f}_1, \mathbf{f}_n) & (\mathbf{f}_2, \mathbf{f}_n) & \cdots & (\mathbf{f}_n, \mathbf{f}_n) \end{bmatrix}^{-1} \begin{bmatrix} (\mathbf{f}, \mathbf{f}_1) \\ (\mathbf{f}, \mathbf{f}_2) \\ \vdots \\ (\mathbf{f}, \mathbf{f}_n) \end{bmatrix}$$

$$\text{where} \quad (\boldsymbol{f}_i, \boldsymbol{f}_j) = \int_a^b \bar{f}_i(t) f_j(t)\, \mathrm{d}t$$

The determinant of the above **Gram matrix** is called the **Gram Determinant**. For linearly independent set of functions it is necessary and sufficient that this Gram determinant be nonzero.

Example 1.19: **Bessel's Inequality**

Let **V** be an inner product space with orthonormal set $\boldsymbol{e}_i$, $i = 1, 2, \ldots\infty$, then

$$\sum_{i=1}^{n} |x_i|^2 = \sum_{i=1}^{n} |(\boldsymbol{x}, \boldsymbol{e}_i)|^2 \le ||\boldsymbol{x}||^2 \qquad \text{(known as Bessel's inequality)}$$

Proof:

$$\mathbf{I} = \left(\left(\boldsymbol{x} - \sum_{i=1}^{n} (\boldsymbol{x}, \boldsymbol{e}_i)\boldsymbol{e}_i \right), \left(\boldsymbol{x} - \sum_{i=1}^{n} (\boldsymbol{x}, \boldsymbol{e}_i)\boldsymbol{e}_i \right) \right) \ge 0$$

Since vectors $\boldsymbol{e}_i$ are orthogonal, it is easy to see

$$\mathbf{I} = ||\boldsymbol{x}||^2 - \left(\sum_{i=1}^{n} (\boldsymbol{x}, \boldsymbol{e}_i)\boldsymbol{e}_i, \boldsymbol{x} \right) - \left(\boldsymbol{x}, \sum_{i=1}^{n} (\boldsymbol{x}, \boldsymbol{e}_i)\boldsymbol{e}_i \right) + \sum_{i=1}^{n} |(\boldsymbol{x}, \boldsymbol{e}_i)|^2$$

$$= ||\boldsymbol{x}||^2 - \sum_{i=1}^{n} |(\boldsymbol{x}, \boldsymbol{e}_i)|^2 \ge 0$$

In fact for an infinite dimensional space,

$$||\boldsymbol{x}||^2 = \sum_{i=1}^{\infty} |x_i|^2 = \sum_{i=1}^{\infty} |(\boldsymbol{x}, \boldsymbol{e}_i)|^2 \ge \sum_{i=1}^{n} |(\boldsymbol{x}, \boldsymbol{e}_i)|^2, n < \infty$$

To summarize:

We are studying signals and operators and their interactions such that they can be analyzed in the setting of a suitable vector space. The vector space structure discussed in this section is an ideal tool for analyzing problems involving control and communication systems. But if a signal is represented as a vector in some space and is expanded along its chosen basis, it is of paramount importance that this expansion converges in the limit to its actual value and therefore the corresponding vector space is "complete." This is a tricky subject and requires further study via the so-called "Cauchy Sequence" and "Completeness" of the Vector Space.

15. **Cauchy Sequence**

If there exists a sequence $\{f_i\}_{i=0}^{\infty}$ such that for every real number $\epsilon > 0$, we can find a natural number $N \in \mathcal{N}$ (integer) such that $d(f_n, f_m) < \epsilon$ whenever $n, m > N$. This sequence is convergent if for a point $f \in \boldsymbol{V}, d(f, f_n) < \epsilon$ for all $n > N$.

Furthermore,

$$\lim_{n,m\to\infty} \mathrm{d}(f_n, f_m) = 0 \qquad \textbf{general metric space} \qquad (1.27)$$

or

$$\lim_{n,m\to\infty} (f_n - f_m, f_n - f_m) = 0 \qquad \textbf{inner product space}$$

Basically the Cauchy sequence implies that we can find a point $f \in \boldsymbol{V}$ that is arbitrarily close to the point f_n for all n.

The limit of this Cauchy sequence is f as $n \to \infty$. A convergent sequence is a Cauchy sequence but vice versa may not be true. As an example, a Cauchy sequence of real numbers converges, but a cauchy sequence of rational numbers may not converge.

The notion of a convergent Cauchy Sequence is fundamental to the approximation and optimization problems.

16. **Complete Vector Space**

The notion of completeness of a space is associated with the quality of goodness of the approximations in that space.

Our space can be finite dimension or infinite dimension. A finite dimensional space that has a convergent **Cauchy Sequence Limit** is called a **Complete Metric Space**. **Infinite dimensional spaces are little trickier**. Suffice it to say that for an infinite dimensional space, the cauchy sequence limit may not be in the space itself. We may have to remove this deficiency by enlarging the space and including the limit. This enlarged infinite dimension completed space is called a **Complete Metric Space**.

Example 1.20:

Consider an infinite sequence of rational numbers

$$2.718, 2.7181, 2.71881, \ldots$$

This is an example of a cauchy sequence which converges not to a **rational number** but to a well-known irrational number e. But since any irrational number can be computed by rational numbers with reasonable accuracy, we might as well consider such a sequence as Cauchy "Complete."

As mentioned earlier, a general metric space is very general and as such its usefulness is restricted. When the notion of "completeness" and "metric" is added, the space becomes useful for "optimal" approximation studies.

17. **Banach Space (metric space with additional structure) L^p or $L^p(a,b)$, $1 \leq p \leq \infty$**

A Banach space is a normed complete space with a metric induced by its norm. The letter "L" stands for the implication that the elements of the space are Lebesgue measurable. This basically implies that the functions in this space are bounded, integrable except for some pathological functions. Here we add the notion of a norm and completeness to the metric space. This notion of the norm endows the space with the ability to measure the "magnitude" of a vector. **A Cauchy Sequence Complete Metric Space with a norm $\|\cdot\|$ is called a Banach Space**. Vector spaces without a norm are not Banach spaces. The metric distance $d(\boldsymbol{u}, \boldsymbol{v})$ between $\boldsymbol{u}$ and $\boldsymbol{v}$ and the norm $\|\boldsymbol{u} - \boldsymbol{v}\|$ are equal in the Banach space, thinking of $\|\boldsymbol{u}\|$ as the distance of the element $\boldsymbol{u}$ from the origin.

Note: Not all metric spaces are normed and therefore not necessarily Banach. **Banach spaces have both metric and norm.**

Example 1.21:

A continuous function $f(t)$, $-\infty < t < \infty$, with a norm $\boldsymbol{L}^p(-\infty, +\infty)$ defined as:

$$\|\boldsymbol{f}\|^p = \left[\int_{-\infty}^{+\infty} |f(t)|^p \, dt \right]^{1/p} \qquad \text{is a Banach space function.}$$

Similarly for $p = \infty$;

$$\|\boldsymbol{f}\|^\infty = \sup_{-\infty<t<\infty} |f(t)| \qquad \text{is a Banach space function.}$$

sup $\Rightarrow$ supremum.

- **Supremum**: is defined as the least upper bound on a set. Thus,

$$\sup(-1, 1) = 1$$

- **Infimum**: is defined as greatest lower bound. Thus,

$$\inf(-1, 1) = -1$$

Example 1.22: Consider a finite dimensional Euclidean space $\boldsymbol{C}^n$ with a vector described as

$$\boldsymbol{v} = \{v_1, v_2, \cdots, v_n\}$$

This is a Banach space with many possible choices of norms. Two important norms are:

$$||\boldsymbol{v}||^p = (|v_1|^p + |v_2|^p + \cdots + |v_n|^p)^{1/p}, \quad 1 \leq p \leq \infty$$

$$||\boldsymbol{v}||^\infty = \max\{|v_1|, |v_2|, \cdots, |v_n|\}$$

$p = 2$ yields the conventional Euclidean Norm.

In general, there are infinitely many norms for $\boldsymbol{C}^n$ that may be of interest depending upon the particular problem. "$\boldsymbol{C}^n$" stands for an n-dimensional space of complex numbers.

In many cases, even the Banach space, being very general, may not be of great use. Some of the most important concepts such as length, angle, energy, and power can be expressed only in terms of inner product.

18. **Hilbert Space H (Inner Product Space)**

A complete normed linear space (or Banach Space) in which norm $|| \cdot ||$ is defined in a very specific manner, called "inner-product, $(\cdot, \cdot)$ " is known as Hilbert Space. This space is populated by a complete set of absolute square integrable or square summable functions.

Examples of l^2 (discrete) and L^2 (continuous)

l^2 spaces (discrete): Consider $\boldsymbol{x}, \boldsymbol{y} \in \boldsymbol{C}^n$

$$\boldsymbol{x} = (x_1, x_2, \cdots, x_n) \qquad \boldsymbol{y} = (y_1, y_2, \cdots, y_n)$$

where $\bar{x}_i$ is the complex conjugate of x_i

Define inner product as:

$$(\boldsymbol{x}, \boldsymbol{y}) = \sum_{i=1}^{n} \bar{x}_i y_i$$

This is a finite dimensional **Hilbert space** also known as Euclidean space. We can extend this notion of Hilbert space to an infinite dimension ($n \to \infty$), provided the summation has a finite value. This infinite dimensional space is denoted as $\boldsymbol{l}^2$, yielding;

$$(\boldsymbol{x}, \boldsymbol{x}) = \sum_{i=1}^{\infty} |x_i|^2 < \infty, \qquad (\boldsymbol{x}, \boldsymbol{y}) = \sum_{i=1}^{\infty} \bar{x}_i y_i < \infty$$

$\boldsymbol{L}^2(a, b)$ spaces (continuous)

$\boldsymbol{L}^2(a, b)$ Hilbert space consists of square integrable Lebesque measurable class of functions $f(t)$, $g(t)$ such that

$$\left.\begin{aligned}(\boldsymbol{f}, \boldsymbol{f}) &= \int_a^b |f(t)|^2 \, \mathrm{d}t < \infty \\ (\boldsymbol{f}, \boldsymbol{g}) &= \int_a^b \bar{f}(t) g(t) \, \mathrm{d}t\end{aligned}\right\} \quad \text{finite time one-dimensional space } \boldsymbol{L}^2(a, b)$$

$$\left.\begin{aligned}(\boldsymbol{f}, \boldsymbol{f}) &= \int_{-\infty}^{+\infty} |f(t)|^2 \, \mathrm{d}t \qquad a = -\infty, \ b = +\infty \\ (\boldsymbol{f}, \boldsymbol{g}) &= \int_{-\infty}^{+\infty} \bar{f}(t) g(t) \, \mathrm{d}t \qquad a = -\infty, \ b = +\infty\end{aligned}\right\} \quad \text{infinite time } \boldsymbol{L}^2(a, b)$$

If $\boldsymbol{f}, \boldsymbol{g}$ are infinite time multi-dimensional Euclidean space-time functions,

$$(\boldsymbol{f}, \boldsymbol{g}) = \int_{-\infty}^{+\infty} \left(\left[\bar{\boldsymbol{f}}(t) \right]^T \boldsymbol{g}(t) \right) \mathrm{d}t = \int_{-\infty}^{+\infty} \left(\sum_{i=1}^{n} \bar{f}_i(t) g_i(t) \right) \mathrm{d}t$$

For a finite multi-dimensional Euclidean Hilbert space in $\boldsymbol{C}^n$, without time dependence

$$(\boldsymbol{f}, \boldsymbol{g}) = \boldsymbol{f} \cdot \boldsymbol{g} = \sum_{i=1}^{n} \bar{f}_i g_i$$

Depending upon the value a and b of the time interval, these spaces are denoted by

$$\boldsymbol{L}^2(0,+\infty), \boldsymbol{L}^2(-\infty,0), \boldsymbol{L}^2(-\infty,+\infty)$$

$$\boldsymbol{L}^{2+} = \boldsymbol{L}^2(0,+\infty), \text{ subspace of } \boldsymbol{L}^2(-\infty,+\infty) \text{ for } t \geq 0$$

$$\boldsymbol{L}^{2-} = \boldsymbol{L}^2(-\infty,0), \text{ subspace of } \boldsymbol{L}^2(-\infty,+\infty) \text{ for } t < 0$$

Instead of time t, the independent variables may be a complex variable such as $j\omega$ or z

$$-\infty \leq \omega \leq +\infty$$

or

$$z = |z|e^{j\theta}, \; 0 \leq \theta \leq 2\pi$$

Example 1.23:

Consider the polynomial space with a positive weighting function $w(t)$ and interval (a, b). Let $\boldsymbol{f}, \boldsymbol{g}$ be the elements of this space such that

$$f(t) = a_0 + a_1 t + \cdots + a_n t^n$$

$$g(t) = b_0 + b_1 t + \cdots + b_m t^m$$

$$w(t) = \text{ a positive weighting function}$$

Then

$$(\boldsymbol{f}, \boldsymbol{g}) = \int_a^b \bar{f}(t) w(t) g(t)\, \mathrm{d}t$$

Example 1.24:

Consider the function $f(t)$

$$f(t) = t^{n/2}e^{-t} \qquad t > 0$$

We can prove that for this function; $f(t) \in \boldsymbol{L}^2(0, \infty)$.

$$(\boldsymbol{f}, \boldsymbol{f}) = \int_0^\infty t^n e^{-2t}\, \mathrm{d}t = \frac{n!}{(2)^{n+1}}$$

Exercise 1.2:

Use the following inequalities:

$$\frac{1}{2}(\alpha + \beta)^2 \le |\alpha|^2 + |\beta|^2$$

$$2|\alpha\beta| \le |\alpha|^2 + |\beta|^2$$

To prove the Cauchy–Schwartz inequality

$$(\boldsymbol{f}, \boldsymbol{g}) \le \|\boldsymbol{f}\| \cdot \|\boldsymbol{g}\|$$

Or

$$\left| \int_a^b \bar{f}(t) g(t)\, \mathrm{d}t \right| \le \left[\int_a^b |f(t)|^2 \right]^{1/2} \left[\int_a^b |g(t)|^2\, \mathrm{d}t \right]^{1/2}$$

Exercise 1.3:

Prove the triangular inequality

$$\|\boldsymbol{f} + \boldsymbol{g}\| \le \|\boldsymbol{f}\| + \|\boldsymbol{g}\|$$

Or

$$\left[\int_a^b |f(t) + g(t)|^2 \,\mathrm{d}t\right]^{1/2} \le \left[\int_a^b |f(t)|^2 \,\mathrm{d}t\right]^{1/2} \left[\int_a^b |g(t)|^2 \,\mathrm{d}t\right]^{1/2}$$

Some Control Theory Concepts

(i) $\boldsymbol{L}^2(j\omega)$ denotes a closed Hilbert space of complex variable functions $F(j\omega)$ and $G(j\omega)$ such that

$$\int_{-\infty}^{+\infty} |F(j\omega)|^2 \,\mathrm{d}\omega < \infty$$

$$\int_{-\infty}^{+\infty} \bar{F}(j\omega) G(j\omega) \,\mathrm{d}\omega < \infty$$

(ii) When the functions are defined **on the unit circle** ∂D, $|z| = 1$, the inner product takes the form

$$(\boldsymbol{f}, \boldsymbol{g}) = \frac{1}{2\pi} \int_{-\pi}^{+\pi} \bar{f}(e^{j\theta}) g(e^{j\theta}) \,\mathrm{d}\theta = \frac{1}{2\pi j} \int_{|z|=1} \bar{f}(z) g(z) \frac{\mathrm{d}z}{z}$$

The function $\boldsymbol{f}$ is square integrable and belongs to $\boldsymbol{L}^2(-\pi, \pi)$ meaning

$$\int_{-\pi}^{+\pi} |f(e^{j\theta}|^2 \,\mathrm{d}\theta < \infty$$

(iii) Consider a function $f(t) \in \boldsymbol{L}^2(-\infty, \infty)$ having a bilateral Laplace

Transform $\boldsymbol{F}_{11}(s) \in \boldsymbol{L}^2(s)$.

$\boldsymbol{H}^2(s)$ denotes a **closed subspace of analytic functions** $\boldsymbol{F}_{11}(s)$ such that

$$\int_{\sigma-j\infty}^{\sigma+j\infty} |\boldsymbol{F}_{11}(\sigma + j\omega)|^2 \, d\omega < 0 \quad \text{for } \sigma > 0, \quad s = \sigma + j\omega$$

$$\text{Re}\, s = \sigma > 0 \qquad \text{(open right-half } s\text{-plane, RHS)}$$

The function $\boldsymbol{F}_{11}$ is said to be analytic at a point s_0 if it has continuous derivatives of all orders at $s = s_0$. Such a function allows a Taylor series representation about the point of analyticity (see Chapter 4).

(iv) $\boldsymbol{H}^{2\perp}(s)$ is the orthogonal complement space of $\boldsymbol{H}^2(s)$ in $\boldsymbol{L}^2(s)$.

$\boldsymbol{H}^{2\perp}(s)$ represents those functions in $\boldsymbol{L}^2(s)$ that are analytic in the open left-half s-plane (LHS). The $\boldsymbol{L}^2(s)$ space in a complex frequency plane and $\boldsymbol{L}^2(-\infty, +\infty)$ in a time domain line are related to each other via bilateral Laplace Transform as in the following:

$$\begin{aligned} \boldsymbol{L}^{2+} &= \boldsymbol{L}^2(0, \infty) \leftrightarrow \boldsymbol{H}^2(s) \\ \boldsymbol{L}^{2-} &= \boldsymbol{L}^2(-\infty, 0) \leftrightarrow \boldsymbol{H}^{2\perp}(s) \\ \boldsymbol{L}^2(-\infty, +\infty) &\leftrightarrow \boldsymbol{L}^2(j\omega) \text{ and } \left(\boldsymbol{H}^2(s) + \boldsymbol{H}^{2\perp}(s)\right) \\ \boldsymbol{H}^2(s) + \boldsymbol{H}^{2\perp}(s) &= \boldsymbol{L}^2(s) \end{aligned}$$

Note: $\boldsymbol{L}^2(s)$ space is a larger space than $\boldsymbol{H}^2(s)$. The nomenclature $\boldsymbol{H}^2$ denotes a Hardy space that is defined later.

If the s-plane is mapped onto a z-plane via bilinear transformation, the $j\omega$ axis gets mapped into $|z| = 1$. The same classification of above spaces applies to the functions $f(z)$.

Thus,

$$L^{\infty}(z) = \max_{|z|=1} |f(z)|$$

If $f(z) \in L^2(z)$ then it implies that $f(z)$ is analytic on $|z| = 1$, yielding a **Fourier** series expansion:

$$f(z) = \sum_{n=-\infty}^{\infty} a_n z^n$$

If $f(z) \in H^2(z)$ then it implies that $f(z)$ is analytic for $|z| \leq 1$, yielding a **Taylor** series expansion about the origin:

$$f(z) = \sum_{n=0}^{\infty} a_n z^n$$

If $f(z) \in H^{2\perp}(z)$ then it implies that $f(z)$ is analytic for $|z| \geq 1$. It can also be written as Taylor series in z^{-1} about the origin:

$$f(z) = \sum_{n=0}^{\infty} a_{-n} \left(z^{-1}\right)^n = \sum_{n=-\infty}^{0} a_n \, (z)^n$$

19. L^{∞} **space** ($-\infty < t < +\infty$ or $|z| = 1$)

L^{∞} denotes the "Banach Space" of essentially Lebesque-measurable functions f such that

$$\|f\|^{\infty} = \lim_{p\to\infty} \left(\int_{-\infty}^{+\infty} |f(t)|^p \, dt \right)^{1/p} = \max_t |f(t)| < \infty$$

or

$$\|f\|^{\infty} = \lim_{p\to\infty} \left[\frac{1}{2\pi} \int_{-\pi}^{+\pi} |f(e^{j\theta})|^p \, d\theta \right]^{1/p} = \sup_{|z|=1} |f(z)| < \infty$$

Note: No integration is performed for computing $||f||^\infty$, rather only the peak value of $|f(t)|$ or $|f(z)|$ is picked out.

As mentioned earlier, instead of variable "t," we may have the frequency variable "ω" or complex variable ("z"). For variables ω or z, it is a measure of the maximum value of its "frequency response magnitude" or the so-called maximum energy amplification. In a z-plane, $f(z) \in \boldsymbol{L}^\infty(z)$ implies that $f(z)$ is analytic and bounded on $|z| = 1$ and can be expanded in a Fourier series:

$$f(z) = \sum_{n=-\infty}^{+\infty} a_n z^n, \quad |z| = 1$$

Thus, if $f(z) \in \boldsymbol{L}^2(z)$ or $\boldsymbol{L}^\infty(z)$, it can always be expanded in a Fourier (or Laurent) series and its eigenvectors are $1, z^{\pm 1}, z^{\pm 2}, \ldots$

20. $\boldsymbol{H}^\infty$ **and** $\boldsymbol{H}^{\perp\infty}$ **space**

$f(z) \in \boldsymbol{H}^\infty$ space implies that $f(z)$ is analytic for $|z| \le 1$ and furthermore,

$$||f||^\infty = \max_{|z|=1} |f(z)| < \infty$$

The eigenvectors in $\boldsymbol{H}^\infty$ space are $1, z, z^2, \ldots$, while the eigenvectors in $\boldsymbol{H}^{\perp\infty}$ are $z^{-1}, z^{-2}, \ldots$ yielding:

$$\boldsymbol{H}^\infty + \boldsymbol{H}^{\perp\infty} = \boldsymbol{L}^\infty$$

Summarizing

- **Every Hilbert Space is a Banach Space, but not vice versa. Similarly, every Banach Space is a Metric Space, but not vice versa.**
- **HILBERT SPACE L^2 is the most useful space for energy associated problems in communication systems.**

Figure 1.34 shows the hierarchy of linear vector spaces in order of their importance in physical applications.

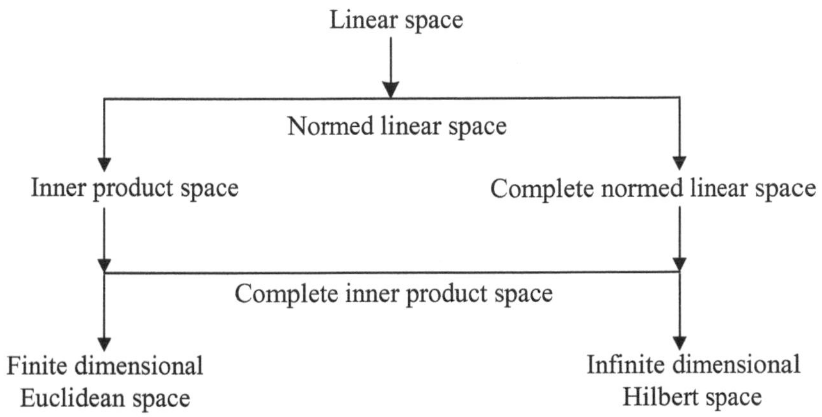

Figure 1.34: Hierarchy of Linear Spaces

21. **Gram-Schmidt Orthonormalization Process of Vectors**

Consider a set of n-linearly independent vectors $\boldsymbol{x}_1, \boldsymbol{x}_2, \cdots, \boldsymbol{x}_n$, (basis vector), spanning the n dimensional space. We can construct a set of n orthonormal unit vectors $\boldsymbol{\varphi}_1, \boldsymbol{\varphi}_2, \cdots, \boldsymbol{\varphi}_n$ spanning the same space and use it as its basis. It is easy to see that none of these vectors $\boldsymbol{\varphi}_i$ are zero. We shall use the following procedure in the construction of the orthonormal set, keeping in mind that each linearly independent vector $\boldsymbol{x}_k$ has new "information," which is not shared with other vectors.

Construction Algorithm:

Let us assume that vectors $\boldsymbol{\varphi}_1, \boldsymbol{\varphi}_2, \cdots, \boldsymbol{\varphi}_{k-1}$ form an orthogonal set with linearly independent vectors $\boldsymbol{x}_1, \boldsymbol{x}_2, \cdots, \boldsymbol{x}_{k-1}$. We are required to find the next vector $\boldsymbol{\varphi}_k$, which is orthonormal to the set $\boldsymbol{\varphi}_1, \boldsymbol{\varphi}_2, \cdots, \boldsymbol{\varphi}_{k-1}$ and involves only the next vector $\boldsymbol{x}_k$. Create a new vector $\boldsymbol{y}_k$ as a linear combination of $\boldsymbol{\varphi}_1, \boldsymbol{\varphi}_2, \cdots, \boldsymbol{\varphi}_{k-1}$ and the next linearly independent vector $\boldsymbol{x}_k$.

Thus,

$$\begin{aligned}
\boldsymbol{y}_1 &= \boldsymbol{x}_1 \\
\boldsymbol{y}_2 &= \boldsymbol{x}_2 - \alpha_{21}\boldsymbol{\varphi}_1 \\
\boldsymbol{y}_3 &= \boldsymbol{x}_3 - \alpha_{31}\boldsymbol{\varphi}_1 - \alpha_{32}\boldsymbol{\varphi}_2 \\
&\vdots \\
\boldsymbol{y}_k &= \boldsymbol{x}_k - \alpha_{k1}\boldsymbol{\varphi}_1 - \alpha_{k2}\boldsymbol{\varphi}_2 - \cdots - \alpha_{k,k-1}\boldsymbol{\varphi}_{k-1} \\
&\vdots \\
\boldsymbol{y}_n &= \boldsymbol{x}_n - \alpha_{n1}\boldsymbol{\varphi}_1 - \alpha_{n2}\boldsymbol{\varphi}_2 - \cdots - \alpha_{n,n-1}\boldsymbol{\varphi}_{n-1}
\end{aligned} \tag{1.28}$$

where,

$$\begin{aligned}
\boldsymbol{\varphi}_1 &= \frac{\boldsymbol{y}_1}{\|\boldsymbol{y}_1\|}, \boldsymbol{\varphi}_2 = \frac{\boldsymbol{y}_2}{\|\boldsymbol{y}_2\|}, \cdots, \boldsymbol{\varphi}_n = \frac{\boldsymbol{y}_n}{\|\boldsymbol{y}_n\|} \\
(\boldsymbol{\varphi}_k, \boldsymbol{\varphi}_k) &= \|\boldsymbol{\varphi}_k\| = 1 \\
(\boldsymbol{x}_k, \boldsymbol{\varphi}_j) &= \alpha_{kj}, \quad k = 1, 2, \cdots, n; \quad j = 1, 2, \cdots, n
\end{aligned}$$

Standard $L^2(a, b)$ Hilbert Space Basis

To every Hilbert Space, we can assign a complete orthonormal basis called its **Span**. An orthonormal sequence $\{\boldsymbol{\varphi}_k\}_{k=-\infty}^{+\infty}$ with a given weighting function $w(t) > 0$ and interval $[a, b]$ is defined as:

$$\int_a^b \varphi_i^*(t)\omega(t)\varphi_j(t)\,\mathrm{d}t = (\boldsymbol{\varphi}_i, \boldsymbol{\varphi}_j) = \delta_{ij}, \quad \boldsymbol{\varphi}_i \perp \boldsymbol{\varphi}_j \tag{1.29}$$

forms a standard basis for a Hilbert Space.

Any function $f(t) \in \boldsymbol{L}^2$ can be written as

$$f(t) = \sum_{k=-\infty}^{+\infty} c_k \varphi_k(t) \quad \text{(also known as general Fourier series)} \tag{1.30}$$

where

$$\sum_{k=-\infty}^{+\infty} |c_k|^2 < \infty$$

Example 1.25: Consider the following sequence $\{f_n(\theta)\}$

$$f_n(\theta) = \frac{1}{\sqrt{2\pi}} e^{jn\theta}$$

$$n = 0, \pm 1, \pm 2, \ldots \qquad w(\theta) = 1, \quad 0 \leq \theta \leq 2\pi$$

$$\begin{aligned}(\boldsymbol{f}_n, \boldsymbol{f}_n) &= \frac{1}{2\pi} \int_0^{2\pi} e^{jn\theta} e^{-jn\theta}\, \mathrm{d}\theta = 1 \\ (\boldsymbol{f}_n, \boldsymbol{f}_m) &= \frac{1}{2\pi} \int_0^{2\pi} e^{jn\theta} e^{-jm\theta}\, \mathrm{d}\theta = 0 \qquad n \neq m\end{aligned}$$

Example 1.26: Kautz, Laguerre, and Legendre Polynomials are typical orthogonal vector sets in the Hilbert Space $\boldsymbol{L}^2(a, b)$. Consider the following sequence of functions generated by a second-order equation:

$$\begin{aligned} f_1(t) &= \sqrt{2p_1} e^{-p_1 t} \\ \sqrt{\frac{p_n}{p_{n+1}}} f_{n+1}(t) &= -f_n(t) + (p_{n+1} + p_n) e^{p_{n+1} t} \int_0^t f_n(\tau) e^{p_{n+1}\tau}\, \mathrm{d}\tau, \quad n = 2, 3, \ldots \end{aligned}$$

Functions $\{f_i(t)\}_{i=1}^{\infty}$ form an orthonormal set defined as

$$(\boldsymbol{f}_i, \boldsymbol{f}_j) = \int_a^b \bar{f}_i(t) w(t) f_j(t)\, \mathrm{d}t = \delta_{ij}$$

where $w(t) = 1$; Interval: $[a, b] = [0, \infty]$.

Depending upon the choice of p_n we shall be able to generate different sets of orthogonal functions.

(i) If $p_n = n$ (an integer), we obtain regular Kautz polynomials, which are useful in optimal approximation. A most interesting recursive relation generating Kautz polynomials in the Laplace transform domain is

$$\mathcal{L}(f_1(t)) = F_1(s) = \sqrt{2p_1}\left(\frac{1}{s+p_1}\right)$$

$$\mathcal{L}(f_{n+1}(t)) = F_{n+1}(s) = \sqrt{\frac{p_{n+1}}{p_n}}F_n(s)\left(\frac{p_n - s}{p_{n+1}+s}\right) \qquad n = 1, 2, \ldots$$

(ii) By letting $p_n = 1$ for all n, well-known Laguerre functions are generated:

$$\begin{aligned} l_1(t) &= \sqrt{2}e^{-t} \\ l_{n+1}(t) &= -l_n(t) + 2e^{-t}\int_0^t l_n(\tau)e^{\tau}\,\mathrm{d}\tau \end{aligned}$$

$$\begin{aligned} \mathcal{L}(l_1(t)) = L_1(s) &= \sqrt{2}\left(\frac{1}{1+s}\right) \\ \mathcal{L}(l_{n+1}(t)) = L_{n+1}(s) &= L_n(s)\left(\frac{1-s}{1+s}\right) \qquad n = 1, 2, \cdots \end{aligned}$$

(iii) If $p_n = n$ (an integer) and $\sqrt{2}e^{-t} = x, \; -1 \le x \le 1$, we obtain Legendre Polynomials:

$$\begin{aligned} p_1(x) &= x \\ p_2(x) &= -3x^2 + 2x \\ p_3(x) &= 10x^3 - 12x^2 + 3x \end{aligned}$$

Example 1.27:

Haar and Walsh Functions

These functions consist of a train of square functions taking only plus or minus 1 values. Both sets of functions satisfy orthogonality relationships and are Cauchy complete. (There is another set known as Rademacher functions, which are the forerunners of these functions but are not Cauchy complete).

Haar Functions Let us define a function $\varphi(t)$

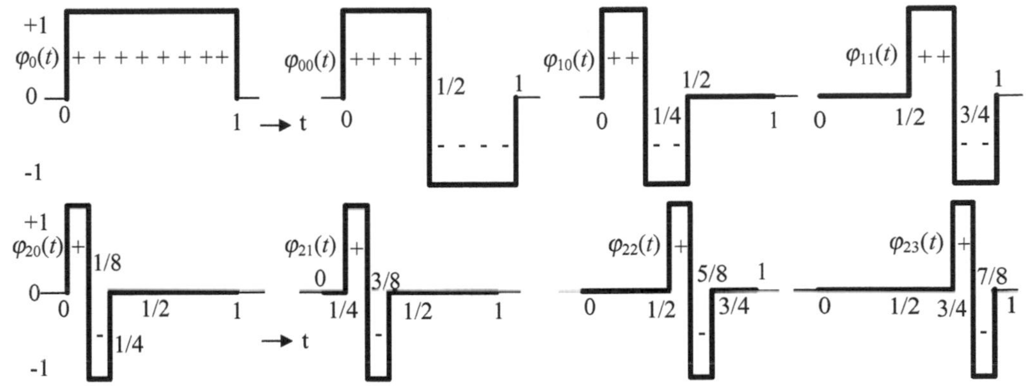

Figure 1.35: The First 8 Haar Functions on a 0 to 1 Interval

$$\varphi(t) = \begin{cases} 1 & 0 \le t \le \frac{1}{2} \\ -1 & \frac{1}{2} \le t \le 1 \\ 0 & \text{otherwise} \end{cases}$$

We generate the following set of functions

$$\varphi_0(t) = 1$$

$$\varphi_{jk}(t) = \varphi(2^j t - k), \qquad j, k = 0, 1, 2, \cdots$$

This is an orthonormal set known as a Haar function.

Figure 1.35 shows the plot of first 8 Haar functions.

Haar functions expansion for $f(t) \in \boldsymbol{L}^2(0, 1)$ yields:

$$f(t) = c_0 + \sum_{j=0}^{\infty} \sum_{k=0}^{2^{j-1}} c_{jk}\varphi_{jk}(t)$$

$$\int_0^1 \varphi_0(t) f(t)\, dt = c_0$$

$$\int_0^1 \varphi(t)\varphi_{jk}(t)\, dt = 0$$

$$\int_0^1 \varphi_{jk}(t)\varphi_{lm}(t)\, dt = 0$$

$$\int_0^1 \varphi_{jk}(t)\varphi_{jk}(t)\, dt = 1$$

Walsh Functions

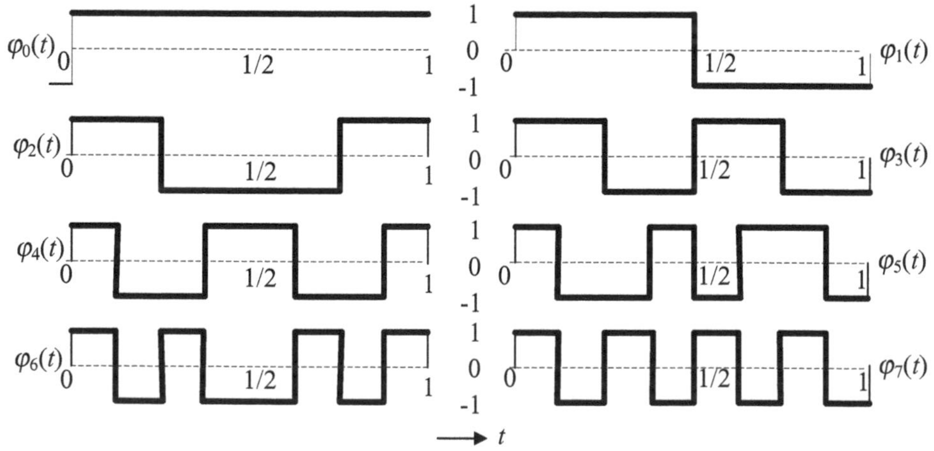

Figure 1.36: The First 8 Walsh Functions on a 0 to 1 Interval.

Let

$$k = \text{ number of zero crossings in the interval 0 to 1.}$$

$$k \text{ also represents the order of the Walsh function.}$$

Define

$$\text{For all } \textbf{odd } k\text{'s}, \quad \varphi_k(t) = \begin{cases} \varphi_{k-1}(t), & 0 \le t \le \frac{1}{2} \\ -\varphi_{k-1}(t), & \frac{1}{2} \le t \le 1 \end{cases}$$

$$\text{For all } \textbf{even } k\text{'s}, \quad \varphi_k(t) = \begin{cases} \varphi_{\frac{k}{2}}(2t), & 0 \le t \le \frac{1}{2} \\ (-1)^{\frac{k}{2}} \varphi_{\frac{k}{2}}(2t-1), & \frac{1}{2} \le t \le 1 \end{cases}$$

$\varphi_k(2t)$ is referred to as a "squeezed half interval" of $\varphi_k(t)$.

Figure 1.36 shows the plot of first 8 Walsh functions.

A complete set of 2^n Walsh functions of order n are given by Hadamard matrix $\boldsymbol{H}_{2n}$. The enteries in the matrix are +1 or -1. Realizing that interval for all Walsh functions is 0 to 1, the Hadamard matrices for 2,4, and 8 Walsh functions are:

$$\boldsymbol{H}_2 = \begin{array}{|c|c|} \hline 1 & 1 \\ \hline -1 & +1 \\ \hline \end{array}$$

$$\boldsymbol{H}_4 = \begin{array}{|c|c|} \hline \boldsymbol{H}_2 & \boldsymbol{H}_2 \\ \hline \boldsymbol{H}_2 & -\boldsymbol{H}_2 \\ \hline \end{array}$$

$$\boldsymbol{H}_8 = \begin{array}{|c|c|} \hline \boldsymbol{H}_4 & \boldsymbol{H}_4 \\ \hline \boldsymbol{H}_4 & -\boldsymbol{H}_4 \\ \hline \end{array}$$

Note that

$$\boldsymbol{H}_n \boldsymbol{H}_n^T = n\boldsymbol{I}_n$$

where $\boldsymbol{I}_n$ is $n \times n$ identity matrix.

We can abbreviate +1 and -1 by + and - respectively.

22. **Least Squares Approximation in Hilbert Space (Projection Theorem)**

Consider an infinite dimensional Hilbert space $\boldsymbol{V}$ with basis $\{\varphi_k(t)\}_{k=-\infty}^{+\infty}$. The time function $f(t)$ is represented as a vector $\boldsymbol{f}, (\boldsymbol{f} \in \boldsymbol{V})$ and can be expanded as:

$$f(t) = \sum_{k=-\infty}^{\infty} c_k\varphi_k(t) = \sum_{k=-N}^{N} c_k\varphi_k(t) + \left(\sum_{k=-\infty}^{-(N+1)} c_k\varphi_k(t) + \sum_{k=N+1}^{\infty} c_k\varphi_k(t) \right)$$

Let

$$W_N(t) = \sum_{k=-N}^{N} c_k\varphi_k(t)$$

$$E_N(t) = \left(\sum_{k=-\infty}^{-(N+1)} c_k\varphi_k(t) + \sum_{k=N+1}^{\infty} c_k\varphi_k(t) \right)$$

Using the notation:

$$f(t) \rightarrow \boldsymbol{f}, \qquad W_N(t) \rightarrow \boldsymbol{W}_N, \qquad E_N(t) \rightarrow \boldsymbol{E}_N$$

yields

$$\boldsymbol{f} = \boldsymbol{W}_N + \boldsymbol{E}_N, \qquad \boldsymbol{W}_N \perp \boldsymbol{E}_N$$

The function space $\boldsymbol{W}_N$ is orthogonal to function space $\boldsymbol{E}_N$. Indeed, $\boldsymbol{f}$, $\boldsymbol{W}_N$, and $\boldsymbol{E}_N$ form right triangle functional subspaces as shown in Figure1.37.

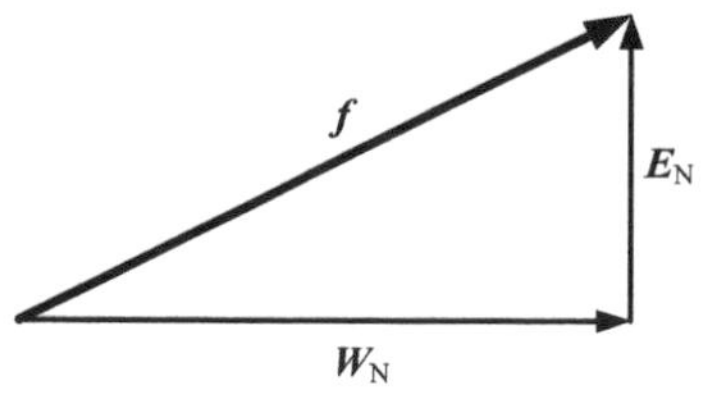

Figure 1.37: Projection of $\boldsymbol{f}$ on Subspaces $\boldsymbol{E}_N$ and $\boldsymbol{W}_N$

If we use the subspace $\boldsymbol{W}_N$ spanned by $\{\varphi_k(t)\}_{k=-N}^{N}$ to represent the signal $f(t)$, then $W_N(t)$ is "the optimal" or least square approximation to $f(t)$. The function $E_N(t)$ is called the **Error Function**. $W_N(t)$ is called the **Projection of f**, onto the approximation subspace $\boldsymbol{W}_N$.

Indeed

$$\|f(t)\|^2 = \|W_N(t)\|^2 + \|E_N(t)\|^2 \quad \text{(Pythagoras or Projection Theorem)}$$

and

$$\sum_{k=-\infty}^{\infty} |c_k|^2 = \sum_{k=-N}^{N} |c_k|^2 + \left(\sum_{k=-\infty}^{-(N+1)} |c_k|^2 + \sum_{k=N+1}^{\infty} |c_k|^2 \right) \tag{1.31}$$

$$\|f(t)\|^2 \geq \|W_N(t)\|^2 \quad \textbf{Bessel's Inequality}$$

In fact:

(a) Subspace $\{\varphi_k(t)\}_{k=-N}^{N}$ is orthogonal to subspace $\left(\{\varphi_k(t)\}_{k=-\infty}^{-(N+1)}, \{\varphi_k(t)\}_{k=N+1}^{\infty}\right)$

(b) $\lim\limits_{N\to\infty} \|E_N\|^2 = 0$, assuring us the set is complete

(c) $W_N(t)$ is referred to as a projection of $f(t)$ to the "useful space $\boldsymbol{W}_N$"

(d) $E_N(t)$ is called the projection of $f(t)$ on to the "error space $\boldsymbol{E}_N$"

Note: Here the L^2 norm is explicitly written as $\|\cdot\|^2$

Example 1.28:

Given: Real functions $\boldsymbol{x}, \boldsymbol{y} \in \boldsymbol{L}^2$,

Determine a constant a such that $\|\boldsymbol{y} - a\boldsymbol{x}\|^2$ is minimized.

This is called the **Least Mean Square** estimate problem.

Solution:

$$\boldsymbol{e} = (\boldsymbol{y} - a\boldsymbol{x}) \qquad \text{error vector}$$

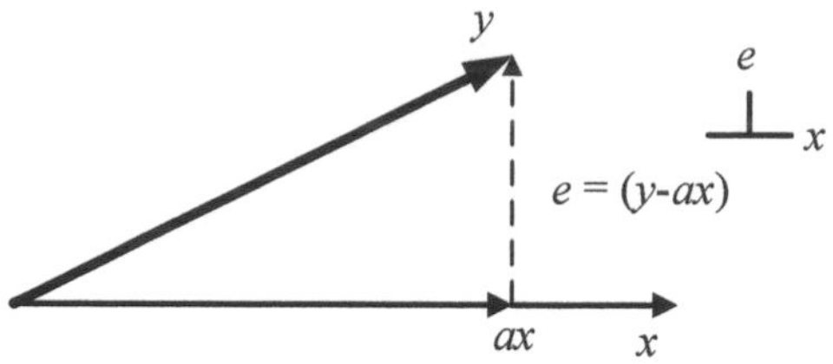

Figure 1.38: Error Vector $\boldsymbol{e}$ and Estimated Vector $a\boldsymbol{x}$ Are Orthogonal

The mean squared error function is:

$$\begin{aligned} I(a) &= \|\boldsymbol{e}\|^2 = (\boldsymbol{e}, \boldsymbol{e}) = ((\boldsymbol{y} - a\boldsymbol{x}), (\boldsymbol{y} - a\boldsymbol{x})) \\ &= (\boldsymbol{y}, \boldsymbol{y}) + a^2(\boldsymbol{x}, \boldsymbol{x}) - 2a|(\boldsymbol{x}, \boldsymbol{y})| \end{aligned}$$

The neccessary condition for the minimum of $I(a)$ w.r.t a is:

$$\frac{1}{2}\frac{\partial I}{\partial a} = a(\boldsymbol{x}, \boldsymbol{x}) - (\boldsymbol{x}, \boldsymbol{y}) = 0$$

or

$$a = \frac{(\boldsymbol{x}, \boldsymbol{y})}{(\boldsymbol{x}, \boldsymbol{x})}$$

"$\boldsymbol{y}$" is often referred to as the observation or the data vector and $a\boldsymbol{x}$ is the optimal estimate. Note that the estimated vector and the error vector are orthogonal to each other, i.e. $a\boldsymbol{x} \perp \boldsymbol{e}$.

Example 1.29:

Given an observation vector $\boldsymbol{y}$ and functions $\{\boldsymbol{x}\}_i \in \boldsymbol{L}^2$, $i = 1$ to n, determine $\{a_i\}_{i=1}^n$ such that $\|\boldsymbol{y} - \sum_{i=1}^n a_i \boldsymbol{x}_i\|^2$ is minimized.

Solution:

$$I(a_1, a_2, \cdots, a_n) = \|\boldsymbol{e}\|^2 = \left((\boldsymbol{y} - \sum_{i=1}^{n} a_i \boldsymbol{x}_i), (\boldsymbol{y} - \sum_{j=1}^{n} a_j \boldsymbol{x}_j) \right)$$

The minimum of I is obtained as:

$$\frac{1}{2}\frac{\partial I}{\partial a_j} = -\left((\boldsymbol{y} - \sum_{i=1}^{n} a_i \boldsymbol{x}_i), \boldsymbol{x}_j\right) = 0, \qquad j = 1, 2, \cdots, n$$

$$\underbrace{\begin{bmatrix} (\boldsymbol{x}_1,\boldsymbol{x}_1) & (\boldsymbol{x}_1,\boldsymbol{x}_2) & \cdots & (\boldsymbol{x}_1,\boldsymbol{x}_n) \\ (\boldsymbol{x}_2,\boldsymbol{x}_1) & (\boldsymbol{x}_2,\boldsymbol{x}_2) & \cdots & (\boldsymbol{x}_2,\boldsymbol{x}_n) \\ \vdots & \vdots & \ddots & \vdots \\ (\boldsymbol{x}_n,\boldsymbol{x}_1) & (\boldsymbol{x}_n,\boldsymbol{x}_2) & \cdots & (\boldsymbol{x}_n,\boldsymbol{x}_n) \end{bmatrix}}_{\text{Gram Matrix}} \begin{bmatrix} a_1 \\ a_2 \\ \vdots \\ a_n \end{bmatrix} = \begin{bmatrix} (\boldsymbol{x}_1,\boldsymbol{y}) \\ (\boldsymbol{x}_2,\boldsymbol{y}) \\ \vdots \\ (\boldsymbol{x}_n,\boldsymbol{y}) \end{bmatrix}$$

$$\boldsymbol{e} \perp \sum_{i=1}^{n} a_i \boldsymbol{x}_i$$

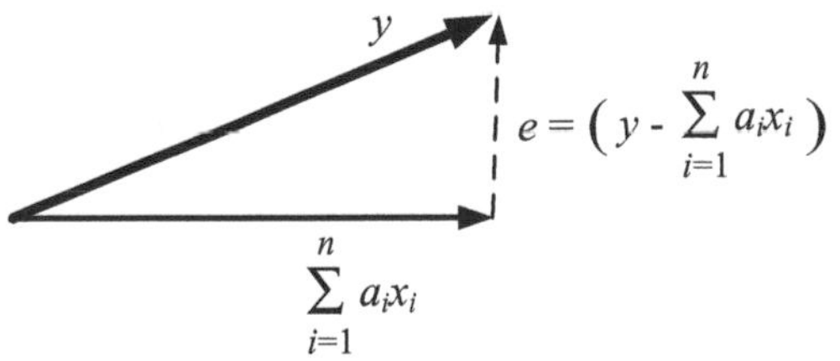

Figure 1.39: Orthogonal Decomposition of $\boldsymbol{y}$ in n-Dimensions

23. **Summary of Important Concepts in Vector Spaces**

Different Independent Variables

Systems engineering involves the study of many intertwined disciplines resulting in system equations with different independent variables. These independent variables can be related to each other via tranformations yielding isometric isomorphism. These variables are:

(a) The time variable t, $-\infty \leq t \leq 0$, $\quad 0 \leq t < \infty$, $\quad$ or $-\infty < t < \infty$

(b) The frequency variable ω, $-\infty < \omega < \infty$

The complex frequency variable $s = \sigma + j\omega$ in "s-plane"

$-\infty < \sigma < \infty$ represents the horizontal axis, $-\infty < \omega < \infty$, the vertical axis. The functions in this space admit transformations between $\boldsymbol{L}^2$ functions in the time domain to $\boldsymbol{L}^2$ functions in the complex frequency plane known as bilateral Laplace transforms.

(c) The complex variable z is defined within a unit disk (as well as on it). We shall denote the open disk with $\boldsymbol{D}$ and the boundary of this unit disk with $\partial\boldsymbol{D}$. The variable z is represented by:

$$z = re^{j\theta} \quad r < 1, \quad 0 \leq \theta \leq 2\pi \quad \text{(inside the disk } \boldsymbol{D}\text{)}.$$

$$z = e^{j\theta} \quad ,0 \leq \theta \leq 2\pi \quad \text{(on } \partial\boldsymbol{D}\text{)}.$$

(d) The variables ω and θ have an interesting relationship (Figure1.40)

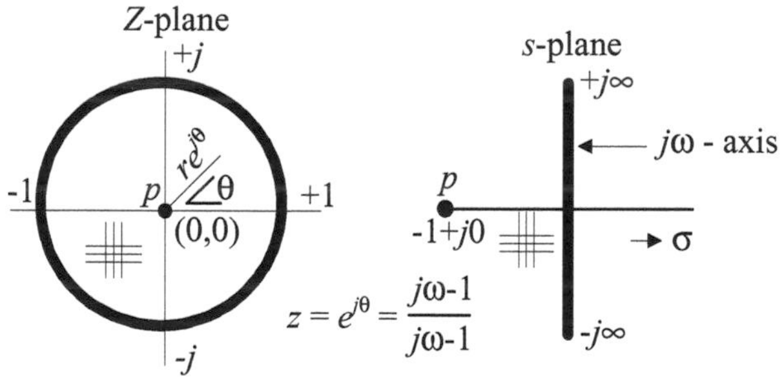

Figure 1.40: Transformation between the Unit Disk and the s-Plane Known as Bilinear Transformation

$$j\omega = \frac{1 + e^{j\theta}}{1 - e^{j\theta}} = \frac{1 + z}{1 - z} \quad , \quad z = e^{j\theta} \quad \text{(point } z \text{ on the unit disk boundary)}$$

or

$$z = e^{j\theta} = \frac{j\omega - 1}{j\omega + 1}, \quad (z \text{ on the unit disk boundary})$$

$$\theta = \pi - 2\tan^{-1}(\omega) \quad (\omega \text{ on the vertical axis in } s\text{-plane})$$

This is the well-known bilinear transformation that transforms the boundary of the unit disk D in the z-plane to the $j\omega$ axis in the s-plane and the inside of the unit disk is transformed to the left-hand side of the s-plane. These tranformations are used to transform the time functions (discrete or continuous) into complex variable functions (variable s) via Laplace, bilateral Laplace, or Fourier transform (variables s or ω) or z-transform.

24. **Examples of various Signal Function Norms**

(a) **Uniform or Chebychev Norm**:

$$\|\boldsymbol{f}\| = \max_{-\infty<t<\infty} |f(t)| \tag{1.32}$$

(b) $\boldsymbol{L^1}$ **Norm** (also called 1-Norm):

$$\|\boldsymbol{f}\|^1 = \int_{-\infty}^{\infty} |f(t)|\,\mathrm{d}t \tag{1.33}$$

For discrete case $\boldsymbol{l^1}$ Norm:

$$\|\boldsymbol{f}\|^1 = \sum_{n=-\infty}^{\infty} |f(n)| \tag{1.34}$$

(c) $\boldsymbol{L^2}$ **Norm** (also called 2-Norm or $\boldsymbol{L^2}(-\infty, \infty)$):

$$\|\boldsymbol{f}\|^2 = \left(\int_{-\infty}^{\infty} |f(t)|^2\,\mathrm{d}t\right)^{\frac{1}{2}} \tag{1.35}$$

For discrete case $\boldsymbol{l}^2$ Norm:

$$||\boldsymbol{f}||^2 = \left(\sum_{n=-\infty}^{\infty} |f(n)|^2 \right)^{\frac{1}{2}} \tag{1.36}$$

This norm is in reality a **Hilbert Space Norm** and is a measure of the generalized "Energy" of a function.

It is important to realize that L^2 norm is a "Hilbert Space Norm" whereas H^2 norm is a Hardy space norm notation discussed below.

The $\boldsymbol{L}^2$ spaces defined in the time domain are analogous to the $\boldsymbol{L}^2$ spaces defined in the frequency domain.

(d) $\boldsymbol{L}^p$ **Norm of a Function** (also called p-Norm):

$$||\boldsymbol{f}||^p = \left(\int_{-\infty}^{\infty} |f(t)|^p \, \mathrm{d}t \right)^{\frac{1}{p}} \tag{1.37}$$

where p is a number ≥ 1. In case $p = 2$ we recover the usual **Euclidean Distance** or Hilbert Space norm.

(e) $\boldsymbol{L}^\infty$ **Norm of a Function** (also called the ∞-Norm):

$$||\boldsymbol{f}||^\infty = \sup_{-\infty<t<\infty} |f(t)| \tag{1.38}$$

For discrete case $\boldsymbol{l}^\infty$ Norm:

$$||\boldsymbol{f}||^\infty = \sup_{-\infty<n<\infty} |f(n)| \tag{1.39}$$

where **sup** stands for the "supremum" of a function. This norm can be interpreted as the largest absolute value of the function $f(t)$ (or $f(n)$) and represents the concept of boundedness, which is essential for stability study. A function $f(t)$ is bounded if

$$\|\boldsymbol{f}\|^{\infty} < \infty$$

Furthermore, $d^{\infty}(\boldsymbol{f}, \boldsymbol{g})$ represents the maximum absolute difference between the two time functions $f(t)$ and $g(t)$. This norm is the Banach space norm.

(f) **Hardy Spaces and Its Norm**

In Lebesque measurable spaces $\boldsymbol{L}^p(-\infty, \infty)$, the independent variable t takes values from $-\infty$ to $+\infty$. For the corresponding transformed frequency functions the whole of the complex plane s including the $j\omega$ axis is involved.

If we are only interested in the signals for $0 < t < \infty$, or only the left-half of the complex plane, or only the $j\omega$ axis, we are dealing with a closed sub-space of $\boldsymbol{L}^p$.

Spaces involving the left-half of the s-plane or the $j\omega$ axis are known as Hardy spaces $\boldsymbol{H}^p$ named after the famous English mathematician G.H. Hardy.

The Hardy space $\boldsymbol{H}^p$ $(0 < p < \infty)$ represents a space of analytic functions, bounded inside and on the open unit disk $\boldsymbol{D}$ such that:

$$\sup_{0<r\leq 1}\left(\frac{1}{2\pi}\int_0^{2\pi}\left[f(re^{j\theta})\right]^p \,d\theta\right)^{\frac{1}{p}} < \infty \quad \text{for} \quad f(re^{j\theta}) \in \boldsymbol{H}^p$$

$$\sup\left(\frac{1}{2\pi}\int_0^{2\pi}\left[f(e^{j\theta})\right]^p \,d\theta\right)^{\frac{1}{p}} < \infty \qquad \text{for} \quad f(e^{j\theta}) \in \boldsymbol{L}^p$$

$$\sup_{|re^{j\theta}|\leq 1} f(re^{j\theta}) \in \boldsymbol{H}^{\infty}$$

In general $\boldsymbol{H}^p$ spaces are subspaces of $\boldsymbol{L}^p$ spaces.

Only when $p = 2, r = 1$, we find that $\boldsymbol{L}^p$ and $\boldsymbol{H}^p$ become the same.

Some Important Facts in L^p and H^p Spaces

- $H^p(-\infty, \infty)$ is a closed subspace of $L^p(-\infty, \infty)(f(t) \in L(-\infty, \infty) : f(t) \equiv 0, t < 0)$.

  H^2 and H^∞ are two important signal spaces.

  H^2 is a Hilbert Space (inner product space) while H^∞ is a Banach space with or without an inner product. Since Banach Space is much larger,

$$H^2 \subseteq H^\infty$$

  Every Hilbert Space is a Banach Space but every Banach Space may not be a Hilbert Space.

- Function $f(t) \in L^2(-\infty, +\infty)$ implies that it is bilateral Laplace transformable.

$$F_{11}(s) = \int_{-\infty}^{+\infty} f(t)e^{-st}\,\mathrm{d}t$$

$$F_{11}(s)|_{s=j\omega} = F(j\omega), \qquad \text{Fourier Transform of f(t)}$$

  Furthermore,

$$L^2(-\infty, +\infty) \cong L^2(j\omega) = \text{Fourier transform on the } j\omega \text{ axis}$$

$$||F(j\omega)|| = ||F_{11}(s)|| = ||f(t)||$$

- Function f(t) $\in L^2(0, \infty)$ implies it is Laplace transformable.

$$F_{11}^{+}(s) = F(s) = \int_0^\infty f(t)e^{-st}\,\mathrm{d}t \quad (F(s) \in H^2)$$

  Thus,

$$L^2(0, \infty) \cong H^2$$

- Function f(t) $\in \boldsymbol{L}^2(-\infty, 0)$ implies f(-t) is Laplace transformable

$$F_{11}^-(s) = F^-(s) = \int_{-\infty}^{0} f(t)e^{-st}\,\mathrm{d}t \qquad (F_{11}^-(s) \in \boldsymbol{H}^{2\perp})$$

$$\boldsymbol{L}^2(-\infty, 0) \cong \boldsymbol{H}^{2\perp}$$

Let us define:

$\boldsymbol{P}^+$ as projection operator for domain $0 < t < \infty$

$\boldsymbol{P}^-$ as projection operator for domain $-\infty < t < 0$

$(\boldsymbol{P}^+)^2 = \boldsymbol{P}^+, \quad (\boldsymbol{P}^-)^2 = \boldsymbol{P}^-, \quad \boldsymbol{P}^+ + \boldsymbol{P}^- = 1$

$\boldsymbol{P}^+\boldsymbol{P}^- = 0, \quad \boldsymbol{P}^-\boldsymbol{P}^+ = 0$

$$\boldsymbol{P}^+\boldsymbol{L}^2(-\infty, +\infty) = \boldsymbol{L}^2(0, \infty), \quad \boldsymbol{P}^-\boldsymbol{L}^2(-\infty, +\infty) = \boldsymbol{L}^2(-\infty, 0)$$

$$\boldsymbol{P}^+\boldsymbol{L}^2(j\omega) = \boldsymbol{H}^2 \qquad \text{Fourier Transform } 0 \le t \le \infty$$

$$\boldsymbol{P}^-\boldsymbol{L}^2(j\omega) = \boldsymbol{H}^{2\perp} \qquad \text{Fourier Transform } -\infty \le t \le 0$$

$$\boldsymbol{H}^2 \oplus \boldsymbol{H}^{2\perp} = \boldsymbol{L}^2, \qquad \boldsymbol{L}^2 \ominus \boldsymbol{H}^{2\perp} = \boldsymbol{H}^2$$

The projection operator implies the projection of a vector of higher dimension into a lower-dimensional subspace, called the **Image of an Idempotent Linear Transformation**

Example 1.30:

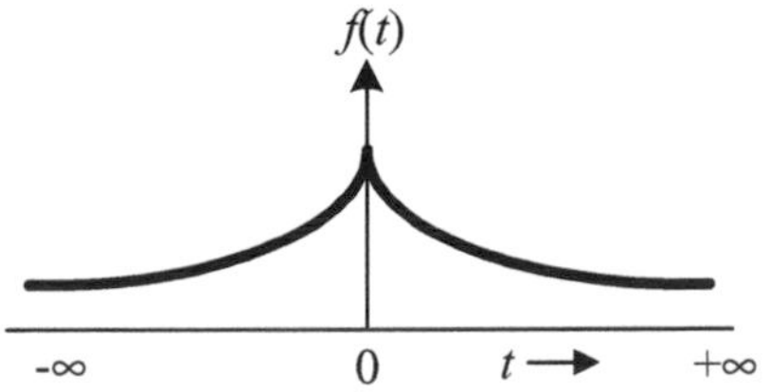

Figure 1.41: Function $f(t)$ in Example 1.21

$$f(t) = e^{-t} + e^{t}, \quad -\infty < t < \infty$$

$$\boldsymbol{P}^{+} f = f^{+}(t) = e^{-t} \quad \text{belongs to} \quad \boldsymbol{H}^{2}, \qquad 0 \le t \le \infty$$

$$\boldsymbol{P}^{-} f = f^{-}(t) = e^{t} \quad \text{belongs to} \quad \boldsymbol{H}^{2\perp}, \qquad -\infty \le t \le 0$$

To Summarize

- Hilbert space is designated as $\boldsymbol{H}$ space.
- $\boldsymbol{L}^2$ space is a Hilbert space $\boldsymbol{H}$.
- $\boldsymbol{L}^2$ is a closed subspace of Banach Space $\boldsymbol{L}^1$.
- Hardy Space $\boldsymbol{H}^2$ is a closed subspace of $\boldsymbol{H}^\infty$, $\boldsymbol{H}^\infty \subseteq \boldsymbol{H}^2$ but $\boldsymbol{H}^\infty$ is not a Hilbert space.
- Hardy Space $\boldsymbol{H}^2$ is a Hilbert space if it is a closed subspace of $\boldsymbol{L}^2$.
- $\|f\|^1 \le \|f\|^2 \le \|f\|^\infty$.
- A complete orthonormal sequence $(z^n)_{n=-\infty}^{\infty}$ forms a basis set for both $\boldsymbol{L}^2$ and $\boldsymbol{L}^\infty$.
- If $f(z) \in \boldsymbol{L}^2$, $(z = e^{j\theta},\ |z| = 1,\ -\pi < \theta < \pi)$

  It is possible to write

$$f(z) = \sum_{n=-\infty}^{+\infty} a_n z^n$$

$$f_1(z) = \boldsymbol{P}^{+} f(z) = \sum_{n=0}^{\infty} a_n z^n \quad \text{analytic part of } \boldsymbol{H}^2 \text{ for } |z| \le 1$$

$$f_2(z) = \boldsymbol{P}^{-} f(z) = \sum_{n=-\infty}^{-1} a_n z^n \quad \text{analytic part of } \boldsymbol{H}^{2\perp} \text{ for } |z| > 1$$

$f_1(z)$ is bounded inside and $f_2(z)$ is bounded outside, the unit circle. On the other hand, $f(z)$ is bounded on the unit circle. From complex variable theory it can be shown that the maximum of $|f(z)|$ occurs on the boundary $|z| = 1$.

$f(z)$ represents Laurent series

$f_1(z)$ represents Taylor series about the origin for $|z| < 1$

$f_2(z)$ represents Taylor series about infinity for $|z| > 1$

The function $f_1(z)$ represents the best approximation to $f(z)$ in the $\boldsymbol{L}^2$ norm by a $\boldsymbol{H}^2$ function analytic inside the disc D.

Example 1.31:

Let

$$f(z) = \frac{1}{1 - 0.5z} + \frac{1}{z - 0.5},$$

$$\boldsymbol{P}^+ f(z) = \left(\frac{1}{1 - 0.5z} \in \boldsymbol{H}^2\right)$$

$$\boldsymbol{P}^- f(z) = \left(\frac{1}{z - 0.5} \in \boldsymbol{H}^{2\perp}\right)$$

Thus, $\left(\frac{1}{z - 0.5}\right)$ represents the best approximation to $f(z)$ in the $\boldsymbol{L}^2$ norm.

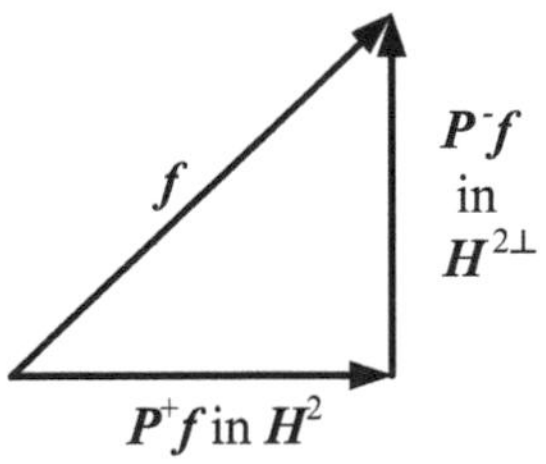

Figure 1.42: Best Approximation from a Function in $\boldsymbol{L}^2$ to $\boldsymbol{H}^2$

Example 1.32:

Let $f(z) \in \boldsymbol{L}^2$ on ∂D. $(z^n)_{n=-\infty}^{+\infty}$ are the basis of $\boldsymbol{L}^2$ in the sense of $\boldsymbol{L}^2$ convergence. This function can be represented by the Fourier series:

$$f(z) = \sum_{-\infty}^{+\infty} a_n z^n$$

$$f_1(z) = \boldsymbol{P}^+ f(z) = \sum_{0}^{\infty} a_n z^n,$$ belongs to $\boldsymbol{H}^2$ and represents analytic part inside the disc D

$$f_2(z) = \boldsymbol{P}^- f(z) = \sum_{-\infty}^{-1} a_n z^n,$$ belongs to $\boldsymbol{H}^{2\perp}$ and represents analytic part outside the disc D

The function $f_1(z)$ represents the best approximation to $f(z)$ in the L^2 norm by a H^2 function analytic inside the disc D.

Bibliography

[Chen, C.T.] Chen, C.T. *Linear System Theory and Design,* New York: Holt, Rinehart and Winston, 1984.

[Chen, W.K.] Chen, W.K. *Linear Network and Systems,* Boston: PWS Publishers, 1994.

[Cooper, G.R.] Cooper, G.R., and Gillan, C.D. *Methods of Signals and Systems Analysis,* New York: Holt, Rinehart and Winston Inc. 1967.

[Kailath, T.] Kailath, T. *Linear Systems,* Englewood Cliffs, NJ: Prentice Hall, 1980.

[Papoulis, A.] Papoulis, A. *Signal Analysis,* New York: McGraw Hill, 1997.

[Poularkis, A.D.] Poularkis, A.D. and Seely S. *Signals and Systems,* Boston: PWS Publishers, 1994.

[Schwartz, L.] Schwartz, L. Sur l'impossibilité de la multiplication des distributions. *Comptes Rendus de L'Academie des Sciences,* Paris, 239, 847–848, 1954.

[Young, N.] Young, N. *Introduction to Hilbert Spaces,* Cambridge University Press, 1988.

Chapter 2

Linear Operators and Matrices

2.1 Introduction

The linear operators (transformations) in Euclidian, Banach, Hilbert, and Hardy spaces play an important role in the communication and control theory. A complete book can justifiably be written on each of these topics. We shall present here some significant concepts from a systems theory viewpoint. Both finite and infinite dimensional matrices and operators are dealt with.

Let,

$$\boldsymbol{x} \in \boldsymbol{V_1}, \quad \boldsymbol{y} \in \boldsymbol{V_2} \quad \text{(Banach, Hilbert, or Euclidian Spaces)}$$

Define $\boldsymbol{A}$ as an operator or a mapping from the space $\boldsymbol{V}_1$ to another space $\boldsymbol{V}_2$, such that, $\boldsymbol{y} = \boldsymbol{A}\boldsymbol{x}$ ($\boldsymbol{y}$ is often called the image of $\boldsymbol{x}$ under $\boldsymbol{A}$)

- **Linear Operator**, $\boldsymbol{A} : \boldsymbol{V}_1 \rightarrow \boldsymbol{V}_2$ implies that,

$$\boldsymbol{y} = A(k_1\boldsymbol{x}_1 + k_2\boldsymbol{x}_2) = k_1\boldsymbol{A}\boldsymbol{x}_1 + k_2\boldsymbol{A}\boldsymbol{x}_2 \qquad \boldsymbol{x}_1, \boldsymbol{x}_2 \in \boldsymbol{V}_1,\ \boldsymbol{y} \in \boldsymbol{V}_2$$

If $\boldsymbol{V}_1 = \boldsymbol{V}_2 = \boldsymbol{V}$, then the operator is onto the space itself.

- **Kernel, Range and Domain of a Linear Operator**

Kernel of an operator $\boldsymbol{A}$(known as Ker $\boldsymbol{A}$) : $\boldsymbol{V}_1 \rightarrow \boldsymbol{V}_2$ is the subspace of $\boldsymbol{V}_1$ such that $\boldsymbol{x} \in \boldsymbol{V}_1$:

$$\boldsymbol{Ax} = \boldsymbol{0} \quad \text{Those elements of } \boldsymbol{V}_1 \text{ which map into origin in } \boldsymbol{V}_2.$$

Domain of an operator $\boldsymbol{A}$ is the set of vectors $\boldsymbol{x} \in \boldsymbol{V}_1$ for which $\boldsymbol{Ax}$ is defined. Other vectors of $\boldsymbol{V}_1$ are of marginal interest as far as the action of the operator is concerned.

Range of an operator $\boldsymbol{A}$ (Ran $\boldsymbol{A}$) is a subspace of the set of vectors $\boldsymbol{y} \in \boldsymbol{V}_2$ such that

$$\boldsymbol{y} = \boldsymbol{Ax}$$

The vector $\boldsymbol{y}$ is called the image of $\boldsymbol{x}$ under $\boldsymbol{A}$. There may be other vectors in $\boldsymbol{V}_2$ which are not in the range of the operator $\boldsymbol{A}$

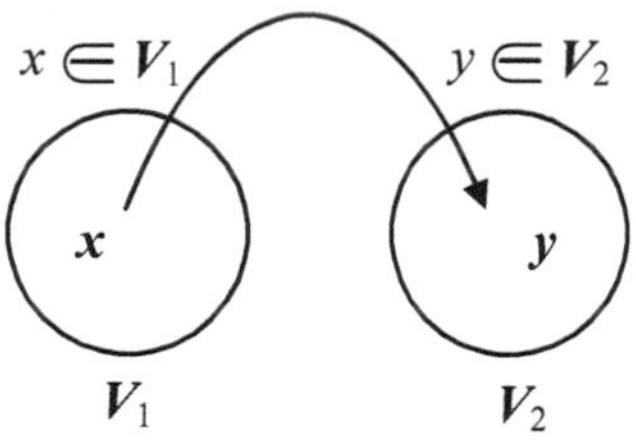

Figure 2.1: Domain and Range Subspaces

Norm of an operator $\boldsymbol{A}$ is denoted by $||\boldsymbol{A}||$, where,

$$||\boldsymbol{A}|| = \sup\left(||Ax|| : x \in \boldsymbol{V}_1, ||\boldsymbol{x}|| \leq 1\right)$$

- **Bounded Operators**: $\boldsymbol{A}$ is bounded if,

$$||\boldsymbol{Ax}|| \leq M||\boldsymbol{x}|| \quad \text{for all } \boldsymbol{x} \in \boldsymbol{V}_1 \quad 0 \leq M < \infty$$

Smallest value of M denotes the **Norm of the operator** $\boldsymbol{A}$ in the sense discussed earlier.

Thus

$$||\boldsymbol{A}|| = \sup_{x \in V} \frac{||\boldsymbol{A}\boldsymbol{x}||}{||\boldsymbol{x}||}$$

The norm $||\boldsymbol{A}||$ can be thought of as the largest factor by which the operator $\boldsymbol{A}$ can magnify any vector.

In fact

$$||\boldsymbol{A}x|| \leq ||\boldsymbol{A}||||x||$$

If there exists a maximizing vector $\boldsymbol{x} \in \boldsymbol{V}_1$ then the operator $\boldsymbol{A}$ attains its norm on that vector $\boldsymbol{x}_{\max}$ such that:

$$||\boldsymbol{A}\,\boldsymbol{x}_{\max}|| = ||\boldsymbol{A}||||\boldsymbol{x}_{\max}||$$

- **Continuous Operator**:

  An operator is continuous in $\boldsymbol{V}$, if for every $\boldsymbol{x} \in \boldsymbol{V}$ there exists a $\hat{\boldsymbol{x}} \in \boldsymbol{V}$ such that for $\epsilon > 0$ and $\delta(\epsilon) > 0$:

$$||\boldsymbol{x} - \hat{\boldsymbol{x}}|| < \delta(\epsilon) \Longrightarrow ||\boldsymbol{A}\boldsymbol{x} - \boldsymbol{A}\hat{\boldsymbol{x}}|| < \epsilon$$

The concept of Boundedness and Continuity are equivalent.

- **Compound Operator**:

  If:

$$\boldsymbol{C} = \boldsymbol{A}\boldsymbol{B} \quad (\boldsymbol{A}\boldsymbol{B} \text{ implies operator } \boldsymbol{B} \text{ followed by } \boldsymbol{A})$$

  Then:

$$||\boldsymbol{C}|| \leq ||\boldsymbol{A}||||\boldsymbol{B}||$$

In general

$$AB \neq BA$$

$(AB - BA)$ is called the **commutator** of A, B and is denoted by:

$$(AB - BA) = [A, B]$$

2.2 Matrix Algebra - Euclidian Vector Space

Matrix Representation of a Vector and a Linear Operator in Finite Dimensional Euclidean Space

Consider a vector $\boldsymbol{x}$ in a Euclidian vector space $\boldsymbol{E}_n$ of dimension n.

If this vector $\boldsymbol{x}$ is expanded in a basis $\{\boldsymbol{e}_i\}_{i=1}^n$, then,

$$\boldsymbol{x} = \sum_{i=1}^{n} x_i \boldsymbol{e}_i$$

Matrix representation of $\boldsymbol{x}$ with $\boldsymbol{e}_i$ basis in $\boldsymbol{E}_n$ is:

$$\boldsymbol{x} = \begin{bmatrix} x_1 \\ x_2 \\ \vdots \\ x_n \end{bmatrix} \quad \text{(called a } n\text{-vector or a column matrix)} \tag{2.1}$$

Note: As is customary, we shall use the same notation for a vector and it's matrix representation. The same is true for the space operator and its matrix representation. Let $\boldsymbol{A}$ be an operator which operates on $\boldsymbol{x}$ in space $\boldsymbol{E}_n$ of dimension n and produces another vector $\boldsymbol{y}$ in the space $\boldsymbol{E}_m$ of dimension m. This is symbolized as:

$$\boldsymbol{A}: \quad \boldsymbol{E}_n \rightarrow \boldsymbol{E}_m$$

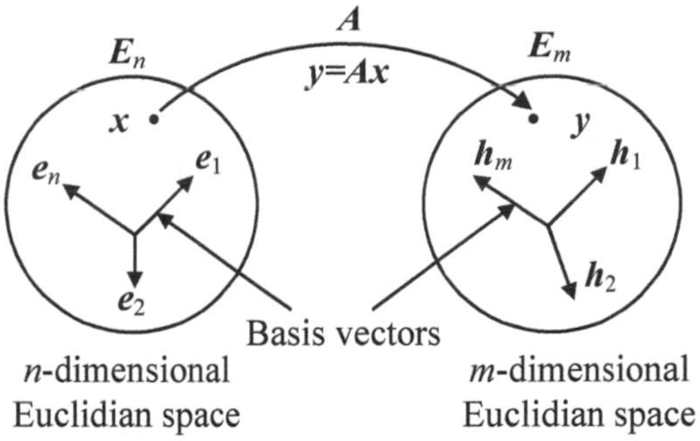

Figure 2.2: Operator Transformation of Vectors from One Space to Another

Let the space $\boldsymbol{E}_m$ have basis $\{\boldsymbol{h}_j\}_{j=1}^{m}$, so the vector $\boldsymbol{y}$ in $\boldsymbol{E}_m$ can be written as:

$$\boldsymbol{y} = \sum_{j=1}^{m} y_j \boldsymbol{h}_j \tag{2.2}$$

Matrix representation of $\boldsymbol{y}$ in $\boldsymbol{E}_m$ is:

$$\boldsymbol{y} = \begin{bmatrix} y_1 \\ y_2 \\ \vdots \\ y_m \end{bmatrix}$$

The operator $\boldsymbol{A}$ operates on the vector $\boldsymbol{x}$ to yield

$$\boldsymbol{Ax} = \sum_{i=1}^{n} x_i \boldsymbol{Ae}_i, \qquad \boldsymbol{Ax} = \boldsymbol{y} = \sum_{j=1}^{m} y_j \boldsymbol{h}_j \tag{2.3}$$

The vector $\boldsymbol{e}_i$ in $\boldsymbol{E}_n$ is acted upon by $\boldsymbol{A}$ to produce a vector $\boldsymbol{Ae}_i$ in $\boldsymbol{E}_m$ which can now be expanded as:

$$\boldsymbol{Ae}_i = \sum_{j=1}^{m} a_{ij} \boldsymbol{h}_j \tag{2.4}$$

From the Eqs. 2.3 and 2.4

$$\boldsymbol{y} = \sum_{j=1}^{m} y_j \boldsymbol{h}_j = \boldsymbol{A}\boldsymbol{x} = \sum_{i=1}^{n} x_i \boldsymbol{A}\boldsymbol{e}_i = \sum_{i=1}^{n}\sum_{j=1}^{m} x_i a_{ij} \boldsymbol{h}_j = \sum_{j=1}^{m}\left(\sum_{i=1}^{n} a_{ij} x_i\right)\boldsymbol{h}_j$$

or,

$$y_j = \sum_{i=1}^{n} a_{ij} x_i \tag{2.5}$$

The above equations can be written as:

$$\begin{bmatrix} y_1 \\ y_2 \\ \vdots \\ y_m \end{bmatrix} = \begin{bmatrix} a_{11} & a_{12} & \cdots & a_{1n} \\ a_{21} & a_{22} & \cdots & a_{2n} \\ \vdots & \vdots & \ddots & \vdots \\ a_{m1} & a_{m2} & \cdots & a_{mn} \end{bmatrix} \begin{bmatrix} x_1 \\ x_2 \\ \vdots \\ x_n \end{bmatrix} \tag{2.6}$$

Or in the matrix form:

$$\boldsymbol{y} = \boldsymbol{A}\,\boldsymbol{x} \tag{2.7}$$

Thus,

$$\underbrace{\boldsymbol{y} = \boldsymbol{A}\boldsymbol{x}}_{\text{space relation}} \quad \leftrightarrow \quad \underbrace{\boldsymbol{y} = \boldsymbol{A}\boldsymbol{x}}_{\text{matrix relation}}$$

Basis vectors need not be specified — Basis vectors of $\boldsymbol{x}$ and $\boldsymbol{y}$ are specified

Thus, $\boldsymbol{x}, \boldsymbol{y}, \boldsymbol{A}$ are the matrix representations of the spaces vectors $\boldsymbol{x}, \boldsymbol{y}$ and the operator $\boldsymbol{A}$ with respect to the basis $\{\boldsymbol{e}_i\}_1^n$ in $\boldsymbol{E}_n$ and the basis $\left\{\boldsymbol{h}_j\right\}_1^m$ in $\boldsymbol{E}_m$. Basis vectors are analogous to the coordinates. **However, the vectors and the operators exist independently of the basis assigned to them**.

A vector space whose vectors belong to some larger space is called a subspace. This concept is useful in developing the canonical forms of a matrix. Let $\boldsymbol{A}$ be an operator (or as is customary to say "mapping") of $\boldsymbol{E}_n$ onto itself. A subspace $\boldsymbol{E}_{n_i}$ of

$\boldsymbol{E}_n$ is invariant with respect to $\boldsymbol{A}$ if $\boldsymbol{Ax} \in \boldsymbol{E}_{n_i}$ implying $\boldsymbol{x} \in \boldsymbol{E}_{n_i}$. The structure of an invariant mapping (matrix) can be very usefully exploited by means of its invariant subspaces:

$$\boldsymbol{E}_n = \sum_{i=1}^{k} \boldsymbol{E}_{n_i} \qquad \left(i = 1, \ldots, k; \quad \sum_{j=1}^{k} n_j = n \right)$$

The basis of $\boldsymbol{E}_{n_i}$ consists of $\boldsymbol{e}_{ij}$ $(i = 1, \ldots, k;\ j = 1, \ldots, n_k)$. In the basis chosen, $\boldsymbol{A}$ can be represented by a quasi-diagonal form,

$$\boldsymbol{A} = \begin{bmatrix} [\boldsymbol{A}_{n_1}] & & & \\ & [\boldsymbol{A}_{n_2}] & & \\ & & \ddots & \\ & & & [\boldsymbol{A}_{n_k}] \end{bmatrix}$$

All other entries besides $[\cdot]$ along the main diagonal are zeros. The multiplication of $\boldsymbol{A}$ and $\boldsymbol{x}$ is interpreted as in the Eq. 2.5. The array a_{11}, a_{12} etc. are called as matrix representation of the operator $\boldsymbol{A}$ which takes a vector $\boldsymbol{x}$ in $\boldsymbol{E}_n$ to a vector $\boldsymbol{y}$ in $\boldsymbol{E}_m$. **The elements a_{ij} of the operator A depend upon the basis of E_n and the basis of E_m and will be different for different choice of the basis.**

Note: We emphasize the fact that as customary, the space vectors $\boldsymbol{x}$, $\boldsymbol{y}$ and the operator $\boldsymbol{A}$ will be represented in the matrix form by the same symbols $\boldsymbol{x}, \boldsymbol{y}$ and $\boldsymbol{A}$. In situations where confusion may arise, the matrix form of $\boldsymbol{x}, \boldsymbol{y}$ and $\boldsymbol{A}$ may be denoted as $\underline{\boldsymbol{x}}, \underline{\boldsymbol{y}}$ and $\underline{\boldsymbol{A}}$. **The Euclidian space of the operator A is designated as $C^{m \times n}$ m rows and n columns.**

Alternate Matrix Representations

Consider a $m \times n$ matrix $\boldsymbol{A}$ of m rows and n columns as shown in Eq.2.6.

Let $\boldsymbol{e}_i$, $i = 1, \cdots, n$ be an orthonormal set of basis column vectors with all the elements of $\boldsymbol{e}_i$ zero except the ith element which takes the value 1.

In terms of basis $\boldsymbol{e}_i$ the matrix $\boldsymbol{A}$ can be written as,

$$\boldsymbol{A} = [\boldsymbol{a}_1, \boldsymbol{a}_2, \cdots, \boldsymbol{a}_n] = \boldsymbol{a}_1\, \boldsymbol{e}_1^T + \boldsymbol{a}_2\, \boldsymbol{e}_2^T + \cdots + \boldsymbol{a}_n\, \boldsymbol{e}_n^T = \sum_{i=1}^{n} \boldsymbol{a}_i\, \boldsymbol{e}_i^T \quad \text{(sum of matrices)}$$

where,

$$\boldsymbol{a}_i = \begin{bmatrix} a_{i1} \\ a_{i2} \\ \vdots \\ a_{im} \end{bmatrix} \qquad (i = 1, 2, \ldots, n)$$

Similarly, an $n \times m$ matrix $\boldsymbol{B}$ can be represented as,

$$\boldsymbol{B} = \begin{bmatrix} \boldsymbol{b}_1^T \\ \boldsymbol{b}_2^T \\ \vdots \\ \boldsymbol{b}_n^T \end{bmatrix} = \boldsymbol{e}_1\, \boldsymbol{b}_1^T + \boldsymbol{e}_2\, \boldsymbol{b}_2^T + \cdots + \boldsymbol{e}_n\, \boldsymbol{b}_n^T = \sum_{j=1}^{n} \boldsymbol{e}_j\, \boldsymbol{b}_j^T \quad \text{(sum of matrices)}$$

where,

$$\boldsymbol{b}_j^T = \begin{bmatrix} b_{1j}\ b_{2j}\ \cdots\ b_{mj} \end{bmatrix} \qquad (j = 1, 2, \ldots, m)$$

Then,

$$\boldsymbol{AB} = \sum_{i=1}^{n} \sum_{j=1}^{n} \boldsymbol{a}_i (\boldsymbol{e}_i^T \boldsymbol{e}_j) \boldsymbol{b}_j^T$$

Since,

$$(\boldsymbol{e}_i^T \boldsymbol{e}_j) = \delta_{ij}$$

Therefore,

$$\boldsymbol{AB} = \sum_{i=1}^{n} \boldsymbol{a}_i\, \boldsymbol{b}_i^T$$

Note:

$\boldsymbol{AB} \neq \boldsymbol{BA}$ in general. (In fact $\boldsymbol{BA}$ may not exist)

2.2.1 Fundamental Concepts

Following is a review of the fundamentals of the matrix theory.

A scalar is a special case of a matrix with one row and one column.

1. **Column matrix (or vector)**

$$\boldsymbol{x} = \begin{bmatrix} x_1 \\ \vdots \\ x_n \end{bmatrix} \qquad (n \times 1 \text{ matrix})$$

2. **Row matrix (or vector)**

$$\boldsymbol{x}^T = \begin{bmatrix} x_1, \ldots, x_n \end{bmatrix} \qquad (1 \times n \text{ matrix})$$

3. **Matrix of order** $m \times n$

   It is an array of elements a_{ij} such that,

$$\boldsymbol{A} = \left((\boldsymbol{A})_{ij}\right) = \left(a_{ij}\right) \qquad (i = 1, \ldots, m; \; j = 1, \ldots, n)$$

   m represents the number of rows and the n the number of columns.

4. **Addition of matrices**

$$\boldsymbol{A} + \boldsymbol{B} = \left((\boldsymbol{A})_{ij} + (\boldsymbol{B})_{ij}\right) = \left(a_{ij} + b_{ij}\right) \qquad (i = 1, \ldots, m; \; j = 1, \ldots, n)$$

5. **Multiplication**

   If $\boldsymbol{A}$ is $m \times p$ and $\boldsymbol{B}$ is $p \times n$ then,

$$\boldsymbol{AB} = \sum_{k=1}^{p} \left((\boldsymbol{A})_{ik} (\boldsymbol{B})_{kj}\right) = \sum_{k=1}^{p} a_{ik} b_{kj} \qquad (i = 1, \ldots, m; \; j = 1, \ldots, n)$$

6. **Inverse**

If A be an $n \times n$ square matrix, its inverse, if it exists, is defined as:

$$A^{-1}A = AA^{-1} = I$$

where I is an identity matrix having ones along the diagonal and zero elsewhere. Later, we shall discuss how to compute A^{-1}, given A.

7. **Transpose, Adjoint**

Let A^T be the transpose of A

$$A^T = (a_{ji}) \qquad \text{(rows and columns of } A \text{ are exchanged)}$$

Then the adjoint matrix of A is

$$A^* = \bar{A}^T$$

"$\bar{A}$" stands for conjugate and "A^*" stands for the transpose conjugate.

A is a **unitary** matrix if,

$$A^{-1} = A^* = \bar{A}^T, \qquad AA^* = I$$

A is **symmetric** matrix if,

$$A^T = A$$

A is **Hermitian** matrix (useful in physics) if,

$$\bar{A}^T = A = A^*$$

This implies that the diagonal terms of a **Hermitian** matrix A are real and its off-diagonal terms are complex conjugates of each other.

Therefore, for a **Hermitian** matrix, a_{ii} = real and $a_{ij} = \bar{a}_{ji}$

8. **Commutator**

The commutator of $\boldsymbol{A}$ and $\boldsymbol{B}$ is

$$\boldsymbol{AB} - \boldsymbol{BA} = [\boldsymbol{A}, \boldsymbol{B}]$$

$[\boldsymbol{A}, \boldsymbol{B}] = 0$ implies that $\boldsymbol{A}$ and $\boldsymbol{B}$ are Hermitian.

9. **Determinant of Matrix $\boldsymbol{A}$**

Let $\boldsymbol{A} = \left(a_{ij}\right)$, $(i, j = 1, \ldots, n)$ be a square matrix.

The determinant of matrix $\boldsymbol{A}$ is a scalar quantity defined as

$$\Delta_A = \det(\boldsymbol{A}) = |\boldsymbol{A}| = \begin{vmatrix} a_{11} & a_{12} & \cdots & a_{1n} \\ a_{21} & a_{22} & \cdots & a_{2n} \\ \vdots & \vdots & \ddots & \vdots \\ a_{n1} & a_{n2} & \cdots & a_{nn} \end{vmatrix} = \sum_{\sigma} \operatorname{sgn}(\sigma)\, a_{1\sigma_1} a_{2\sigma_2} \cdots a_{n\sigma_n}$$

The index σ in the above sum varies over all the permutations of $[1, \ldots, n]$. The symbol $\operatorname{sgn}(\sigma)$ denotes the parity of the permutation σ and is +1 for the even and -1 for the odd permutations. There are n! terms in summation defining the determinant.

10. **Cofactors of a_{ij}**

Let $\boldsymbol{A} = \left(a_{ij}\right)$ be a square matrix. Then $\boldsymbol{A}^{ij}$ is defined as the co-factor of a_{ij} and is computed as the determinant of the matrix $\boldsymbol{A}$, after striking the i-th row and the j-th column and multiplied by $(-1)^{i+j}$.

Hence,

$$\Delta_A = \det(\boldsymbol{A}) = \sum_{j=1}^{n} a_{ij} (\boldsymbol{A})^{ij} = \sum_{i=1}^{n} a_{ij} (\boldsymbol{A})^{ij}$$

11. **Exterior Algebra and the Determinant Computation**

Exterior Algebra, also known as the 'Grassman Algebra,' has many applications in physics. Here we will introduce it only as an algorithm to compute the determinant of a matrix. Consider a n-dimensional space $\boldsymbol{L}$, with elements $\boldsymbol{x}_1, \boldsymbol{x}_2, \ldots$, etc, over a field $\boldsymbol{R}$ of real numbers $a, b, \cdots$. We shall construct a new vector space $\wedge^2$ which consists of sums of the form $\sum_i \sum_j a_i b_j (\boldsymbol{x}_i \wedge \boldsymbol{x}_j)$ subject to the following rules.

$$\left.\begin{aligned}(a_1\boldsymbol{x}_1 + a_2\boldsymbol{x}_2) \wedge \boldsymbol{x}_3 &= a_1\boldsymbol{x}_1 \wedge \boldsymbol{x}_3 + a_2\boldsymbol{x}_2 \wedge \boldsymbol{x}_3 \\ \boldsymbol{x}_3 \wedge (a_1\boldsymbol{x}_1 + a_2\boldsymbol{x}_2) &= a_1\boldsymbol{x}_3 \wedge \boldsymbol{x}_1 + a_2\boldsymbol{x}_3 \wedge \boldsymbol{x}_2 \\ \boldsymbol{x}_1 \wedge \boldsymbol{x}_2 &= -\boldsymbol{x}_2 \wedge \boldsymbol{x}_1 \\ \boldsymbol{x}_1 \wedge \boldsymbol{x}_1 &= -\boldsymbol{x}_1 \wedge \boldsymbol{x}_1 = 0 \\ \boldsymbol{x}_1 \wedge \boldsymbol{x}_2 \wedge \boldsymbol{x}_3 &= \boldsymbol{x}_1 \wedge (\boldsymbol{x}_2 \wedge \boldsymbol{x}_3) = (\boldsymbol{x}_1 \wedge \boldsymbol{x}_2) \wedge \boldsymbol{x}_3\end{aligned}\right\} \tag{2.8}$$

If $\boldsymbol{x}_2 = a\boldsymbol{x}_1$, then $\boldsymbol{x}_1 \wedge \boldsymbol{x}_2 = \boldsymbol{x}_1 \wedge a\boldsymbol{x}_1 = a\boldsymbol{x}_1 \wedge \boldsymbol{x}_1 = 0$

otherwise

$$\boldsymbol{x}_2 \wedge \boldsymbol{x}_1 \neq 0$$

$\boldsymbol{x}_1 \wedge \boldsymbol{x}_2$ is referred to as the "wedge product" or exterior product of $\boldsymbol{x}_1$ with $\boldsymbol{x}_2$. Let $\boldsymbol{e}_1, \boldsymbol{e}_2, \ldots, \boldsymbol{e}_n$ be a basis of space $\boldsymbol{L}$ represented as an n-tuple,

$$(1, 0, 0, \ldots, 0), (0, 1, 0, \ldots, 0), \ldots, (0, 0, 0, \ldots, 1).$$

Let $\boldsymbol{A}$ be a linear transformation on the space $\boldsymbol{L}$ onto itself with representation,

$$\boldsymbol{A} = \begin{bmatrix} a_{11} & a_{12} & & a_{1n} \\ a_{21} & a_{22} & \cdots & a_{2n} \\ \vdots & \vdots & & \vdots \\ a_{n1} & a_{n2} & & a_{nn} \end{bmatrix} = \begin{bmatrix} \boldsymbol{a}_1 & \boldsymbol{a}_2 & \cdots & \boldsymbol{a}_n \end{bmatrix} \tag{2.9}$$

The determinant of matrix $\boldsymbol{A}$ (a scalar number) can be written as a wedge product "volume,"

$$\begin{aligned} \det \boldsymbol{A} &= \boldsymbol{a}_1 \wedge \boldsymbol{a}_2 \wedge \cdots \wedge \boldsymbol{a}_n \\ 1 &= \boldsymbol{e}_1 \wedge \boldsymbol{e}_2 \wedge \cdots \wedge \boldsymbol{e}_n \end{aligned} \tag{2.10}$$

where,

$$\boldsymbol{a}_i = (a_{1i}\,\boldsymbol{e}_1 + a_{2i}\,\boldsymbol{e}_2 + \cdots + a_{ni}\,\boldsymbol{e}_n) \tag{2.11}$$

Wedge products can be taken two at a time and hence provide a very **efficient algorithm** for computing the determinants of the large matrices, particularly when they are sparse. Following example illustrates the algorithm.

Example 2.1:

$$\boldsymbol{A} = \begin{bmatrix} a_{11} & a_{12} \\ a_{21} & a_{22} \end{bmatrix} = \begin{bmatrix} \boldsymbol{a}_1 & \boldsymbol{a}_2 \end{bmatrix}$$

$$\det \boldsymbol{A} = \boldsymbol{a}_1 \wedge \boldsymbol{a}_2 = (a_{11}\,\boldsymbol{e}_1 + a_{21}\,\boldsymbol{e}_2)\,(a_{12}\,\boldsymbol{e}_1 + a_{22}\,\boldsymbol{e}_2)$$

$$\det \boldsymbol{A} = \boldsymbol{a}_{11}a_{12}\ \boldsymbol{e}_1 \wedge \boldsymbol{e}_1 + a_{11}a_{22}\ \boldsymbol{e}_1 \wedge \boldsymbol{e}_2 + a_{21}a_{12}\ \boldsymbol{e}_2 \wedge \boldsymbol{e}_1 + a_{21}a_{22}\ \boldsymbol{e}_2 \wedge \boldsymbol{e}_2$$

But,

$$\begin{aligned} &\boldsymbol{e}_1 \wedge \boldsymbol{e}_1 = \boldsymbol{e}_2 \wedge \boldsymbol{e}_2 = 0 \qquad \text{(repeated indices in the wedge product)} \\ &\boldsymbol{e}_2 \wedge \boldsymbol{e}_1 = -\boldsymbol{e}_1 \wedge \boldsymbol{e}_2 \\ &\boldsymbol{e}_1 \wedge \boldsymbol{e}_2 = 1 \end{aligned}$$

Thus,

$$\det \boldsymbol{A} = (a_{11}a_{22} - a_{21}a_{12}) \tag{2.12}$$

Example 2.2:

$$\boldsymbol{A} = \begin{bmatrix} a_{11} & a_{12} & a_{13} \\ a_{21} & a_{22} & a_{23} \\ a_{31} & a_{32} & a_{33} \end{bmatrix} = \begin{bmatrix} \boldsymbol{a}_1 & \boldsymbol{a}_2 & \boldsymbol{a}_3 \end{bmatrix}$$

$$\det \boldsymbol{A} = \boldsymbol{a}_1 \wedge \boldsymbol{a}_2 \wedge \boldsymbol{a}_3$$

$$\boldsymbol{a}_1 \wedge \boldsymbol{a}_2 = (a_{11}\boldsymbol{e}_1 + a_{21}\boldsymbol{e}_2 + a_{31}\boldsymbol{e}_3) \wedge (a_{12}\boldsymbol{e}_1 + a_{22}\boldsymbol{e}_2 + a_{32}\boldsymbol{e}_3)$$

$$= (a_{11}a_{22} - a_{21}a_{12})\boldsymbol{e}_1 \wedge \boldsymbol{e}_2 + (a_{11}a_{32} - a_{31}a_{12})\boldsymbol{e}_1 \wedge \boldsymbol{e}_3 + (a_{21}a_{32} - a_{31}a_{22})\boldsymbol{e}_2 \wedge \boldsymbol{e}_3$$

$$= \alpha_{11}\boldsymbol{e}_1 \wedge \boldsymbol{e}_2 + \alpha_{13}\boldsymbol{e}_1 \wedge \boldsymbol{e}_3 + \alpha_{23}\boldsymbol{e}_2 \wedge \boldsymbol{e}_3$$

Now,

$$\det \boldsymbol{A} = \boldsymbol{a}_1 \wedge \boldsymbol{a}_2 \wedge \boldsymbol{a}_3 = (\alpha_{11}\boldsymbol{e}_1 \wedge \boldsymbol{e}_2 + \alpha_{13}\boldsymbol{e}_1 \wedge \boldsymbol{e}_3 + \alpha_{23}\boldsymbol{e}_2 \wedge \boldsymbol{e}_3)(a_{13}\boldsymbol{e}_1 + a_{23}\boldsymbol{e}_2 + a_{33}\boldsymbol{e}_3)$$

Note:

$$\boldsymbol{e}_1 \wedge \boldsymbol{e}_2 \wedge \boldsymbol{e}_1 = \boldsymbol{e}_1 \wedge \boldsymbol{e}_2 \wedge \boldsymbol{e}_2 = \boldsymbol{e}_1 \wedge \boldsymbol{e}_3 \wedge \boldsymbol{e}_1 = \boldsymbol{e}_1 \wedge \boldsymbol{e}_3 \wedge \boldsymbol{e}_3 = \boldsymbol{e}_2 \wedge \boldsymbol{e}_3 \wedge \boldsymbol{e}_2 = \boldsymbol{e}_2 \wedge \boldsymbol{e}_3 \wedge \boldsymbol{e}_3 = 0$$

We obtain,

$$\det \boldsymbol{A} = \alpha_{11}a_{33}\ \boldsymbol{e}_1 \wedge \boldsymbol{e}_2 \wedge \boldsymbol{e}_3 + \alpha_{13}a_{23}\ \boldsymbol{e}_1 \wedge \boldsymbol{e}_3 \wedge \boldsymbol{e}_2 + \alpha_{23}a_{13}\ \boldsymbol{e}_2 \wedge \boldsymbol{e}_3 \wedge \boldsymbol{e}_1$$

$$\boldsymbol{e}_1 \wedge \boldsymbol{e}_3 \wedge \boldsymbol{e}_2 = -\boldsymbol{e}_1 \wedge \boldsymbol{e}_2 \wedge \boldsymbol{e}_3$$

$$\boldsymbol{e}_2 \wedge \boldsymbol{e}_3 \wedge \boldsymbol{e}_1 = -\boldsymbol{e}_2 \wedge \boldsymbol{e}_1 \wedge \boldsymbol{e}_3 = (-1)^2\boldsymbol{e}_1 \wedge \boldsymbol{e}_2 \wedge \boldsymbol{e}_3$$

Thus,

$$\det \boldsymbol{A} = (\alpha_{11}a_{33} - \alpha_{13}a_{23} + \alpha_{23}a_{13})$$

$$\det \boldsymbol{A} = (a_{11}a_{22} - a_{21}a_{12})\,a_{33} - (a_{11}a_{32} - a_{31}a_{12})\,a_{23} + (a_{21}a_{32} - a_{31}a_{22})\,a_{13} \quad (2.13)$$

Note: Any time the indices of any two elements repeat in a wedge product, it yields a zero term. **Geometric Product and Planar Geometric Algebra**

In this section, $\boldsymbol{x}_i, \boldsymbol{e}_i, i = 1, 2, \ldots$ are vectors and $\boldsymbol{e}_0, a_i, b_i, c_i, i = 1, 2, \ldots$ are scalar numbers. Geometric product of two vectors $\boldsymbol{x}_1$ and $\boldsymbol{x}_2$ is defined in terms of "inner product" and "exterior product" as:

$$\boldsymbol{x}_1\boldsymbol{x}_2 = \boldsymbol{x}_1 \cdot \boldsymbol{x}_2 + \boldsymbol{x}_1 \wedge \boldsymbol{x}_2 \qquad \textbf{Definition}$$

$$\begin{aligned} \boldsymbol{x}_1 \cdot \boldsymbol{x}_2 &= \frac{1}{2}(\boldsymbol{x}_1\boldsymbol{x}_2 + \boldsymbol{x}_2\boldsymbol{x}_1) && \text{Scalar product} \\ \boldsymbol{x}_1 \wedge \boldsymbol{x}_2 &= \frac{1}{2}(\boldsymbol{x}_1\boldsymbol{x}_2 - \boldsymbol{x}_2\boldsymbol{x}_1) && \text{Exterior or Wedge product} \end{aligned}$$

Obviously,

$$\begin{aligned} \boldsymbol{x}_1 \cdot \boldsymbol{x}_2 &= \boldsymbol{x}_2 \cdot \boldsymbol{x}_1 \\ \boldsymbol{x}_1 \wedge \boldsymbol{x}_2 &= -\boldsymbol{x}_2 \wedge \boldsymbol{x}_1 \\ (\boldsymbol{x}_1 + \boldsymbol{x}_2)^2 &= \boldsymbol{x}_1^2 + \boldsymbol{x}_2^2 + \boldsymbol{x}_1\boldsymbol{x}_2 + \boldsymbol{x}_2\boldsymbol{x}_1 \end{aligned}$$

Two-dimensional space:

Let $\boldsymbol{e}_1$ and $\boldsymbol{e}_2$ be unit vectors and

$\boldsymbol{e}_0$ as a unit scalar ($\boldsymbol{e}_0 = 1$, any multiplication with $\boldsymbol{e}_0$ leaves the vector as it is)

$$\begin{aligned} &\boldsymbol{e}_1 \cdot \boldsymbol{e}_1 = 1\,, \quad \boldsymbol{e}_1 \wedge \boldsymbol{e}_1 = 0 \\ &\boldsymbol{e}_2 \cdot \boldsymbol{e}_2 = 1\,, \quad \boldsymbol{e}_2 \wedge \boldsymbol{e}_2 = 0 \\ &\boldsymbol{e}_1 \cdot \boldsymbol{e}_2 = \boldsymbol{e}_2 \cdot \boldsymbol{e}_1 = 0\,, \quad \boldsymbol{e}_1 \wedge \boldsymbol{e}_2 = -\boldsymbol{e}_2 \cdot \boldsymbol{e}_1 \end{aligned}$$

$$\boldsymbol{e}_0\boldsymbol{e}_1 = \boldsymbol{e}_1\,, \quad \boldsymbol{e}_0\boldsymbol{e}_1 = \boldsymbol{e}_2$$

If $\quad \boldsymbol{x}_1 = a_1\boldsymbol{e}_1 + a_2\boldsymbol{e}_2\,, \quad \boldsymbol{x}_2 = b_1\boldsymbol{e}_1 + b_2\boldsymbol{e}_2$

Thus, $\quad \boldsymbol{x}_1\boldsymbol{x}_2 = (a_1b_1 + a_2b_2)\boldsymbol{e}_0 + (a_1b_2 - a_2b_1)\boldsymbol{e}_1 \wedge \boldsymbol{e}_2$

If:

$$\boldsymbol{x}_1 \text{ is oriented along } \boldsymbol{e}_1 \text{ implying } a_2 = 0$$
$$\boldsymbol{x}_2 \text{ is oriented along } \boldsymbol{e}_2 \text{ implying } b_1 = 0$$

Then,

$$\boldsymbol{x}_1\boldsymbol{x}_2 = \boldsymbol{x}_1 \wedge \boldsymbol{x}_2 = a_1 b_2 \boldsymbol{e}_1 \wedge \boldsymbol{e}_2$$

$$\begin{aligned}(\boldsymbol{x}_1 \wedge \boldsymbol{x}_2)^2 &= (a_1\boldsymbol{e}_1 \wedge b_2\boldsymbol{e}_2) \wedge (a_1\boldsymbol{e}_1 \wedge b_2\boldsymbol{e}_2) = -(a_1\boldsymbol{e}_1 \wedge b_2\boldsymbol{e}_2 \wedge b_2\boldsymbol{e}_2 \wedge a_1\boldsymbol{e}_1)\\ &= -(a_1 b_2)^2\end{aligned}$$

Clifford Algebra

Introducing the notion of "multivectors" we define the elements of a multivector algebra as:

$$\begin{aligned}\boldsymbol{x}_1 &= a_0\boldsymbol{e}_0 + a_1\boldsymbol{e}_1 + a_2\boldsymbol{e}_2 + a_3\boldsymbol{e}_1 \wedge \boldsymbol{e}_2\\ \boldsymbol{x}_2 &= b_0\boldsymbol{e}_0 + b_1\boldsymbol{e}_1 + b_2\boldsymbol{e}_2 + b_3\boldsymbol{e}_1 \wedge \boldsymbol{e}_2\end{aligned}$$

It is easy to see that:

$$\begin{aligned}\boldsymbol{x}_1 + \boldsymbol{x}_2 &= (a_2 + b_2)\boldsymbol{e}_0 + (a_1 + b_1)\boldsymbol{e}_1 + (a_2 + b_2)\boldsymbol{e}_2 + (a_3 + b_3)\boldsymbol{e}_1 \wedge \boldsymbol{e}_2\\ \boldsymbol{x}_1\boldsymbol{x}_2 &= c_0\boldsymbol{e}_0 + c_1\boldsymbol{e}_1 + c_2\boldsymbol{e}_2 + c_3\boldsymbol{e}_1 \wedge \boldsymbol{e}_2\end{aligned}$$

$$\begin{aligned}c_0 &= a_0b_0 + a_1b_1 + a_2b_2 - a_3b_3\\ c_1 &= a_0b_1 + a_1b_0 + a_3b_2 - a_2b_3\\ c_2 &= a_0b_2 + a_2b_0 + a_1b_3 - a_3b_1\\ c_3 &= a_0b_3 + a_3b_0 + a_1b_2 - a_2b_1\end{aligned}$$

12. **Adjugate of matrix A**

The matrix formed by $(\boldsymbol{A})^{ij}$ (the co-factors of the elements a_{ij} of $\boldsymbol{A}$) is defined as Adj $(\boldsymbol{A})$

13. **Inverse matrix A^{-1}**

We shall assume the knowledge of the determinant of a matrix and its elementary properties.

Let Δ_A be the determinant of $\boldsymbol{A}$, where $\boldsymbol{A}$ is an $n \times n$ square matrix.

$(\boldsymbol{A})^{ij}$ be the ij cofactor of a_{ij}, that is, the determinant of the matrix $\boldsymbol{A}$ after striking out the ith row and jth column, multiplied by $(-1)^{i+j}$.

From Laplace's expansion

$$\sum_{j=1}^{n} a_{ij}(A)^{ij} = \Delta_A$$

$$\sum_{j=1}^{n} a_{ij}(A)^{kj} = \Delta_A \delta_{ik}$$

$$\sum_{j=1}^{n} a_{ij} \frac{(A)^{kj}}{\Delta_A} = \delta_{ik} \qquad i, k = 1, 2, \ldots, n \tag{2.14}$$

$$\boldsymbol{A}\boldsymbol{A}^{-1} = \boldsymbol{A}^{-1}\boldsymbol{A} = \boldsymbol{I} \qquad \text{Identity matrix}$$

Elements of the matrix $\boldsymbol{A}\boldsymbol{A}^{-1}$ are,

$$(\boldsymbol{A}\boldsymbol{A}^{-1})_{ik} = \sum_{j=1}^{n} a_{ij}(\boldsymbol{A}^{-1})_{jk} = \delta_{ik} \tag{2.15}$$

Comparing Eqs. 2.14 and 2.15,

$$(\boldsymbol{A}^{-1})_{jk} = \frac{(\boldsymbol{A})^{kj}}{\Delta_A}$$

Hence,

$$\boldsymbol{A}^{-1} = \left(\frac{1}{\Delta_A}\right)(\text{Adj}\,(\boldsymbol{A}))^T \tag{2.16}$$

Recursive Inverse Algorithm for Large Matrices

For large values of n, $\boldsymbol{A}^{-1}$ is a very computationally extensive task.

Following algorithm can be used to compute $\boldsymbol{A}^{-1}$ recursively for large matrices.

Given:

$$\boldsymbol{A} = \boldsymbol{A}_n = \begin{bmatrix} a_{11} & a_{12} & \cdots & a_{1n} \\ a_{21} & a_{22} & \cdots & a_{2n} \\ \vdots & \vdots & & \\ a_{n1} & a_{n2} & \cdots & a_{nn} \end{bmatrix}, \quad \boldsymbol{A}_k = \begin{bmatrix} a_{11} & a_{12} & \cdots & a_{1k} \\ a_{21} & a_{22} & \cdots & a_{2k} \\ \vdots & \vdots & & \\ a_{k1} & a_{k2} & \cdots & a_{kk} \end{bmatrix}, \quad \text{we assume } \boldsymbol{A}_k^{-1} \text{ is known}$$

$$\boldsymbol{A}_{k+1} = \left[\begin{array}{c|c} & a_{1,k+1} \\ \boldsymbol{A}_k & \vdots \\ & a_{k,k+1} \\ \hline a_{k+1,1} \; a_{k+1,2} \; \cdots & a_{k+1,k+1} \end{array}\right] = \left[\begin{array}{c|c} \boldsymbol{A}_k & \boldsymbol{x}_{k+1} \\ \hline \boldsymbol{y}_{k+1}^T & a_{k+1,k+1} \end{array}\right]$$

Furthermore, we can write $\boldsymbol{A}_{k+1}^{-1} = \left[\begin{array}{c|c} \boldsymbol{W}_{k+1} & \boldsymbol{u}_{k+1} \\ \hline \boldsymbol{v}_{k+1}^T & \alpha_{k+1} \end{array}\right]$

Given $\boldsymbol{A}_k^{-1}, \boldsymbol{x}_{k+1}, \boldsymbol{y}_{k+1}^T, a_{k+1,k+1}, k = 2, 3, \ldots, n-1$, we are required to compute: $\boldsymbol{W}_{k+1}, \boldsymbol{u}_{k+1}, \boldsymbol{v}_{k+1}$ and α_{k+1}, and hence $\boldsymbol{A}_{k+1}^{-1}$

Solution:

$$\boldsymbol{A}_{k+1}\boldsymbol{A}_{k+1}^{-1} = \boldsymbol{I}_{k+1} = \left[\begin{array}{c|c} \boldsymbol{I}_k & \boldsymbol{0} \\ \hline \boldsymbol{0} & 1 \end{array}\right]$$

Thus

$$\left[\begin{array}{c|c} \boldsymbol{A}_k & \boldsymbol{x}_{k+1} \\ \hline \boldsymbol{y}_{k+1}^T & a_{k+1,k+1} \end{array}\right] \left[\begin{array}{c|c} \boldsymbol{W}_{k+1} & \boldsymbol{u}_{k+1} \\ \hline \boldsymbol{v}_{k+1}^T & \alpha_{k+1,k+1} \end{array}\right] = \left[\begin{array}{c|c} \boldsymbol{I}_k & \boldsymbol{0} \\ \hline \boldsymbol{0} & 1 \end{array}\right] \tag{2.17}$$

From Eq 2.17, it is easy to compute:

$$
\begin{aligned}
\alpha_{k+1} &= -\left(a_{k+1,k+1} - \boldsymbol{y}_{k+1}^T \boldsymbol{A}_k^{-1} \boldsymbol{x}_{k+1}\right)^{-1} \\
\boldsymbol{v}_{k+1}^T &= \alpha_{k+1} \boldsymbol{y}_{k+1}^T \boldsymbol{A}_k^{-1} \\
\boldsymbol{u}_{k+1} &= -\alpha_{k+1} \boldsymbol{A}_k^{-1} \boldsymbol{x}_{k+1} \\
\boldsymbol{W}_{k+1} &= \boldsymbol{A}_k^{-1} \left(\boldsymbol{I}_k - \boldsymbol{x}_{k+1} \boldsymbol{v}_{k+1}^T\right)
\end{aligned}
$$

14. **The determinant of the product of two square matrices *A* and *B***

$$\det(\boldsymbol{AB}) = \det(\boldsymbol{A}) \det(\boldsymbol{B}) = \Delta_A \Delta_B$$

15. **A matrix *A* is singular if**

$$\det(\boldsymbol{A}) = \Delta_A = 0$$

16. **Minors of a matrix**

Choose any k rows and any k columns from the matrix $\boldsymbol{A}$ and form a sub-matrix. The determinant of this sub-matrix, with rows and columns in their natural order, is called the k-th order minor of the matrix $\boldsymbol{A}$.

17. **Trace of a square matrix *A***

When $m = n$, we have a square matrix. Sum of all the diagonal elements of a square matrix is an important parameter in stability study of a differential equation and is called the Trace.

$$\text{Trace } \boldsymbol{A} = \sum_{i=1}^{n} a_{ii} = \text{Tr}\,(\boldsymbol{A})$$

Some books use the word "spure" for the trace (German word).

18. **Projection matrices**

If $\boldsymbol{P}$ is Hermitian and $\boldsymbol{P}^n = \boldsymbol{P}$, $(n = 2, \ldots)$, then $\boldsymbol{P}$ is called a projection matrix. Any arbitrary vector $\boldsymbol{x}$ can be decomposed into two vectors $\boldsymbol{y}$ and $\boldsymbol{z}$ such that

$$\boldsymbol{x} = \boldsymbol{y} + \boldsymbol{z}$$

$$\boldsymbol{y} = \boldsymbol{P}\boldsymbol{x} \quad , \quad \boldsymbol{z} = (\boldsymbol{I} - \boldsymbol{P})\boldsymbol{x} \quad , \quad \boldsymbol{P}\boldsymbol{z} = \boldsymbol{0}$$

Any vector space $\boldsymbol{V}$ can be decomposed into $\boldsymbol{W}$ and $\boldsymbol{W}^{\perp}$, where

$$\boldsymbol{y} = \boldsymbol{P}\boldsymbol{x} \in \boldsymbol{W} \text{ and } \boldsymbol{z} = (\boldsymbol{I} - \boldsymbol{P})\boldsymbol{x} \in \boldsymbol{W}^{\perp}, \quad \boldsymbol{W} + \boldsymbol{W}^{\perp} = \boldsymbol{V}$$

19. **Rank of a matrix**.

The rank is a very useful concept in the solution of simultaneous equations. It can be defined in many different (but equivalent) ways. In particular, it is the largest order of a nonvanishing minor, and it is also the maximum number of linearly independent rows (or of linearly independent columns of the matrix). Thus, given a $m \times n$ matrix $\boldsymbol{A}$, its rank $r \leq m, n$.

20. **Kernel, Range and Nullity of a matrix**

Let $\boldsymbol{A}$ be a matrix transformation from $\boldsymbol{E}_n$ on $\boldsymbol{E}_m$.

The kernel of the matrix $\boldsymbol{A}$ is the totality of $\boldsymbol{x} \in \boldsymbol{E}_n$ for which $\boldsymbol{A}\boldsymbol{x} = \boldsymbol{0}$.

The range of $\boldsymbol{A}$ is the totality of all the vectors $\boldsymbol{A}\boldsymbol{x} \in \boldsymbol{E}_m$.

These are denoted as ker $\boldsymbol{A}$ and rng $\boldsymbol{A}$ respectively.

Let dim stand for the dimension. Then dim (ker $\boldsymbol{A}$) is also known as the nullity of $\boldsymbol{A}$. Furthermore, dim (rng $\boldsymbol{A}$) is the rank of $\boldsymbol{A}$.

Sylvester's law of nullity states that

$$\dim(\ker \boldsymbol{A}) + \dim(\text{rng}\, \boldsymbol{A}) = \dim(\boldsymbol{E}_n)$$

21. **Representation of $C^{m\times n}$ Euclidean Space Operator as a Matrix Norm**

Let the Bounded LTI operator be defined by $m \times n$ matrix $\underline{A}$:

$$\underline{y} = \underline{A}\,\underline{x}$$

$\underline{x}$ is a m vector and $\underline{y}$ is a n vector and $\underline{A}$ is in Euclidean space $C^{m\times n}$. The "magnitude" of matrices can be determined by any vector norm on $C^{m\times n}$. The norm of a matrix $\boldsymbol{A}$ is called Norm $||\boldsymbol{A}||$ and corresponds to the "greatest stretching" of the vectors under its mapping.

Common matrix norms are:

(a) **Frobenius Norm** $||\underline{A}||_F$ (or Hilbert-Schmidt Norm)

$$||\underline{A}||_F^2 = \sum_{i=1}^{n}\sum_{j=1}^{n} |a_{ij}|^2 = \sum_{i=1}^{n} ||\underline{a}_i||^2 = \sum_{j=1}^{n} ||\underline{a}_j^*||^2 = \text{Trace}(\underline{A}^*\underline{A})$$

Example 2.3:

$$\underline{A} = \begin{bmatrix} 1 & 2 \\ 3 & -4 \end{bmatrix}, \quad ||\underline{A}||_F = [1^2 + 2^2 + 3^2 + (-4)^2]^{\frac{1}{2}} = (30)^{\frac{1}{2}}$$

(b) **Matrix p-Norm** $||\underline{A}||^p = \max_{||\boldsymbol{x}||^p=1} ||\underline{\boldsymbol{A}\boldsymbol{x}}||^p, \quad 1 < p \leq \infty$

(c) **Maximum Absolute Column Sum Norm** $||\underline{A}||^1$

$$||\underline{A}||^1 = \max_j \sum_{i=1}^{n} |a_{ij}|$$

Example 2.4:

$$\underline{A} = \begin{bmatrix} 1 & 2 \\ 3 & -4 \end{bmatrix}, \quad ||\underline{A}||^1 = [|3| + |-4|] = 7$$

22. **Few other Norms of Matrices and Vectors**

(a) $||\boldsymbol{A}||^m = \max_i \sum_j |a_{ij}|, \quad ||\boldsymbol{x}||_m = \max_i |\boldsymbol{x}_i| \quad (m\text{-norm})$

(b) $||\boldsymbol{A}||^l = \max_j \sum_i |a_{ij}|, \quad ||\boldsymbol{x}||_l = \sum_j |\boldsymbol{x}_j| \quad (l\text{-norm})$

(c) $||\boldsymbol{A}||^k = \left[\sum_{ij} |a_{ij}^2|\right]^{1/2}, ||\boldsymbol{x}||_k = \left[\sum_j |\boldsymbol{x}_j|^2\right]^{1/2} \quad (k\text{-norm})$

(d) $||\underline{\boldsymbol{A}}||^2 = (\text{Maximum Eigenvalue of } \underline{\boldsymbol{A}}^* \underline{\boldsymbol{A}})^{\frac{1}{2}}$ **Spectral Norm**

Eigenvalue defined later.

(e) $||\underline{\boldsymbol{A}}||^\infty = \max_i \sum_{j=1}^m |a_{ij}|$

Furthermore;

- $||\underline{\boldsymbol{A}}||^2 \leq ||\underline{\boldsymbol{A}}||^1 ||\underline{\boldsymbol{A}}||^\infty$
- If $\underline{\boldsymbol{A}}$ is a square matrix with Eigenvalues $\lambda_1, \lambda_2, \cdots, \lambda_n$, then

$$\frac{1}{||\underline{\boldsymbol{A}}^{-1}||} \leq |\lambda_i| \leq ||\underline{\boldsymbol{A}}||, \quad i = 1, 2, \cdots, n$$

- If $\underline{\boldsymbol{C}} = \underline{\boldsymbol{A}}\,\underline{\boldsymbol{B}}$, then

$$||\underline{\boldsymbol{C}}|| = ||\underline{\boldsymbol{A}}|| ||\underline{\boldsymbol{B}}||$$

23. **Hadamard's inequality (bound on the magnitude of a determinant)**

We want to obtain a bound on the determinant $\det(\boldsymbol{A}) = \Delta_A$.

Let

$$\boldsymbol{A} = \begin{bmatrix} \boldsymbol{a}_1^T \\ \boldsymbol{a}_2^T \\ \vdots \\ \boldsymbol{a}_n^T \end{bmatrix}, \quad \boldsymbol{A}^T = \begin{bmatrix} \boldsymbol{a}_1 & \boldsymbol{a}_2 & \cdots & \boldsymbol{a}_n \end{bmatrix}$$

Thus,

$$AA^T = A^TA = \begin{bmatrix} a_1^T a_1 & a_1^T a_2 & \cdots & a_1^T a_n \\ a_2^T a_1 & a_2^T a_2 & \cdots & a_2^T a_n \\ \vdots & & & \\ a_n^T a_1 & a_n^T a_2 & \cdots & a_n^T a_n \end{bmatrix}$$

$$\det(AA^T) = |\Delta_A|^2 \le \prod_{i=1}^{n} \|a_i\|^2 = \left(\sum_{j=1}^{n} |a_{1j}|^2\right)\left(\sum_{j=1}^{n} |a_{2j}|^2\right)\cdots\left(\sum_{j=1}^{n} |a_{nj}|^2\right)$$

If every element a_{ij} of the matrix A is such that

$$|a_{ij}| < K \qquad i, j = 1, 2, \ldots, n$$

Then,

$$|\Delta_A| \le \left(n^{n/2}\right)(K^n)$$

This is known as Hadamard's inequality.

24. **Matrix Functions - Geometric Series and Convergence Conditions**

$$I + A + A^2 + \cdots = \sum_{k=1}^{\infty} A^k = (I - A)^{-1}$$

$\|A\| < 1$ is the convergence condition for the above geometric series.

$$\|(I - A)^{-1}\| \le (1 - \|A\|)^{-1}$$

Let

$$f(\lambda) = \lambda^n + a_1\lambda^{n-1} + \ldots + a_n \quad \text{(scalar polynomial)}$$

We can define a polynomial matrix function:

$$f(A) = A^n + a_1 A^{n-1} + \ldots + a_n I$$

25. **Exponential, Trignometric Matrix Functions**

As in scalar functions, the exponential, logarithmic and trignometric matrix functions can be considered as definitions (convergence conditions satisfied).

$$
\begin{aligned}
e^{\boldsymbol{A}t} &= \sum_{n=0}^{\infty} \frac{(\boldsymbol{A}t)^n}{n!} = \boldsymbol{I} + \boldsymbol{A}t + \frac{1}{2}(\boldsymbol{A}t)^2 + \cdots + \frac{1}{n!}(\boldsymbol{A}t)^n + \cdots, \qquad n \to \infty \\
e^{\boldsymbol{A}} e^{\boldsymbol{B}} &= e^{(\boldsymbol{A}+\boldsymbol{B})} \text{ if } \boldsymbol{AB} = \boldsymbol{BA} \\
\frac{\mathrm{d}}{\mathrm{d}t}\left(e^{\boldsymbol{A}t}\right) &= \boldsymbol{A}e^{\boldsymbol{A}t} \qquad 0 \le t \le \infty \\
\frac{\mathrm{d}}{\mathrm{d}t}\operatorname{Tr}\left(e^{\boldsymbol{A}t}\right) &= \operatorname{Tr}\left(\boldsymbol{A}e^{\boldsymbol{A}t}\right) \\
\det e^{\boldsymbol{A}t} &= e^{\operatorname{Tr}(\boldsymbol{A}t)} \\
\ln(\boldsymbol{I} + \boldsymbol{A}) &= \sum_{n=1}^{\infty} (-1)^{n+1} \frac{\boldsymbol{A}^n}{n} = \boldsymbol{A} - \frac{1}{2}\boldsymbol{A}^2 + \cdots + \frac{(-1)^{n+1}}{n}\boldsymbol{A}^n + \cdots, \; n \to \infty \\
\sin(\boldsymbol{A}) &= \sum_{n=0}^{\infty} \frac{(-1)^n \boldsymbol{A}^{2n+1}}{(2n+1)!} \qquad \text{odd powers} \\
\cos(\boldsymbol{A}) &= \sum_{n=0}^{\infty} \frac{(-1)^n \boldsymbol{A}^{2n}}{(2n)!} \qquad \text{even powers}
\end{aligned}
$$

26. **Summation of Finite Series**

$$
\boldsymbol{I} + \boldsymbol{A} + \boldsymbol{A}^2 + \cdots + \boldsymbol{A}^n = \left(\boldsymbol{I} - \boldsymbol{A}^{n+1}\right)(\boldsymbol{I} - \boldsymbol{A})^{-1}
$$

27. **Taylor Series expansion for vector functions**

$$
\boldsymbol{f}(\boldsymbol{x} + \boldsymbol{h}) = \boldsymbol{f}(\boldsymbol{x}) + (\boldsymbol{\nabla}_x \boldsymbol{f}(\boldsymbol{x}))^T \boldsymbol{h} + \frac{1}{2}(\boldsymbol{\nabla}_{xx} \boldsymbol{f}(\boldsymbol{x}))\,\boldsymbol{h} + \text{ HOT}
$$

where

$$
\boldsymbol{\nabla}_x = \begin{bmatrix} \dfrac{\partial}{\partial x_1} \\ \vdots \\ \dfrac{\partial}{\partial x_n} \end{bmatrix}, \qquad
\boldsymbol{\nabla}_{xx} = \begin{bmatrix} \dfrac{\partial^2}{\partial x_1 \partial x_1} & \dfrac{\partial^2}{\partial x_1 \partial x_2} & \cdots & \dfrac{\partial^2}{\partial x_1 \partial x_n} \\ \vdots & & & \\ \dfrac{\partial^2}{\partial x_n \partial x_1} & \dfrac{\partial^2}{\partial x_n \partial x_2} & \cdots & \dfrac{\partial^2}{\partial x_n \partial x_n} \end{bmatrix}
$$

28. **Derivative of the Determinant**

$$\frac{\partial}{\partial x}\boldsymbol{A}(x) = [\det(\boldsymbol{A}(x))]\ \mathrm{Tr}\left[\boldsymbol{A}^{-1}\frac{\partial}{\partial x}\boldsymbol{A}(x)\right]$$

$$\frac{\mathrm{d}}{\mathrm{d}x}(\alpha\boldsymbol{A}(x)) = \alpha\frac{\mathrm{d}}{\mathrm{d}x}\boldsymbol{A}(x), \quad \alpha \text{ is a scalar.}$$

$$\frac{\mathrm{d}}{\mathrm{d}x}\boldsymbol{A}^{-1}(x) = -\boldsymbol{A}^{-1}\left(\frac{\mathrm{d}}{\mathrm{d}x}\boldsymbol{A}(x)\right)\boldsymbol{A}^{-1}$$

$$\nabla_x\left(\boldsymbol{x}^T\boldsymbol{A}\boldsymbol{x}\right) = \left(\boldsymbol{A} + \boldsymbol{A}^T\right)\boldsymbol{x}$$

$$\nabla_x\left(\boldsymbol{x}^T\boldsymbol{A}\boldsymbol{x} + \boldsymbol{b}\boldsymbol{x}\right) = \left(\boldsymbol{A} + \boldsymbol{A}^T\right)\boldsymbol{x} + \boldsymbol{b}$$

$$\nabla_{xx}\left(\boldsymbol{x}^T\boldsymbol{A}\boldsymbol{x} + \boldsymbol{b}\boldsymbol{x}\right) = \boldsymbol{A} + \boldsymbol{A}^T$$

29. **Structured Matrices**

(1) **Vandermonde Matrix**

$$\boldsymbol{V} = V(\lambda_1, \lambda_2, \ldots, \lambda_n) = \begin{bmatrix} 1 & \lambda_1 & \lambda_1^2 & \cdots & \lambda_1^{n-1} \\ 1 & \lambda_2 & \lambda_2^2 & \cdots & \lambda_2^{n-1} \\ \vdots & \vdots & \vdots & & \vdots \\ 1 & \lambda_n & \lambda_n^2 & \cdots & \lambda_n^{n-1} \end{bmatrix} \det\boldsymbol{V} = \prod_{i>j}^{n}\left(\lambda_i - \lambda_j\right)$$

The matrix $\boldsymbol{V}$ can be associated with a characteristic polynomial

$$P_n(\lambda) = a_1 + a_2\lambda + a_3\lambda^2 + \cdots + a_n\lambda^{n-1}$$

Given n distinct points $\lambda_1, \lambda_2, \ldots, \lambda_n$ and n values $w_1, w_2, \ldots, w_n$, there exists unique polynomial $P_n(\lambda)$ for which

$$P_n(\lambda_i) = w_i$$

or

$$\begin{bmatrix} 1 & \lambda_1 & \lambda_1^2 & \cdots & \lambda_1^{n-1} \\ \vdots & \vdots & \vdots & & \vdots \\ 1 & \lambda_n & \lambda_n^2 & \cdots & \lambda_n^{n-1} \end{bmatrix} \begin{bmatrix} a_1 \\ \vdots \\ a_n \end{bmatrix} = \begin{bmatrix} w_1 \\ \vdots \\ w_n \end{bmatrix}$$

(2) **Toeplitz Matrix, Finite Dimension**

Toeplitz Matix $\boldsymbol{T}$ is a $n \times n$ square matrix whose entries along the main diagonal and diagonals parallel to it are same, yielding,

$$\boldsymbol{T} = \begin{bmatrix} t_{11} & t_{12} & \cdots & t_{1n} & \\ t_{21} & \ddots & \ddots & \vdots & \\ \vdots & \ddots & \ddots & t_{12} & \\ t_{n1} & \cdots & t_{21} & \ddots & t_{11} \end{bmatrix} = \begin{bmatrix} t_0 & t_1 & \cdots & t_{n-1} \\ t_{-1} & \ddots & \ddots & \vdots \\ \vdots & \ddots & \ddots & t_1 \\ t_{-(n-1)} & \cdots & t_{-1} & t_0 \end{bmatrix}$$

A typical i-th row of Toeplitz Matrix is represented by a row vector (or sequence) $t_{-i}, t_{-i+1}, \ldots, t_{-i+(n-1)}$ $i = 0, 1, 2, \ldots, n-1$. A general entry in matrix $\boldsymbol{T}$ takes the form on the diagonals

$$t_{ij} = t_{j-i} \qquad i, j = 0, 1, 2, \ldots, n-1$$

Finite size Toeplitz matrix has some computation advantages. Toeplitz matrices of infinite dimension play an important role in the construction of multiplication operators in $\boldsymbol{H}^p$ spaces. Toeplitz matrix can be written as $\boldsymbol{T} = \boldsymbol{T}_U + \boldsymbol{T}_L$, $\boldsymbol{T}_U$ is upper diagonal and $\boldsymbol{T}_L$ is lower diagonal matrix.

Observation about Toeplitz Matrices:

Consider the multiplication of two polynomials:

$$P(z) = (a_0 + a_1 z + a_2 z^2), \quad Q(z) = (b_0 + b_1 z + b_2 z^2)$$

$$R(z) = P(z)Q(z) = Q(z)P(z) = (c_0 + c_1 z + c_2 z^2 + c_3 z^3 + c_4 z^4)$$

The above polynomial multiplication can be written as:

$$\boldsymbol{T}_p \cdot \boldsymbol{q} = \boldsymbol{r}, \quad \text{or} \quad \boldsymbol{T}_q \cdot \boldsymbol{p} = \boldsymbol{r}$$

where

$$\boldsymbol{T}_p = \begin{bmatrix} a_0 & 0 & 0 \\ a_1 & a_0 & 0 \\ a_2 & a_1 & a_0 \\ 0 & a_2 & a_1 \\ 0 & 0 & a_2 \end{bmatrix}, \ \boldsymbol{T}_q = \begin{bmatrix} b_0 & 0 & 0 \\ b_1 & b_0 & 0 \\ b_2 & b_1 & b_0 \\ 0 & b_2 & b_1 \\ 0 & 0 & b_2 \end{bmatrix}, \ \boldsymbol{p} = \begin{bmatrix} a_0 \\ a_1 \\ a_2 \end{bmatrix}, \ \boldsymbol{q} = \begin{bmatrix} b_0 \\ b_1 \\ b_2 \end{bmatrix}, \ \boldsymbol{r} = \begin{bmatrix} c_0 \\ c_1 \\ c_2 \\ c_3 \\ c_4 \end{bmatrix}$$

$\boldsymbol{T}_p$ and $\boldsymbol{T}_q$ are the Toeplitz matrix representation of the polynomials $P(z)$ and $Q(z)$ respectively. The main thrust of the above example is that polynomial multiplication which is very common in Signal processing can be accomplished via Toeplitz matrices.

(3) **Hankel Matrices, Finite Dimensional**

A $n \times n$ square Hankel matrix takes the form

$$\boldsymbol{H} = \begin{bmatrix} h_{11} & h_{12} & \cdots & h_{1,n-1} & h_{1n} \\ h_{12} & & \ddots & \ddots & h_{1,n+1} \\ \vdots & \ddots & \ddots & \ddots & \vdots \\ h_{1,n-1} & \ddots & \ddots & & \\ h_{1n} & h_{1,n+1} & & & h_{1,2n-1} \end{bmatrix} = \begin{bmatrix} h_0 & h_{-1} & h_{-2} & \cdots & h_{-n+2} & h_{-n+1} \\ h_{-1} & & & \ddots & \ddots & \ddots \\ \vdots & \ddots & \ddots & \ddots & & \\ h_{-n+2} & \ddots & \ddots & & & \\ h_{-n+1} & h_{-n} & & \cdots & & h_{-2n+2} \end{bmatrix}$$

The entries in the matrix are constant along the diagonals which are perpendicular to the main diagonal and

$$h_{ij} = h_{-i-j} \qquad i, j = 0, 1, \ldots, n-1$$

Hankel matrices just like Toeplitz matrices, also have a polynomial multiplication interpretation. Let us again consider the multiplication of two polynomials except in one of the polynomials, the variable z is replaced with variable z^{-1}. Thus,

$$P(z) = (a_0 + a_1 z + a_2 z^2)$$

$$Q(z) = (b_0 z^{-2} + b_1 z^{-1} + b_2) = (b_2 + b_1 z^{-1} + b_0 z^{-2})$$

$$R(z) = P(z)Q(z) = Q(z)P(z) = (c_0 z^{-2} + c_1 z^{-1} + c_2 + c_3 z + c_4 z^2)$$

The above polynomial multiplication can be represented as:

$$\boldsymbol{H}_p.\hat{\boldsymbol{q}} = \hat{\boldsymbol{r}} \quad \text{or} \quad \boldsymbol{H}_q.\hat{\boldsymbol{p}} = \hat{\boldsymbol{r}}$$

$$\boldsymbol{H}_p = \begin{bmatrix} 0 & 0 & a_0 \\ 0 & a_0 & a_1 \\ a_0 & a_1 & a_2 \\ a_1 & a_2 & 0 \\ a_2 & 0 & 0 \end{bmatrix}, \ \boldsymbol{H}_q = \begin{bmatrix} 0 & 0 & b_0 \\ 0 & b_0 & b_1 \\ b_0 & b_1 & b_2 \\ b_1 & b_2 & 0 \\ b_2 & 0 & 0 \end{bmatrix}, \ \hat{\boldsymbol{p}} = \begin{bmatrix} a_2 \\ a_1 \\ a_0 \end{bmatrix}, \ \hat{\boldsymbol{q}} = \begin{bmatrix} b_2 \\ b_1 \\ b_0 \end{bmatrix}, \ \boldsymbol{r} = \begin{bmatrix} c_0 \\ c_1 \\ c_2 \\ c_3 \\ c_4 \end{bmatrix}$$

Just as Toeplitz matrix, Hankel matrix is useful for polynomial multiplication provided z is replaced with z^{-1}.

(4) **Circulant Matrices**

A $n \times n$ square matrix $\boldsymbol{C}$ has the following structure

$$\boldsymbol{C} = \begin{bmatrix} c_0 & c_{-1} & \cdots & c_{1-n} \\ c_1 & c_0 & \ddots & \\ \vdots & \ddots & \ddots & \\ c_{n-1} & c_{n-2} & \cdots & c_0 \end{bmatrix}, \qquad \begin{matrix} c_{ij} = c_{j-i}, & i, j = 0, 1, \ldots, n-1 \\ c_{-k} = c_{n-k}, & 1 < k < n-1 \end{matrix}$$

(5) **Stochastic Matrix *S***

All the entries of this square matrix are non-negative and row sum is unity. Thus,

$$\boldsymbol{S} = \begin{bmatrix} s_{11} & s_{12} & \cdots & s_{1n} \\ s_{21} & s_{22} & \cdots & s_{2n} \\ & & \vdots & \\ s_{n1} & s_{n2} & \cdots & s_{nn} \end{bmatrix}, \qquad \begin{array}{ll} s_{ij} \geq 0 & i, j = 1, 2, \ldots, n \\ \sum_{i=1}^{n} s_{ij} = 1 & j = 1, 2, \ldots, n \end{array}$$

The entries s_{ij} may be functions of time $t \geq 0$

(6) ***Q*-matrix**

Q-matrix is $n \times n$ symmetric matrix with following properties:

$$\boldsymbol{Q} = \begin{bmatrix} q_{11} & q_{12} & \cdots & q_{1n} \\ q_{12} & q_{22} & \cdots & q_{2n} \\ & & \vdots & \\ q_{1n} & q_{2n} & \cdots & q_{nn} \end{bmatrix}, \qquad \begin{array}{l} 0 \leq -q_{ii} = -q_i, \quad q_i > 0 \\ 0 \leq q_{ij}, \quad q_{ij} = q_{ji}, \quad i, j = 1, 2, \ldots, n \end{array}$$

$$\sum_{j=1}^{n} q_{ij} = 0$$

Fact: Given a ***Q***-matrix, we can always generate a Stochastic matrix

$$\boldsymbol{S}(t) = e^{\boldsymbol{Q}t}$$

30. **Nonsingular Matrices**

For a $n \times n$ matrix $\boldsymbol{A}$ to be nonsingular (invertable), its determinant Δ_A must be nonzero, which implies that all its rows (or columns) are linearly independent. This further means the rank of $\boldsymbol{A}$ is n. For $\boldsymbol{A}$ to be invertable, $\boldsymbol{Ax} = \boldsymbol{0}$, implying no linearly independent solutions besides $\boldsymbol{x} = \boldsymbol{0}$.

2.3 Systems of Linear Algebraic Equations

Let $\boldsymbol{A}$ be $m \times n$ matrix, $\boldsymbol{x}$ be an $n \times 1$ matrix, and $\boldsymbol{b}$ be an $m \times 1$ matrix forming a system of equations

$$\boldsymbol{A}\boldsymbol{x} = \boldsymbol{b} \tag{2.18}$$

$\boldsymbol{B} = [\boldsymbol{A}, \boldsymbol{b}]$ be the $m \times (n+1)$ augmented matrix.

This system has a solution only if both $\boldsymbol{A}$ and $\boldsymbol{B}$ have same rank r. Only r of the n components of $\boldsymbol{x}$ can be uniquely determined, and other $(n-r)$ components can be assigned at will. Let us look at this problem from a geometrical point of view.

2.3.1 Geometric Interpretation of Simultaneous Equations

(a) **Homogenous System of Equations**

Consider a system of homogenous equations

$$\boldsymbol{A}\boldsymbol{x} = \boldsymbol{0} \tag{2.19}$$

$$\begin{bmatrix} \boldsymbol{a}_1^T \\ \boldsymbol{a}_2^T \\ \vdots \\ \boldsymbol{a}_n^T \end{bmatrix} [x] = \boldsymbol{0} \tag{2.20}$$

The Eq. 2.20 implies

$$(\boldsymbol{x}, \boldsymbol{a}_i) = 0 \qquad i = 1, 2, \ldots, n \tag{2.21}$$

The Eq. 2.21 states that the vector $\boldsymbol{x}$ is orthogonal to all the n-vectors, $\boldsymbol{a}_1, \boldsymbol{a}_2, \ldots, \boldsymbol{a}_n$

Two cases arise depending upon whether $\Delta_A = |\boldsymbol{A}| = \det(\boldsymbol{A})$, vanishes or not.

(1) Case 1: $\det(\boldsymbol{A}) = \Delta_A \neq 0$, Rank $(\boldsymbol{A}) = n$ (Trivial Solution)

Rank $(\boldsymbol{A}) = n$ implies the vectors $\boldsymbol{a}_i$ are linearly independent and fully span the n-dimensional space. Since no vector can be orthogonal to the n-linearly independent vectors spanning an n-dimensional space, the only solution for Eq. 2.20 is, $\boldsymbol{x} = 0$

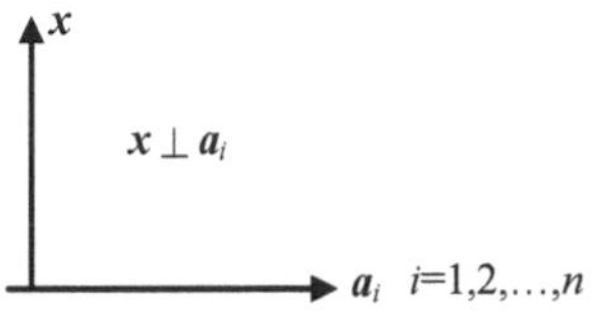

Figure 2.3: The Vectors Spanning the Matrix $\boldsymbol{A}$ Are Orthogonal to the Vector $\boldsymbol{x}$

(2) Case 2: $\det(\boldsymbol{A}) = \Delta_A = 0$, Rank $(\boldsymbol{A}) < n$ (Nontrivial Solution)

Let Rank $(\boldsymbol{A}) = k < n$. Then the Eq 2.19 has infinite set of non trivial solutions. We can easily rearrange the rows of the matrix $\boldsymbol{A}$, such that the first k rows represent vectors which are linearly independent. Let us designate these vectors as,

$$\boldsymbol{a}_j, \quad j = 1, 2, \ldots, k \text{ (linearly independent)} \tag{2.22}$$

The remaining $(n - k)$ row vectors of $\boldsymbol{A}$ are linearly dependent and be designated as $\boldsymbol{a}_{k+1}, \boldsymbol{a}_{k+2}, \ldots, \boldsymbol{a}_n$. These vectors can be expressed as:

$$\boldsymbol{a}_{k+i} = \sum_{i=1}^{k} \beta_{ij}\boldsymbol{a}_j \qquad i = 1, 2, \ldots, (n-k) \tag{2.23}$$

From the Eqs. 2.21 and 2.23, it is clear that if,

$$\boldsymbol{x} \perp \boldsymbol{a}_j \qquad (j = 1, 2, \ldots, k)$$

then, $\boldsymbol{x} \perp \boldsymbol{a}_i \qquad (i = 1, 2, \ldots, (n-k))$

Thus, it is sufficient to solve for only k equations

$$\begin{bmatrix} \boldsymbol{a}_1^T \\ \boldsymbol{a}_2^T \\ \cdots \\ \boldsymbol{a}_k^T \end{bmatrix} \begin{bmatrix} x \end{bmatrix} = \boldsymbol{0} \tag{2.24}$$

and obtain $(n-k)$ linearly independent solutions of $\boldsymbol{x}$. Call these solutions as $\boldsymbol{x}_1, \boldsymbol{x}_2, \ldots, \boldsymbol{x}_{n-k}$. This leads us to the following important conclusion:

> The n-dimensional space represented by the vectors of matrix $\boldsymbol{A}$ can be decomposed into two orthogonal sub-spaces $\boldsymbol{L}_k$and $\boldsymbol{M}_{n-k}$. The subspace $\boldsymbol{L}_k$ is of dimension k and is spanned by basis vectors $\boldsymbol{a}_1, \boldsymbol{a}_2, \ldots, \boldsymbol{a}_k$. The subspace $\boldsymbol{M}_{n-k}$ is of dimension $(n-k)$ and is orthogonal to the subspace $\boldsymbol{L}_k$ and is also spanned by linearly independent vectors $\boldsymbol{x}_1, \boldsymbol{x}_2, \ldots, \boldsymbol{x}_{n-k}$.

The n-dimensional space is spanned by vectors $\boldsymbol{a}_1, \boldsymbol{a}_2, \ldots, \boldsymbol{a}_k, \boldsymbol{x}_1, \boldsymbol{x}_2, \ldots, \boldsymbol{x}_{n-k}$. Thus any vector $\boldsymbol{x}$ in the n-dimensional space of vectors of the matrix $\boldsymbol{A}$ can be represented by

$$\boldsymbol{x} = \sum_{j=1}^{k} \alpha_j \boldsymbol{a}_j + \sum_{i=1}^{n-k} \beta_i \boldsymbol{x}_i \tag{2.25}$$

The essential point to be realized is that, only the vectors $\boldsymbol{a}_1, \boldsymbol{a}_, \ldots, \boldsymbol{a}_k$ have the "necessary information" while $\boldsymbol{a}_{k+1}, \boldsymbol{a}_{k+2}, \ldots, \boldsymbol{a}_n$ are really "redundant" as far as the system structure is concerned. Different choice of α_i and β_i result in different infinite solutions. Algorithm for computation of $\boldsymbol{x}_1, \boldsymbol{x}_2, \ldots, \boldsymbol{x}_k$ is as follows:

Computation Algorithm for $x_1, x_2, \ldots, x_k (M_{n-k}$ space vectors)

Consider the Eq. 2.24 for $x = x_i$,

$$\begin{bmatrix} a_1^T \\ a_2^T \\ \vdots \\ a_k^T \end{bmatrix} \begin{bmatrix} x_i \end{bmatrix} = \mathbf{0}, \quad i = 1, 2, \ldots, (n-k)$$

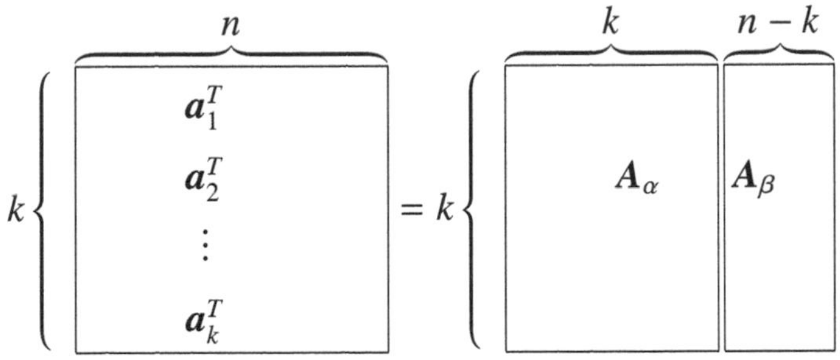

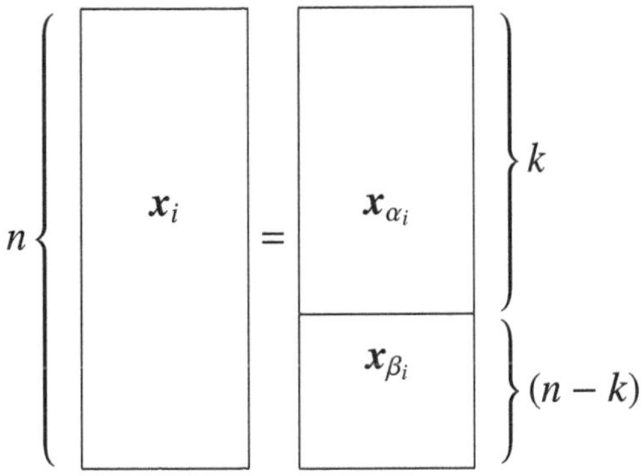

$$A_\alpha x_{\alpha_i} + A_\beta x_{\beta_i} = \mathbf{0}$$

$$x_{\alpha_i} = -A_\alpha^{-1} A_\beta x_{\beta_i}$$

Let,

$$x_{\beta_1} = \begin{bmatrix} c_1 \\ 0 \\ \vdots \\ 0 \end{bmatrix}, x_{\beta_2} = \begin{bmatrix} 0 \\ c_2 \\ \vdots \\ 0 \end{bmatrix}, x_{\beta_{n-k}} \begin{bmatrix} 0 \\ 0 \\ \vdots \\ c_{n-k} \end{bmatrix} \qquad c_1, c_2, \ldots, c_{n-k} \text{ are arbitrary constants}$$

This yields the solutions $\boldsymbol{x}_i,\ i = 1, 2, \ldots, k$.

To summarize: The vector solution $\boldsymbol{x}$ for the Eq. 2.20 is made up of sum of two parts

$$\begin{aligned} \boldsymbol{x} &= \boldsymbol{y} + \boldsymbol{z} \\ \boldsymbol{y} &= \sum_{j=1}^{k} \alpha_j \boldsymbol{a}_j \\ \boldsymbol{z} &= \sum_{i=1}^{n-k} \beta_i \boldsymbol{x}_i \end{aligned} \tag{2.26}$$

$\boldsymbol{y}$ is called a projection of $\boldsymbol{x}$ onto space $\boldsymbol{L}_k$

$\boldsymbol{z}$ is called the projection of $\boldsymbol{x}$ onto the orthogonal complimentary space $\boldsymbol{M}_{n-k}$

(b) **Inhomogenous System of Equations** Consider the inhomogenous system

$$\boldsymbol{A}\boldsymbol{x} = \boldsymbol{b} \tag{2.27}$$

$$\boldsymbol{A} = \begin{bmatrix} \boldsymbol{a}_1^T \\ \boldsymbol{a}_2^T \\ \vdots \\ \boldsymbol{a}_n^T \end{bmatrix}, \ \boldsymbol{b} = \begin{bmatrix} b_1 \\ b_2 \\ \vdots \\ b_n \end{bmatrix}, \boldsymbol{x} = \begin{bmatrix} x_1 \\ x_2 \\ \vdots \\ x_n \end{bmatrix},$$

The solution vector $\boldsymbol{x}$ satisfies

$$(\boldsymbol{x}, \boldsymbol{a}_i) = b_i, \quad i = 1, 2, \ldots, n \tag{2.28}$$

If the matrix $\boldsymbol{A}$ has full rank ($\det(\boldsymbol{A}) \neq 0$), then

$$\boldsymbol{x} = \boldsymbol{A}^{-1}\boldsymbol{b} \qquad \text{(Cramer's rule)} \tag{2.29}$$

The vector $\boldsymbol{b}$ ranges over the whole n-dimensional space and unique values of

$\boldsymbol{x}$ can be computed. This unique solution of the Eq. 2.29 $\boldsymbol{x} = (x_1, x_2, \ldots, x_n)$ is,

$$x_i = \frac{\Delta_i}{\Delta_A}, \qquad i = 1, 2, \ldots, n \tag{2.30}$$

where Δ_i is determined from Δ_A by replacing it's elements of the i-th column by the corresponding components of $\boldsymbol{b}$. The interesting situation occurs when $\boldsymbol{A}$ is singular in the equation $\boldsymbol{Ax} = \boldsymbol{b}$.

(c) **Singular Inhomogenous System**

Consider

$$\boldsymbol{Ax} = \boldsymbol{b}, \qquad \Delta_A = \det(\boldsymbol{A}) = 0 \tag{2.31}$$

Let us find the condition under which the Eq. 2.31 has a non-trivial solution. Multiplying the Eq. 2.31 with the matrix Adj $(\boldsymbol{A})$ on both sides,

$$[\text{Adj}\,(\boldsymbol{A})]\,[\boldsymbol{Ax}] = [\text{Adj}\,(\boldsymbol{A})]\,\boldsymbol{b}$$

$$[\text{Adj}\,(\boldsymbol{A})]\,\boldsymbol{A} = \Delta_A \boldsymbol{I} = \boldsymbol{0}$$

Hence

$$[\text{Adj}\,(\boldsymbol{A})]\,\boldsymbol{b} = \boldsymbol{0} \tag{2.32}$$

But Eq. 2.32 implies $\Delta_j = 0,\ \ j = 1, 2, \ldots, n$ (Δ_j defined in the Eq. 2.30)

We arrive at the following conclusion:

If Δ_A vanishes then a necessary and sufficient condition for Eq. 2.31,

$$\boldsymbol{Ax} = \boldsymbol{b},$$

to have a non-trivial solution is that vector $\boldsymbol{b}$ be orthogonal to all the row vectors of Adj $(\boldsymbol{A})$

Let us summarize the general solution of Eq. 2.31

$$\left.\begin{aligned} \boldsymbol{Ax} &= \boldsymbol{b} \\ \text{Rank } \boldsymbol{A} &= k < n \end{aligned}\right\} \tag{2.33}$$

The general solution $\boldsymbol{x}$ is made of two vectors

$$\boldsymbol{x} = \boldsymbol{x}_h + \boldsymbol{x}_p$$

The vector $\boldsymbol{x}_h$ represents the solution corresponding to the homogenous system and involves $(n - k)$ arbitrary constants and is a solution to the equation.

$$\boldsymbol{Ax}_h = 0, \quad \text{Rank } (\boldsymbol{A}) = k$$

The vector $\boldsymbol{x}_p$ represents a particular solution of the Eq. 2.32 and is given by

$$\boldsymbol{Ax}_p = \boldsymbol{b}$$

$$[\text{Adj}\,(\boldsymbol{A})]\,\boldsymbol{b} = 0$$

2.3.2 Eigenvalues and Eigenvectors of Matrices

Consider a $n \times n$ square matrix $\boldsymbol{A}$. For a vector $\boldsymbol{x} \in \boldsymbol{E}_n$,

If,

$$\boldsymbol{Ax} = \lambda \boldsymbol{x}$$

Then λ is known as the eigenvalue (scalar) and $\boldsymbol{x}$ is referred to as the eigenvector of the matrix $\boldsymbol{A}$ belonging to the eigenvalue λ.

Thus,

$$(\lambda \boldsymbol{I} - \boldsymbol{A})\boldsymbol{x} = \boldsymbol{A}(\lambda)\boldsymbol{x} = 0 \tag{2.34}$$

implying

$$\sum_{j=1}^{n}(\lambda\delta_{ij} - a_{ij})\boldsymbol{x}_j = 0 \qquad (i = 1, \ldots, n)$$

If the rank of the matrix $\boldsymbol{A}(\lambda)$ is $r \leq n$, then it has r linearly independent nontrivial eigenvectors and a maximum of r distinct eigenvalues. For at least one nontrivial solution of the Eq. 2.34, it is necessary that $\boldsymbol{A}(\lambda)$ is non-invertable, implying:

$$\det(\boldsymbol{A}(\lambda)) = \Delta_{\boldsymbol{A}}(\lambda) = |(\lambda\boldsymbol{I} - \boldsymbol{A})| = 0 \qquad \text{(scalar equation)} \tag{2.35}$$

$\Delta_{\boldsymbol{A}}(\lambda)$ is called the **Characteristic Polynomial** of the matrix $\boldsymbol{A}$. It is a polynomial of degree n in λ and can be written as,

$$\Delta_{\boldsymbol{A}}(\lambda) = P(\lambda) = \begin{vmatrix} (\lambda - a_{11}) & \cdots & -a_{1n} \\ \vdots & & \\ -a_{n1} & \cdots & (\lambda - a_{nn}) \end{vmatrix} = \lambda^n + a_1\lambda^{n-1} + \cdots + a_n \tag{2.36}$$

$P(\lambda) = 0$ is known as the **Characteristic Equation** of the matrix $\boldsymbol{A}$

According to the fundamental theorem of linear algebra, the polynomial $P(\lambda)$ has n roots $\lambda_1, \ldots, \lambda_n$, not all necessarily **distinct** known as the eigenvalues of $\boldsymbol{A}$.

Some Important Concepts in Polynomials

(i) **Greatest Common Divisor**

The greatest common divisor (GCD) of polynomials $P(\lambda)$ and $Q(\lambda)$ is the highest degree polynomial $D(\lambda)$ such that $D(\lambda)$ is a divisor of both $P(\lambda)$ and $Q(\lambda)$. It is denoted as $\text{GCD}(P(\lambda), Q(\lambda))$. This GCD polynomial $D(\lambda)$ is also the smallest-degree polynomial which can be written as a linear combination of $P(\lambda)$ and $Q(\lambda)$ such that, $D(\lambda) = P(\lambda)R(\lambda) + Q(\lambda)S(\lambda)$. A common way to find GCD of two polynomials is to simply factor the two polynomials. Then take the product of all the common simple factors as GCD.

Example 2.5:

Find the GCD of $(\lambda^3 + 6\lambda^2 + 11\lambda + 2)$ and $(\lambda^2 + 7\lambda + 10)$

$$P(\lambda) = \lambda^3 + 6\lambda^2 + 11\lambda + 6 = (\lambda + 1)(\lambda + 2)(\lambda + 3)$$

$$Q(\lambda) = \lambda^2 + 7\lambda + 10 = (\lambda + 2)(\lambda + 5)$$

Thus $(\lambda + 2)$ is GCD$(P(\lambda), Q(\lambda))$. In some cases the synthetic division or Euclidean algorithm yields the result.

(ii) **Elementary Symmetric Functions of a Matrix *A***

$$|\lambda \boldsymbol{I} - \boldsymbol{A}| = P(\lambda) = a_0\lambda^n + a_1\lambda^{n-1} + a_2\lambda^{n-2} + \cdots + a_n = \prod_{i=1}^{n}(\lambda - \lambda_i) \quad (2.37)$$

Equating coefficients for various powers of λ

$$a_0 = 1, \qquad (-1)a_1 = \sum_{i=1}^{n} \lambda_i$$

$$(-1)^2 a_2 = \frac{1}{2!}\sum_{\substack{i,j=1\\ i\neq j}}^{n'}(\lambda_i\lambda_j)$$

$$(-1)^m a_m = \frac{1}{m!}\sum_{\substack{i_1,i_2,\ldots,i_m\\ i_1\neq i_2\ldots\neq i_m}}^{n'}(\lambda_{i_1}\lambda_{i_2}\ldots\lambda_{i_m})$$

$$(-1)^n a_n = \prod_{i=1}^{n}\lambda_i$$

Prime on the summation implies the sum only over the distinct subscripts.

Two important parameters associated with A are the **trace** and **determinant**.

$$\textbf{Trace of } \boldsymbol{A} = \text{Tr}\,(\boldsymbol{A}) = \sum_{i=1}^{n} a_{ii} = \sum_{i=1}^{n}\lambda_i \quad \text{also called "spure } \boldsymbol{A}\text{"}$$

$$\textbf{determinant } \boldsymbol{A} = \det(\boldsymbol{A}) = \Delta_{\boldsymbol{A}} = \Delta_{\boldsymbol{A}}(0) = \prod_{i=1}^{n}\lambda_i$$

(iii) **Annihilating Polynomial $N(A)$, and the Minimal Polynomial $M(A)$**

Given a $n \times n$ matrix $\boldsymbol{A}$, the **annihilating polynomial** of the matrix $\boldsymbol{A}$ is defined as a polynomial $N(\lambda)$ such that when λ is replaced by $\boldsymbol{A}$, the resulting polynomial matrix $N(\boldsymbol{A})$ is:

$$N(\boldsymbol{A}) = \mathbf{0}$$

In general there are more than one **annihilating polynomials** of A. One of the **annihilating polynomials** of $\boldsymbol{A}$ is the characteristic polynomial $P(\lambda)$:

$$P(\lambda) = \Delta_A(\lambda) = |\lambda \boldsymbol{I} - \boldsymbol{A}|$$

$$P(\boldsymbol{A}) = \mathbf{0}$$

This is the well known **Cayley-Hamilton** theorem which we shall formally prove later. But as an observation, we can see that

$$(\lambda \boldsymbol{I} - \boldsymbol{A})^{-1} = \frac{1}{\Delta_A(\lambda)} \left[\text{Adj } (\lambda \boldsymbol{I} - \boldsymbol{A})^T \right]$$

or

$$\Delta_A(\lambda)\boldsymbol{I} = [\lambda \boldsymbol{I} - \boldsymbol{A}] \left[\text{Adj } (\lambda \boldsymbol{I} - \boldsymbol{A})^T \right]$$

Now, if we substitute $\boldsymbol{A}$ for λ, we obtain,

$P(\boldsymbol{A}) = \Delta_A(\boldsymbol{A}) = \mathbf{0}$ **Cayley-Hamilton Theorem**

As stated, there are many annihilating polynomials of a matrix. The particular annihilating polynomial of $\boldsymbol{A}$, having the lowest degree in λ and the coefficient of the highest degree in λ as unity is called the Minimal Polynomial $M(\lambda)$ of the matrix $\boldsymbol{A}$. This minimal polynomial $M(\lambda)$ is a factor of any other annihilating polynomial $N(\lambda)$ of $\boldsymbol{A}$.

Given two annihilating polynomials $P(\lambda)$ and $M(\lambda)$ of a square matrix $\boldsymbol{A}$,

$$P(\lambda) = Q(\lambda)M(\lambda) + R(\lambda) \qquad \text{degree } R(\lambda) < \text{degree } M(\lambda)$$

$$P(\lambda)\boldsymbol{I} = Q(\lambda)M(\lambda)\boldsymbol{I} + R(\lambda)\boldsymbol{I} \tag{2.38}$$

since $P(\boldsymbol{A}) = M(\boldsymbol{A}) = \boldsymbol{0}, \quad R(\boldsymbol{A}) = \boldsymbol{0}$

Implying, Both $P(\lambda)$ and $M(\lambda)$ are annihilating polynomials of $\boldsymbol{A}$.

$$P(\lambda) = Q(\lambda)M(\lambda)$$

(a) **Algorithm to compute Minimal Polynomial $M(\lambda)$ given a matrix A**

$$\boldsymbol{A} = (a_{ij}) \qquad (i, j = 1, 2, \ldots, n) \qquad \text{be a square matrix}$$

$$P(\lambda) = \Delta_{\boldsymbol{A}}(\lambda) = |\lambda\boldsymbol{I} - \boldsymbol{A}| = \text{Characteristic Polynomial of degree } n \text{ in } \lambda$$

$D(\lambda)$ = GCD of the n^2 square elements of Adj $(\lambda\boldsymbol{I} - \boldsymbol{A})$,

and of degree less than n

$M(\lambda)$ = Minimal polynomial of the matrix $\boldsymbol{A}$ of degree m, $m \leq n$

Show,

$$M(\lambda) = \frac{P(\lambda)}{D(\lambda)} = \textbf{Minimal Polynomial}$$

or,

$$P(\lambda) = M(\lambda)D(\lambda), \quad Q(\lambda) = D(\lambda)$$

Proof:

$$P(\boldsymbol{A}) = \boldsymbol{0} \qquad \text{implies} \qquad P(\lambda)\boldsymbol{I} = (\lambda \boldsymbol{I} - \boldsymbol{A})\,\boldsymbol{Q}(\lambda, \boldsymbol{A}) \tag{2.39}$$

$$M(\boldsymbol{A}) = \boldsymbol{0} \qquad \text{implies} \qquad M(\lambda)\boldsymbol{I} = (\lambda \boldsymbol{I} - \boldsymbol{A})\,\boldsymbol{R}(\lambda, \boldsymbol{A}) \tag{2.40}$$

where $\boldsymbol{Q}(\lambda, \boldsymbol{A})$ and $\boldsymbol{R}(\lambda, \boldsymbol{A})$ are polynomial matrices in λ and $\boldsymbol{A}$

Let

$$[\text{Adj}\,(\lambda \boldsymbol{I} - \boldsymbol{A})]^T = [D(\lambda)\boldsymbol{B}(\lambda, \boldsymbol{A})]$$

where GCD of $B(\lambda, \boldsymbol{A})$ is unity polynomial.

Thus

$$P(\lambda)\boldsymbol{I} = (\lambda \boldsymbol{I} - \boldsymbol{A})\,[\text{Adj}\,(\lambda \boldsymbol{I} - \boldsymbol{A})]^T = (\lambda \boldsymbol{I} - \boldsymbol{A})\,(D(\lambda)\boldsymbol{B}(\lambda, \boldsymbol{A})) \tag{2.41}$$

From Eq. 2.41

$$\frac{P(\lambda)}{D(\lambda)}\boldsymbol{I} = (\lambda \boldsymbol{I} - \boldsymbol{A})\,\boldsymbol{B}(\lambda, \boldsymbol{A}) = \Psi(\lambda)\boldsymbol{I}$$

The GCD of $\boldsymbol{B}(\lambda, \boldsymbol{A})$ is unity polynomial.

From above,

$$\Psi(\boldsymbol{A}) = 0$$

Let,

$$\Psi(\lambda) = \beta(\lambda)M(\lambda) + \alpha(\lambda) \tag{2.42}$$

where $\beta(\lambda)$ is quotient polynomial and $\alpha(\lambda)$ is the remainder.

Substituting $\boldsymbol{A}$ for λ in Eq. 2.41,

$$\Psi(\boldsymbol{A}) = \beta(\boldsymbol{A})M(\boldsymbol{A}) + \alpha(\boldsymbol{A})$$

But,

$$\Psi(\boldsymbol{A}) = 0 \text{ and } M(\boldsymbol{A}) = 0$$

Therefore

$$\alpha(\boldsymbol{A}) = 0 \qquad \text{implying} \qquad \alpha(\lambda) = 0$$

Thus,

$$\Psi(\lambda) = \beta(\lambda)M(\lambda)$$

or,

$$\Psi(\lambda)\boldsymbol{I} = \beta(\lambda)M(\lambda)\boldsymbol{I}$$

From the Eqs. 2.40 and 2.41

$$(\lambda \boldsymbol{I} - \boldsymbol{A})\,\boldsymbol{B}(\lambda, \boldsymbol{A}) = \beta(\lambda)\,(\lambda \boldsymbol{I} - \boldsymbol{A})\,\boldsymbol{R}(\lambda, \boldsymbol{A})$$

or

$$\boldsymbol{B}(\lambda, \boldsymbol{A}) = \beta(\lambda)\boldsymbol{R}(\lambda, \boldsymbol{A})$$

since GCD of $\boldsymbol{B}(\lambda, \boldsymbol{A})$ is unity,

$$\beta(\lambda) = 1$$

Implying

$$\boldsymbol{B}(\lambda, \boldsymbol{A}) = \boldsymbol{R}(\lambda, \boldsymbol{A})$$

Hence

$$\boxed{\Psi(\lambda) = \frac{P(\lambda)}{D(\lambda)} = M(\lambda) = \text{ Minimal Polynomial}}$$

Example 2.6

Compute the characteristic and the minimal polynomial of the matrix

$$\boldsymbol{A} = \begin{bmatrix} 0 & 0 & 1 \\ 0 & 0 & 0 \\ 0 & 0 & 1 \end{bmatrix}$$

Solution:

$$(\lambda \boldsymbol{I} - \boldsymbol{A}) = \begin{bmatrix} \lambda & 0 & -1 \\ 0 & \lambda & 0 \\ 0 & 0 & \lambda - 1 \end{bmatrix}$$

$$P(\lambda) = \Delta_A(\lambda) = \lambda^2(\lambda - 1) \qquad \textbf{(characteristic polynomial)}$$

To compute minimal polynomial $M(\lambda)$

$$[\text{Adj}\, \boldsymbol{A}] = \begin{bmatrix} \lambda(\lambda - 1) & 0 & 0 \\ 0 & \lambda(\lambda - 1) & 0 \\ \lambda(\lambda - 1) & 0 & \lambda^2 \end{bmatrix}$$

The GCD of the elements of [Adj $\boldsymbol{A}$] is λ. Thus,

$$D(\lambda) = \lambda$$

$$M(\lambda) = \frac{\lambda^2(\lambda - 1)}{\lambda} = \lambda(\lambda - 1), \quad M(\boldsymbol{A}) = \boldsymbol{A}(\boldsymbol{A} - \boldsymbol{I}) = \boldsymbol{0}$$

Example 2.7

Let us compute the minimal polynomial of the following matrix $\boldsymbol{A}$

$$\boldsymbol{A} = \begin{bmatrix} \lambda_1 & 1 & & \\ & \lambda_1 & & \\ & & \lambda_2 & \\ & & & \lambda_2 \end{bmatrix}$$

$$(\lambda \boldsymbol{I} - \boldsymbol{A}) = \begin{bmatrix} (\lambda - \lambda_1) & -1 & & \\ & (\lambda - \lambda_1) & & \\ & & (\lambda - \lambda_2) & \\ & & & (\lambda - \lambda_2) \end{bmatrix}$$

$$\frac{1}{(\lambda - \lambda_2)}[\text{Adj}\,\boldsymbol{A}] = \begin{bmatrix} (\lambda - \lambda_1)(\lambda - \lambda_2) & & & \\ (\lambda - \lambda_2) & (\lambda - \lambda_1)(\lambda - \lambda_2) & & \\ & & (\lambda - \lambda_1)^2 & \\ & & & (\lambda - \lambda_1)^2 \end{bmatrix}$$

The GCD of the elements of Adj $\boldsymbol{A}$ is $(\lambda - \lambda_2)$

$$D(\lambda) = (\lambda - \lambda_2), \quad P(\lambda) = \Delta_{\boldsymbol{A}}(\lambda) = (\lambda - \lambda_1)^2(\lambda - \lambda_2)^2$$

$$M(\lambda) = \frac{P(\lambda)}{D(\lambda)} = (\lambda - \lambda_1)^2(\lambda - \lambda_2)$$

Exercise 2.1

Given $M(\lambda)$ as the minimal polynomial of $n \times n$ matrix $\boldsymbol{A}$, such that

$$M(\lambda) = \sum_{i=0}^{m} a_{m-i}\,\lambda^i, \qquad a_0 = 1, \quad m \leq n$$

show,

$$(\lambda \boldsymbol{I} - \boldsymbol{A})^{-1} = \frac{\sum\limits_{j=0}^{m-1} \lambda^j \left(\sum\limits_{i=j+1}^{m} a_{m-i}\,\boldsymbol{A}^{i-j-1} \right)}{M(\lambda)}$$

$$(\lambda \boldsymbol{I} - \boldsymbol{A})^{-1} = \frac{\sum\limits_{j=0}^{n-1} \lambda^j \left(\sum\limits_{i=j+1}^{n} a_{n-i}\,\boldsymbol{A}^{i-j-1} \right)}{\Delta_{\boldsymbol{A}}(\lambda)}, \qquad n = m$$

Hint:

$$M(\lambda)\boldsymbol{I} = (\lambda \boldsymbol{I} - \boldsymbol{A})\left[\boldsymbol{R}_0(\boldsymbol{A}) + \boldsymbol{R}_1(\boldsymbol{A})\lambda + \boldsymbol{R}_2(\boldsymbol{A})\lambda^2 + \cdots + \boldsymbol{R}_{m-1}(\boldsymbol{A})\lambda^{m-1}\right]$$

(b) **Computation Algorithm for the matrix functions** $f(\boldsymbol{A})$

We shall deal only with analytic functions of matrix $\boldsymbol{A}$ represented by a Taylor series and which are regular except at the zeroes of the annihilating polynomial $N(\lambda)$.

Given an analytic function $f(\lambda)$ of degree n and an annihilating polynomial $N(\lambda)$ of a matrix $\boldsymbol{A}$ of degree m, $n \geq m$.

We can write

$$f(\lambda) = d(\lambda)N(\lambda) + g(\lambda), \qquad \text{degree } g(\lambda) < \text{degree } N(\lambda)$$

Then

$$f(\boldsymbol{A}) = d(\boldsymbol{A})N(\boldsymbol{A}) + g(\boldsymbol{A})$$

$$\text{But } N(\boldsymbol{A}) = 0$$

Therefore

$$f(\boldsymbol{A}) = g(\boldsymbol{A})$$

where degree of $g(\boldsymbol{A})$ is less than degree of $f(\boldsymbol{A})$ and also $\leq m$

Note: $f(\boldsymbol{A})$ is a function of $\boldsymbol{A}$ whereas $g(\boldsymbol{A})$ is polynomial in $\boldsymbol{A}$.

In fact $g(\lambda)$ can be written as

$$g(\lambda) = g_0 + g_1\lambda + g_2\lambda^2 + \ldots + g_{n-1}\lambda^{k-1} \qquad k \leq m$$

Example 2.8

Consider an annihilating polynomial $N(\lambda)$ of the matrix $\boldsymbol{A}$ with distinct roots $\lambda_1, \lambda_2, \ldots, \lambda_m$.

Then,

$$g(\lambda_i) = f(\lambda_i)$$

or,

$$\begin{bmatrix} 1 & \lambda_1 & \lambda_1^2 & \cdots & \lambda_1^{m-1} \\ 1 & \lambda_2 & \lambda_2^2 & \cdots & \lambda_2^{m-1} \\ \vdots & & & & \\ 1 & \lambda_m & \lambda_m^2 & \cdots & \lambda_m^{m-1} \end{bmatrix} \begin{bmatrix} g_0 \\ g_1 \\ \vdots \\ g_{m-1} \end{bmatrix} = \begin{bmatrix} f(\lambda_1) \\ f(\lambda_2) \\ \vdots \\ f(\lambda_m) \end{bmatrix}$$

or,

$$g(\lambda) = \sum_{j=1}^{m} \prod_{\substack{k=1 \\ k \neq j}}^{m} \left(\frac{\lambda - \lambda_k}{\lambda_j - \lambda_k} \right) f(\lambda_j)$$

Exercise 2.2

Given

$$\boldsymbol{A} = \begin{bmatrix} 2 & 1 & 4 \\ 0 & 2 & 0 \\ 0 & 3 & 1 \end{bmatrix}$$

show,

$$\boldsymbol{e}^{\boldsymbol{A}t} = \begin{bmatrix} e^{2t} & 12e^{t} - 12e^{2t} - 13te^{2t} & -4e^{t} + 4e^{2t} \\ 0 & e^{2t} & 0 \\ 0 & -3e^{t} + 3e^{-2t} & e^{t} \end{bmatrix}$$

(c) **The minimal polynomial matrix of a vector $\boldsymbol{b}$ w.r.t a $n \times n$ matrix A**

Consider an equation of the form:

$$\boldsymbol{C}(\boldsymbol{A})\boldsymbol{b} = \left(\boldsymbol{A}^{k} + c_1 \boldsymbol{A}^{k-1} + \ldots + c_0 \boldsymbol{I}\right) \boldsymbol{b} = 0 \qquad k \leq n \tag{2.43}$$

The minimum value of k ($k = m$) for which the Eq. 2.43 is true is called the **minimal polynomial matrix of the vector $\boldsymbol{b}$** with respect to the matrix $\boldsymbol{A}$. The degree of the polynomial matrix $\boldsymbol{C}(\boldsymbol{A})$ is called the grade of the vector $\boldsymbol{b}$. Thus for the vector $\boldsymbol{b}$ to be of grade n, the minimal polynomial and the characteristic polynomial are the same.

Krylov Spaces

Let $\boldsymbol{b}_0$ be a vector of grade n, and define

$$\boldsymbol{b}_k = \boldsymbol{A}^k \boldsymbol{b}_0 \qquad k = 0, 1, \ldots, n$$

These vectors are called Krylov vectors and span an n dimensional space with basis $\boldsymbol{b}_0, \boldsymbol{A}\boldsymbol{b}_0, \ldots, \boldsymbol{A}^{n-1}\boldsymbol{b}_0$ and satisfy the equation below:

$$\boldsymbol{b}_n + a_1 \boldsymbol{b}_{n-1} + \ldots + a_n \boldsymbol{b}_0 = 0$$

where $a_1, a_2, \ldots, a_n$ are the co-efficients of the characteristic polynomial of the vector $\boldsymbol{b}$ with respect to matrix $\boldsymbol{A}$. Note that $\boldsymbol{b}_0$ is a vector of grade n. If $\boldsymbol{b}_0$ is a vector less than grade n, then the minimal polynomial of the vector $\boldsymbol{b}$ with respect to $\boldsymbol{A}$ will be of degree less than n. In most situations, we do not know in advance whether $\boldsymbol{b}_0$ is a vector of grade less than n. In order to ensure that $\boldsymbol{b}_0$ is of grade n, we can choose

$$\boldsymbol{b}_0 = \begin{bmatrix} b_{01} \\ b_{02} \\ \vdots \\ b_{0n} \end{bmatrix} = \begin{bmatrix} 1 \\ 1 \\ \vdots \\ 1 \end{bmatrix}$$

In this case, the degree of minimal polynomial of $\boldsymbol{A}$ is same as degree of minimal polynomial of $\boldsymbol{A}$ with respect to the vector $\boldsymbol{b}_0$

2.3.3 Generalized Eigenvectors—Matrix of Multiplicity k

If,

$$\begin{aligned} (\lambda \boldsymbol{I} - \boldsymbol{A})^k \boldsymbol{x} &= \boldsymbol{0} \\ (\lambda \boldsymbol{I} - \boldsymbol{A})^{k-1} &\neq \boldsymbol{0} \quad (k \leq n) \end{aligned} \tag{2.44}$$

The vector $\boldsymbol{x}$ is called a generalized eigenvector of the matrix $\boldsymbol{A}$ of multiplicity k. As a result of the generalized eigenvector $\boldsymbol{x}_i$ of the order k_i belonging to the eigenvalue λ_i, a chain of k_i generalized eigenvectors

$$\begin{Bmatrix} \boldsymbol{x}_i \\ (\lambda_i \boldsymbol{I} - \boldsymbol{A})\boldsymbol{x}_i \\ \vdots \\ (\lambda_i \boldsymbol{I} - \boldsymbol{A})^{k_i - 1}\boldsymbol{x}_i \end{Bmatrix}$$

and can be utilized as k_i **linearly independent eigenvectors of matrix $\boldsymbol{A}$**.

Observation:

(1) Generalized eigenvectors of a matrix corresponding to different eigenvalues are linearly independent.

(2) **Hermitian Matrix**- Eigenvalues of a Hermitian matrix are real and the eigenvectors of a Hermitian matrix for different eigenvalues are orthogonal. This result is important in physics, particularly in Quantum mechanics.

- **Eigenvalues of the function of the matrix A**

  If λ_i is an eigenvalue of $\boldsymbol{A}$ (designated as $\boldsymbol{A} \to \lambda_i$), then

$$\boldsymbol{A} \to \lambda_i, \quad \boldsymbol{A}^{-1} \to \lambda_i^{-1}$$

$$\boldsymbol{A}^T \to \lambda_i, \quad \boldsymbol{A}^k \to \lambda_i^k$$

$$f(\boldsymbol{A}) \to f(\lambda_i)$$

- **Sylvester Theorem 1**

  For a quadratic form $\left[\boldsymbol{x}^T \boldsymbol{A}\boldsymbol{x}\right]$ to be **positive definite**, it is necessary and sufficient that all the **principle minors** (main diagonal) of $\boldsymbol{A}$ are **positive**.

2.4 Diagonalization—Eigenvalue Decomposition

If a matrix $\boldsymbol{A}$ has n linearly independent vectors, only then can a matrix be diagonalized. This is always possible if all the n eigenvalues of the matrix $\boldsymbol{A}$ are distinct. Thus

$$\boldsymbol{A}\boldsymbol{x}_i = \lambda_i \boldsymbol{x}_i \quad (i = 1, \ldots, n)$$

$\boldsymbol{x}_i$ is the eigenvector belonging to λ_i,

Then

$$\boldsymbol{A}\,[\boldsymbol{x}_1 \cdots \boldsymbol{x}_n] = [\boldsymbol{x}_1 \cdots \boldsymbol{x}_n] \begin{bmatrix} \lambda_1 & & \\ & \ddots & \\ & & \lambda_n \end{bmatrix}$$

$$\boldsymbol{P} = [\boldsymbol{x}_1 \cdots \boldsymbol{x}_n] \qquad \text{(modal matrix)}$$

$$\boldsymbol{\Lambda} = \begin{bmatrix} \lambda_1 & & \\ & \ddots & \\ & & \lambda_n \end{bmatrix} \qquad \text{(a diagonal matrix)}$$

$$\boldsymbol{A}\boldsymbol{P} = \boldsymbol{P}\boldsymbol{\Lambda} \qquad \text{or} \qquad \boldsymbol{A} = \boldsymbol{P}^{-1}\boldsymbol{\Lambda}\boldsymbol{P}$$

Furthermore, $\boldsymbol{A}^k = \boldsymbol{P}^{-1}\boldsymbol{\Lambda}^k\boldsymbol{P}$

This is also known as the eigenvalue Decomposition of the matrix $\boldsymbol{A}$.

Two matrices $\boldsymbol{A}$ and $\boldsymbol{B}$ are called **similar** if one can find a non-singular matrix $\boldsymbol{T}$,

$$\boldsymbol{A} = \boldsymbol{T}^{-1}\boldsymbol{B}\boldsymbol{T}$$

$\boldsymbol{T}$ is known as the Similarity Transformation.

Similarity Transformation preserves the eigenvalues and the characteristic polynomial to be the same for the original and the transformed matrices.

- Similar matrices $\boldsymbol{A}$ and $\boldsymbol{B}$ have same eigenvalues, equal determinants, and same characteristic polynomials. Thus,

$$\boldsymbol{A} \to \lambda_i \text{ implies } \boldsymbol{B} \to \lambda_i, \quad \Delta_A = \Delta_B, \quad \Delta_A(\lambda) = \Delta_B(\lambda)$$

- Every Hermitian matrix is diagonalizable, and its modal matrix $\boldsymbol{P}$ is unitary,

$$\left(\boldsymbol{P}^T\right)^* \boldsymbol{P} = \boldsymbol{I}$$

- A companions matrix, $\boldsymbol{A}_c$ is defined as, then

$$\boldsymbol{A}_c = \begin{bmatrix} 0 & 1 & 0 & \cdots & 0 \\ 0 & 0 & 1 & \cdots & 0 \\ 0 & 0 & 0 & \cdots & 1 \\ -a_n & -a_{n-1} & -a_{n-2} & \cdots & -a_1 \end{bmatrix},$$

$$\Delta_{\boldsymbol{A}_c}(\lambda) = a_0\lambda^n + a_1\lambda^{n-1} + a_2\lambda^{n-2} + \cdots = a_n = \prod_{i=1}^{n}(\lambda - \lambda_i)$$

If the eigenvalues of the companion matrix $\boldsymbol{A}_c$ are distinct, then

$$\boldsymbol{A} = \boldsymbol{P}^{-1}\boldsymbol{\Lambda}\boldsymbol{P}$$

$$\boldsymbol{P} = \boldsymbol{V}_n = \begin{bmatrix} 1 & & 1 \\ \lambda_1 & \cdots & \lambda_n \\ \vdots & & \\ \lambda_1^{n-1} & & \lambda_n^{n-1} \end{bmatrix}$$

Matrix $\boldsymbol{V}_n$ is known as the **Vandermonde** matrix, which is non-singular.

In fact,

$$\det(\boldsymbol{V}_n) = \prod_{i=2}^{n}\left(\prod_{j=1}^{i-1}(\lambda_i - \lambda_j)\right)$$

Sylvester's Theorem 2

If all the eigenvalues ($\lambda_i,\ i = 1, 2, \ldots, n$) of a $n \times n$ matrix $\boldsymbol{A}$ are distinct then

$$f(\boldsymbol{A}) = \sum_{i=1}^{n} f(\lambda_i) \frac{\prod_{i \neq k}(\boldsymbol{A} - \lambda_k \boldsymbol{I})}{\prod_{i \neq k}(\lambda_i - \lambda_k)} \tag{2.45}$$

$$f(\boldsymbol{A}) = \frac{1}{\Delta} \sum_{i=1}^{n} \Delta_i \boldsymbol{A}^{i-1}$$

$$\Delta = \det(\boldsymbol{V}_n) = \begin{vmatrix} 1 & 1 & \cdots & 1 \\ \lambda_1 & \lambda_2 & & \lambda_n \\ \vdots & \vdots & & \vdots \\ \lambda_1^{n-1} & \lambda_2^{n-1} & & \lambda_n^{n-1} \end{vmatrix} \quad \text{Vandermonde matrix determinant}$$

and Δ_i is the determinant obtained from Δ by replacing the i-th row $\left(\lambda_1^{i-1} \lambda_2^{i-1} \cdots \lambda_n^{i-1}\right)$ of $\boldsymbol{V}_n$ with the row $(f(\lambda_1) f(\lambda_2) \cdots f(\lambda_n))$.

Proof is left as an exercise for the reader.

2.4.1 Solution to the Equation $Ax = b$, Revisited

(1) Case 1: $\boldsymbol{A}$ is non-singular

Let minimal polynomial of matrix $\boldsymbol{A}$ be

$$M(\lambda) = \lambda^m + m_1 \lambda^{m-1} + \cdots + m_m$$

Then

$$M(\boldsymbol{A}) = \boldsymbol{A}^m + m_1 \boldsymbol{A}^{m-1} + \cdots + m_m \boldsymbol{I} = \boldsymbol{0}$$

or

$$\boldsymbol{A}^{-1} = -\frac{1}{m_m} \left[\sum_{j=0}^{m-1} m_{m-j-1} \boldsymbol{A}^j \right]$$

(2) Case 2: $\boldsymbol{A}$ is singular

Generalized Inverse of a Matrix (also called Pseudo-Inverse)

Upto now, we have been dealing with only square matrices.

Let us consider,

$$\boldsymbol{Ax} = \boldsymbol{b}$$

where $\boldsymbol{A}$ is not a square matrix or not invertable. Obviously the matrix $\boldsymbol{A}$ has no inverse. In this case the solution for $\boldsymbol{x}$ in the above equation is not so easy. There may be no solutions or many solutions. One very useful result is the least mean square error solution which is given as ,

$$\boldsymbol{x} = \left[\left[\bar{\boldsymbol{A}}^T \boldsymbol{A}\right]^{-1} \bar{\boldsymbol{A}}^T\right]\boldsymbol{b} = \boldsymbol{A}^{\uparrow}\boldsymbol{b}$$

where,

$$\boldsymbol{A}^{\uparrow} = \left[\left[\bar{\boldsymbol{A}}^T \boldsymbol{A}\right]^{-1} \bar{\boldsymbol{A}}^T\right] = \text{Generalized Inverse of } \boldsymbol{A} \tag{2.46}$$

In general $\boldsymbol{A}^{\uparrow}$ is not easy to compute. However, if we can diagonalize $\left[\bar{\boldsymbol{A}}^T \boldsymbol{A}\right]$ then the computation of $\boldsymbol{A}^{\uparrow}$ is very easy. This is what is accomplished by Singular Value Decomposition (SVD) discussed below. Generalized Inverse is also referred to as Pseudo Inverse. Matrix $\bar{\boldsymbol{A}}$ is obtained from $\boldsymbol{A}$ by replacing elements of $\boldsymbol{A}$ by its complex conjugate.

2.4.2 Singular Value Decomposition of a Matrix (SVD)

Square matrices are decomposed via eigenvalues. On the other hand, the rectangular matrices are decomposed via singular value discussed below. Associated with the matrix $\boldsymbol{A}$ are the vectors $\boldsymbol{u}_i$ and $\boldsymbol{v}_i$

$$\begin{aligned} \boldsymbol{A}\boldsymbol{v}_i &= \sigma_i \boldsymbol{u}_i \\ \bar{\boldsymbol{A}}^T \boldsymbol{u}_i &= \sigma_i \boldsymbol{v}_i \end{aligned} \qquad i = 1, 2, \ldots \tag{2.47}$$

A is rectangular matrix $\quad n \times m, \quad n > m$

$\bar{A}^T$ = Hermitian Transpose of A

v_i = Right singular vector of A belonging to the singular value σ_i

u_i = Left singular vector of A belonging to the singular value σ_i

σ_i = Singular value (a non-negative scalar)

The pair (v_i, u_i) is called the **Schmidt Pairs of A** and is usually normalized to yield,

$$\|u_i\| = \|v_i\| = 1 \qquad \text{(unity Euclidean length)}$$

From the Eq. 2.47

$$\bar{A}^T A v_i = \sigma_i^2 v_i, \qquad A\bar{A}^T u_i = \sigma_i^2 u_i \tag{2.48}$$

Thus, $\bar{A}^T A$ and $A\bar{A}^T$ is a positive definite or positive semi-definite matrix and has countably many positive eigenvalues $\sigma_1^2 \geq \sigma_2^2 \geq \ldots \geq \sigma_k^2$, $k \leq n$. There are some parallels between eigenvalues and singular values of a matrix but they are far from being the **same** quantities. Eigenvalues play an important role when a matrix is a transformation from a vector space onto itself. On the other hand singular values are of importance when the matrix is a transformation from one vector space to a different vector space of different dimension, resulting in a non-existing inverse.

Comparison between Eigenvalue and Singular Value Decomposition

a) **Eigenvalue Decomposition (EVD):**

Mapping of the n dimensional space onto itself resulting in a square matrix. Let us take the simple case where $n \times n$ matrix A has eigenvalues $\lambda_1, \lambda_2, \ldots, \lambda_n$ and corresponding set of eigenvectors $p_1, p_2, \ldots, p_n$ which are linearly independent. Define

$$\Lambda = [\text{Diag } [\lambda_1, \lambda_2, \ldots, \lambda_n]]$$

$$P = \begin{bmatrix} p_1 & p_2 & \cdots & p_n \end{bmatrix}, \text{ (since } p_1, p_2, \ldots, p_n \text{ are linearly independent, } P^{-1} \text{ exists)}$$

Thus,

$$A\,[\boldsymbol{p}_1\ \boldsymbol{p}_2\ \cdots\ \boldsymbol{p}_n] = [\boldsymbol{p}_1\ \boldsymbol{p}_2\ \cdots\ \boldsymbol{p}_n]\ [\text{Diag}\ [\lambda_1, \lambda_2, \ldots, \lambda_n]]$$

or,

$$\boldsymbol{AP} = \boldsymbol{P\Lambda}$$

Thus,

$$\boldsymbol{A} = \boldsymbol{P\Lambda P}^{-1}$$

$$\boldsymbol{A}^k = \boldsymbol{P\Lambda}^k\boldsymbol{P}^{-1}, \qquad k = 0, 1, \ldots, n \tag{2.49}$$

Matrix $\boldsymbol{P}$ is a similarity transformation which decomposes the matrix $\boldsymbol{A}$ to a diagonal form (or Jordan-Canonical form discussed later). This is only possible because $\boldsymbol{A}$ is a square matrix. When the matrix is not square, the above decomposition is not possible so we look for another transformation called singular value decomposition.

b) **Singular Value Decomposition of a singular matrix (SVD)**

Let $\boldsymbol{A}$ mapping from m dimensional space onto another n dimensional space. Define

$$\begin{aligned} \boldsymbol{U} &= [\boldsymbol{u}_1\ \boldsymbol{u}_2\ \ldots\ \boldsymbol{u}_n] \\ \boldsymbol{V} &= [\boldsymbol{v}_1\ \boldsymbol{v}_2\ \ldots\ \boldsymbol{v}_m] \end{aligned} \tag{2.50}$$

$\boldsymbol{\Sigma} = n \times m$ matrix, same size as $\boldsymbol{A}$, with all entries zero, except along the main diagonal some of which may also be zero. The singular vectors can always be choosen to be orthonormal.

Thus,

$$\bar{\boldsymbol{U}}^T\boldsymbol{U} = \boldsymbol{I}_n, \qquad \bar{\boldsymbol{V}}^T\boldsymbol{V} = \boldsymbol{I}_m \tag{2.51}$$

The matrices $\boldsymbol{U}$ and $\boldsymbol{V}$ are unitary if complex and orthonormal if real.

Eq. 2.47 can be re-written as,

$$\boldsymbol{AV} = \boldsymbol{U\Sigma}, \qquad \bar{\boldsymbol{A}}^T\boldsymbol{U} = \boldsymbol{V}\bar{\boldsymbol{\Sigma}}^T \tag{2.52}$$

Thus the matrix $\boldsymbol{A}$ can be decomposed as,

$$\boldsymbol{A} = \boldsymbol{U\Sigma}\bar{\boldsymbol{V}}^T \qquad \text{(compared to } \boldsymbol{A} = \boldsymbol{P\Lambda P}^{-1} \text{ for a square matrix.)} \tag{2.53}$$

Notice that when $\boldsymbol{A}$ is non-singular, $\boldsymbol{U}$ and $\bar{\boldsymbol{V}}^T$ are replaced with $\boldsymbol{P}$ and $\boldsymbol{P}^{-1}$. The matrix $\boldsymbol{P}$ and $\boldsymbol{P}^{-1}$ represent basis in the n dimensional space of $\boldsymbol{A}$, so that the transformed matrix $\boldsymbol{\Lambda}$ in the same n dimensional space is diagonal. The singular value decomposition is relevant when $n \times m$ **matrix $\boldsymbol{A}$ represents a mapping from m space onto n space.** $\boldsymbol{U}$ and $\boldsymbol{V}$ are orthonormal or unitary and hence preserve lengths. For $n > m$

$$\boldsymbol{\Sigma} = \left.\overbrace{\begin{bmatrix} \sigma_1 & & & \\ & \sigma_2 & & \\ & & \ddots & \\ & & & \sigma_m \\ \hline & \boldsymbol{0} & & \end{bmatrix}}^{m}\right\} n = \left.\overbrace{\begin{bmatrix} \boldsymbol{\Sigma}_1 \\ \hline \boldsymbol{0} \end{bmatrix}}^{m}\right\} n \tag{2.54}$$

So, we ignore the last $(n - m)$ columns of $\boldsymbol{U}$ and do not compute them.

$$\boldsymbol{A} = n\left\{ \overbrace{\begin{bmatrix} \boldsymbol{u}_1 & \boldsymbol{u}_2 & \cdots & \boldsymbol{u}_m \end{bmatrix}}^{m} \right. \overbrace{\begin{bmatrix} \sigma_1 & & & \\ & \sigma_2 & & \\ & & \ddots & \\ & & & \sigma_m \end{bmatrix}}^{m} \overbrace{\begin{bmatrix} \boldsymbol{v}_1 & \boldsymbol{v}_2 & \cdots & \boldsymbol{v}_m \end{bmatrix}}^{m} \tag{2.55}$$

Summary:

For an $n \times m$ matrix $\boldsymbol{A}$,

$$\boldsymbol{A}\bar{\boldsymbol{A}}^T = \boldsymbol{S}_n = \boldsymbol{U}\boldsymbol{\Sigma}\bar{\boldsymbol{\Sigma}}\boldsymbol{U}^T \qquad n \times n \text{ symmetric matrix}$$

$$\bar{\boldsymbol{A}}^T\boldsymbol{A} = \boldsymbol{S}_m = \boldsymbol{V}^T\boldsymbol{\Sigma}\bar{\boldsymbol{\Sigma}}^T\boldsymbol{V} \qquad m \times m \text{ symmetric matrix}$$

SVD consists of finding the eigenvalues and eigenvector of $\boldsymbol{A}\bar{\boldsymbol{A}}^T$ of $\bar{\boldsymbol{A}}^T\boldsymbol{A}$.

SVD Algorithm:

Given: An $n \times m$ rectangular matrix $\boldsymbol{A}$, $n \geq m$

(1) Compute $\boldsymbol{A}\bar{\boldsymbol{A}}^T = \boldsymbol{S}_n$

- Compute eigenvalues of $\boldsymbol{S}_n$ as $\sigma_1^2, \sigma_2^2, \ldots, \sigma_i^2$, $\quad (i \leq n)$

  (Singular values of $\boldsymbol{A}$ corresponding to $\boldsymbol{A}\bar{\boldsymbol{A}}^T$ are $\sigma_1, \sigma_2, \ldots, \sigma_i$)
- Compute the eigenvectors of $\boldsymbol{A}\bar{\boldsymbol{A}}^T$ with respect to eigenvalues $\sigma_1^2, \sigma_2^2, \ldots, \sigma_i^2$. These are the left singular vectors of $\boldsymbol{A}$. **Normalize** these vectors to yield $\boldsymbol{U} = [\boldsymbol{u}_1\ \boldsymbol{u}_2\ \ldots\ \boldsymbol{u}_n]$

(2) Compute $\bar{\boldsymbol{A}}^T\boldsymbol{A} = \boldsymbol{S}_m$

- Eigenvalues of $\boldsymbol{S}_m$ are the same as of $\boldsymbol{S}_n$ and therefore the same singular values $\sigma_1, \sigma_2, \ldots, \sigma_i$
- Compute the eigenvectors of $\bar{\boldsymbol{A}}^T\boldsymbol{A}$ with respect to eigenvalues $\sigma_1^2, \sigma_2^2, \ldots, \sigma_i^2$. These are the right singular vectors of $\boldsymbol{A}$. Normalize these vectors to obtain $\boldsymbol{V}$, where:

$$\boldsymbol{V} = [\boldsymbol{v}_1\ \boldsymbol{v}_2\ \ldots\ \boldsymbol{v}_m]$$

Note: Both $\boldsymbol{A}\bar{\boldsymbol{A}}^T$ and $\bar{\boldsymbol{A}}^T\boldsymbol{A}$ are symmetric real matrices. Hence their eigenvalues and eigenvectors are real and orthogonal.

2.5 Multiple Eigenvalues—Jordan Canonical Form

An $n \times n$ matrix can be diagonalized if it has n distinct eigenvalues and therefore n linearly independent eigenvectors. On the other hand, if the matrix has eigenvalues of multiplicity greater than 1 than diagonalization is still possible if n linearly independent eigenvectors can be found. Otherwise, the similarity transformation produces not the diagonal form but the Jordan form. In this case, there may exist many different transformed Jordan forms. We discuss the following two cases.

Case #1: Elementary Divisors.

Let,

$$\Delta_A(\lambda) = |(\lambda \boldsymbol{I} - \boldsymbol{A})| = \prod_{i=1}^{r} (\lambda - \lambda_i)^{k_i}, \qquad \sum_{i=1}^{r} k_i = n$$

The matrix $\boldsymbol{A}$ can be transformed to matrix $\boldsymbol{J}$ with Canonical super boxes $\boldsymbol{J}_i\,(i = 1, \ldots, r)$,

$$\boldsymbol{J} = \sum_{i=1}^{r} \text{Diag}\, \boldsymbol{J}_i$$

These super boxes $\boldsymbol{J}_i$ are further divided into boxes $\boldsymbol{J}_{ij}\,(j = 1, \ldots, r_i;\ r_i \leq k_i)$,

$$\boldsymbol{J}_i = \sum_{j=1}^{r_i} \text{Diag}\, \boldsymbol{J}_{ij}$$

$$\boldsymbol{J} = \begin{bmatrix} \boldsymbol{J}_1 & & & \\ & \ddots & \boldsymbol{0} & \\ & & \boldsymbol{J}_i & \\ & \boldsymbol{0} & & \ddots \\ & & & & \boldsymbol{J}_r \end{bmatrix} = \sum_{i=1}^{r} \text{Diag}\, \boldsymbol{J}_i, \quad \boldsymbol{J}_i = \begin{bmatrix} \boldsymbol{J}_{i1} & & & \\ & \ddots & & \\ & & \boldsymbol{J}_{ij} & \\ & & & \ddots & \\ & & & & \boldsymbol{J}_{ir_i} \end{bmatrix} = \sum_{j=1}^{r_i} \text{Diag}\, \boldsymbol{J}_{ij}$$

Dimension of the box $\boldsymbol{J}_i$ is $k_i \times k_i$, the multiplicity of λ_i, $\left(\sum_{i=1}^{r} k_i = n \right)$.

Dimention of the ij-th box within $\boldsymbol{J}_i$ is $l_{ij} \times l_{ij}$, $\left(\sum_{j=1}^{r_i} l_{ij} = k_i \right)$

Following procedure is used to determine

$$\lambda_i, \qquad (i = 1, \ldots, r)$$

$$k_i, \qquad (i = 1, \ldots, r)$$

$$l_{ij}, \qquad (i = 1, \ldots, r; j = 1, \ldots, r_i)$$

Step 1: Determine the characteristic polynomial $\Delta_{\boldsymbol{A}}(\lambda)$ of $\boldsymbol{A}$ as the determinant of $(\lambda \boldsymbol{I} - \boldsymbol{A})$ and its roots $\lambda_i \quad (i = 1, \ldots, r)$

Step 2: Determine the multiplicity index $k_i\,(i = 1, \ldots, r)$ such that $(\lambda - \lambda_i)^{k_i}$ is a factor of $\Delta_{\boldsymbol{A}}(\lambda)$ but $(\lambda - \lambda_i)^{k_i+1}$ is not.

Step 3: Consider all the minors of the order $(n - j)$ of the matrix $(\lambda_i \boldsymbol{I} - \boldsymbol{A})$ $(i = 1, \ldots r; j = 1, \ldots r_i)$. If the greatest common divisor (GCD) of any one of these minors contains a factor $(\lambda - \lambda_i)^{k_i - l_{ij}}$ but not the factor $(\lambda - \lambda_i)^{k_i - l_{i,j} - 1}$ then $(\lambda - \lambda_i)^{l_{ij}}$ are the elementary divisors of the matrix $\boldsymbol{A}$ such that,

$$\Delta_{\boldsymbol{A}}(\lambda) = \prod_{i=1}^{r} \prod_{j=1}^{r_i} (\lambda - \lambda_i)^{l_{ij}} \quad \text{(elementary divisor product)}$$

The minors of the order $(n - k_i - 1)$ of the matrix $(\lambda \boldsymbol{I} - \boldsymbol{A})$ contain no factor $(\lambda - \lambda_i)$. Each Jordan sub box $\boldsymbol{J}_{ij}$ appears as:

$$\boldsymbol{J}_{ij} = \begin{bmatrix} \lambda_i & 1 & 0 & \cdots & 0 \\ 0 & \lambda_i & 1 & & 0 \\ \vdots & & & & \\ & & & \ddots & 1 \\ 0 & & & & \lambda_i \end{bmatrix}, \quad l_{ij} \times l_{ij} \text{ matrix}$$

In reality we have used the method of " elementary divisors" to arrive at the structure of Jordan Cannonical form. We transformed $\boldsymbol{A}$ into Jordanform $\boldsymbol{J}$ via similarity transformation $\hat{\boldsymbol{P}}$

$$\boldsymbol{A} = \hat{\boldsymbol{P}}\boldsymbol{J}\hat{\boldsymbol{P}}^{-1}, \quad \boldsymbol{A}^k = \hat{\boldsymbol{P}}\boldsymbol{J}^k\hat{\boldsymbol{P}}^{-1}.$$

The model matrix $\hat{\boldsymbol{P}}$ is made up of the chain of generalized eigenvectors

$$\boldsymbol{x}_{ij}, \ (\lambda_i\boldsymbol{I} - \boldsymbol{A})\boldsymbol{x}_{ij}, \cdots, (\lambda_i\boldsymbol{I} - \boldsymbol{A}^{l_{ij}-1})\boldsymbol{x}_{ij} \quad (i = 1, \ldots r, \ j = 1, \ldots, r_i)$$

$$(\lambda_i\boldsymbol{I} - \boldsymbol{A})^{l_{ij}}\boldsymbol{x}_{ij} = \boldsymbol{0}$$

These are $\sum_{i=1}^{r} r_i$ independent vectors.

Every non-singular square matrix $\boldsymbol{A}$ can be transformed into **Jordan Form**.

Minimal polynomial of $\boldsymbol{J}$ (or A)

The minimal polynomial is

$$P_m(\lambda) = \prod_{i=1}^{r_i} (\lambda - \lambda_i)^{l_{i1}}$$

while l_{i1} the largest size Jordan sub box associated with λ_i

Notation: Using "Dg" for diagonal,

$$\boldsymbol{J} = \text{Dg}\,[\boldsymbol{J}_1, \boldsymbol{J}_2, \ldots, \boldsymbol{J}_i, \ldots, \boldsymbol{J}_r]$$
$$\boldsymbol{J}_i = \text{Dg}\left[\boldsymbol{J}_{i1}, \ldots, \boldsymbol{J}_{ij}, \ldots, \boldsymbol{J}_{ir_i}\right] \qquad (i = 1, \ldots, r)$$

$(\lambda - \lambda_i)^{l_{ij}}$ are known as **elementary divisors** of $\boldsymbol{A}$ $(i = 1, \ldots, r, \ j = 1, \ldots, r_i)$

$(\lambda_i\boldsymbol{I} - \boldsymbol{A})$ acts as "elevator matrix". It raises an eigenvector to the next higher eigenvectors till the last vector in the chain is reached and then it gets annihilated.

Case #2: Generalized Eigenvectors

When the matrix $\boldsymbol{A}$ is nonderogatory such as the companion matrix, its minimal and characteristic polynomials are the same. In this case, we avoid determining elementary divisors and hence less computing is needed. We determine, one independent eigenvector for each distinct eigenvalue of the matrix $\boldsymbol{A}$. Therefore, only r independent eigenvectors are needed. Let us define these vectors as $\boldsymbol{x}_{11}, \boldsymbol{x}_{21}, \cdots, \boldsymbol{x}_{r1}$. Consider the eigenvector $\boldsymbol{x}_{i1}$, the matrix $(\lambda_i \boldsymbol{I} - \boldsymbol{A})^{k_i}$ annihilates the eigenvector $\boldsymbol{x}_{i1}$. Thus,

$$(\lambda_i \boldsymbol{I} - \boldsymbol{A})^{k_i} \boldsymbol{x}_{i1} = \boldsymbol{0}$$

$$\begin{aligned}
(\lambda_i \boldsymbol{I} - \boldsymbol{A}) \boldsymbol{x}_{i1} &= \boldsymbol{x}_{i2} \\
(\lambda_i \boldsymbol{I} - \boldsymbol{A}) \boldsymbol{x}_{i2} &= \boldsymbol{x}_{i3} \\
&\vdots \\
(\lambda_i \boldsymbol{I} - \boldsymbol{A}) \boldsymbol{x}_{ik_i} &= \boldsymbol{0}
\end{aligned}$$

$$\begin{bmatrix} \boldsymbol{A} \end{bmatrix} \begin{bmatrix} \boldsymbol{x}_{i1} & \boldsymbol{x}_{i2} & \cdots & \boldsymbol{x}_{ik_i} \end{bmatrix} = \begin{bmatrix} \boldsymbol{x}_{i1} & \boldsymbol{x}_{i2} & \cdots & \boldsymbol{x}_{ik_i} \end{bmatrix} \begin{bmatrix} \lambda_i & 1 & 0 & \cdots & 0 \\ 0 & \lambda_i & 0 & & 0 \\ \vdots & \vdots & \vdots & & 1 \\ 0 & 0 & 0 & \cdots & \lambda_i \end{bmatrix}$$

$$\begin{bmatrix} \boldsymbol{A} \end{bmatrix} \begin{bmatrix} \hat{\boldsymbol{P}}_i \end{bmatrix} = \begin{bmatrix} \hat{\boldsymbol{P}}_i \end{bmatrix} \begin{bmatrix} \boldsymbol{J}_i \end{bmatrix}$$

yielding,

$$\begin{bmatrix} \boldsymbol{A} \end{bmatrix} \begin{bmatrix} \hat{\boldsymbol{P}}_1 & \hat{\boldsymbol{P}}_2 & \cdots & \hat{\boldsymbol{P}}_r \end{bmatrix} = \begin{bmatrix} \boldsymbol{J}_1 & & & \\ & \boldsymbol{J}_2 & & \\ & & \ddots & \\ & & & \boldsymbol{J}_r \end{bmatrix} \begin{bmatrix} \hat{\boldsymbol{P}}_1 & \hat{\boldsymbol{P}}_2 & \cdots & \hat{\boldsymbol{P}}_r \end{bmatrix}$$

2.5.1 Cayley-Hamilton Theorem

This remarkable theorem states,

"A matrix satisfied its own Characteristic Equation"

Specifically

$$\Delta_A(\lambda) = p(\lambda) = |(\lambda \boldsymbol{I} - \boldsymbol{A})| = \lambda_n + a_1\lambda^{n-1} + \cdots + a_n$$

$\boldsymbol{A}$ is $n \times n$ matrix.

If $\quad \boldsymbol{x}, \boldsymbol{Ax} \in \boldsymbol{E}_n$

Then

$$p(\boldsymbol{A})\boldsymbol{x} = \boldsymbol{0}$$

implying

$$p(\boldsymbol{A}) \equiv \boldsymbol{A}^n + a_1\boldsymbol{A}^{n-1} + \cdots + a_n\boldsymbol{I} = \boldsymbol{0}$$

Cayley-Hamilton Theorem

Proof:

$$(\boldsymbol{A}(\lambda))^{-1} = (\lambda \boldsymbol{I} - \boldsymbol{A})^{-1} = \left(\frac{1}{\Delta_A(\lambda)}\right)(\text{Adj}\boldsymbol{A}(\lambda)) = \frac{1}{p(\lambda)}\boldsymbol{B}(\lambda) \tag{2.56}$$

$$\boldsymbol{B}(\lambda) = \text{Adj}\boldsymbol{A}(\lambda) = \text{ polynomial matrix in } \lambda \text{ of degree } n-1$$

$$\boldsymbol{B}(\lambda) = \boldsymbol{B}_1\lambda^{n-1} + \boldsymbol{B}_2\lambda^{n-2} + \cdots + \boldsymbol{B}_n = \sum_{i=0}^{n-1} \boldsymbol{B}_{n-i}\lambda^i \tag{2.57}$$

From the Eqs. 2.56 and 2.57,

$$(\lambda \boldsymbol{I} - \boldsymbol{A})\boldsymbol{B}(\lambda) = p(\lambda)\boldsymbol{I}$$

Equating powers of λ on both sides

$$\begin{aligned}
\mathbf{0} - \boldsymbol{A}\boldsymbol{B}_n &= a_n \boldsymbol{I} \\
\boldsymbol{B}_n - \boldsymbol{A}\boldsymbol{B}_{n-1} &= a_{n-1} \boldsymbol{I} \\
&\vdots \\
\boldsymbol{B}_2 - \boldsymbol{A}\boldsymbol{B}_1 &= a_1 \boldsymbol{I} \\
\boldsymbol{B}_1 - \mathbf{0} &= \boldsymbol{I}
\end{aligned} \tag{2.58}$$

Multiplying these equations with $\mathbf{1}, \boldsymbol{A}, \ldots, \boldsymbol{A}^n$ respectively and adding,

$$\mathbf{0} \equiv \boldsymbol{A}^n + a_1 \boldsymbol{A}^{n-1} + \cdots + a_n \boldsymbol{I} \equiv p(\boldsymbol{A}) \tag{2.59}$$

Cayley-Hamilton Theorem.

This theorem is significant in System theory for it implies that all matrices $\boldsymbol{A}^k$ $(k \geq n)$ can be expressed as a linear combination of matrices $\boldsymbol{A}^j$ $(j < n)$.

2.6 Co-efficients of Characteristic Polynomial

Consider a method due to A. N. Krylov for finding the co-efficients of the characteristic polynomial $\Delta_{\boldsymbol{A}}(\lambda)$ of a $n \times n$ matrix $\boldsymbol{A}$. This is also a good introduction to Krylov spaces which will be discussed later.

Given: A $n \times n$ square matrix $\boldsymbol{A}$, we are required to find the co-efficients $a_1, a_2, \ldots, a_n$ of the characteristic polynomial,

$$\Delta_{\boldsymbol{A}}(\lambda) = |\lambda \boldsymbol{I} - \boldsymbol{A}| = \lambda^n + a_1 \lambda^{n-1} + \cdots + a_n \tag{2.60}$$

From the Caley-Hamilton theorem,

$$\boldsymbol{A}^n + a_1 \boldsymbol{A}^{n-1} + \cdots + a_n \boldsymbol{I} = \mathbf{0} \tag{2.61}$$

Now choose an **arbitrary vector** $\boldsymbol{b}_0$ of grade n and postmultiply with Eq. 2.61, yielding,

$$\boldsymbol{A}^n\boldsymbol{b}_0 + a_1\boldsymbol{A}^{n-1}\boldsymbol{b}_0 + \cdots + a_n\boldsymbol{b}_0 = \boldsymbol{0} \tag{2.62}$$

Let,

$$\begin{aligned} \boldsymbol{y}_0 &= \boldsymbol{b}_0 \\ \boldsymbol{y}_k &= \boldsymbol{A}\boldsymbol{y}_{k-1} = \boldsymbol{A}^k\boldsymbol{b}_0 \qquad k = 1, 2, \ldots, n \qquad \text{(Krylov vectors)} \end{aligned}$$

or,

$$\begin{bmatrix} \boldsymbol{y}_1 \; \boldsymbol{y}_2 \cdots \boldsymbol{y}_n \end{bmatrix} \begin{bmatrix} a_1 \\ a_2 \\ \cdots \\ a_n \end{bmatrix} = -\begin{bmatrix} \boldsymbol{y}_n \end{bmatrix}$$

Note: notice that we have avoided calculating the powers of $\boldsymbol{A}$ but still need to invert matrix made up of Krylov vectors and a clever choice of vector $\boldsymbol{b}_0$.

2.7 Computation of Matrix Polynomial Function

$$F(\boldsymbol{A}) = \sum c_k\boldsymbol{A}^k \qquad m \geq n,\; \lambda_i(i = 1, \ldots, n) \text{ are eigenvalues of } \boldsymbol{A}. \tag{2.63}$$

$$\frac{F(\lambda)}{\Delta_{\boldsymbol{A}}(\lambda)} = Q(\lambda) + \frac{R(\lambda)}{\Delta_{\boldsymbol{A}}(\lambda)} \quad \text{(long division)} \tag{2.64}$$

Polynomial $R(\lambda)$ less than degree n.

$$\begin{aligned} F(\lambda) &= Q(\lambda)\Delta_{\boldsymbol{A}}(\lambda) + R(\lambda) \\ F(\lambda_i) &= R(\lambda_i) \qquad \Delta_{\boldsymbol{A}}(\lambda_i) = 0, \quad i = 1, \ldots, n \end{aligned} \tag{2.65}$$

Compute the coefficients of $R(\lambda_i)$ from $F(\lambda_i)$.

If λ_i is an eigenvalue of the multiplicity m_i, then not only $\Delta_{\boldsymbol{A}}(\lambda_i) = 0$, but also the first $(m_i - 1)$ derivatives of $\Delta_A(\lambda)$ w.r.t. λ at $\lambda = \lambda_i$ vanish, resulting in

$$\left.\frac{\mathrm{d}^k}{\mathrm{d}\lambda^k}F(\lambda)\right|_{\lambda=\lambda_i} = \left.\frac{\mathrm{d}^k}{\mathrm{d}\lambda^k}R(\lambda)\right|_{\lambda=\lambda_i} \qquad (k = 0, 1, \ldots, m_i - 1)$$

Matrix exponential:

$$\boldsymbol{e}^{\boldsymbol{A}t} = \sum_{k=0}^{\infty} \frac{\boldsymbol{A}^k t^k}{k!} \quad \text{(not generally recommended for computing)}$$

$$\boldsymbol{e}^{\boldsymbol{A}t} = \sum_{i=0}^{n-1} \alpha_i(t)\boldsymbol{A}^i, \quad \alpha_0(0) = 1, \quad \alpha_i(0) = 0, \ (i = 2, \ldots)$$

$$\boldsymbol{e}^{(\boldsymbol{A}+\boldsymbol{B})t} = \left(\boldsymbol{e}^{\boldsymbol{A}t}\right)\left(\boldsymbol{e}^{\boldsymbol{B}t}\right), \quad \boldsymbol{AB} = \boldsymbol{BA}$$

Convergent Series:

$$g(\lambda) = \sum_{k=0}^{\infty} g_k \lambda^k, \quad |\lambda| \le r \le 1 \quad \text{implies convergence}$$

$$g(\boldsymbol{A}) = \sum_{k=0}^{\infty} g_k \boldsymbol{A}^k, \ \boldsymbol{A} \text{ with eigenvalues } \lambda_i, \ |\lambda_i| \le r \le 1 \text{ implies convergence}$$

Complex Integration:

$$\boldsymbol{e}^{\boldsymbol{A}t} = \frac{1}{2\pi j} \oint_c (\lambda \boldsymbol{I} - \boldsymbol{A})^{-1} \boldsymbol{e}^{\lambda t}\, \mathbf{d}\lambda, \qquad |\lambda_i| \le \text{radius } c$$

Riccati Equation Solution:

$$\boldsymbol{AS} + \boldsymbol{SA}^T = -\boldsymbol{Q}, \quad \boldsymbol{S}, \boldsymbol{Q} \text{ are symmetric}$$

then,

$$\boldsymbol{S} = \int_0^{\infty} \boldsymbol{e}^{\boldsymbol{A}t} \boldsymbol{Q} \boldsymbol{e}^{\boldsymbol{A}^T t}\, \mathbf{d}t$$

Functions of Jordan matrices J and diagonal matrix Λ

$$J = \begin{bmatrix} \lambda_1 & 1 & & \cdot \\ & \lambda_1 & 1 & \\ & & \lambda_1 & \\ & & & \cdot \end{bmatrix}, \; f(J) = \begin{bmatrix} f(\lambda_1) & \frac{f'(\lambda_1)}{1!} & \frac{f''(\lambda_1)}{2!} & \cdot \\ & f(\lambda_1) & \frac{f'(\lambda_1)}{1!} & \\ & & f(\lambda_1) & \\ & & & \cdot \end{bmatrix}, \; e^{Jt} = \begin{bmatrix} e^{\lambda_1 t} & te^{\lambda_1 t} & t^2 e^{\lambda_1 t} & \cdot \\ & e^{\lambda_1 t} & te^{\lambda_1 t} & \\ & & e^{\lambda_1 t} & \\ & & & \cdot \end{bmatrix}$$

$$\Lambda = \text{Diag}\ [\lambda, \lambda 2, \cdots, \lambda_n]$$

$$f((\Lambda)) = \text{Diag}\ [f(\lambda_1), f(\lambda_2), \cdots, f(\lambda_n)]$$

$$e^{\Lambda t} = \text{Diag}\ \left[e^{\lambda_1 t}, e^{\lambda_2 t}, \cdots, e^{\lambda_n t}\right]$$

$$A = P^{-1}\Lambda P$$

$$f(A) = P^{-1} f(\Lambda) P$$

$$A = S^{-1} A S$$

$$f(A) = S^{-1} f(J) S$$

If a $n \times n$ matrix A has **minimal polynomial of degree** $m < n$, then

$$e^{At} = \alpha_0(t) I + \alpha_1(t) A + \cdots + \alpha_{m-1}(t) A^{m-1}$$

Coefficients $\alpha_j(t)$ $(j = 0, \ldots, m-1)$ can be computed from eigenvalues, distinct or multiple. Matrix A is called "stable" if the real part of all its eigenvalues λ_i $(i = 1, \ldots, n)$ are **negative**.

Nilpotent matrix:

Given a $n \times n$ matrix A, it is called Nilpotent matrix if

$A^k = 0$ for some positive integer k.

2.8 S-N Decomposition of a Non-singular Matrix

Any non-singular $n \times n$ matrix $\boldsymbol{A}$ can be decomposed into two unique matrices $\boldsymbol{S}$ and $\boldsymbol{N}$ such that:

$$\boldsymbol{A} = \boldsymbol{S} + \boldsymbol{N}, \quad \boldsymbol{SN} = \boldsymbol{NS}, \quad \boldsymbol{N}^n = \boldsymbol{0} \quad \text{(Nilpotent)}$$

Proof:

Let

$$\det(\lambda \boldsymbol{I} - \boldsymbol{A}) = P(\lambda) = \prod_{i=1}^{m} (\lambda - \lambda_i)^{r_i}, \quad r_1 + r_2 + \cdots + r_m = n$$

Using partial fraction:

$$\frac{1}{P(\lambda)} = \sum_{i=1}^{m} \frac{n_i(\lambda)}{(\lambda - \lambda_i)^{r_i}}$$

Define:

$$f_i(\lambda) = n_i(\lambda) \prod_{\substack{i \neq j \\ j=1}}^{m} (\lambda - \lambda_j)^{r_j}, \quad i = 1, 2, \cdots, m$$

Clearly:

$$\sum_{i=1}^{m} f_i(\lambda) = 1, \quad \sum_{i=1}^{m} f_i(\boldsymbol{A}) = \boldsymbol{I}$$

Let

$$f_i(\boldsymbol{A}) = n_i(\boldsymbol{A}) \prod_{\substack{j \neq i \\ j=1}}^{m} (\boldsymbol{A} - \lambda_i \boldsymbol{I})^{r_j}$$

$$f_i(\boldsymbol{A}) f_j(\boldsymbol{A}) = \boldsymbol{0} \quad i \neq j, \quad f_i^2(\boldsymbol{A}) = f_i(\boldsymbol{A}) \quad i = 1, \cdots, m$$

Furthermore, from Cayley-Hamilton Theorem:

$$(\boldsymbol{A} - \lambda_i \boldsymbol{I})^{r_i} f_i(\boldsymbol{A}) = f_i(\boldsymbol{A})(\boldsymbol{A} - \lambda_i \boldsymbol{I})^{r_i} = n_i(\boldsymbol{A}) \prod_{j=1}^{m} (\boldsymbol{A} - \lambda_j \boldsymbol{I})^{r_j} = \boldsymbol{0} \quad i \neq j$$

$$\boldsymbol{A} f_i(\boldsymbol{A}) = f_i(\boldsymbol{A}) \boldsymbol{A}$$

Letting:

$$S = \sum_{i=1}^{m} \lambda_i f_i(\boldsymbol{A})$$

$$\boldsymbol{N} = (\boldsymbol{A} - \boldsymbol{S}) = \left(\boldsymbol{A} - \sum_{i=1}^{m} \lambda_i f_i(\boldsymbol{A})\right) = \left(\boldsymbol{A} \sum_{i=1}^{m} f_i(\boldsymbol{A}) - \sum_{i=1}^{m} \lambda_i f_i(\boldsymbol{A})\right)$$

$$= \sum_{i=1}^{m} (\boldsymbol{A} - \lambda_i \boldsymbol{I})\, f_i(\boldsymbol{A})$$

Clearly:

$$\boldsymbol{N}^n = (\boldsymbol{A} - \boldsymbol{S})^n = \sum_{i=1}^{m} (\boldsymbol{A} - \lambda_i \boldsymbol{I})^i\, f_i^i(\boldsymbol{A}) = \boldsymbol{0} \qquad \text{Nilpotent Matrix}$$

Also,

$$\boldsymbol{SA} = \boldsymbol{AS}$$

implying

$$\boldsymbol{NS} = \boldsymbol{SN}$$

$$\boldsymbol{A} = \sum_{i=1}^{m} (\lambda_i \boldsymbol{I} + \boldsymbol{N})\, f_i(\boldsymbol{A})$$

$$\boldsymbol{A}^k = \sum_{i=1}^{m} (\lambda_i \boldsymbol{I} + \boldsymbol{N})^k\, f_i^k(\boldsymbol{A})$$

$f_i(\boldsymbol{A})$ are also known as projection matrices.

Example 2.9:

Let

$$P(\lambda) = (\lambda - \lambda_1)(\lambda - \lambda_2)$$

Then

$$N_1(\lambda) = \frac{1}{(\lambda_1 - \lambda_2)}, \quad N_2(\lambda) = -\frac{1}{(\lambda_1 - \lambda_2)}$$

$$f_1(\lambda) = N_1(\lambda)(\lambda - \lambda_2), \quad f_2(\lambda) = N_2(\lambda)(\lambda - \lambda_1)$$

$$f_1(\lambda) + f_2(\lambda) = \frac{1}{(\lambda_1 - \lambda_2)}(\lambda - \lambda_2 - \lambda + \lambda_1) = 1$$

$$f_1(\boldsymbol{A}) + f_2(\boldsymbol{A}) = \frac{1}{(\lambda_1 - \lambda_2)}(\boldsymbol{A} - \lambda_2\boldsymbol{I} - \boldsymbol{A} + \lambda_1\boldsymbol{I}) = \boldsymbol{I}$$

$$f_1(\boldsymbol{A})f_2(\boldsymbol{A}) = \frac{1}{(\lambda_1 - \lambda_2)^2}(\boldsymbol{A} - \lambda_2\boldsymbol{I})(\boldsymbol{A} - \lambda_1\boldsymbol{I}) = \boldsymbol{0}$$

$$f_1^2(\boldsymbol{A}) = \frac{1}{(\lambda_1 - \lambda_2)^2}(\boldsymbol{A} - \lambda_2\boldsymbol{I})^2 = f_1(\boldsymbol{A})$$

$$f_2^2(\boldsymbol{A}) = \frac{1}{(\lambda_2 - \lambda_1)^2}(\boldsymbol{A} - \lambda_1\boldsymbol{I})^2 = f_2(\boldsymbol{A})$$

$$\begin{aligned} \boldsymbol{S} &= \lambda_1 f_1(\boldsymbol{A}) + \lambda_2 f_2(\boldsymbol{A}) = \frac{\lambda_1}{\lambda_1 - \lambda_2}(\boldsymbol{A} - \lambda_2\boldsymbol{I}) - \frac{\lambda_2}{\lambda_1 - \lambda_2}(\boldsymbol{A} - \lambda_1\boldsymbol{I}) \\ &= \frac{1}{\lambda_1 - \lambda_2}(\lambda_1\boldsymbol{A} - \lambda_2\boldsymbol{A}) = \boldsymbol{A} \end{aligned}$$

$$\boldsymbol{N} = 0$$

2.9 Computation of A^n without Eigenvectors

Consider a $n \times n$ matrix $\boldsymbol{A}$. We are required to compute:

$$\boldsymbol{A}^N, \quad N \geq n$$

This task can be accomplished using the similarity transformations, yielding: Either

$$\boldsymbol{A}^N = \boldsymbol{P}\boldsymbol{\Lambda}^N\boldsymbol{P}^{-1}, \quad \boldsymbol{\Lambda} = \boldsymbol{P}^{-1}\boldsymbol{A}\boldsymbol{P} \qquad \text{Diagonal}$$

or

$$\boldsymbol{A}^N = \hat{\boldsymbol{P}}\boldsymbol{J}^N\hat{\boldsymbol{P}}^{-1}, \quad \boldsymbol{J} = \hat{\boldsymbol{P}}^{-1}\boldsymbol{A}\hat{\boldsymbol{P}} \qquad \text{Jordan form}$$

In either case, we are required to compute the eigenvectors or the generalized eigenvectors which is a computationally expensive task. Following method suggested by Elaydi and Harris [Elaydi, S.N.]. The proof of this method is clearly related to the fundamental solutions of the n-th order difference equations discussed in chapter 3. Here we state the algorithm to compute $\boldsymbol{A}^N$ without proof.

Algorithm for Computing $\boldsymbol{A}^N$, $N \geq n$

(i) Compute the eigenvalues $\lambda_1, \lambda_2, \ldots, \lambda_n$ (not necessarily distinct) as the roots of the characteristic polynomial $P(\lambda)$:

$$P(\lambda) = \det(\lambda \boldsymbol{I} - \boldsymbol{A}) = \lambda^n + a_1 \lambda^{n-1} + \cdots + a_n$$

(ii) Compute the matrices:

$$\boldsymbol{M}(j) = (-1)^j \prod_{i=1}^{j} (\lambda_i \boldsymbol{I} - \boldsymbol{A}), \quad \boldsymbol{M}(0) = \boldsymbol{I}, \quad j = 1, 2, \ldots, n-1$$

Note:

$$\boldsymbol{M}(n) = \prod_{i=1}^{n} (\lambda_i \boldsymbol{I} - \boldsymbol{A}) = \boldsymbol{0} \quad \text{(Caley-Hamilton Theorem.)}$$

(iii) Compute the scalar fundamental recursive functions $\varphi_j(N)$:

$$\varphi_1(N) = \lambda_1^N$$

$$\varphi_{j+1}(N) = \sum_{i=0}^{N-1} (\lambda_{j+1})^{N-i-1} \varphi_j(i), \quad j = 1, 2, \ldots, n-1 \tag{2.66}$$

(iv) For $N \geq n$, compute:

$$\boldsymbol{A}^N = \sum_{j=0}^{n-1} \varphi_{j+1}(N) \boldsymbol{M}(j)$$

2.10 Companion Matrices

This is a special matrix of the form

$$A_c = \begin{bmatrix} 0 & 1 & 0 & \cdots & 0 \\ 0 & 0 & 1 & \cdots & 0 \\ \vdots & & & & \\ 0 & 0 & 0 & \cdots & 1 \\ -a_n & -a_{n-1} & -a_{n-2} & \cdots & -a_1 \end{bmatrix},$$

$$\Delta_{A_c}(\lambda) = |\lambda \boldsymbol{I} - \boldsymbol{A}_c| = \lambda^n + \boldsymbol{a}_1\lambda^{n-1} + \cdots + \boldsymbol{a}_n = p(\lambda)$$

The polynomial $\Delta_{A_c}(\lambda)$ can be associated with the companion matrix $\boldsymbol{A}_c$. Following special properties are associated with companion matrices:

1. If λ_i is an eigenvalue of multiplicity one (distinct), the associated eigenvector is,

$$\boldsymbol{p}_i^T = \left[\left\{1 \ \lambda_i \ \lambda_i^2 \ \cdots \ \lambda_i^{n-1}\right\}\right]$$

2. If λ_i is an eigenvalue of the multiplicity $k_i \leq n$, namely, $((\lambda - \lambda_i)^{k_i}$ is a factor of $\Delta_{A_c}(\lambda)$ but $(\lambda - \lambda_i)^{k_i+1}$ is not) then this eigenvalue has k_i "**generalized eigenvectors**" and one and only one Jordan block of size $k_i \times k_i$ belonging to the eigenvalue λ_i. This implies that companion matrix is nonderogatory. The corresponding k_i eigenvectors of the eigenvalue λ_i are:

$$\begin{aligned} \boldsymbol{p}_{i1}^T &= \left[1 \ \lambda_i \ \lambda_i^2 \ \cdots \ \lambda_i^{n-1}\right] \\ \boldsymbol{p}_{i2}^T &= \left[0 \ 1 \ 2\lambda_i \ \cdots \ (n-1)\lambda_i^{n-2}\right] \\ &\vdots \\ \boldsymbol{p}_{ik_i}^T &= \left[0 \ 0 \ 0 \cdots \left(\prod_{j=1}^{k_i-1}(n-j)\lambda_i^{n-k_i}\right)\right] \end{aligned}$$

3. An n-th order Linear Differential Equation

$$x^{(n)} + a_1 x^{(n-1)} + \cdots + a_{n-1}\dot{x} + a_n x = 0$$

can be written as:

$$\dot{\boldsymbol{x}} = \boldsymbol{A}_c \boldsymbol{x}\,, \text{ where } \boldsymbol{A}_c \text{ is companion matrix}$$

4. **Important :** A matrix $\boldsymbol{A}$ is similar to the companion matrix $\boldsymbol{A}_c$ if and only if the **minimal** and the **characteristic** polynomial of $\boldsymbol{A}$ and $\boldsymbol{A}_c$ are the same. This implies $\boldsymbol{A}$ being nonderogatory.

2.11 Choletsky Decomposition (*LU* Decomposition)

This is a convenient scheme for machine computation of,

$$\boldsymbol{A}\boldsymbol{x} = \boldsymbol{b}$$

$$\boldsymbol{A} \text{ is } n \times n \text{ of rank } n,\ \boldsymbol{b} \text{ is } n \times 1$$

Write $\boldsymbol{A}$ as :

$$\boldsymbol{A} = \boldsymbol{L}\boldsymbol{U}, \text{ where } \boldsymbol{L} \text{ is } \textbf{lower triangular} \text{ and } \boldsymbol{U} \text{ is } \textbf{upper triangular}$$

$$\boldsymbol{L} = \begin{bmatrix} l_{11} & 0 & \cdots & 0 \\ l_{21} & l_{22} & \cdots & 0 \\ \vdots & & & \\ l_{n1} & l_{n2} & \cdots & l_{nn} \end{bmatrix} \quad \boldsymbol{U} = \begin{bmatrix} 1 & c_{12} & \cdots & c_{1n} \\ 0 & 1 & \cdots & c_{2n} \\ \vdots & & & \\ 0 & 0 & \cdots & 1 \end{bmatrix}, \quad \boldsymbol{A} = (a_{ij})$$

l_{ij} and u_{ij} are computed as

$$l_{i1} = a_{i1}, \quad u_{1j} = \frac{a_{1j}}{l_{11}}, \qquad (i = 1, \ldots, n;\ j = 1, \ldots, n)$$

$$l_{ij} = a_{ij} - \sum_{k=1}^{j-1} (l_{ik} u_{kj}); \qquad (i \geq j > 1)$$

$$u_{ij} = \frac{1}{l_{ii}} \left(a_{ij} - \sum_{k=1}^{i-1} l_{ik} u_{kj} \right); \qquad (j > i > 1),\ u_{ii} = 1$$

Knowing $\boldsymbol{L}$ and $\boldsymbol{U}$ matrices, solve the two sets of equations

$$\boldsymbol{U}\boldsymbol{x} = \boldsymbol{y}, \qquad \boldsymbol{L}\boldsymbol{y} = \boldsymbol{b}$$

when $\boldsymbol{A}$ is symmetric. The computation of $\boldsymbol{U}$ is simplified as

$$u_{ij} = \frac{1}{l_{ii}} l_{ji}, \qquad (i \leq j)$$

2.12 Jacobi and Gauss-Seidel Methods

When all the diagonal elements of $\boldsymbol{A}$ are non-zero, we can decompose $\boldsymbol{A}$ as:

$$\boldsymbol{A} = \boldsymbol{L} + \boldsymbol{D} + \boldsymbol{U}$$

$\boldsymbol{U} =$ Upper triangle with zero on the diagonal

$\boldsymbol{L} =$ Lower triangle with zero on the diagonal

$\boldsymbol{D} =$ Diagonal matrix

The iterative schemes for solving $\boldsymbol{A}\boldsymbol{x} = \boldsymbol{b}$, with initial guess $\boldsymbol{x}^{(0)}$ are:

$$\boldsymbol{x}^{(i+1)} = \boldsymbol{D}^{-1}\boldsymbol{b} - \boldsymbol{D}^{-1}(\boldsymbol{L} + \boldsymbol{U})\boldsymbol{x}^{(i)} \quad (i = 0, 1, 2, 3 \ldots) \qquad \text{(Jacobi)}$$

$$\boldsymbol{x}^{(i+1)} = (\boldsymbol{L} + \boldsymbol{D})^{-1}\boldsymbol{b} - (\boldsymbol{L} + \boldsymbol{D})^{-1}\boldsymbol{U}\boldsymbol{x}^{(i)} \qquad \text{(Gauss-Seidel)}$$

2.13 Least Squares (Pseudo Inverse Problem)

Given

$$\boldsymbol{Ax} = \boldsymbol{b},$$

subject to condition $\boldsymbol{Bx} = 0$,

$\boldsymbol{A}$ is $n \times p$, Rank $\boldsymbol{A} = p$;

$\boldsymbol{B}$ is $r \times p$, Rank $\boldsymbol{B} = r$, $\quad (r \leq p \leq n)$

Compute the $\boldsymbol{x}$.

Define

$$\left(\boldsymbol{A}^T\boldsymbol{A}\right)^{-1}\boldsymbol{A}^T = \boldsymbol{A}^+ \text{ (pseudo inverse)}$$
$$\left(\boldsymbol{A}^T\boldsymbol{A}\right)^{-1}\boldsymbol{B}^T = \boldsymbol{B}_1$$

The least square solution is:

$$\hat{\boldsymbol{x}} = \left[\boldsymbol{A}^+ - \boldsymbol{B}_1(\boldsymbol{B}\boldsymbol{B}_1)^{-1}\boldsymbol{B}\boldsymbol{A}^+\right]\boldsymbol{b}$$

2.14 Hermitian Matrices and Definite Functions

1. $\boldsymbol{A}$ is Hermitian, then for all $\boldsymbol{x}$

$\boldsymbol{x}^T\boldsymbol{Ax} = \boldsymbol{x}^*\boldsymbol{Ax} > 0$ implies A is **positive definite matrix**

$\boldsymbol{x}^*\boldsymbol{Ax} \leq 0$ implies $\boldsymbol{A}$ is **positive semi-definite matrix**

If for some $\boldsymbol{x}$, $\boldsymbol{x}^*\boldsymbol{Ax} > 0$ and for other $\boldsymbol{x}^*\boldsymbol{Ax} < 0$ implies $\boldsymbol{A}$ is **indefinite matrix**.

2. Hermitian (or symmetric real) matrices have **distinct eigenvalues** and their eigenvectors are mutually orthogonal. If in addition the matrix is **positive definite**, then all its eigenvalues are necessarily positive. If λ_1 is the largest and λ_n is the smallest eigenvalue of $\boldsymbol{A}$, then

$$\lambda_n(\boldsymbol{x}^*\boldsymbol{x}) \le (\boldsymbol{x}^*\boldsymbol{A}\boldsymbol{x}) \le \lambda_1(\boldsymbol{x}^*\boldsymbol{x})$$

In fact any Hermitian (or real symmetric) matrix can be diagonalized by similarity transformation $\boldsymbol{P}$ in which all its columns are mutually orthonormal (called as unitary matrix). All the eigenvalues of a Hermitian (or symmetric real) positive definite matrix are strictly positive. The coefficients of the characteristic polynomial $|(\lambda\boldsymbol{I} - \boldsymbol{A})|$ of a positive definite matrix alternate in sign yielding a necessary and sufficient condition for positive definiteness. Looking at the **principle diagonal minors** of the **determinant of a positive definite Hermitian matrix, they should be strictly positive**. If two Hermitian matrices **commute**, then they can be **simultaneously diagonalized**.

3. **Simultaneous diagonalization** of two real matrices $\boldsymbol{R} > 0$ and $\boldsymbol{Q} \le 0$.
Select a nonsingular $\boldsymbol{W}$, such that $\boldsymbol{R} = \boldsymbol{W}^T\boldsymbol{W}$ **square root matrix of** $\boldsymbol{R}$.
Choose a **orthogonal matrix** $\boldsymbol{O}$, such that :

$$\boldsymbol{O}^T\boldsymbol{W}^T\boldsymbol{Q}\boldsymbol{W}\boldsymbol{O} = \boldsymbol{D} \quad (\boldsymbol{D} \ge 0 \text{ is a diagonal matrix})$$

4. **Liapunov Stability Theorem:**
Given a $n \times n$ real matrix $\boldsymbol{A}$ with eigenvalues λ_i, if there exists a matrix $\boldsymbol{S} \ge 0$, such that:

$$\left(\boldsymbol{A}^T\boldsymbol{S} + \boldsymbol{S}\boldsymbol{A}\right) \le 0, \text{then, } \mathrm{Re}(\lambda_i) < 0, \quad (i = 1, \ldots, n)$$

2.15 Summary of Useful Facts and Identities

$\boldsymbol{e}_i$ denotes the i-th column vector, i-th entry being unity and zero everywhere else.

Then

$$\begin{aligned} \boldsymbol{A}\boldsymbol{e}_i &= \quad i\text{-th column of } \boldsymbol{A} = \boldsymbol{a}_i \\ \boldsymbol{e}_j^T \boldsymbol{A} &= \quad j\text{-th row of } \boldsymbol{A} = \boldsymbol{a}_j^T \end{aligned}$$

1. $\left(\boldsymbol{A}^{-1} - \boldsymbol{B}^{-1}\right)^{-1} = \boldsymbol{A} - \boldsymbol{C}\boldsymbol{A}$, where $\boldsymbol{C} = (\boldsymbol{A} - \boldsymbol{B})^{-1}$

2. $(\boldsymbol{I} - \boldsymbol{A}\boldsymbol{B})^{-1} = \boldsymbol{I} - \boldsymbol{A}(\boldsymbol{I} + \boldsymbol{B}\boldsymbol{A})^{-1}\boldsymbol{B}$, $\quad$ $\boldsymbol{B}\boldsymbol{A}$ non-singular (Woodbury's form).

If $\boldsymbol{B} = \boldsymbol{x}$, $n \times 1$ vector and $\boldsymbol{A} = \boldsymbol{y}^T$, $1 \times n$ column vector,

then, the associated **Sherman-Morrison Formula is**:

$$\left.\begin{aligned} (\boldsymbol{I} + \boldsymbol{x}\boldsymbol{y}^T)^{-1} &= \boldsymbol{I} - \frac{1}{\beta}\left(\boldsymbol{x}\boldsymbol{y}^T\right); \ \beta = (1 + \boldsymbol{x}^T\boldsymbol{y}) \\ \left(\boldsymbol{A} + \boldsymbol{x}\boldsymbol{y}^T\right)^{-1} &= \boldsymbol{A}^{-1} - \frac{1}{\alpha}\boldsymbol{A}^{-1}\boldsymbol{x}\boldsymbol{y}^T\boldsymbol{A}^{-1}; \ \alpha = 1 + \ \text{trace}\left(\boldsymbol{x}\boldsymbol{y}^T\boldsymbol{A}^{-1}\right) \end{aligned}\right\}$$

Proof:

$$\left(\boldsymbol{A} + \boldsymbol{x}\boldsymbol{y}^T\right)^{-1} = (\boldsymbol{C} + \boldsymbol{D})$$

$$\boldsymbol{I} = (\boldsymbol{C} + \boldsymbol{D})\left(\boldsymbol{A} + \boldsymbol{x}\boldsymbol{y}^T\right) = \boldsymbol{C}\boldsymbol{A} + \boldsymbol{C}\boldsymbol{x}\boldsymbol{y}^T + \boldsymbol{D}\boldsymbol{A} + \boldsymbol{D}\boldsymbol{x}\boldsymbol{y}^T$$

Let

$$\boldsymbol{C} = \boldsymbol{A}^{-1}$$

Then:

$$\boldsymbol{0} = \boldsymbol{A}^{-1}\boldsymbol{x}\boldsymbol{y}^T + \boldsymbol{D}\boldsymbol{A} + \boldsymbol{D}\boldsymbol{x}\boldsymbol{y}^T$$

or

$$-A^{-1}xy^T A^{-1} = D\left(I + xy^T A^{-1}\right)$$

Let

$$\left(1 + \text{Trace}\,(xy^T A^{-1})\right) = \alpha$$

Then

$$D = -\frac{1}{\alpha}A^{-1}xy^T A^{-1}$$

yielding:

$$\left(A + xy^T\right)^{-1} = A^{-1} - \frac{1}{\alpha}A^{-1}xy^T A^{-1}, \quad \alpha = 1 + \text{Trace}\,(xy^T A^{-1})$$

3. **Inverses in Filtering**

$$\left(P^{-1} + B^T R^{-1} B\right)^{-1} B^T R^{-1} = PB^T\left(BP^{-1}B^T + R\right)^{-1} \quad P > 0, \quad R > 0$$

$$(A + BC)^{-1} = A^{-1} - A^{-1}B\left(I + CA^{-1}B\right)CA^{-1}$$

$$(I + AB)^{-1} = I - A\,(I + BA)^{-1}\,B$$

4. **Eigenvalues**

(i) eig(A) stands for eigenvalues of a square matrix A

(ii) rank(A) = r, implies that the matrix A has r linearly independent rows (or columns) and at the most r non-zero eigenvalues.

(iii) When $A = A^T$, its eigenvalues λ_i are real, eigenvectors v_i are orthogonal. Let us form a matrix of orthogonal vectors:

$$V = \left[v_1 \; v_2 \; \cdots \; v_n\right]$$

$$Av_i = \lambda_i v_i, \quad \lambda_i \text{ is real} \quad (i = 1, 2, \cdots, n)$$

$$VV^T = I$$

$$\text{Trace}\left(A^k\right) = \sum_{i=1}^{n} \lambda_i^K$$

$$\text{eig}\left(A^{-1}\right) = \lambda_i^{-1}$$

$$\text{eig}\,(I + \alpha A) = I + \alpha\lambda_i$$

5. **Matrix Functions**

$$f(A) = \sum_{n=0}^{\infty} c_n A^n, \quad \sum_{n=0}^{\infty} c_n x^n < \infty, \quad |x| < 1$$

(i) $A = T^{-1}\hat{A}T$ implies $f(A) = T^{-1}f(\hat{A})T$

$\lim_{n\to\infty} A^n \to 0, \quad |A| < 1$

(ii) if $AB = BA$, then $e^A e^B = e^{A+B}$

(iii) A be $n \times m$, (tall) $(n > m)$ rank$(A) = m$, then

$$Ax = b \quad \text{yields (if there exists a solution)}$$

$$x = A^+ b$$

$$A^+ = \left(A^T A\right)^{-1} A^T b, \quad A^+ \text{ stands for Pseudo-inverse of } A$$

6. Let A be a $n \times n$ real matrix. It can be decomposed as:

$$A = \sum_{i=1}^{n} \lambda_i x_i y_i^T, \quad x_i y_i^T \text{ being a matrix of rank 1}$$

λ_i $(i = 1, \ldots, n)$ are distinct eigenvalues of A. x_i the corresponding eigenvector of A, y_i the corresponding eigenvector of A^T.

Furthermore, if A is Hermitian, then $y_i^T = \bar{x}_i^T = x_i^*$

7. $\boldsymbol{A} = \boldsymbol{x}\boldsymbol{y}^T$ implies that the matrix $\boldsymbol{A}$ is of rank one.

8. **Gerschgorin Circles**

   Given a $n \times n$ nonsingular matrix $\boldsymbol{A} = (a_{ij})$ with eigenvalues $\lambda_k\,(k = 1, \ldots, n)$ then

$$|a_{ij}| > \sum_{i \neq j} |a_{ij}| \qquad (i = 1, \ldots, n)$$

$$|\lambda_k - a_{ii}| \leq \sum_{i \neq j} (a_{ij}) \qquad (\text{for at least one } k), \quad \text{i} = 1, \ldots, \text{n}$$

9. **Bordering matrices:** Matrices discussed below, are useful in sequential filtering algorithms and system realization problems.

   Given:

$$\tilde{\boldsymbol{A}} = \begin{bmatrix} \boldsymbol{A} & \boldsymbol{x} \\ \boldsymbol{y}^T & \alpha \end{bmatrix}, \ \boldsymbol{A} \text{ is } n \times n \text{ and } \boldsymbol{A}^{-1} \text{ exists}$$

$\boldsymbol{x}$ and $\boldsymbol{y}$ are $n \times 1$ vectors

Then,

$$\tilde{\boldsymbol{A}}^{-1} = \begin{bmatrix} \left(\boldsymbol{A} - \dfrac{1}{\alpha}\boldsymbol{x}\boldsymbol{y}^T\right)^{-1} & -\dfrac{1}{\beta}\boldsymbol{A}^{-1}\boldsymbol{x} \\ -\dfrac{1}{\beta}\boldsymbol{y}^T\boldsymbol{A}^{-1} & -\dfrac{1}{\beta} \end{bmatrix}, \qquad \beta = \alpha - \boldsymbol{y}^T\boldsymbol{A}^{-1}\boldsymbol{x}$$

Proof:

Let,

$$\tilde{\boldsymbol{A}}^{-1} = \begin{bmatrix} \boldsymbol{C} & \boldsymbol{u} \\ \boldsymbol{v}^T & \beta_1 \end{bmatrix}$$

Then,

$$\tilde{\boldsymbol{A}}^{-1}\tilde{\boldsymbol{A}} = \begin{bmatrix} \boldsymbol{C} & \boldsymbol{u} \\ \boldsymbol{v}^T & \beta_1 \end{bmatrix} \begin{bmatrix} \boldsymbol{A} & \boldsymbol{x} \\ \boldsymbol{y}^T & \alpha \end{bmatrix} = \begin{bmatrix} \left(\boldsymbol{C}\boldsymbol{A} + \boldsymbol{u}\boldsymbol{y}^T\right) & (\boldsymbol{C}\boldsymbol{x} + \alpha\boldsymbol{u}) \\ \boldsymbol{v}^T\boldsymbol{A} + \beta_1\boldsymbol{y}^T & \alpha\beta_1 + \boldsymbol{v}^T\boldsymbol{x} \end{bmatrix} = \left[\boldsymbol{I}_{n+1}\right]$$

Thus,

$$\boldsymbol{C}\boldsymbol{A} + \boldsymbol{u}\boldsymbol{y}^T = \boldsymbol{I}_n, \quad \boldsymbol{C}\boldsymbol{x} + \alpha\boldsymbol{u} = \boldsymbol{0}$$
$$\boldsymbol{v}^T\boldsymbol{A} + \beta_1\boldsymbol{y}^T = \boldsymbol{0}, \quad \alpha\beta_1 + \boldsymbol{v}^T\boldsymbol{x} = 1$$

From the above equations:

$$\alpha\boldsymbol{u} = -\boldsymbol{C}\boldsymbol{x} \quad \text{and} \quad \alpha\boldsymbol{C}\boldsymbol{A} - \boldsymbol{C}\boldsymbol{x}\boldsymbol{y}^T = \alpha\boldsymbol{I}_n$$

or

$$\boldsymbol{C} = \left(\boldsymbol{A} - \alpha^{-1}\boldsymbol{x}\boldsymbol{y}^T\right)^{-1}$$
$$\boldsymbol{u} = -\frac{1}{\alpha}\left(\boldsymbol{A} - \alpha^{-1}\boldsymbol{x}\boldsymbol{y}^T\right)\boldsymbol{x} = -\left(\alpha - \boldsymbol{y}^T\boldsymbol{A}^{-1}\boldsymbol{x}\right)^{-1}\boldsymbol{A}^{-1}\boldsymbol{x}$$

Similarly

$$\boldsymbol{v}^T = -\beta_1\boldsymbol{y}^T\boldsymbol{A}^{-1}, \quad \beta_1 = -\left(\alpha - \boldsymbol{y}^T\boldsymbol{A}^{-1}\boldsymbol{x}\right)^{-1}$$

In confirmity with the given identity, let $\beta_1 = -\frac{1}{\beta}$. Summarizing the above proof:

$$\tilde{\boldsymbol{A}}^{-1} = \begin{bmatrix} \boldsymbol{C} & \boldsymbol{u} \\ \boldsymbol{v}^T & \beta_1 \end{bmatrix} = \begin{bmatrix} \left(\boldsymbol{A} - \frac{1}{\alpha}\boldsymbol{x}\boldsymbol{y}^T\right)^{-1} & -\frac{1}{\beta}\boldsymbol{A}^{-1}\boldsymbol{x} \\ -\frac{1}{\beta}\boldsymbol{y}^T\boldsymbol{A}^{-1} & -\frac{1}{\beta} \end{bmatrix}$$

If $\boldsymbol{A}$ is Hermitian (meaning diagonalizable, $\boldsymbol{A} = \boldsymbol{U}\boldsymbol{\Lambda}\boldsymbol{U}^*$, $\boldsymbol{U}$ is unitary) and $\boldsymbol{y} = \boldsymbol{x}$, then the eigenvalues $\tilde{\lambda}$ of $\tilde{\boldsymbol{A}}$ are computed from

$$\boldsymbol{x}^*\boldsymbol{U}\left(\tilde{\lambda}\boldsymbol{I} - \boldsymbol{\Lambda}\right)^{-1}\boldsymbol{U}^*\boldsymbol{x} = 0, \qquad (\tilde{\boldsymbol{A}} \text{ is also Hermitian})$$

If $\boldsymbol{A} > 0$ and $\alpha > \boldsymbol{y}^T\boldsymbol{A}^{-1}\boldsymbol{x}$, $\boldsymbol{y} = \boldsymbol{x}$, then $\tilde{\boldsymbol{A}} > 0$ (positive definite matrix).

10. **Kronecker Product:** Let $\boldsymbol{A}$ be $m \times n$ and $\boldsymbol{B}$ be $p \times q$.

Then the Kronecker product "⊗" is defined as:

(a) $\boldsymbol{A} \otimes \boldsymbol{B} = \begin{bmatrix} a_{11}\boldsymbol{B} & \cdots & a_{1n}\boldsymbol{B} \\ \vdots & & \\ a_{m1}\boldsymbol{B} & \cdots & a_{mn}\boldsymbol{B} \end{bmatrix}$, $= mp \times np$ matrix, called **Kronecker Product**

(b) $(\boldsymbol{A} \otimes \boldsymbol{B})(\boldsymbol{C} \otimes \boldsymbol{D}) = (\boldsymbol{AC} \otimes \boldsymbol{BD})$ provided $\boldsymbol{AC}$ and $\boldsymbol{BD}$ exists.

(c) $(\boldsymbol{A} \otimes \boldsymbol{B})^T = (\boldsymbol{A}^T \otimes \boldsymbol{B}^T)$

(d) $(\boldsymbol{A} \otimes \boldsymbol{B})^{-1} = (\boldsymbol{A}^{-1} \otimes \boldsymbol{B}^{-1})$

(e) Let us express the Liapunov matrix equation (all matrices are $n \times n$)

$$\boldsymbol{AS} + \boldsymbol{SA}^T = \boldsymbol{Q} \tag{2.67}$$

in Kronecker product form. $\boldsymbol{S}$ and $\boldsymbol{Q}$ are symmetric. We leave $\boldsymbol{A}$ alone and express $\boldsymbol{Q}$ and $\boldsymbol{S}$ as representation of n vectors each, yielding

$$\boldsymbol{Q} = \begin{bmatrix} \boldsymbol{q}_1 \mid \cdots \mid \boldsymbol{q}_n \end{bmatrix}, \quad \boldsymbol{S} = \begin{bmatrix} \boldsymbol{s}_1 \mid \cdots \mid \boldsymbol{s}_n \end{bmatrix}, \boldsymbol{q} = \begin{bmatrix} \boldsymbol{q}_1 \\ \vdots \\ \boldsymbol{q}_n \end{bmatrix} \quad \boldsymbol{s} = \begin{bmatrix} \boldsymbol{s}_1 \\ \vdots \\ \boldsymbol{s}_n \end{bmatrix}$$

Dimensions of vectors $\boldsymbol{q}$ and $\boldsymbol{s}$ being n^2 each. The Matrix equation, Eq. 2.67, takes the form

$$(\boldsymbol{I} \otimes \boldsymbol{A} + \boldsymbol{A} \otimes \boldsymbol{I})\boldsymbol{s} = \boldsymbol{q}$$

11. **Hadamard Product $\boldsymbol{H}$:**

$\boldsymbol{A}$ and $\boldsymbol{B}$ are $n \times n$. Their Hadamard Product is defined as:

$$\boldsymbol{H} = \boldsymbol{A} * \boldsymbol{B},$$

$$\boldsymbol{H} = \left(h_{ij}\right) = \left(a_{ij}b_{ij}\right) \quad (i, j = 1, \ldots, n)$$

12. **Tridiagonal Form, Cholesky-decomposition:**

If a $n \times n$ matrix $\boldsymbol{A}$ is symmetric, it can be transformed via similarity transformation into a **Tridiagonal** form having non-zero entries only directly below or directly above the **main diagonal** as well as non-zero entries along the main diagonal. When $\boldsymbol{A}$ is positive definite there exists a special Cholesky-decomposition, $\boldsymbol{A} = \boldsymbol{B}^T\boldsymbol{B}$, where $\boldsymbol{B}$ is upper triangular matrix.

13. **Binet-Cauchy Theorem:**

Binet-Cauchy theorem is a useful theorem in electrical network theory. It states the algorithm for computing the determinant of the product $\boldsymbol{AB}$ where $\boldsymbol{A}$ is $m \times n$ and $\boldsymbol{B}$ is $n \times m$, $m < n$. Define Major of $\boldsymbol{A}$ (or of $\boldsymbol{B}$) as the determinant of the sub-matrix of maximum order (in this case m). By **Binet-Cauchy Theorem.**

$$\det(\boldsymbol{AB}) = \sum_{\text{all majors}} (\text{product of corresponding majors of } \boldsymbol{A} \text{ and } \boldsymbol{B})$$

14. **Lancasters Formula**:

Let x be a n vector and:

$$p(\boldsymbol{x}) = e^{-f(\boldsymbol{x})}, \quad f(\boldsymbol{x}) = \frac{1}{2}\boldsymbol{x}^T\boldsymbol{R}^{-1}\boldsymbol{x} > 0, \quad R \text{ is } n \times n \text{ positive definite}$$

$$\int_{-\infty}^{\infty} p(\boldsymbol{x})\,\mathrm{d}\boldsymbol{x} = (2\pi)^{-n/2}\Delta_{\boldsymbol{R}} \qquad \Delta_{\boldsymbol{R}} = \det R = \text{determinant of } \boldsymbol{R}$$

$$\int_{-\infty}^{+\infty} e^{-(ax^2+bx+c)}\,\mathrm{d}x = \left(\sqrt{\pi a^{-1}}\right) e^{[b^2-4ac][4a]^{-1}}, \qquad \int_{-\infty}^{+\infty} e^{-(x-\mu)^2/2\sigma^2}\,\mathrm{d}x = \sqrt{2\pi\sigma^2}$$

$$\int_{-\infty}^{+\infty} f(\boldsymbol{x})\delta(\boldsymbol{y} - \boldsymbol{A}\boldsymbol{x})\,\mathrm{d}x = (\Delta_A)^{-1} f(\boldsymbol{A}^{-1}\boldsymbol{y}), \quad \Delta_A = \det(\boldsymbol{A})$$

15. **Singular Value Decomposition and Solution of Linear Equations:**

$\boldsymbol{A}$ is $n \times m$ real matrix with $n > m$ (tall), with rank $r \leq m$.

We can decompose $\boldsymbol{A}$ as,

$$\boldsymbol{A} = \boldsymbol{U\Sigma V}$$

where,

$$\boldsymbol{U} = \begin{bmatrix} \boldsymbol{e}_1 \; \boldsymbol{e}_2 \cdots \boldsymbol{e}_n \end{bmatrix} = \text{ Eigenvectors of } \boldsymbol{AA}^T, \quad n \times n \text{ matrix}$$
$$\boldsymbol{V} = \begin{bmatrix} \boldsymbol{f}_1 \; \boldsymbol{f}_2 \cdots \boldsymbol{f}_m \end{bmatrix} = \text{ Eigenvectors of } \boldsymbol{A}^T\boldsymbol{A}, \quad m \times m \text{ matrix}$$
$$\boldsymbol{\Sigma} = \sqrt{\text{diag}\left(\boldsymbol{AA}^T\right)} = \text{ diag}\left(\sigma_1, \sigma_2, \cdots, \sigma_r, 0, \cdots, 0\right), \quad r \leq m$$
$$\sigma_1 > \sigma_2 > \cdots > \sigma_r \quad \text{are all non-negative.}$$

If $\boldsymbol{A}$ is symmetric, then

$$\boldsymbol{A} = \boldsymbol{V}^T\boldsymbol{\Sigma V}, \quad \boldsymbol{V} \text{ is orthogonal, its column vectors being eigenvectors of } \boldsymbol{A}$$

The solution to the linear equation $\boldsymbol{Ax} = \boldsymbol{b}$ is:

$$\boldsymbol{x} = \boldsymbol{x}_a + \boldsymbol{x}_b,$$
$$\boldsymbol{x}_a = \sum_{i=1}^{r} \left(\boldsymbol{e}_i^T \boldsymbol{b}\right) \sigma_i^{-1} \boldsymbol{f}_i$$
$$\boldsymbol{x}_b = \sum_{i=r+1}^{m} c_i \boldsymbol{f}_i \quad (c_i \text{ is arbitrary}).$$

$\boldsymbol{x}_b$ represents the auxillary (arbitrary) part of $\boldsymbol{x}$ which can be taken as zero.

16. **Schur-Cohen Criteria:**

In order that the roots of a polynomial

$$p(\lambda) = a_0\lambda^n + a_1\lambda^{n-1} + \cdots + a_n,$$

Lie within the unit circle in complex λ-plane, it is necessary and sufficient that the following conditions are satisfied:

$$\left.\begin{aligned}(-1)^n p(-1) &> 0\\ p(1) &> 0\\ \det(\boldsymbol{X}_i + \boldsymbol{Y}_i) &> 0\\ \det(\boldsymbol{X}_i - \boldsymbol{Y}_i) &> 0\end{aligned}\right]$$

$$\boldsymbol{X}_i = \begin{bmatrix} a_0 & a_1 & \cdots & a_{i-1} \\ & a_0 & \cdots & a_{i-2} \\ 0 & & \ddots & \\ & & & a_0 \end{bmatrix}, \quad \boldsymbol{Y}_i = \begin{bmatrix} & & a_n \\ 0 & a_n & a_{n-1} \\ & & \vdots \\ a_n & \cdots & a_{n-i+1} \end{bmatrix}$$

$$i = 1, 2, \cdots, n$$

17. **Positive Definite and Semidefinite Matrices**

When $\boldsymbol{A}$ is positive definite matrix denote by $\boldsymbol{A} > 0$, $\boldsymbol{x}^T\boldsymbol{A}\boldsymbol{x} > 0$ for all $\boldsymbol{x}$

When $\boldsymbol{A}$ is positive semidefinite denoted by $\boldsymbol{A} \geq 0$, $\boldsymbol{x}^T\boldsymbol{A}\boldsymbol{x} \geq 0$ for all $\boldsymbol{x}$

$\mathrm{eig}(\boldsymbol{A}) > 0$ when $\boldsymbol{A} > 0$

$\mathrm{eig}(\boldsymbol{A}) \geq 0$ when $\boldsymbol{A} \geq 0$

$\mathrm{Trace}(\boldsymbol{A}) > 0$ when $\boldsymbol{A} > 0$

$\mathrm{Trace}(\boldsymbol{A}) \geq 0$ when $\boldsymbol{A} \geq 0$

$\boldsymbol{A} = \boldsymbol{B}\boldsymbol{B}^T$ when $\boldsymbol{A} > 0$, implies $\boldsymbol{B}^{-1}$exists.

$\boldsymbol{A} = \boldsymbol{B}\boldsymbol{B}^T$ when $\boldsymbol{A} \geq 0$, implies both $\boldsymbol{A}$ and $\boldsymbol{B}$ have rank r

$\boldsymbol{A} - t\boldsymbol{B} > 0$, $\boldsymbol{B} = \boldsymbol{B}^T$ implies$\boldsymbol{A} > 0$, when t is sufficiently small

$(\boldsymbol{A} + t\boldsymbol{B})^{-1} \approx \boldsymbol{A}^{-1} - t\boldsymbol{A}^{-1}\boldsymbol{B}\boldsymbol{A}^{-1}$, when t is sufficiently small

18. **Derivatives and Gradients**

$$\frac{\partial}{\partial\alpha}(\det \boldsymbol{Y}(\alpha)) = (\det \boldsymbol{Y}(\alpha))\,\mathrm{Trace}\left[\boldsymbol{Y}^{-1}(\alpha)\frac{\partial}{\partial\alpha}\boldsymbol{Y}(\alpha)\right]$$

$$\frac{\partial}{\partial\alpha}\left(\boldsymbol{Y}^{-1}(\alpha)\right) = -\boldsymbol{Y}^{-1}(\alpha)\frac{\partial \boldsymbol{Y}(\alpha)}{\partial\alpha}\boldsymbol{Y}^{-1}(\alpha)$$

$$\boldsymbol{\nabla}_x\left(\boldsymbol{x}^T\boldsymbol{b}\right) = \boldsymbol{b}$$

$$\boldsymbol{\nabla}_x\left(\boldsymbol{x}^T\boldsymbol{A}\boldsymbol{x}\right) = \left(\boldsymbol{A} + \boldsymbol{A}^T\right)\boldsymbol{x}$$

$$\boldsymbol{\nabla}_x\boldsymbol{s}^T(\boldsymbol{x})\boldsymbol{A}\boldsymbol{s}(\boldsymbol{x}) = [\boldsymbol{\nabla}_x s(x)]^T\left[(\boldsymbol{A} + \boldsymbol{A}^T)\right]\boldsymbol{s}(\boldsymbol{x})$$

If, $f = \boldsymbol{x}^T\boldsymbol{A}\boldsymbol{x} + \boldsymbol{b}^T\boldsymbol{x}$ then,

$$\boldsymbol{\nabla}_x f = \left(\boldsymbol{A} + \boldsymbol{A}^T\right)\boldsymbol{x} + \boldsymbol{b}$$

$$\boldsymbol{\nabla}_{xx} f = \left(\boldsymbol{A} + \boldsymbol{A}^T\right)$$

Consider differentiating a scalar function $f(\boldsymbol{A}), \boldsymbol{A} = \{a_{ij}\}_{i,j=1}^{n}$

$$\frac{\mathrm{d}f}{\mathrm{d}a_{ij}} = \sum_{k=1}^{n}\sum_{l=1}^{n}\frac{\partial f}{\partial a_{kl}}\frac{\partial a_{kl}}{\partial a_{ij}} = \mathrm{Trace}\left[\left[\frac{\partial f}{\partial \boldsymbol{A}}\right]^T\left[\frac{\partial \boldsymbol{A}}{\partial a_{ij}}\right]\right]$$

$$\frac{\partial a_{kl}}{\partial a_{ij}} = \delta_{ik}\delta_{lj}$$

19. **Swaping Rows and Columns**

Let $\boldsymbol{e}_i$ denote column vectors of dimension m, such that it's i-th entry is 1 and zero else where. Similarly $\boldsymbol{e}_j$ denote column vectors of dimension p, which has 1 on the j-th entry and zero else where. Define:

$$\boldsymbol{J}^{ij} = \boldsymbol{e}_i\boldsymbol{e}_j^T, \quad \text{a } m \times p \text{ matrix}$$

$$\left(\boldsymbol{e}_i\boldsymbol{e}_j^T\right)_{i,j} = \delta_{ij}, \quad i = 1, 2, \cdots, m, \quad j = 1, 2, \cdots, p$$

$$\text{If} \quad \boldsymbol{A} = \begin{bmatrix} \boldsymbol{a}_1 \ \boldsymbol{a}_2 \cdots \boldsymbol{a}_i \cdots \boldsymbol{a}_m \end{bmatrix}, \quad n \times m \text{ matrix}$$

$$\text{Then} \quad \boldsymbol{A}\boldsymbol{J}^{ij} = \boldsymbol{A}\boldsymbol{e}_i\boldsymbol{e}_j^T = \begin{bmatrix} 0\ 0 \cdots a_{1i} \cdots 0\ 0 \\ 0\ 0 \cdots a_{2i} \cdots 0\ 0 \\ \vdots \\ 0\ 0 \cdots a_{ni} \cdots 0\ 0 \end{bmatrix}, \quad n \times p \text{ matrix}$$

thus $\boldsymbol{A}\boldsymbol{J}^{ij}$ is a $n \times p$ matrix of zeroes except that its j-th column is represented by the vector $\boldsymbol{a}_i$. Similarly

$$\boldsymbol{J}^{ij}\boldsymbol{A} = \boldsymbol{e}_i\boldsymbol{e}_j^T\boldsymbol{A} = \begin{bmatrix} 0 & \cdots & & 0 \\ \vdots & & & \\ 0 & \cdots & & 0 \\ a_{j1} & a_{j2} & \cdots & a_{jp} \\ 0 & & \cdots & 0 \\ \vdots & & & \\ 0 & 0 & \cdots & 0 \end{bmatrix} \quad \rightarrow i$$

In this case, the j-th row of matrix $\boldsymbol{A}$ replaces the i-th row of $\boldsymbol{J}^{ij}\boldsymbol{A}$, all other entries being zero. This allows one to replace columns and rows of a matrix with its other columns and rows.

20. **Elementary Operation on Matrices**

Given a matrix $\boldsymbol{A}$, number of operations can be performed on it resulting in a related transformed matrix $\boldsymbol{B}$. When the matrix is square, we have seen the importance of such operations as the Similarity Transformations. Even when the matrix is not square, there are many operations resulting in simplified transformed matrix. A number of operations of great importance and simplicity are called as **Elementary operations** performed via **Elementary Transformations**.

Following **three elementary operations** on a matrix $\boldsymbol{A}$ are quite useful.

1. Interchange of any two rows (or any two columns) of $\boldsymbol{A}$.
2. Multiplication of each element of a row (or column) of $\boldsymbol{A}$ by a scalar constant.
3. Addition of the elements of a row (or column) of $\boldsymbol{A}$ multiplied by a constant to another row (or column) of the matrix $\boldsymbol{A}$.

These operations (or transformations) do not change the intrinsic properties of the matrix $\boldsymbol{A}$. In case the matrix is square:

(i) Transformation #1 simply changes the sign of the determinant, $\Delta_A \neq 0$,

(ii) Second transformation changes the determinant of $\boldsymbol{A}$ by a constant.

(iii) Third transformation leaves the determinant Δ_A unchanged.

These operations on $\boldsymbol{A}$ (square or rectangular) can be carried out via certain simple, nonsingular matrices known as **elementary** or **unimodular matrices**.
Definition of the Left Elementary Matrix $\boldsymbol{L}$: Given $\boldsymbol{A}$, define $\boldsymbol{LA} = \boldsymbol{A}_L$
The left elementary matrix $\boldsymbol{L}$ operating on the left side of $\boldsymbol{A}$, performs any of the above mentioned three elementary row operations on the matrix, $\boldsymbol{A}$.
Definition of the Right Elementary matrix $\boldsymbol{R}$: Given $\boldsymbol{A}$, define $\boldsymbol{AR} = \boldsymbol{A}_R$
The right elementary matrix $\boldsymbol{R}$ operating on right side of $\boldsymbol{A}$ performs any of the three column operations mentioned above on the matrix, $\boldsymbol{A}$.

- An elementary matrix, not necessarily square, is referred to as **totally unimodular matrix**.
- Any elementary matrix is nonsingular.
- Product of any number of elementary matrices is nonsingular.
- Any nonsingular matrix can be decomposed into product of a finite number of elementary matrices (Proof is left to the reader).

L-R Elementary Transformation

Example 2.10:

Given:

$$A = \begin{bmatrix} a_{11} & a_{12} & a_{13} & a_{14} \\ a_{21} & a_{22} & a_{23} & a_{24} \\ a_{31} & a_{32} & a_{33} & a_{34} \end{bmatrix}$$

Required:

(i) Find matrix $\boldsymbol{L}$ which adds the 3rd row to the 2nd row

(ii) Find matrix $\boldsymbol{R}$ which adds the 3rd column to the 2nd column

Solution:

$$\boldsymbol{L} = \begin{bmatrix} 1 & 0 & 0 \\ 0 & 1 & 1 \\ 0 & 0 & 1 \end{bmatrix} \rightarrow \begin{bmatrix} r_3 \text{ is added to } r_2 \text{ in Identity matrix} \\ r_3 + r_2 \rightarrow r_2 \end{bmatrix}$$

$$\boldsymbol{R} = \begin{bmatrix} 1 & 0 & 0 & 0 \\ 0 & 1 & 0 & 0 \\ 0 & 1 & 1 & 0 \\ 0 & 0 & 0 & 1 \end{bmatrix} \; to \; \begin{bmatrix} c_3 \text{ is added to } c_2 \text{ in Identity matrix} \\ c_3 + c_2 \rightarrow c_2 \end{bmatrix}$$

"r" stands for row, "c" stands for column.

$$\boldsymbol{LA} = \begin{bmatrix} 1 & 0 & 0 \\ 0 & 1 & 1 \\ 0 & 0 & 1 \end{bmatrix} \begin{bmatrix} a_{11} & a_{12} & a_{13} & a_{14} \\ a_{21} & a_{22} & a_{23} & a_{24} \\ a_{31} & a_{32} & a_{33} & a_{34} \end{bmatrix} = \begin{bmatrix} a_{11} & a_{12} & a_{13} & a_{14} \\ (a_{21} + a_{31}) & (a_{22} + a_{32}) & (a_{23} + a_{33}) & (a_{24} + a_{34}) \\ a_{31} & a_{32} & a_{33} & a_{34} \end{bmatrix}$$

$$AR = \begin{bmatrix} a_{11} & a_{12} & a_{13} & a_{14} \\ a_{21} & a_{22} & a_{23} & a_{24} \\ a_{31} & a_{32} & a_{33} & a_{34} \end{bmatrix} \begin{bmatrix} 1 & 0 & 0 & 0 \\ 0 & 1 & 0 & 0 \\ 0 & 1 & 1 & 0 \\ 0 & 0 & 0 & 1 \end{bmatrix} = \begin{bmatrix} a_{11} & (a_{12}+a_{13}) & a_{13} & a_{14} \\ a_{21} & (a_{22}+a_{23}) & a_{23} & a_{24} \\ a_{31} & (a_{32}+a_{33}) & a_{33} & a_{34} \end{bmatrix}$$

Equivalent Matrices

Consider a set of nonsingular left and right elementary matrices $\boldsymbol{L}_i$ and $\boldsymbol{R}_j$ respectively such that:

$$\boldsymbol{L} = \prod_{i=1}^{r_1} \boldsymbol{L}_{r_1+i-i}, \quad \boldsymbol{R} = \prod_{j=2}^{r_2} \boldsymbol{R}_{r_2+1-j}$$

(i) Two matrices $\boldsymbol{A}$ and $\boldsymbol{B}$ are equivalent **if and only if**:

$$\boldsymbol{B} = \boldsymbol{LAR}, \quad \boldsymbol{A} = \boldsymbol{L}^{-1}\boldsymbol{BR}^{-1}$$

Both $\boldsymbol{A}$ and $\boldsymbol{B}$ have same rank and same order.

(ii) If $\boldsymbol{A}$ is nonsingular (square), then one can find $\boldsymbol{L}$ and $\boldsymbol{R}$ matrices such that:

$$\boldsymbol{LAR} = \boldsymbol{I} \quad \text{(Identity matrix)}, \qquad \boldsymbol{A} = \boldsymbol{L}^{-1}\boldsymbol{R}^{-1}$$

(iii) If $\boldsymbol{A}$ is rectangular matrix of the order $(n \times m)$ and rank r, then it can be reduced to the form:

$$\boldsymbol{B} = \boldsymbol{LAR} = \begin{bmatrix} \boldsymbol{D}_r & \boldsymbol{0} \\ \boldsymbol{0} & \boldsymbol{0} \end{bmatrix}, \boldsymbol{D}_r = \begin{bmatrix} d_1 & & & \boldsymbol{0} \\ & d_2 & & \\ & & \ddots & \\ \boldsymbol{0} & & & d_r \end{bmatrix}$$

with a proper choice of $\boldsymbol{L}$ and $\boldsymbol{R}$, the reduced form is:

$$\boldsymbol{B} = \boldsymbol{LAR} = \begin{bmatrix} \boldsymbol{I}_r & \boldsymbol{0} \\ \boldsymbol{0} & \boldsymbol{0} \end{bmatrix}$$

Hermite Form A_h (or Row-Reduced Echelon Form)

Hermite matrix mean a row-reduced echelon matrix in which some parts of the matrix can be divided into upper and lower parts via a "staircase". Every corner of this "staircase" stars with 1. Elements above the staircase are arbitrary and below are zero. Thus the first nonzero element of each row is 1 and the column in which this "1" appears is a column of the identity matrix. If a row has all zeros it appears in the end. The definition applies equally to the column-reduced echelon form.

Example 2.11:

$$A_h = \begin{bmatrix} 1 & 3 & 0 & 5 & 1 \\ 0 & 0 & 1 & 2 & -2 \\ 0 & 0 & 0 & 1 & 1 \\ 0 & 0 & 0 & 0 & 0 \end{bmatrix}$$

Notice the non-zero first element of each row is 1. The columns where this first "1" appears, form an indentity matrix. Every matrix can be reduced to Hermite form via Elementary operations.

Example 2.12: Given:

$$A = \begin{bmatrix} 1 & 2 & 3 & 1 \\ 2 & 4 & 4 & 4 \\ 3 & 6 & 7 & 7 \end{bmatrix}$$

Convert A to A_h (row echelon matrix) via elementary operations. Procedure:

$$\begin{Bmatrix} r_1 \\ r_2 \\ r_3 \end{Bmatrix} \begin{bmatrix} 1 & 2 & 3 & 1 \\ 2 & 4 & 4 & 4 \\ 3 & 6 & 7 & 7 \end{bmatrix} \to \begin{bmatrix} 1 & 2 & 3 & 1 \\ 0 & 0 & -2 & 2 \\ 0 & 0 & -2 & 4 \end{bmatrix} \begin{Bmatrix} r_1 \to r_1 \\ r_2 - 2r_1 \to r_2 \\ r_3 - 3r_1 \to r_3 \end{Bmatrix} \to \begin{bmatrix} 1 & 2 & 3 & 1 \\ 0 & 0 & 1 & -1 \\ 0 & 0 & 0 & 1 \end{bmatrix} \begin{Bmatrix} r_1 \\ -r_1/2 \to r_2 \\ (r_3 - r_2)/2 \to r_3 \end{Bmatrix}$$

$$A_h = \begin{bmatrix} 1 & 2 & 3 & 1 \\ 0 & 0 & 1 & -1 \\ 0 & 0 & 0 & 1 \end{bmatrix}$$

2.16 Finite and Infinite Dimensional Operators

In this section we are dealing with operators in finite and infinite dimensions and not matrices, even though we use the same notation.

- **Projection Operator:**

  Projection operator $\boldsymbol{P}$ is defined as:

  (i) $(\boldsymbol{f}, \boldsymbol{Pg}) = (\boldsymbol{Pf}, \boldsymbol{g})$ for $\boldsymbol{f}, \boldsymbol{g} \in \boldsymbol{V}$

  (ii) $\boldsymbol{P}^2 \equiv \boldsymbol{P}, \qquad \boldsymbol{P}^* = \boldsymbol{P}, \qquad ||\boldsymbol{P}|| = 1$

  (iii) Projection operator can split an arbitrary vector $\boldsymbol{f}$ in $\boldsymbol{V}$ into $\boldsymbol{W}_N$ and $\boldsymbol{E}_N$ such that $\boldsymbol{f} = \boldsymbol{W}_N + \boldsymbol{E}_N, \quad \boldsymbol{W}_N \perp \boldsymbol{E}_N$

$$f(t) = W_N(t) + E_N(t)$$

$$\boldsymbol{Pf} = \boldsymbol{PW}_N = \boldsymbol{W}_N, \quad \boldsymbol{PE}_N = \boldsymbol{0}$$

Thus, the projection operator $\boldsymbol{P}$ divides the vector space into and onto space $\boldsymbol{W}_N$ and space $\boldsymbol{E}_N$ which is the orthogonal compliment of $\boldsymbol{W}_N$.

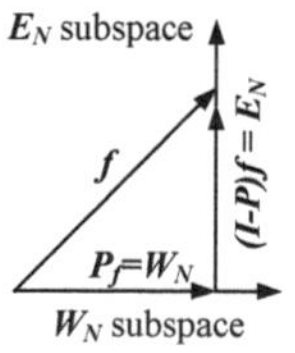

Figure 2.4: Projection Operator.

Example 2.13:

In n-dimensional Euclidean Space:

$$\boldsymbol{x} = (x_1, x_2, \cdots, x_n), \quad \boldsymbol{Px} = (x_1, x_2, 0, \cdots, 0)$$

The operator $\boldsymbol{P}$ defines a special projection operator which retains only the first two components.

- **Identity Operator:**

$$\boldsymbol{I}\boldsymbol{x} = \boldsymbol{x} \quad \forall \boldsymbol{x} \in \boldsymbol{V}$$

- **Non Singular Inverse Operator:**

$\boldsymbol{A}^{-1}$ is called the inverse of $\boldsymbol{A}$ if:

$$\boldsymbol{A}^{-1}\boldsymbol{A} = \boldsymbol{A}\boldsymbol{A}^{-1} = \boldsymbol{I}$$

If $\boldsymbol{A}^{-1}$ does not exist, we call $\boldsymbol{A}$ a singular operator.

It is easy to show that:

$$(\boldsymbol{A}_1\boldsymbol{A}_2)^{-1} = \boldsymbol{A}_2^{-1}\boldsymbol{A}_1^{-1}$$

- **Adjoint Operator:**

$\boldsymbol{A}^*$ is called as the adjoint operator of $\boldsymbol{A}$ and is defined as

$$(\boldsymbol{A}\boldsymbol{x}, \boldsymbol{x}) = (\boldsymbol{x}, \boldsymbol{A}^*\boldsymbol{x})$$

$$(\boldsymbol{A}_1\boldsymbol{A}_2)^* = \boldsymbol{A}_2^*\boldsymbol{A}_1^*$$

If

$$\boldsymbol{A}^* = \boldsymbol{A} \text{ then } \boldsymbol{A} \text{ is called or self adjoint operator}$$

Furthermore,

$$(\boldsymbol{A}^*)^* = \boldsymbol{A}$$

For Hermitian operator:

$$(\boldsymbol{x}, \boldsymbol{A}\boldsymbol{x}) = \overline{(\boldsymbol{x}, \boldsymbol{A}\boldsymbol{x})} \quad \text{(a real number)}, \quad \overline{(\cdot)} \text{ stands for conjugate.}$$

- **Unitary Operator:**

  If $\boldsymbol{A}^{-1} = \boldsymbol{A}^*$, operator is said to be unitary.

  In fact for a unitary operator

$$\boldsymbol{AA}^* = \boldsymbol{A}^*\boldsymbol{A} = \boldsymbol{I}, \quad (\boldsymbol{x}, \boldsymbol{y}) = (\boldsymbol{Ax}, \boldsymbol{Ay}) \quad \text{(isometric property)}$$

- **Eigenvalues of an operator**

$$\boldsymbol{Ax} = \lambda \boldsymbol{x} \quad \text{(for all } \boldsymbol{x} \text{ in } \boldsymbol{V}\text{)}$$

Such a vector $\boldsymbol{x}$ and the corresponding λ is called the eigenvector and the eigenvalue of the operator $\boldsymbol{A}$ belonging to eigenvector $\boldsymbol{x}$.

Example 2.14:

Consider a space $\boldsymbol{V}$ of exponential functions:

$$f(t) = \sum_{k=-\infty}^{\infty} a_k e^{jk\omega t}, \qquad \boldsymbol{f} \in \boldsymbol{V}$$

Then, the operator $\boldsymbol{A} = \mathrm{d}^2/\mathrm{d}t^2$ yields:

$$\frac{\mathrm{d}^2 f}{\mathrm{d}t^2} = \sum_{k=-\infty}^{+\infty} (-k^2\omega^2)\, f(t)$$

In operator notation,

$$\boldsymbol{f} = \sum_{k=-\infty}^{+\infty} a_k \boldsymbol{e}_k, \quad \boldsymbol{f} \to f(t), \quad \boldsymbol{e}_k \to \boldsymbol{e}_k(t)$$

$$\boldsymbol{Af} = \sum_{k=-\infty}^{+\infty} a_k \boldsymbol{A} \boldsymbol{e}_k$$

$$\boldsymbol{Ae}_k = \lambda_k \boldsymbol{e}_k, \quad e_k(t) = e^{jk\omega t}, \quad \lambda_k = -k^2\omega^2$$

Hence the operator $\frac{\mathrm{d}^2}{\mathrm{d}t^2}$ has infinitely many independent eigenvectors $e^{jk\omega t}$ with eigenvalues $-k^2\omega^2$. Eigenvalues of a Hermitian operator are real; If a Hermitian operator has two distinct Eigenvalues λ_1 and λ_2 belonging to distinct eigenvectors $\boldsymbol{x}_1$ and $\boldsymbol{x}_2$, then the two vectors $\boldsymbol{x}_1$ and $\boldsymbol{x}_2$ are orthogonal to each other.

- **Integral Operator**:
  Define $\boldsymbol{K}$: $\boldsymbol{L}^2(a, b) \to \boldsymbol{L}^2(c, d)$

$$(\boldsymbol{K}\boldsymbol{x})(t) = \int_a^b k(t, \tau)x(\tau)\,\mathrm{d}\tau, \quad c < t < d$$

$$|(\boldsymbol{K}\boldsymbol{x})(t)|^2 \leq \left[\int_a^b |k(t, \tau)|^2\,\mathrm{d}\tau\right]^{1/2} \left[\int_a^b |x(\tau)|^2\,\mathrm{d}\tau\right]^{1/2}, \quad c < t < d$$

- **Differential Operator:**
  Define $\boldsymbol{D} : \boldsymbol{L}^2(-\infty, \infty)$

$$(\boldsymbol{D}^n\boldsymbol{x})(t) = \frac{\mathrm{d}^n}{\mathrm{d}t^n}x, \quad n = 1, 2, \cdots$$

- **Shift Operator $\boldsymbol{S}$** (forward shift):

$$\boldsymbol{S}(x_1, x_2, x_3, \cdots) = (0, x_1, x_2, x_3, \cdots), \quad ||\boldsymbol{S}|| = 1$$

The matrix of the shift operator is:

$$\boldsymbol{S} = \begin{bmatrix} 0\,0\,0\,0 \cdots \\ 1\,0\,0\,0 \cdots \\ 0\,0\,1\,0 \cdots \\ \cdot\ \ \cdot\ \ \cdot\ \ \cdot\ \ \cdot \end{bmatrix}$$

- **Backward Shift Operator:**

$$S^*(x_1, x_2, x_3, \cdots) = (x_2, x_3, x_4, \cdots), \quad \|S^*\| = 1$$

The matrix of the backward shift operator is:

$$S^* = \begin{bmatrix} 0 & 1 & 0 & 0 & \cdots \\ 0 & 0 & 1 & 0 & \cdots \\ 0 & 0 & 0 & 0 & \cdots \\ \cdot & \cdot & \cdot & \cdot & \cdot \end{bmatrix}$$

- **Flip Operator $\hat{J}$:**

$$\hat{J}(x_1, x_2, x_3, \cdots, x_n) = (x_n, x_{n-1}, x_{n-2}, \cdots, x_1)$$

If,

$$f(e^{j\theta}) = \sum_{n=0}^{\infty} a_n e^{jn\theta}, \quad e^{j\theta} \text{ treated as a vector,}$$

then,

$$\hat{J}f(e^{j\theta}) = \sum_{n=0}^{\infty} a_n e^{-jn\theta}$$

If,

$$f(z) = \sum_{n=0}^{\infty} a_n z^n$$

then,

$$\hat{J}f(z) = \sum_{n=0}^{\infty} a_n \bar{z}^n$$

- **Involution Operator J:**

$$Jf(z) = z^{-1} f(z^{-1})$$

2.16.1 Operators and Matrices in Infinite Dimensional Space

(Multiplication, Hankel, and Toeplitz operators)

Motivation for this section

In Control Theory, a transfer function is written as:

$$G(z) = \frac{N(z)}{D(z)} = \frac{\prod_{i=1}^{m}(z - z_i)}{\prod_{k=1}^{n}(z - z_k)}, \qquad n \geq m$$

z_i are zeros and z_k are poles of the above transfer function. For stable systems, the poles are within the unit circle. $G(z)$ is a rational function of the variable z. If we are given a series function:

$$h(z) = \sum_{n=-1}^{-\infty} a_n z^{-n}$$

How can one decide that $h(z)$ represents a rational function of $G(z)$. The answer to this question lies in the theory of Hankel operator. A theorem by Kronecker asserts that if the Hankel operator associated with $h(z)$ is bounded and has a rank n, then there exists rational function $G(z)$ which has exactly n poles. This is a very significant result. Theory of Hankel operators is important for model reduction problem which can be stated as follows:

Model Reduction Problem:

How to simplify a relatively complicated stable transfer function with a simplified lower order stable model transfer function without the loss of its essential properties? This is of importance in control system theory. The answer to the above problem, can be obtained via infinite dimensional operator theory involving Fourier series. We shall give only the rudimentary picture and fundamental concepts associated with infinite dimensional operators. For greater details, the reader should refer such excellent references as [Young, N.], [Peller, V.V.] and [Khrushev, S.].

Summary of Some Useful Facts about Infinite Vector Spaces

1. We consider only $\boldsymbol{L}^p$ and $\boldsymbol{H}^p$ spaces ($p = 2$ and ∞)

2. D represents a unit disk centered at the origin in the z-plane and ∂D as this unit disc boundary surrounding the unit disc D,

$$\boldsymbol{z} \in D \quad \text{implies} \quad z = re^{j\theta}, \ \|r\| \leq 1, \ 0 \leq \theta \leq 2\pi$$

$$\boldsymbol{z} \in \partial D \quad \text{implies} \quad z = e^{j\theta}, \ 0 \leq \theta \leq 2\pi$$

Notation: The variable $\boldsymbol{z}$ will be treated as a vector or variable z as a scalar as the need arises. This should not represent any conflict.

3. $\boldsymbol{L}^2$ and $\boldsymbol{L}^\infty$ spaces are spanned by a complete orthonormal basis $\{\boldsymbol{z}^n\}_{n=-\infty}^{n=+\infty}$. Any function $\boldsymbol{\varphi} \in \boldsymbol{L}^2$ or $\boldsymbol{L}^\infty$ implies

$$\boldsymbol{\varphi}(\boldsymbol{z}) = \sum_{n=-\infty}^{n=+\infty} a_n \boldsymbol{z}^n, \ \sum_{n=-\infty}^{n=+\infty} |a_n|^2 < \infty \quad \text{implying } \boldsymbol{L}^2 \text{ convergence.}$$

$$(\boldsymbol{\varphi}, \boldsymbol{z}^n) = \frac{1}{2\pi} \int_{-\pi}^{+\pi} \varphi(e^{j\theta}) e^{-jn\theta} \, \mathrm{d}\theta, \qquad n = -\infty \text{ to } +\infty \ (integer)$$

$$\|\boldsymbol{\varphi}\|^2 = (\boldsymbol{\varphi}, \boldsymbol{\varphi}) = \frac{1}{2\pi} \int_{-\pi}^{+\pi} \varphi(e^{j\theta}) \varphi(e^{-j\theta}) \, \mathrm{d}\theta = \frac{1}{2\pi j} \int_{\partial D} \varphi(z)\varphi(\bar{z}) \frac{\mathrm{d}z}{z}$$

$$\|\boldsymbol{\varphi}\|^\infty = \underset{|z|=1}{\text{ess}} \sup |\varphi(z)|$$

4. $\boldsymbol{H}^2$ and $\boldsymbol{H}^\infty$ spaces are spanned by a complete orthonormal set $\{\boldsymbol{z}^n\}_{n=0}^{\infty}$.

$$\boldsymbol{f} \in \boldsymbol{H}^2 \text{ and } \boldsymbol{H}^\infty \text{ implies } f(z) = \sum_{n=0}^{\infty} a_n z^n, \ \sum_{n=0}^{\infty} |a_n|^2 < \infty$$

where

$$a_n = (f, z^n) = \frac{1}{2\pi}\int_{-\pi}^{+\pi} f(e^{j\theta})e^{-jn\theta}\,d\theta, \qquad n = 0, 1, \ldots, \infty$$

5. $\boldsymbol{L}^2$ is a complete subspace of $\boldsymbol{L}^\infty$

   $\boldsymbol{H}^\infty$ is a complete subspace of $\boldsymbol{L}^\infty$

   $\boldsymbol{H}^2$ is a complete subspace of $\boldsymbol{L}^2$

   Thus

$$\boldsymbol{L}^\infty \subseteq \boldsymbol{H}^\infty \subseteq \boldsymbol{H}^2$$

$$\boldsymbol{L}^\infty \subseteq \boldsymbol{L}^2 \subseteq \boldsymbol{H}^2$$

   Furthermore,

   $\boldsymbol{H}^2$ is a Hilbert space, $\boldsymbol{H}^\infty$ is a Banach space.

   Also,

   $\boldsymbol{H}^\infty = \boldsymbol{L}^\infty \cap \boldsymbol{H}^2$ is a space of bounded analytic functions on the unit disc.

$$\left.\begin{array}{ll} \boldsymbol{H}^p & \text{basis } \{z^n\}_{n=0}^{\infty} \\ \boldsymbol{L}^p \ominus \boldsymbol{H}^p & \text{basis } \{z^n\}_{n=-\infty}^{-1} \end{array}\right\} p = 2, \infty$$

- It is customary to specify complex analytic functions in $\boldsymbol{L}^2$ space with bounded functions on the unit circle ∂D. Thus, $\boldsymbol{f} \in \boldsymbol{L}^2$, implies

$$f(z)|_{|z|=1} = f(e^{j\theta}), \qquad |f(e^{j\theta})| \le M < \infty\ ,\ 0 \le \theta \le 2\pi$$

$$f(z) = \sum_{\substack{n=-\infty \\ |z|=1}}^{+\infty} a_n z^n = \sum_{-\infty}^{+\infty} a_n e^{jn\theta} \quad \text{(Fourier Series.)}$$

Fourier series is the **important** tool here.

- Bounded analytic functions **inside as well as on the the unit disk** are considered as belonging to Hilbert space, $\boldsymbol{H}^2$. That is, $\boldsymbol{f} \in \boldsymbol{H}^2$ implies

$$|f(z)|_{|z|\leq 1} = |f(re^{j\theta})| = |M| < \infty, \quad r \leq 1$$

$$f(z) = \sum_{\substack{n=0 \\ |z|\leq 1}}^{\infty} a_n z^n = \sum_{n=0}^{\infty} a_n f(re^{j\theta}), \quad r \leq 1, \quad 0 \leq \theta \leq 2\pi$$

$$||\boldsymbol{f}|| = \sum_{n=0}^{\infty} |a_n|^2 < \infty \quad (convergence)$$

$\boldsymbol{zH}^2$ is a subspace resulting from multiplying every function in $\boldsymbol{H}^2$ with $\boldsymbol{z}$.

$\boldsymbol{f}$ is analytic if $\boldsymbol{f} \in \boldsymbol{L}^2, \boldsymbol{H}^2$, $\boldsymbol{f}$ is coanalytic if $\boldsymbol{f} \in \boldsymbol{L}^2, \boldsymbol{zH}^2$.

$\boldsymbol{P}^+, \boldsymbol{P}^-$ Projection Operator in $\boldsymbol{L}^2, \boldsymbol{L}^\infty$.

Let $\boldsymbol{P}^+$ is the orthogonal projection from $\boldsymbol{L}^2 \to \boldsymbol{H}^2$ or from $\boldsymbol{L}^\infty \to \boldsymbol{H}^\infty$

$$\boldsymbol{P}^+ : \boldsymbol{L}^2 \to \boldsymbol{H}^2, \quad \boldsymbol{L}^\infty \to \boldsymbol{H}^\infty$$

$\boldsymbol{P}^-$ is the orthogonal projection from $\boldsymbol{L}^2 \to \boldsymbol{L}^2 \ominus \boldsymbol{H}^2$ or from $\boldsymbol{L}^\infty \to \boldsymbol{L}^\infty \ominus \boldsymbol{H}^\infty$

$$\boldsymbol{P}^- : \boldsymbol{L}^2 \to \boldsymbol{L}^2 \ominus \boldsymbol{H}^2, \quad \boldsymbol{L}^\infty \to \boldsymbol{L}^\infty \ominus \boldsymbol{H}^\infty$$

$$\text{If, } \boldsymbol{f}(z) \in \boldsymbol{L}^2, \boldsymbol{L}^\infty, \quad |z| = 1, \text{ then } f(z) = \sum_{n=-\infty}^{+\infty} a_n z^{-n}$$

$$f_1(z) = f^+(z) = \boldsymbol{P}^+ \sum_{n=-\infty}^{+\infty} a_n z^{+n} = \sum_{n=0}^{+\infty} a_n z^{+n} \quad (\boldsymbol{H}^2, \boldsymbol{H}^\infty) \quad |z| \leq 1$$

$$f_2(z) = f^-(z) = \boldsymbol{P}^- \sum_{n=-\infty}^{+\infty} a_n z^{+n} = \sum_{n=-\infty}^{-1} a_n z^{+n} \quad (\boldsymbol{H}^{2\perp}, \boldsymbol{H}^{\infty\perp}) \quad |z| \geq 1$$

$f_1(z)$ is analytic **on** and **inside** the unit circle.

$f_2(z)$ is analytic **on** and **outside** the unit circle.

Projection operator is a mathematical equivalent for the truncation of a series representation of a function.

Note: Reader should be very careful about the notation anomaly. Given a sequence $\{a_n\}_{n=0}^{\infty}$, the conventional Z-tranfer function used in Control Engineering is given by:

$$G(z) = \sum_{n=0}^{\infty} a_n z^{-n}$$

But in our $\boldsymbol{H}^2$ space, we associate with the above sequence $\{a_n\}_{n=0}^{\infty}$ the function

$$f(z) = \sum_{n=0}^{\infty} a_n z^n$$

It is obvious that $G(z)$ and $f(z)$ can be obtained from each other by replacing z with z^{-1} and should not cause any further confusion.

Example 2.15:

$$\boldsymbol{f} \rightarrow f(t) = \sum_{n=-\infty}^{\infty} a_n e^{j\omega_0 nt}, \quad \omega_0 = \frac{1}{T} \quad \text{(Fourier Series)}$$

The operator $\boldsymbol{P}^+$ and $\boldsymbol{P}^-$ associates with vector $\boldsymbol{f} \rightarrow f(t)$

$$\boldsymbol{P}^+\boldsymbol{f} = \sum_{n=0}^{\infty} a_n e^{j\omega_0 nt}$$

$$\boldsymbol{P}^-\boldsymbol{f} = \sum_{n=-\infty}^{-1} a_n e^{j\omega_0 nt}$$

$$\boldsymbol{P}^- = \boldsymbol{I} - \boldsymbol{P}$$

Toeplitz operators and Matrices (Infinite Dimensional)

Let $\boldsymbol{\varphi} \in \boldsymbol{L}^\infty$, its fourier series is:

$$\boldsymbol{\varphi}(z) = \sum_{-\infty}^{+\infty} a_n z^n$$

The orthonormal basis in the **domain** are $\{z^j\}_{j=-\infty}^{+\infty}$, j being the column index. We define a Toeplitz operator T_φ :

$$T_\varphi : L^2, L^\infty \to L^2, L^\infty$$

such that,

$$T_\varphi z = \varphi(z)z$$

$$\text{Matrix } \{\tau_{ij}\}_{i,j=-\infty}^{+\infty} = \left(T_\varphi z^i, z^j\right) = a_{i-j}, \qquad i, j = (-\infty, \cdots, -1, 0, +1, \ldots, \infty)$$

The matrix T_φ of the operator T_φ, with a symbol φ, w.r.t these orthonormal basis has same entries along each diagonal parallel to the main diagonal. This matrix with no borders is

$$T_\varphi = \begin{bmatrix} \cdots & \cdot & \cdot & \cdot & \cdots \\ \cdots & a_0 & a_{-1} & a_{-2} & \cdots \\ \cdots & a_1 & a_0 & a_{-1} & \cdots \\ \cdots & a_2 & a_1 & a_0 & \cdots \\ \cdots & \cdot & \cdot & \cdot & \cdots \end{bmatrix} \downarrow i, -\infty \text{ to } +\infty$$

$$\to j, -\infty \text{ to } +\infty$$

The basis in the **codomain** representing the domain of the inverse of the operator are $\{z^{-i}\}_{-\infty}^{+\infty}$

Proof

$$\varphi(z) = \sum_{-\infty}^{+\infty} a_n z^n$$

$$T_\varphi z^j = \varphi(z)z^j = \sum_{n=-\infty}^{+\infty} a_n z^{n+j} = \sum_{m=-\infty}^{+\infty} a_{m-j} z^m$$

$$(T_\varphi z^j, z^i) = \sum_{m=-\infty}^{+\infty} a_{m-j}(z^m, z^i) = \sum_{m=-\infty}^{+\infty} a_{m-j}\delta_{m-i} = a_{i-j}$$

$i, j = (-\infty, \ldots, -1, 0, 1, \ldots, \infty)$, i, j represents the row the column of the matrix $\boldsymbol{T}_\varphi$, φ is known as the "symbol" of $\boldsymbol{T}_\varphi$

$\boldsymbol{T}_\varphi$ is Toeplitz if and only if,

$$\boldsymbol{S}^*\boldsymbol{T}_\varphi\boldsymbol{S} = \boldsymbol{T}_\varphi, \qquad \boldsymbol{T}_\varphi : \boldsymbol{L}^2 \to \boldsymbol{L}^2$$

Furthermore,

$$\boldsymbol{S}\boldsymbol{T}_\varphi = \boldsymbol{T}_\varphi\boldsymbol{S} \qquad \boldsymbol{T}_\varphi \text{ is analytic Toeplitz}$$

$$\boldsymbol{S}^*\boldsymbol{T}_\varphi = \boldsymbol{T}_\varphi\boldsymbol{S}^* \qquad \boldsymbol{T}_\varphi \text{ is coanalytic Toeplitz}$$

$\boldsymbol{S}$ is the forward shift operator.

$\boldsymbol{S}^*$ is backward shift operator.

Inner Function:

A function $\boldsymbol{u} \in \boldsymbol{H}^2$ is called an inner function, if

$$|u(z)| = 1 \qquad \text{for} \quad |z| = 1$$

A typical example of an inner function is :

$$\varphi(z) = u(z) = \left(\frac{z+\alpha}{z-\alpha}\right)^n, \qquad |\alpha| < 1$$

When $\alpha = 1$, it is called a singular **inner function**. This function represents an all pass "filter".

Blaschke Product:

A function $\boldsymbol{b} \in \boldsymbol{H}^\infty$ and $|b(z)| < 1$ for $|z| < 1$ is defined as:

$$b(z) = z^m \prod_{j=1}^{n} \frac{|\alpha_j|}{\alpha_j}\left(\frac{\alpha_j - z}{1 - \overline{\alpha}_j z}\right), \qquad |z| < 1, \ |\alpha_j| > 1 \ \boldsymbol{b} \in \boldsymbol{H}^\infty$$

This is also an inner function. The zeroes of $b(z)$ consist of one zero of multiplicity m at the origin and n zeros at α_j inside the unit circle and n poles outside the unit circle at $(\bar{\alpha}_j)^{-1}$.

Multiplication Operator:

As the name implies, the multiplication operator represents multiplication of two vectors in z or "frequency" domain (convolution in time domain). Let $\boldsymbol{M}$ an operator on $\boldsymbol{L}^2(-\infty, \infty)$ or $\boldsymbol{L}^\infty$ such that,

$$(\boldsymbol{M}\boldsymbol{x})(t) = \varphi(t)x(t), \quad \boldsymbol{\varphi}(t) \in \boldsymbol{L}^2(-\infty, \infty) \text{ or } \boldsymbol{L}^\infty \text{ and }, \quad \boldsymbol{x}(t) \in \boldsymbol{L}^2(-\infty, +\infty)$$

Then,

$$\|\boldsymbol{M}\boldsymbol{x}\| = \left[\int_{-\infty}^{+\infty} |\varphi(t)|^2 |x(t)|^2 \, \mathrm{d}t\right]^{1/2} \leq \|\boldsymbol{\varphi}\|^\infty \|\boldsymbol{x}\|$$

This concept of multiplication operator plays a role in Control Theory.

If $\boldsymbol{\varphi} \in (\boldsymbol{L}^\infty \text{or } \boldsymbol{L}^2)$ and $\boldsymbol{f} \in \boldsymbol{L}^2$, then $\boldsymbol{\varphi f} \in \boldsymbol{L}^2$

How to compute $(\boldsymbol{\varphi f})(z)$? We can perform this computation in two ways.

(i) Direct representation of $\boldsymbol{\varphi f}$

Fourier series of $(\boldsymbol{\varphi})(z)$

$$\varphi(z) = \sum_{n=-\infty}^{+\infty} a_n z^n$$

Fourier series of $\boldsymbol{f}$

$$f(z) = \sum_{n=-\infty}^{+\infty} c_n z^n$$

Fourier series of $\boldsymbol{\varphi f}$

$$(\boldsymbol{\varphi f})(z) = \varphi(z)f(z) = \left(\sum_{n=-\infty}^{+\infty} a_n z^n\right)\left(\sum_{m=-\infty}^{+\infty} c_m z^m\right)$$

$$= \sum_{i=-\infty}^{+\infty}\left(\sum_{j=-\infty}^{+\infty} a_{i-j} c_j\right) z^{+i}$$

(ii) $(\boldsymbol{\varphi f})(z)$ can be computed via multiplication operator $\boldsymbol{M}_\varphi$ operating on f.

Operator $\boldsymbol{M}_{\varphi} : \boldsymbol{L}^2 \to \boldsymbol{L}^2$ is defined as,

$$\boldsymbol{M}_{\varphi} z^j = \varphi(z) z^j = \left(\sum_{n=-\infty}^{+\infty} a_n z^n \right) z^j = \sum_{n=-\infty}^{+\infty} a_n z^{n+j}$$

$$(\boldsymbol{M}_{\varphi} \boldsymbol{f})(z) = \boldsymbol{M}_{\varphi} \sum_{j=-\infty}^{+\infty} c_j z^j = \sum_{j=-\infty}^{+\infty} c_j \boldsymbol{M}_{\varphi} z^j = \sum_{j=-\infty}^{+\infty} c_j \sum_{n=-\infty}^{+\infty} a_n z^{n+j}$$

$$= \sum_{i=-\infty}^{+\infty} \left(\sum_{j=-\infty}^{+\infty} a_{i-j} c_j \right) z^j$$

Both ways we arrive at the same result.

Example 2.16:

$$\text{For } \boldsymbol{\varphi}, \boldsymbol{f} \in \boldsymbol{L}^2 \quad \varphi(z) = \sum_{n=-\infty}^{+\infty} a_n z^n \quad , f(z) = \sum_{n=-\infty}^{+\infty} c_n z^n \quad , z = e^{j\theta}$$

Solution:

$$|\varphi(z)| = \sum_{n=-\infty}^{+\infty} |a_n| < \infty, \qquad |f(z)| = \sum_{n=-\infty}^{+\infty} |c_n| < \infty$$

$$||\varphi(z)|| = \sum_{n=-\infty}^{+\infty} |a_n|^2, \qquad ||f(z)|| = \sum_{-\infty}^{+\infty} |c_n|^2$$

Compute the fourier series expansion of $(\boldsymbol{\varphi f})$ and show its convergence.

$$\text{Let} \quad \varphi(z) = \lim_{m\to\infty} \varphi_m(z) = \lim_{m\to\infty} \sum_{n=-m}^{+m} a_n z^n \quad \text{(Truncated } \varphi(z))$$

$$b(z) = (\varphi_m(z) f(z)) = \sum_{n=-\infty}^{+\infty} b_n z^n \quad \text{(Truncated } \varphi(z)) f(z)$$

where

$$b_n = \frac{1}{2\pi} \int_{-\pi}^{+\pi} b(e^{j\theta}) e^{-jn\theta} \, \mathrm{d}\theta = \frac{1}{2\pi} \int_{-\pi}^{+\pi} \varphi_m(e^{j\theta}) f(e^{j\theta}) e^{-jn\theta} \, \mathrm{d}\theta$$

or

$$b_n = \frac{1}{2\pi}\int_{-\infty}^{+\infty}\left(\sum_{k=-m}^{+m} a_k e^{jk\theta}\right)\left(\sum_{i=-\infty}^{\infty} c_i e^{ji\theta}\right)e^{-jn\theta}\,\mathrm{d}\theta$$

Therefore

$$b_n = 0 \qquad \text{when } i \neq n-k$$

$$b_n = \sum_{k=-m}^{+m} a_k c_{n-k} \qquad i = n-k$$

Thus

$$\varphi_m(z)f(z) = \sum_{n=-\infty}^{+\infty}\left(\sum_{k=-m}^{+m} a_k c_{n-k}\right)z^n$$

Consider

$$I_m = \frac{1}{2\pi}\int_{-\pi}^{+\pi} |f(e^{j\theta})|\left|\varphi(e^{j\theta}) - \sum_{n=-m}^{+m} a_n e^{jn\theta}\right|\mathrm{d}\theta$$

$$I_m \leq |f(e^{j\theta})|\left|\sum_{n=-\infty}^{+\infty}|a_n| - \sum_{n=-m}^{+m}|a_n|\right|$$

$$\lim_{m\to\infty} I_m = 0$$

$$\lim_{m\to\infty}\frac{1}{2\pi}\int_{-\pi}^{+\pi}\varphi_m(e^{j\theta})f(e^{j\theta})e^{-jn\theta}\,\mathrm{d}\theta = \frac{1}{2\pi}\int_{-\pi}^{+\pi}\varphi(e^{j\theta})f(e^{j\theta})e^{-jn\theta}\,\mathrm{d}\theta$$

Hankel Operator and Matrices (Infinite Dimension)

Two kinds of bounded Hankel operators arise in $\boldsymbol{H}^2$ spaces.

The first kind operates on $\boldsymbol{\varphi}$ in $\boldsymbol{H}^2$ and transports it to $\boldsymbol{L}^2 \ominus \boldsymbol{H}^2$.

This is called as the Hankel operator $\boldsymbol{H}_\varphi$ with the symbol φ and is:

$$\boldsymbol{H}_\varphi : \boldsymbol{H}^2 \to \boldsymbol{L}^2 \ominus \boldsymbol{H}^2, \quad \boldsymbol{L}^2 = \boldsymbol{H}^2 \oplus \boldsymbol{H}^{2\perp}, \quad \boldsymbol{L}^2 \ominus \boldsymbol{H}^2 = \boldsymbol{H}^{2\perp}$$

The second kind of Hankel operator acts on the function φ in $\boldsymbol{H}^2$ and maps these functions (vectors) into $\boldsymbol{H}^2$ itself. This is defined as Hankel operator $\boldsymbol{\Gamma}_\varphi$ with the "symbol φ" and written as:

$$\boldsymbol{\Gamma}_\varphi : \boldsymbol{H}^2 \rightarrow \boldsymbol{H}^2$$

Both the operators $\boldsymbol{H}_\varphi$ and $\boldsymbol{\Gamma}_\varphi$ are bounded if and only if the symbol function $\varphi(z)$ is bounded on the unit circle ∂D.

Proof:

The non analytic part of $\varphi(z)$ denoted by $\varphi_2(z)$ (also known as "noncausal" or anti-analytic function) which is in $\boldsymbol{L}^2 \ominus \boldsymbol{H}^2$ and **outside the unit circle**, is represented by the complex sequence $\{a_n\}_{-\infty}^{-1}$. Associated with this function $\varphi_2(z)$ is a Hankel operator $\boldsymbol{H}_\varphi : \boldsymbol{H}^2 \rightarrow \boldsymbol{L}^2 \ominus \boldsymbol{H}^2$ and it's matrix representation is given by:

$$h_{ij} = a_{-i-j} = (\boldsymbol{H}_\varphi z^i, z^j), \qquad i = -1, -2, -3, \cdots, \quad j = 0, +1, +2, \cdots,$$

The span of basic vectors are:

$$1, z, z^2, \cdots \qquad \text{in } \boldsymbol{H}^2$$

and

$$z^{-1}, z^{-2}, z^{-3}, \cdots \quad \text{in } \boldsymbol{L}^2 \ominus \boldsymbol{H}^2$$

Where

$$z^m = z^m \boldsymbol{e}_m = e^{jm\theta} \boldsymbol{e}_m \qquad m = 0, \pm 1, \pm 2, \cdots$$

The vectors $\boldsymbol{e}_m$ are orthonormal unit vectors defined earlier.

The Hankel operator $\boldsymbol{H}_\varphi$ transports a vector from space $\boldsymbol{H}^2$ to space $\boldsymbol{L}^2 \ominus \boldsymbol{H}^2$.

Computation of Hankel Matrix H_φ.

Let $\varphi \in L^\infty, f \in L^2$, we define a multiplication operator M_φ and the Hankel operator H_φ as follows:

$$(M_\varphi f)(z) = (\varphi f)(z) = \varphi(z)f(z)$$

$$(H_\varphi f)(z) = P^-(M_\varphi f)(z) = P^-(\varphi f)(z) \in L^2 \ominus H^2$$

Important:

If $\varphi \in L^\infty$ and $f \in L^2$ then $\varphi f \in L^2$.

If $\varphi \in L^\infty$ and $f \in H^2$ then $\varphi f \in H^2$.

The matrix H_φ associated with the operator H_φ of the symbol φ is:

$$H_\varphi = \begin{bmatrix} a_{-1} & a_{-2} & a_{-3} & \cdots \\ a_{-2} & a_{-3} & a_{-4} & \cdots \\ a_{-3} & a_{-4} & a_{-5} & \cdots \\ . & . & . & \cdots \end{bmatrix}, \qquad \varphi(z) = \sum_{n=-\infty}^{+\infty} a_n z^n$$

Proof

Let,

$$f = z^j, \quad \varphi = \sum_{n=-\infty}^{+\infty} a_n z^n$$

$$M_\varphi z^j = \sum_{n=-\infty}^{+\infty} a_n z^{n+j} = \sum_{n=-\infty}^{+\infty} a_{m-j} z^m$$

$$H_\varphi z^j = P^-\left[M_\varphi z^j\right] = P^-\left[\sum_{m=-\infty}^{+\infty} a_{m-j} z^m\right]$$

$$H_\varphi z^j = \sum_{n=-\infty}^{-1} a_{m-j} z^m$$

The above truncation of $M_\varphi z^j$ yields the vector in $L^2 \ominus H^2$.

Taking inner product of $\boldsymbol{H}_\varphi z^j$ with z^i

$$(z^i, \boldsymbol{H}_\varphi z^j) = \sum_{m=-\infty}^{-1} a_{m-j}(z^i, z^m)$$

But,

$$(z^i, z^m) = \frac{1}{2\pi}\int_{-\pi}^{+\pi} e^{ij\theta}e^{-mj\theta}\,\mathrm{d}\theta = \delta_{m-i}$$

Hence,

$$h_{ij} = (z^i, \boldsymbol{H}_\varphi z^j) = \sum_{m=-\infty}^{-1} a_{m-j}\delta_{m-i} = a_{-i-j}$$

$$i = 1, 2, 3, \ldots \quad j = 0, 1, 2, \ldots$$

If

$$f(z) = \sum_{k=0}^{\infty} b_k z^k$$

$$\varphi(z) = \sum_{n=-\infty}^{+\infty} a_n z^n$$

$$\boldsymbol{P}^{-}\left[(\boldsymbol{M}_\varphi \boldsymbol{f})(z)\right] = \sum_{m=-\infty}^{-1} c_m z^m$$

In the literature, φ is known as the symbol of the operator $\boldsymbol{H}_\varphi$.

Example 2.17:

Given $\boldsymbol{f}(z)$, $\boldsymbol{\varphi}(z)$, and $\boldsymbol{P}^{-1}\left[(\boldsymbol{M}_\varphi \boldsymbol{f})(z)\right]$, show

$$\boldsymbol{H}_\varphi \boldsymbol{b} = \boldsymbol{c}$$

$$\boldsymbol{b} = \begin{bmatrix} b_0 \\ b_1 \\ b_2 \\ \vdots \end{bmatrix}, \quad \boldsymbol{c} = \begin{bmatrix} c_{-1} \\ c_{-2} \\ c_{-3} \\ \vdots \end{bmatrix}$$

Proof:

$$\varphi(z) = \sum_{n=-3}^{+3} a_n z^n \qquad \boldsymbol{\varphi} \in \boldsymbol{L}^2 \text{ or } \boldsymbol{L}^\infty$$

$$f(z) = \sum_{k=0}^{+3} b_k z^k \qquad \boldsymbol{f} \in \boldsymbol{H}^2$$

$$\begin{aligned}\left(\boldsymbol{M}_\varphi \boldsymbol{f}\right)(z) &= (a_{-3}b_0)z^{-3} + (a_{-2}b_0 + a_{-3}b_1)z^{-2} + (a_{-3}b_2 + a_{-2}b_1 + a_{-1}b_0)z^{-1} \\ &\quad + (a_0b_0) + (a_0b_1 + a_1b_0)z + (a_2b_0 + a_1b_1 + a_0b_2)z^2 \\ &\quad + (a_1b_2 + a_2b_1 + a_3b_0)z^3\end{aligned}$$

$$\boldsymbol{P}^- \left[(\boldsymbol{M}_\varphi \boldsymbol{f})(z)\right] = (a_{-3}b_0)z^{-3} + (a_{-2}b_0 + a_{-3}b_1)z^{-2} + (a_{-3}b_2 + a_{-2}b_1 + a_{-1}b_0)z^{-1}$$

Thus,

$$\boldsymbol{H}_\varphi = \begin{bmatrix} a_{-1} & a_{-2} & a_{-3} \\ a_{-2} & a_{-3} & 0 \\ a_{-3} & 0 & 0 \end{bmatrix} \qquad \text{Hankel Matrix}$$

and,

$$\begin{bmatrix} a_{-1} & a_{-2} & a_{-3} \\ a_{-2} & a_{-3} & 0 \\ a_{-3} & 0 & 0 \end{bmatrix} \begin{bmatrix} b_0 \\ b_1 \\ b_2 \end{bmatrix} = \begin{bmatrix} c_{-1} \\ c_{-2} \\ c_{-3} \end{bmatrix}$$

or

$$\boldsymbol{H}_\varphi \boldsymbol{b} = \boldsymbol{c}$$

Computation of Hankel Operator $\Gamma_\varphi : \boldsymbol{H}^2 \rightarrow \boldsymbol{H}^2$

Often, we need a Hankel operator from $\boldsymbol{H}^2$ to $\boldsymbol{H}^2$. This Hankel operator is denoted with the symbol $\boldsymbol{\Gamma}_\varphi$. We show the relationship between $\boldsymbol{H}_\varphi$ and $\boldsymbol{\Gamma}_\varphi$: Let $\boldsymbol{\varphi} \in \boldsymbol{L}^2$ or $\boldsymbol{L}^\infty$, $\boldsymbol{f} \in \boldsymbol{H}^2$, and the involution operator $\boldsymbol{J}$, such that

$$\boldsymbol{J}(z^j) = z^{-j-1}, \quad |z| = 1$$

$$J(f(z)) = z^{-1}f(z^{-1}), |z| = 1$$

Define,

$$\mathbf{\Gamma}_{\varphi} = \boldsymbol{J}\boldsymbol{H}_{\varphi} : \boldsymbol{H}^2 \to \boldsymbol{H}^2$$

$$\mathbf{\Gamma}_{\varphi} = \boldsymbol{P}^{-}(\boldsymbol{J}\boldsymbol{M}_{\varphi})(z)$$

Example 2.18:

Given,

$$\boldsymbol{\varphi} \in \boldsymbol{L}^{\infty}, \boldsymbol{L}^2 \qquad \varphi(z) = \sum_{n=-\infty}^{+\infty} a_n z^n, \quad \text{show}$$

$$\mathbf{\Gamma}_{\varphi} = \begin{bmatrix} a_{-1} & a_{-2} & a_{-3} & \cdots \\ a_{-2} & a_{-3} & a_{-4} & \cdots \\ a_{-3} & a_{-4} & a_{-5} & \cdots \\ . & . & . & \cdots \end{bmatrix}$$

Proof:

$$f(z) = z^j, \quad \boldsymbol{\varphi} \in \boldsymbol{L}^{\infty}, \boldsymbol{L}^2$$

$$\begin{aligned} \boldsymbol{H}_{\varphi}\boldsymbol{z}^j &= \boldsymbol{P}^{-}(\boldsymbol{\varphi}\boldsymbol{z}^j) = P^{-}\left(\sum_{n=-\infty}^{+\infty} a_n z^{n+j}\right) \\ &= \boldsymbol{P}^{-}\left(\sum_{m=-\infty}^{+\infty} a_{m-j} z^m\right) = \sum_{m=-\infty}^{-1} a_{m-j} z^m = \sum_{m=1}^{\infty} a_{-m-j} z^{-m} \end{aligned}$$

$$\mathbf{\Gamma}_{\varphi}\boldsymbol{z}^j = \boldsymbol{J}\left[\sum_{m=-\infty}^{-1} a_{m-j} z^{-m}\right] = \sum_{m=-\infty}^{-1} a_{m-j} z^{m+1} = \sum_{k=0}^{\infty} a_{-j-k-1} z^k$$

Taking inner product of $\mathbf{\Gamma}_{\varphi}\boldsymbol{z}^j$ with $\boldsymbol{z}^i$

$$(z^i, \mathbf{\Gamma}_{\varphi}\boldsymbol{z}^j) = \sum_{k=0}^{\infty} a_{-j-k-1}(z^i, z^k) = a_{-j-i-1}, \quad i, j = 0, 1, 2, \ldots$$

Note: Matrix representation of both $\boldsymbol{H}_{\varphi}$ and $\mathbf{\Gamma}_{\varphi}$ is the same.

Few basic facts about Hankel and Toeplitz Operators:

1. $\boldsymbol{T}_\varphi : \boldsymbol{L}^2, \boldsymbol{L}^\infty \to \boldsymbol{L}^2, \boldsymbol{L}^\infty, \boldsymbol{T}_\varphi \boldsymbol{z}^j = \boldsymbol{\varphi}(z)\boldsymbol{z}^j$
2. $\boldsymbol{H}_\varphi : \boldsymbol{L}^2, \boldsymbol{L}^\infty \to \boldsymbol{L}^2 \ominus \boldsymbol{H}^2, \boldsymbol{H}_\varphi : \boldsymbol{P}^- \boldsymbol{M}_\varphi \longrightarrow \boldsymbol{H}_\varphi \boldsymbol{z}^j = P^{-1}\left(\varphi(z)\, z^j\right)$
3. $\left(\boldsymbol{P}^- \boldsymbol{M}_\varphi g, h\right) = \left(H_\varphi g, h\right) = \frac{1}{2\pi} \int\limits_0^{2\pi} \overline{\varphi}(e^{j\theta})\overline{g}(e^{j\theta})h(e^{j\theta})\, \mathrm{d}\theta$
4. $\boldsymbol{H}_{a_1\varphi_1 + a_2\varphi_2} = a_1 \boldsymbol{H}_{\varphi_1} + a_2 \boldsymbol{H}_{\varphi_2}$
5. If $\widehat{\varphi} \in \boldsymbol{H}^\infty$ and $f \in \boldsymbol{H}^2$ then $\widehat{\varphi} f \in \boldsymbol{H}^2$

   Therefore,

   $$\boldsymbol{P}^-(\widehat{\varphi} f) = \boldsymbol{H}_{\widehat{\varphi}} = 0$$

   Hence

   $$\boldsymbol{H}_{\varphi - \widehat{\varphi}} = \boldsymbol{H}_\varphi - \boldsymbol{H}_{\widehat{\varphi}} = \boldsymbol{H}_\varphi$$

   and

   $$||\boldsymbol{H}_{\varphi - \widehat{\varphi}}|| = ||\boldsymbol{H} s_\varphi|| \le ||\varphi - \widehat{\varphi}||^\infty$$

6. If $\boldsymbol{H} = \boldsymbol{H}_\varphi$, then $\boldsymbol{H}_\varphi^* = \boldsymbol{H}_{\varphi^*}$
7. **Hankel operator is not unique**

   Consider a function $\varphi_1 - \varphi_2 \in \boldsymbol{H}^2$

   $$\boldsymbol{P}^-(\boldsymbol{M}_{\varphi_1 - \varphi_2} f) = 0 \longrightarrow \boldsymbol{H}_{\varphi_1 - \varphi_2} = 0$$

   Thus,

   $$\boldsymbol{H}_{\varphi_1} = \boldsymbol{H}_{\varphi_2} \qquad \text{if and only if} \qquad \varphi_1 - \varphi_2 \in \boldsymbol{H}^2$$

8. The Hankel operator $\boldsymbol{H}_\varphi$ is bounded.
9. An operator $\boldsymbol{H}_\varphi$ is Hankel if and only if

   $$\boldsymbol{S}^* \boldsymbol{H}_\varphi = \boldsymbol{H}_\varphi \boldsymbol{S}$$

where $\boldsymbol{S}$ is a unilateral forward shift operator. $\boldsymbol{S}^*$ is unilateral backward shift operator. Shift operator is one of the most important operators in Disk Algebra.

10. The self adjoint flip operator $J : \boldsymbol{L}^2 \to \boldsymbol{L}^2$ is defined as:

$$\boldsymbol{J}\boldsymbol{f} = \tilde{\boldsymbol{f}}$$

$$\tilde{f}(z) = f(\bar{z}), \quad \bar{z} \text{ is conjugate of } z$$

$$z = e^{j\theta}, \quad \bar{z} = z^{-1} = e^{-j\theta}$$

$$\boldsymbol{\Gamma}_{\varphi} = \boldsymbol{J}\boldsymbol{H}_{\varphi}$$

11. $\boldsymbol{H}_{\varphi}$ is never invertable.

Kronecker's Theorem for Hankel Matrices

Consider the infinite Hankel matrix

$$\boldsymbol{\Gamma}_{\varphi} = \{a_{-i-j-1}\} \qquad i, j = 0, 1, 2, \cdots, \infty$$

Kronecker Theorem states:

1. $\boldsymbol{\Gamma}_{\varphi}$ is a finite rank matrix if and only if

$$\boldsymbol{P}^{-}\varphi(z) = \sum_{n=-1}^{-\infty} a_n\, z^n \quad \text{is rational}$$

2. $\boldsymbol{\Gamma}_{\varphi}$ is a bounded operator if the poles of $\boldsymbol{P}^{-}\varphi(z) = \sum_{n=-1}^{-\infty} a_n\, z^n$ are all inside the unit circle $|z| < 1$

In general, Hilbert space is useful when we are interested in the signal error functional minimization and the best approximation problems. However, there are situations in engineering systems, particularly Control Technology, where Hilbert norm may not be the best norm. In fact in many problems in systems engineering we are interested in bounds on the maximum absolute error of a functional. Such problems belong to $\boldsymbol{L}^{\infty}$ or $\boldsymbol{H}^{\infty}$ norm.

Approximation problems in infinite dimensional spaces

The most popular criterion for optimization and approximation is the "least integral squared error" type. We look for an error function whose Hilbert space norm is as small as possible, taking advantage of Hilbert space geometry. The best approximation and the error function are orthogonal in the Hilbert space. Therefore the best approximation of a function in $\boldsymbol{L}^2$-norm by a function in $\boldsymbol{H}^2$ space is $(\boldsymbol{P}^{+}\boldsymbol{\varphi})(z)$ where $(\boldsymbol{P}^{+}\boldsymbol{\varphi})(z)$ belongs to $\boldsymbol{H}^2$ and the error $(\boldsymbol{P}^{-}\boldsymbol{\varphi})(z)$ belongs to $\boldsymbol{H}^{2\perp}$.

Best Approximation of an L^2 function in H^2 (Least Square Optimization)

Let $\varphi \in \boldsymbol{L}^2$ and $h \in \boldsymbol{H}^2$ such that,

$$||\boldsymbol{\varphi} - \boldsymbol{h}|| = \inf_{\boldsymbol{h}' \in \boldsymbol{H}^2} ||\boldsymbol{\varphi} - \boldsymbol{h}'||$$

This vector $\boldsymbol{h}$ is the best approximation of $\boldsymbol{\varphi}$ in $\boldsymbol{H}^2$ space with respect to the $\boldsymbol{L}^2$ norm. That is to say $\boldsymbol{h}$ in the subspace $\boldsymbol{H}^2$ is nearest to $\boldsymbol{\varphi} \in \boldsymbol{L}^2$. Thus,

$$\boldsymbol{\varphi} = \boldsymbol{h} + \boldsymbol{e}, \ \boldsymbol{e} \text{ being the error vector.}$$

$$\boldsymbol{h} = \boldsymbol{P}^{+}(\boldsymbol{\varphi}), \quad \boldsymbol{e} = \boldsymbol{P}^{-}(\boldsymbol{\varphi}), \quad ||\boldsymbol{\varphi}||^2 = ||\boldsymbol{h}||^2 + ||\boldsymbol{e}||^2, \qquad \boldsymbol{e} \perp \boldsymbol{H}^2$$

Least Square Optimization

$\boldsymbol{\varphi} = \boldsymbol{h} + \boldsymbol{e}, \quad \boldsymbol{h}$ is the projection of $\boldsymbol{\varphi}$ on $\boldsymbol{H}^2$

Example 2.19:

$$\varphi(z) = \frac{4z+3}{(2z-1)(z+2)} = \frac{1}{(z-1/2)} + \frac{1}{(z+2)}, \quad |z| = 1$$

$$(\boldsymbol{P}^{+}\boldsymbol{\varphi})(z) = \left(\frac{1}{z+2}\right) \quad \text{Analytic} \in \boldsymbol{H}^2$$

$$(\boldsymbol{P}^{-}\boldsymbol{\varphi})(z) = \left(\frac{1}{z-1/2}\right) \quad \text{Non-Analytic} \in \boldsymbol{L}^2 \ominus H^2$$

Hence $(1/(z+2)$ is the best approximation to $\varphi(z)$ which minimizes $||\varphi(z) - (1/(z-1/2))||^2$. This is the least square minimization problem. We shall discuss this aspect of the problem in details in Chapter on Transforms(Chapter 5). The best approximation problem with respect to non-Hilbert norm as mentioned earlier has no satisfactory answer in general. Very often, we are interested in the peak value of a function and its norm. In the system theory, we deal with the input functions, the output functions and so called "system impulse responses". The output functions can be considered as some "multiplication(convolution)" operation between the input function and the impulse response function. $\boldsymbol{L}^{\infty}$-norm of a function is not the same as the Hilbert norm but the $\boldsymbol{L}^{\infty}$-norm of a function, $\boldsymbol{\varphi}$ is the same as the operator norm of multiplication by $\boldsymbol{\varphi}$ with functions in $\boldsymbol{L}^2$ space.

Therefore the operator theory in $\boldsymbol{L}^{\infty}$-norm optimization problem is needed.

Minimization Problem in $\boldsymbol{H}^{\infty}$ (NEHARI's PROBLEM)

Nehari's Problem states:

Question: Given a complex sequence $\{a_n\}_{n=-\infty}^{n=-1}$, does there exist a bounded function $\varphi(z)$ on the unit circle i.e. $\varphi \in \boldsymbol{L}^{\infty}$,

such that

$$(\boldsymbol{\varphi}, \boldsymbol{z}^n) = \frac{1}{2\pi}\int_{-\pi}^{+\pi} \varphi(e^{j\theta})e^{-jn\theta}\, d\theta = a_n, \quad n = -1, -2, \ldots, -\infty$$

$$||\varphi||^{\infty} \leq 1$$

Nehari's Theorem states that the answer is Yes. Such a function $\varphi(z)$ exists if and only if the Hankel operator $\mathbf{\Gamma}_{\varphi}$ defined by the sequence $\{a_n\}_{n=-\infty}^{n=-1}$ is bounded and furthermore,

$$||\mathbf{\Gamma}_{\varphi}|| = 1$$

For the minimization problem, we are seeking a Hankel matrix $\mathbf{\Gamma}_{\hat{\varphi}}$ such that

$$||\mathbf{\Gamma}_{\varphi} - \mathbf{\Gamma}_{\hat{\varphi}}|| = \text{Minimum}$$

Problem is solved via singular value decomposition, keeping only the largest singular values [Adamjan, V.M.].

Bibliography

[Adamjan, V.M.] Adamjan, V.M., Arov, D.Z. and Krein, M.G. *Mat USSR sbornik,* (English Translation) (Original treatment of H^{∞} Optimization.) 15, pp 31–78, 1971.

[Aitken, A.] Aitken, A.C. *Determinants and Matrices,* New York: Interscience Inc., 1942. This is an excellent reading.

[Aplevich, J.D.] Aplevich J.D. Direct Computation of Canonical Forms for Linear Systems by ELementary Matrix operation, *IEEE Trans, Autom Control,* AC-18, No 3, pp 306–307, June 1973.

[Barnet, S.] Barnet, S, Matrices, *Methods and Applications,* New York, Oxford Applied Mathematics and Computing Science Series, Clarendon Press, 2003.

[Bellman, R.] Bellman R. *Introduction to Matrix Algebra,* New York: McGraw Hill Co., 1960. Very comprehensive coverage of timely applications, very readable, lots of references.

[Brooks, M.] Brooks, M. *Matrix Reference Manual* http://www.ee.ic.ac. uk/hp/staff/dmb/matrix/intro.html, May 20, 2004.

[Desoer, C.A.] Desoer C.A. and Schulman, J.D. Zeros and Poles of Matrix Transfer Function and their Dynamical interpretation, *IEEE Trans. Circuits Syst.* CAS-21, pp 3–8, 1974.

[Elaydi, S.N.] Elaydi, S.N and Harris W.A. On the Computation of A^N, *Siam Rev.* Vol 40, No. 4, pp 965–971, December 1998.

[Gantmacher, F.R.] Gantmacher, F.R. *The Theory of Matrices, Vols 1, 2,* Translated from Russian by K.A. Hirsh Chelsea, N.Y., 1959. This is a very complete work.

[Horn, R.A.] Horn, R. A. and Johnson, C.R. *Matrix Analysis,* London: Cambridge University Press, 1993. A must reading for Hermitian and Symmetric matrices.

[Kailath, T.] Kailath, T. *Linear Systems,* New Jersey: Prentice Hall, This is a very complete collection of matrix theory for the study of Linear Dynamical Systems.

[Khrushev, S.] Khrushev, S. *Hankel Operators, best approximation and stationary processes,* Translation from Russian: Russian Math Surveys 37, pp 61–144, 1982.

[Marcus, M.] Marcus, M. and Minc, Henryk, *A survey of Matrix Theory and Matrix Inequalities,* New York: Dover Publications Inc, 1964. A very good survey, very precise and concise.

[Myskis, A.D.] Myskis, A. D. *Advanced Mathematics For Engineers,* Special Course Translated from the Russian by V. M. Volosov and I. G. Volosova, Mir Publishers Moscow, 1975.

[Noble, B.] Noble, B. *Applied Linear Algebra,* New Jersey: Prentice-Hall Inc., 1969.

[Pontryagin, L.S.] Pontryagin, L.S. *Ordinary Differential Equations,* Translated from the Russian by L. Kacinskas and W.B. Counts, New York: Addision-Wesley Company Inc., 1962.

[Peller, V.V.] Peller, V.V. *Hankel Operators and Their Applications,* New York: Springer Verlag, 2003.

[Petersen, K.B] Petersen, K.B, Petersen, M.K. *The Matrix Cookbook*

[Puri, N.N.] Puri, N.N. Linear Algebra, *Wiley Encyclopedia of Electrical and Electronics Engineering* Edited by John G. Webster, Vol 11, 2006.

[Smirnov, V.I.] Smirnov, V.I. *Linear Algebra and Group Theory,* Translated from the Russian and revised by R. A. Silverman, New York: McGraw Hill, 1961.

[Strang, G.] Strang, G. *Linear Algebra and its Applications,* New York: Academic press, 1976.

[Wilf, H.S.] Wilf, H.S. *Mathematics For The Physical Sciences,* New York: John Wiley and Sons, 1962.

[Wilkinson, J.H.] Wilkinson, J.H. *The Algebraic Eigenvalue Problem,* London: Oxford University Press, 1965.

[Young, N.] Young, N. *An Introduction to Hilbert Space,* London: Cambridge University Press, 1988.

Chapter 3

Ordinary Differential and Difference Equations

3.1 Introduction

Linear differential and difference equations with constant coefficients play a very important part in engineering problems. The solution of these equations is reasonably simple and most system textbooks treat the subject in a gingerly fashion. In reality, the thought process involved in the solution of these equations is of fundamental importance. The parallelism between differential and difference equations is emphasized. Matrix notation is introduced for its conciseness. The treatment of matrix differential (or difference) equations is presented here in greater detail. Furthermore, the stability of differential and difference equations has been studied via second method of Liapunov including an extensive table of various differential equations and conditions under which the systems representing these equations are stable.

3.2 System of Differential and Difference Equations

3.2.1 First Order Differential Equation Systems

Ideas developed here are later applied to higher order systems. Consider

$$\dot{x} + ax = f(t), \quad x(t)\,|_{t=0} = x_0, \quad \text{a constant} \tag{3.1}$$

where $f(t)$ is a known continuous function of time (forcing function) and x is a system response, sometimes denoted as $x(t)$. Equation 3.1 is called **linear nonhomogeneous** because the left-hand of the equation is a function of independent variable t and all the terms are linear in the dependent variable x.

Method of Solution

First consider the homogeneous equation

$$\dot{x} + ax = 0 \tag{3.2}$$

We seek a solution of the form

$$x = x(t) = e^{\lambda t}k \tag{3.3}$$

$$\dot{x} = \lambda e^{\lambda t}k = \lambda x$$

where k and λ are unknown constants. From Eqs. 3.2 and 3.3,

$$(\lambda + a)x = 0 \tag{3.4}$$

For a nontrivial solution,

$$P(\lambda) = (\lambda + a) = 0 \quad \Rightarrow \quad \lambda = -a \tag{3.5}$$

The polynomial $P(\lambda)$, is referred to as the characteristic polynomial of the homogeneous part of Eq. 3.1. Thus,

$$x(t) = e^{-at}k \qquad (f(t) \equiv 0) \tag{3.6}$$

is the solution of the homogeneous part.

The constant k is determined from the **Initial Conditions**,

$$x(0) = (e^{-at}|_{t=0})k = k \tag{3.7}$$

Thus, the solution to the homogeneous differential Eq. 3.1 is

$$\dot{x} + ax = 0 \qquad \text{is} \qquad x(t) = e^{-at}x(0) \tag{3.8}$$

The function $[e^{\lambda t}]|_{\lambda=-a}$ is called the **Fundamental Solution** $\phi(t)$ of the homogeneous differential equation. In general, a n-th order differential equation has n fundamental solutions and are determined from the roots of the polynomial $P(\lambda)$. We shall fully discuss this point later.

The Solution of the Nonhomogeneous Eq. 3.1

Let

$$x(t) = \phi(t)c(t) = e^{-at}c(t) \tag{3.9}$$

where

$\phi(t)$ = fundamental solution of the homogeneous part of the Eq. 3.1 = e^{-at}

$c(t)$ = unknown time function

This method of solution is called the **Method of Variation of Parameters** because the unknown parameter k is replaced with an unknown time function $c(t)$.

Differentiating Eq. 3.9,

$$\begin{aligned} \dot{x} &= \dot{\phi}c + \phi\dot{c} \\ &= -ax + e^{-at}\dot{c} \end{aligned} \tag{3.10}$$

Comparing Eqs. 3.1 and 3.10,

$$e^{-at}\dot{c} = f(t)$$

$$\dot{c} = e^{+at}f(t) \tag{3.11}$$

Integrating

$$c(t) = c(0) + \int_0^t e^{a\tau} f(\tau)\,\mathrm{d}\tau$$

Applying initial conditions to Eq. 3.9,

$$c(0) = x(0) \tag{3.12}$$

In summary,

$$\dot{x} + ax = f(t) \tag{3.13}$$

has a solution

$$\boxed{x(t) = e^{-at}\left[x(0) + \int_0^t e^{a\tau} f(\tau)\,\mathrm{d}\tau\right]}$$

The solution $x(t)$ is made of two parts. The first part [$e^{-at}x(0)$] is often called the Complimentary Function or Transient Response. The second part involving the forcing function under the integral sign is referred to as the Particular Integral (PI) or the Forced Function. The complete response may be thought of as the superposition of the following two responses:

1. Response due to initial conditions $x(0)$ only, with $f(t) = 0$.
2. Response due to forcing function $f(t) \neq 0$ and with initial condition $x(0) = 0$ (no initial energy).

The initial condition as an alternative can be considered as an additional forcing function. In fact, $x(t)$ can be rewritten as:

$$x(t) = \left[\int_0^t e^{-a(t-\tau)} \{x(0)\delta(\tau) + f(\tau)\}\, d\tau\right] \tag{3.14}$$

Expression (3.14) can be interpreted as a convolution of the system **impulse response**, $h(t - \tau) = e^{-a(t-\tau)}$ with the input function $[x(0)\delta(\tau) + f(t)]$.

Important Fact: The fundamental solution $\phi(t)$ of the homogeneous linear differential Eq. 3.2 satisfies the differential equation itself.

3.2.2 First Order Difference Equation

Consider the difference equation:

$$x(n + 1) + ax(n) = f(n), \quad f(0) \text{ given} \tag{3.15}$$

For the homogeneous case, $f(n) = 0$, yielding

$$x(n + 1) + ax(n) = 0 \tag{3.16}$$

Let

$$x(n) = \lambda^n k.$$

Substituting this into Eq. 3.16,

$$(\lambda + a)x(n) = 0 \tag{3.17}$$

which implies that either

$$x(n) = 0, \qquad \text{which is a trivial solution}$$

or

$$(\lambda + a) = 0, \Rightarrow \qquad \lambda = -a$$

Thus

$$x(n) = (-a)^n k, \quad k = x(0) \tag{3.18}$$

Important Fact: The fundamental solution $\phi(n) = (-a)^n$ satisfies the homogeneous Eq. 3.16

$$\phi(n + 1) + a\phi(n) = 0 \tag{3.19}$$

Solution of Eq. 3.15 :Assume

$$x(n) = \phi(n)c(n) \qquad \text{where } c(n) \text{ is an unknown function.}$$

Substituting the above expression into Eq. 3.15,

$$\phi(n + 1)c(n + 1) + a\phi(n)c(n) = f(n) \tag{3.20}$$

Define

$$c(n + 1) = c(n) + \Delta c(n)$$

Thus,

$$[\phi(n + 1) + a\phi(n)]\, c(n) + \phi(n + 1)\Delta c(n) = f(n) \tag{3.21}$$

Equation 3.21 can be further simplified via Eq. 3.19 to yield

$$\Delta c(n) = c(n+1) - c(n) = \phi^{-1}(n+1)f(n)$$

Summing $\Delta c(n)$ from 0 to k-1

$$\sum_{n=0}^{k-1} \Delta c(n) = c(k) - c(0) = \sum_{n=0}^{k-1} \left[\phi^{-1}(n+1)f(n)\right] = \sum_{i=0}^{k-1} \phi^{-1}(i+1)f(i) \tag{3.22}$$

Thus,

$$\begin{aligned} c(n) &= c(0) + \sum_{i=0}^{n-1} \phi^{-1}(i+1)f(i) \\ c(0) &= x(0) \end{aligned} \tag{3.23}$$

Thus, the solution of the nonhomogeneous Eq. 3.15

$$x(n+1) + ax(n) = f(n), \qquad f(0) \text{ given}$$

is:

$$\boxed{x(n) = \phi(n)[x(0) + \sum_{i=0}^{n-1} \phi^{-1}(i+1)f(i)]}$$

Note: There is a one-to-one parallel between the solutions of the differential and difference equations, the only difference being that integration is replaced with the Summation and vice versa.

3.2.3 n-th Order Constant Coefficient Differential Equation

Method of variation of parameters

Given

$$x^{(n)} + \sum_{i=1}^{n} a_i x^{(n-i)} = f(t) \tag{3.24}$$

$$x^{(n-i)} = \frac{d^{n-i}x}{dt^{n-i}},$$

Initial conditions $x^{(i)}(0) \quad (i = 0, 1, \ldots, n-1)$ are given.

The following steps lead to the solution of Eq. 3.24:

- **Homogeneous part:**

$$x^{(n)} + \sum_{i=1}^{n} a_i x^{(n-i)} = 0 \tag{3.25}$$

- **Characteristic equation:**

$$P(\lambda) = \lambda^n + \sum_{i=1}^{n} a_i \lambda^{n-i} = 0 \tag{3.26}$$

- **Characteristic polynomial:**

$$P(\lambda) = \lambda^n + \sum_{i=1}^{n} a_i \lambda^{n-i} \tag{3.27}$$

$$P(\lambda) = (\lambda - \lambda_1)^{r_1}(\lambda - \lambda_2)^{r_2} \ldots (\lambda - \lambda_k)^{r_k} \tag{3.28}$$

where $\lambda_1, \lambda_2, \ldots, \lambda_k$ are the roots of the characteristic equation with respective multiplicities $r_1, r_2, \ldots, r_k$ and $r_1 + r_2 + \ldots + r_k = n$.

The n fundamental solutions are:

$$\begin{array}{llll} \phi_1(t) = e^{\lambda_1 t} & \phi_{r_1+1}(t) = e^{\lambda_2 t} & \ldots & \phi_{r_1+\ldots+r_{k-1}+1}(t) = e^{\lambda_k t} \\ \phi_2(t) = te^{\lambda_1 t} & \phi_{r_1+2}(t) = te^{\lambda_2 t} & \ldots & \phi_{r_1+\ldots+2r_{k-1}}(t) = te^{\lambda_k t} \\ \vdots & \vdots & \vdots & \vdots \\ \phi_{r_1}(t) = t^{r_1-1}e^{\lambda_1 t} & \phi_{r_1+r_2}(t) = t^{r_2-1}e^{\lambda_2 t} & \ldots & \phi_{r_1+\ldots+r_k}(t) = t^{r_k-1}e^{\lambda_k t} \end{array} \tag{3.29}$$

Eigenfunctions of a Differential Operator

These linearly independent fundamental solutions $\phi_1(t), \phi_2(t), \ldots, \phi_n(t)$ of Eq. 3.25 can also be considered as the eigenfunctions of the differential operator T:

$$T = \left(\frac{d^n}{dt^n} + \sum_{i=1}^{n} a_i \frac{d^{n-i}}{dt^{n-i}} \right) \qquad \text{Linear Time Invariant Operator}$$

Thus, the solution of the equation $Tx = 0$ are the eigenfunctions of T, given by $\phi_1(t), \phi_2(t), \ldots, \phi_n(t)$, and are computed from the characteristic equation:

$$\lambda^n + \sum_{i=1}^{n} a_i \lambda^{n-i} = 0$$

In the case of the time varying differential operator:

$$T(t) = \left(\frac{d^n}{dt^n} + \sum_{i=1}^{n} a_i(t) \frac{d^{n-i}}{dt^{n-i}} \right),$$

there are still n fundamental, linearly independent solutions $\phi_1(t), \phi_2(t), \ldots, \phi_n(t)$ given by:

$$T(t)x = 0$$

But there is no general way to compute these solutions, except for lower order cases.

Important Fact: Fundamental solutions $\phi_j(t),\ j = 1, 2, \ldots, n$ satisfy the homogeneous part of differential Eq. 3.24.

$$x^{(n)} + \sum_{i=1}^{n} a_i x^{(n-i)} = f(t) = 0$$

are

$$\phi_j^{(n)}(t) + \sum_{i=1}^{n-1} a_i \phi_j^{(n-i)}(t) = 0, \quad j = 1, 2, \ldots, n \tag{3.30}$$

For the nonhomogenous Eq. 3.24, we seek a solution of the form

$$x^{(m)}(t) = \sum_{j=1}^{n} \phi_j^{(m)}(t) c_j(t), \quad m = 0, 1, \ldots, n-1, \qquad (\phi_j^{(0)}(t) = \phi_j(t)) \tag{3.31}$$

The constraints imposed by Eq. 3.31 result in having to solve the n **first order equations only**.

Differentiating Eq. 3.31,

$$x^{(m+1)}(t) = \sum_{j=1}^{n} \phi_j^{(m+1)}(t) c_j(t) + \sum_{j=1}^{n} \phi_j^{(m)}(t) \dot{c}_j(t) \tag{3.32}$$

In order for Eqs. 3.31 and 3.32 to be compatible, the unknown function $c_i(t)$ is constrained as:

$$\sum_{j=1}^{n} \phi_j^{(m)}(t) \dot{c}_j(t) = 0, \quad m = 0, 1, \ldots, n-2 \tag{3.33}$$

The last n-th equation is,

$$x^{(n)}(t) = \sum_{j=1}^{n} \phi_j^{(n)}(t) c_j(t) + \sum_{j=1}^{n} \phi_j^{(n-1)}(t) \dot{c}_j(t) \tag{3.34}$$

Substituting Eqs. 3.33 and 3.34 into 3.24,

$$\sum_{j=1}^{n}\left[\phi_j^{(n)}(t)+\sum_{i=1}^{n-1}a_i\phi_j^{(n-i)}(t)\right]c_j(t)+\sum_{j=1}^{n}\phi_j^{(n-1)}(t)\dot{c}_j(t)=f(t) \tag{3.35}$$

Realizing that the first term in the above expression vanishes due to Eq. 3.30.

The differential equations for unknown functions $c_j(t)$ are:

$$\sum_{j=1}^{n}\phi_j^{(m)}(t)\dot{c}_j(t)=0,\quad m=0,1,\ldots,n-2\quad (n-1\text{ equations.}) \tag{3.36}$$

$$\sum_{j=1}^{n}\phi_j^{(n-1)}(t)\dot{c}_j(t)=f(t)\qquad (n\text{-th equation}). \tag{3.37}$$

In the matrix notation,

$$\begin{bmatrix}\phi_1(t) & \phi_2(t) & \cdots & \phi_n(t)\\ \dot{\phi}_1(t) & \dot{\phi}_2(t) & \cdots & \dot{\phi}_n(t)\\ \vdots & \vdots & \ddots & \vdots\\ \phi_1^{(n-1)}(t) & \phi_2^{(n-1)}(t) & \cdots & \phi_n^{(n-1)}(t)\end{bmatrix}\begin{bmatrix}\dot{c}_1\\ \dot{c}_2\\ \vdots\\ \dot{c}_n\end{bmatrix}=\begin{bmatrix}0\\ 0\\ \vdots\\ f(t)\end{bmatrix} \tag{3.38}$$

From Cramer's Rule,

$$\dot{c}_j(t)=\frac{D_j(t)}{D(t)},\quad j=1,2,\ldots,n \tag{3.39}$$

where

$$D(t)=\det\begin{bmatrix}\phi_1(t) & \phi_2(t) & \cdots & \phi_n(t)\\ \dot{\phi}_1(t) & \dot{\phi}_2(t) & \cdots & \dot{\phi}_n(t)\\ \vdots & \vdots & \ddots & \vdots\\ \phi_1^{(n-1)}(t) & \phi_2^{(n-1)}(t) & \cdots & \phi_n^{(n-1)}(t)\end{bmatrix} \tag{3.40}$$

$$D_j(t) = \det \begin{bmatrix} \phi_1(t) & \cdots & 0 & \cdots & \phi_n(t) \\ \dot{\phi}_1(t) & \cdots & 0 & \cdots & \dot{\phi}_n(t) \\ \vdots & \vdots & \vdots & \cdots & \vdots \\ \phi_1^{(n-1)}(t) & \cdots & f(t) & \cdots & \phi_n^{(n-1)}(t) \end{bmatrix} \tag{3.41}$$

$$\downarrow$$

j-th Column.

The determinants $D_j(t)$ are obtained from the determinant $D(t)$ by replacing its j-th column with $[0, 0, \cdots, 0, f(t)]^T$.

Integrating Eq. 3.39,

$$c(t) = c(0) + \int_0^t \frac{D_j(\tau)}{D(\tau)} \, d\tau \tag{3.42}$$

The unknown $c(0)$ is computed from Eq. 3.31 as

$$\begin{bmatrix} \phi_1(0) & \phi_2(0) & \cdots & \phi_n(0) \\ \dot{\phi}_1(0) & \dot{\phi}_2(0) & \cdots & \dot{\phi}_n(0) \\ \vdots & \vdots & \ddots & \vdots \\ \phi_1^{(n-1)}(0) & \phi_2^{(n-1)}(0) & \cdots & \phi_n^{(n-1)}(0) \end{bmatrix} \begin{bmatrix} c_1(0) \\ c_2(0) \\ \vdots \\ c_n(0) \end{bmatrix} = \begin{bmatrix} x(0) \\ \dot{x}(0) \\ \vdots \\ x^{(n-1)}(0) \end{bmatrix} \tag{3.43}$$

Special Case

For the case of simple roots of $P(\lambda)$,

$$k = n, \qquad r_i = 1, \qquad i = 1, \ldots, n$$

$$\phi_i(t) = e^{\lambda_i t}, \ i = 1, 2, \ldots, n \tag{3.44}$$

Example 3.1:

$$\ddot{x} + 3\dot{x} + 2x = e^{-3t}, \quad x(0) = 1, \dot{x} = 0$$

$$P(\lambda) = \lambda^2 + 3\lambda + 2 = (\lambda + 1)(\lambda + 2)$$

$$\lambda = -1, \quad \lambda_2 = -2, \quad f(t) = e^{-3t}$$

$$\phi_1(t) = e^{-t}, \phi_2(t) = e^{-2t}$$

From Eq. 3.39,

$$\begin{aligned} \dot{c_1} &= \frac{-\phi_2(t)f(t)}{\phi_1(t)\dot{\phi_2}(t) - \dot{\phi_1}(t)\phi_2(t)} = e^{-2t} \\ \dot{c_2} &= \frac{-\phi_1(t)f(t)}{\phi_1(t)\dot{\phi_2}(t) - \dot{\phi_1}(t)\phi_2(t)} = -e^{-t} \end{aligned} \tag{3.45}$$

Integrating,

$$\begin{aligned} c_1(t) &= \frac{1}{2}\left(1 - e^{-2t}\right) + c_1(0) \\ c_2(t) &= -\left(1 - e^{-t}\right) + c_2(0) \end{aligned} \tag{3.46}$$

From Eq. 3.42,

$$x(0) = 1 = c_1(0) + c_2(0)$$

$$\dot{x}(0) = 0 = -c_1(0) - 2c_2(0)$$

Thus,

$$c_1(0) = 2, \quad c_2(0) = -1$$

The resulting solution for the above example is:

$$\begin{aligned} x(t) &= \phi_1(t)c_1(t) + \phi_2(t)c_2(t) \\ &= \frac{5}{2}e^{-t} - 2e^{-2t} + \frac{1}{2}e^{-3t} \end{aligned} \tag{3.47}$$

Example 3.2:

$$\ddot{x} + 2\dot{x} + x = t,$$

$$x(0) = 1, \dot{x}(0) = 0$$

$$P(\lambda) = \lambda^2 + 2\lambda + 1 = (\lambda + 1)^2$$

$$\lambda_1 = \lambda_2 = -1, \quad f(t) = t$$

$$\phi_1(t) = e^{-t}, \quad \phi_2(t) = te^{-t}$$

Following the steps discussed earlier,

$$\begin{aligned} & c_1(0) = 1, \quad c_2(0) = 1 \\ & c_1(t) = 3 - \left(t^2 - 2t + 2\right)e^t, \\ & c_2(t) = 2 + (t-1)e^t \\ & x(t) = t - 2 + 3e^{-t} + 2te^{-t} \end{aligned} \tag{3.48}$$

Exercise 3.1:

Given:

$$\ddot{x}(t) + x(t) = \tan t, \quad x(0) = A_1, \ \dot{x}(0) = A_2$$

Show that the solution is:

$$x(t) = A_1 \cos t + A_2 \sin t + \sin t - \cos t \log(\sec t + \tan t)$$

3.2.4 k-th Order Difference Equations

Same kind of reasoning developed in previous Section is used in this section for the discrete (or difference) equations.

$$x(n+k) + \sum_{i-1}^{k} a_i x(n+k-i) = f(n) \tag{3.49}$$

The initial conditions $x(i)$, $\quad i = 0, 1, \ldots, k-1$ are given.

The characteristic polynomial of the above equation is:

$$P(\lambda) = \lambda^k + \sum_{i=1}^{k} a_i \lambda^{k-i} \tag{3.50}$$

- (Later on we shall see that λ is replaced with z because of the introduction of z-transforms to solve the difference equations.)

For multiple roots, the characteristic Eq. 3.49 is given by:

$$\begin{aligned} &P(\lambda) = \lambda^k + \sum_{i=1}^{k} a_i \lambda^{k-i} = (\lambda - \lambda_1)^{r_1} \cdots (\lambda - \lambda_m)^{r_m} = 0 \\ &\text{where} \quad r_1 + r_2 + \cdots + r_m = k \end{aligned} \tag{3.51}$$

The fundamental solutions are:

$$\begin{array}{cccc} \phi_1(n) = \lambda_1^n & \phi_{r_1+1}(n) = \lambda_2^n & \cdots & \phi_{r_1+\ldots+r_{m-1}+1}(n) = \lambda_m^n \\ \phi_2(n) = n\lambda_1^n & \phi_{r_1+2}(n) = n\lambda_2^n & \cdots & \phi_{r_1+r_2+\ldots+r_{m-1}+2}(n) = n\lambda_m^n \\ \vdots & \vdots & \vdots & \vdots \\ \phi_{r_1}(n) = n^{r_1-1}\lambda_1^n & \phi_{r_1+r_2}(n) = n^{r_2-1}\lambda_2^n & \cdots & \phi_{r_1+r_2+\ldots+r_{m-1}+r_m}(n) = n^{r_m-1}e^{\lambda_m^n} \end{array}$$

Important Fact: Fundamental solutions that satisfy the homogeneous part of the difference equation.

$$x(n+k) + \sum_{i=1}^{k} a_i x(n+k-i) = 0$$

are:

$$\phi_j(n+k) + \sum_{i=1}^{k} a_i \phi_j(n+k-i) = 0, \quad j = 1, 2, \ldots, k \tag{3.52}$$

Solution of the Homogenous Eq. 3.49

Continuing with the solution of Eq. 3.49, we seek a solution of the form:

$$\boxed{x(n+k-i) = \sum_{j=1}^{k} \phi_j(n+k-i) c_j(n), \quad i = k, k-1, \ldots, 1} \tag{3.53}$$

Replacing n with $n + 1$,

$$x(n+1+k-i) = \sum_{j=1}^{k} \phi_j(n+1+k-i) c_j(n+1), \quad \text{where } i = k, k-1, \ldots, 1 \tag{3.54}$$

Replacing i with $i - 1$,

$$x(n+k-i+1) = \sum_{j=1}^{k} \phi_j(n+k-i+1) c_j(n), \quad \text{where } i = k, k-1, \ldots, 1 \tag{3.55}$$

But Eq. 3.54 can be written as:

$$x(n+1+k-i) = \sum_{i=1}^{k} \phi_j(n+1+k-i) \left[c_j(n) + \Delta c_j(n) \right], \quad i = k, k-1, \ldots, 1 \tag{3.56}$$

Comparing Eqs. 3.54 and 3.55,

$$\boxed{\sum_{j=1}^{k} \phi_j(n+1+k-i) \Delta c_j(n) = 0, \quad i = k, k-1, \ldots, 2} \tag{3.57}$$

Furthermore, for $i = 1$, in Eq. 3.54

$$\boxed{x(n+k) = \sum_{j=1}^{k} \phi_j(n+k) c_j(n) + \sum_{j=1}^{k} \phi_j(n+k) \Delta c_j(n)} \tag{3.58}$$

Substituting Eqs. 3.58 and 3.57 into 3.49,

$$\sum_{j=1}^{k}\left\{\left[\phi_j(n+k)+\sum_{i=1}^{k-1}a_i\phi_j(n+k-i)\right]c_i(n)\right\}+\sum_{j=1}^{k}\left[\phi_j(n+k)\Delta c_j(n)\right]=f(n) \quad (3.59)$$

Since the first term in Eq. 3.59 vanishes due to Eq. 3.52, Eq. 3.59 yields:

$$\sum_{j=1}^{k}\phi_j(n+k)\Delta c_j(n)=f(n) \quad (3.60)$$

Combining $(k-1)$ equations from 3.57 with 3.60 yields,

$$\begin{bmatrix}\phi_1(n+1) & \phi_2(n+1) & \cdots & \phi_k(n+1)\\ \phi_1(n+2) & \phi_2(n+2) & \cdots & \phi_k(n+2)\\ \vdots & \vdots & \ddots & \vdots\\ \phi_1(n+k) & \phi_2(n+k) & \cdots & \phi_k(n+k)\end{bmatrix}\begin{bmatrix}\Delta c_1(n)\\ \Delta c_2(n)\\ \vdots\\ \Delta c_k(n)\end{bmatrix}=\begin{bmatrix}0\\ \vdots\\ 0\\ f(n)\end{bmatrix} \quad (3.61)$$

Eq. 3.61 is solved to compute $\Delta c_i(n), \quad i=1,2,\ldots,k.$

The functions $c_i(n)$ are:

$$c_i(n)=c_i(0)+\sum_{l=0}^{n-1}\Delta c_i(n) \quad (3.62)$$

The initial values $c_i(0)$ are obtained from Eq. 3.53 for $n=0$, yielding

$$\begin{bmatrix}\phi_1(0) & \phi_2(0) & \cdots & \phi_k(0)\\ \phi_1(1) & \phi_2(1) & \cdots & \phi_k(1)\\ \vdots & \vdots & \ddots & \vdots\\ \phi_1(k-1) & \phi_2(k-1) & \cdots & \phi_k(k-1)\end{bmatrix}\begin{bmatrix}c_1(0)\\ c_2(0)\\ \vdots\\ c_k(0)\end{bmatrix}=\begin{bmatrix}x(0)\\ x(1)\\ \vdots\\ x(k-1)\end{bmatrix} \quad (3.63)$$

Computation of $c_i(n)$ along with $\phi_i(n)$ yields the resultant solution:

$$x(n)=\sum_{j=1}^{k}\phi_j(n)c_j(n)$$

This completes the solution of the n-th order nonhomogenous difference equation.

Example 3.3:

$$x(n+2) + 3x(n+1) + 2x(n) = e^{-n^2},$$
$$x(0) = 0, \quad x(1) = 0, \quad f(n) = e^{-n^2}$$
$$P(\lambda) = \lambda^2 + 3\lambda + 2 \qquad \Rightarrow \lambda_1 = -1, \lambda_2 = -2,$$
$$\phi_1(n) = (-1)^n, \quad \phi_2(n) = (-2)^n$$

From Eq. 3.61,

$$\begin{bmatrix} (-1)^{n+1} & (-2)^{n+1} \\ (-1)^{n+2} & (-2)^{n+2} \end{bmatrix} \begin{bmatrix} \Delta c_1(n) \\ \Delta c_2(n) \end{bmatrix} = \begin{bmatrix} 0 \\ f(n) \end{bmatrix} \tag{3.64}$$

From Eq. 3.63,

$$\begin{bmatrix} 1 & 1 \\ -1 & -2 \end{bmatrix} \begin{bmatrix} c_1(0) \\ c_2(0) \end{bmatrix} = \begin{bmatrix} 1 \\ 0 \end{bmatrix} \tag{3.65}$$

Using Cramer's Rule,

$$\begin{aligned} \Delta c_1(n) &= (-1)^{n+1} e^{-n^2}, \quad c_1(0) = 2 \\ \Delta c_2(n) &= \left(\frac{1}{2}\right)^{n+1} e^{-n^2}, \quad c_2(0) = -1 \end{aligned} \tag{3.66}$$

Thus,

$$\begin{aligned} c_1(n) &= 2 + \sum_{i=0}^{n-1} (-1)^{i+1} e^{-i^2} \\ c_2(n) &= -1 - \sum_{i=0}^{n-1} \left(\frac{1}{2}\right)^{i+1} e^{-i^2} \end{aligned} \tag{3.67}$$

Hence,

$$x(n) = (-1)^n \left[2 + \sum_{i=0}^{n-1} (-1)^{i+1} e^{-i^2}\right] + (-2)^n \left[-1 - \sum_{i=0}^{n-1} \left(\frac{1}{2}\right)^{i+1} e^{-i^2}\right] \tag{3.68}$$

Example 3.4:

$$x(n+2) + 2x(n+1) + x(n) = (0.5)^{-2n},$$

$$x(0) = 0, \quad x(1) = 0, \quad f(n) = (0.5)^{-2n}$$

$$P(\lambda) = \lambda^2 + 2\lambda + 1 \qquad \Rightarrow \lambda_1 = \lambda_2 = -1,$$

$$\phi_1(n) = (-1)^n, \quad \phi_2(n) = n(-1)^n$$

$$\begin{bmatrix} (-1)^{n+1} & (n+1)(-1)^{n+1} \\ (-1)^{n+2} & (n+2)(-1)^{n+2} \end{bmatrix} \begin{bmatrix} \Delta c_1(n) \\ \Delta c_2(n) \end{bmatrix} = \begin{bmatrix} 0 \\ (0.5)^{-2n} \end{bmatrix}$$

$$\begin{bmatrix} 1 & 0 \\ -1 & -1 \end{bmatrix} \begin{bmatrix} c_1(0) \\ c_2(0) \end{bmatrix} = \begin{bmatrix} 1 \\ 0 \end{bmatrix}$$

From Cramer's Rule,

$$\Delta c_1(n) = (n+1)(-4)^n, \quad c_1(0) = 1$$

$$\Delta c_2(n) = (-4)^n, \qquad c_2(0) = -1$$

Thus,

$$c_1(n) = \frac{24}{25} + \left(\frac{1}{25} + \frac{n}{5}\right)(-4)^n$$

$$c_2(n) = -\frac{4}{5} - \frac{1}{5}(-4)^n$$

Hence,

$$x(n) = \frac{1}{25}\left[4^n + (24 - 20n)(-1)^n\right] \tag{3.69}$$

At this point, we would like to introduce the compact notation of matrix algebra **All matrices are distinguished by boldface cap letters.**

3.3 Matrix Formulation of the Differential Equation

Consider the n-th order nonhomogeneous differential equation

$$x^{(n)}(t) + a_1 x^{(n-1)}(t) + \cdots + a_n x(t) = f(t), \qquad x^{(i)}(t) = \frac{d^i}{dt^i} \text{ (i-th derivative)}$$
$$x^{(i)}(0) \text{ are known}, \quad (i = 0, 1, \ldots, n-1), \tag{3.70}$$

Let

$$\left.\begin{aligned} x(t) = x^{(0)}(t) &= x_1(t) \\ \dot{x}(t) = x^{(1)}(t) = \dot{x}_1(t) &= x_2(t) \\ \ddot{x}(t) = x^{(2)}(t) = \dot{x}_2(t) &= x_3(t) \\ &\vdots \\ x^{(n-1)}(t) = \dot{x_{n-1}}(t) &= x_n(t) \end{aligned}\right\} \quad (n) \text{ equations} \tag{3.71}$$

Substituting Eqs. 3.71 into 3.70, we obtain

$$\frac{d}{dt}x_n = x^{(n)}(t) = -a_n x_1(t) - a_{n-1}x_2(t) - \cdots - a_1 x_n(t) + f(t) \tag{3.72}$$

$$\begin{bmatrix} \dot{x}_1(t) \\ \dot{x}_2(t) \\ \vdots \\ \dot{x}_{n-1}(t) \\ \dot{x}_n(t) \end{bmatrix} = \begin{bmatrix} 0 & 1 & 0 & \cdots & 0 \\ 0 & 0 & 1 & \cdots & 0 \\ \vdots & \vdots & \vdots & \vdots & \vdots \\ 0 & 0 & 0 & \cdots & 1 \\ -a_n & -a_{n-1} & -a_{n-2} & \cdots & -a_1 \end{bmatrix} \begin{bmatrix} x_1(t) \\ x_2(t) \\ \vdots \\ x_{n-1}(t) \\ x_n(t) \end{bmatrix} + \begin{bmatrix} 0 \\ 0 \\ \vdots \\ 0 \\ 1 \end{bmatrix} f(t) \tag{3.73}$$

In the matrix form, the above equation is written as

$$\dot{\boldsymbol{x}} = \boldsymbol{A}_E\boldsymbol{x} + \boldsymbol{b}f(t) \tag{3.74}$$

Matrix $\boldsymbol{A}_E$ is a special form of matrix and is called the **Companion Matrix**. All the elements of vector $\boldsymbol{b}$ except the last one are zero. The last element being unity. The function $f(t)$ is a scalar.

An n-th order system of equations with m general inputs $u_1(t), u_2(t), \ldots, u_m(t)$ (referred to as a **normal system**) can be written as:

$$\dot{\boldsymbol{x}} = \boldsymbol{A}\boldsymbol{x} + \boldsymbol{B}\boldsymbol{u}(t)$$

$$\boldsymbol{x} = \begin{bmatrix} x_1 \\ \vdots \\ x_n \end{bmatrix}, \boldsymbol{A} = \begin{bmatrix} a_{11} & \cdots & a_{1n} \\ \vdots & \ddots & \vdots \\ a_{n1} & \ldots & a_{nn} \end{bmatrix}, \boldsymbol{B} = \begin{bmatrix} b_{11} & \cdots & b_{1n} \\ \vdots & \ddots & \vdots \\ b_{m1} & \cdots & b_{mn} \end{bmatrix}, \boldsymbol{u}(t) = \begin{bmatrix} u_1(t) \\ \vdots \\ u_m(t) \end{bmatrix} \tag{3.75}$$

3.3.1 Solution of Equation $\dot{\boldsymbol{x}} = \boldsymbol{A}_E\boldsymbol{x} + \boldsymbol{b}f(t)$

Consider the equation:

$$\dot{\boldsymbol{x}} = \boldsymbol{A}_E\boldsymbol{x} + \boldsymbol{b}f(t) \tag{3.76}$$

$$\boldsymbol{A}_E = \begin{bmatrix} 0 & 1 & 0 & \cdots & 0 \\ 0 & 0 & 1 & \cdots & 0 \\ \vdots & \vdots & \ddots & \vdots & \\ 0 & 0 & 0 & \cdots & 1 \\ -a_n & -a_{n-1} & -a_{n-2} & \cdots & -a_1 \end{bmatrix}, \quad \boldsymbol{b} = \begin{bmatrix} 0 \\ 0 \\ \vdots \\ 0 \\ 1 \end{bmatrix},$$

$$P(\lambda) = |\lambda\boldsymbol{I} - \boldsymbol{A}_E| = \begin{vmatrix} \lambda & -1 & 0 & \cdots & 0 & 0 \\ 0 & \lambda & -1 & \cdots & 0 & 0 \\ \vdots & \vdots & \ddots & \ddots & \vdots & \vdots \\ 0 & 0 & 0 & \cdots & \lambda & -1 \\ a_n & a_{n-1} & a_{n-2} & \cdots & a_2 & (\lambda + a_1) \end{vmatrix} = (\lambda^n + a_1\lambda^{n-1} + \cdots + a_n)$$

where $P(\lambda)$ is the characteristic polynomial of the matrix $\boldsymbol{A}_E$.

Let $\phi_1(t), \phi_2(t), \ldots, \phi_n(t)$ be independent fundamental solutions of the equation:

$$\phi_i^{(n)}(t) + a_1\phi_i^{(n-1)}(t) + a_2\phi_i^{(n-2)}(t) + \cdots + a_n\phi_i(t) = 0, \quad \text{for } i = 1, 2, \ldots, n \tag{3.77}$$

In order to solve Eq. 3.76, assume a solution of the form:

$$\boldsymbol{x}(t) = \boldsymbol{W}(t)\boldsymbol{c}(t) \tag{3.78}$$

$$\boldsymbol{W}(t) = \begin{bmatrix} \boldsymbol{w}_1(t) \ \boldsymbol{w}_2(t) \ \cdots \ \boldsymbol{w}_n(t) \end{bmatrix} = \begin{bmatrix} \phi_1(t) & \phi_2(t) & \cdots & \phi_n(t) \\ \dot{\phi}_1(t) & \dot{\phi}_2(t) & \cdots & \dot{\phi}_n(t) \\ \vdots & \vdots & \cdots & \vdots \\ \phi_1^{(n-1)}(t) & \phi_2^{(n-1)}(t) & \cdots & \phi_n^{(n-1)}(t) \end{bmatrix}, \quad \boldsymbol{c}(t) = \begin{bmatrix} c_1(t) \\ c_2(t) \\ \vdots \\ c_n(t) \end{bmatrix}$$

The determinant of $\boldsymbol{W}(t)$ is called the Wronskian of the functions, $\phi_i(t)$'s and is nonzero because of the linear independence of the columns.

From Eqs. 3.77 and 3.78, it is clear that:

$$\dot{\boldsymbol{W}} = \boldsymbol{A}_E\boldsymbol{W} \tag{3.79}$$

Differentiating Eq. 3.78,

$$\dot{\boldsymbol{x}} = \dot{\boldsymbol{W}}\boldsymbol{c}(t) + \boldsymbol{W}\dot{\boldsymbol{c}}(t) = \boldsymbol{A}_E\boldsymbol{c}(t) + \boldsymbol{W}\dot{\boldsymbol{c}}(t) = \boldsymbol{A}_E\boldsymbol{x} + \boldsymbol{W}\dot{\boldsymbol{c}}(t) \tag{3.80}$$

Comparing Eqs. 3.80 and 3.76,

$$\boldsymbol{W}\dot{\boldsymbol{c}}(t) = \boldsymbol{b}f(t)$$

Integrating

$$\boldsymbol{c}(t) = \boldsymbol{c}(0) + \int_0^t \boldsymbol{W}^{-1}(\tau)\boldsymbol{b}f(\tau)\,\mathrm{d}\tau \tag{3.81}$$

$c(0)$ is obtained from Eq. 3.78 by substituting $t = 0$.

$$c(0) = W^{-1}(0)x(0) \tag{3.82}$$

Summarizing,

Solution of $\dot{x} = A_E x + b f(t)$, $x(0)$ known.

$\dot{W} = A_E W$, $A_E =$ **Companion Matrix**

$$x(t) = W(t)\left[W^{-1}(0)x(0) + \int_0^t W^{-1}(\tau) b f(\tau)\,d\tau\right]$$

It is easy to verify that

$$W(t)W^{-1}(\tau) = W(t-\tau)W^{-1}(0) \tag{3.83}$$

Solution of the general equation:

$$\dot{x} = Ax + Bu(t) \tag{3.84}$$

The fundamental solutions associated with a general matrix A may not be as easy to compute as for the companion matrix A_E, which can be obtained via transformation (see Chapter 7).

For the above equation, we seek a solution of the form:

$$x(t) = e^{At}c(t) \tag{3.85}$$

where

$$e^{At} = I + At + \frac{A^2t^2}{2!} + \cdots + \frac{A^kt^k}{k!}, \quad k \to \infty \tag{3.86}$$

Differentiating Eq. 3.85,

$$\dot{\boldsymbol{x}} = \boldsymbol{A}e^{\boldsymbol{A}t}\boldsymbol{c}(t) + e^{\boldsymbol{A}t}\dot{\boldsymbol{c}}(t) = \boldsymbol{A}\boldsymbol{x}(t) + e^{\boldsymbol{A}t}\dot{\boldsymbol{c}}(t) \tag{3.87}$$

Comparing Eqs. 3.84 and 3.87,

$$e^{\boldsymbol{A}t}\dot{\boldsymbol{c}}(t) = \boldsymbol{B}\boldsymbol{u}(t) \tag{3.88}$$

Integrating both sides,

$$\dot{\boldsymbol{c}}(t) = \boldsymbol{c}(0) + \int_0^t e^{-\boldsymbol{A}\tau}\boldsymbol{B}\boldsymbol{u}(\tau)\,\mathrm{d}\tau \tag{3.89}$$

Note:

$$e^{at}e^{-at} = 1 \qquad \text{(scalar)}$$

Similarly,

$$e^{\boldsymbol{A}t}e^{-\boldsymbol{A}t} = \boldsymbol{I} \qquad \left(e^{\boldsymbol{A}t}\right)^{-1} = e^{-\boldsymbol{A}t} \qquad \text{(matrix)} \tag{3.90}$$

Summarizing: The solution of the equation

$$\dot{\boldsymbol{x}} = \boldsymbol{A}\boldsymbol{x} + \boldsymbol{B}\boldsymbol{u}(t), \quad \boldsymbol{x}(0) \text{ given}$$

is:

$$\begin{aligned} \boldsymbol{x}(t) &= e^{\boldsymbol{A}t}\left[\boldsymbol{x}(0) + \int_0^t e^{-\boldsymbol{A}\tau}\boldsymbol{B}\boldsymbol{u}(\tau)\,\mathrm{d}\tau\right] \\ &= e^{\boldsymbol{A}t}\boldsymbol{x}(0) + \int_0^t e^{\boldsymbol{A}(t-\tau)}\boldsymbol{B}\boldsymbol{u}(\tau)\,\mathrm{d}\tau \end{aligned} \tag{3.91}$$

$$\left. e^{\boldsymbol{A}t}\right|_{t=0} = \boldsymbol{I}$$

Matrix $e^{\boldsymbol{A}t}$ is called the Transition Matrix and has many other properties. The computation of $e^{\boldsymbol{A}t}$ is of fundamental importance and we will discuss it in detail in this chapter as well as in other places, such as the chapters on Laplace

Transforms and the State Space. It is easy to see that if the fundamental solutions associated with a matrix can be determined then the corresponding fundamental solution matrix $\boldsymbol{W}(t)$ and the transition matrix $e^{\boldsymbol{A}t}$ are related to each other.

$$e^{\boldsymbol{A}t} = \boldsymbol{W}(t)\boldsymbol{W}^{-1}(0) \tag{3.92}$$

3.4 Matrix Formulation of the Difference Equation

Consider

$$\begin{aligned} &x(n+k) + a_1 x(n+k-1) + \cdots + a_k x(n) = f(n), \\ &x(i) \text{ is given,} \quad i = 0, \ldots, k-1 \end{aligned} \tag{3.93}$$

$$\begin{aligned} x(n+i-1) &= x_i(n) \quad i = 1, 2, \ldots, k \\ x(n+k) &= x_k(n+1) \end{aligned} \tag{3.94}$$

Substituting the above equation in Eq. 3.93,

$$x_k(n+1) = x(n+k) = -a_k x_1(n) - a_{k-1} x_2(n) - \cdots - a_1 x_k(n) + f(n) \tag{3.95}$$

$$\begin{aligned} x_1(n+1) &= x(n+1) = x_2(n) \\ x_2(n+1) &= x(n+2) = x_3(n) \\ &\vdots \\ x_{k-1}(n+1) &= x(n+k-1) = x_k(n) \\ x_k(n+1) &= -a_k x_1(n) - a_{k-1} x_2(n) - \cdots - a_1 x_k(n) + f(n) \end{aligned} \tag{3.96}$$

In matrix form, the above equations are:

$$\boldsymbol{x}(n+1) = \boldsymbol{A}_E \boldsymbol{x}(n) + \boldsymbol{b} f(n) \tag{3.97}$$

The characteristic polynomial of the matrix $\boldsymbol{A}_E$ is:

$$|\lambda \boldsymbol{I} - \boldsymbol{A}_E| = p(\lambda) = \lambda^k + a_1\lambda^{k-1} + a_2\lambda^{k-2} + \cdots + a_k \tag{3.98}$$

As discussed in Section 2.2.4, $\phi_1(n), \phi_2(n), \ldots, \phi_k(n)$ are the fundamental, linearly independent solutions depending on the roots of the characteristic polynomial $P(\lambda)$. The fundamental matrix is defined as:

$$\boldsymbol{W}(n) = \begin{bmatrix} \phi_1(n) & \phi_2(n) & \cdots & \phi_k(n) \\ \phi_1(n+1) & \phi_2(n+1) & \cdots & \phi_k(n+1) \\ \vdots & \vdots & \cdots & \vdots \\ \phi_1(n+k-1) & \phi_2(n+k-1) & \cdots & \phi_k(n+k-1) \end{bmatrix} \tag{3.99}$$

and satisfies the homogeneous part of Eq. 3.93, yielding

$$\boldsymbol{W}(n+1) = \boldsymbol{A}_E \boldsymbol{W}(n) \tag{3.100}$$

Assume the solution of Eq. 3.93 is:

$$\begin{aligned} \boldsymbol{x}(n) &= \boldsymbol{W}(n)\boldsymbol{c}(n) \\ \boldsymbol{x}(n+1) &= \boldsymbol{W}(n+1)\boldsymbol{c}(n+1) \\ &= \boldsymbol{W}(n+1)\boldsymbol{c}(n) + \boldsymbol{W}(n+1)\Delta\boldsymbol{c}(n) \end{aligned} \tag{3.101}$$

From Eq. 3.100,

$$\begin{aligned} \boldsymbol{x}(n+1) &= \boldsymbol{A}_E\boldsymbol{W}(n)\boldsymbol{c}(n) + \boldsymbol{W}(n+1)\Delta\boldsymbol{c}(n) \\ &= \boldsymbol{A}_E\boldsymbol{x}(n) + \boldsymbol{W}(n+1)\Delta\boldsymbol{c}(n) \end{aligned} \tag{3.102}$$

Comparing Eqs. 3.97 and 3.102,

$$\boldsymbol{W}(n+1)\Delta\boldsymbol{c}(n) = \boldsymbol{b}f(n) \tag{3.103}$$

or,

$$\boldsymbol{c}(n) = \boldsymbol{c}(0) + \sum_{j=0}^{n-1} \boldsymbol{W}^{-1}(j+1)\boldsymbol{b}f(j) \tag{3.104}$$

The initial value of $\boldsymbol{c}(0)$ is obtained from 3.101 as:

$$\boldsymbol{c}(0) = \boldsymbol{W}^{-1}(0)\boldsymbol{x}(0) \tag{3.105}$$

Important Fact: The solution of the equation

$$\boldsymbol{x}(n+1) = \boldsymbol{A}_E\boldsymbol{x} + \boldsymbol{b}f(n)$$

is:

$$\boldsymbol{x}(n) = \boldsymbol{W}(n)\boldsymbol{W}^{-1}(0)\boldsymbol{x}(0) + \sum_{j=0}^{n-1} \boldsymbol{W}(n)\boldsymbol{W}^{-1}(j+1)\boldsymbol{b}f(j)$$

$$\boldsymbol{W}(n+1) = \boldsymbol{A}_E\boldsymbol{W}(n)$$

$$\boldsymbol{W}(n)\boldsymbol{W}^{-1}(j+1) = \boldsymbol{W}(n-j-1)\boldsymbol{W}^{-1}(0)$$

In case we have a general matrix $\boldsymbol{A}$ instead of $\boldsymbol{A}_E$, and a vector function $\boldsymbol{f}(n)$, then the difference equation takes the form:

$$\boldsymbol{x}(n+1) = \boldsymbol{A}\boldsymbol{x}(n) + \boldsymbol{B}\boldsymbol{f}(n) \tag{3.106}$$

We assume a solution of the form: $\boldsymbol{x}(n) = \boldsymbol{A}^n\boldsymbol{c}(n)$.

Then

$$\boldsymbol{x}(n+1) = \boldsymbol{A}^{n+1}\boldsymbol{c}(n+1) = \boldsymbol{A}^{n+1}\boldsymbol{c}(n) + \boldsymbol{A}^{n+1}\Delta\boldsymbol{c}(n) = \boldsymbol{A}\boldsymbol{x}(n) + \boldsymbol{A}^{n+1}\Delta\boldsymbol{c}(n) \tag{3.107}$$

Comparing Eq. 3.106 and 3.107,

$$\boldsymbol{A}^{n+1}\Delta\boldsymbol{c}(n) = \boldsymbol{B}\boldsymbol{f}(n)$$

or

$$\boldsymbol{x}(n) = \boldsymbol{A}^n \boldsymbol{x}(0) + \sum_{j=0}^{n-1} \boldsymbol{A}^{n-j-1} \boldsymbol{B} \boldsymbol{f}(j)$$

Summarizing: The solution of the equation

$$\boldsymbol{x}(n+1) = \boldsymbol{A}\boldsymbol{x}(n) + \boldsymbol{B}\boldsymbol{f}(n), \quad \boldsymbol{x}(0) \text{ given}$$

is:

$$\boldsymbol{x}(n) = \boldsymbol{A}^n \boldsymbol{x}(0) + \sum_{j=0}^{n-1} \boldsymbol{A}^{n-j-1} \boldsymbol{B} \boldsymbol{f}(j)$$

3.5 Time Varying Linear Differential Equations

Consider

$$\dot{x}_i = \sum_{j=1}^{n} a_{ij}(t) x_j + f_i(t) \quad x_i(t_0) = x_{i0}, \quad i = 1, 2, \ldots, n \tag{3.108}$$

In matrix form,

$$\dot{\boldsymbol{x}} = \boldsymbol{A}(t)\boldsymbol{x} + \boldsymbol{f}(t), \quad \boldsymbol{x}(t_0) = \boldsymbol{x}_0 \tag{3.109}$$

Let us first study the homogeneous system

$$\dot{\boldsymbol{x}} = \boldsymbol{A}(t)\boldsymbol{x}, \quad \boldsymbol{x}(t_0) = \boldsymbol{x}_0 \tag{3.110}$$

Let $\boldsymbol{\phi}_1(t, t_0), \ldots, \boldsymbol{\phi}_n(t, t_0)$ be n fundamental (linearly independent) solutions of the Eq. 3.110. There is no systematic way to find these solutions except in simple cases. Furthermore, if the differential equation is nonlinear as well as time varying then the problem becomes harder.

The fundamental matrix formed from the fundamental solutions is:

$$\boldsymbol{\Phi}(t,t_0) = \begin{bmatrix} \boldsymbol{\phi}_1(t,t_0) & \boldsymbol{\phi}_2(t,t_0) & \cdots & \boldsymbol{\phi}_n(t,t_0) \end{bmatrix}$$

or

$$\boldsymbol{\Phi}(t,t_0) = \begin{bmatrix} \phi_{11}(t,t_0) & \phi_{12}(t,t_0) & \cdots & \phi_{1n}(t,t_0) \\ \phi_{21}(t,t_0) & \phi_{22}(t,t_0) & \cdots & \phi_{2n}(t,t_0) \\ \vdots & \vdots & \cdots & \vdots \\ \phi_{n1}(t,t_0) & \phi_{n2}(t,t_0) & \cdots & \phi_{nn}(t,t_0) \end{bmatrix} \tag{3.111}$$

The matrix $\boldsymbol{\Phi}(t,t_0)$ (**Fundamental Matrix**) satisfies the equation:

$$\dot{\boldsymbol{\Phi}}(t,t_0) = \boldsymbol{A}(t)\boldsymbol{\Phi}(t,t_0), \quad \boldsymbol{\Phi}(t_0,t_0) = \boldsymbol{I} \tag{3.112}$$

The solution of the homogenous Eq. 3.108 can be written as

$$\boldsymbol{x}(t) = \boldsymbol{\Phi}(t,t_0)\boldsymbol{x}(0) \tag{3.113}$$

Matrix $\boldsymbol{\Phi}(t,t_0)$ is nonsingular, meaning: $\det(\boldsymbol{\Phi}(t,t_0)) = |\boldsymbol{\Phi}(t,t_0)| \neq 0$.

The reader can easily verify that

$$\begin{aligned} \boldsymbol{\Phi}(t_1,t_2) &= \boldsymbol{\Phi}(t_1,t_0)\boldsymbol{\Phi}(t_0,t_2) \\ \boldsymbol{\Phi}^{-1}(t_1,t_2) &= \boldsymbol{\Phi}(t_2,t_1) \end{aligned} \tag{3.114}$$

For the solution of nonhomogenous Eq. 3.109, let

$$\begin{aligned} \boldsymbol{x}(t) &= \boldsymbol{\Phi}(t,t_0)\boldsymbol{c}(t) \\ \dot{\boldsymbol{x}} &= \dot{\boldsymbol{\Phi}}(t,t_0)\boldsymbol{c}(t) + \boldsymbol{\Phi}(t,t_0)\dot{\boldsymbol{c}}(t) = \boldsymbol{A}(t)\boldsymbol{\Phi}(t,t_0)\boldsymbol{c}(t) + \boldsymbol{\Phi}(t,t_0)\dot{\boldsymbol{c}}(t) \\ &= \boldsymbol{A}(t)\boldsymbol{x}(t) + \boldsymbol{\Phi}(t,t_0)\dot{\boldsymbol{c}}(t) \end{aligned} \tag{3.115}$$

Comparing Eqs. 3.109 and 3.115,

$$\boldsymbol{\Phi}(t, t_0)\dot{\boldsymbol{c}}(t) = \boldsymbol{f}(t)$$

Integrating,

$$\begin{aligned}\boldsymbol{c}(t) &= \boldsymbol{c}(0) + \int_{t_0}^{t} \boldsymbol{\Phi}^{-1}(\tau, t)\boldsymbol{f}(\tau)\,\mathrm{d}\tau \\ \boldsymbol{x}(t) &= \boldsymbol{\Phi}(t, t_0)\boldsymbol{x}(0) + \int_{t_0}^{t} \boldsymbol{\Phi}(t, \tau)\boldsymbol{f}(\tau)\,\mathrm{d}\tau\end{aligned} \tag{3.116}$$

For $\boldsymbol{A}(t) = \boldsymbol{A}$, a constant matrix, $\boldsymbol{\Phi}(t, t_0) = \boldsymbol{\Phi}(t - t_0) = e^{\boldsymbol{A}(t-t_0)}$ (3.117)

We can generate $\boldsymbol{\Phi}(t, t_0)$ as an iterative solution of Eq. 3.112, yielding:

$$\boldsymbol{\Phi}(t, t_0) = \boldsymbol{I} + \int_{t_0}^{t} \boldsymbol{A}(\tau)\boldsymbol{\Phi}(\tau, t_0)\,\mathrm{d}\tau \tag{3.118}$$

Reiterating,

$$\begin{aligned}\boldsymbol{\Phi}(t, t_0) = \boldsymbol{I} &+ \int_{t_0}^{t} \boldsymbol{A}(\tau_1)\,\mathrm{d}\tau_1 + \left[\int_{t_0}^{t} \boldsymbol{A}(\tau_1)\,\mathrm{d}\tau_1 \int_{t_0}^{\tau_1} \boldsymbol{A}(\tau_2)\,\mathrm{d}\tau_2\right] \\ &+ \left[\int_{t_0}^{t} \boldsymbol{A}(\tau_1)\,\mathrm{d}\tau_1 \int_{t_0}^{\tau_1} \boldsymbol{A}(\tau_2)\,\mathrm{d}\tau_2 \int_{t_0}^{\tau_3} \boldsymbol{A}(\tau_3)\,\mathrm{d}\tau_3\right] + \cdots\end{aligned} \tag{3.119}$$

- **Interesting relationship regarding $|\boldsymbol{\Phi}(t, t_0)|$**

  We shall prove:

$$\frac{\mathrm{d}}{\mathrm{d}t}|\boldsymbol{\Phi}(t, t_0)| = \operatorname{trace}(\boldsymbol{A}(t))\,|\boldsymbol{\Phi}(t, t_0)|, \quad |\boldsymbol{\Phi}(t, t_0)| = \det[\boldsymbol{\Phi}(t, t_0)]$$

$$\int_{t_0}^{t} \operatorname{trace}(\boldsymbol{A}(\tau))\,\mathrm{d}\tau = e^{|\boldsymbol{\Phi}(t,t_0)|}$$

  Furthermore, if $\boldsymbol{A}(t) = \boldsymbol{A}$, Then

$$|\boldsymbol{\Phi}(t, t_0)| = \left|e^{\boldsymbol{A}(t-t_0)}\right| = e^{(\operatorname{trace}\boldsymbol{A}(t-t_0))} \tag{3.120}$$

Proof:

Taking the derivative of $\det(\mathbf{\Phi}) = |\mathbf{\Phi}| = |\mathbf{\Phi}^T|$ yields:

$$\frac{\mathrm{d}}{\mathrm{d}t}|\mathbf{\Phi}| = \begin{vmatrix} \dot{\phi}_{11} & \dot{\phi}_{12} & \cdots & \dot{\phi}_{1n} \\ \phi_{21} & \phi_{22} & \cdots & \phi_{2n} \\ \vdots & \vdots & \cdots & \vdots \\ \phi_{n1} & \phi_{n2} & \cdots & \phi_{nn} \end{vmatrix} + \begin{vmatrix} \phi_{11} & \phi_{12} & \cdots & \phi_{1n} \\ \dot{\phi}_{21} & \dot{\phi}_{22} & \cdots & \dot{\phi}_{2n} \\ \vdots & \vdots & \cdots & \vdots \\ \phi_{n1} & \phi_{n2} & \cdots & \phi_{nn} \end{vmatrix} + \cdots + \begin{vmatrix} \phi_{11} & \phi_{12} & \cdots & \phi_{1n} \\ \phi_{21} & \phi_{22} & \cdots & \phi_{2n} \\ \vdots & \vdots & \cdots & \vdots \\ \dot{\phi}_{n1} & \dot{\phi}_{n2} & \cdots & \dot{\phi}_{nn} \end{vmatrix} \tag{3.121}$$

$$= D_1 + D_2 + \ldots + D_n$$

Consider the first determinant D_1, from Eq. 3.112:

$$\dot{\phi}_{1i} = \sum_{j=1}^{n} a_{1j}(t)\phi_{ij}$$

Substituting the above expression in D_1,

$$D_1 = \begin{vmatrix} \sum_{j=1}^{n} a_{1j}(t)\phi_{j1} & \sum_{j=1}^{n} a_{1j}(t)\phi_{j2} & \cdots & \sum_{j=1}^{n} a_{1j}(t)\phi_{jn} \\ \phi_{21} & \phi_{22} & \cdots & \phi_{2n} \\ \vdots & \vdots & \cdots & \vdots \\ \phi_{n1} & \phi_{n2} & \cdots & \phi_{nn} \end{vmatrix} \tag{3.122}$$

Multiply every element in the second row of D_1 with a_{12}, third row with a_{13}, $\cdots$, n-th row with a_{1n}, and subtracting from the first row of D_1, yielding:

$$\boxed{D_1 = \begin{vmatrix} a_{11}(t)\phi_{11} & a_{11}(t)\phi_{12} & \cdots & a_{11}(t)\phi_{1n} \\ \phi_{21} & \phi_{22} & \cdots & \phi_{2n} \\ \vdots & \vdots & \cdots & \vdots \\ \phi_{n1} & \phi_{n2} & \cdots & \phi_{nn} \end{vmatrix} = a_{11}(t)\,|\mathbf{\Phi}|} \tag{3.123}$$

All the other determinants are simplified in the same manner.

Thus,

$$D_i = a_{ii}(t)\,|\boldsymbol{\Phi}|, \quad i = 1, 2, \ldots, n.$$

Therefore,

$$\begin{aligned}\frac{\mathrm{d}}{\mathrm{d}t}|\boldsymbol{\Phi}(t)| = \sum_{i=1}^{n} D_i &= \left(\sum_{i=1}^{n} a_{ii}(t)\right)|\boldsymbol{\Phi}(t)| \\ &= (\text{Trace})\,|\boldsymbol{\Phi}(t)|\end{aligned}$$

Integrating

$$\begin{aligned}|\boldsymbol{\Phi}(t)| &= \left[e^{\int_{t_0}^{t}\sum_{i=1}^{n} a_{ii}(\tau)\,\mathrm{d}\tau}\right]|\boldsymbol{\Phi}(t_0)| \qquad (3.124)\\ &= \left[e^{\int_{t_0}^{t}\text{trace }(A(\tau))\,\mathrm{d}\tau}\right]|\boldsymbol{\Phi}(t_0)| \\ |\boldsymbol{\Phi}(t_0)| &= \boldsymbol{I}\end{aligned}$$

The various methods of solution presented above are referred to as "classical" or "time domain" techniques for the solution of ordinary linear differential and difference equations. For time invariant systems, the transform methods are more convenient because the differential and difference equations get transformed to algebraic equations, which are much easier to solve. However, these methods have one important prerequisite, namely the complex variable theory. In order to fully appreciate these transform methods and their inversion, the theory of complex variables and contour integration in the complex plane is needed. With this as the basis, the chapter on transform techniques starts with just enough complex variable theory information needed to understand the transform methods.

Summary

1. **Scalar Differential Equation**

> $$Tx(t) = f(t), \qquad T = \sum_{i=1}^{n} a_i(t)x^{(n-i)}(t), \qquad a_0(t) = 1$$
>
> The initial condition $x^{(n-i)}(0)$ and parameters $a_i(t), (i = 1, \ldots, n)$ and a function $f(t)$ are given.

- We seek a solution of the form

$$\begin{aligned}
x(t) &= [\boldsymbol{\phi}(t)]^T [\boldsymbol{c}(t)] \quad \text{scalar} \\
[\boldsymbol{\phi}(t)]^T &= [\phi_1(t),\ \phi_2(t),\ \cdots,\ \phi_n(t)] \quad \text{row vector} \\
[\boldsymbol{c}(t)]^T &= [c_1(t),\ c_2(t),\ \cdots,\ c_n(t)] \quad \text{row vector}
\end{aligned}$$

- The functions $\phi_i(t), (i = 1, 2, \ldots, n)$ are determined by: $T\phi_i(t) = 0$
- The functions $c_i(t), (i = 1, 2, \ldots, n)$ are determined by integrating:

$$\left.\begin{aligned}
\left[\boldsymbol{\phi}^{(n-i)}\right]^T [\dot{\boldsymbol{c}}] &= 0, \\
\left[\boldsymbol{\phi}^{(n-1)}\right]^T [\dot{\boldsymbol{c}}] &= f(t)
\end{aligned}\right\} \qquad (i = n, \ldots, 2)$$

$$\boldsymbol{c}(t) = \boldsymbol{c}(0) + \int_0^t \dot{\boldsymbol{c}}(t)\,\mathrm{d}t$$

and the initial conditions $\boldsymbol{c}(0)$ are computed as:

$$\boldsymbol{c}(0) = \begin{bmatrix} \left[\boldsymbol{\phi}(0)^{(0)}\right]^T \\ \vdots \\ \left[\boldsymbol{\phi}(0)^{(n-1)}\right]^T \end{bmatrix}^{-1} \begin{bmatrix} x(0) \\ \vdots \\ x^{(n-1)}(0) \end{bmatrix}$$

2. **Scalar Difference Equation**

$$Tx(n) = \sum_{i=0}^{k} a_i x(n+k-i) = f(n), \qquad a_0 = 1$$

The initial condition $x(k-i)$, parameters a_i, $(i = 1, 2, \ldots, k)$, and the forcing function $f(n)$ are given.

- We seek a solution of the form

$$x(n) = [\boldsymbol{\phi}(n)]^T\,[\boldsymbol{c}(n)] \quad \text{scalar equation}$$

$$[\boldsymbol{\phi}(n)]^T = [\phi_1(n),\ \phi_2(n),\ \cdots,\ \phi_k(n)] \quad \text{row vector}$$

$$[\boldsymbol{c}(n)]^T = [c_1(n),\ c_2(n),\ \cdots,\ c_k(n)] \quad \text{row vector}$$

- The functions $\phi_i(n)$ are determined by: $T\phi_i(n) = 0,\ (i = 1, 2, \ldots, k)$
- The functions $c_i(n), (i = 1, 2, \ldots, k)$ are computed as following:

$$\left.\begin{aligned} [\boldsymbol{\phi}(n+i-1)]^T\,[\Delta\boldsymbol{c}(n)] &= 0 \\ [\boldsymbol{\phi}(n+k)]^T\,[\Delta\boldsymbol{c}(n)] &= f(n) \end{aligned}\right\}(i = k, \ldots, 2)$$

$$\boldsymbol{c}(n) = \boldsymbol{c}(0) + \sum_{j=1}^{n} \Delta\boldsymbol{c}(j)$$

$$\boldsymbol{c}(0) = \begin{bmatrix} [\boldsymbol{\phi}(0)]^T \\ [\boldsymbol{\phi}(1)]^T \\ \vdots \\ [\boldsymbol{\phi}(k-1)]^T \end{bmatrix}^{-1} \begin{bmatrix} x(0) \\ x(1) \\ \vdots \\ x(k-1) \end{bmatrix}$$

3.6 Computing e^{At}, A^N, $f(A)$ without Determination of Eigenvectors

In the section on Matrix Algebra in Chapter 2, we discussed how to compute the different functions of the matrix $\boldsymbol{A}$ such as e^{At}, $\boldsymbol{A}^N$, and polynomial functions of the matrix $\boldsymbol{A}$. Such a task was accomplished via Similarity Transformations, which invariably require the computation of the eigenvectors (and the eigenvalues). The computation of the eigenvectors, is rather a tedious task. The following method is suggested by Elaydi and Harris [Elaydi, S.N.], which is based on the "linearly independent fundamental functions" of the matrix $\boldsymbol{A}$ computed in a recursive manner.

(a) **Algorithm for computation of e^{At}, $\boldsymbol{A}$ is nonsingular $k \times k$ matrix**

There are many methods for computing e^{At}. Suggested below is one of the methods which requires the following four steps for the computation of e^{At}

(i) Compute the eigenvalues $\lambda_1, \lambda_2, \cdots, \lambda_k$, counting their algebraic multiplicity, as roots of the polynomial $p(\lambda)$:

$$p(\lambda) = \det(\lambda \boldsymbol{I} - \boldsymbol{A}) = \lambda^k + a_1\lambda^{k-1} + \cdots + a_k = \prod_{i=1}^{k}(\lambda - \lambda_i)$$

(ii) Compute the recursive functions $\alpha_j(t)$:

$$\begin{aligned}
\dot{\alpha}_1(t) &= \lambda_1\alpha_1(t) & \alpha_1(0) &= 1 \\
\dot{\alpha}_2(t) &= \lambda_2\alpha_2(t) + \alpha_1(t) & \alpha_2(0) &= 0 \\
&\vdots \\
\dot{\alpha}_k(t) &= \lambda_k\alpha_k(t) + \alpha_{k-1}(t) & \alpha_k(0) &= 0
\end{aligned} \tag{3.125}$$

Recursive solutions of these equations yield:

$$\alpha_1(t) = e^{\lambda_1(t)}$$

$$\alpha_j(t) = e^{\lambda_j(t)}\left[\int_0^t e^{-\lambda_j\tau}\alpha_{j-1}(\tau)\,\mathrm{d}\tau\right], \qquad j = 2,\cdots,k$$

(iii) Compute the matrices $\boldsymbol{M}(j)$ recursively:

$$\boldsymbol{M}(j) = (\lambda_j\boldsymbol{I} - \boldsymbol{A})\boldsymbol{M}(j-1), \qquad \boldsymbol{M}(0) = \boldsymbol{I},\; j = 1,2,\cdots,k \tag{3.126}$$

or $\quad \boldsymbol{M}(j) = \prod_{i=1}^{j}(\lambda_i\boldsymbol{I} - \boldsymbol{A}), \quad \boldsymbol{M}(k) = \boldsymbol{0} \qquad$ Cayley-Hamilton Theorem

(iv) Compute $e^{\boldsymbol{A}t}$

$$e^{\boldsymbol{A}t} = \sum_{j=0}^{k-1}\alpha_{j+1}(t)\boldsymbol{M}(j) \tag{3.127}$$

Proof:

If Eq. 3.127 is true, then

$$\boldsymbol{S} = \left[\frac{\mathrm{d}}{\mathrm{d}t}\left(e^{\boldsymbol{A}t}\right) - \boldsymbol{A}e^{\boldsymbol{A}t}\right] = \sum_{j=0}^{k-1}\left[\dot{\alpha}_{j+1}(t)\boldsymbol{M}(j) - \alpha_{j+1}(t)\boldsymbol{A}\boldsymbol{M}(j)\right]$$

Also from Eq. 3.126, by substituting $j+1$ for j:

$$\boldsymbol{A}\boldsymbol{M}(j) = \lambda_{i+1}\boldsymbol{M}(j) - \boldsymbol{M}(j+1)$$

Thus,

$$\boldsymbol{S} = \sum_{j=0}^{k-1}\left(\dot{\alpha}_{j+1}(t) - \lambda_{j+1}\right)\boldsymbol{M}(j) - \sum_{j=0}^{k-1}\alpha_{j+1}(t)\boldsymbol{M}(j+1) \tag{3.128}$$

Now

$$\sum_{j=0}^{k-1} \alpha_{j+1}(t) \boldsymbol{M}(j+1) = \sum_{j=1}^{k-1} \alpha_j(t) \boldsymbol{M}(j), \quad \boldsymbol{M}(k) = 0$$

Hence,

$$\begin{aligned} \boldsymbol{S} &= (\dot{\alpha}_1(t) - \lambda_1 \alpha_1(t))\, \boldsymbol{M}(0) \\ &\quad + \sum_{j=1}^{k-1} \left(\dot{\alpha}_{j+1}(t) - \lambda_{j+1} \alpha_{j+1}(t) - \alpha_j(t) \right) \boldsymbol{M}(j) \end{aligned} \tag{3.129}$$

From Eqs. 3.125 and 3.129, each term in the bracket on the left-hand side vanishes, resulting in: $\boldsymbol{S} = \boldsymbol{0}$ which is true, justifying the algorithm.

(b) **Algorithm for computing $\boldsymbol{A}^N$, $\boldsymbol{A}$ is nonsingular $k \times k$ matrix, $N \geq k$**

Following the same line of reasoning as for e^{At}, with the only difference that we are dealing with is the discrete rather than continuous functions, we state the following algorithm:

(i) Compute the eigenvalues $\lambda_1, \lambda_2, \cdots, \lambda_k$, counting their algebraic multiplicity, as the roots of the characteristic polynomial $P(\lambda)$:

$$P(\lambda) = \det(\lambda \boldsymbol{I} - \boldsymbol{A}) = \lambda^k + a_1 \lambda^{k-1} + \cdots + a_n = \prod_{i=1}^{k} (\lambda - \lambda_i) \tag{3.130}$$

(ii) Compute the matrices in a recursive manner:

$$\boldsymbol{M}(j) = \prod_{i=1}^{j} (\lambda_i \boldsymbol{I} - \boldsymbol{A}), \quad \boldsymbol{M}(0) = \boldsymbol{I}, \quad j = 1, 2, \cdots, k-1$$

$$\boldsymbol{M}(k) = \prod_{i=1}^{k} (\lambda_i \boldsymbol{I} - \boldsymbol{A}) = \boldsymbol{0} \qquad \text{Cayley-Hamilton Theorem}$$

(iii) Compute the fundamental recursive functions

$$
\begin{aligned}
&\alpha_1(n+1) = \lambda_1\alpha_1(n) && \alpha_1(0) = 1 \\
&\alpha_2(n+1) = \lambda_2\alpha_2(n) + \alpha_1(n) && \alpha_2(0) = 0 \\
&\vdots \\
&\alpha_k(n+1) = \lambda_k\alpha_k(n) + \alpha_{k-1}(n) && \alpha_k(0) = 0
\end{aligned}
\tag{3.131}
$$

The resulting functions are:

$$
\left.
\begin{aligned}
&\alpha_1(N) = \lambda_1^N \\
&\alpha_{j+1}(N) = \sum_{i=0}^{N-1}\left[\left(\lambda_{j+1}\right)^{N-i+1}\alpha_j(i)\right], \quad j = 1,\cdots,k-1 \\
&\alpha_{j+1}(0) = 0
\end{aligned}
\right\}
\tag{3.132}
$$

(iv) For $N \geq k$, an integer

$$
\boldsymbol{A}^N = \sum_{j=0}^{k-1}\alpha_{j+1}(N)\boldsymbol{M}(j)
$$

The proof is exactly the same as for the computation of e^{At}.

(c) **Algorithm for computation of e^{At} via generalized eigenvectors of A**

- Let $\boldsymbol{A}$ be real $k \times k$ matrix
- Define

$$
\begin{aligned}
\boldsymbol{A}(\lambda) &= (\lambda\boldsymbol{I} - \boldsymbol{A}) \\
\det\boldsymbol{A}(\lambda) &= P(\lambda) = \prod_{i=1}^{m}(\lambda-\lambda_i)^{r_i}, \qquad \sum_{i=1}^{m} r_i = k
\end{aligned}
$$

The λ_i are k eigenvalues, counting their algebraic multiplicity. These eigenvalues may be real or complex conjugate pairs.

- For a real eigenvalue λ_i of multiplicity r_i, there exists r_i real linearly independent cigenvectors $\boldsymbol{u}_1(i)$, $\boldsymbol{u}_2(i), \cdots, \boldsymbol{u}_{r_i}(i)$.

  For a complex conjugate pair of eigenvalues λ_j, λ_{j+1}, there exists $2r_j \left(r_j = r_{j+1}\right)$ linearly independent eigenvectors $\boldsymbol{u}_1(j)$, $\boldsymbol{u}_2(j), \cdots, \boldsymbol{u}_{r_j}(j)$ and their complex conjugates $\bar{\boldsymbol{u}}_1(j)$, $\bar{\boldsymbol{u}}_2(j), \cdots, \bar{\boldsymbol{u}}_{r_j}(j)$. Each of these vectors are called **Generalized Eigenvectors of** $\boldsymbol{A}$. Following two steps result in computation of e^{At}

(i) Computation of **Generalized Eigenvectors of** $\boldsymbol{A}$

$$\left.\begin{aligned} (\boldsymbol{A} - \lambda_i \boldsymbol{I})^{r_i}\, \boldsymbol{u}_{r_i}(i) &= 0 \\ (\boldsymbol{A} - \lambda_i \boldsymbol{I})^{r_i - 1}\, \boldsymbol{u}_{r_i}(i) = \boldsymbol{u}_{r_i - 1}(i) &\neq 0 \\ &\vdots \\ (\boldsymbol{A} - \lambda_i \boldsymbol{I})\, \boldsymbol{u}_{r_i}(i) = \boldsymbol{u}_1(i) &\neq 0 \end{aligned}\right\} \quad i = 1, 2, \ldots, m$$

Same procedure applies to real or complex eigenvalues and corresponding real or complex eigenvectors. We need to choose $\boldsymbol{u}_1(i)$ carefully so that subsequent vectors are nonzero.

(ii) Fundamental solution of e^{At} (via generalized eigenvcctors)

   (a) Case1: Eigenvalues of $\boldsymbol{A}$ are real

   Let λ_i be the real eigenvalue of multiplicity r_i. We have already computed real generalized eigenvectors $\boldsymbol{u}_1(i)$, $\boldsymbol{u}_2(i), \cdots, \boldsymbol{u}_{r_i}(i)$. Corresponding to each real generalized eigenvectors, we obtain a real fundamental solution vector, $\boldsymbol{\phi}$, for e^{At}, given by:

$$e^{At}\boldsymbol{u}_p(i) = e^{\lambda_i t}\left[e^{(A - \lambda_i t)}\right]\boldsymbol{u}_p(i) = \boldsymbol{\phi}_p(i, t)$$

Thus,

$$\boldsymbol{\phi}_p(i,t) = e^{\lambda_i t}\left[\sum_{n=0}^{r_i-1} (\boldsymbol{A} - \lambda_i \boldsymbol{I})^n \frac{t^n}{n!}\right]\boldsymbol{u}_p(i)$$

$$p = 1, 2, \ldots r_i$$

Note that as a result of the definition of the vectors $\boldsymbol{u}_p(i)$

$$e^{(\boldsymbol{A}-\lambda_i)t}\boldsymbol{u}_p(i) = \left[\sum_{n=0}^{r_i-1} (\boldsymbol{A} - \lambda_i \boldsymbol{I})^n \frac{t^n}{n!}\right]\boldsymbol{u}_p(i)$$

The terms of order higher than $(r_i - 1)$ in this series will contribute to zero (Cayley-Hamilton Theorem).

The generated fundamental solutions for the real eigenvalues (and real eigenvectors) are:

$$\begin{aligned}
\boldsymbol{\phi}_1(i,t) &= e^{\lambda_i t}\left[\boldsymbol{u}_1(i) + t\boldsymbol{u}_2(i) + \cdots + \frac{t^{r_i-1}}{(r_i-1)!}\boldsymbol{u}_{r_i}(i)\right] \\
\boldsymbol{\phi}_2(i,t) &= e^{\lambda_i t}\left[\qquad\quad \boldsymbol{u}_2(i) + \cdots + \frac{t^{r_i-2}}{(r_i-1)!}\boldsymbol{u}_{r_i}(i)\right] \\
&\vdots \\
\boldsymbol{\phi}_{r_i}(i,t) &= e^{\lambda_i t}\left[\qquad\qquad\qquad\qquad \boldsymbol{u}_{r_i}(i)\right]
\end{aligned}$$

The above equations can be represented as:

$$\left[\boldsymbol{\phi}_1(i,t)\ \boldsymbol{\phi}_2(i,t)\ \cdots\ \boldsymbol{\phi}_{r_i}(i,t)\right] = \left[\boldsymbol{u}_1(i)\ \boldsymbol{u}_2(i)\ \cdots\ \boldsymbol{u}_{r_i}(i)\right]\left[e^{\boldsymbol{J}_i t}\right]$$

(b) Case2: Eigenvalues of $\boldsymbol{A}$ are Complex Conjugate pairs.

Let λ_j and $\lambda_{j+1} = \bar{\lambda}_j$ be a complex conjugate pair of eigenvalues with multiplicity $r_j = r_{j+1}$. Just as in real cases, let us construct r_j, complex fundamental vector solutions as following:

$$\boldsymbol{\psi}_l(j,t) = e^{\lambda_j t}\left[e^{(\boldsymbol{A}-\lambda_j \boldsymbol{I})t}\,\boldsymbol{u}_l(j)\right], \qquad l = 1,2,\ldots,r_j$$

The resulting $2r_j$ real fundamental solution vectors are:

$$\left.\begin{aligned} \text{Re }(\boldsymbol{\psi}_l(j,t)) &= \boldsymbol{\phi}_l(j,t) \\ \text{Im }(\boldsymbol{\psi}_l(j,t)) &= \boldsymbol{\phi}_{l+r_j}(j,t) \end{aligned}\right\} \quad , \; l = 1,2,\ldots,r_j$$

(c) Combining the real and complex conjugate fundamental solution vectors in lexical order:

$$\begin{aligned} e^{At}\begin{bmatrix}\boldsymbol{u}_1 & \boldsymbol{u}_2 & \cdots & \boldsymbol{u}_k\end{bmatrix} &= \begin{bmatrix}\boldsymbol{\phi}_1(t) & \boldsymbol{\phi}_2(t) & \cdots & \boldsymbol{\phi}_k(t)\end{bmatrix} \\ &= \begin{bmatrix}\boldsymbol{u}_1 & \boldsymbol{u}_2 & \cdots & \boldsymbol{u}_k\end{bmatrix}\begin{bmatrix}e^{Jt}\end{bmatrix} \end{aligned}$$

where $\boldsymbol{J}$ is known as the Jordan Canonical form. The model matrix, formed by column vectors $\boldsymbol{u}_1, \boldsymbol{u}_2, \cdots, \boldsymbol{u}_k$, known as Similarity Transformation.

In fact:

$$\begin{bmatrix}A\end{bmatrix}\begin{bmatrix}\boldsymbol{u}_1 & \boldsymbol{u}_2 & \cdots & \boldsymbol{u}_k\end{bmatrix} = \begin{bmatrix}\boldsymbol{u}_1 & \boldsymbol{u}_2 & \cdots & \boldsymbol{u}_k\end{bmatrix}[\boldsymbol{J}]$$

$$\begin{bmatrix}e^{At}\end{bmatrix}\begin{bmatrix}\boldsymbol{u}_1 & \boldsymbol{u}_2 & \cdots & \boldsymbol{u}_k\end{bmatrix} = \begin{bmatrix}\boldsymbol{u}_1 & \boldsymbol{u}_2 & \cdots & \boldsymbol{u}_k\end{bmatrix}\begin{bmatrix}e^{Jt}\end{bmatrix}$$

$$\boldsymbol{J} = \begin{bmatrix} \boldsymbol{J}_1 & & & \\ & \boldsymbol{J}_2 & & \\ & & \ddots & \\ & & & \boldsymbol{J}_m \end{bmatrix}$$

As a special case when all the eigenvalues are distinct, $\boldsymbol{J}$ takes the form $\boldsymbol{\Lambda}$ where:

$$\boldsymbol{\Lambda} = \begin{bmatrix} \lambda_1 & & & \\ & \lambda_2 & & \\ & & \ddots & \\ & & & \lambda_k \end{bmatrix}, \qquad e^{\boldsymbol{\Lambda} t} = \begin{bmatrix} e^{\lambda_1 t} & & & \\ & e^{\lambda_2 t} & & \\ & & \ddots & \\ & & & e^{\lambda_k t} \end{bmatrix}$$

3.7 Stability of Autonomous Differential Equations

Consider a n-dimensional differential equation:

$$\dot{\boldsymbol{y}}(t) = \boldsymbol{g}(\boldsymbol{y}(t)) \tag{3.133}$$

If at any point $\boldsymbol{y}^*$,

$$\boldsymbol{g}(\boldsymbol{y}^*) = \boldsymbol{0}$$

Then $\boldsymbol{y}^*$ is called the **Equilibrium Point** or the singular points of Eq. 3.133. There may be many equilibrium points $\boldsymbol{y}^*$ for which

$$\boldsymbol{g}(\boldsymbol{y}^*) = \boldsymbol{0} \text{ and } \dot{\boldsymbol{y}}^* = 0$$

For a particular $\boldsymbol{y}^*$, define

$$\boldsymbol{x}(t) = \boldsymbol{y}(t) - \boldsymbol{y}^*, \qquad \text{Perturbation Equation}$$

The perturbation equations for the system Eq. 3.133 are:

$$\dot{\boldsymbol{x}}(t) = \dot{\boldsymbol{y}}(t) = \boldsymbol{g}(\boldsymbol{y}(t)) = \boldsymbol{g}(\boldsymbol{x}(t) + \boldsymbol{y}^*) = \boldsymbol{f}(\boldsymbol{x}(t)) \qquad \boldsymbol{f}(\boldsymbol{0}) = \boldsymbol{0} \tag{3.134}$$

Equation 3.134 is known as having a null equilibrium solution.

For each different equilibrium point of Eq. 3.133, we have different perturbed system equations. Each of these perturbed systems have a **Null Equilibrium** solutions given by

$$\dot{\boldsymbol{x}}(t) = \boldsymbol{f}(\boldsymbol{x}(t)), \quad \boldsymbol{f}(\boldsymbol{0}) = \boldsymbol{0}, \quad \boldsymbol{x}(t_0) = \boldsymbol{x}_0$$

We like to emphasize the fact that most dynamical systems are described by their null equilibrium solution perturbation equations, particularly for studying stability. **For each equilibrium point there is a different perturbation differential equation**. In what follows, we shall study the stability of a perturbed, null equilibrium, differential equation

$$\dot{\boldsymbol{x}}(t) = \boldsymbol{f}(\boldsymbol{x}(t)), \quad \boldsymbol{f}(\boldsymbol{0}) = \boldsymbol{0}, \quad \boldsymbol{x}(t_0) = \boldsymbol{x}_0 \tag{3.135}$$

Various Definitions of Stability:

- The null equilibrium point of Eq. 3.135 is **stable**,

  $$\text{If for each } \epsilon > 0, \text{ there exists a } \delta(\epsilon) > 0 \text{ such that: } \|\boldsymbol{x}(t_0)\| < \delta(\epsilon)$$

  $$\text{then} \quad \|\boldsymbol{x}(t) - \boldsymbol{x}(t_0)\| < \epsilon \quad \text{for all t} > t_0$$

- If the system Eq. 3.135 is stable **and furthermore,**

  $$\lim_{t \to \infty} x(t) = 0$$

  then the system is called **Asymptotically Stable**.

  These definitions are in the sense of "Liapunov" and play an important part in the control systems study. The notion of Asymptotic Stability is more stringent than the Stability Only and are illustrated via figure 3.1.

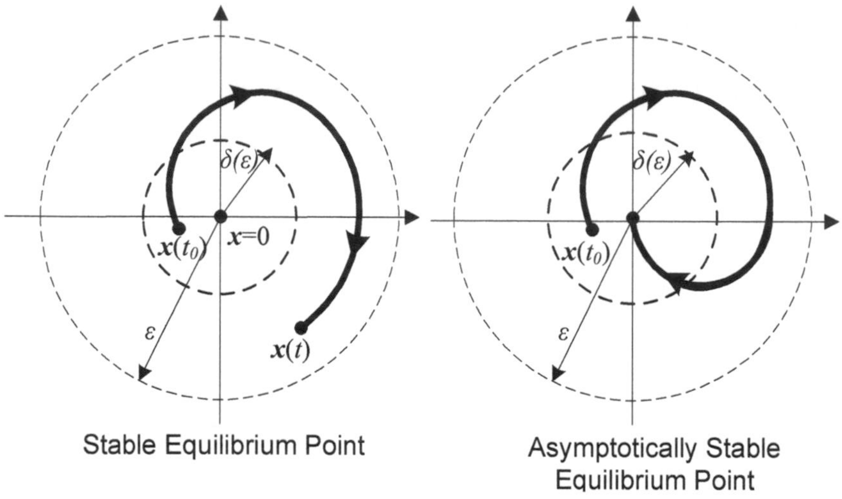

Figure 3.1: Stable and Asymptotically Stable Equilibrium Points

- The null equilibrium solution of the linearized version of Eq. 3.135, can be written as:

$$\dot{\boldsymbol{x}} = \boldsymbol{A}\boldsymbol{x},$$

This solution is **Liapunov Stable** and **Liapunov Asymptotically Stable** if there exists a constant K and a positive number α, such that:

$$||e^{\boldsymbol{A}t}\boldsymbol{x}(0)|| \leq Ke^{-\alpha t}||\boldsymbol{x}(0)|| \quad \text{for all} \quad t \geq 0\,, \boldsymbol{x}(t)|_{t=0} = \boldsymbol{x}(0)$$

- **Periodic Solutions and Limit Cycles**

Except for harmonic equations, a periodic solution of a differential equation is a result of nonlinearities present in the system. Consider

$$\dot{\boldsymbol{x}} = f(t, \boldsymbol{x})\,, \quad \boldsymbol{x}(t_0) = \boldsymbol{x}(0)$$

with a **Periodic Solution** $\varphi(t, t_0)$. This solution is **Orbitally Stable** if for each $\epsilon > 0,\ t \geq t_0 \geq 0$ there exists a $\delta(\epsilon, t_0) > 0$ such that

$$\|\boldsymbol{x}(t_0) - \boldsymbol{\varphi}(t_0)\| \leq \delta(\epsilon, t_0) \text{ implies } \|\boldsymbol{x}(t, t_0; \boldsymbol{x}(t_0)) - \boldsymbol{\varphi}(t, t_0)\| < \epsilon$$

Poincare-Benedixon Theory: Limit Cycle Theorem

If for $t \geq t_0$, a trajectory of the system $\dot{\boldsymbol{x}} = \boldsymbol{f}(t, \boldsymbol{x})$ does not approach any of its singular points (where $\boldsymbol{f}(t, \boldsymbol{x}) = \boldsymbol{0}$) and $\|\boldsymbol{x}(t)\| < \infty$ (bounded), then it is either a limit cycle or reaches a limit cycle as $t \to \infty$.

Benedixon Limit Cycle Criterion (Autonomous Case)

Consider a two-dimensional nonlinear system:

$$\begin{aligned} \dot{x}_1 &= f_1(x_1, x_2) \\ \dot{x}_2 &= f_2(x_1, x_2) \end{aligned} \qquad \text{or} \qquad \dot{\boldsymbol{x}} = \boldsymbol{f}(\boldsymbol{x})$$

D be simply connected region in (x_1, x_2) plane and $\boldsymbol{f}(\boldsymbol{x})$ is continously differentiable in D. The following test confirms the **absence of a limit cycle** inside the region D for this nonlinear system.

Theorem

$$\text{Let} \qquad I = \text{div}_{\boldsymbol{x}} \cdot \boldsymbol{f} = (\nabla_{\boldsymbol{x}})^T \boldsymbol{f} = \begin{bmatrix} \partial/\partial x_1 & \partial/\partial x_2 \end{bmatrix} \begin{bmatrix} f_1(x_1, x_2) \\ f_2(x_1, x_2) \end{bmatrix} = \left(\frac{\partial f_1}{\partial x_1} + \frac{\partial f_2}{\partial x_2} \right) \neq 0 \tag{3.136}$$

If I is of the same sign throughout the region D, then no limit cycle can exist in D.

Proof:

$$\begin{aligned} \dot{x}_1 &= f_1(x_1, x_2) \\ \dot{x}_2 &= f_2(x_1, x_2) \end{aligned} \qquad \text{or} \qquad \frac{\mathrm{d}x_1}{\mathrm{d}x_2} = \frac{f_1(x_1, x_2)}{f_2(x_1, x_2)}$$

Thus,

$$f_1(x_1, x_2)\,dx_2 - f_2(x_1, x_2)\,dx_1 = 0$$

or

$$\oint_D [f_1(x_1, x_2)\,dx_2 - f_2(x_1, x_2)\,dx_1] = 0$$

From Green's Theorem:

$$\begin{aligned}\oint_D (f_1(x_1, x_2)\,dx_2 - f_2(x_1, x_2)\,dx_1) &= \iint_0 \left(\frac{\partial f_1}{\partial x_1} + \frac{\partial f_2}{\partial x_2}\right) dx_1\,dx_2 \\ &= \iint_0 (\text{div}_x \cdot \boldsymbol{f})\,dx_1\,dx_2 = 0\end{aligned}$$

This is only possible if $\text{div}_x \cdot \boldsymbol{f} = 0$ otherwise a change of sign occurs in the domain *D*. This only means that there are no limit cycles of the system Eq. 3.136 inside the region *D*. **This does not mean that there are no limit cycles outside the region *D* for the system Eq. 3.136**. A further consequence of the above result is that if a trajectory remains inside a domain and does not approach a critical point (singular), then the trajectory is a stable limit cycle or approaches a stable limit cycle.We shall conclude this discussion with the famous van der Pol equation.

Example 3.5:

The van der Pol differential equation is the forerunner of most of the nonlinear studies, particularly a damped harmonic oscillator, stated as:

$$\ddot{x} - 2\epsilon\omega_0\left(1 - \beta x^2\right)\dot{x} + \omega_0^2 x = 0\,, \quad \beta > 0, \epsilon > 0$$

1. When $\beta = 0$,

   This is the equation of an unstable oscillator.

2. When $\beta \neq 0$,

For any positive value of β, however small, the system is still negatively damped as long as $\beta x^2 \leq 1$. As x further increases, the damping becomes positive resulting in the reduction of the amplitude of x until βx^2 again becomes less than or equal to 1 at which point again the damping becomes negative resulting in an oscillation. The state variable form of the van der Pol equation takes the form

$$\dot{x}_1 = x_2 \quad = f_1(x_1, x_2)$$

$$\dot{x}_2 = 2\epsilon\omega_0\left(1 - \beta^2 x_1\right)x_2 - \omega_0^2 x_1 = f_2(x_1, x_2)\,, \quad \beta \neq 0$$

$$(\nabla_x)^T f = \frac{\partial}{\partial x_1}(f_1) + \frac{\partial}{\partial x_2}(f_2) = 2\epsilon\omega_0\left(1 - \beta x_1^2\right)$$

$$\text{if } |x_1| < (\beta)^{-1/2} \quad \text{then} \quad (\nabla_x)^T f > 0$$

Hence, for $-\beta^{-1/2} < |x| < \beta^{-1/2}$ there can be no limit cycle.

Indeed for other postive values of β, there will be a limit cycle. On the other hand, for negative values of β the system is Liapunov unstable.

- **Classfication of Singular Points (Critical Points)**

This classification is useful in fluid dynamics. Let us study a two-dimensional system behavior about the neighborhood of its stationary solutions represented by its singular points from a phase-portrait point of view.

$$\begin{bmatrix} \dot{x}_1 \\ \dot{x}_2 \end{bmatrix} = \begin{bmatrix} a_{11} & a_{12} \\ a_{21} & a_{22} \end{bmatrix} \begin{bmatrix} x_1 \\ x_2 \end{bmatrix} \tag{3.137}$$

$$\dot{\boldsymbol{x}} = \boldsymbol{A}\boldsymbol{x}\,, \quad \det\boldsymbol{A} \neq 0\,, \quad \boldsymbol{A} \text{ real nonsingular matrix}$$

$$\boldsymbol{0} = \boldsymbol{A}\boldsymbol{x} \quad \text{at the equilibrium point}$$

$$\boldsymbol{x} = \boldsymbol{A}^{-1}\boldsymbol{0} = \boldsymbol{0} \text{ is the only equilibrium point.}$$

Let $\boldsymbol{T}$ be a nonsingular transformation such that:

$$\hat{\boldsymbol{A}} = \boldsymbol{T}^{-1}\boldsymbol{A}\boldsymbol{T}, \quad \boldsymbol{y} = \boldsymbol{T}\boldsymbol{x}$$

$$\dot{\boldsymbol{y}} = \hat{\boldsymbol{A}}\boldsymbol{y}$$

This linear transformation does not alter the **qualitative behavior** of system Eq. 3.137 about the critical points. Hence, we shall study the transformed equations in Eq. 3.137. The characteristic polynomial of $\boldsymbol{A}$ or $\hat{\boldsymbol{A}}$ is:

$$\begin{aligned} P(\lambda) &= \det[\lambda \boldsymbol{I} - \boldsymbol{A}] = \det\left(\lambda \boldsymbol{I} - \boldsymbol{T}\hat{\boldsymbol{A}}\boldsymbol{T}^{-1}\right) = \det\left(\lambda \boldsymbol{T}\boldsymbol{T}^{-1} - \boldsymbol{T}\hat{\boldsymbol{A}}\boldsymbol{T}^{-1}\right) \\ &= \det\left(\lambda \boldsymbol{I} - \hat{\boldsymbol{A}}\right) = \lambda^2 + a_1\lambda + a_2 \end{aligned}$$

$$P(\lambda) = (\lambda - \lambda_1)(\lambda - \lambda_2), \quad -(\lambda_1 + \lambda_2) = a_1, \quad \lambda_1\lambda_2 = a_2$$

Classification of Singular Points

Depending upon a_1 and a_2, the transformed matrix $\hat{\boldsymbol{A}}$ takes the following four forms and yields the system behavior about the critical points (singular points) and their classification. The critical points are classified as:

(i) Node (ii) Saddle points (iii) Focus points (iv) Center points

(i) **Node**

The equilibrium point looks like a "node," which acts as an "attractor" or "repeller" of the trajectories. In general, there are three classifications of nodes, depending upon the eigenvalues of the matrix $\boldsymbol{A}$ or $\hat{\boldsymbol{A}}$.

(a)

$$\hat{\boldsymbol{A}} = \begin{bmatrix} \lambda_1 & 0 \\ 0 & \lambda_2 \end{bmatrix},$$

$$p(\lambda) = \lambda^2 + a_1\lambda + a_2$$

$\lambda_1 \neq \lambda_2$ and both real λ_1 and λ_2 have the same sign.

$$\dot{y}_1 = \lambda_1 y_1, \quad y_1(t) = K_1 e^{\lambda_1 t}$$

$$\dot{y}_2 = \lambda_2 y_2, \quad y_2(t) = K_2 e^{\lambda_2 t}$$

$\lambda_1 > 0, \lambda_2 > 0$ System has unstable node (repeller or source)

$\lambda_1 < 0, \lambda_2 < 0$ System has stable node (attractor or sink)

Phase Portrait

The phase portrait in the $y_1 - y_2$ plane is given by

$$\frac{\mathrm{d}y_1}{\mathrm{d}y_2} = \left(\frac{\lambda_2}{\lambda_1}\right)\left(\frac{y_1}{y_2}\right)$$

$$y_1 = K\,(y_2)^{|\lambda_1/\lambda_2|}$$

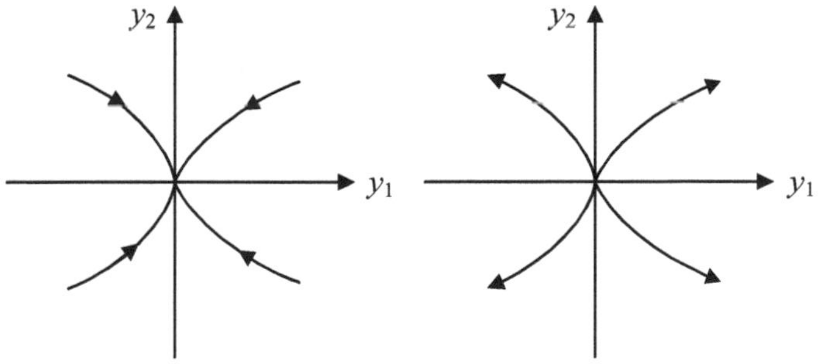

(a) $\lambda_1 < 0, \lambda_2 < 0$ sink (b) $\lambda_1 > 0, \lambda_2 > 0$ source

Figure 3.2: Phase Portrait: Both λ_1, λ_2 Real and of the Same Sign

(b)

$$\hat{\boldsymbol{A}} = \begin{bmatrix} \lambda_1 & 0 \\ 0 & \lambda_1 \end{bmatrix}, \quad \lambda_1 \text{ is real}$$

$$y_1 = K_1 e^{\lambda_1 t}, \quad y_2 = K_2 e^{\lambda_1 t} \quad \Rightarrow \; y_1 = K y_2$$

$\lambda_1 > 0$ system unstable, $\lambda_1 < 0$ system unstable.

Figure 3.3 shows the phase portrait

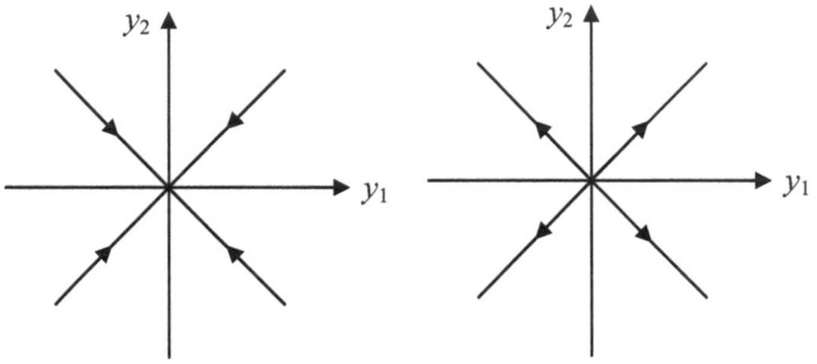

(a) $\lambda_1 < 0$ sink (b) $\lambda_1 > 0$ source

Figure 3.3: Phase Portrait: Diagonal Matrix

(c)

$$\hat{\mathbf{A}} = \begin{bmatrix} \lambda_1 & 1 \\ 0 & \lambda_1 \end{bmatrix}, \quad \lambda_1 \text{ is real}$$

$$y_1(t) = K_1 e^{\lambda_1 t}, \quad y_2(t) = K_2 t e^{\lambda_1 t}$$

The phase portrait is time dependent.

$$\lambda_1 > 0 \text{ unstable}, \quad \lambda_1 < 0 \text{ stable}.$$

(ii) **Saddle Points**

$$\hat{\mathbf{A}} = \begin{bmatrix} \lambda_1 & 0 \\ 0 & \lambda_2 \end{bmatrix}, \lambda_1, \lambda_2 \text{ are real and different signs.}$$

In this case

$$y_1 = K\,(y_2)^{-|\lambda_1/\lambda_2|}$$

There are stable as well as unstable solutions and the phase portrait, shown in figure 3.4, looks like a camel's saddle.

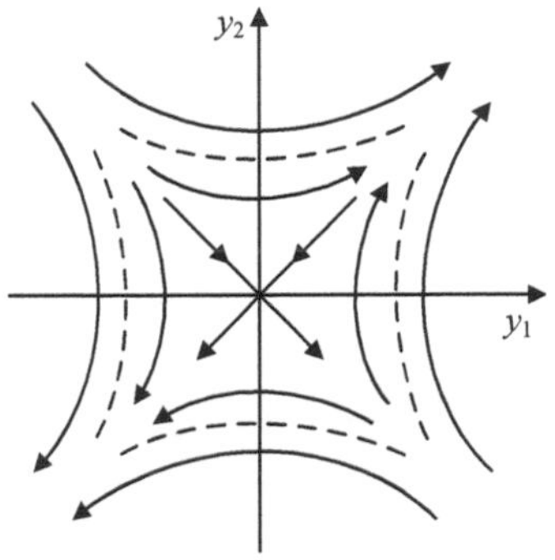

Figure 3.4: Saddle Point λ_1, λ_2 Real and Different Signs

(iii) **Focus Points**

$$A = \begin{bmatrix} \lambda_1 & 0 \\ 0 & \bar{\lambda}_1 \end{bmatrix} \quad \text{where} \quad \lambda = \alpha + j\beta, \quad \bar{\lambda}_1 = \alpha - j\beta$$

$$y_1(t) = e^{\alpha t}\left[k_1 \cos\beta t + k_2 \sin\beta t\right], y_2(t) = e^{\alpha t}\left[k_3 \cos\beta t + k_4 \sin\beta t\right]$$

The phase portrait in the $y_1 - y_2$ plane as shown in figure 3.5 is **spiraling in** or **spiraling out** of the equilibrium point acting as a "**Focus.**"

$\alpha < 0$ stable focus, $\alpha > 0$ unstable focus

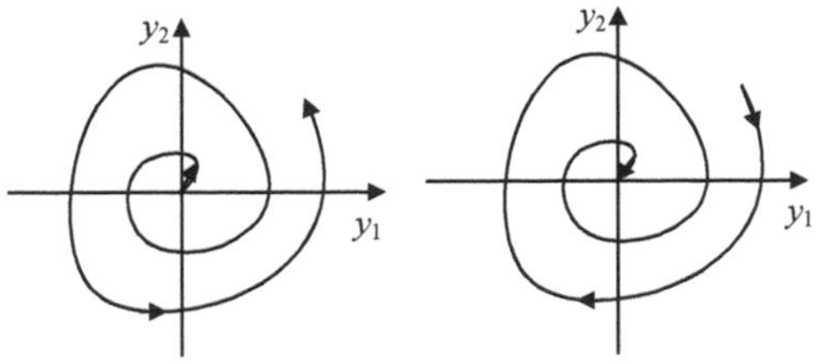

(a) Unstable Focus (b) Stable Focus
Re $\lambda_1 > 0$ Re $\lambda_1 < 0$

Figure 3.5: Focus: λ_1, λ_2 Complex Conjugates

(iv) **Center Points**

$$A = \begin{bmatrix} j\beta & 0 \\ 0 & -j\beta \end{bmatrix}$$

The phase portraits in Figure 3.6 look like ellipses (or circles) around this singularity, hence the name.

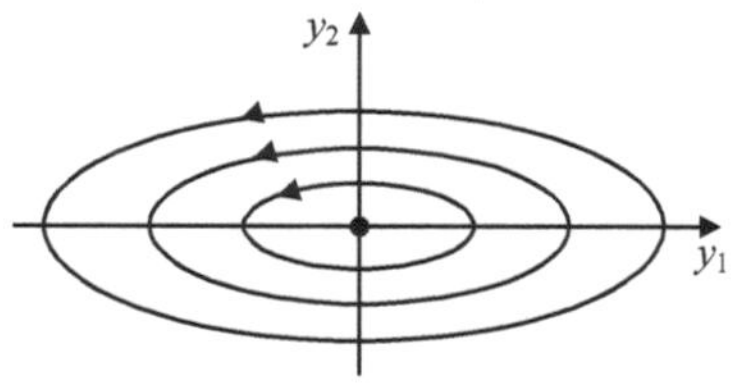

Figure 3.6: Center Point Singularity

- **The stability of nonlinear (null equilibrium) system** can be studied by the construction of "Liapunov Functions" via so-called "**Second Method of Liapunov.**" However, the stability of the linearized version of Eq. 3.135 about its null equilibrium point tells us a lot about the behavior of the system around its equilibrium.

- **The stability of the linearized equation of the nonlinear Eq. 3.135**
  This method is known as Liapunov's First Method.
  The Taylor Series expansion of $\boldsymbol{f}(\boldsymbol{x}(t))$ in Eq. 3.135 yields,

$$\frac{\mathrm{d}}{\mathrm{d}t}\boldsymbol{x}(t) = \boldsymbol{f}(\boldsymbol{x}(t)) = \boldsymbol{f}(\boldsymbol{0}) + \left\{\left[\nabla_x \boldsymbol{f}^T(\boldsymbol{x})\right]|_{x=0}\right\}\boldsymbol{x} + \text{higher order terms}$$

Neglecting higher order terms and realizing that $\boldsymbol{f}(\boldsymbol{0}) = \boldsymbol{0}$, yields:

$$\dot{\boldsymbol{x}} = \left\{\left[\nabla_x \boldsymbol{f}^T(\boldsymbol{x})\right]|_{x=0}\right\}\boldsymbol{x} = \boldsymbol{A}\boldsymbol{x}$$

We state the following observation:

- The behavior of the solution of Eq. 3.135 is essentially governed by the real part of the eigenvalues of the matrix $\boldsymbol{A}$.

- Liapunov's First Method

  Given the nonlinear system

$$\dot{\boldsymbol{x}} = \boldsymbol{f}(\boldsymbol{x}), \qquad \boldsymbol{f}(\boldsymbol{0}) = \boldsymbol{0}$$

  Linearized verion of the above:

$$\dot{\boldsymbol{x}} = \left\{\left[\nabla_x \boldsymbol{f}^T(\boldsymbol{x})\right] |_{x=0}\right\} \boldsymbol{x} = \boldsymbol{A}\boldsymbol{x}$$

  Let **each of the eigenvalues** of $\boldsymbol{A}$ have a nonzero real part. Then the behavior of the nonlinear system is similar to the behavior of the linear system in the neighborhood of the equilibrium point.

  Furthermore:

  1. If all the eigenvalues of $\boldsymbol{A}$ have their real part negative, then the null equilibrium point of the nonlinear system is stable.
  2. If at least one of the eigenvalues of $\boldsymbol{A}$ have its real part positive, then the null equilibrium point of the nonlinear system is unstable.
  3. If at least one of the eigenvalues of $\boldsymbol{A}$ has its real part as zero, then it is not possible to arrive at any conclusion about the stability of the null equilibrium solution of the nonlinear system. The only way to study the stability of the null equilibrium solution of the nonlinear system is via the "Second Method of Liapunov" involving the nonlinearities.

3.7.1 Liapunov's Second Method (Direct Method)

Let us study the stability of the equilibrium solution of a n-dimension dynamic system

$$\dot{\boldsymbol{x}} = \boldsymbol{f}(\boldsymbol{x}), \qquad \boldsymbol{f}(\boldsymbol{0}) = \boldsymbol{0} \tag{3.138}$$

The **direct** or the **second method** of Liapunov is a sufficient condition method to study the stability of the equilibrium of the dynamical system without any explicit knowledge of the actual solution. In fact, this is the only general way to study the equilibrium solution study of a general nonlinear and time varying system. In this second method we generate a "general energy-type function" $V(\boldsymbol{x})$, known as the **Liapunov Function**. In a nutshell, the Liapunov's main theorem states the following:

Given the system of Eq. 3.138

(i) If there exists a scalar function $V(\boldsymbol{x})$ which is positive definite, i.e.,

$$V(\boldsymbol{x}) > 0 \text{ for all } \boldsymbol{x} \quad (\, V(\boldsymbol{x}) \text{ going to zero only at the origin})$$

(ii) Its time derivative $\dot{V}(\boldsymbol{x})$ (taking into account the Eq. 3.138),

$$\dot{V}(\boldsymbol{x}) \;=\; (\nabla V)^T \cdot \boldsymbol{f}(x) \;=\; \sum_{i=1}^{n} \frac{\partial V}{\partial x_i} f_i(x) \leq 0 \tag{3.139}$$

Furthermore,

$\dot{V}(\boldsymbol{x})$ does not vanish along any solution of Eq. 3.138 except at $\boldsymbol{x} = 0$.

If such a function satisfying the above conditions can be found, then the dynamical system Eq. 3.138 is asymptotically stable. **Liapunov functions are not unique**. For the same system, many Liapunov functions can be found. There are a number of Liapunov theorems, dealing with different aspects of stabilty, which we shall not discuss here. We illustrate the method with some examples. At the end of this section, we present a table of different differential equations, their associated Liapunov functions, and the conditions of stabiliy on the differential equation parameters.

Example 3.6:

Consider a mass-spring system:

$$\begin{aligned} &m\ddot{x} + b\dot{x} + kx = 0 \\ &x = x_1, \qquad \dot{x} = x_2, \qquad m > 0,\ b > 0,\ k > 0 \end{aligned} \tag{3.140}$$

Choose an "energy-like function" $V(x_1, x_2)$ (kinetic plus twice the potential energy):

$$\boxed{V(x_1, x_2) = \frac{1}{2}m\dot{x}^2 + \frac{1}{2}kx^2 = \frac{1}{2}mx_2^2 + \frac{1}{2}kx_1^2}$$

Taking its time derivative:

$$\dot{V}(x_1, x_2) = mx_2\dot{x}_2 + kx_1\dot{x}_1 = (m\ddot{x} + kx)\,\dot{x}$$

From Eq. 3.140:

$$\boxed{\dot{V}(x_1, x_2) = -b\dot{x}^2 = -bx_2^2}$$

1. $\dot{V} < 0$ and does not take the value zero along the solution of Eq. 3.138 except at the origin.

2. $V(x_1, x_2) > 0$ in the whole $x_1 - x_2$ plane and only vanishes when $x_1 = x_2 = 0$.

Conclusion:

The equilibrium solution of system Eq. 3.140 is asymptotically stable as long as the damping parameter $b > 0$, mass $m > 0$, and spring constant $k > 0$.

Example 3.7:

Consider a system

$$\ddot{x} + \dot{x} + x^3 = 0 \tag{3.141}$$

or

$$(\ddot{x} + x^3) = -\dot{x}$$

Multiplying both sides with $\dot{x}$

$$\dot{x}(\ddot{x} + x^3) = -\dot{x}^2 = \dot{V}$$

Integrating the above expression and realizing $x = x_1, \dot{x} = x_2$, yields:

$$V(x_1, x_2) = \frac{x_1^4}{4} + \frac{1}{2}x_2^2$$
$$\dot{V}(x_1, x_2) = -x_2^2$$

Note: $\dot{V}$ goes to zero only at the origin since $x_2 = \dot{x} = 0 \rightarrow \ddot{x} = 0$, otherwise it is negative.

Taking Eq. 3.141 into consideration $x \rightarrow 0$ as $t \rightarrow \infty$.

Hence, the system represented by Eq. 3.141 is asymptotically stable.

Example 3.8:

Consider:

$$\dddot{x} + a_1\ddot{x} + a_2\dot{x} + a_3x = 0 \tag{3.142}$$

Let

$$x = x_1, \ \dot{x} = x_2, \ \ddot{x} = x_3, \ \dddot{x} = x_4$$

(Only three coordinates are necessary to describe the equation in state space.)

We can rewrite Eq. 3.142 as

$$(x_4 + a_2x_2) = -(a_1x_3 + a_3x_1)$$

Multiplying both sides with $(a_1x_3 + a_3x_1)$,

$$(x_4 + a_2x_2)(a_1x_3 + a_3x_1) = -(a_1x_3 + a_3x_1)^2 = \dot{V}(x_1, x_2, x_3) \tag{3.143}$$

Integrating Eq. 3.143

$$V(x_1, x_2, x_3) = \frac{a_1x_3^2}{2} + a_3 \int x_4x_1 \, \mathrm{d}t + \frac{a_1a_2}{2}x_2^2 + \frac{a_2a_3}{2}x_1^2$$

Now,

$$\int x_4x_1 \, \mathrm{d}t = x_1x_3 - \int x_2x_3 \, \mathrm{d}t = x_1x_3 - \frac{x_2^2}{2}$$

Thus,

$$V(x_1, x_2, x_3) = \frac{a_1x_3^2}{2} + a_3x_1x_3 - \frac{a_3x_2^2}{2} + \frac{a_1a_2}{2}x_2^2 + \frac{a_2a_3}{2}x_1^2$$

or,

$$2V(x_1, x_2, x_3) = \begin{bmatrix} x_1 & x_2 & x_3 \end{bmatrix} \begin{bmatrix} a_2a_3 & 0 & a_3 \\ 0 & (a_1a_2 - a_3) & 0 \\ a_3 & 0 & a_1 \end{bmatrix} \begin{bmatrix} x_1 \\ x_2 \\ x_3 \end{bmatrix}$$

$V(x_1, x_2, x_3)$ is positive definite, provided

$a_1 > 0,\ a_2 > 0,\ a_3 > 0$ and $(a_1a_2 - a_3) > 0$

$$\dot{V}(x_1, x_2, x_3) = -(a_1x_3 + a_3x_1)^2$$

Note:

$$\int x_ix_j \, \mathrm{d}t = x_ix_{j-1} - \int x_{i+1}x_{j-1} \, \mathrm{d}t \quad i < j$$

If $(j - i)$ = odd integer,

$$\int x_i x_j \, \mathrm{d}t = x_i x_{j-1} - x_{i+1} x_{j-2} + x_{i+2} x_{j-3} \cdots + (-1)^{(j-i+1)/2} \int x^2_{(i+j-1)/2} \, \mathrm{d}t$$

If $(j - i)$ = even integer,

$$\int x_i x_j \, \mathrm{d}t = x_i x_{j-1} - x_{i+1} x_{j-2} dt + x_{i+2} x_{j-3} \cdots + (-1)^{(j-i)/2} \int x^2_{(i+j)/2} \, \mathrm{d}t$$

When x_i is integrated, it leads to x_{i-1}, but when x_i is differentiated, it results in x_{i+1}.

3.7.2 Stability Studies for Nonlinear Difference Equations Using Liapunov's Second Method

Consider the following nonlinear, nonautonomous difference equations:

$$x(n+2) + a_1 x(n+1) + a_2 x(n) + f\,[x(n), x(n+1), n] = 0 \tag{3.144}$$

and

$$x(n+3) + a_1 x(n+2) + a_2 x(n+1) + a_3 x(n) + f\,[x(n), x(n+1), x(n+2), n] = 0 \tag{3.145}$$

where a_1, a_2, and a_3 are real constants, and f is a nonlinear function of its arguments.

The domain of a definition for n is any non-negative integer. Essentially, our object is to establish some sufficient conditions for the asymptotic stability of the equilibrium solutions in Eqs. 3.144 and 3.145. For the linear case, $f \equiv 0$, the sufficient conditions derived here coincide with those pertinent to the Schur-Cohen criterion [Jury, I.E.].

The method of analysis used to study asymptotic stability is the discrete analog of Liapunov's Direct Method. This procedure consists of finding a scalar function, $V(n)$, which is positive definite and has a first difference,

$$\Delta V(n) = V(n+1) - V(n) \tag{3.146}$$

which is negative semidefinite. $V(n)$ is not identically zero on any trajectory of Eq. 3.144 [Hahn, W.]. If such a function exists, the equilibrium solution of the system is asymptotically stable.

(a) **Linear Case Corresponding to Eq. 3.144**

Linearized version of Eq. 3.144, given by

$$x(n+2) + a_1 x(n+1) + a_2 x(n) = 0. \tag{3.147}$$

This equation can be written in the matrix form

$$\boldsymbol{x}(n+1) = \boldsymbol{A}\boldsymbol{x}(n), \tag{3.148}$$

where

$$x_1(n) = x(n) \tag{3.149}$$

$$x_2(n) = x(n+1) \tag{3.150}$$

$$\boldsymbol{x}(n) = \begin{bmatrix} x_1(n) \\ x_2(n) \end{bmatrix}, \tag{3.151}$$

and

$$\boldsymbol{A} = \begin{bmatrix} 0 & 1 \\ -a_2 & -a_1 \end{bmatrix} \tag{3.152}$$

We now seek a transformation defined by a real triangular matrix $\boldsymbol{Q}$ such that

$$\boldsymbol{y}(n) = \boldsymbol{Q}\boldsymbol{x}(n) \tag{3.153}$$

Applying this transformation to Eq. 3.148 gives

$$\boldsymbol{y}(n+1) = \boldsymbol{R}\boldsymbol{y}(n), \tag{3.154}$$

where

$$\boldsymbol{R} = \boldsymbol{Q}\boldsymbol{A}\boldsymbol{Q}^{-1} \tag{3.155}$$

Let $\boldsymbol{R}$ be of the following form [Puri, N.N.],

$$\boldsymbol{R} = \begin{bmatrix} r_2/r & r_1 \\ -r_1/r & r_2 \end{bmatrix}, \tag{3.156}$$

$$r = \sqrt{r_1^2 + r_2^2}$$

From Eq. 3.155, we see that $\boldsymbol{R}$ and $\boldsymbol{A}$ are similar matrices and their characteristic determinants are identical. This results in the relationship

$$\lambda^2 + a_1\lambda + a_2 = \lambda^2 - \left(r_2 + \frac{r_2}{r}\right)\lambda + r, \tag{3.157}$$

where λ is an arbitrary variable. Equating the coefficients of equal powers of λ

$$a_2 = r > 0, \tag{3.158}$$

$$a_1 = -r_2\left[\frac{r+1}{r}\right]. \tag{3.159}$$

The corresponding $\boldsymbol{Q}$ obtained from Eqs. 3.155, 3.158, and 3.159 in terms of r_1, r_2, and r, is

$$Q = \begin{bmatrix} 1 & 0 \\ -r_2/rr_1 & 1/r_1 \end{bmatrix} \tag{3.160}$$

where the first element as one was chosen arbitrarily.

We are now in a position to choose a Liapunov function. Consider a Liapunov function defined by

$$V(n) = \boldsymbol{y}^T(n)\boldsymbol{y}(n), \tag{3.161}$$

where $\boldsymbol{y}^T$ denotes the transpose of $\boldsymbol{y}$. Combining Eqs. 3.153, 3.160, and 3.161

$$V(n) = x_1^2(n) + \frac{1}{r_1^2}\left[x_2(n) - \frac{r_2}{r}x_1(n)\right]^2 \geq 0. \tag{3.162}$$

It is easily seen from Eq. 3.162 that $V(n)$ is zero if and only if $x_1(n) = x_2(n) = 0$. Thus, $V(n)$ is positive definite. We now form the first difference of $V(n)$,

$$\Delta V(n) = \boldsymbol{y}^T(n+1)\boldsymbol{y}(n+1) - \boldsymbol{y}^T(n)\boldsymbol{y}(n). \tag{3.163}$$

Eqs. 3.154 and 3.163 yield

$$\Delta V(n) = \boldsymbol{y}^T(n)\left[\boldsymbol{R}^T\boldsymbol{R} - \boldsymbol{I}\right]\boldsymbol{y}(n), \tag{3.164}$$

where $\boldsymbol{I}$ is the unity matrix. From Eqs. 3.156 and 3.164:

$$\boldsymbol{R}^T\boldsymbol{R} - \boldsymbol{I} = \begin{bmatrix} 0 & 0 \\ 0 & (r^2 - 1) \end{bmatrix} \tag{3.165}$$

Combining Eqs. 3.164 and 3.165:

$$\Delta V(n) = -\left(1 - r^2\right)y_2^2(n) \tag{3.166}$$

The first difference, $\Delta V(n)$, is negative semidefinite, If

$$1 - r^2 = 1 - a_2^2 > 0. \qquad \text{first stability condition.} \tag{3.167}$$

Since $y_2(n) = 0$ on a trajectory of Eq. 3.144, it implies that $x_1(n) = x_2(n) = 0$. Thus the equilibrium solution, $x_1(n) = x_2(n) = 0$, of Eq. 3.144 is asymptotically stable. We have assumed that $\boldsymbol{R}$ is a real matrix, thus the coefficients a_1 and a_2 must satisfy the additional condition

$$(1 + a_2)^2 - a_1^2 > 0 \qquad \text{second stability condition.} \tag{3.168}$$

Summarizing, the conditions of asymptotic stability of the system, Eq. 3.147, are precisely the same as yielded by the Schur-Cohen conditions [Jury, I.E.].

(b) **Nonlinear Case Given by Eq. 3.144**

The matrix formulation of Eq. 3.144 is

$$\boldsymbol{x}(n+1) = \boldsymbol{A}\boldsymbol{x}(n) - \boldsymbol{b}f\left[x_1(n), x_2(n), n\right], \tag{3.169}$$

$$\boldsymbol{b} = \begin{bmatrix} 0 \\ 1 \end{bmatrix} \tag{3.170}$$

and where $\boldsymbol{x}$ and $\boldsymbol{A}$ are defined by Eqs. 3.149 – 3.151. Applying the transformation given by Eqs. 3.153 – 3.169 produces the following equation:

$$\boldsymbol{y}(n+1) = \boldsymbol{R}\boldsymbol{y}(n) - \boldsymbol{Q}\boldsymbol{b}f\left[x_1(n), x_2(n), n\right]. \tag{3.171}$$

We assume that the linear system corresponding to Eq. 3.169 is asymptotically stable at $x_1(n) = x_2(n) = 0$. This leads to the following choice for $V(n)$,

$$V(n) = \boldsymbol{y}^T(n)\boldsymbol{y}(n). \tag{3.172}$$

As in Eq. 3.162, $V(n)$ is positive definite. The first difference of $V(n)$, from Eqs. 3.171 and 3.172, is:

$$\Delta V(n) = \boldsymbol{y}^T(n)\boldsymbol{y}(n) - 2\boldsymbol{b}^T\boldsymbol{Q}^T\boldsymbol{R}\boldsymbol{y}(n) + \boldsymbol{b}^T\boldsymbol{Q}^T\boldsymbol{Q}\boldsymbol{b}f^2 \tag{3.173}$$

By using Eqs. 3.153, 3.156, 3.160, and 3.170, Eq. 3.173 can be rewritten as

$$\Delta V(n) = -K + \frac{2}{r}x_1 f(x_1(n), x_2(n), n) + \frac{1-r_1^2}{r_1^2(1-r^2)}f^2(x_1(n), x_2(n), n) \tag{3.174}$$

where

$$K = \left(1-r^2\right)\left(-\frac{r_2}{r_1 r}x_1(n) + \frac{1}{r_1}x_2(n) + \frac{r_2 f(x_1(n), x_2(n), n)}{r_1\left(1-r^2\right)}\right)^2 \geqq 0. \tag{3.175}$$

We assume that the only solution of $f(x_1(n), x_2(n), n) = 0$ is the equilibrium solution, $x_1(n) = x_2(n) = 0$. Thus, whenever $x_1(n)$ and $x_2(n)$ are not zero, the negative definiteness of $\Delta V(n)$ requires

$$\frac{2r_1^2\left(1-r^2\right)}{r\left(1-r_1^2\right)} \leqq \frac{f\left[x_1(n), x_2(n), n\right]}{x_1(n)} < 0, \quad \text{for all } n \geq 0. \tag{3.176}$$

The relationships between the r's and the a's are given by Eqs. 3.158 and 3.159. The results involving the sufficient conditions of stability for Eqs. 3.144 and 3.145 are summarized in Theorem I.

Theorem I: Given the difference equation

$$x(n+2) + a_1 x(n+1) + a_2 x(n) + f\left[x(n), x(n+1), n\right] = 0$$

assume that

(i) $1 - a_2^2 > 0$

(ii) $\left(1 + a_2^2\right)^2 - a_1^2 > 0$

(iii) $f[x(n), x(n+1), n] \neq 0$ if $x(n) \neq 0$ and $x(n+1) \neq 0$

(iv) $f(0, 0, n) = 0$

(v) $-\dfrac{2r_1^2\left(1 - r^2\right)}{r\left(1 - r_1^2\right)} \leqq \dfrac{f[x(n), x(n+1), n]}{x(n)} < 0$

If all the above conditions are satisfied, then the equilibrium solution is asymptotically stable. When f is identically zero, the above conditions coincide with those yielded by the Schur-Cohen criterion.

Example 3.9: **Linear, Nonautonomous System**

Consider the system given by

$$x(n+2) + x(n+1) + \left[\frac{1}{2} - \frac{c}{1+n^2}\right] x(n) = 0, \quad c \text{ is a constant.}$$

The condition for stability yields:

$$a_1 = 1, \quad a_2 = \frac{1}{2}, \quad f = -\frac{cx(n)}{1+n^2}$$

$$1 - a_2^2 = \frac{3}{4} > 0,$$

$$(1 + a_2)^2 - a_1^2 = \frac{5}{4} > 0,$$

$$0 < \frac{c}{1+n^2} \leq \frac{15}{31}.$$

The system is asymptotically stable at its equilibrium solution if $0 < c \leq 15/31$

Example 3.10: **Nonlinear, Nonautonomous System**

Consider the system given by

$$x(n+2) + x(n+1) + \frac{1}{2}x(n) = [1 + c\sin(n)]\, x^3(n) + x(n)x^2(n+1),$$

where c is a constant. The condition for stability is:

$$0 < [1 + c\sin(n)]\, x^2(n) + x^2(n+1) \leqq \frac{15}{31}$$

This relationship places a restriction on the constant c and also gives a region of guaranteed asymptotic stability for the above system.

Sufficient Conditions for Stability of the Third Order Eq. 3.145

The matrix equation corresponding to Eq. 3.145 is

$$\boldsymbol{x}(n+1) = \boldsymbol{A}\boldsymbol{x}(n) - \boldsymbol{b}f(x_1(n), x_2(n), n) \tag{3.177}$$

$$x_1(n) = x(n)$$
$$x_2(n) = x(n+1)$$
$$x_3(n) = x(n+2),$$

$$\boldsymbol{x}(n) = \begin{bmatrix} x_1(n) \\ x_2(n) \\ x_3(n) \end{bmatrix} \tag{3.178}$$

$$\boldsymbol{A} = \begin{bmatrix} 0 & 1 & 0 \\ 0 & 0 & 1 \\ -a_3 & -a_2 & -a_1 \end{bmatrix}, \qquad \boldsymbol{b} = \begin{bmatrix} 0 \\ 0 \\ 1 \end{bmatrix} \tag{3.179}$$

The linear case, $f(x_1(n), x_2(n), n) \equiv 0$, is analyzed in the same way as the second order case, thus only the "highlights" will be given here. The matrix $\boldsymbol{R}$, Eq. 3.154, is chosen such that

$$\boldsymbol{R}^T\boldsymbol{R} = \begin{bmatrix} 1 & 0 & 0 \\ 0 & 1 & 0 \\ 0 & 0 & r^2 \end{bmatrix}, \qquad r = \left(r_1^2 + r_2^2 + r_3^2\right)^{1/2} \tag{3.180}$$

The resulting $\boldsymbol{R}$ and $\boldsymbol{Q}$ matrices are

$$\boldsymbol{R} = \begin{bmatrix} \dfrac{-r_1 r_3}{\rho r} & \dfrac{r_2}{\rho} & r_1 \\ \dfrac{-r_2 r_3}{\rho r} & \dfrac{-r_1}{\rho} & r_2 \\ \dfrac{\rho}{r} & 0 & r_3 \end{bmatrix} \tag{3.181}$$

$$\boldsymbol{Q} = \begin{bmatrix} (r^2/\rho) & (\dfrac{r_1 r_3 r}{\rho^2} + \dfrac{r_1 r^2}{\rho^2} - \dfrac{r_3}{\rho}) & (-\dfrac{r_1 r}{\rho^2} - \dfrac{r_1 r_3}{\rho^2}) \\ (-\dfrac{r_1 r^2}{r_2 \rho}) & (\dfrac{r_2 r^2}{\rho^2} + \dfrac{r_1 r_3}{r_2 \rho} - \dfrac{r_1^2 r_3 r}{r_2 \rho^2}) & (\dfrac{r_1^2 r}{r_2 \rho^2} - \dfrac{r_2 r_3}{\rho^2}) \\ 0 & 0 & 1 \end{bmatrix} \tag{3.182}$$

$$\rho = \sqrt{r_1^2 + r_2^2}$$

By equating the characteristic determinants of $\boldsymbol{A}$ and $\boldsymbol{R}$,

$$a_1 = \frac{r_1}{\rho}\left[1 + \frac{r_3}{r}\right] - r_3, \tag{3.183}$$

$$a_2 = \frac{r_3}{r} - \frac{r_1}{\rho}\left[r + r_3\right], \tag{3.184}$$

$$a_3 = -r < 0. \tag{3.185}$$

The resulting conditions for asymptotic stability of the equilibrium solution for the linear case are

$$\left.\begin{aligned} a_3 &< 0 \\ b_1 &> 0 \\ b_1^2 - b_2^2 &> 0 \\ (b_1 + b_2)^2 - b_3^2 &> 0 \end{aligned}\right\} \tag{3.186}$$

$$b_1 = 1 - a_3^2,$$

$$b_2 = a_2 - a_1 a_3,$$

$$b_3 = a_1 - a_2 a_3.$$

These conditions, Eqs. 3.186, are similar to those yielded by the Schur-Cohen conditions [Jury, I.E.].

The additional condition for asymptotic stability of the nonlinear system is obtained in the same way as in the second order case and is given by

$$x_1 f\left(x_1, x_2, x_3, n\right) + c_1 x_2 f\left(x_1, x_2, x_3, n\right) \geqq \frac{c_2}{2r} f^2\left(x_1, x_2, x_3, n\right), \tag{3.187}$$

where

$$c_1 = \frac{\left(1 + a_2\right) b_2}{a_3\left(b_1 + b_2\right)}, \tag{3.188}$$

and

$$c_2 = 1 + \frac{b_1 b_3^2}{\left(b_1^2 - b_2^2\right)\left[\left(b_1 + b_2\right)^2 - b_3^2\right]}. \tag{3.189}$$

In summary, the sufficient conditions for asymptotic stability of the equilibrium of Eq. 3.177 are Eqs. 3.186 and 3.187, and the requirement that $f \equiv 0$ if and only if $x_1(n) = x_2(n) = x_3(n) = 0$.

Example 3.11: **Nonlinear, Nonautonomous System**

Consider the following equation:

$$x(n+3) - \frac{1}{2}x(n+2) + \frac{1}{2}x(n+1) - \frac{1}{2}x(n) + c(n)\left[x^3(n) - \frac{3}{4}x^2(n)x(n+1)\right] = 0.$$

Thus, we have

$$a_1 = -\frac{1}{2}, \qquad a_2 = \frac{1}{2}, \qquad a_3 = -\frac{1}{2},$$

$$b_1 = \frac{3}{4}, \qquad b_2 = \frac{1}{4}, \qquad b_3 = -\frac{1}{4},$$

$$c_1 = -\frac{3}{4}, \qquad \text{and} \qquad \frac{1}{2r}c_2 = \frac{11}{10}.$$

The inequality in Eq. 3.187 becomes

$$c(n)\left[x^4(n) - \frac{3}{2}x^3(n)x(n+1) + \frac{9}{16}x^2(n)x^2(n+1)\right]$$
$$+ -\frac{11}{10}c^2(n)\left[x^6(n) - \frac{3}{2}x^5(n)x(n+1) + \frac{9}{16}x^4(n)x^2(n+1)\right] \geqq 0 \quad (3.190)$$

or

$$\left[c(n)x^2(n) - \frac{11}{10}c^2(n)x^4(n)\right]\left[x(n) - \frac{3}{4}x(n+1)\right]^2 \geqq 0.$$

Therefore, we see that the above system is asymptotically stable if

$$c(n) - \frac{11}{10}c^2(n)x^2(n) \geqq 0,$$

which is a restriction on the function $c(n)$ and also gives a region of asymptotic stability.

Conclusion

The stability of a class of second and third order nonlinear, nonautonomous difference equations was analyzed by a discrete analogue of Liapunov's Direct Method. The stability conditions obtained for the corresponding linear systems were related to the Schur-Cohen criteria for linear systems. The additional stability conditions for the nonlinear systems are new results.

The following tables represent Liapunov functions for many important differential equations that arise in control system stability studies.

Table 3.1: Method Used: Routh Canonical Transformation

Differential Equation $x = x_1, \dot{x} = x_2, \frac{d}{dt}(x_n) = x_{n+1}$	Liapunov Function V	Derivative $\dot{V}$	Conditions of Global Asymptotic Stability
$\ddot{x} + \dot{x} + Kx^3 = 0$	$V = (x_1 + x_2)^2 + (K/2)x_1^4$	$\dot{V} = -2Kx_1^4$	$K > 0$
$\ddot{x} + f(x)\dot{x} + x = 0$	$V = x_1^2 + x_2^2$	$\dot{V} = -2f(x)x_2^2$	$f(x) > 0$
$\ddot{x} + a\dot{x} + bf(x) = 0$	$V = 2\int_0^{x_1} f(x_1)\, dx_1 + x_2^2$	$\dot{V} = -ax_2^2$	$\int_0^{x_1} f(x_1)\, dx_1 > 0$
$\ddot{x} + a_1(x, \dot{x})x + a_2(x, \dot{x})x = 0$	$V = \left[x_2 + \int_0^{x_1} a_1(x_1, x_2)\, dx_1\right]^2 + \int_0^{x_1} a_2(x_1, x_2)x_1\, dx_1$	$\dot{V} = -a_2(x_1, x_2)x_1 \int_0^{x_1} a_1(x_1, x_2)\, dx_1$	$a_1(x_1, x_2) > 0$ $a_2(x_1, x_2) > 0$
$\dddot{x} + a_1(x, \dot{x}, \ddot{x})\ddot{x} + a_2(x, \dot{x}, \ddot{x})\dot{x} + a_3 x = 0$ $a_1(x, \dot{x}, \ddot{x}) = a_{1o} + a_{1n}(x, \dot{x}, \ddot{x})$ $a_2(x, \dot{x}, \ddot{x}) = a_{2o} + a_{2n}(x, \dot{x}, \ddot{x})$	$V = a_3^2 x_1^2 + 2a_{2o} \int_o^{x_2} a_2(x_1, x_2, x_3)x_2\, dx_2$ $+ 2a_3 \int_o^{x_2} a_1(x_1, x_2, x_3)x_2\, dx_2$ $+ 2a_{2o}a_3 x_1 x_2 + 2a_3 x_2 x_3 + a_2 x_3^2$	$\dot{V} = -2a_3 a_{2n} x_2^2$ $- 2(a_1 a_{2o} - a_3)x_3^2$	$a_3 > 0,\ a_{2o} > 0$ $a_{2n} \geq 0,\ a_1 \geq \frac{a_3}{a_{2o}}$
$\ddot{x} + \left[f(x) + xf'(x)\right]\dot{x} + \beta f(x) = 0$ $f'(x) = \frac{d}{dx}f(x)$	$V = (x_1 + x_2)^2$ $+ 2\int_o^{x_1} \left[\beta f(x_1) + \frac{d}{dx_1}\left(x_1 f' x_1\right)\right] x_1\, dx_1$	$\dot{V} = -2\beta x_1^2 f(x_1)$ $- 2x_2^2 \frac{d}{dx_1}[x_1 f(x_1)]$	$\frac{d}{dx_1}[x_1 f(x_1)] > 0$ $\beta f(x_1) > 0$

continued on next page

Differential Equation	Liapunov Function V	Derivative $\dot{V}$	Conditions of Global Asymptotic Stability
$\dddot{x} + a_1\ddot{x} + a_2\dot{x} + a_1 a_2 x + K x_2^{2m+1}$	$V = a_2(a_1x_1 + x_2)^2 + (a_1x_2 + x_3)^2 + \frac{K}{m+1}x_2^{2m+2}$	$\dot{V} = -2a_1 K x_2^{2(m+1)}$	$K > 0,\ a_1 > 0,\ a_2 > 0$
$\dddot{x} + a_1\ddot{x} + a_2 f(x) = 0$ $f(x) = K_1x + K_2\dot{x} - K_2\int_o^x F(\alpha)\,\mathrm{d}\alpha$	$V = (a_1x_2 + x_3)^2 + \frac{2a_1a_2}{K_1}\int_o^x F(\alpha)\,\mathrm{d}\alpha + \left(\frac{K_2a_1}{K_1} - 1\right)x_3^2$	$\dot{V} = -2\left(\frac{K_2}{K_1}a_1 - 1\right)x_3^2 - 2\frac{a_1a_2K_3}{K_1}F^2(x)$	$K_1 > 0,\ K_3 > 0$ $\frac{K_2a_1}{K_1} - 1 > 0$ $a_1 > 0$ $\int_o^x F(\alpha)\,\mathrm{d}\alpha > 0$
$\dddot{x} + a_1\ddot{x} + a_2(\dot{x})\dot{x} + a_3x = 0$ $a_2(\dot{x}) = a_{2o} + a_{2n}(\dot{x})$	$V = \left(a_{2o} - \frac{a_3}{a_1}\right)x_3^2 + (a_3x_1 + a_{2o}x_2)^2 + \frac{a_3}{a_1}(a_1x_2 + x_3)^2 + 2\int_o^{x_2} a_2a_{2n}x_2\,\mathrm{d}x_2$	$\dot{V} = -2(a_1a_{2o} - a_3)x_3^2 - 2a_3a_{2n}x_2^2$	$a_1 > 0,\ a_3 > 0$ $a_{2n} > 0$ $a_1a_{2o} - a_3 > 0$
$\dddot{x} + a_1(x,\dot{x})\ddot{x} + a_2\dot{x} + a_3x = 0$ $a_1(x,\dot{x}) = a_{1o} + a_{1n}(x,\dot{x})$	$V = \left(a_2 - \frac{a_3}{a_1}\right)x_3^2 + (a_3x_1 + a_2x_2)^2 + \frac{a_3}{a_{1o}}(a_{1o}x_2 + x_3)^2 + 2a_3\int_o^{x_2} a_{1n}x_2\,\mathrm{d}x_2$	$\dot{V} = -2(a_1a_2 - a_3)x_3^2$	$a_1(x,\dot{x})a_2 - a_3 > 0$ $a_{1o} > 0,\ a_2 > 0$ $a_{1o}a_2 - a_3 > 0$ $\int_o^{x_2} a_{1n}x_2\,\mathrm{d}x_2 > 0$

continued on next page

Differential Equation	Liapunov Function V	Derivative $\dot V$	Conditions of Global Asymptotic Stability
$\dddot{x} + a_1(x,\dot x)\ddot x + a_2(x,\dot x)\dot x + a_3 x = 0$ $a_1(x,\dot x) = a_{1o} + a_{1n}(x,\dot x)$ $a_2(x,\dot x) = a_{2o} + a_{2n}(x,\dot x)$	$V = \left(a_{2o} - \frac{a_3}{a_{10}}\right)x_3^2 + (a_3x_1 + a_{2o}x_2)^2$ $+ \frac{a_3}{a_{10}}(a_{10}x_2 + x_3)^2$ $+ 2\int_0^{x_2} [a_3a_{1n} + a_{2o}a_{2n}]\, x_2\, dx_2$	$\dot V = -2(a_1a_{2o} - a_3)\, x_3^2$ $- 2a_3a_{2n}x_2^2$	$a_{10}a_{2o} - a_3 > 0$ $a_{10} > 0,\ a_{2o} > 0$ $a_3a_{1n} + a_{2o}a_{2n} \geq 0$
$\ddddot{x} + a_1\dddot{x} + a_2(\dot x)\ddot x + a_3\dot x + a_4 x = 0$ $a_2(\dot x) = a_{2o} + a_{2n}(\dot x)$	$\frac{a_1}{a_1a_{2o} - a_3}\left(\frac{a_3}{a_1}x_2 + x_4\right)^2$ $+ \left(\frac{a_1a_4}{a_1a_{2o} - a_3}x_1 + x_3\right)^2$ $+ \left(\frac{a_3}{a_1} - \frac{a_1a_4}{a_1a_{2o} - a_3}\right)x_2^2$ $+ \left[\frac{a_1a_4}{a_1a_{2o} - a_3}\right]\left[\frac{a_3}{a_1} - \frac{a_1a_4}{a_1a_{2o} - a_3}\right]x_1^2$ $+ \left(\frac{2a_3}{a_1a_{2o} - a_3}\right)\int_o^{x_2} a_{2n}x_2\, dx_2$ $+ \frac{2a_1}{a_1a_2 - a_3}\int_o^{x_3} a_{2n}x_3\, dx_3$	$\dot V = -\gamma(a_3x_2 + a_1x_4)^2$ $\gamma = \frac{2a_1^2}{a_1a_{2o} - a_3}$	$a_1 > 0$ $a_{2o} > 0$ $a_1a_{2o} - a_3 > 0$ $\frac{a_3}{a_1} - \frac{a_1a_4}{a_1a_{2o} - a_3} > 0$ $\int_o^{x_2} a_{2n}(x_2)x_2\, dx_2 \geq 0$ $\int_o^{x_2} a_{2n}(x_2)x_3\, dx_3 \geq 0$

Table 3.2: Method Used: Integration by Parts

Differential Equation	Liapunov Function V	Derivative $\dot{V}$	Conditions of Global Asymptotic Stability
$\ddot{x} + a_1\dot{x} + a_2(x)x = 0$	$V = x_2^2 + 2\int_o^{x_1} a_2(x_1)x_1\,\mathrm{d}x_1$	$\dot{V} = -2a_1x_2^2$	$a_1 > 0$ $a_2(x_1) > 0$
$\ddot{x} + a_1(x,\dot{x})\dot{x} + a_2(x)\dot{x} = 0$	$V = x_2^2 + 2\int_o^{x_1} a_2(x_1)x_1\,\mathrm{d}x_1$	$\dot{V} = -2a_1(x_1,x_2)x_2^2$	$a_1(x_1,x_2) > 0$ $a_2(x_1) > 0$
$\ddot{x} + a_1(x)\dot{x} + a_2(x)x = 0$	$V = \left[x_2 + \int_o^{x_1} a_1(x_1)\,\mathrm{d}x_1\right]$ $+ 2\int_o^{x_2} a_2(x_1)x_1\,\mathrm{d}x_1$	$\dot{V} = -2a_2(x_1)\int_o^{x_1} a_1(x_1)\,\mathrm{d}x_1$	$a_1(x_1) > 0$ $a_2(x_1) > 0$
$\ddot{x} + a_1\dot{x} + a_2(1 - x^2)x = 0$	$V = x_2^2 + a_2x_1^2\left(1 - \frac{x_1^2}{2}\right)$	$\dot{V} = -2a_1x_2^2$	$a_1 > 0$ $a_2\left(1 - x_1^2\right) > 0$
$\dddot{x} + a_1\ddot{x} + a_2(x)\dot{x} + a_3x = 0$	$V = (a_1x_2 + x_3)^2 + \left(a_2(x) - \frac{a_3}{a_1}\right)x_2^2$ $+ \frac{a_3}{a_1}(a_1x_1 + x_2)^2$	$\dot{V} = -2\Big(a_1a_2(x) - a_3$ $- \frac{\mathrm{d}}{\mathrm{d}x_1}a_2(x_1)\Big)x_2^2$	$a_1 > 0$ $a_2 > 0$ $a_1a_2(x) - a_3 > \frac{\mathrm{d}}{\mathrm{d}x_1}a_2(x_1) > 0$

continued on next page

Differential Equation	Liapunov Function V	Derivative $\dot{V}$	Conditions of Global Asymptotic Stability
$\dddot{x} + a_1\ddot{x} + a_2(\dot{x})\dot{x} + a_3x = 0$	$V = (a_1x_2 + x_3)^2 + \frac{a_3}{a_1}(a_1x_1 + x_2)^2 + \int_o^{x_2}\left(a_2(x_2) - \frac{a_3}{a_1}\right)x_2\,\mathrm{d}x_2$	$\dot{V} = -2(a_1a_2(\dot{x}) - a_3)x_2^2$	$a_1 > 0$ $a_3 > 0$ $a_1a_2(\dot{x}) - a_3 > 0$
$\dddot{x} + a_1(\dot{x})\ddot{x} + a_2\dot{x} + a_3x = 0$ $a_1(\dot{x}) = a_{10} + a_{1n}(\dot{x})$	$V = (a_3x_1 + a_2x_2)^2 + \left(a_2 - \frac{a_3}{a_{10}}\right)x_3^2 + \frac{a_3}{a_{10}}(a_{10}x_2 + x_3)^2 + a_3\int_o^{x_2} a_{1n}(x_2)\,\mathrm{d}x_2$	$\dot{V} = -2(a_1(\dot{x}) - a_3)x_3^2$	$a_{10} > 0$ $a_2 > 0$ $a_{10}a_2 - a_3 > 0$ $a_{1n} \geq 0$

Table 3.3: Method Used: Variable Multiplier

Differential Equation	Liapunov Function $V, \dot{V}$	Stability Conditions
$\dddot{x} + a_1\ddot{x} + (a_2\dot{x} + a_3x)^{2n+1} = 0$	$V = \frac{a_3}{a_1a_2 - a_3}\left\{\frac{(a_3x_1 + a_2x_2)^{2n+2}}{n+1} + a_1x_3^2 + 2x_2x_3 + \frac{a_2}{a_3}x_3^2\right\}$ $\dot{V} = -x_3^2$ $a_1a_2 - a_3 > 0$, asymptotic stability of origin $a_1a_2 - a_3 < 0$, asymptotic instability of the origin (Chetaev's Theorem)	$a_1a_2 - a_3 > 0$ $a_1 > 0$ $a_2 > 0$
$\dddot{x} + a_1(\dot{x})\ddot{x} + a_2\dot{x} + a_3x = 0$	$V = \frac{1}{a_3}(a_3x_1 + a_2x_2)^2 + \frac{1}{a_2a_3}(a_2x_3 + a_1x_2)^2 + \int_o^{x_2}\left[a_2(x_2) - \frac{a_3}{a_1}\right]x_2\,dx_2$ $\dot{V} = -\frac{a_3}{a_2}\left(a_1(x_2) - \frac{a_3}{a_1}\right)x_3^2$	$a_2 > 0$ $a_3 > 0$ $a_1(x_2) - \frac{a_3}{a_1} > 0$
$\ddot{x} + a_1(t)\dot{x} + a_2(t)x = 0$ $\lim_{t\to\infty} a_1(t), a_2(t) \to 0$	$V = x_1^2 + \frac{1}{a_2(t)}x_2^2, \quad \lvert a_1(t)\rvert < M_1,\ \lvert a_2(t)\rvert < M_2$ $\dot{V} = \frac{1}{a_2}\left(2a_1 - \frac{\dot{a}_2}{a_2}\right)x_2^2$	$a_2(t) > 0$ $2a_1 - \frac{\dot{a}_2}{a_2} > 0$

continued on next page

Differential Equation	Liapunov Function $V, \dot{V}$	Stability Conditions
$\dddot{x} + a_1\ddot{x} + a_2\dot{x} + a_3x^3$ $a_4\dot{x}\left(\sqrt{3}x + c_2x\right)^2 = 0$	$V = (a_1^2 + a_2 + c_2a_3)x_2^2 + 2a_1x_1x_3 + x_3^2 + a_1a_4\frac{x_1^4}{2} + 2a_4x^3x_2 + 3c_2a_4x_1^2x_2^2 + 2a_4c_2^2x_1x_2^3 + a_4c_2^2\frac{x_2^4}{2}$ $\dot{V} = -\left\{a_1a_2 + a_4(a_1a_2 - a_3x^2) + \left(\sqrt{3}x_1 + c_2x_2\right)^2\right\}x_2^2$	$V > 0$ $\dot{V} \leq 0$
$a_0\dddot{x} + a_1\ddot{x} + a_2\dot{x} + a_3x = 0$	$V = x'Sx$, $x' =$ Transpose x $S = \frac{1}{\Delta a_3}\begin{bmatrix} a_3^3 & a_3^2a_2 & 0 \\ a_3^2a_2 & a_3^2a_1 + a_3a_2^2 & a_3^2a_0 \\ 0 & a_3^2a_0 & a_3a_2a_0 \end{bmatrix}$ $\dot{V} = -(2/\Delta)x_3^2$	$a_i > 0$ $i = 1, 2, 3, 4$ $V > 0$ $\Delta = a_1a_2 - a_0a_3 > 0$ $\dot{V} \leq 0$

continued on next page

Differential Equation	Liapunov Function $V, \dot{V}$	Stability Conditions
$\dddot{x} + g_1(\ddot{x})\ddot{x} + a_2\dot{x} + a_3x = 0$	$V = x'S_2x + \frac{2a_1}{\Delta}\int_o^{x_2} [g_1(x_3) - a_1]\, x_2\, \mathrm{d}x_2$ $S_2 = \begin{bmatrix} a_1a_3 & a_3 & 0 \\ a_3 & a_1^2 + a_2 & a_1 \\ 0 & a_1 & 1 \end{bmatrix}$ $\dot{V} = -2x_2^2 - (2/\Delta)\,[g(x_3) - a_3]\, x_3^2$	$a_i > 0 \ i = 1, 2, 3$ $\Delta = a_1a_2 - a_3 > 0$
$\dddot{x} + g_1(\ddot{x})\ddot{x} + g_2(\dot{x})\dot{x} + g_3(x)x = 0$	$V = x'S_2x + \frac{2}{\Delta}(g_3 - a_3)x_1x_2$ $+ \frac{2a_1}{\Delta}\int_o^{x_1} (g_3 - a_3)x_1\, \mathrm{d}x_1$ $+ \frac{2}{\Delta}\int_o^{x_2} [(g_2 - a_2) + a_1(g_1 - a_1)]\, x_2\, \mathrm{d}x_2$ $\dot{V} = -\frac{2}{\Delta}\left(a_1g_2 - g_3 - \frac{\mathrm{d}g_3}{\mathrm{d}x_1}x_1\right)x_2^2$ $-\frac{2}{\Delta}(g_1 - a_1)x_3^2$	$a_i > 0 \ i = 1, 2, 3$ $a_1a_2 - a_3 > 0$ $g_i - a_i > 0 \ i = 1, 2, 3$ $a_1a_2 - g_3 - \frac{\mathrm{d}g_3}{\mathrm{d}x_1}x_1 \geq 0$

continued on next page

Differential Equation	Liapunov Function $V, \dot{V}$	Stability Conditions
$\dddot{x} + g_1(\ddot{x})\ddot{x} + g_2(\dot{x})\dot{x} + g_3(x)x = p(t)$ $S_2 = \begin{bmatrix} a_1a_3 & g_3 & 0 \\ g_3 & a_1^2 + a_2 & a_1 \\ 0 & a_1 & 1 \end{bmatrix}$ $\lvert p(t)\rvert < M < \infty$ $\lim_{t\to\infty} p(t) = 0$	$V = x'S_2x + \frac{2a_1}{\Delta}\int_o^{x_1}(g_3 - a_3)x_1\,\mathrm{d}x_1$ $+ \frac{2a_1}{\Delta}\int_o^{x_1}(g_3 - a_3)x_1\,\mathrm{d}x_1$ $+ \frac{2}{\Delta}\int_o^{x_2}[(g_2 - a_2) + a_1(g_1 - a_1)]\,x_2\,\mathrm{d}x_2$ $+ 2\left\{k_1 - \frac{1}{\Delta}\int_o^{t}\left[\frac{a_1}{g_2 - a_2} + \frac{1}{(g_1 - a_1)}\right]p^2(t)\,\mathrm{d}t\right\}$ $\dot{V} = -2\left[a_1a_2 - g_3 - \frac{\mathrm{d}g_3}{\mathrm{d}x_1}x_1\right]\frac{x_2^2}{\Delta}$ $- \frac{2a_1}{\Delta}(g_2 - a_2)\left(x_2 - \frac{p(t)}{g_2 - a_2}\right)^2$ $- \frac{2}{\Delta}(g_1 - a_1)\left(x_3 - \frac{p(t)}{g_1 - a_1}\right)^2$	$a_i > 0$ $g_i - a_i > 0, \quad i = 1, 2, 3$ $g_2 - a_2 \geq k_2 > 0$ $g_1 - a_1 \geq k_1 > 0$ $\frac{1}{\Delta}\left[\frac{a_1}{k_2} + \frac{1}{k_1}\right]\int_o^{t} p^2(t)\,\mathrm{d}t \leq k_3$ $a_1a_2 - g_3 - \frac{\mathrm{d}g_3}{\mathrm{d}x_1}x_1 > 0$

Table 3.4: Method Used: Variable Gradient

Differential Equation	Liapunov Function V	Derivative $\dot{V}$	Conditions of Global Asymptotic Stability
$\dddot{x} + a_1\ddot{x} + a_2\dot{x} + a_3(x) = 0$	$V = 2a_1 \int_o^{x_1} a_3(x_1)\,\mathrm{d}x_1 + 2a_3(x_1)x_2 + a_2x_2^2 + (a_3x_2 + x_3)^2$	$\dot{V} = -2\left[a_2a_3 - \frac{\mathrm{d}}{\mathrm{d}x_1}a_3(x_1)\right]x_2^2$	$a_1 > 0$ $a_2 > 0$ $a_1a_2 - \frac{\mathrm{d}}{\mathrm{d}x_1}a_3(x) > 0$
$\dddot{x} + a_1\ddot{x} + a_2(x)\dot{x} + a_3 = 0$	$V = \frac{a_3}{a_1}(a_1x_1 + x_2)^2 + \left(a_2(x_1) - \frac{a_3}{a_1}\right)x_2^2 + (a_1x_2 + x_3)^2$	$\dot{V} = -2\left[a_1a_2(x_1) - a_3 - \frac{1}{2}x_2\frac{\mathrm{d}}{\mathrm{d}x_1}a_2(x_1)\right]x_2^2$	$a_1 > 0$ $a_2 > 0$ $a_1a_2(x_1) - a_3 > 0$ $\left[a_1a_2(x_1) - a_3 - \frac{1}{2}x_2\frac{\mathrm{d}}{\mathrm{d}x_1}a_2(x_1)\right] > 0$
$\dddot{x} + a_1\ddot{x} + a_2(\dot{x}) + a_3x = 0$	$V = \frac{a_3}{a_1}(a_1x_1 + x_2)^2 + (a_1x_2 + x_3)^2 + 2\int_o^{x_2}\left(\frac{a_2(x_2)}{x_2} - \frac{a_3}{a_1}\right)x_2\,\mathrm{d}x_2$	$\dot{V} = -2\left[a_1\frac{a_2(x_2)}{x_2} - a_3\right]x_2^2$	$a_1 > 0$ $a_3 > 0$ $a_1\frac{a_2(x_2)}{x_2} - a_3 > 0$

continued on next page

Differential Equation	Liapunov Function V	Derivative $\dot{V}$	Conditions of Global Asymptotic Stability
$\dddot{x} + a_1\ddot{x} + a_2(\dot{x}) + a_3(x) = 0$ $\int_o^{x_1} a_3(x_1)\,\mathrm{d}x_1 = A_3$ $\int_o^{x_2} a_2(x_2)\,\mathrm{d}x_2 = A_2$	$V = 2a_1 \int_o^{x_1} a_3(x_1)\,\mathrm{d}x_1$ $+ 2a_3(x_1)x_2$ $+ 2\int_o^{x_2} a_2(x_2)\,\mathrm{d}x_2$ $+ (a_1x_2 + x_3)^2$	$\dot{V} = -2\left[a_1\frac{a_2(x_2)}{x_2} - \frac{\mathrm{d}}{\mathrm{d}x_1}a_3(x_1)\right]x_2^2$	$a_1 > 0$ $A_2 > 0$ $A_3 > 0$ $\frac{a_1a_2(x_2)}{x_2} - \frac{\mathrm{d}}{\mathrm{d}x_1}a_3(x_1) > 0$ $4a_1A_2A_3 - a_3^2(x_1)x_2^2 > 0$
$\ddot{x} + \dot{x} + \alpha x^3 = 0$	$V = \boldsymbol{x}'\boldsymbol{S}\boldsymbol{x} = \alpha\frac{x_1^4}{2} + x_2^2$ $\boldsymbol{S} = \begin{pmatrix} \alpha x_1^2/2 & 0 \\ 0 & 1 \end{pmatrix}$	$\dot{V} = \boldsymbol{x}'\boldsymbol{T}_e\boldsymbol{x} = -2x_2^2$ $T_e = 2\begin{pmatrix} 0 & 0 \\ 0 & 1 \end{pmatrix}$	$\alpha > 0$
$\ddot{x} + \dot{x}\left[1 + \frac{\mathrm{d}}{\mathrm{d}x}(f_1(x_1)x_1)\right]$ $+ \beta x_1 f(x_1) = 0$	$V = 2\int_0^{x_1} \beta f(x_1)x_1\,\mathrm{d}x_1 + x_2^2$	$\dot{V} = -2\left[1 + \frac{\mathrm{d}}{\mathrm{d}x}[f(x_1)x_1]\right]x_2^2$	$\beta f(x_1) > 0$ $\frac{\mathrm{d}}{\mathrm{d}x_1}[x_1 f(x_1)] > -1$

continued on next page

Differential Equation	Liapunov Function V	Derivative $\dot{V}$	Conditions of Global Asymptotic Stability
$\dddot{x} + 3\ddot{x} + \left(2 + 3\dot{x}^2\right)\dot{x}$ $+\beta x^3 = 0$	$V = 3\beta\frac{x_1^4}{2} + 2\beta x_1^3 x_2 + 11x_2^2$ $+ \frac{3x_2^4}{2} + 6x_2x_3 + x_3^2$ $V = \frac{3\beta}{2}\left(x_1^2 + \frac{2x_2}{3}\right)^2 + \frac{3x_2^4}{2}$ $+ (3x_2 + x_3)^2 + 2x_2^2\left(1 - \frac{\beta}{3}\right)$	$\dot{V} = -6\left(2 + 3x_2^2 - \beta x_1^2\right)x_2^2$	$\beta < 3$ $2 + 3x_2^2 > \beta x_1^2$
$\dddot{x} + 3\ddot{x} + \left(2 + 3x^2\right)\dot{x}$ $+\beta x^3 = 0$	$V = \boldsymbol{x'Sx}$ $= x_1^6 + \left(1 + \frac{\beta^2}{2}\right)x_1^4$ $+ 2(1+\beta)x_1^3x_2 + 2x_1^3x_3$ $+ (2+3\beta)x_2^2 + 2\beta x_2x_3 + x_3^2$	$\dot{V} = -2K_1x_3^2 - 2K_2x_2^2$ $K_1 = (3-\beta)$ $K_2 = (2\beta - 3)$	$\frac{3}{2} \leq \beta \leq 3$
$\dddot{x} + a_1\dot{x} + a_2(\dot{x})\dot{x}$ $+ a_3(x)x = 0$	$V = 2a_1\int_0^{x_1} a_3(x_1)x_1\,\mathrm{d}x_1$ $+ 2a_3(x_1)x_1x_2$ $+ a_1^2x_2^2 + 2\int_0^{x_2} a_2(x_2)x_2\,\mathrm{d}x_2$ $+ 2a_1x_2x_3 + x_3^2$	$\dot{V} = 2x_2^2\left[a_1a_2(x_2) - a_3(x_1) - x_1\frac{\mathrm{d}a_3(x_1)}{\mathrm{d}x_1}\right]$	$a_1 > 0$ $a_2(x_2) > 0$ $a_3(x_1) > 0$ $a_1a_2(x_2) - a_3(x_1) - x_1\frac{\mathrm{d}}{\mathrm{d}x_1}a_3(x_1) > 0$

continued on next page

Differential Equation	Liapunov Function V	Derivative $\dot{V}$	Conditions of Global Asymptotic Stability
$\dddot{x} + a_1(x,\dot{x})\ddot{x} + a_2\dot{x} + a_3x = 0$	$V = a_2(a_3x_1 + a_2x_2)^2 + (a_3x_2 + a_2x_3)^2 + 2a_2a_3I$ $I = \int_0^{x_2}\left(a_1(x_1,x_2) - \frac{a_3}{a_2}\right)x_2\,\mathrm{d}x_2$	$\dot{V} = 2a_2^2\left[a_2(x_1x_2) - \frac{a_3}{a_2}\right]x_3^2 + 2a_2a_3x_2\int_0^{x_2}\frac{\partial a_1}{\partial x_1}x_2\,\mathrm{d}x_2$	$a_1(x_1,x_2)a_2 - a_3 = \alpha > 0$ $\alpha x_3^2 > a_3x_2\int_0^{x_2}\frac{\partial a_1}{\partial x_1}x_2\,\mathrm{d}x_2$
$\ddddot{x} + a_1\dddot{x} + a_2\ddot{x} + a_3\dot{x} + a_4(x) = 0$	$V = (x_4 + a_1x_3 + \Delta_2x_2)^2 + \frac{a_3}{a_1}\left(x_3 + a_1x_2 + \frac{a_1}{a_3}a_4(x_1)\right)^2 + \left(\frac{a_3}{a_1}\Delta_2 - \frac{\mathrm{d}}{\mathrm{d}x}a_4(x)\right)x_2^2 + 2\Delta_2\int_0^{x_1}a_4(x_1)\,\mathrm{d}x_1 + \frac{a_1}{a_3}(a_4(x_1))^2$ $\Delta_2 = a_2 - \frac{a_3}{a_1}$	$\dot{V} = -2\left[\left(a_2 - \frac{a_3}{a_1}\right)a_3 - a_1\frac{\mathrm{d}a_4(x)}{\mathrm{d}x} + \frac{1}{2}\frac{\mathrm{d}^2a_4(x)}{\mathrm{d}x^2}x_2\right]x_2^2$	$a_1 > 0$ $a_2 > 0$ $a_3 > 0$ $a_1a_2 - a_3 > 0$ $\left[\Delta_2a_3 - a_1\frac{\mathrm{d}a_4(x)}{\mathrm{d}x} + \frac{1}{2}\frac{\mathrm{d}^2a_4(x)}{\mathrm{d}x^2}x_2\right] > 0$ $\int_0^{x_1}a_4(x_1)\,\mathrm{d}x_1 > \alpha$ $\alpha = \frac{a_1}{2a_3\Delta_2}a_4^2(x_1)$

continued on next page

Differential Equation	Liapunov Function V	Derivative $\dot{V}$	Conditions of Global Asymptotic Stability
$\ddddot{x} + a_1 \dddot{x} + a_2(x, \dot{x})\ddot{x} + a_3\dot{x} + a_4 x = 0$ $a_4 = a_4(x, \dot{x})$	$V = \frac{a_3}{a_1}\left[\frac{a_1 a_4}{a_3}x_1 + a_1 x_2 + x_3\right]^2 + \left[\frac{a_1 a_4}{a_3}x_2 + a_1 x_3 + x_4\right]^2 + \left[a_2(x_1, x_2) - \frac{a_3}{a_1} - \frac{a_1 a_4}{a_3}\right]x_3^2 + \left[2a_1 A_2 - \frac{a_4}{a_3}\left(a_3 + \frac{a_1^2 a_4}{a_3}\right)x_2^2\right]$ $A_2 = \frac{2a_1 a_4}{a_3}\int_0^{x_2} a_2(x_1, x_2)x_2\,\mathrm{d}x_2$	$\dot{V} = -2\left[a_1 a_2(x_1 x_2) - a_3 - \frac{a_1^2 a_4}{a_3} - \frac{1}{2}\left(x_2\frac{\partial}{\partial x_1}a_4 + x_3\frac{\partial}{\partial x_2}a_4\right)\right]x_3^2 + \frac{a_1 a_4}{a_3}x_2\int_0^{x_2}\frac{\partial a_4}{\partial x_1}x_2\,\mathrm{d}x_2$	$V > 0$ $\dot{V} \leq 0$

continued on next page

Differential Equation	Liapunov Function V	Derivative $\dot{V}$	Conditions of Global Asymptotic Stability
$\ddddot{x} + a_1\dddot{x} + a_2(x,\dot{x})\ddot{x} + a_3(\dot{x}) + a_4(x) = 0$ $\int_0^{x_1} a_4(x_1)\,\mathrm{d}x_1 = A_4$ $\int_0^{x_2} a_3(x_2)\,\mathrm{d}x_2 = A_3$ $\int_0^{x_2}\left[a_1^2 a_2(x_1,x_2) - \frac{\mathrm{d}a_4}{\mathrm{d}x_1}\right]x_2\,\mathrm{d}x_2 = A_2$	$V = 2a_1^2A_4 + 2a_1a_4(x_1)x_4 + 2a_4(x_1)x_3 + 2a_1A_3 + 2x_3a_3(x_2) + a_2(x_1,x_2)x_3^2 + 2a_1^3x_2x_3 + 2a_1^2x_2x_4 + 2a_1x_3x_4 + 2A_2$	$\dot{V} = -2a_1\left[a_1\frac{a_3(x_2)}{x_2} - \frac{\mathrm{d}}{\mathrm{d}x_1}a_4(x_1)\right]x_2^2 - 2\left[a_1a_2(x_1,x_2) - \frac{\mathrm{d}}{\mathrm{d}x_2}a_3(x_2) - \frac{1}{2}\frac{\mathrm{d}}{\mathrm{d}t}a_2(x_1x_2) - a_1^3\right]x_3^2 + 2x_4\int_0^{x_2}\left[a_1^2\frac{\partial}{\partial x_1}a_2(x_1,x_2) - \frac{\mathrm{d}^2}{\mathrm{d}x_1^2}a_4(x_1)\right]x_2\,\mathrm{d}x_2$	$V > 0$ $\dot{V} \le 0$

Bibliography

[Elaydi, S.N.] Elaydi, S.N and Harris, W.A. On the Computation of A^N, *Siam Rev.* 40(4), 965–971, December 1998.

[Hahn, W.] Hahn, W. *Theory and Application of Liapunov's Direct Method,* Englewood Cliffs, NJ: Prentice-Hall Inc., 146–150, 1963.

[Jury, I.E.] Jury, I.E. *Sampled Data Control Systems,* New York: John Wiley and Sons, Chap. 1, 34–35, 1958.

[Puri, N.N.] Ku, Y.H. and Puri, N.N. Liapunov Functions for High Order Systems, *Journal of the Franklin Institute,* Vol. 276(5), 349–365, November 1963.

[Puri, N.N.] Puri, N.N. and Weygandt, C.N. Second method of Liapunov and Routh's Canonical Form, *Journal of the Franklin Institute,* 276(5), 365–384, November 1963.

[Puri, N.N.] Puri, N.N. and Drake, R.L. Stability studies for a class of nonlinear difference equations using second method of Liapunov, *Journal of the Franklin Institute,* Vol. 279, 209–217, March 1965.

Chapter 4

Complex Variables for Transform Methods

4.1 Introduction

The Fourier series is a very powerful tool in solving various problems in engineering. It is nonetheless restricted to the periodic functions. In Chapter 5, we generalize the Fourier series tool to the nonperiodic functions. Digital, continuous, and resultant transforms are referred to as Fourier transforms, Laplace transforms, Z-transforms, or discrete Fourier transforms.

Fourier transforms are useful for determining the steady state response of a network and frequency contents of the signals and various levels of energy bands of different frequencies. However, the Fourier transform is not well suited for the time dependent transients and the stability studies. Instead, the Laplace transform is ideally suited for this purpose and represents a powerful tool for the study of the linear time invariant systems. The digital systems are studies via the Z-transform.

When numerical computations have to be performed, the Discrete Fourier Transforms (DFT) and Fast Fourier Transforms (FFT) have a comparative advantage. In order to understand the true importance of these transforms, a rudimentary (maybe more) knowledge of the complex plane integration is absolutely essential. For this reason, this chapter starts with the complex variable and complex plane integration and lays the foundations for the transform theory. Inspite of many wonderful books on analytical theory, this chapter provides only the essentials, fitting it into the larger frame of an engineering mathematical structure.

4.2 Complex Variables and Contour Integration

4.2.1 Definition of a Complex Variable

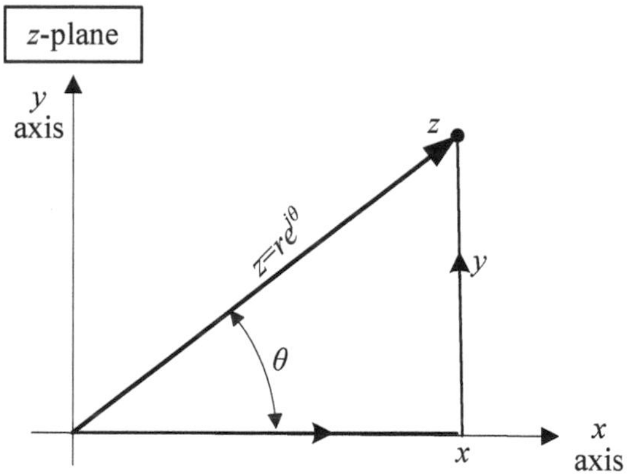

Figure 4.1: Complex Plane

Figure 4.1 shows a complex plane z, complex number is represented as a point.

$$z = x + jy = \sqrt{x^2 + y^2} \angle \arctan \frac{y}{x} = re^{j\theta} \tag{4.1}$$

x, y are called Rectangular Coordinates, whereas, r, θ are designated as polar coordinates. We shall extend the calculus of functions of a real variable to the complex variable z, involving ideas of differentiation, integration, power series, etc.

Consider a simple function

$$f(z) = z^2 = (x + jy)^2 = (x^2 - y^2) + j2xy$$

In general,

$$f(z) = f(x + jy) = u(x, y) + jv(x, y) \tag{4.2}$$

where u and v are real functions of the variables x and y. The function $f(z)$ can also be considered as a transformation, where every point z in the z-plane is transformed to a corresponding point $f(z)$ in the $f(z)$-plane as shown in Figure 4.2.

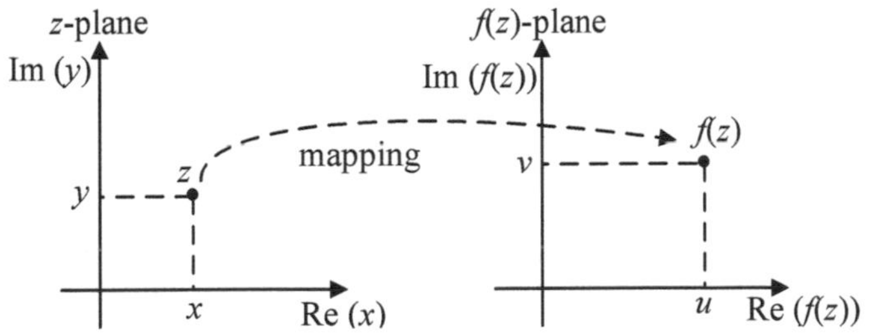

Figure 4.2: Mapping from Complex Plane z to Complex Plane $f(z)$

Unless otherwise stated, we shall deal with only the single valued functions.

Definition of a Simple Curve

A simple curve simply means that it does not cross itself, yielding a single valued function only. Very often it is also referred to as Jordan curve. We shall deal with only simple curves.

4.2.2 Analytic Function

A function $f(z)$ is called **Analytic** (or regular) in a **region** or a domain, of a complex Z-plane if it has a unique derivative at all the points in that region. The statement "$f(z)$ is analytic at a point"(called Regular Point) means that the function $f(z)$ has a unique continuous derivative at every point inside a small region surrounding the point known as the neighborhood. A curious result of the existence of the first derivative of a complex function at the point is that it guarantees the existence of the derivatives of all higher orders. This is an entirely different situation from the real variable case where the existence of first derivative is no guarantee for the existence of higher derivatives. As a result of this we can define Analyticity in another way:

> If a function $f(z)$, is analytic at a point $z = z_0$, then it can be expanded in a Taylor series in the neighborhood of the point z_0. "Analytic" and "Regular" are often used synonymously.

In the literature, analytic function is often referred to as a "holomorphic" function. When a function is analytic everywhere, it is called an **Entire Function**. A region or a point where the function is not analytic is known as "singular." We shall see that a function that is analytic everywhere is really not very exciting. In fact, it is the singularities of a function that are of greater importance and provide some exciting results (discussed later in this chapter).

The whole theory of complex variables can be developed from the above powerful definition of analyticity. Its significance will be apparent later on.

4.2.3 Derivative of Complex Variable at a Point

The derivative of a function is defined as:

$$f'(z_0) = \left.\frac{\mathrm{d}f(z)}{\mathrm{d}z}\right|_{z=z_0} = \lim_{\substack{\Delta z \to 0 \\ z \to z_0}} \left[\frac{(f(z_0 + \Delta z) - f(z_0)}{\Delta z}\right]$$

$$\Delta z = (\Delta^2 x + \Delta^2 y)^{1/2} \angle \arctan \frac{\Delta y}{\Delta x}$$

$f(z)$ can be written in terms of its rectangular coordinates as:

$$f(z) = u(x, y) + jv(x, y)$$

For the above definition of the derivative to have a meaning, it should have a unique value regardless of the direction in which $\Delta z \rightarrow 0$ as the point z_0 is approached. Let us approach the point z_0 from two different directions. This will result in certain conditions to be satisfied by the analytic functions $f(z)$.

1. Approaching z_0 horizontally $\Delta z = \Delta x, \Delta y = 0$

$$f'(z_0) = \left.\frac{\mathrm{d}f(z)}{\mathrm{d}z}\right|_{z=z_0} = \lim_{\substack{\Delta x \to 0 \\ z \to z_0}} \left[\frac{(f(z + \Delta x) - f(z)}{\Delta x}\right] \tag{4.3}$$

2. Approaching z_0 vertically $\Delta z = j\Delta y, \Delta x = 0$

$$f'(z_0) = \left.\frac{\mathrm{d}f(z)}{\mathrm{d}z}\right|_{z=z_0} = \lim_{\substack{\Delta y \to 0 \\ z \to z_0}} \left[\frac{(f(z + \Delta y) - f(z)}{\Delta y}\right] \tag{4.4}$$

For analytic functions, equating the expressions in the above Eqs. 4.3 and 4.4 because of the uniqueness of $\left.\frac{\mathrm{d}f(z)}{\mathrm{d}z}\right|_{z=z_0}$, yields

$$\begin{aligned} \frac{\partial u}{\partial x} &= \frac{\partial v}{\partial y} \\ \frac{\partial u}{\partial y} &= -\frac{\partial v}{\partial x} \end{aligned} \qquad \text{at } x = x_0, y = y_0 \tag{4.5}$$

Furthermore,

$$\frac{\partial f}{\partial z} = \frac{\mathrm{d}f}{\mathrm{d}z} = \frac{\partial f}{\partial x} = \frac{1}{j}\frac{\partial f}{\partial y}$$

Known as Cauchy-Reimann conditions of analyticity at a point $z = z_0$.

Example 4.1:

$f(z) = z^2$ is analytic in the whole z-plane. This can be verified as the following:

$$f(z) = (x + jy)^2 = x^2 - y^2 + j2xy$$
$$u(x, y) = x^2 - y^2$$
$$v(x, y) = 2xy$$
$$\frac{\partial u}{\partial x} = \frac{\partial v}{\partial y} = 2x$$
$$\frac{\partial u}{\partial y} = -\frac{\partial v}{\partial x} = -2y$$

Furthermore, these Cauchy-Reimann conditions are satisfied at every point, implying

$$\frac{\mathrm{d}f}{\mathrm{d}z} = \frac{\partial f}{\partial x} = \frac{1}{j}\frac{\partial f}{\partial y} = \frac{\partial u}{\partial x} + j\frac{\partial v}{\partial y} = 2x + j2y = 2z$$

Differentiating Eq. 4.5, yields

$$\frac{\partial^2 u}{\partial x^2} + \frac{\partial^2 u}{\partial y^2} = \frac{\partial^2 v}{\partial x^2} + \frac{\partial^2 v}{\partial y^2} = 0$$

which is known as the Laplace equation in two dimensions. A three-dimensional version of this equation is very common in field theory. Thus,

> If, $f(z) = u + jv$ is Analytic in a region, then u and v satisfy the Laplace equation. Knowing u, one can determine v, and vice versa. The pair u, v is called a **Conjugate** or **Harmonic Function**.

4.2.4 Path of Integration

Figure 4.3 shows C as a simple oriented curve (does not cross itself).

By definition

$$\int_{z_1 C}^{z_2} f(z)\,\mathrm{d}z = \lim_{\Delta z_k \to 0} \sum_{k=0}^{\infty} f(z_k)\Delta z_k \tag{4.6}$$

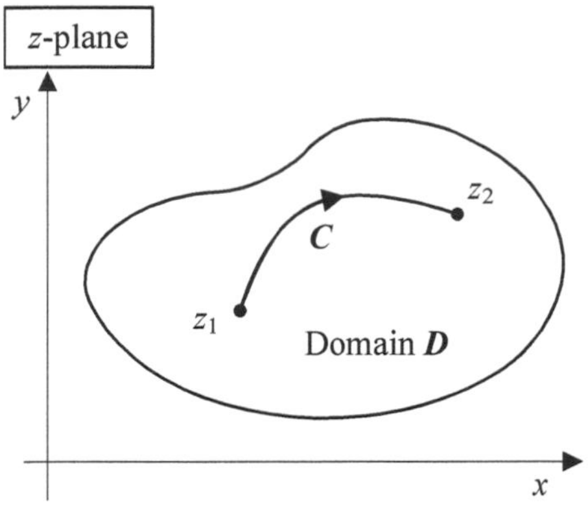

Figure 4.3: Complex Function Integration Path

In the case of real variable x, $\int_{x_1}^{x_2} f(x)dx$ depends only on the endpoints x_1 and x_2. But the complex integral in Eq. 4.6 is more complicated. It does not depend only on the endpoints z_1 and z_2 but also on the path of integration taken to arrive from z_1 to z_2.

Question:

> Under what circumstances does $\int_{z_1}^{z_2} f(z)dz$ depend only on the endpoints z_1 and z_2, independent of the path of integration?

This is an important question and the answer is as following:

If the paths connecting z_1 and z_2 lies in a domain (D) of the complex z-plane where $f(z)$ is analytic everywhere, then the integral $\int_{z_1}^{z_2} f(z)\,\mathrm{d}z$ is independent of paths of integration lying in the domain (D).

Proof: For the above statement to be true, (see Figure 4.4)

$$\int_{z_1 C_i}^{z_2} f(z)\,\mathrm{d}z = \int_{z_1}^{z_2} \mathrm{d}F(z) = F(z_2) - F(z_1) \qquad i = 1, \ldots, n \tag{4.7}$$

From Eq. 4.7,

$$F(z) = \int f(z)\,\mathrm{d}z = \int \frac{\mathrm{d}F(z)}{\mathrm{d}z}\,\mathrm{d}z$$

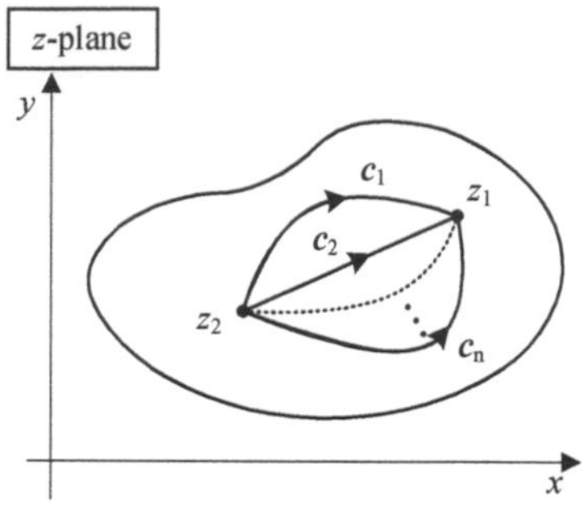

Figure 4.4: Integration Independent of the Path

$$F(z) = P(x, y) + jQ(x, y) = P + jQ$$

$$f(z) = u(x, y) + jv(x, y) = u + jv$$

$$\mathrm{d}z = \Delta x + j\Delta y$$

$$\mathrm{d}F(z) = f(z)\,\mathrm{d}z \tag{4.8}$$

Also,

$$\mathrm{d}F(z) = \frac{\partial F(z)}{\partial x}\Delta x + \frac{\partial F(z)}{\partial y}\Delta y = \left(\frac{\partial P}{\partial x} + j\frac{\partial Q}{\partial x}\right)\Delta x + \left(\frac{\partial P}{\partial y} + j\frac{\partial Q}{\partial y}\right)\Delta y \tag{4.9}$$

Equating the expressions in Eqs. 4.8 and 4.9, we get

$$\left(\frac{\partial P}{\partial x}\Delta x + \frac{\partial P}{\partial y}\Delta y\right) + j\left(\frac{\partial Q}{\partial x}\Delta x + \frac{\partial Q}{\partial y}\Delta y\right) = (u + jv)(\Delta x + j\Delta y)$$

Separating the real and imaginary parts,

$$\left(\frac{\partial P}{\partial x}\Delta x + \frac{\partial P}{\partial y}\Delta y\right) = u\Delta x - v\Delta y$$

$$\left(\frac{\partial Q}{\partial x}\Delta x + \frac{\partial Q}{\partial y}\Delta y\right) = u\Delta x + v\Delta y$$

or

$$u = \frac{\partial P}{\partial x} = \frac{\partial Q}{\partial y}$$
$$v = \frac{\partial Q}{\partial x} = -\frac{\partial P}{\partial y}$$

Taking partial differentials of u and v,

$$\begin{aligned}
\frac{\partial u}{\partial x} &= \frac{\partial^2 P}{\partial x^2} = \frac{\partial^2 Q}{\partial x \partial y} \\
\frac{\partial u}{\partial y} &= \frac{\partial^2 Q}{\partial y^2} = \frac{\partial^2 P}{\partial y \partial x} \\
\frac{\partial v}{\partial x} &= \frac{\partial^2 Q}{\partial x^2} = -\frac{\partial^2 P}{\partial x \partial y} \\
\frac{\partial v}{\partial y} &= -\frac{\partial^2 P}{\partial y^2} = \frac{\partial^2 Q}{\partial y \partial x}
\end{aligned} \tag{4.10}$$

Fulfillment of the above Eq. 4.10 implies,

$$\frac{\partial u}{\partial x} = \frac{\partial v}{\partial y}, \qquad \frac{\partial u}{\partial y} = -\frac{\partial v}{\partial x}$$
$$\frac{\partial^2 u}{\partial x \partial y} = \frac{\partial^2 u}{\partial y \partial x}, \qquad \frac{\partial^2 v}{\partial x \partial y} = \frac{\partial^2 v}{\partial y \partial x}$$

These are Cauchy-Reimann conditions.

Therefore, for the integral of a complex function on a simple curve to be independent of the path of integration lying in a domain (D), the following two conditions must be satisfied.

1. Function is analytic in (D).
2. Function is single valued in (D).

The above results can be obtained from the Green's theorem stated below.

Green's Theorem

If $\int(Pdx + Qdy)$ is independent of the path of integration in D, then there exists a function:

$$F(z) = F(x + jy) = P(x, y) + jQ(x, y)$$

such that

$$\frac{\partial F}{\partial x} = P(x, y), \quad \frac{\partial F}{\partial y} = Q(x, y)$$

Converse of the above theorem is also true yielding Cauchy's Integral Theorem.

4.2.5 Useful Facts about Complex Variable Differentiation

i.

$$f(z) = u + jv,\ z = x + jy,\ z^* = x - jy,\ x = \frac{1}{2}(z + z^*),\ y = \frac{j}{2}(z - z^*)$$

ii.

$$f(z) = f(x + jy) = f(x, y) = f(\frac{(z + z^*)}{2}, \frac{(z - z^*)}{2j})$$

Hence, $\frac{\mathrm{d}f(z)}{\mathrm{d}z}$ and $\frac{\partial f(z)}{\partial z}$ will be used interchangeably.

iii.

$$\frac{\mathrm{d}f(z)}{\mathrm{d}z} = \frac{\partial u}{\partial x} + j\frac{\partial v}{\partial x} = \frac{1}{j}\left(\frac{\partial u}{\partial y} + j\frac{\partial v}{\partial y}\right)$$

iv.

$$\frac{\partial f(z)}{\partial y} = j\frac{\partial f(z)}{\partial x}$$

v.

$$\frac{\mathrm{d}f(z)}{\mathrm{d}z} = \frac{\partial f(z)}{\partial z} = \frac{1}{2}\left(\frac{\partial f(z)}{\partial x} - j\frac{\partial f(z)}{\partial y}\right)$$

vi.

$$\frac{\mathrm{d}f(z)}{\mathrm{d}z^*} = \frac{\partial f(z)}{\partial z^*} = \frac{1}{2}\left(\frac{\partial f(z)}{\partial x} + j\frac{\partial f(z)}{\partial y}\right)$$

vii.

$$\frac{\mathrm{d}f(z)}{\mathrm{d}z} \neq \frac{\partial f(z)}{\partial x} + j\frac{\partial f(z)}{\partial y}$$

viii.

$$\nabla_z f(z) = \frac{\partial f(z)}{\partial x} + j\frac{\partial f(z)}{\partial y} = z\frac{\mathrm{d}f(z)}{\mathrm{d}z^*}$$

ix. **If a function $f(z)$ is nonanalytic, then**

$$\frac{\partial f(z)}{\partial x} = \frac{\partial f(z)}{\partial z}\frac{\partial z}{\partial x} + \frac{\partial f(z)}{\partial z^*}\frac{\partial z^*}{\partial x}$$

$$\frac{\partial g(f(z))}{\partial x} = \left(\frac{\partial g(f(z))}{\partial f(z)}\right)\left(\frac{\partial f(z)}{\partial x}\right) + \left(\frac{\partial g(f(z^*))}{\partial f(z^*)}\right)\left(\frac{\partial f(z^*)}{\partial x}\right)$$

The last term involving $\frac{\partial f(z)}{\partial z^*}$ becomes zero when $f(z)$ is analytic.

4.2.6 Cauchy's Integration Theorem

Let C be a simple Jordan closed curve in the z-plane such that a function $f(z)$ is analytic **everywhere inside as well as on it.**

Cauchy's Integral Theorem states:

$$\oint_C f(z)\,\mathrm{d}z = \oint_C f(z)\,\mathrm{d}z = 0$$

The arrows show the direction of integration, one being clockwise and the other being counterclockwise.

Proof:

Let C be the closed curve and two points z_1 and z_2 lying on it as shown in the above Figure 4.5. C_1 and C_2 are the two segments of C joining z_1 and z_2. "↺" stands for travel "Around the closed curve in the counterclockwise direction."

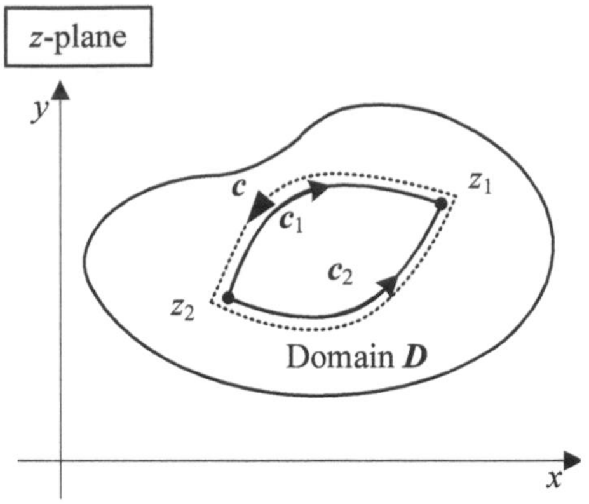

Figure 4.5: Cauchy's Integral Theorem

The function $f(z)$ is analytic, therefore (from the previous section)

$$\circlearrowright \int_{z_1 C_1}^{z_2} f(z)\,\mathrm{d}z = \circlearrowright \int_{z_1 C_2}^{z_2} f(z)\,\mathrm{d}z = -\circlearrowright \int_{z_1 C_2}^{z_2} f(z)\,\mathrm{d}z$$

$$\circlearrowright \int_{z_1 C_1}^{z_2} f(z)\,\mathrm{d}z + \circlearrowright \int_{z_2 C_2}^{z_1} f(z)\,\mathrm{d}z = 0$$

$$\oint_{C} f(z)\,\mathrm{d}z = 0 \tag{4.11}$$

This is the famous Cauchy's Integral Theorem.

4.2.7 Modified Cauchy's Integral Theorem

Consider two closed curves C_1 and C_2 in the z-plane as shown in the Figure 4.6. The function $f(z)$ is analytic everywhere on C_1 and C_2 as well as inside the annular region. We can prove the following Modified Cauchy's Integral Theorem

$$\oint_{C_1} f(z)\,\mathrm{d}z = \oint_{C_2} f(z)\,\mathrm{d}z$$

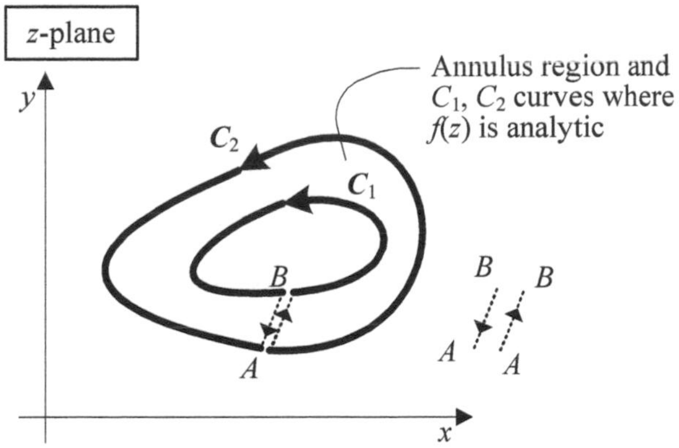

Figure 4.6: Modified Cauchy's Integral Theorem

Proof:

Make a cut between C_1 and C_2 along the path A, B. Pictorially, they are two separate edges but in reality they coincide.

Let us choose the following path of integration: from A to B, along C_1 in counterclockwise fashion, then B to A and then along C_2 in the clockwise fashion. This results in a complete closed path. Keep in mind the following three points:

1. The integrals along the opposite directions AB and BA cancel each other.
2. The integral along the curve in the clockwise direction is negative of the integral along the same curve in the counterclockwise direction.
3. The function to be integrated, $f(z)$, is analytic along the closed curve as well as the enclosed region.

Thus,

$$\oint_{C_1} f(z)\,\mathrm{d}z + \oint_{C_2} f(z)\,\mathrm{d}z = 0$$

or,

$$\oint_{C_1} f(z)\,\mathrm{d}z = -\oint_{C_2} f(z)\,\mathrm{d}z = \oint_{C_2} f(z)\,\mathrm{d}z \tag{4.12}$$

4.2.8 Taylor Series Expansion and Cauchy's Integral Formula

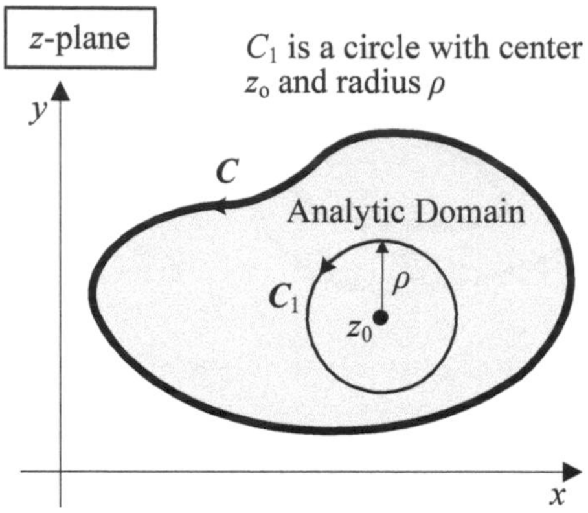

Figure 4.7: Taylor Series Expansion about Analytic Point

Figure 4.7 shows a closed curve C in the z-plane. The function, $f(z)$, is analytic everywhere on as well as inside this closed curve C. **Cauchy's Integral Formula** is a very remarkable result, which states that if we know $f(z)$ at **every point on such a curve**, the function is completely known and computable **everywhere inside** the region inside the curve C. Thus, if z_0 is any point in the region **inside** C, then via Cauchy's Integral Formula, we are able to compute the value of the function and all its derivatives on the point z_0.

This is accomplished through the following equation:

$$\left.\frac{\mathrm{d}^n f(z)}{\mathrm{d}z^n}\right|_{z=z_0} = f^{(n)}(z_0) = \frac{n!}{2\pi j}\oint \frac{f(z)}{(z-z_0)^{n+1}}\,\mathrm{d}z, \qquad n = 0, 1, \cdots \tag{4.13}$$

Proof:

The proof follows from the Taylor Series Expansion. Since $f(z)$ is analytic at $z = z_0$, the Taylor series expansion yields:

$$f(z) = \sum_{k=0}^{\infty} \frac{f^k(z_0)}{k!}(z-z_0)^k$$

Dividing both sides by $(z - z_0)^{n+1}$,

$$\phi(z) = \frac{f(z)}{(z - z_0)^{n+1}} = \sum_{k=0}^{\infty} \frac{f^k(z_0)}{k!}(z - z_0)^{k-n-1} \tag{4.14}$$

Since the above function $\phi(z)$ is analytic in the region surrounded by C_1 and C as well as on both the curves, utilizing Eq. 4.12,

$$\oint_{C_1} f(z)\,\mathrm{d}z = \oint_{C} f(z)\,\mathrm{d}z \tag{4.15}$$

From Eqs. 4.14 and 4.15

$$\oint_{C} \frac{f(z)}{(z - z_0)^{n+1}}\,\mathrm{d}z = \sum_{k=0}^{\infty} \frac{f^k(z_0)}{k!} \oint_{C_1} (z - z_0)^{k-n-1}\,\mathrm{d}z \tag{4.16}$$

Along the circle C_1, $\quad z - z_0 = \rho e^{j\theta}, \quad \mathrm{d}z = j\rho e^{j\theta}\,\mathrm{d}\theta,$

$$\oint_{C_1} (z - z_0)^{k-n-1}\,\mathrm{d}z = j\rho^{k-n} \int_0^{2\pi} e^{j(k-n)\theta}\,\mathrm{d}\theta$$

$$\int_0^{2\pi} e^{j(k-n)\theta} d\theta = \begin{cases} 2\pi & n = k \\ 0 & n \neq k \end{cases} \tag{4.17}$$

From Eqs. 4.15, 4.16, and 4.17, we get

$$\frac{n!}{2\pi j} \oint_{C} \frac{f(z)}{(z - z_0)^{n+1}}\,\mathrm{d}z = f^{(n)}(z_0), \qquad n = 0, 1, 2, \cdots \tag{4.18}$$

This is **Cauchy's Integral Formula**.

4.2.9 Classification of Singular Points

A function may not be analytic at a point but still may have some most interesting properties. Let us consider a function $f(z)$, which can be expanded in the "Laurent" series about a point $z = z_0$, yielding

$$f(z) = \sum_{n=-\infty}^{\infty} a_n(z - z_0)^n = \sum_{n=0}^{\infty} a_n(z - z_0)^n + \sum_{n=1}^{\infty} a_{-n}(z - z_0)^{-n} \tag{4.19}$$

1. **Analytic Function**

$$a_{-n} \equiv 0 \text{ for all positive integers } n$$

2. **Function with pole of order** m

$$a_{-n} = 0, \qquad \text{for all integers } n > m, \qquad m < \infty$$

We call it a function having a pole of finite order m.

3. **Essential Singularity**

When $m \to \infty$, we call it a function having **essential singularity**.

4. **Branch Point**

Consider

$$f(z) = \sqrt{(z)}$$

This equation has a singular point at $z = z_0$ called a Branch Point. This kind of singularity appears in partial differential equations and series summation. This is further elaborated in a later section.

4.2.10 Calculation of Residue of $f(z)$ at $z = z_0$

When a function is not analytic at a point, it cannot be expanded in a Taylor series. However, such a function can be expanded about a singular point in the "Laurent" series. Let the function $f(z)$ have a pole of order m at $z = z_0$, then the Laurent series expansion yields:

$$f(z) = \sum_{n=0}^{\infty} a_n(z - z_0)^n + \sum_{n=1}^{m} a_{-n}(z - z_0)^{-n}$$

Multiplying both sides by $(z - z_0)^m$,

$$(z - z_0)^m f(z) = \sum_{n=0}^{\infty} a_n(z - z_0)^{n+m} + a_{-1}(z - z_0)^{m-1} + \cdots + a_{-m} \tag{4.20}$$

We shall see that a_{-1} is the only important parameter in the above equation and is known as the **residue** of the function $f(z)$ for reasons explained in the section involving the computation of complex integrals. To find the **residue**, a_{-1}, the expression in the Eq. 4.20 is differentiated $(m - 1)$ times and computed at $z = z_0$. Successive differentiations eliminate a_{-2}, a_{-3}, ..., a_{-n} and the substitution of $z = z_0$ removes the positive powers of the series, yielding,

$$\frac{d^{m-1}}{dz^{m-1}}\Big[(z - z_0)^m f(z)\Big]_{z=z_0} = (m - 1)!a_{-1}$$

$$a_{-1} = \text{Residue of } f(z)|_{z=z_0} = \frac{1}{(m - 1)!}\frac{d^{m-1}}{dz^{m-1}}[(z - z_0)^m f(z)]|_{z=z_0} \tag{4.21}$$

For a simple pole, $a_{-1} = [(z - z_0)f(z)]\Big|_{z=z_0}$.

Obviously, m has to be finite integer for this concept to be useful.

4.2.11 Contour Integration

Application of Residue Theorem to Isolated Singularities

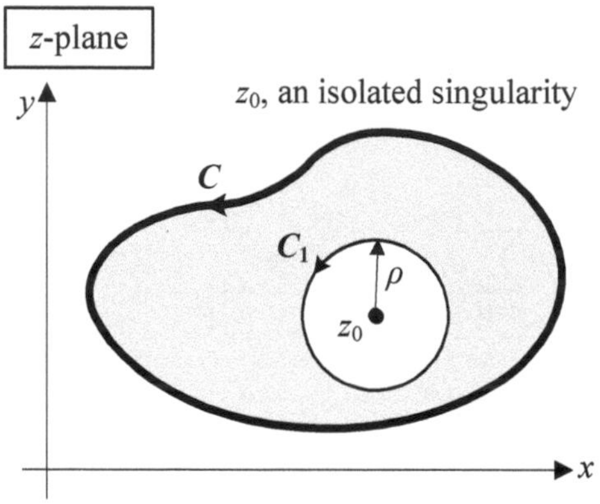

Figure 4.8: Isolated Singularity of $f(z)$ at z_0

Figure 4.8 shows a closed curve C surrounding an isolated singular point z_0 of the function $f(z)$. Function $f(z)$ is analytic everywhere inside and on the curve C, except at the point z_0. Let us surround the point z_0 with a small curve of radius ρ. Via Cauchy's integral formula

$$\oint_C f(z)\,\mathrm{d}z = \oint_{C_1} f(z)\,\mathrm{d}z$$

Substituting from Eq. 4.19 to the right-hand side of the above equation

$$\oint_C f(z)\,\mathrm{d}z = \sum_{n=0}^{\infty} a_n \oint_{C_1} (z-z_0)^n\,\mathrm{d}z + \sum_{n=1}^{\infty} a_{-n}(z-z_0)^{-n}\,\mathrm{d}z \tag{4.22}$$

$$z - z_0 = \rho e^{j\theta}, \quad \mathrm{d}z = j\rho e^{j\theta} d\theta$$

$$\oint_C f(z)\,\mathrm{d}z = \sum_{n=0}^{\infty} a_n\rho^{n+1} \int_0^{2\pi} e^{j(n+1)\theta}\,\mathrm{d}\theta + \sum_{n=1}^{\infty} a_{-n}\rho^{-n+1} \int_0^{2\pi} e^{j(-n+1)\theta}\,\mathrm{d}\theta$$

$$\int_0^{2\pi} e^{j(n+1)\theta}\, \mathrm{d}\theta = \left[\frac{e^{j(n+1)\theta}}{j(n+1)}\right]_0^{2\pi} = 0, \quad n > 0$$

$$\int_0^{2\pi} e^{j(-n+1)\theta}\, \mathrm{d}\theta = \begin{cases} 0 & n \neq 1 \\ 2\pi & n = 1 \end{cases}$$

Thus, Eq. 4.22 is simplified as

$$\oint_C f(z)\, \mathrm{d}z = 2\pi j a_{-1} = 2\pi j[\text{Residue of } f(z) \text{ at the singular point } z_0]$$

Only the term involving a_{-1} has not vanished upon integration. It is called the "Residue of the function $f(z)$ at $z = z_0$ (upon integration)" defined as Res $[f(z)]_{z=z_0}$.

4.2.12 Contour Integral Computation

Residue Theorem—Several Isolated Singularities

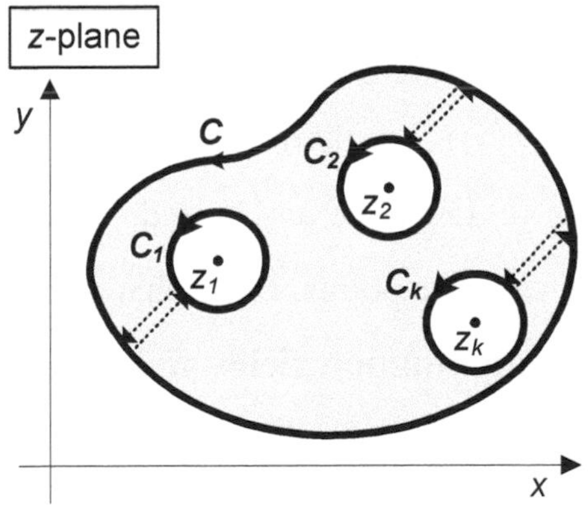

Figure 4.9: Several Isolated Singularities

Figure 4.9 shows the closed curve C surrounding the singular points $z_1, z_2, \ldots, z_k$ of the function $f(z)$. Except for these points, the function $f(z)$ is analytic everywhere inside and on the curve C. Let us enclose each point $z_i (i = 1, 2, \ldots, k)$ with a small circle of radius ρ centered at z_i. Introducing cuts along C and C_i, $(i = 1, 2, \ldots, k)$ as shown in the Figure 4.9. As discussed earlier, in the section on modified Cauchy's Integral Theorem, the integrals along the cuts cancel each other, resulting in

$$\text{Near } z = z_i, \quad f(z) = \sum_{n=0}^{\infty} a_n(i)(z - z_i)^n + \sum_{n=1}^{\infty} a_{-n}(i)(z - z_i)^{-n}$$

$$\oint_C f(z)\,\mathrm{d}z = \sum_{i=1}^{k} \oint_{C_i} f(z)\,\mathrm{d}z = 2\pi j \sum_{i=1}^{k} a_{-1}(i)$$

$= 2\pi j$ [Sum of Residues of the singularities of the function inside C]

Important: No singularity is allowed on the contour of integration C, which can be modified to include simple singularities on the contour C.

4.2.13 Summary on Evaluation of Residues—Special Results

- **Functions with no singularity at infinity** (∞)

  Consider a function $f(z)$ that has finite distance singular points z_i, $i = 1, 2, \ldots, k$. Everywhere else in the z-plane, this function is analytic including infinity ($z \to \infty$). Furthermore, the function $f(z)$ is of the order $z^{-(2+\epsilon)}$, as $z \to \infty$, $\epsilon \geq 0$. Let C be any closed curve in the z-plane that surrounds all the above singular points. Note that $f(z)$ is analytic outside this curve C. **This particular situation yields an interesting result**:

$$I = \oint_C f(z)\,\mathrm{d}z = 2\pi j \sum_{i=1}^{k} \text{Res}\,[f(z)]_{z=z_i} = 0$$

$$\sum_{i=1}^{k} \text{Res}\,[f(z)]_{z=z_i} = 0 \tag{4.23}$$

Proof:

Let

$$z = \frac{1}{\lambda}, \qquad dz = -\frac{1}{\lambda^2}\, d\lambda \tag{4.24}$$

Furthermore, let C_1 be a circle in the z-plane with a radius R and centered at the origin, R being large enough to surround the curve C. The point z on this circle (C_1) is defined as:

$$z = Re^{j\theta}, \quad \lambda = \frac{1}{R}e^{-j\theta}$$

From Cauchy's Integral Formula,

$$I = \oint_C f(z)\, dz = \oint_{C_1} f(z)\, dz$$

From Eq. 4.24,

$$I = \oint_{C_1} f(z)\, dz = \oint_\Gamma \left[-\frac{1}{\lambda^2} f\left(\frac{1}{\lambda}\right)\right] d\lambda = \oint_\Gamma \left[\frac{1}{\lambda^2} f\left(\frac{1}{\lambda}\right)\right] d\lambda \tag{4.25}$$

where Γ is a circle centered at the origin in the λ-plane with the radius $\frac{1}{R}$. Realizing that $f\left(\frac{1}{\lambda}\right)$ has no poles inside the circle Γ, the kernel $\frac{1}{\lambda^2} f\left(\frac{1}{\lambda}\right)$ contributes zero to the integral I, yielding

$$I = \oint_\Gamma \left[\frac{1}{\lambda^2} f\left(\frac{1}{\lambda}\right)\right] d\lambda = 0 = \oint_C f(z)\, dz \tag{4.26}$$

Hence,

$$\sum_{i=1}^{k} \text{Res}\,[f(z)]_{z=z_i} = 0$$

- **Functions with singularity at infinity** (∞)

  In this case, the function $f(z)$ has isolated finite distance singularities at z_i, $i = 1, 2, \ldots, k$ as well as a singularity at infinity (∞). Otherwise, the function is analytic everywhere else in the z-plane.

  Furthermore, the function $f(z)$ is of the order $z^{-(2+\epsilon)}$, as $z \to \infty$, $\epsilon \geq 0$. Following the same line of reasoning as above, we shall show:

$$\text{Res}\,[f(z)]_{z\to\infty} = -\sum_{i=1}^{k} \text{Res}\,[f(z)]_{z=z_i} = -\text{Res}\left[\frac{1}{z^2} f\left(\frac{1}{z}\right)\right]_{z=0} \tag{4.27}$$

The above result can be arrived at as following:

The circle C_1 is enlarged with $R \to \infty$. As a result C_1 encloses all the finite distance singularities as well as the singularity at infinity, yielding

$$[I]_{R\to\infty} = \oint_{C_1} f(z)\, \mathrm{d}z = 2\pi j \left(\sum_{i=1}^{k} \text{Res}\,[f(z)]_{z=z_i} + \text{Res}\,[f(z)]_{z\to\infty} \right) = 0$$

$$\text{Res}\,[f(z)]_{z\to\infty} = -\sum_{i=1}^{k} \text{Res}\,[f(z)]_{z=z_i} = -\left(\frac{1}{2\pi j}\right) \oint_{C} f(z)\, \mathrm{d}z \tag{4.28}$$

From Eq. 4.25, the above equation can be also be written as

$$\oint_{C} f(z)\, \mathrm{d}z = \oint_{\Gamma} \left[\frac{1}{\lambda^2} f\left(\frac{1}{\lambda}\right)\right] \mathrm{d}\lambda = \text{Res}\left[\frac{1}{\lambda^2} f\left(\frac{1}{\lambda}\right)\right]_{\lambda\to 0} \tag{4.29}$$

Combining Eqs. 4.28 and 4.29, the final result is:

$$\text{Res}\,[f(z)]_{z\to\infty} = -\sum_{i=1}^{k} \text{Res}\,[f(z)]_{z=z_i} = -\text{Res}\left[\frac{1}{z^2} f\left(\frac{1}{z}\right)\right]_{z=0}$$

Note: If the residue at infinity is zero, the above result simplifies as

$$\sum_{i=1}^{k} \text{Res}\,[f(z)]_{z=z_i} = 0$$

- If z_0 is a regular point of $f(z)$, then

$$\text{Res}\,[f(z)]_{z\to z_0} = 0, \qquad z_0 \neq \infty$$

- If z_0 is a pole of order n for the function $f(z)$, then

$$\text{Res}\,[f(z)]_{z\to z_0} = \frac{1}{(n-1)!}\lim_{z\to z_0}\left[\frac{\mathrm{d}^{n-1}}{\mathrm{d}z^{n-1}}(z-z_0)^n f(n)\right], \qquad z_0 \neq \infty$$

- If $\lim z f(z)|_{z\to\infty} = A_\infty\ (\neq \infty)$, then

$$\text{Res}\,[f(z)]_{z\to\infty} = -A_\infty$$

- If $\lim f(z)|_{z\to z_0} = A_0$, then

$$\text{Res}\,[f(z)]_{z\to z_0} = A_0$$

- Consider

$$f(z) = \frac{n(z)}{d(z)}, \qquad f(z_0) \neq 0$$

a. If $z = z_0$ is a simple zero of $d(z)$, namely a simple pole of $f(z)$, then

$$\text{Res}\,[f(z)]_{z\to z_0} = \frac{n(z_0)}{d'(z_0)}$$

b. If $z = z_0$ is a second order zero of $d(z)$, namely a double pole of $f(z)$

then

$$\text{Res}\,[f(z)]_{z\to z_0} = \frac{6n'(z_0)d''(z_0) - 2n(z_0)d'''(z_0)}{3[d''(z_0)]^2}$$

" ′ " stands for derivative wrt the variable z.

Example 4.2:

Consider a meromorphic function $f(z)$ with no singularities other than the poles in the entire z-plane. Let

$$f(z) = \frac{h(z)}{g(z)}$$

where $h(z)$ and $g(z)$ are analytic or regular at $z = z_0$. We can write

$$f(z) = \frac{\sum_{n=0}^{\infty} c_n(z - z_0)^n}{\sum_{n=0}^{\infty} d_n(z - z_0)^n} = \sum_{n=0}^{\infty} e_n(z - z_0)^n$$

$$\sum_{n=0}^{\infty} c_n(z - z_0)^n = \left(\sum_{n=0}^{\infty} d_n(z - z_0)^n\right)\left(\sum_{n=0}^{\infty} e_n(z - z_0)^n\right)$$

Equating powers of $(z - z_0)$ we obtain

$$c_i = \sum_{k=0}^{i} d_{i-k}e_k \qquad i = 1, 2, \cdots$$

The residue of this function at $z = z_0$ is 0.

Example 4.3:

Consider a meromorphic function $f(z)$ having a pole of order r_k at $z = z_k$.

$$f(z) = \frac{1}{(z - z_k)^{r_k}}\frac{h(z)}{g(z)}$$

where $h(z)$ and $g(z)$ are regular at $z = z_k$. Since $h(z)$ and $g(z)$ are regular at $z = z_k$

$$\frac{h(z)}{g(z)} = \frac{\sum_{n=0}^{\infty} c_n(z - z_0)^n}{\sum_{n=0}^{\infty} d_n(z - z_0)^n} = \sum_{n=0}^{\infty} e_n(z - z_0)^n \qquad (4.30)$$

$$f(z) = \frac{1}{(z-z_k)^{r_k}} \frac{h(z)}{g(z)} = \frac{1}{(z-z_k)^{r_k}} \left[\sum_{n=0}^{\infty} e_n (z-z_0)^n \right] \tag{4.31}$$

In the above expansion, only $(z-z_k)^{r_k-1}$ in the numerator contributes to the residue

$$\text{Res}\,[f(z)]_{z=z_k} = e_{r_k-1} \tag{4.32}$$

From Eq. 4.30,

$$\sum_{n=0}^{\infty} c_n (z-z_0)^n = \left[\sum_{n=0}^{\infty} d_n (z-z_0)^n \right] \left[\sum_{n=0}^{\infty} e_n (z-z_0)^n \right]$$

$$\begin{aligned} c_0 &= d_0\, e_0 \\ c_1 &= d_1\, e_0 + d_0\, e_1 \\ &\vdots \\ c_{r_k-1} &= d_{r_k-1}\, e_0 + d_{r_k-2}\, e_1 + \cdots + d_0\, e_{r_k-1} \end{aligned}$$

$$\text{Res}\,[f(z)]_{z=z_k} = e_{r_k-1} = \frac{\Delta}{(d_0)^{r_k}}$$

$$\Delta = \begin{vmatrix} d_0 & 0 & 0 & \cdots & c_0 \\ d_1 & d_0 & 0 & \cdots & c_1 \\ d_2 & d_1 & d_0 & \cdots & c_2 \\ \vdots & & & & \\ d_{r_k-1} & d_{r_k-2} & d_{r_k-3} & \cdots & c_{r_k-1} \end{vmatrix}$$

4.2.14 Laurent Series Expansion of a Function $f(z)$

Figure 4.10 shows two closed curves C_1 and C_2, with a point z_0 inside C_1. The function $f(z)$ is analytic in annulus region D between the two curves, as well as on C_1 and C_2. Let us choose a curve C through the point z and lying in annulus region D between C_1 and C_2 as shown in Figure 4.10.

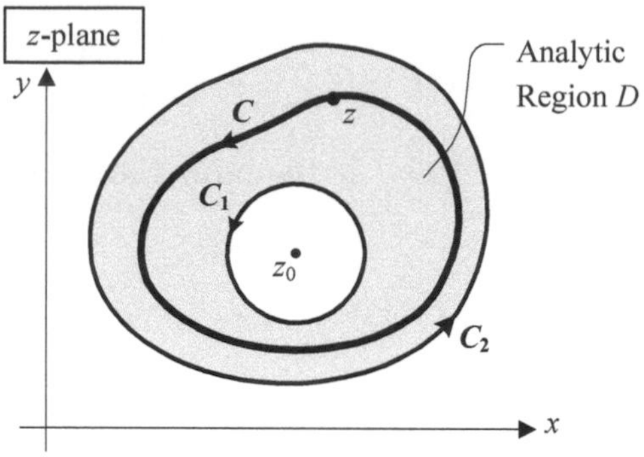

Figure 4.10: Laurent Series

Such a function $f(z)$ can be expanded in the Laurent series as following:

$$f(z) = \sum_{n=0}^{+\infty} a_n(z - z_0)^n + \sum_{n=1}^{+\infty} a_{-n}(z - z_0)^{-n} = \sum_{n=-\infty}^{+\infty} a_n(z - z_0)^n$$

$$a_n = \frac{1}{2\pi j} \oint_C \frac{f(z)}{(z - z_0)^{(n+1)}} \, \mathrm{d}z \quad n = 0, 1, 2, \ldots$$

$$a_{-n} = \frac{1}{2\pi j} \oint_C \frac{f(z)}{(z - z_0)^{(-n+1)}} \, \mathrm{d}z \quad n = 1, 2, \ldots$$

Here, D is the region of convergence, the a_n series converges when the point z is inside the curve C. The "a_{-n}" series is an inverse power series and converges when the point z is outside the curve C. It is important to note that the point z is always in the annulus region D. Both the series converge on the curve C.

- If all $a_{-n} = 0$ then the function $f(z)$ is analytic (or regular) at z_0.
- If $n = k$, $a_{-n} = a_{-k} \neq 0$ but all subsequent a_{-n} are zero, then $f(z)$ is said to have a pole of order k, thereby implying that $f(z)(z - z_0)^k$ is analytic at $z = z_0$.
- For $k = 1$, $f(z)$ has a simple pole.

- Poles are referred to as a nonessential singularity.

- If there are infinite number of a_{-n} different from zero, then $f(z)$ is said to have an essential singularity.

- When a function has nonessential singularity, namely pole at $z = z_0$, then a_{-1} is called the **Residue** of $f(z)$ at $z = z_0$. As explained earlier, the **residue** of a function at a singular point plays an essential role in complex integration. This is one of the most important concepts in the study of complex variables.

- If $C_1 = C_2 = C$ are circles of radius unity, then $f(z)$ defines an analytic function on a unit disk. Letting $z = e^{j\theta}$, we realize that such an analytic function has a Fourier series expansion, and all the other nice properties that go with it.

Proof of Laurent Series

Let us draw a small circle Γ with the center at any general point z in the region D.

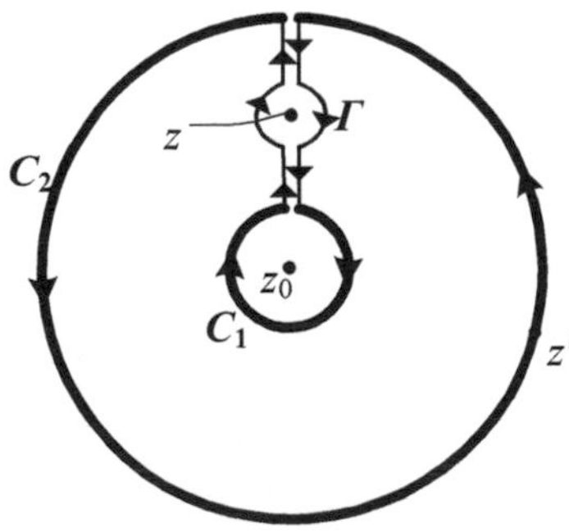

Figure 4.11: Proof of Laurent Series

Consider the contour involving C_1, Γ, and C_2. Let $z^{'}$ be a point on this contour.

Since the function $\frac{f(z')}{z'-z}$ has no singularity inside the above contour,

$$I(z) = \oint_{C_1,\Gamma,C_2} \frac{f(z')}{z'-z}\,dz' = 0 \tag{4.33}$$

Expanding the above integral

$$I(z) = \oint_{C_1} \frac{f(z')}{z'-z}\,dz' + \oint_{\Gamma} \frac{f(z')}{z'-z}\,dz' + \oint_{C_2} \frac{f(z')}{z'-z}\,dz' = 0$$

But,

$$\oint_{\Gamma} \frac{f(z')}{z'-z}dz' = -2\pi j f(z)$$

Thus,

$$2\pi j f(z) = \oint_{C_1} \frac{f(z')}{z'-z}\,dz' + \oint_{C_2} \frac{f(z')}{z'-z}\,dz'$$

$$2\pi j f(z) = I_1(z) + I_2(z) \tag{4.34}$$

where,

$$I_1(z) = \oint_{C_1} \frac{f(z')}{z'-z}\,dz', \qquad I_2(z) = \oint_{C_2} \frac{f(z')}{z'-z}\,dz'$$

Computation of $I_2(z)$

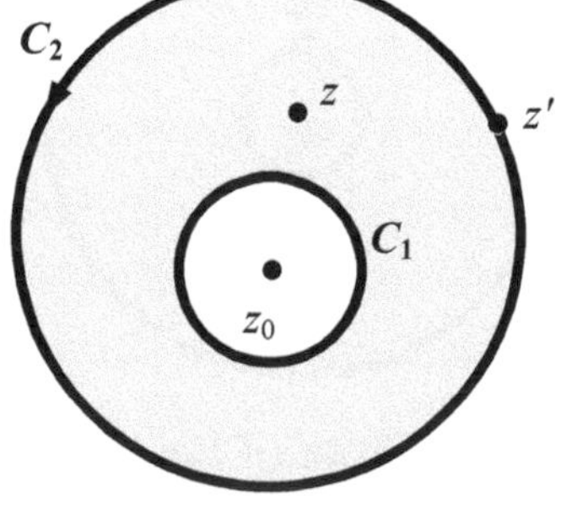

Figure 4.12: Computation of $I_2(z)$

The points z and z_0 are inside the curve C_2 shown in Figure 4.12. Therefore,

$$\left|\frac{z - z_0}{z' - z_0}\right| < 1$$

$$\frac{1}{z' - z} = \frac{1}{(z' - z_0) - (z - z_0)} = \frac{1}{(z' - z_0)}\left(\frac{1}{1 - (z - z_0)(z' - z_0)^{-1}}\right)$$

$$\frac{1}{1 - x} = \sum_{0}^{\infty} x^n, \qquad |x| < 1, \quad x = \frac{(z - z_0)}{(z' - z_0)}$$

$$\frac{1}{z' - z} = \frac{1}{z' - z_0}\sum_{n=0}^{\infty}\frac{(z - z_0)^n}{(z' - z_0)^n} = \sum_{n=0}^{\infty}\frac{(z - z_0)^n}{(z' - z_0)^{n+1}}$$

$$I_2(z) = \oint_{C_2} f(z')\left[\sum_{n=0}^{\infty}\frac{(z - z_0)^n}{(z' - z_0)^{n+1}}\right] dz' = \sum_{n=0}^{\infty}\left[\oint_{C_2}\frac{f(z')}{(z' - z_0)^{(n+1)}}\, dz'\right](z - z_0)^n$$

Define

$$\oint_{C_2}\frac{f(z')}{(z' - z_0)^{(n+1)}}\, dz' = 2\pi j a_n, \qquad n = 0, 1, 2, \ldots \tag{4.35}$$

Thus,

$$I_2(z) = 2\pi j\sum_{n=0}^{+\infty} a_n(z - z_0)^n$$

Computation of $I_1(z)$

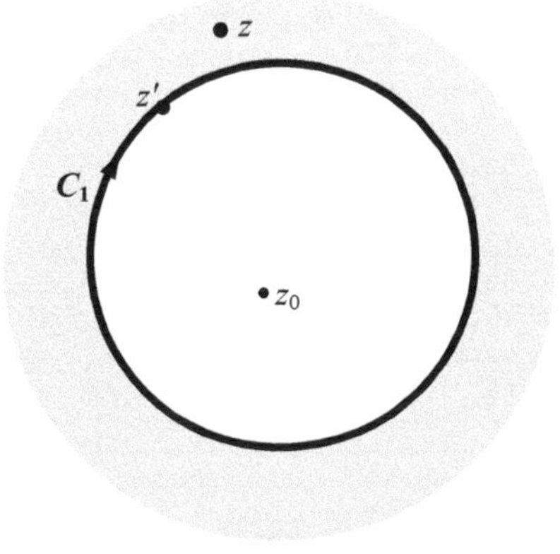

Figure 4.13: Computation of $I_1(z)$

The point z is outside the C_1, z_0 is inside the C_1, and z' is on C_1. Therefore,

$$\left|\frac{z' - z_0}{z - z_0}\right| < 1$$

$$\frac{1}{z' - z} = \frac{-1}{(z - z_0) - (z' - z_0)} = \frac{-1}{(z - z_0)} \sum_{n=0}^{\infty} \left[\frac{z' - z_0}{z - z_0}\right]^n$$

$$I_1(z) = \oint_{C_1} \frac{f(z')}{(z' - z)} \mathrm{d}z' = -\oint_{C_1} f(z') \sum_{n=0}^{\infty} \frac{(z' - z_0)^n}{(z - z_0)^{n+1}}$$

$$= \oint_{C_1} f(z') \left[\sum_{n=1}^{\infty} \frac{(z' - z_0)^{n-1}}{(z - z_0)^n}\right] \mathrm{d}z' = 2\pi j \sum_{n=1}^{\infty} \frac{b_n}{(z - z_0)^n}$$

where

$$2\pi j b_n = \oint_{C_1} f(z')(z' - z_0)^{n-1} \,\mathrm{d}z', \qquad n = 1, 2, \ldots \tag{4.36}$$

Let us define $b_n = a_{-n}$.

Combine Eqs. 4.34, 4.35, and 4.36 and realize that the integrations around the contour C_1, C_2, and C are the same (C resides in the analytic region).

$$f(z) = \sum_{n=0}^{\infty} a_n(z - z_0)^n + \sum_{n=1}^{\infty} a_{-n}(z - z_0)^{-n} = \sum_{n=-\infty}^{\infty} a_n(z - z_0)^n$$

$$\oint_{C} \frac{f(z')}{(z' - z_0)^{n+1}} \mathrm{d}z' = \oint_{C_1} \frac{f(z')}{(z' - z_0)^{n+1}} \mathrm{d}z' = \oint_{C_2} \frac{f(z')}{(z' - z_0)^{n+1}} \mathrm{d}z'$$

$$a_n = \frac{1}{2\pi j} \oint_{C} \frac{f(z')}{(z' - z_0)^{n+1}} \mathrm{d}z' \qquad n = 0, \pm 1, \pm 2, \cdots$$

The coefficients a_n are computed on C_2, while coefficients a_{-n} are computed on C_1. But if we choose a curve C lying between C_1 and C_2, then both a_n and a_{-n} can be computed on C as shown above.

4.2.15 Evaluation of Real Integrals by Residues

Cauchy's Principle Value of a Definite Integral

While studying the Fourier transform, one comes across the integrals of the type

$$I = \int_{-\infty}^{+\infty} f(x)\,\mathrm{d}x \tag{4.37}$$

These integrals have a value in common mathematical sense only when $f(x)$ is finite everywhere for $-\infty < x < \infty$. What happens when $f(x)$ has singularities along the line of integration $-\infty < x < \infty$? For all type of singularities we have no general method for computing Eq. 4.37. However, if thc singularities are isolated and simple, then the integral I can be computed as a limit via the contour integration in terms of residues of these singularities. To emphasize this use of residues, the value so obtained is called the **Cauchy's Principle Value** and designated as "PV." Thus,

$$I = \mathrm{PV}\int_{-\infty}^{+\infty} f(x)\,\mathrm{d}x, \text{ when computed via residues} \tag{4.38}$$

As an illustration, consider: $I = \int_{a}^{b} \frac{\mathrm{d}x}{x-c}, \qquad b > c > a$

Since the function $(x-c)^{-1}$ goes to $\pm\infty$ at $x = c^+, x = c^-$, we remove a small symmetric interval of $2r$ centered at $x = c$ and in the limit let $r \to 0$. Thus,

$$\begin{aligned} I &= \int_{a}^{c-r} \frac{\mathrm{d}x}{x-c} + \int_{c+r}^{b} \frac{\mathrm{d}x}{x-c} = [\ln(x-c)]\,|_a^{c-r} + [\ln(x-c)]\,|_{c+r}^{b} \\ &= \ln(r) - \ln(c-a) + \ln(b-c) - \ln(r) \\ &= \ln\left(\frac{b-c}{c-a}\right) = \left(\mathrm{PV}\int_{a}^{b} \frac{\mathrm{d}x}{x-c}\right) \end{aligned}$$

In reality as x crosses the point c, an infinite height positive area right around c is canceled with an infinite negative height area as a result of the introduction of symmetric cut $2r$. Essentially, the same result is accomplished by the introduction of contour distortion around singularity. This process of computing the proper integrals is referred to as **Cauchy's Principle Value** (PV $\int_{-\infty}^{+\infty} f(x)\,\mathrm{d}x$). Very often, "PV" will be omitted as being understood and only repeated to emphasize. We shall evaluate these real integrals using integration in the complex plane.

Let

$$z = jx \qquad \mathrm{d}z/j = \mathrm{d}x$$

$$x = \pm\infty \quad \Rightarrow \quad z = \pm j\infty$$

Thus,

$$I = \int_{-\infty}^{+\infty} f(x)\,\mathrm{d}x = -j\int_{-j\infty}^{+j\infty} f(z/j)\,\mathrm{d}z$$

Note that the above integral is performed along the j axis in the z-plane from $z = -j\infty$ to $z = +j\infty$. The contour of integration is only a line and therefore open. But Cauchy's Residue Theorem applies only to the closed contours. Therefore, we need to close the path from $+j\infty$ to $-j\infty$ over some other curve C. **How to choose this path such that contribution from this chosen curve C is either zero or easily computable is a question answered by Jordan's Lemmas**. Thus, before we proceed with the computation of the above integral, it is necessary to study these Jordan's Lemmas.

The following facts in conjunction with these lemmas are very useful:

- If L_C is the length of a simple curve C, then $\int_C f(z)\,\mathrm{d}z \le \max_{z\in C} |f(z)| L_c$
- If C is a semicircular arc of radius r centered at the origin, then

$$L_C = \pi R\,, \quad L_C \to \infty \quad \text{as} \quad R \to \infty$$

Furthermore,

$$\lim_{R\to\infty}\int_C f(z)\,\mathrm{d}z = 0, \quad \text{if } f(z) \text{ varies as } \frac{1}{z^p}, \quad p > 1$$

$$\lim_{R\to\infty}\int_C f(z)\,\mathrm{d}z = \text{constant}, \quad \text{if } f(z) \text{ varies as } \frac{1}{z} \text{ on the curve } C$$

- If C is a curve of radius R then

$$\circlearrowleft \int_{C_1}^{C_2} \frac{\mathrm{d}z}{z} = -j(\theta_2 - \theta_1), \quad \theta \in [\theta_1, \theta_2] \quad z = Re^{j\theta}, \; z_1 = Re^{j\theta_1}, \; z_2 = Re^{j\theta_2}$$

- $$\lim_{|x|\to 0} x \ln x = 0$$

- $$\lim_{|x|\to 0} x \ln(1 + x) = x$$

- If

$$f(z) = \ln\left(\frac{z-a}{z+a}\right), \quad z = Re^{j\theta}, \quad \theta \in [-\pi/2, +\pi/2]$$

Then

$$I_R = \lim_{R\to\infty}\int_C f(z)\,\mathrm{d}z = (\pi j a)$$

Jordan's Lemmas

Figure 4.14 shows z_0 to be an isolated singularity of $f(z)$, such that $f(z)(z - z_0)^{-1}$ is analytic at $z = z_0$. Let C_ρ and C_R be the arcs of circles with radius ρ and R such that $\rho \to 0$ and $R \to \infty$ as shown in the figure.

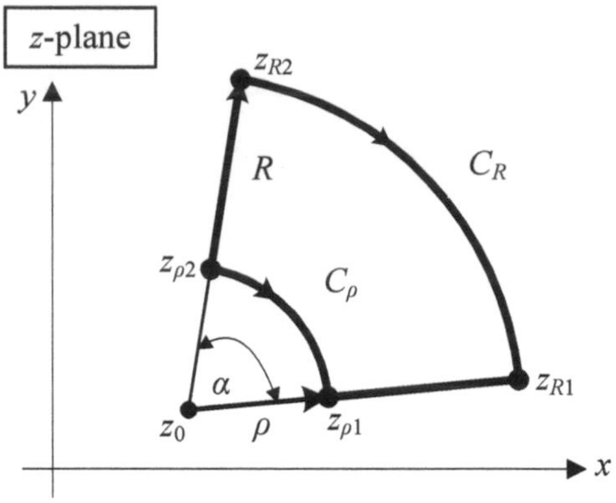

Figure 4.14: Explanation for Jordan Lemma 1

Thus,

$$z_{\rho_1} - z_0 = \rho e^{j\theta_1}$$

$$z_{\rho_2} - z_0 = \rho e^{j(\theta_1+\alpha)}$$

$$z_{\mathbb{R}_1} - z_0 = Re^{j\theta_1}$$

$$z_{\mathbb{R}_2} - z_0 = Re^{j(\theta_1+\alpha)}$$

Jordan's Lemma 1

If the function $f(z)$ meets the following two conditions:

1. $\operatorname{Res}\,[f(z)]_{z\to z_0} = \lim_{z\to z_0}(z - z_0)f(z) = a_0$ ($f(z)$ has a simple pole at z_0)

2. $\lim zf(z)|_{z\to\infty} = A_0$

Then Jordan's Lemma 1 states

$$\lim_{\rho\to 0} \circlearrowright \int_{C_\rho} f(z)\,\mathrm{d}z = -j\alpha a_0$$

$$\lim_{R\to\infty} \circlearrowright \int_{C_R} f(z)\,\mathrm{d}z = -j\alpha A_0$$

Proof:

Let us prove the first of the above equations.

The function $f(z)$ has a simple pole at $z = z_0$, therefore $(z - z_0)f(z)$ is analytic at z_0 and can be expanded in the Taylor series, yielding

$$(z - z_0)f(z) = \sum_{k=0}^{\infty} a_k(z - z_0)^k$$

Let

$$z - z_0 = \rho e^{j\theta}, \qquad \mathrm{d}z = j\rho e^{j\theta}\,\mathrm{d}\theta$$

Thus,

$$\lim_{\rho\to 0} \circlearrowleft \int_{C_\rho} f(z)\,\mathrm{d}z = \lim_{\rho\to 0}\left[\sum_{k=0}^{\infty} a_k\rho^k \int_{\theta_1+\alpha}^{\theta_1} je^{k\theta}\,\mathrm{d}\theta\right] = -j\alpha a_0 \tag{4.39}$$

To prove the second equation, let us expand $(z - z_0)f(z)$ in the Taylor series about powers of $(z - z_0)^{-1}$, yielding

$$(z - z_0)f(z) = \sum_{k=0}^{\infty} \frac{A_k}{(z - z_0)^k}$$

or

$$f(z) = \sum_{k=0}^{\infty} \frac{A_k}{(z - z_0)^{k+1}}$$

$$z - z_0 = Re^{j\theta}, \qquad \mathrm{d}z = jRe^{j\theta}\,\mathrm{d}\theta$$

$$\lim_{R\to\infty} \circlearrowleft \int_{C_R} f(z)\,\mathrm{d}z = \lim_{R\to\infty}\left[\sum_{k=0}^{\infty} \frac{A_k}{R^k} \int_{\theta_1+\alpha}^{\theta_1} je^{k\theta}\,\mathrm{d}\theta\right] = -j\alpha A_0$$

Jordan's Lemma 2

If the function $f(z)$ satisfies the following two conditions:

1. $f(z)$ is analytic except for a finite number of isolated singularities.

2. $\lim f(z)|_{z\to\infty} = 0$.

Then the Jordan's Lemma 2 states

$$\lim_{R\to\infty} \int_{C_{1R}} f(z)e^{zt}\, dz = 0 \qquad t > 0$$

$$\lim_{R\to\infty} \int_{C_{2R}} f(z)e^{zt}\, dz = 0 \qquad t < 0$$

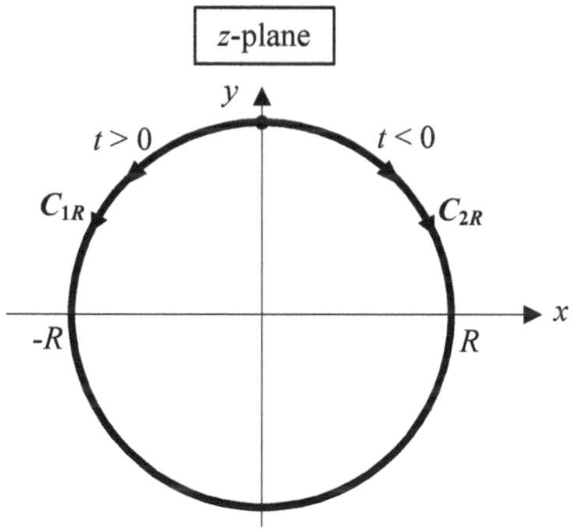

Figure 4.15: Explanation for Jordan Lemma 2

Proof:

From the conditions of the Lemma and the kind of functions we are interested in, we conclude:

$$\text{For } |z| = R, \quad |f(z)| \le M_R \qquad \lim_{R\to\infty} M_R = 0$$

Thus, for $t > 0$

$$\int_{C_{1R}} f(z)e^{zt}\, dz = \left[\int_{\pi/2}^{3\pi/2} f\left(Re^{j\theta}\right) e^{R(\cos\theta + j\sin\theta)t} \left(jRe^{j\theta}\right) d\theta \right] \tag{4.40}$$

Let

$$\theta_1 = \theta - \pi/2, \qquad \cos\theta = -\sin\theta_1$$

Using the fact that

$$\frac{\sin\theta_1}{\theta_1} > \left(\frac{\sin\pi/2}{\pi/2} = \frac{2}{\pi}\right) \qquad (0 < \theta_1 < \pi/2)$$

Eq. 4.40 can be simplified to yield

$$\left| \circlearrowleft \int_{C_{1R}} f(z)e^{zt}\,\mathrm{d}z \right| \leq \left[2RM_R \int_0^{\pi/2} e^{(-\frac{2}{\pi}Rt\theta_1)}\,\mathrm{d}\theta_1 = \frac{\pi}{t} M_R\left(1 - e^{-Rt}\right) \right] \tag{4.41}$$

Thus, for $t > 0$,

$$\lim_{R\to\infty} \left| \circlearrowleft \int_{C_{1R}} f(z)e^{zt}\,\mathrm{d}z \right| \leq \lim_{R\to\infty} \frac{\pi}{t} M_R\left(1 - e^{-Rt}\right) = 0$$

Which implies

$$\lim_{R\to\infty} \circlearrowleft \int_{C_{1R}} f(z)e^{zt}\,\mathrm{d}z = 0, \qquad t > 0$$

In an analogous manner, it is easy to show that

$$\lim_{R\to\infty} \circlearrowright \int_{C_{2R}} f(z)e^{zt}\,\mathrm{d}z = 0, \qquad t < 0$$

In what follows, we shall see the importance of these Lemmas in computing real integrals and inverse Laplace and Fourier transforms also. **With the help of Jordan's two Lemmas, we are now in a position to use contour integration and evaluate the principle value of the definite integrals for the appropriate functions.**

Application of Jordan's Lemmas

- Application of Jordan's Lemma 1.

  Evaluation of the integral

$$I = \int_{-\infty}^{+\infty} f(x)\,\mathrm{d}x = -j \int f(z/j)\,\mathrm{d}z$$

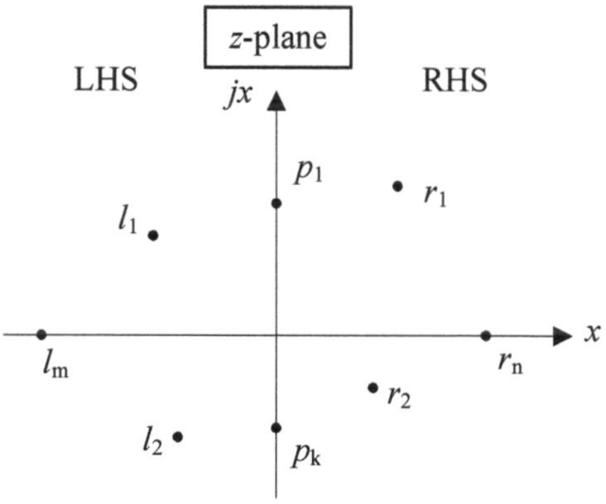

Figure 4.16: Singularities of Function $f(z)$

If the following conditions are satisfied by function $f(z/j)$:

1. It has isolated singularities in the LHS of z-plane denoted by l_i, $i = 1, 2, \ldots, m$ and in the RHS of the z-plane denoted by r_k, $k = 1, 2, \ldots, n$.
2. On the imaginary axis $z = jx$, the function $f(z/j)$ has simple poles $p_1, p_2, \ldots, p_v$ (analytic everywhere else).
3. $\lim zf(z)|_{z\to\infty} = A_0$

Then

$$\mathrm{PV} \int_{-\infty}^{+\infty} f(x)\,\mathrm{d}x = -j \int_{-j\infty}^{+j\infty} f(z/j)\,\mathrm{d}z \tag{4.42}$$

$$= -2\pi j(-j)\left[\sum_{v=1}^{m} \text{Res}\,[f(z)]_{z=l_v} + \frac{1}{2}\sum_{v=1}^{k} \text{Res}\,[f(z)]_{z=p_v} - \frac{1}{2}A_0\right] \tag{4.43}$$

$$= -2\pi j(-j)\left[-\sum_{v=1}^{n} \text{Res}\,[f(z)]_{z=r_v} + \frac{1}{2}\sum_{v=1}^{k} \text{Res}\,[f(z)]_{z=p_v} - \frac{1}{2}A_0\right] \tag{4.44}$$

Proof:

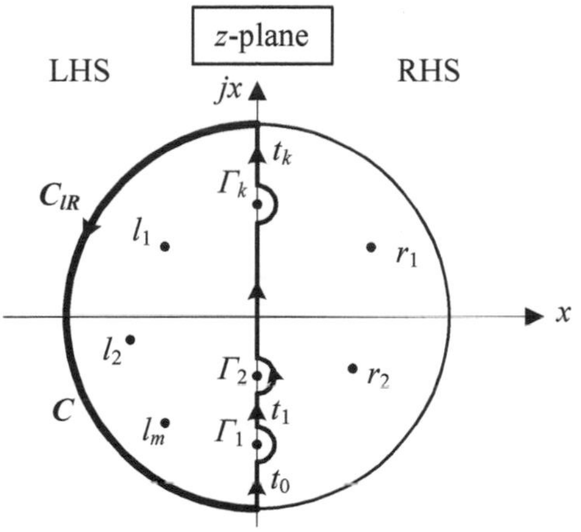

Figure 4.17: Complex Integration Contour

Proof follows from Jordan's Lemma.

Let us choose a closed contour C as shown in Figure 4.17. Contour C consists of semicircles Γ_v of radius ρ, centered at p_v, $(v = 1, 2, \ldots, k)$, segments t_v on the imaginary axis $(v = 1, 2, \ldots, k)$ and semicircle C_{lR} of radius R (large enough to enclose all singularities in left hand side of z-plane (LHS)).

According to Cauchy's Residue Theorem, integration along various segments of C,

$$\oint_C f\left(\frac{z}{j}\right) \mathrm{d}z = z\pi j\left[\sum_{v=1}^{m} \text{Res}\,[f(z)]_{z=l_v} + \sum_{v=1}^{k} \text{Res}\left[f\left(\frac{z}{j}\right)\right]_{z=p_v}\right] \tag{4.45}$$

$$\oint_C f\left(\frac{z}{j}\right) \mathrm{d}z = \circlearrowleft \int_{C_{lR}} f\left(\frac{z}{j}\right) \mathrm{d}z + \sum_{v=0}^{k} \int_{t_v} f\left(\frac{z}{j}\right) \mathrm{d}z + \sum_{v=1}^{k} \int_{\Gamma_v} f\left(\frac{z}{j}\right) \mathrm{d}z \tag{4.46}$$

According to Jordan's Lemma

$$\lim_{\rho\to 0} \circlearrowleft \int_{\Gamma_v} f\left(\frac{z}{j}\right) \mathrm{d}z = j\pi \mathrm{Res}\left[f\left(\frac{z}{j}\right)\right]_{z=p_v}$$

$$\lim_{R\to\infty} \circlearrowleft \int_{C_{lR}} f\left(\frac{z}{j}\right) \mathrm{d}z = j\pi A_0$$

Furthermore,

$$\lim_{\rho\to 0} \sum_{v=0}^{k} \int_{t_v} f\left(\frac{z}{j}\right) \mathrm{d}z = \int_{-j\infty}^{+j\infty} f\left(\frac{z}{j}\right) \mathrm{d}z$$

Equating Eq. 4.45 to Eq. 4.46 and using the above simplifications,

$$2\pi j\left[\sum_{v=1}^{m} \mathrm{Res}_{\,z=l_v}[f(z)] + \frac{1}{2}\sum_{v=1}^{k} \mathrm{Res}_{\,z=p_v}\left[f\left(s\frac{z}{j}\right)\right]\right]$$
$$= j\pi A_0 + \int_{-j\infty}^{+j\infty} f\left(\frac{z}{j}\right) \mathrm{d}z + j\pi \sum_{v=1}^{k} \mathrm{Res}_{\,z=p_v}[f(z)]$$

or

$$\int_{-j\infty}^{+j\infty} f\left(\frac{z}{j}\right) \mathrm{d}z = 2\pi j\left[\sum_{v=1}^{m} \mathrm{Res}_{\,z=l_v}\left[f\left(\frac{z}{j}\right)\right] + \frac{1}{2}\sum_{v=1}^{k} \mathrm{Res}\left[f\left(\frac{z}{j}\right)\right] - \frac{1}{2}A_0\right]$$

This proves Eq. 4.43. The Eq. 4.44 can be proven in the same way, replacing C_{rR} with C_{lR} and realizing that C encloses only poles at $z = r_v$, $v = 1, 2, \ldots, n$ and none of the poles on the vertical axis.

Example 4.4:

Evaluate

$$I = \int_{0}^{+\infty} \frac{\mathrm{d}x}{1+x^2}$$

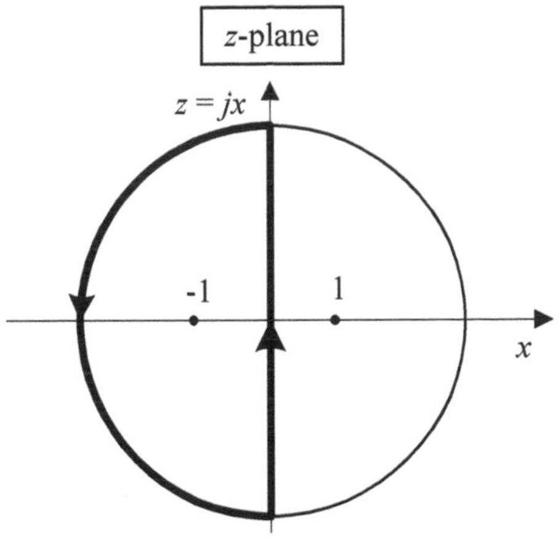

Figure 4.18: Computation of $I = \int_0^{+\infty} \frac{dx}{1+x^2}$

Solution: Since

$$f(x) = \frac{1}{1+x^2} \quad \text{is an even function,}$$

Therefore,

$$I = \frac{1}{2}\int_{-\infty}^{+\infty} \frac{dx}{1+x^2} = -\frac{j}{2}\int_{-j\infty}^{+j\infty} \frac{dz}{1-z^2} = \frac{j}{2}\int_{-j\infty}^{+j\infty} \frac{dz}{(z-1)(z+1)}$$

Furthermore, from Eq. 4.43

$$I = \frac{j}{2}\oint \frac{dz}{z^2-1} = \frac{j}{2}(2\pi j)\left(\text{Res}_{z=-1}\left[\frac{1}{z^2-1}\right]\right) = -\pi\left(-\frac{1}{2}\right) = \frac{\pi}{2}$$
$$I = \frac{j}{2}\oint \frac{dz}{z^2-1} = \frac{j}{2}(-2\pi j)\left(\text{Res}_{z=1}\left[\frac{1}{z^2-1}\right]\right) = \pi\left(\frac{1}{2}\right) = \frac{\pi}{2}$$

Exercise 4.1:

Evaluate

$$I = \int_{-\infty}^{+\infty} \frac{1}{(x^2+a^2)(x^2+b^2)}\,dx \quad a>0, b>0, \qquad \text{Answer: } I = \frac{\pi}{ab(a+b)}$$

Exercise 4.2:

Evaluate

$$I = \int_{-\infty}^{+\infty} \sin^4 x \,\mathrm{d}x, \qquad \text{Answer } I = \frac{3\pi}{4}$$

Example 4.5:

Evaluate

$$I = \int_{-\infty}^{+\infty} \frac{\mathrm{d}x}{1 + x^4}$$

Solution

$$I = \int_{-\infty}^{+\infty} \frac{\mathrm{d}x}{1 + x^4} = -j \int_{-j\infty}^{+j\infty} \frac{\mathrm{d}z}{1 + z^4}$$

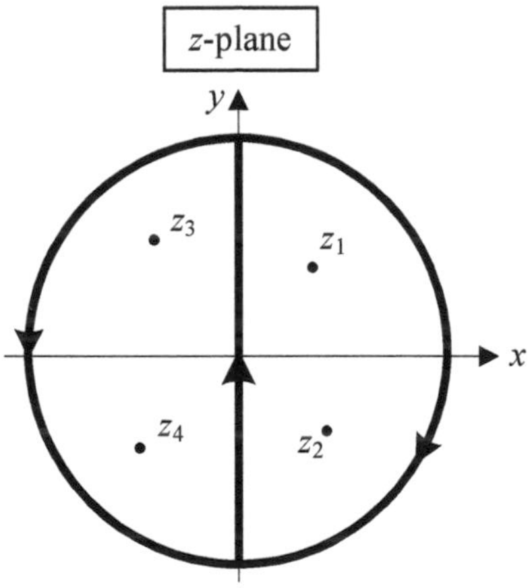

Figure 4.19: Computation of $\int_{-\infty}^{+\infty} \frac{dx}{1 + x^4}$

Function $(1 + z^4)^{-1}$ has four poles obtained by solving $1 + z^4 = 0$,

$$z = (-1)^{1/4} = e^{j(\pi + 2n\pi)/4}, \qquad n = 0, -1, 1, 2$$

$$z_1 = e^{j\pi/4}, \qquad z_2 = e^{-j\pi/4}$$

$$z_3 = e^{j3\pi/4}, \qquad z_4 = e^{j5\pi/4}$$

From Eq. 4.43

$$I = (-j)(-2\pi j)\left[\text{Res}_{z=z_1}\left[\frac{1}{z^4+1}\right] + \text{Res}_{z=z_2}\left[\frac{1}{z^4+1}\right]\right]$$

$$\text{Res}\left[\frac{1}{z^4+1}\right]_{z=z_1} = \lim_{z\to z_1}\left[\frac{z-z_1}{z^4+1}\right] = \lim_{z\to z_1}\left[\frac{1}{4z^3}\right] \qquad \text{(use L'hospital's Rule)}$$

$$= \frac{1}{4}e^{-j3\pi/4} = \frac{1}{4}\left(-\frac{1}{\sqrt{2}} - \frac{j}{\sqrt{2}}\right) = \text{Res}\,[z_1] = R_1$$

Similarly

$$\text{Res}\left[\frac{1}{z^4+1}\right]_{z=z_2} = \frac{1}{4z^3} = \frac{1}{4}\left(-\frac{1}{\sqrt{2}} + \frac{j}{\sqrt{2}}\right) = \text{Res}\,[z_2] = R_2$$

Thus,

$$I = (R_1 + R_2) = (-j)(-2\pi j)\left(-\frac{1}{2\sqrt{2}}\right) = \frac{\pi}{\sqrt{2}}$$

$$\int_{-\infty}^{+\infty} \frac{\mathrm{d}x}{1+x^4} = \frac{\pi}{\sqrt{2}}$$

When $f(x)$ is an odd function of x, the substitution of $z = jx$ may not improve the situation. We illustrate this point with the following example.

Example 4.6:

Evaluate

$$I = \int_0^{\infty} \frac{dx}{1+x^3}$$

Solution

Let

$$z = xe^{j\theta}; \quad z^3 = x^3\,e^{j3\theta}, \quad \text{choose } 3\theta = 2\pi$$

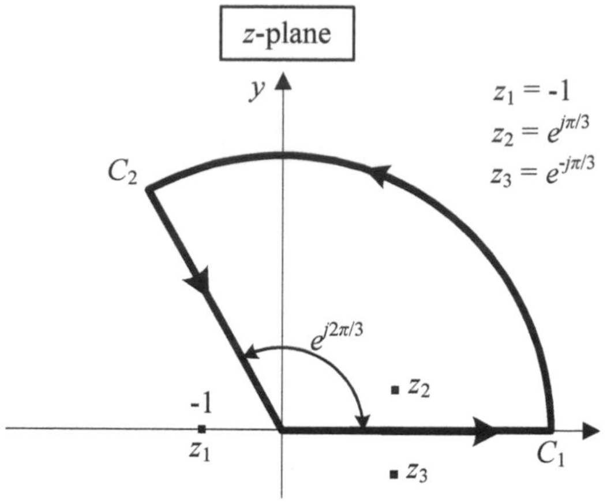

Figure 4.20: Integration $I = \int_0^\infty \frac{dx}{1+x^3}$

$$z^3 = x^3; \quad dz = e^{j2\pi/3}\,dx; \quad f(z) = \frac{1}{1+z^3}$$

The function $f(z)$ has poles at $(1 + z^3 = 0)$. These are

$$z_1 = -1, \quad z_2 = e^{j\pi/3}, \quad z_3 = e^{-j\pi/3}$$

$$I = \oint_{o-c_1-c_2-o} \frac{dz}{1+z^3} = 2\pi j \,\text{Res} f(z)|_{z=z_2}$$

$$\text{Res} f(z)|_{z=z_2} = 2\pi j \left[\frac{(z-z_2)}{(z-z_1)(z-z_2)(z-z_3)} \right] \Bigg|_{z=z_2} = \frac{1}{3} e^{\frac{-j\pi}{3}}$$

From Jordan's Lemma as $R \to \infty$

$$I = \int_0^\infty \frac{dx}{1+x^3} + \int_{c_2-o} \frac{dz}{1+z^3} = \int_0^\infty \frac{dx}{1+x^3} + e^{j2\pi/3} \int_\infty^0 \frac{dx}{1+x^3}$$

$$= (1 - e^{j\pi/3}) \int_0^\infty \frac{dx}{1+x^3}$$

or

$$I = \frac{2}{3}\left(\frac{j\pi e^{-j\pi/3}}{1 - e^{j2\pi/3}}\right) = \frac{2\pi\sqrt{3}}{9}$$

The technique applied in this example can also be used to evaluate integrals of the form

$$I = \int_0^\infty \frac{\mathrm{d}x}{1 + x^{(2n+1)}}, \quad n = 1, 2, \cdots$$

Example 4.7:

Evaluate the integrals of the type

$$I = \int_0^{2\pi} f(\theta)\,\mathrm{d}\theta$$

Cauchy's Residue theorem is used to solve these types of integral.

As an example

$$I = \int_0^{2\pi} \frac{\mathrm{d}\theta}{a + b\cos\theta}, \qquad a > b > 0$$

Let

$$z = e^{j\theta}$$

$$\mathrm{d}z = je^{j\theta}\,\mathrm{d}\theta = jz\,\mathrm{d}\theta$$

$$\cos\theta = \frac{1}{2}[e^{j\theta} + e^{-j\theta}] = \frac{1}{2}\left(z + \frac{1}{z}\right) = \frac{z^2 + 1}{2z}$$

Thus,

$$I = \frac{2}{jb}\oint_C \frac{\mathrm{d}z}{z^2 + 2(a/b)z + 1}$$

The poles of the integrand are

$$z_1 = -\frac{a}{b} + \sqrt{\frac{a^2 - b^2}{b^2}}, \qquad z_2 = -\frac{a}{b} - \sqrt{\frac{a^2 - b^2}{b^2}}$$

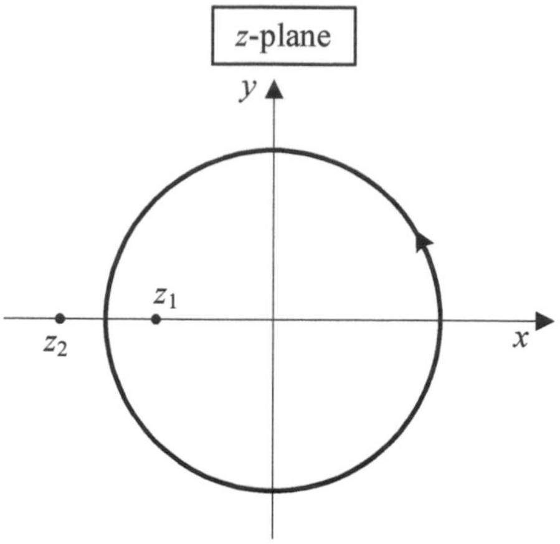

Figure 4.21: Integration $I = \int_0^{2\pi} \frac{d\theta}{a + b\cos\theta}, a > b > 0$

For $a > b$, only z_1 is inside the curve C. Thus

$$I = \frac{2}{jb}(2\pi j)\left[\text{Res}\left[\frac{1}{(z-z_1)(z-z_2)}\right]_{z=z_1}\right] = \frac{4\pi}{b}\frac{1}{z_1 - z_2} = \frac{2\pi}{\sqrt{a^2 - b^2}}$$

Similarly, we can show that

$$\int_0^{2\pi} \left(\frac{\cos 3\theta}{5 - 4\cos\theta}\right) d\theta = \frac{\pi}{2}$$

$$\int_0^{2\pi} \left(\frac{1}{(5 - 3\sin\theta)^2}\right) d\theta = \frac{5\pi}{32}$$

Example 4.8:

Minimum Mean Squared Error Computation via Contour Integration

In Control Systems theory, we are often required to compute and minimize the mean squared error.

$$I = \int_0^{\infty} e^2(t)\, dt \tag{4.47}$$

where $e(t)$ is the error between the desired and the actual output and represents a stable function involving:

$$\lim_{t\to\infty} e(t) = 0$$

From the Laplace tranforms and the Parseval's relation:

$$I = \int_0^{\infty} e^2(t)\,\mathrm{d}t = \frac{1}{2\pi j}\int_{c-j\infty}^{c+j\infty} E(s)E(-s)\,\mathrm{d}s \tag{4.48}$$

Let

$$E(s) = \frac{N(s)}{D(s)} \tag{4.49}$$

where

$$N(s) = c_0 s^{n-1} + c_1 s^{n-2} + \cdots + c_{n-1}$$

$$D(s) = a_0 s^n + a_1 s^{n-1} + \cdots\cdots + a_n$$

The stability of $e(t)$ implies that $D(s)$ is a Hurwitz polynomial that has roots only in the LHS of s-plane.

$$B(s^2) = N(s)N(-s) = \left(\sum_{k=0}^{n-1} c_k s^{n-1-k}\right)\left(\sum_{i=0}^{n-1} c_i (-s)^{n-1-i}\right) = \sum_{j=0}^{n-1} b_j s^{2(n-1-i)}$$

Thus,

$$I = \frac{1}{2\pi j}\int_{c-j\infty}^{c+j\infty}\left[\frac{b_0 s^{2(n-1)} + b_1 s^{2(n-2)} + \cdots + b_{n-1}}{D(s)D(-s)}\right]\mathrm{d}s \tag{4.50}$$

We shall prove that

$$I = \frac{(-1)^{n-1}}{2a_0}\left(\frac{N}{D}\right) \tag{4.51}$$

where

$$D = \begin{vmatrix} a_1 & a_0 & 0 & 0 & \cdots \\ a_3 & a_2 & a_1 & a_0 & \cdots \\ a_5 & a_4 & a_3 & a_2 & \cdots \\ a_7 & a_6 & a_5 & a_4 & \cdots \\ \vdots & \vdots & \vdots & \vdots & \cdots \end{vmatrix}, \qquad N = \begin{vmatrix} b_0 & a_0 & 0 & 0 & \cdots \\ b_1 & a_2 & a_1 & a_0 & \cdots \\ b_2 & a_4 & a_3 & a_2 & \cdots \\ b_3 & a_6 & a_5 & a_4 & \cdots \\ \vdots & \vdots & \vdots & \vdots & \cdots \end{vmatrix} \tag{4.52}$$

Note: That the determinant N can be obtained from the determinant D by replacing the first column of D with the coefficients, b's.

Proof:

$$I = \int_{c-j\infty}^{c+j\infty} \frac{G(s^2)}{D(s)D(-s)} \mathrm{d}s$$

Let

$$\frac{G(s^2)}{D(s)D(-s)} = \frac{A(s)}{D(s)} + \frac{A(-s)}{D(-s)} \tag{4.53}$$

Thus,

$$A(s)D(-s) + A(-s)D(s) = G(s^2) \tag{4.54}$$

using Jordan's Lemma

$$I = \frac{1}{2\pi j} \oint_C \left[\frac{A(s)}{D(s)} + \frac{A(-s)}{D(-s)} \right] \mathrm{d}s$$

the closed curve c goes from $c - j\infty$ to $c + j\infty$ and infinite semicircle in the left-half of the s-plane.

Thus,

$$I = \frac{1}{2\pi j} \oint_C \frac{A(s)}{D(s)} \mathrm{d}s + \frac{1}{2\pi j} \oint_C \frac{A(-s)}{D(-s)} \mathrm{d}s$$

But

$$\oint_C \frac{A(-s)}{D(-s)}\,\mathrm{d}s = 0$$

(all the poles of the integrand are outside the contour of integration C).

$$I = \frac{1}{2\pi j}\oint_C \frac{A(s)}{D(s)}\,\mathrm{d}s$$

$$\frac{A(s)}{D(s)} = \frac{\sum\limits_{r=1}^{n-1} d_r(s)^{n-r}}{\sum\limits_{r=1}^{n} a_r(s)^{n-r}} = \frac{\alpha_1}{s} + \frac{\alpha_2}{s^2} + \cdots \tag{4.55}$$

where

$$\alpha_1 = (-1)^{n-1}\frac{d_1}{2a_0}$$

The other α's are unnecessary to determine because α_1 is the residue and the only parameter that contributes to the integration.

From the Eq. 4.55

$$I = \frac{1}{2\pi j}\oint_C \left(\frac{\alpha_1}{s} + \frac{\alpha_2}{s^2} + \cdots\right)\mathrm{d}s = \alpha_1$$

Hence,

$$I = (-1)^{n-1}\frac{\alpha_1}{2a_0}$$

The coefficient α_1 is determined from Eq. 4.54 by equating powers of s^2,

$$\begin{bmatrix} b_0 \\ b_1 \\ b_2 \\ b_3 \\ \vdots \end{bmatrix} = 2\begin{bmatrix} a_1 & a_0 & 0 & 0 & \cdots \\ a_3 & a_2 & a_1 & a_0 & \cdots \\ a_5 & a_4 & a_3 & a_2 & \cdots \\ a_7 & a_6 & a_5 & a_4 & \cdots \\ \vdots & \vdots & \vdots & \vdots & \cdots \end{bmatrix}\begin{bmatrix} \alpha_1 \\ \alpha_2 \\ \alpha_3 \\ \alpha_4 \\ \vdots \end{bmatrix}$$

$$\alpha_1 = \frac{\begin{vmatrix} b_0 & a_0 & 0 & 0 & \cdots \\ b_1 & a_2 & a_1 & a_0 & \cdots \\ b_2 & a_4 & a_3 & a_2 & \cdots \\ b_3 & a_6 & a_5 & a_4 & \cdots \\ \vdots & \vdots & \vdots & \vdots & \cdots \end{vmatrix}}{\begin{vmatrix} a_1 & a_0 & 0 & 0 & \cdots \\ a_3 & a_2 & a_1 & a_0 & \cdots \\ a_5 & a_4 & a_3 & a_2 & \cdots \\ a_7 & a_6 & a_5 & a_4 & \cdots \\ \vdots & \vdots & \vdots & \vdots & \cdots \end{vmatrix}} = \frac{N}{2D}$$

$$I = \frac{(-1)^{n-1}}{2a_0}\left(\frac{N}{D}\right)$$

It should be noted that D is the Hurwitz's determinant of the polynomial $D(s)$ because the roots of $D(s)$ lie in the left-hand side of the s-plane.

- **Application of Jordan's Lemma 2 to the Evaluation of $\int_{-\infty}^{+\infty} f(x)e^{jxt}\,\mathrm{d}x$**

We shall evaluate the integrals of the type

$$\int_{-\infty}^{+\infty} f(x)e^{jxt}\,\mathrm{d}x = (-j)\int_{-j\infty}^{+j\infty} f\left(\frac{z}{j}\right)e^{zt}\,\mathrm{d}z$$

Consider a function $f(z/j)$ satisfying the following conditions:

1. $\lim_{|z|\to\infty} f(z) = 0$
2. It has isolated singularities denoted by $l_1, \ldots, l_m$ in the LHS and denoted by $r_1, \ldots, r_n$ in the RHS of z-plane.
3. On the imaginary axis $z = jx$, the function $f(z/j)$ has simple poles $p_1, p_2, \ldots, p_k$. (It is analytic everywhere else.)

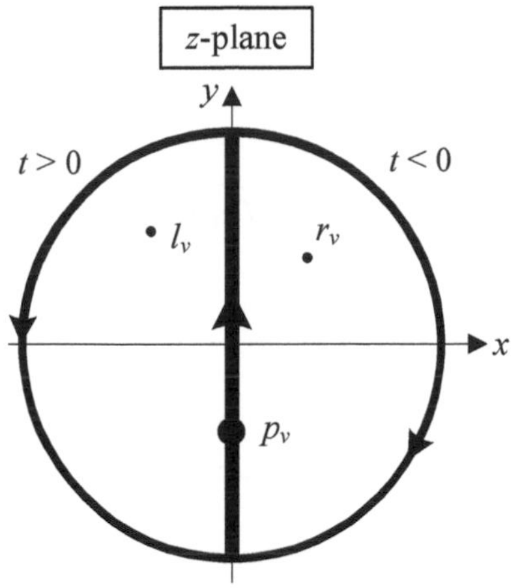

Figure 4.22: Singularities of the Function $f(z/j)e^{zt}$

Then

$$I = \text{PV}\int_{-\infty}^{+\infty} f(x)e^{jxt}\,\mathrm{d}x = -j\int_{-j\infty}^{+j\infty} f\Big(\frac{z}{j}\Big)e^{zt}\,\mathrm{d}z \tag{4.56}$$

$$\begin{aligned} I = (2\pi j)(-j)\Big\{&\sum_{\nu=1}^{m}\text{Res}\Big[f\Big(\frac{z}{j}\Big)e^{zt}\Big]_{z=l_\nu} \\ &+\frac{1}{2}\sum_{\nu=1}^{k}\text{Res}\Big[f\Big(\frac{z}{j}\Big)e^{zt}\Big]_{z=p_\nu}\Big\},\quad t>0 \end{aligned} \tag{4.57}$$

$$\begin{aligned} I = (-2\pi j)(-j)\Big\{&\sum_{\nu=1}^{m}\text{Res}\Big[f\Big(\frac{z}{j}\Big)e^{zt}\Big]_{z=r_\nu} \\ &+\frac{1}{2}\sum_{\nu=1}^{k}\text{Res}\Big[f\Big(\frac{z}{j}\Big)e^{zt}\Big]_{z=p_\nu}\Big\},\quad t<0 \end{aligned} \tag{4.58}$$

Proof of this integral is analogous to the previous section.

Example 4.9

Evaluate

$$I = \int_{-\infty}^{+\infty} \frac{e^{jx}}{x}\,\mathrm{d}x$$

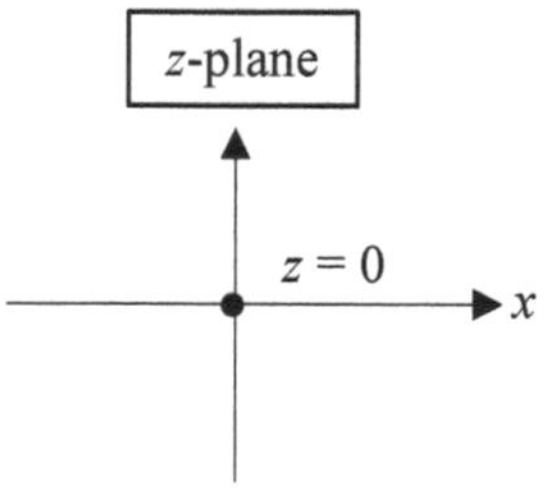

Figure 4.23: $I = \int_{-j\infty}^{+j\infty} \frac{e^z}{z} dz$

$$I = \int_{-\infty}^{+\infty} \frac{e^{jx}}{x}\,\mathrm{d}x = -j\int_{-j\infty}^{+j\infty} \frac{e^{z}}{z/j}\,\mathrm{d}z = \int_{-j\infty}^{+j\infty} \frac{e^{z}}{z}\,\mathrm{d}z$$

According to Eq. 4.56

$$I = (2\pi j)\left(\frac{1}{2}\right)\left\{\mathrm{Res}\left[\frac{e^{zt}}{z}\right]_{z=0}\right\} = j\pi$$

Then

$$\int_{-\infty}^{+\infty} \frac{e^{jx}}{x}\,\mathrm{d}x = j\pi$$

Equating real and imaginary parts of the above integral

$$\int_{-\infty}^{+\infty} \frac{\cos x}{x}\,\mathrm{d}x = 0, \qquad \int_{-\infty}^{+\infty} \frac{\sin x}{x}\,\mathrm{d}x = \pi$$

Note: It is important to realize that in order to evaluate these integrals, $f(z/j)$ is allowed to have only simple poles on the imaginary axis in the z-plane.

Example 4.10:

Evaluate

$$\int_{-\infty}^{+\infty} \frac{e^{jax}}{x^2+b^2}\,\mathrm{d}x, \qquad b>0$$

Solution

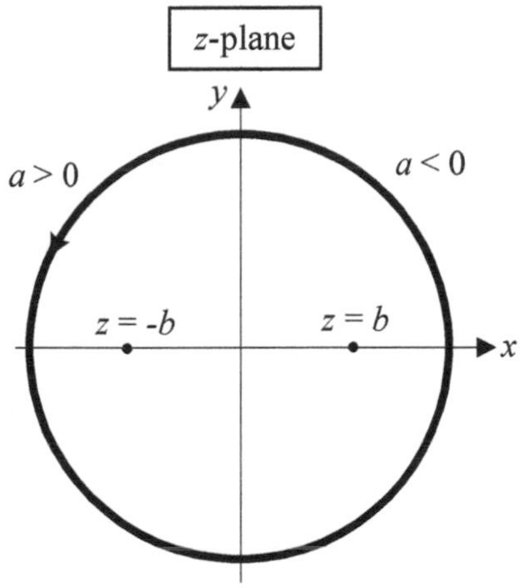

Figure 4.24: $\int_{-\infty}^{+\infty} \frac{e^{jax}}{x^2+b^2}\,\mathrm{d}x, \quad b>0$

$$I = \int_{-\infty}^{+\infty} \frac{e^{jax}}{x^2+b^2}\,\mathrm{d}x = -j\int_{-j\infty}^{+j\infty} \frac{-e^{az}}{z^2-b^2}\,\mathrm{d}z$$

From Eq. 4.57

$$I = 2\pi j(-j)\left\{\mathrm{Res}\left[\frac{-e^{az}}{(z+b)(z-b)}\right]_{z=-b}\right\} = 2\pi\left(\frac{-e^{-ab}}{-2b}\right) = \frac{\pi}{b}e^{-ab}, \qquad a>0$$

$$I = (-2\pi j)(-j)\left\{\mathrm{Res}\left[\frac{-e^{az}}{(z+b)(z-b)}\right]_{z=b}\right\} = -2\pi\left(\frac{e^{-ab}}{-2b}\right) = \frac{\pi}{b}e^{ab}, \qquad a<0$$

Thus,

$$I = \frac{\pi}{b}e^{-b|a|}$$

Example 4.11:

Evaluate

$$I = \int_a^b \frac{f(x)}{x - x_0 \pm j\epsilon}\,\mathrm{d}x, \qquad \epsilon \to 0,\ b > x_0 > a$$

It is easy to show that

$$I = \int_a^b \frac{f(x)}{x - x_0 \pm j\epsilon}\,\mathrm{d}x = \mathrm{PV}\int_a^b \frac{f(x)}{x - x_0}\,\mathrm{d}x \mp j\pi f(x_0)$$

4.2.16 Branch Points—Essential Singularities

So far, the functions considered involved only nonessential singularities or poles. We now look at functions involving so-called "branch points" or essential singularities. The difficulty associated with branch points can be illustrated by the following example. Consider a function,

$$f(z) = z^{-1/2}$$

which is a mapping of the points from the z-plane to the $f(z)$-plane.

Consider a point

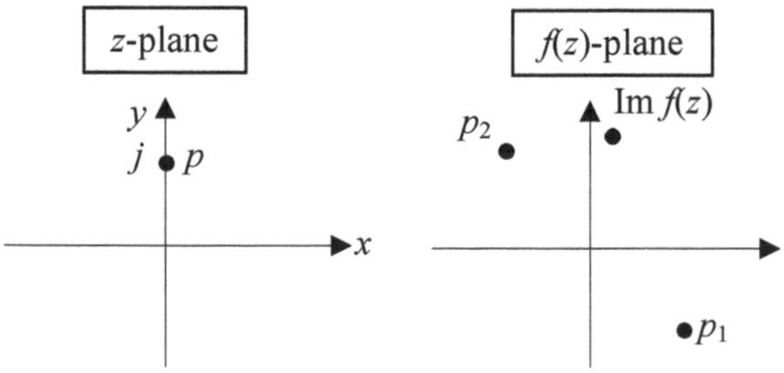

Figure 4.25: Branch Point Singularity

$$z = j = e^{j\pi/2} = e^{j(2\pi+\pi/2)}$$

then,

for $z = e^{j\pi/2}$ (point p in the z-plane),

$f(z) = e^{-j\pi/4}$ (point p in the z-plane maps to point p_1 in $f(z)$-plane)

For $z = e^{j(2\pi+\pi/2)}$ (same point p in the z-plane),

$f(z) = e^{-j(\pi+\pi/4)}$ (point p in the z-plane maps to point p_2 in $f(z)$-plane)

Therefore, the same point $z = p$ has been mapped into two points p_1 and p_2. This is due to the fact that the origin here is a "branch point." As we move around the origin from the point p and back to p in the z-plane, we go from point p_1 in the $f(z)$-plane to point p_2 in the $f(z)$-plane yielding a multivalued function. To avoid this multivaluedness we agree from the start that point p in z-plane maps to point p_1 in the $f(z)$-plane only and under no circumstances maps to point p_2 in the $f(z)$-plane. This boils down to avoiding the total encirclement of the singular point $z = 0$ in the Z-plane. **To ensure this we make a barrier by drawing a so-called "branch cut" or branch line from $z = 0$ to $z = \infty$ and agree not to cross it**. Essentially, we agree to restrict the value of θ from 0^+ to $2\pi^-$ in evaluating the contour integral. A point that cannot be encircled without crossing a branch cut is referred to as "**Branch Point**."

Example 4.12:

Evaluate

$$I = \int_0^\infty \frac{x^{p-1}}{1+x} \mathrm{d}x \qquad 0 < p < 1$$

Consider

$$J = \oint_C \frac{z^{p-1}}{1+z} \mathrm{d}z \qquad 0 < p < 1$$

"C" represents a branch cut contour shown below. J is computed first from the residue theorem on the smaller circle and the value of the function along the various parts of the integration contour.

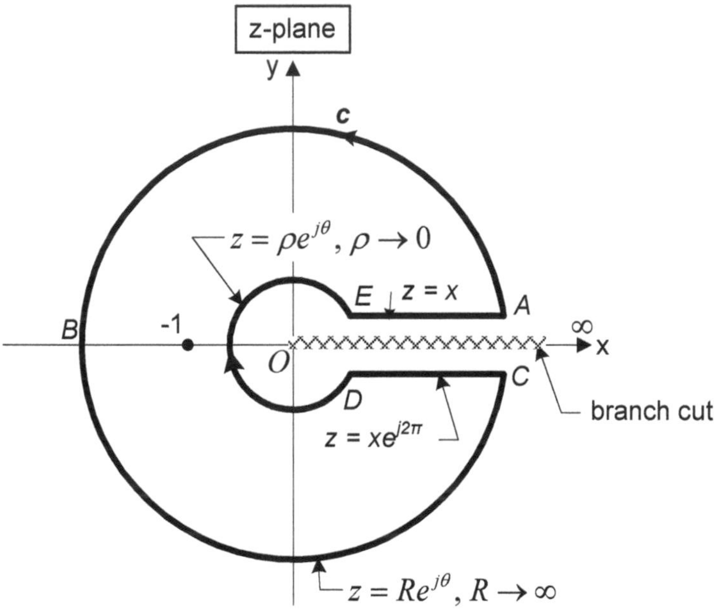

Figure 4.26: Branch Cut

1. From residue theorem

$$J = 2\pi j\left\{\text{Res}\left[\frac{z^{p-1}}{1+z}\right]\right\}_{z=-1=e^{j\pi}} = 2\pi j e^{j\pi(p-1)}$$

2. From contour integration, letting $\rho \to 0$, $R \to \infty$

$$J = J_1 + J_2 + J_3 + J_4$$

where

$$J_1 = \int_c^d \frac{z^{p-1}}{1+z}\,dz = \int_R^\rho \frac{(xe^{j2\pi})^{p-1}}{1+xe^{j2\pi}}\,dx \simeq -e^{j2\pi(p-1)} \int_0^\infty \frac{x^{p-1}}{1+x}\,dx = -e^{j2\pi(p-1)} I$$

$$J_2 = \oint \frac{z^{p-1}}{1+z}\,dz = j\int_{2\pi}^0 \frac{\rho^p e^{-jp\theta}}{1+\rho e^{j\theta}}\,d\theta \simeq 0 \quad \rho \to 0$$

$$J_3 = \int_E^A \frac{z^{p-1}}{1+z}\,dz = \int_\rho^R \frac{x^{p-1}}{1+x}\,dx \simeq \int_0^\infty \frac{x^{p-1}}{1+x}\,dx = I$$

$$J_4 = \oint \frac{z^{p-1}}{1+z}\,dz = j\int_0^{2\pi} \frac{(R)^p e^{jp\theta}}{1+Re^{j\theta}}\,d\theta \simeq 0, \quad R \to \infty$$

Thus,

$$2\pi j\left(e^{j\pi(p-1)}\right) = \left(1 - e^{j2\pi(p-1)}\right) I$$

or

$$I = \frac{2\pi j e^{j\pi(p-1)}}{1 - e^{j2\pi(p-1)}} = \frac{-2\pi j}{e^{-j\pi p} - e^{j\pi p}} = \frac{\pi}{\sin \pi p}$$

$$I = \pi \qquad \text{when} \qquad p = \frac{1}{2}$$

Exercise 4.3:

Show that for $1 < p < 2$

$$I = \int_0^\infty \frac{x^{p-1}}{x^2+1}\,dx = \frac{\pi}{2}\csc\left(\frac{\pi P}{2}\right)$$

Hint: Use the Branch cut.

Example 4.13:

Evaluate

$$I = \int_{-\infty - jr}^{\infty - jr} e^{-\alpha z^2}\, dz$$

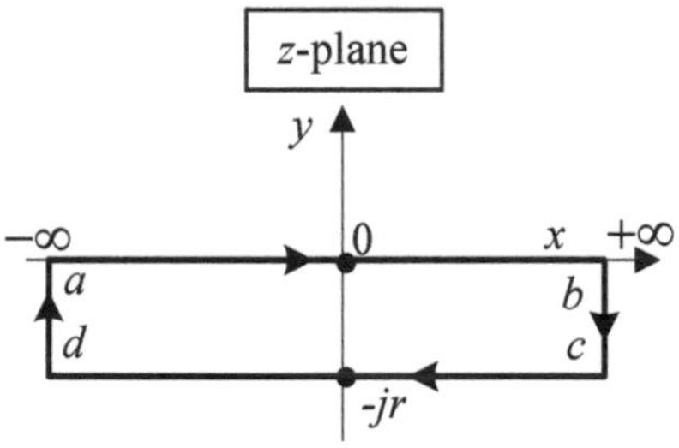

Figure 4.27: $I = \int_{-\infty - jr}^{\infty - jr} e^{-\alpha z^2}\, dz$

Let

$$e^{-\alpha z^2} = f(z)$$

Consider the following contour *abcd* of integration. There are no poles of $f(z)$ inside this contour.

Therefore

$$\hat{I} = \oint_{abcd} f(z)\, dz = I_1 + I_2 + I_3 + I_4 = 0$$

where

$$I = I_1 = \int_a^b f(z)\, dz \qquad I_2 = \int_b^c f(z)\, dz$$

$$I_3 = \int_c^d f(z)\, dz \qquad I_4 = \int_d^a f(z)\, dz$$

I_2 and I_4 are equal and opposite, canceling each other. Thus,

$$I = -I_3$$

Therefore

$$I = \int_{-\infty-jr}^{\infty-jr} e^{-\alpha z^2}\,\mathrm{d}z = -\int_{c}^{d} f(z)\,\mathrm{d}z = \int_{d}^{c} f(z)\,\mathrm{d}z = \int_{a}^{b} f(z)\,\mathrm{d}z$$
$$= \int_{-\infty}^{+\infty} f(z)\,\mathrm{d}z = \int_{-\infty}^{+\infty} e^{-\alpha z^2}\,\mathrm{d}z = \int_{-\infty}^{+\infty} e^{-\alpha x^2}\,\mathrm{d}x$$

This integral belongs to a general class of integrals shown below.

Evaluate

$$I_k = \int_{0}^{+\infty} e^{-x^{2k}}\,\mathrm{d}x, \quad 2k \geq 1, \quad k \text{ a positive integer.}$$

Solution:

Let C be the contour of integration given by $ADEBA$ as shown in the figure below.

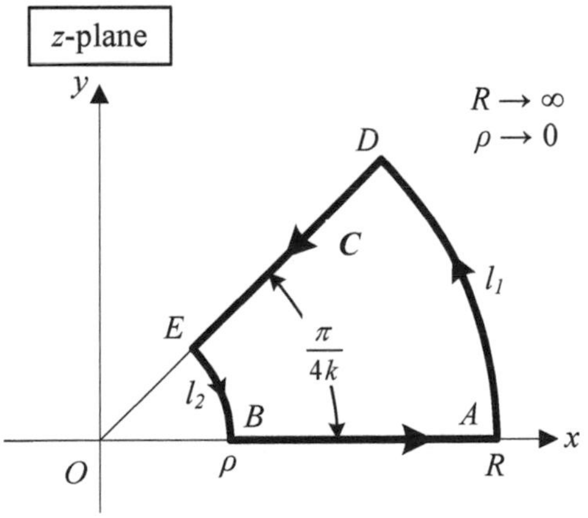

Figure 4.28: $\int_{0}^{+\infty} e^{-x^{2k}}\,\mathrm{d}x,\ 2k \geq 1$

$e^{-z^{2k}}$ is analytic everywhere inside C, hence,

$$0 = \oint\limits_C e^{-z^{2k}}\,dz = \int\limits_\rho^R e^{-x^{2k}}\,dx + \int\limits_{l_1} e^{-z^{2k}}\,dz + \int\limits_{DE} e^{-z^{2k}}\,dz + \int\limits_{l_2} e^{-z^{2k}}\,dz$$

As $R \to \infty$ and $\rho \to 0$ and $k > 1/2$, it can be easily verified that both the second and the fourth terms tend to zero, yielding

$$I_k = \int\limits_{\rho\to 0}^{R} e^{-x^{2k}}\,dx = \int\limits_0^\infty e^{-x^{2k}}\,dx = -\lim_{\substack{R\to\infty\\ \rho\to 0}} \int\limits_{DO} e^{-z^{2k}}\,dz = \lim_{\substack{R\to\infty\\ \rho\to 0}} \int\limits_{OD} e^{-z^{2k}}\,dz$$

Along OD, $z = re^{j\pi/4k}$, $dz = dre^{j\pi/4k}$,

thus,

$$I_k = \int\limits_0^\infty e^{-x^{2k}}\,dx = \int\limits_0^\infty e^{-(re^{j\pi/4k})^{2k}} e^{j\pi/4k}\,dr = e^{j\pi/4k} \int\limits_0^\infty e^{-jr^{2k}}\,dr$$

Multiplying both sides with $e^{-j\pi/4k}$

$$e^{-j\pi/4k} I_k = \int\limits_0^\infty e^{-jr^{2k}}\,dr$$

or

$$\left(\cos\frac{\pi}{4k} - j\sin\frac{\pi}{4k}\right) I_k = \int\limits_0^\infty \left(\cos r^{2k} - j\sin r^{2k}\right) dr$$

or

$$I_k = \frac{1}{\cos\frac{\pi}{4k}} \int\limits_0^\infty \cos r^{2k}\,dr = \frac{1}{\sin\frac{\pi}{4k}} \int\limits_0^\infty \sin r^{2k}\,dr$$

For $k = 1$, probability integral

$$I = I_1 = \frac{1}{\sqrt{2}} \int\limits_0^\infty \cos x^2\,dx = \frac{1}{\sqrt{2}} \int\limits_0^\infty \sin x^2\,dx = \frac{1}{2} \int\limits_{-\infty}^{+\infty} e^{-x^2}\,dx$$

Example 4.14:

$$I_1 = \oint \frac{e^{jz^2}}{\sin\sqrt{\pi}z}\,\mathrm{d}z$$

Let us now consider the closed contour $ABCDA$, shown in Figure 4.29 The inte-

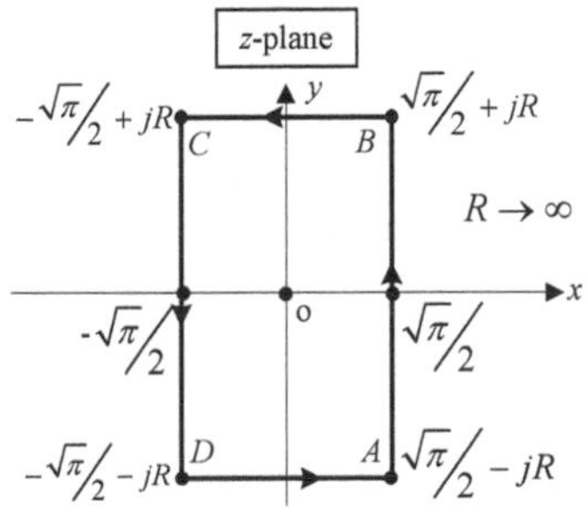

Figure 4.29: $I_1 = \oint \frac{e^{jz^2}}{\sin\sqrt{\pi}z}$

grand has only one pole at $z = 0$, thus,

$$I_1 = 2\pi j\mathrm{Res}\left[\frac{e^{jz^2}z}{\sin\sqrt{\pi}z}\right]_{z\to 0} = \frac{2\pi j}{\sqrt{\pi}} = 2j\sqrt{\pi}$$

Furthermore, going around the contour

$$I_1 = \int_{AB} \frac{e^{-jz^2}}{\sin\sqrt{\pi}z}\,\mathrm{d}z + \int_{BC} \frac{e^{-jz^2}}{\sin\sqrt{\pi}z}\,\mathrm{d}z + \int_{CD} \frac{e^{-jz^2}}{\sin\sqrt{\pi}z}\,\mathrm{d}z + \int_{DA} \frac{e^{-jz^2}}{\sin\sqrt{\pi}z}\,\mathrm{d}z$$

It is easy to see

$$\lim_{R\to\infty} \int_{BC} \frac{e^{-jz^2}}{\sin\sqrt{\pi}z}\,\mathrm{d}z = -\int_{DA} \frac{e^{-jz^2}}{\sin\sqrt{\pi}z}\,\mathrm{d}z$$

Along AB,

$$z = \frac{\sqrt{\pi}}{2} + jy, \qquad \mathrm{d}z = j\,\mathrm{d}y$$

Along CD,

$$z = -\frac{\sqrt{\pi}}{2} + jy, \qquad \mathrm{d}z = j\,\mathrm{d}y$$

Thus, along AB,

$$\left.\frac{e^{-jz^2}}{\sin\sqrt{\pi}z}\right|_{z=\frac{\sqrt{\pi}}{2}+jy} = \frac{e^{j(\frac{\pi}{4}-y^2)}e^{\sqrt{\pi}y}}{\cos j\sqrt{\pi}y} = \frac{e^{j(\frac{\pi}{4}-y^2)}e^{\sqrt{\pi}y}}{\cos h\sqrt{\pi}y}$$

Along CD,

$$\left.\frac{e^{-jz^2}}{\sin\sqrt{\pi}z}\right|_{z=-\frac{\sqrt{\pi}}{2}+jy} = -\frac{e^{j(\frac{\pi}{4}-y^2)}e^{-\sqrt{\pi}y}}{\cos j\sqrt{\pi}y} = -\frac{e^{j(\frac{\pi}{4}-y^2)}e^{-\sqrt{\pi}y}}{\cos h\sqrt{\pi}y}$$

Hence

$$I_1 = j\int_{-\infty}^{+\infty}\frac{e^{j(\frac{\pi}{4}-y^2)}e^{\sqrt{\pi}y}}{\cos h\sqrt{\pi}y}\,\mathrm{d}y - j\int_{+\infty}^{-\infty}\frac{e^{j(\frac{\pi}{4}-y^2)}e^{-\sqrt{\pi}y}}{\cos h\sqrt{\pi}y}\,\mathrm{d}y \tag{4.59}$$

$$= j\int_{-\infty}^{+\infty}\frac{e^{j(\frac{\pi}{4}-y^2)}\left[e^{\sqrt{\pi}y}+e^{-\sqrt{\pi}y}\right]}{\cos h\sqrt{\pi}y}\,\mathrm{d}y \tag{4.60}$$

$$= 2j\int_{-\infty}^{+\infty}e^{j(\frac{\pi}{4}-y^2)}\,\mathrm{d}y \tag{4.61}$$

From Eqs. 4.2.16 and 4.61

$$\int_{-\infty}^{+\infty}e^{j(\frac{\pi}{4}-y^2)}\,\mathrm{d}y = \sqrt{\pi}$$

or

$$\int_{0}^{+\infty}e^{j(\frac{\pi}{4}-y^2)}\,\mathrm{d}y = \frac{\sqrt{\pi}}{2}$$

Expanding these integrals

$$\int_0^{\infty}\cos x^2\,\mathrm{d}x = \int_0^{\infty}\sin x^2\,\mathrm{d}x = \sqrt{2}\int_0^{\infty}e^{-x^2}\,\mathrm{d}x = \frac{1}{2}\sqrt{\frac{\pi}{2}} \tag{4.62}$$

4.3 Poisson's Integral on Unit Circle (or Disk)

Cauchy's integral formula makes a very powerful statement. Essentially, it states that if a function $f(z)$ is analytic everywhere inside as well as on a closed curve, we can find the value of the function $f(z)$ at any point inside this closed curve by knowing the function only all along the curve. We do not need to know the function explicitly inside the closed curve. When this curve is a unit circle, we can get the same result via the Poisson's integral.

Poisson's Kernel and Integral on the Unit Circle (or Disk)

The function $f(z)$ is analytic on and inside a unit circle or a unit disk with the center at the origin.

Consider a point

$$z_0 = re^{j\theta}, \quad 0 \leq r < 1$$

The Poisson's integral formula states that

$$f(z_0) = \frac{1}{2\pi} \int_0^{2\pi} P_r(\theta - t) f(e^{jt})\, \mathrm{d}t = \frac{1}{2\pi} \int_{-\pi}^{+\pi} P_r(\theta - t) f(e^{jt})\, \mathrm{d}t$$

where

$$P_r(\theta - t) = \frac{1 - r^2}{1 - 2r\cos(\theta - t) + r^2} = \text{Poisson's Kernel.}$$

It can be considered as an alternate form of the Cauchy Integral Formula.

Proof:

From Cauchy's Integral formula

$$f(z_0) = \frac{1}{2\pi j} \oint_C \frac{f(z)}{(z - z_0)}\, \mathrm{d}z \tag{4.63}$$

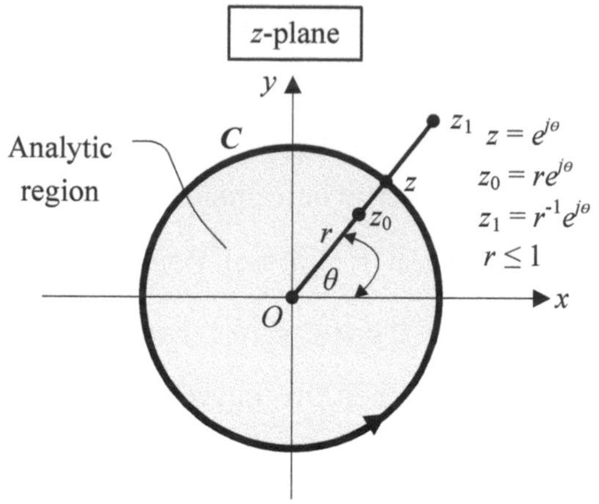

Figure 4.30: Poisson's Integral

$$z_0 = re^{j\theta}, \quad |r| < 1 \quad \text{inside the unit circle } C.$$

$$z_1 = \frac{1}{r}e^{j\theta}, \quad \frac{1}{r} > 1 \quad \text{outside the unit circle } C.$$

From Cauchy's integral theorem:

$$\frac{1}{2\pi j}\oint_C \frac{f(z)}{z - z_1} = 0 \quad (z_1, \text{ outside the circle } C) \tag{4.64}$$

Combining Eqs. 4.63 and 4.64 and letting $z = e^{jt}$

$$f(z_0) = \frac{1}{2\pi}\int_0^{2\pi}\left[\frac{1}{e^{jt} - re^{j\theta}} - \frac{r}{re^{jt} - e^{j\theta}}\right] f(e^{jt})e^{jt}\,\mathrm{d}t$$

$$f(z_0) = \frac{1}{2\pi}\int_0^{2\pi}\left[\frac{(1 - r^2)}{1 - 2r\cos(\theta - t) + r^2}\right] f(e^{jt})\,\mathrm{d}t = \frac{1}{2\pi}\int_0^{2\pi} P_r(\theta - t) f(e^{jt})\,\mathrm{d}t \tag{4.65}$$

where

$$P_r(\theta - t) = \frac{1 - r^2}{1 - 2r\cos(\theta - t) + r^2} = \text{Poisson's Kernel.}$$

Poisson's integral can be extended to the case where a function $g(z)$ is analytic only outside the unit circle. In that case, we can define:

$$\xi = \frac{1}{z}$$

$$g(z) = f(\xi)\Big|_{\xi=\frac{1}{z}}$$

Now $f(\xi)$ is analytic inside the circle $|\xi| = 1$.

Let

$$\xi_0 = \frac{1}{z_0} = r^{-1}e^{-j\theta}, \quad r^{-1} < 1$$

It is easy to see

$$f(\xi_0^{-1}) = g(z_0) = g(re^{j\theta}) = \frac{1}{2\pi}\int_0^{2\pi}\left[\frac{(r^2-1)}{r^2 - 2r\cos(\theta - t) + 1}\right]g(e^{jt})\,\mathrm{d}t \qquad (4.66)$$

where z_0 is outside the unit circle.

Poisson's Integral for Functions with Simple Zeroes Inside the Open Disk

Consider $f(z)$ with following properties:

1. $f(z)$ is analytic for $|z| \leq 1$ (closed disk).
2. $f(z)$ has no zeroes on $|z| = 1$ but has simple zeroes inside the open disk at points $\alpha_1, \alpha_2, \cdots, \alpha_n$.
3. The point $z_0 = re^{j\theta}, \quad r < 1$.

Then

$$\ln|f(z_0)| = \sum_{i=1}^{n}\ln\left|\frac{(z_0 - \alpha_i)}{(1 - \overline{\alpha}_i z_0)}\right| + \frac{1}{2\pi}\int_0^{2\pi} P_r(\theta - t)\ln|f(e^{jt})|\,\mathrm{d}t \qquad (4.67)$$

Proof:

Define a new function $g(z)$:

$$g(z) = f(z) \prod_{i=1}^{n} \frac{(1 - \overline{\alpha}_i z)}{(z - \alpha_i)}$$

Compared to $f(z)$ this new function $g(z)$ so obtained, is analytic inside as well as on the closed disk $|z| \leq 1$. Taking the natural log (ln) on both sides

$$\ln g(z) = \ln f(z) + \sum_{i=1}^{n} \ln\left(\frac{1 - \overline{\alpha}_i z}{z - \alpha_i}\right)$$

$$\ln g(z_0) = \ln f(z_0) + \sum_{i=1}^{n} \ln\left(\frac{1 - \overline{\alpha}_i z_0}{z_0 - \alpha_i}\right) = \frac{1}{2\pi} \int_0^{2\pi} P_r(\theta - t) \ln\left(g(e^{it})\right) dt$$

$$z_0 = re^{j\theta}, \ \ r \leq 1$$

$$\ln x = \ln |x| + j\angle x, \text{ where } x \text{ is a complex number}$$

Thus for $x = g(e^{j\omega})$,

$$\ln |g(e^{j\omega})| = \ln |f(e^{j\omega})|, \text{ because the factor } \left(\frac{1 - \overline{\alpha}_i z_0}{z_0 - \alpha_i}\right) \text{ has a magnitude} = 1.$$

Furthermore,

$$\ln\left(\frac{1 - \overline{\alpha}_i z_0}{z_0 - \alpha_i}\right) = -\ln\left(\frac{z_0 - \alpha_i}{1 - \overline{\alpha}_i z_0}\right)$$

Taking real part of Eq. 4.3 and applying the above simplifications we obtain

$$\ln |f(z_0)| = \sum_{i=1}^{n} \ln\left|\frac{(z_0 - \alpha_i)}{(1 - \overline{\alpha}_i z_0)}\right| + \frac{1}{2\pi} \int_0^{2\pi} P_r(\theta - t) \ln |f(e^{jt})| \, dt \qquad (4.68)$$

$$\ln |f(0)| = \sum_{i=1}^{n} \ln |\alpha_i| + \frac{1}{2\pi} \int_0^{2\pi} \ln \left|f(e^{jt})\right| dt$$

Poisson's Integral for Functions with Simple Poles and Simple Zeros Inside the Unit Disk

Consider a function $h(z)$ with the following properties:

1. $h(z)$ is analytic outside as well as on the unit disk, $|z| \geq 1$.
2. $h(z)$ has simple zeroes inside the open disk at $\beta_1, \beta_2, \ldots, \beta_n$.
3. $h(z)$ has simple poles inside the open disk at $\alpha_1, \alpha_2, \ldots, \alpha_m$.
4. There might be zero at the origin of order k ($k \in \mathcal{N}$).

But none of the poles and zeros lie on the unit circle.

Let

$$h(z) = z^k \frac{f_1(z)}{g_1(z)} \tag{4.69}$$

where

$$f_1(z) = \prod_{i=1}^{n} \left(\frac{z - \beta_i}{1 - \overline{\beta}_i z} \right)$$
$$g_1(z) = \prod_{j=1}^{m} \left(\frac{z - \alpha_j}{1 - \overline{\alpha}_j z} \right)$$

Then

$$\int_0^{2\pi} \ln |h(j\omega)| \, d\omega = \ln \left| \frac{f_1(0)}{g_1(0)} \right| - \ln \left| \frac{\prod_{i=1}^{n} \beta_i}{\prod_{j=1}^{m} \alpha_i} \right|$$

Proof:

Taking natural log (ln) of the absolute value on both sides of Eq. 4.69

$$\ln |h(z)| = k \ln |z| + \ln |f_1(z)| - \ln |g_1(z)|$$

Now applying Poisson's formula to $f_1(z)$ and $g_1(z)$ at $z = z_0$, we obtain

$$P_1(\theta - t) = 0$$

$$P_0(\theta - t) = 1$$

$$\ln\left|\frac{z_0 - \alpha_j}{1 - z_0\overline{\alpha}_j}\right|_{z_0=0} = \ln\left|\alpha_j\right|$$

$$\ln\left|\frac{z_0 - \beta_i}{1 - z_0\overline{\beta}_i}\right|_{z_0=0} = \ln|\beta_i|$$

$$\int_0^{2\pi} \ln|e^{j\omega}|\, d\omega = 0$$

Thus,

$$\ln|f_1(0)| = \sum_{i=1}^{n} \ln|\beta_i| + \frac{1}{2\pi}\int_0^{2\pi} \ln|f_1(j\omega)|\, d\omega$$

$$\ln|g_1(0)| = \sum_{j=1}^{n} \ln|\alpha_j| + \frac{1}{2\pi}\int_0^{2\pi} \ln|g_1(j\omega)|\, d\omega$$

Therefore,

$$\int_0^{2\pi} \ln|h(j\omega)|\, d\omega = \ln\left|\frac{f_1(0)}{g_1(0)}\right| - \ln\left|\frac{\prod_{i=1}^{n}\beta_i}{\prod_{j=1}^{m}\alpha_j}\right|$$

In fact without the pole at the origin,

$$h(z) = \frac{K\prod_{i=1}^{n}(z - \beta_i)}{\prod_{j=1}^{m}(z - \alpha_j)}, \quad h(0) = K\left|\frac{\prod_{i=1}^{n}\beta_i}{\prod_{j=1}^{m}\alpha_j}\right|$$

Poisson's Integral along Imaginary Axis–Analytic in RHS with No Zeros

Very often we know a bounded function $f(z)$ all along the imaginary axis $z = j\omega, -\infty \leq \omega \leq \infty$ and we want to extend the knowledge of this function to the rest of the z-plane. This is exactly the situation of obtaining the "Laplace Transforms" from the knowledge of the Fourier transforms. This is not always possible. But for a very large class of functions satisfying some restrictions mentioned below, we can accomplish this task via the following theorem.

Theorem:

Let $f(z)$ be a function of complex variable z such that

1. $\lim_{z\to\infty} |z|^{-1}|f(z)| = 0$

2. $f(z)$ is analytic for all z in the right hand side of the z-plane, such that, $z = \alpha + j\omega,\ \alpha \geq 0,\ -\infty \leq \omega \leq \infty$

$$f(z_0) = \frac{1}{2\pi}\int_{-\infty}^{+\infty}\left[\frac{1}{z_0 - j\omega}\right] f(j\omega)\, d\omega, \text{ or} \tag{4.70}$$

$$f(z_0) = \frac{1}{2\pi}\int_{-\infty}^{+\infty}\left[\frac{\alpha_0}{\alpha_0^2 + (\alpha_0 - \omega)^2}\right] f(j\omega)\, dw, \quad z_0 = \alpha_0 + j\omega_0, \quad \alpha_0 > 0$$

Proof:

Consider the closed contour C (Figure 4.31) made up of $AOBRA \equiv AB + C_1$. Applying Cauchy's Integral theorem to contour C_1

$$-2\pi j f(z_0) = \oint_C \frac{f(z)}{z - z_0}\, dz = j\int_{-j\infty}^{+j\infty} \frac{f(j\omega)}{j\omega - z_0}\, d\omega + \circlearrowright \int_{C_1} \frac{f(z)}{(z - z_0)}\, dz \tag{4.71}$$

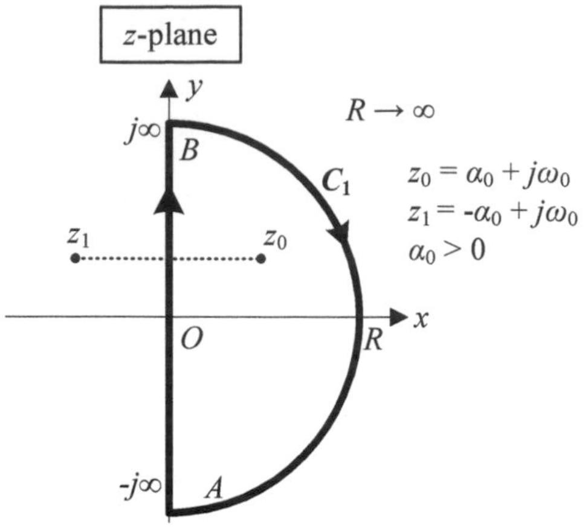

Figure 4.31: Poisson-Jensen's Integral

The second term in the above equation tends to 0 due to condition 1 and Jordan's lemma. Therefore,

$$f(z_0) = +\frac{1}{2\pi j}\int_{-j\infty}^{+j\infty}\left[\frac{1}{z_0 - j\omega}\right] f(j\omega)\,\mathrm{d}\omega \tag{4.72}$$

Furthermore,

$z_1 = -\alpha_0 + j\omega_0$, $\alpha_0 > 0$ is a point outside the contour C, yielding:

$$0 = \frac{1}{2\pi j}\oint \frac{f(z)}{(z_1 - z)}\,\mathrm{d}z = \frac{1}{2\pi}\int_{-\infty}^{+\infty}\left[\frac{1}{z_1 - j\omega}\right] f(j\omega)\,\mathrm{d}\omega \tag{4.73}$$

Adding and subtracting the Eqs. 4.72 and 4.73

$$f(z_0) = \frac{1}{2\pi}\int_{-\infty}^{+\infty}\left[\frac{1}{(z_0 - j\omega)} - \frac{1}{(z_1 - j\omega)}\right] f(j\omega)\,\mathrm{d}\omega$$

$$f(z_0) = \frac{1}{2\pi}\int_{-\infty}^{+\infty}\left[\frac{1}{(z_0 - j\omega)} + \frac{1}{(z_1 - j\omega)}\right] f(j\omega)\,\mathrm{d}\omega$$

Further simplification with substitution of z_0 and z_1 yields

$$\begin{aligned} f(z_0) &= -\frac{1}{\pi}\int_{-\infty}^{+\infty}\left[\frac{\alpha_0}{\alpha_0^2+(\omega_0-\omega)^2}\right]f(j\omega)\,\mathrm{d}\omega \\ &= \frac{1}{\pi}\int_{-\infty}^{+\infty}\left[\frac{(\omega_0-\omega)}{\alpha_0^2+(\omega_0-\omega)^2}\right]f(j\omega)\,\mathrm{d}\omega \end{aligned} \tag{4.74}$$

Poisson-Jensen Formula for Analytic Functions with Zeros in RHS

Consider a function $f(z)$ such that

1. $f(z)$ is analytic in the RHS of the z-plane except for simple zeros at $\alpha_1, \alpha_2, \cdots, \alpha_n$
2. No zeros on the $j\omega$ axis
3. $\lim\limits_{|z|\to\infty}\dfrac{\ln|f(z)|}{|z|} = 0$

Then, for a point z_0:

$$\ln|f(z_0)| = \sum_{i=1}^{n}\ln\left|\frac{\alpha_i - z_0}{\overline{\alpha}_i + z_0}\right| + \frac{1}{\pi}\int_{-\infty}^{+\infty}\frac{x_0}{x_0^2+(\omega-y_0)^2}\ln|f(j\omega)|\,\mathrm{d}\omega$$

$$z_0 = x_0 + jy_0, \quad x_0 > 0 \quad (z_0 \text{ in RHS})$$

Proof

The proof follows the same outline as in the previous section. Let

$$f_1(z) = f(z)\prod_{i=1}^{n}\left(\frac{z+\overline{\alpha}_i}{z-\alpha_i}\right) \tag{4.75}$$

$f_1(z)$ is analytic within the closed contour involving the $j\omega$ axis and the semicircle (in LHS or RHS). Taking the natural log (ln) of both sides of Eq. 4.75 and applying Poisson's formula at $z = z_0$:

$$\begin{aligned} \ln f_1(z_0) &= \ln f(z_0) + \sum_{i=1}^{n} \ln \frac{z_0 + \overline{\alpha}_i}{z_0 - \alpha_i} \\ &= \frac{1}{\pi} \int_{-\infty}^{+\infty} \left[\frac{x_0}{x_0^2 + (\omega - y_0)^2} \right] \ln f_1(j\omega)\, d\omega \end{aligned} \tag{4.76}$$

Realizing that

$$\begin{aligned} x &= |x| e^{j\theta} \qquad x \text{ being a complex number} \\ \ln x &= \ln |x| + j\theta \\ \ln |f_1(j\omega)| &= \ln |f(j\omega)| \end{aligned}$$

and equating real parts on both sides of Eq. 4.76, yields

$$\ln |f(z_0)| = \sum_{i=1}^{n} \ln \left| \frac{z_0 - \alpha_i}{z_0 + \overline{\alpha}_i} \right| + \frac{1}{\pi} \int_{-\infty}^{+\infty} \frac{x_0}{x_0^2 + (\omega - y_0)^2} \ln |f(j\omega)|\, d\omega \tag{4.77}$$

Fundamental Theorem of Algebra

Theorem:

Every real or complex polynomial $P_n(z)$ of degree n, $P_n(z) = z^n + a_1 z^{n-1} + \cdots + a_n$ has n finite roots and therefore can be written as:

$$P_n(z) = \sum_{i=1}^{r} (z - z_i)^{m_i}, \quad m_1 + m_2 + \cdots + m_r = n \tag{4.78}$$

It is customary to take this theorem for granted. However, it's proof has a long history. To prove this theorem via Cauchy's residue is illuminating.

Proof:

We shall use the argument that $P_n(z)$ has at least one finite root in the z-plane, otherwise as shown later, a contradiction takes place.

Let

$$\psi(z^2) = \frac{1}{P_n(z)P_n(-z)} = \frac{1}{(z^{2n} + b_1 z^{2(n-1)} + \cdots + b_n)}$$

where b_i, $(i = 1 \cdots n)$ are real.

Consider the closed contour C_1 involving real axis from $+R$ to $-R$ and the semicircle C shown in the figure 4.32. Let us assume $P_n(z)$ has no roots in the z-plane and therefore no zeros inside as well as on C_1, which means $\psi(z^2)$ has no poles inside as well as on C_1. We will show that this will lead to a contradiction as follows.

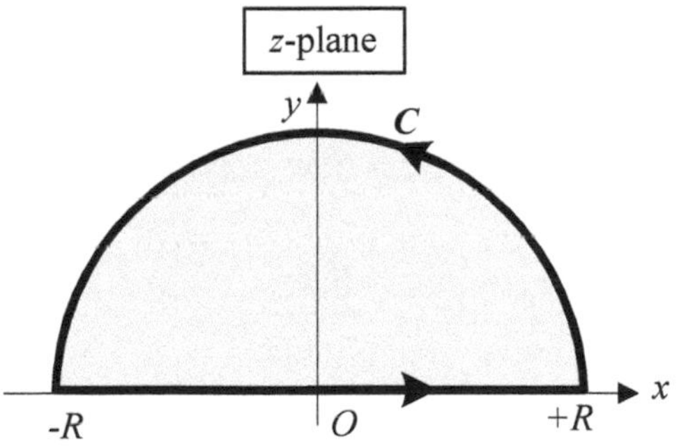

Figure 4.32: Proof of Fundamental Theorem of Algebra

Since $\psi(z^2)$ has no poles inside the contour C_1

$$I_1 = \oint \psi(z^2)\, dz = 0 \tag{4.79}$$

By Jordan's lemma, the integral of the infinite semicircular arc C is zero, yielding

$$I_1 = I = \int_{-R}^{+R} \psi(x^2)\, dx = 0$$

In the case of a zero of $P_n(z)$ on the real axis

$$I_1 = I = \mathrm{PV}\int_{-R}^{+R} \psi(x^2)\,\mathrm{d}x + \frac{j\pi}{b_n} = 0 \tag{4.80}$$

Since there are no zeros of $P_n(z)$ in the z-plane,

$$I = I_1 = 0 \quad \text{implies} \quad \int_{-R}^{+R} \psi(x^2)\,\mathrm{d}x = 0 \tag{4.81}$$

But for Eq. 4.81 to be true, $\psi(x^2)$ must change sign as x moves along the real axis from $-R$ to $+R$. To do so, $\psi(x^2)$ must go through the zero value, which is a contradiction to our assumption.

Conclusion: $P_n(z)$ has least one root.

Let this root be at $z = z_1$. We can re-write a new polynomial

$$P_{n-1}(z) = \frac{P_n(z)}{z - z_1}$$

$P_{n-1}(z)$ is a polynomial of degree $(n - 1)$, which again must have at least one root. Extending the argument further, we have proved that $P_n(z)$ has exactly n roots (counting multiplicity).

Maximum-Minimum Modulus Principle

1. **Maximum Principle**

   If $f(z) \in D$ (closed contour) is analytic and M_A is its maximum absolute value on the boundary ∂D, then $|f(z)| < M_A$ everywhere inside D, unless $f(z)$ is a constant of absolute value M_A The statement implies, that the maximum absolute value of an analytic function $f(z) \in D$ occurs at the boundary.

2. **Minimum Principle**

   If $f(z) \in D$ is analytic and M_I is its minimum absolute value on the boundary ∂D, then $|f(z)| > M_I$ everywhere in D, unless $f(z) \equiv 0$ or a constant. This principle implies that the minimum absolute value also occurs at the boundary.

3. **Mean Value Theorem**

$$f(z) \in D \text{ is analytic.}$$

   Its value at a point z_0 in D is the mean value of its integral on any circle centered at z_0 and radius less than the distance from z_0 to the boundary of domain D. This theorem is analogous to the Cauchy's integral formula and takes the form:

$$f(z_0) = \frac{1}{2\pi} \int_0^{2\pi} f(z_0 + re^{j\theta})\, \mathrm{d}\theta$$

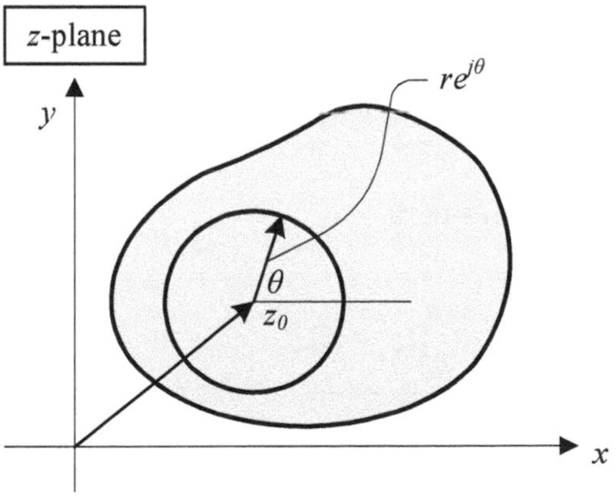

Figure 4.33: Mean Value Theorem

Meromorphic Functions and Theorem on Principle of the Argument

We shall refer to a function $f(z)$ in the domain D as meromorphic if it has no singularities other than poles in the domain D. We shall only deal with rational meromorphic functions with finite number of poles and zeros.

Preliminaries

(i) $z = re^{j\theta}$, a point on curve C in the z- plane

(ii) $f(z) = |f(z)|\angle\varphi_f(z)$, a mapping from $f(z)$ to z-plane.

(iii) $\Delta\ \varphi_f(z)\Big]_{z_1}^{z_2}$ = change in angle $\varphi_f(z)$ as z changes from z_1 to z_2

(iv) $\Delta_C\varphi_f(z)$ = change in angle $\varphi_f(z)$ as z moves along the closed contour C

(v) $(z - z_0) = |z - z_0|\angle\varphi_{z_0}(z)$

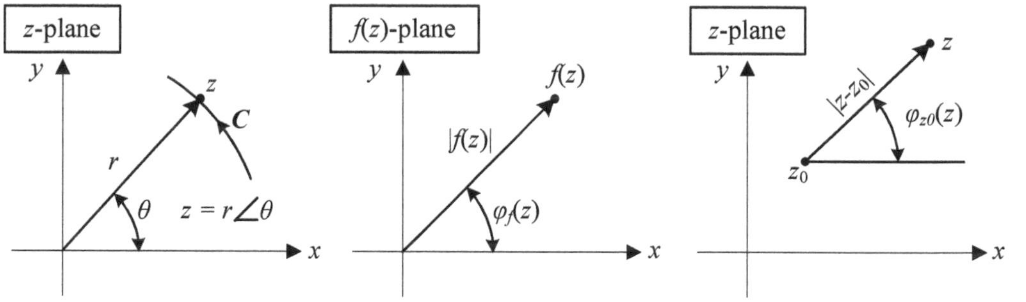

Figure 4.34: Preliminaries

Principle of the Argument Theorem

Let

$$f(z) = K\frac{\prod_{i=1}^{m}(z - z_i)}{\prod_{j=1}^{n}(z - p_j)} = |f(z)|\angle\varphi_f(z) \tag{4.82}$$

Consider a countour C enclosing the zeros $(z_1, z_2, \cdots, z_r)$ and poles $(p_1, p_2, \cdots, p_k)$ but excluding the zeros $(z_{r+1}, z_{r+2}, \cdots, z_m)$ and the poles $(p_{k+1}, p_{k+2}, \cdots, p_n)$. Then the total change in the argument of $f(z)$ as we move along the contour C is

$$\Delta_C\varphi_f(z) = 2\pi(r - k) \tag{4.83}$$

This theorem essentially results in the celebrated Nyquist criterion for stability theory in Control Systems. Essentially, the theorem states that if the closed contour C encloses r zeros and k poles of a function then the total change in the argument of the function as the point z moves along this curve is $2\pi(r - k)$.

Proof:

Let

$$(z - z_i) = |z - z_i| \angle \varphi_{z_i}(z)$$
$$(p - p_j) = |p - p_j| \angle \varphi_{p_j}(z)$$

From Eq. 4.82

$$\varphi_f(z) = \sum_{i=1}^{n} \varphi_{z_i}(z) - \sum_{j=1}^{n} \varphi_{p_j}(z)$$
$$\Delta_C \varphi_f(z) = \sum_{i=1}^{n} \Delta_C \varphi_{z_i}(z) - \sum_{j=1}^{n} \Delta_C \varphi_{p_j}(z)$$

But

$$\Delta_C \varphi_{z_i}(z) = 0 \quad \text{if } z_i \text{ is outside } C$$
$$= 2\pi, \quad z_i \text{ is inside } C$$

Similarly,

$$\Delta_C \varphi_{p_j}(z) = 0 \quad \text{if } p_j \text{ is outside } C$$
$$= 2\pi, \quad p_j \text{ is inside } C$$

Hence,

$$\Delta_C \varphi_f(z) = 2\pi(r - k)$$

4.4 Positive Real Functions

Theory of positive real functions plays an important role in design of Networks. A **Positive Real Function** $F(s)$ is an analytic function of a complex variable $s = \sigma + j\omega$, having the following properties:

1. $F(s)$ is analytic in the RHS of the s-plane
2. $\mathrm{Re}\,[F(s)] \geq 0$ for $\mathrm{Re}\,(s) \geq 0$ (RHS)
3. $F(\sigma)$ is real

Positive Real Rational Function

A rational function $F(s)$ with a real coefficient is **Positive Real** (PR), if and only if,

1. $F(s)$ is analytic for $\mathrm{Re}\,(s) > 0$
2. The poles on $j\omega$-axis are simple with positive residues
3. $\mathrm{Re}\,[F(j\omega)] \geq 0$ for all ω except for the poles of $F(s)$ on the $j\omega$ axis
4. A real rational function $F(s)$ is positive real if and only if

$$|\varphi_F(s)| \leq \varphi(s) \quad (0 \leq \varphi(s) \leq \pi/2), \quad s = |s|\angle\varphi(s), \quad F(s) = |F(s)|\angle\varphi_F(s)$$

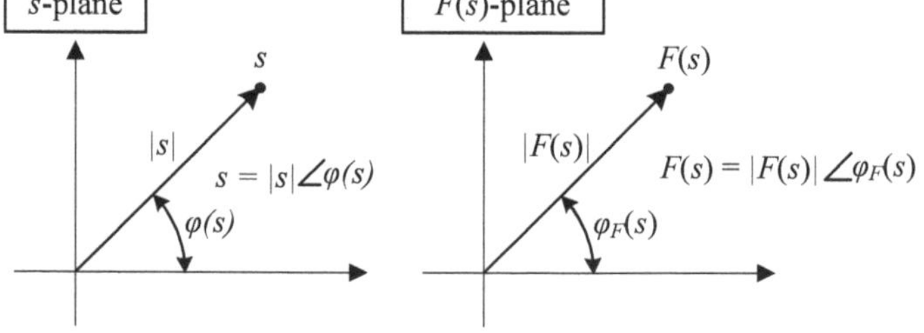

Figure 4.35: Representation of s and $F(s)$

There are many properties of positive real functions that are useful in network synthesis as well as in system stability studies. In systems analysis we deal with either continuous time signals that are transformed to s-plane or digital signals that are transformed to z-plane. Therefore, mapping from s-plane to z-plane and vice versa is of great importance. Particularly, PR functions are important for mapping analog filters into digital filters and vice versa.

4.4.1 Bilinear Transformation

Consider the following bilinear transformation:

$$s = \alpha\frac{z-1}{z+1}, \quad z = \frac{\alpha+s}{\alpha-s}, \quad \alpha > 0, \quad \alpha \text{ can be usually taken as } 1 \tag{4.84}$$

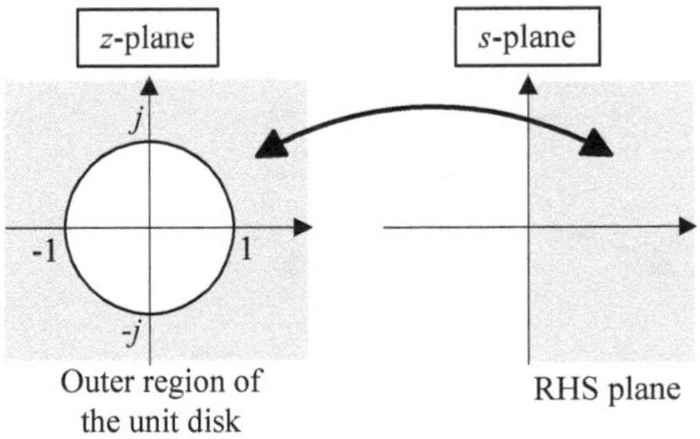

Figure 4.36: Mapping of Outer Region of Unit Disk in z-Plane to RHS in s-Plane

This change of variables provides a mapping from the z-plane to the s-plane transforming the region $|z| \geq 1$ to the region $\mathrm{Re}\,(s) \geq 0$, leading to the following results:

1. If $\mathrm{Re}\,(f(z)) \geq 0$ for $|z| \geq 1$, then $\mathrm{Re}\,(f(\frac{\alpha+s}{\alpha-s})) \geq 0$ for $\mathrm{Re}\,(s) \geq 0$ (RHS), and vice versa

2. if $f(z)$ is PR for $|z| \geq 1$, then $f(\frac{\alpha+s}{\alpha-s})$ is PR for $\mathrm{Re}\,(s) \geq 0$ and $\alpha > 0$

3. $z = e^{sT}$, $T > 0$ maps the outside of the unit disk in z-plane to RHS in s-plane

4. If $f(z)$ is PR for $|z| \geq 1$, then $f(e^{sT})$ is PR for $\mathrm{Re}\,(s) \geq 0$ and $T > 0$

5. The Following is an all-pass transformation from a unit disk in the z-plane to a unit disc in the w-plane:

$$z = e^{j\theta}\left(\frac{w - w_0}{\overline{w_0}w - 1}\right), \quad |w_0| < 1$$

$\theta =$ angle of pure rotation, $\quad w_0 =$ zero of the all pass function.

6. A positive real function of a positive real function is itself a positive real function. Thus if $F_1(s)$ is PR, $F_2(s)$ is PR, then $F_1[F_2(s)]$ is PR

Bounded Positive Real Functions

Using bilinear transformation we can relate one bounded PR function to another bounded PR function.

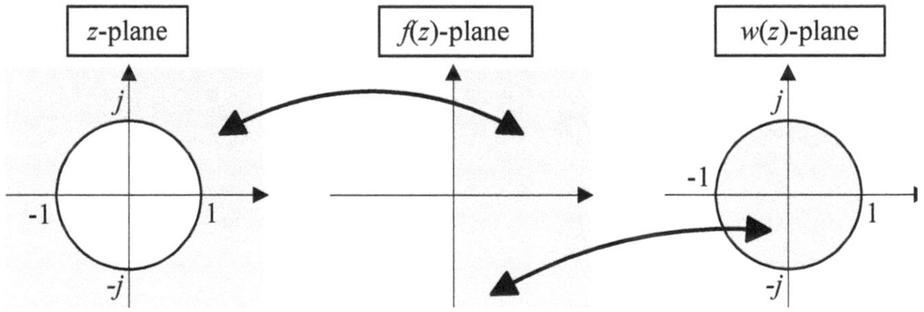

Figure 4.37: Transformation of Bounded PR Function to a Bounded PR Function

Consider the following transformation

$$w(z) = \frac{1 - f(z)}{1 + f(z)}, \quad f(z) = \frac{1 - w(z)}{1 + w(z)}, \quad |z| \geq 1$$

If $f(z)$ is PR, then $w(z)$ is also PR.

$|w(z)| \leq 1$ for $\quad |z| \leq 1$

Function of Matrices

Given a function of complex variable $f(z)$ and a matrix $\boldsymbol{A}$, we shall define $f(\boldsymbol{A})$ by substituting the variable z by the variable $\boldsymbol{A}$, denoting

$$f(z) \to f(\boldsymbol{A})$$

$$\begin{aligned} f(z) &= a_0 z^2 + a_1 z + a_2 \ \to f(\boldsymbol{A}) = a_0 \boldsymbol{A}^2 + a_1 \boldsymbol{A} + a_2 \boldsymbol{I} \\ f(z) &= \frac{a_0 z + a_1}{b_0 z + b_1} \ \to f(\boldsymbol{A}) = (a_0 \boldsymbol{A} + a_1 \boldsymbol{I})\,(b_0 \boldsymbol{A} + b_1 \boldsymbol{I})^{-1} \\ &= (b_1 \boldsymbol{A} + b_1 \boldsymbol{I})^{-1}\,(a_0 \boldsymbol{A} + a_1 \boldsymbol{I}) \end{aligned}$$

$$e^z = \sum_{k=0}^{\infty} \frac{z^k}{k!} \ \to e^{\boldsymbol{A}} = \sum_{k=0}^{\infty} \frac{\boldsymbol{A}^k}{k!}$$

$$e^{jz} = \cos(z) + j\sin(z) \ \to e^{j\boldsymbol{A}} = \cos(\boldsymbol{A}) + j\sin(\boldsymbol{A})$$

The functions $f(\boldsymbol{A})$ can be computed via Cauchy's Residue Theorem.

Computation of Matrix Functions via Residues

If $f(A)$ is an analytic function defined on the spectrum of a square matrix A, then we can write

$$f(A) = \frac{1}{2\pi j} \oint_C f(z)\,(z\boldsymbol{I} - \boldsymbol{A})^{-1}\ \mathrm{d}z \tag{4.85}$$

where C is a closed contour enclosing all the eigenvalues of A (spectrum of A.) Let

$$\left((z\boldsymbol{I} - \boldsymbol{A})^{-1}\right) = \frac{1}{P(z)}\,\boldsymbol{B}(\boldsymbol{A}, z) \tag{4.86}$$

$$P(z) = \det(z\boldsymbol{I} - \boldsymbol{A}) = z^n + a_1 z^{n-1} + a_2 z^{n-2} + \cdots + a_n \tag{4.87}$$

The coefficients $a_i,\ i = 1, 2, \cdots, n$ are determined from elements of the matrix $\boldsymbol{A}$.

Furthermore,

$$\boldsymbol{B}(\boldsymbol{A},z) = \left(\boldsymbol{B}_0(\boldsymbol{A})z^{n-1} + \boldsymbol{B}_1(\boldsymbol{A})z^{n-2} + \cdots + \boldsymbol{B}_{n-1}(\boldsymbol{A})\right) = \sum_{k=0}^{n-1} \boldsymbol{B}_k(\boldsymbol{A})\, z^{n-k-1} \tag{4.88}$$

From Eq. 4.86

$$P(z)\boldsymbol{I} = (z\boldsymbol{I} - \boldsymbol{A})(\boldsymbol{B}(\boldsymbol{A},z)) \tag{4.89}$$

Substituting Eqs. 4.87 and 4.88 into Eq. 4.89 and equating the powers of z

$$\begin{aligned}
\boldsymbol{B}_0(\boldsymbol{A}) &= \boldsymbol{I} \\
\boldsymbol{B}_1(\boldsymbol{A}) &= \boldsymbol{A} + a_1\boldsymbol{I} \\
\boldsymbol{B}_2(\boldsymbol{A}) &= \boldsymbol{A}^2 + a_1\boldsymbol{A} + a_2\boldsymbol{I} \\
&\vdots \\
\boldsymbol{B}_{n-1}(\boldsymbol{A}) &= \boldsymbol{A}^{n-1} + a_1\boldsymbol{A}^{n-2} + a_2\boldsymbol{A}^{n-3} + \cdots + a_{n-1}\boldsymbol{I} \\
0 &= A_n + a_1\boldsymbol{A}^{n-2} + a_2\boldsymbol{A}^{n-3} + \cdots + a_{n-1}\boldsymbol{A} + a_n\boldsymbol{I}
\end{aligned} \tag{4.90}$$

Using Eq. 4.88, Eq. 4.85 can be rewritten as

$$f(\boldsymbol{A}) = \frac{1}{2\pi j}\oint_C \left(\frac{f(z)}{P(z)}\right)\left(\sum_{k=0}^{n-1} \boldsymbol{B}_k(\boldsymbol{A})\, z^{n-k-1}\right) \mathrm{d}z = \sum_{k=0}^{n-1} \boldsymbol{B}_k(\boldsymbol{A}) \left[\frac{1}{2\pi j}\oint_C \frac{f(z)z^{n-k-1}}{P(z)}\,\mathrm{d}z\right]$$

$$P(z) = \prod_{i=1}^{m} (z - z_k)^{r_i}, \qquad r_1 + r_2 + \cdots + r_m = n, \quad r_i \in N$$

Using residue theorem

$$\frac{1}{2\pi j}\oint_C \frac{f(z)z^{n-k-1}}{\prod_{i=1}^{m}(z - z_k)^{r_i}} = \sum_{i=1}^{m}\left[\frac{1}{(r_i - 1)!}\right]\left[\frac{\mathrm{d}^{r_i-1}}{\mathrm{d}z^{r_i-1}}\left(f(z)(z - z_k)^{r_i}\, z^{n-k-1}\right)\right]$$

$$f(\boldsymbol{A}) = \sum_{k=0}^{n-1} \boldsymbol{B}_k(\boldsymbol{A}) \left[\sum_{i=1}^{m} \frac{1}{(r_i - 1)!}\left(\frac{\mathrm{d}^{r_i-1}}{\mathrm{d}z^{r_i-1}}\left(f(z)(z - z_k)^{r_i} z^{n-k-1}\right)\right)\right]$$

Bibliography

[Ahlfors, L.V.] Ahlfors, L.V. *Complex Analysis,* New York: McGraw-Hill, 1979.

[Churchill, R.V.] Churchill, R.V. *Introduction to Complex Variables and Applications,* New York: McGraw Hill, 1948.

[Doetsch, G.] Doetsch, G. *Theory Unt Anwendung der Laplace=Transformation,* New York: Dover Publication, 1943.

[Krantz, S.G.] Krantz, S.G. *Handbook of Complex Variables,* Boston: Birkhäuser, 1999.

[Lange, R.] Lange, R., Walsh, R.A. A Heuristic for the Poisson Integral for the Half Plane and Some Caveats, *The American Mathematical Monthly,* 92(5), 356–358, May 1985.

[Papoulis, A.] Papoulis, A. *Fourier Integral and Its Application,* New York: McGraw Hill, 1962.

[Titchmarsh, E.C.] Titchmarsh, E.C. *The Theory of Functions,* London: Oxford University Press, 1939.

[Whittaker, E.T.] Whittaker, E.T., Watson, G.N. *A Course in Modern Analysis,* 4th ed. Cambridge, England: Cambridge University Press, 1990.

Chapter 5

Integral Transform Methods

5.1 Introduction

Is it reasonable to ask why one should study various integral transforms and their properties. The most important reason is the so-called "translation invariance" of the linear time-invariant systems. This property of these systems can be described in terms of complex exponentials, which are eigenfunctions of time-invariant or space invariant systems. These simple "eigenfunctions" facilitate the analysis of the response of a linear time-invariant system to any input signal. If our independent variable takes values from 0 to ∞, then the eigenfunctions are such that they result in single-sided Laplace transforms. On the other hand, if the independent variable varies from $-\infty$ to $+\infty$, then the eigenfunctions result in the double-sided Laplace transforms or the Fourier transforms. It is customary to omit the word "single-sided" when it is obvious.

This chapter is devoted to understanding the derivation of various transform methods and their applications. Fourier transforms are useful in computing the amplitude and the frequency of the modulated communication systems. The starting point for the study of the Fourier transforms is the Fourier series and applying appropriate limits the transform relationships are derived. But not all signals are Fourier transformable and its application to systems involving the transient study of dynamical systems (involving differential equations) is tedious. Transient studies problems are best dealt with using Laplace transforms. Fourier transforms are extended to Laplace transforms by the application of appropriate conditions. Discrete transforms play the same role in the study of the discrete systems as the Fourier and the Laplace transforms in the continuous (or analog) systems.

We shall define the Fourier Transform first and derive the Laplace Transform as an extension. Since Laplace transforms are relatively straightforward, a detailed treatment of the Fourier transforms will be presented after the Laplace transforms and their applications.

5.2 Fourier Transform Pair Derivation

Consider a periodic signal $f(t)$, with a period T and expand it in the Fourier series with a complete set of basis functions $e^{j\omega_0 kt}, k = 0, \pm 1, \pm 2, \ldots$, orthonormal in the interval $[-T/2, T/2]$, yielding:

$$f(t) = \frac{1}{T} \sum_{k=-\infty}^{+\infty} F(jk\omega_0) e^{j\omega_0 kt}, \quad \omega_0 = 2\pi/T \tag{5.1}$$

where

$$F(jk\omega_0) = \int_{-T/2}^{T/2} f(t)\, e^{-j\omega_0 kt}\, \mathrm{d}t \tag{5.2}$$

Let

$$k\omega_0 = \omega_k$$

$$\Delta\omega_k = (k+1)\omega_0 - k\omega_0 = 2\pi/T$$

As $T \to 0, \omega_k \to \omega$ (continuous variable) and in the limit, summation is replaced by integration. Eqs. 5.1 and 5.2 become

$$f(t) = \frac{1}{2\pi}\int_{-\infty}^{+\infty} F(j\omega)e^{+j\omega t}\,\mathrm{d}w \qquad \text{Fourier transform of } f(t)$$

$$F(j\omega) = \int_{-\infty}^{+\infty} f(t)e^{-j\omega t}\,\mathrm{d}t \qquad \text{Fourier inverse of } F(j\omega) \qquad (5.3)$$

In order for these integrals to converge, the following Fourier transformability condition must be satisfied:

Conditions for Fourier Transformability

1. $f(t)$ has a finite number of discontinuities and a finite number of maxima and minima in any finite interval.

2. $\int_{-\infty}^{+\infty} |f(t)|\,\mathrm{d}t < \infty$

Even if the **first condition** is violated, the Fourier transform can be defined. The second condition is a must.

It is easy to see that

$$f(t) = \frac{1}{2\pi}\int_{-\infty}^{+\infty}\left[\int_{-\infty}^{+\infty} f(t)e^{-j\omega t}\,\mathrm{d}t\right]e^{j\omega t}\,\mathrm{d}\omega = \frac{1}{\sqrt{2\pi}}\int_{-\infty}^{+\infty}\left[\frac{1}{\sqrt{2\pi}}\int_{-\infty}^{+\infty} f(t)e^{-j\omega t}\,\mathrm{d}t\right]e^{j\omega t}\,\mathrm{d}\omega$$

The factor $\frac{1}{2\pi}$ can be distributed as $\frac{1}{\sqrt{2\pi}}$ to each part of the pair, yielding a normalized transform pair.

5.3 Another Derivation of Fourier Transform

This derivation makes use of Delta function properties.

Consider $f(t)$ as a convolution of $f(t)$ and $\delta(t)$, yielding

$$f(t) = \int_{\infty}^{+\infty} f(\tau)\delta(t-\tau)\,\mathrm{d}\tau \tag{5.4}$$

where

$$\delta(t-\tau) = \frac{1}{2\pi}\int_{-\infty}^{+\infty} e^{j\omega(t-\tau)}\,\mathrm{d}\omega \tag{5.5}$$

From Eqs. 5.4 and 5.5

$$\begin{aligned} f(t) &= \int_{-\infty}^{+\infty} f(\tau)\left[\frac{1}{2\pi}\int_{-\infty}^{+\infty} e^{j\omega(t-\tau)}\,\mathrm{d}\omega\right]\mathrm{d}\tau \\ &= \frac{1}{2\pi}\int_{-\infty}^{+\infty}\left[\int_{-\infty}^{+\infty} f(\tau)e^{-j\omega t}\,\mathrm{d}\tau\right]e^{j\omega t}\,\mathrm{d}\omega \end{aligned} \tag{5.6}$$

Eq. 5.6 can be interpreted as:

Fourier Transform Pair

$$F[f(t)] = F(j\omega) = \int_{-\infty}^{+\infty} f(\tau)e^{-j\omega\tau}\,\mathrm{d}\tau = \int_{-\infty}^{+\infty} f(t)e^{-j\omega t}\,\mathrm{d}t$$

(t and τ are both dummy variables)

$$F^{-1}[F(j\omega)] = f(t) = \frac{1}{2\pi}\int_{-\infty}^{+\infty} F(j\omega)e^{-j\omega t}\,\mathrm{d}\omega$$

5.4 Derivation of Bilateral Laplace Transform L_b

The main condition of the Fourier transformability is:

$$\int_{-\infty}^{+\infty} |f(t)|\, \mathrm{d}t < \infty \tag{5.7}$$

This condition is a most stringent requirement and is violated by many common functions found in physics and engineering. To overcome this difficulty, bilateral Laplace transform, denoted by L_b, is introduced.

Let the function $f(t)$ be such that

$$|f(t)| \leq Me^{\alpha t} \qquad 0 \leq t < \infty$$

$$|f(t)| \leq Me^{\beta t} \qquad -\infty \leq t < 0$$

Where $M, \alpha,$ and β are some constants.

If we select some value σ (as shown in Figure 5.1) such that

$$-\infty < \alpha < \sigma < \beta < +\infty$$

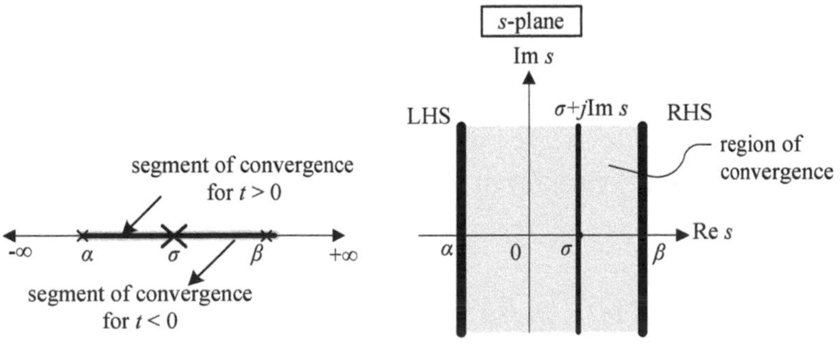

Figure 5.1: Double-Sided Laplace Transform Region of Convergence

Then

$$\lim_{t \to \pm\infty} \left| f(t)e^{-\sigma t} \right| \to 0$$

Furthermore,

$$\int_{-\infty}^{+\infty} \left| f(t)e^{-\sigma t} \right| \, \mathrm{d}t < \infty$$

In light of the above results

$$F[f(t)e^{-\sigma t}] = F(\sigma, j\omega) = F(\sigma + j\omega) = \int_{-\infty}^{+\infty} f(t)e^{-\sigma t}e^{-j\omega t} \, \mathrm{d}t$$

$$\begin{aligned} F^{-1}[F(\sigma, j\omega)] = F^{-1}[F(\sigma + j\omega)] &= \frac{1}{2\pi j} \int_{-j\infty}^{+j\infty} F(\sigma, j\omega)e^{j\omega t} \, \mathrm{d}\omega \\ &= \frac{1}{2\pi j} \int_{-j\infty}^{+j\infty} F(\sigma + j\omega)e^{j\omega t} \, \mathrm{d}\omega \end{aligned}$$

Let

$$\sigma + j\omega = s$$

$$F(j\omega, \sigma) = F_b(s) = L_b[f(t)]$$

$$\mathrm{d}\omega = \frac{1}{j} \mathrm{d}s, \qquad \sigma = \mathrm{Re}\, s$$

Thus the **Bilateral Laplace Transform Pair** can be defined as:

$$\begin{aligned} L_b[f(t)] \equiv F_b(s) &= \int_{-\infty}^{+\infty} f(t)e^{-st} \, \mathrm{d}t \\ L_b^{-1}[f(t)] &= \frac{1}{2\pi j} \int_{\sigma - j\infty}^{\sigma + j\infty} F_b(s)e^{st} \, \mathrm{d}s \end{aligned} \tag{5.8}$$

The vertical strip $\alpha < \operatorname{Re} s = \sigma < \beta$, $-\infty < \operatorname{Im} s < +\infty$ is called the Region of Convergence (ROC). There are no singularities or poles of $F_b(s)$ inside this ROC strip. The subscript "b" stands for bilateral.

The function $F_b(s)$ is analytic,

For $t > 0$, $\quad \alpha < \sigma < \infty$

For $t < 0$, $\quad -\infty < \sigma < \beta$

Summary

$$|f(t)| < \begin{cases} M_1 e^{\alpha t} & t > 0 \\ M_2 e^{\beta t} & t < 0 \end{cases}$$

$$\int_{-\infty}^{+\infty} |f(t)e^{-\sigma t}|\,\mathrm{d}t \text{ is finite,} \qquad -\infty < \alpha < \sigma < \beta$$

$f(t)$ has a bilateral or double-sided Laplace Transform denoted by $F_b(s)$.

Furthermore,

$$\lim_{t\to\infty} e^{\alpha t} f(t) = 0$$

$$\lim_{t\to-\infty} e^{\beta t} f(t) = 0$$

5.5 Another Bilateral Laplace Transform Derivation

Let

$$f(t) = \int_{-\infty}^{+\infty} f(\tau)\delta(t-\tau)\,\mathrm{d}\tau \tag{5.9}$$

where

$$\delta(t-\tau) = \frac{1}{2\pi j}\int_{\sigma-j\infty}^{\sigma+j\infty} e^{s(t-\tau)}\,\mathrm{d}s, \qquad s = \sigma + j\omega \tag{5.10}$$

$$-\infty < \alpha < \sigma < \beta < +\infty$$

Thus,

$$f(t) = \int_{-\infty}^{+\infty} f(\tau) \left[\frac{1}{2\pi j} \int_{\sigma-j\infty}^{\sigma+j\infty} e^{s(t-\tau)}\, \mathrm{d}s \right] \mathrm{d}\tau$$

$$= \frac{1}{2\pi j} \int_{\sigma-j\infty}^{\sigma+j\infty} \left[\int_{-\infty}^{+\infty} f(\tau) e^{-s\tau}\, \mathrm{d}\tau \right] e^{st}\, \mathrm{d}s$$

The above equation yields the transform pair,

$$F_b(s) = \int_{-\infty}^{+\infty} f(\tau) e^{-s\tau}\, \mathrm{d}\tau = \int_{-\infty}^{+\infty} f(t) e^{-st}\, \mathrm{d}t$$

$$f(t) = \frac{1}{2\pi j} \int_{\sigma-j\infty}^{\sigma+j\infty} F_b(s)\, e^{st}\, \mathrm{d}s$$

5.6 Single-Sided Laplace Transform

Single-sided Laplace Transform is referred to as "Laplace Transform".

Consider the causal function $f(t)$ satisfying the following conditions:

1. $f(t) \equiv 0, \quad -\infty < t < 0$

2. $|f(t)| < Me^{\alpha t}, \quad 0 < t < \infty$

3. $\lim_{t \to \infty} e^{\alpha t} f(t) = 0$

Laplace transform of $f(t)$ yields:

$$L[f(t)] \equiv F(s) = \int_0^\infty f(t)e^{-st}\,\mathrm{d}t \qquad \mathrm{Re}\, s = \sigma > \alpha$$

$$L^{-1}[F(s)] \equiv f(t) = \frac{1}{2\pi j}\int_{\sigma-j\infty}^{\sigma+j\infty} F(s)\,e^{st}\,\mathrm{d}s, \quad \alpha < \sigma < \infty \tag{5.11}$$

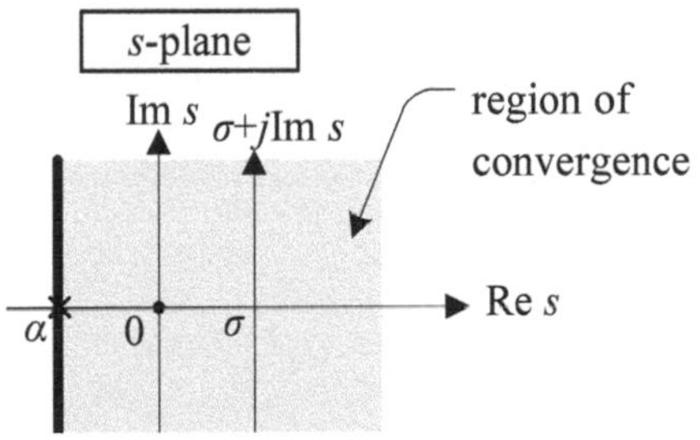

Figure 5.2: Single Laplace Transform Region of Convergence

The functions satisfying the conditions discussed earlier are called Functions of the Exponential Order. **Functions such as e^{t^2} grow more rapidly than the exponential order functions and therefore violate the conditions of convergence and are not Laplace transformable.**

Notation: The following notation is often used:

1. $f(t) \leftrightarrow F(s)$ implying the "Laplace transform of $f(t)$ is $F(s)$."

2. $F(s) \leftrightarrow f(t)$ implying the "Laplace inverse of $F(s)$ is $f(t)$."

3. $u(t)$ is a step function, such that

$$u(t) = \begin{cases} 0 & -\infty < t < 0 \\ 1 & 0 < t < \infty \\ \frac{1}{2} & \text{at } t = 0 \end{cases}$$

Thus, a causal $f(t)$ can be defined for $-\infty < t < \infty$, as

$$f(t)u(t) \equiv 0, \qquad -\infty < t < 0$$
$$f(t)u(t) \equiv f(t), \qquad 0 < t < \infty$$

In what follows, unless otherwise stated, $f(t)$ is always causal when dealing with the single-sided Laplace transform and can be written as $f(t)$ instead of $f(t)u(t)$, realizing $f(t) \equiv 0$ for negative values of time.

Note

1. The Laplace transform "L" is a linear operator. Thus, if $f_1(t)$ and $f_2(t)$ are two time functions, then

$$L[a_1 f_1(t)] = a_1 L[f_1(t)]$$
$$L[a_1 f_1(t) + a_2 f_2(t)] = a_1 L[f_1(t)] + a_2 L[f_2(t)]$$

   Where a_1 and a_2 are constants.

2. Many formulas in this chapter can be easily derived by the use of integration by parts, namely:

$$\int_a^b f_1(t) f_2(t) = \left[f_1(t) \int f_2(t)\,\mathrm{d}t \right]_a^b - \int_a^b \left(\frac{\mathrm{d}f_1(t)}{\mathrm{d}t} \right) \left(\int f_2(t)\,\mathrm{d}t \right) \mathrm{d}t$$

5.7 Summary of Transform Definitions

Table 5.1 summarizes the **Transform Pairs**. Conditions under which these Transform Pairs exist have already been discussed in the previous section.

Table 5.1: Transform Definitions

Fourier Transform Pair	$F[f(t)] \equiv F(j\omega) = \int_{-\infty}^{+\infty} f(t)e^{-j\omega t}\,\mathrm{d}t$	Direct
	$F^{-1}[F(j\omega)] \equiv f(t) = \frac{1}{2\pi} \int_{-\infty}^{+\infty} F(j\omega)e^{+j\omega t}\,\mathrm{d}\omega$	Inverse
Laplace Transform (single sided)	$L[f(t)] \equiv F(s) = \int_{0}^{+\infty} f(t)e^{-st}\,\mathrm{d}t$	Direct
	$L^{-1}[F(s)] \equiv f(t) = \frac{1}{2\pi j} \int_{c-j\infty}^{c+j\infty} F(s)e^{st}\,\mathrm{d}s$	Inverse
Bilateral Laplace Transform (double sided)	$L_b[f(t)] \equiv F_b(s) = \int_{-\infty}^{+\infty} f(t)e^{-st}\,\mathrm{d}t$	Direct
	$L_b^{-1}[F_b(s)] \equiv f(t) = \frac{1}{2\pi j} \int_{c-j\infty}^{c+j\infty} F_b(s)e^{st}\,\mathrm{d}s$	Inverse
The Z-Transform Pair	$Z[f(n)] \equiv F(z) = \sum_{n=0}^{\infty} f(n)z^{-n}$	Direct
	$Z^{-1}[F(z)] \equiv f(n) = \frac{1}{2\pi j} \oint F(z)z^{n-1}\,\mathrm{d}z$	Inverse
Discrete Fourier Transform (DFT)	$D[f(nT)] \equiv F(jk\Omega) = \sum_{n=0}^{N-1} f(nT)e^{-j\Omega nkT}$	Direct
	$D^{-1}[F(k\Omega)] \equiv f(nT) = \frac{1}{N} \sum_{k=0}^{N-1} F(jk\Omega)e^{j\Omega nkT}$	Inverse
	$\Omega = 2\pi/NT$	
General Discrete Transforms	$F(k) = \sum_{n=0}^{N-1} f(n)\,g(n,k,N)$	Direct
	$f(n) = \sum_{k=0}^{N-1} F(k)h(n,k,N)$	Inverse
	$\sum_{n=0}^{N-1} \sum_{k=0}^{N-1} g(n,k,N)h(n,k,N) = \delta(n,k)$	
	$\delta(n,k) = 1$ when $n = k$, 0 otherwise	

5.8 Laplace Transform Properties

The following table lists a number of Laplace transform properties

Table 5.2: Laplace Transform Properties

Property	Time Function $f(t)$	Laplace Transform $F(s)$	(ROC) $\text{Re}\, s > \sigma$
1. Linearity	$f_i(t), i = 1, \cdots, N$ $f(t) = \sum_{i=1}^{N} a_i f_i(t)$	$F_i(s), i = 1, \cdots, N$ $F(s) = \sum_{i=1}^{N} a_i F_i(s)$	$\text{Re}\, s > \sigma_i$ $\text{Re}\, s > \text{Max}\, \sigma_{i0}$
2. Scaling	$f(ct),\ c > 0$	$\frac{1}{c} F\left(\frac{s}{c}\right)$	$\text{Re}\, s > c\sigma_i$
3. Time-delay	$f(t-T)u(t-T)$	$F(s)e^{-sT}$	$\text{Re}\, s > \sigma$
4. Frequency Shift	$e^{-\lambda t} f(t)$	$F(s+\lambda)$	$\text{Re}\, s > \sigma - \text{Re}\,\lambda$
5. Time Derivative	$\frac{\mathrm{d}}{\mathrm{d}t} f(t)$	$sF(s) - f(0)$	$\text{Re}\, s > \sigma$
6. Higher Time Derivative	$\frac{\mathrm{d}^n}{\mathrm{d}t^n} f(t) = f^{(n)}(t)$	$s^n F(s) - \sum_{i=0}^{n-1} s^{n-i-1} f^{(i)}(0)$	$\text{Re}\, s > \sigma$
7. Differentiation of Transform	$(-t)^n f(t)$	$\frac{\mathrm{d}^n}{\mathrm{d}s^n} F(s) = F^{(n)}(s)$	$\text{Re}\, s > \sigma$
8. Integration of Time Function	$\int_0^t f(\tau)\, \mathrm{d}\tau$	$\frac{1}{s} F(s)$	$\text{Re}\, s > \sigma$
9. Integration of Transform Function	$\frac{f(t)}{t}$	$\int_s^\infty F(\lambda)\, \mathrm{d}\lambda$	$\text{Re}\, s > \sigma$
10. Initial Value	$\lim_{t\to 0} f(t) = \lim_{s\to\infty} sF(s)$		
11. Final Value	$\lim_{t\to\infty} f(t) = \lim_{s\to 0} sF(s)$		
12. Time Convolution	$\int_0^t f_1(\tau) f_2(t-\tau)\, \mathrm{d}\tau$	$F_1(s) F_2(s)$	$\text{Re}\, s > \sigma_1 \geq \sigma_2$

(continued)

Property	Time Function $f(t)$	Laplace Transform $F(s)$	(ROC) $\text{Re}\, s > \sigma$
13. Frequency Convolution	$f_1(t)\, f_2(t)$ $\lvert f_i(t)\rvert \le M\, e^{\alpha_i t}$ $0 \le t < \infty$ $\lvert f_i(t)\rvert \le M\, e^{\beta_i t}$ $-\infty \le t < 0$ $i = 1, 2$	$F(\lambda, s) = F_1(\lambda) \times F_2(s-\lambda)$ $\frac{1}{2\pi j}\int\limits_{c-j\infty}^{c+j\infty} F(\lambda, s)\, \mathrm{d}\lambda$	$\alpha_1 < c < \beta_1$ $\sum\limits_{i=1}^{2} \alpha_i < \sigma < \sum\limits_{i=1}^{2} \beta_i$

We shall illustrate some of these properties by the following examples.

Example 5.1:

$$f(t) = u(t) \quad \text{implying ROC } \alpha = 0$$

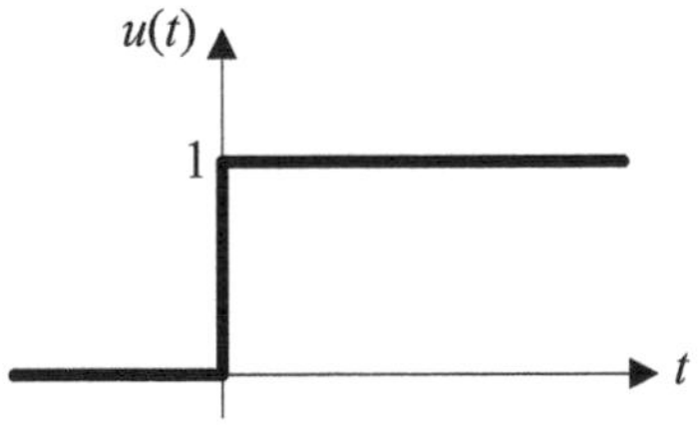

Figure 5.3: Step Function

$$F(s) = \int\limits_0^\infty u(t)e^{-st}\, \mathrm{d}t = \int\limits_0^\infty e^{-st}\, \mathrm{d}t = \left.\frac{e^{-st}}{-s}\right]_0^\infty = \left. e^{-st}\right]_{t=\infty} + \frac{1}{s}$$

$$\alpha = 0 < \text{Re}\, s < \infty \quad \text{implying } \lim_{t\to\infty} e^{-st} = 0$$

Thus,

$$L[u(t)] = \frac{1}{s}$$

or

$$u(t) \leftrightarrow \frac{1}{s}$$

Example 5.2:

$$f(t) = t, \qquad \text{ROC } \alpha > 0$$

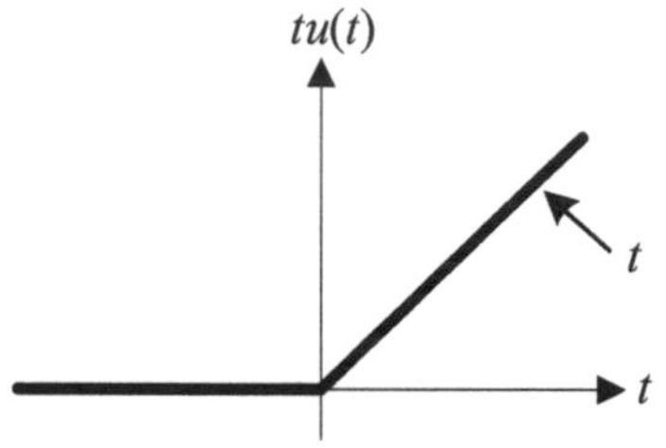

Figure 5.4: Ramp Function

$$F(s) = \int_0^\infty te^{-st}\,dt$$

$$f_1(t) = t$$

$$f_2(t) = e^{-st}$$

Integrating by parts,

$$F(s) = \left.\frac{te^{-st}}{-s}\right]_0^\infty - \int_0^\infty (1)\left(\frac{e^{-st}}{-s}\right) dt = \frac{1}{s}\lim_{t\to 0}(te^{-st}) + \frac{1}{s^2} = \frac{1}{s^2}$$

Computing this transform another way,

$$F(s) = \int_0^\infty te^{-st}\,dt = -\frac{d}{ds}\int_0^\infty e^{-st}\,dt = -\frac{d}{ds}\left(\frac{1}{s}\right) = \frac{1}{s^2}$$

Example 5.3:

$$f(t) = t^n \quad \text{ROC } \alpha > 0$$

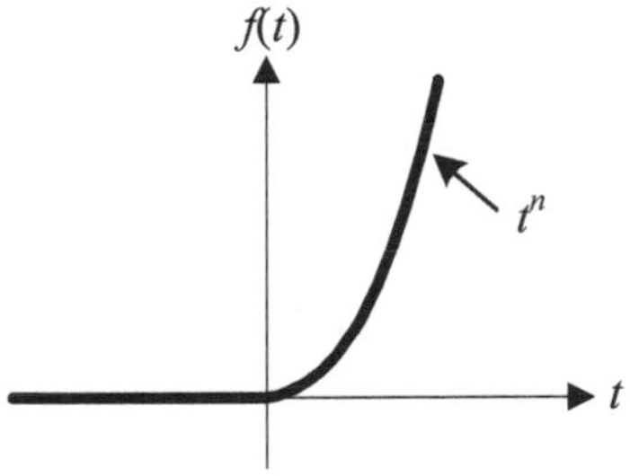

Figure 5.5: $f(t) = t^n,\ n = 0, 1, 2, \ldots$

$$F(s) = \int_0^{\infty} t^n e^{-st}\,\mathrm{d}t$$

Integrating by parts and repeating it n times

$$F(s) = \frac{n!}{s^{n+1}}$$

Another way

$$F(s) = \int_0^{\infty} t^n e^{-st}\,\mathrm{d}t = (-1)^n \frac{\mathrm{d}^n}{\mathrm{d}s^n}\frac{1}{s} = \frac{n!}{s^{n+1}}$$

Example 5.4:

$$f(t) = e^{-\alpha_1 t}, \qquad (\alpha_1 > 0) \qquad \alpha > -\alpha_1$$

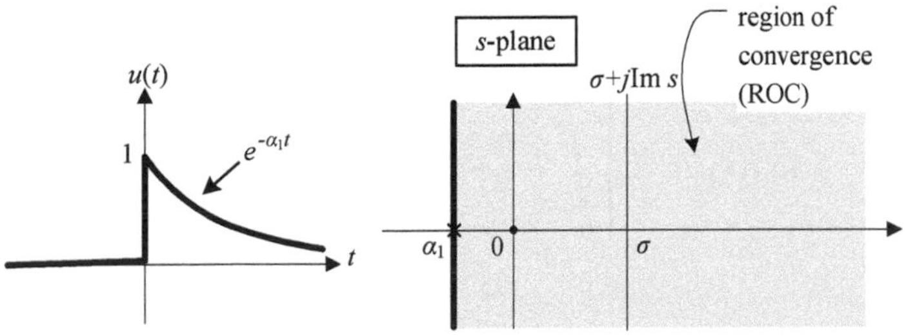

Figure 5.6: Time Function $e^{-\alpha t}u(t)$ and ROC

$$F(s) = \int_0^\infty e^{-\alpha_1 t} e^{-st}\, dt = \int_0^\infty e^{-(s+\alpha_1)t}\, dt = \frac{1}{s+\alpha_1}$$

Function $F(s)$ is analytic in Re $s > -\alpha_1$ and its singularities lie in the shaded region.

Example 5.5:

Evaluate the Laplace transforms of:

a. $\dfrac{\sin t}{t}$

b. $\mathrm{Si}(t) = \int_0^t \dfrac{\sin \tau}{\tau}\, d\tau$

Solution: Let

$$\sin t = \frac{e^{jt} - e^{-jt}}{2j}$$

From Property 1 in Table 5.2

$$L[\sin t] = \frac{1}{2j} L[e^{jt}] - \frac{1}{2j} L[e^{-jt}] = \frac{1}{2j}\left[\frac{1}{s-j} - \frac{1}{s+j}\right] = \frac{1}{s^2+1}$$

From Property 9 in Table 5.2

$$L\left[\frac{\sin t}{t}\right] = \int_s^\infty \frac{d\lambda}{1+\lambda^2} = \frac{\pi}{2} - \tan^{-1} s$$

From Property 8 in Table 5.2

$$L\,[S\,i(t)] = L\left[\int_0^t \frac{\sin \tau}{\tau}\, d\tau\right] = \frac{1}{s}\left[\frac{\pi}{2} - \tan^{-1} s\right]$$

Example 5.6:

Prove the initial and final value properties i.e., items 10 and 11 in Table 5.2 of the Laplace transform.

Proof:

$$L\left[\frac{df}{dt}\right] = \int_0^\infty e^{-st}\frac{df}{dt}\,dt = sF(s) - f(0)$$

Taking the limit of both sides as $s \to 0$

$$\lim_{s\to 0}\left[L\frac{df}{dt}\right] = \lim_{s\to 0}[sF(s) - f(0)] = \lim_{s\to 0}\int_0^\infty e^{-st}\frac{df}{dt}\,dt = \int_0^\infty df(t) = f(\infty) - f(0)$$

$$\lim_{s\to 0} sF(s) = f(\infty)$$

Similarly taking the limit of $L\left[\frac{df}{dt}\right]$ as $s \to \infty$

$$\lim_{s\to\infty}\left[L\frac{df}{dt}\right] = \lim_{s\to\infty}\left[\int_0^\infty e^{-st}\frac{df}{ds}\,dt\right] = \lim_{s\to\infty}[sF(s) - f(0)] = 0$$

$$\lim_{s\to\infty} sF(s) = f(0)$$

Convolution Theorems

1. **Time Domain Convolution Theorem**

$$\mathcal{L}^{-1}[F_1(s)F_2(s)] = \int_0^t f_1(\tau)f_2(t-\tau) = \int_0^t f_1(t-\tau)f_2(\tau)\,d\tau = f_1(t) * f_2(t)$$

where "$*$" stands for **convolution symbol** (**not multiplicity**) **Proof**:

From the definition

$$F_1(s) = \int_0^\infty f_1(t)e^{-st}\,dt = \int_0^\infty f_1(u)e^{-su}\,du$$

$$F_2(s) = \int_0^\infty f_2(t)e^{-st}\,dt = \int_0^\infty f_2(v)e^{-sv}\,dv$$

Thus,

$$F_1(s)F_2(s) = \int_0^\infty \int_0^\infty f_1(u) f_2(v) e^{-s(u+v)}\, du\, dv, \qquad u \geq 0, v \geq 0 \tag{5.12}$$

Let $u + v = t, \quad u = \tau$

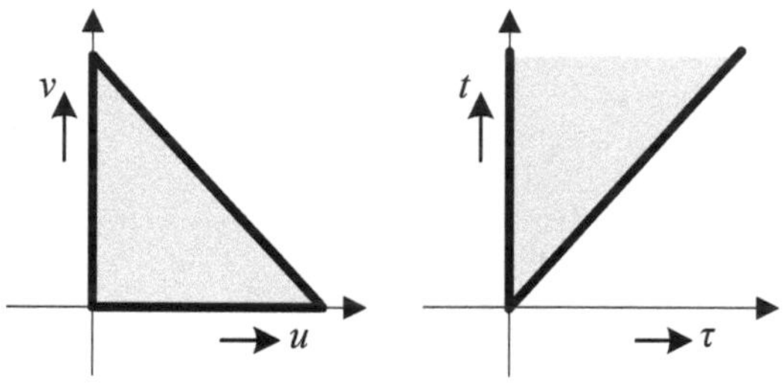

Figure 5.7: Region of Integration

The variables u, v are related to new variables t and τ through the Jacobian of the transformation,

$$du\, dv = \begin{vmatrix} \partial u/\partial t & \partial u/\partial \tau \\ \partial v/\partial t & \partial v/\partial \tau \end{vmatrix} dt\, d\tau \tag{5.13}$$

or

$$du\, dv = dt\, d\tau$$

$$t = \tau + v, \quad v \geq 0 \quad \text{implies} \quad \tau \leq t$$

Thus,

$$F_1(s)F_2(s) = \int_0^\infty \left[\int_0^t f_1(\tau) f_2(t-\tau)\, d\tau \right] e^{-st}\, dt \tag{5.14}$$

or

$$\mathcal{L}^{-1}[F_1(s)F_2(s)] = \int_0^t f_1(\tau) f_2(t-\tau)\, d\tau = f_1(t) * f_2(t)$$

2. **Frequency Domain Convolution Theorem**

If

$$L[f_1(t)] = F_1(s), \quad \sigma_1 = \text{ abcissa of convergence of } F_1(s)$$

$$L[f_2(t)] = F_2(s), \quad \sigma_2 = \text{ abcissa of convergence of } F_2(s)$$

Then

$$L[f_1(t)f_2(t)] = \frac{1}{2\pi j}\int_{c-j\infty}^{c+j\infty} F_1(\lambda)F_2(s-\lambda)\,\mathrm{d}\tau = [F_1(s) * F_2(s)] \tag{5.15}$$

$$c > \sigma_1 + \sigma_2$$

Proof:

By definition

$$L[f_1(t)f_2(t)] = \int_0^{\infty} f_1(t)f_2(t)e^{-st}\,\mathrm{d}t \tag{5.16}$$

Let

$$f_1(t) = \frac{1}{2\pi j}\int_{c-j\infty}^{c+j\infty} F_1(\lambda)e^{\lambda t}\,\mathrm{d}t, \quad \mathrm{Re}\,\{\lambda\} > \sigma_1 \tag{5.17}$$

Substituting Eq. 5.17 into Eq. 5.16

$$L\,[f_1(t)f_2(t)] = \int_0^{+\infty}\left[\frac{1}{2\pi j}\int_{c-j\infty}^{c+j\infty} F_1(\lambda)\,e^{\lambda t}\,\mathrm{d}\lambda\right] f_2(t)\,e^{-st}\,\mathrm{d}t$$

Interchanging the order of integration

$$L\,[f_1(t)f_2(t)] = \frac{1}{2\pi j}\int_{c-j\infty}^{c+j\infty} F_1(\lambda)\left[\int_0^{+\infty} f_2(t)e^{-(s-\lambda)t}\,\mathrm{d}t\right]\mathrm{d}\lambda$$

$$\text{Re}\{s-\lambda\} > \sigma_2,$$

Thus,

$$\text{Re}\{s\} > \sigma_2 + \sigma_1$$

or

$$L[f_1(t)f_2(t)] = \frac{1}{2\pi j}\int_{c-j\infty}^{c+j\infty} F_1(\lambda)F_2(s-\lambda)\,d\lambda = \frac{1}{2\pi j}F_1(s) * F_2(s)$$

$$\text{Re}\{s\} > \sigma_2 + \sigma_1$$

Notation:

The following notation will be used whenever necessary:

$$f_1(t) \leftrightarrow F_1(s)$$

$$f_2(t) \leftrightarrow F_2(s)$$

$$f_1(t)f_2(t) \leftrightarrow F_1(s) * F_2(s)$$

$$F_1(s)F_2(s) \leftrightarrow f_1(t) * f_2(t)$$

5.9 Recovery of the Original Time Function

The Laplace transform is one of the most extensively studied topics. In order to make use of the Laplace transform technique it is important that the transformation is carried out from the t-domain to the s-plane and the function $f(t)$ is recovered back from $F(s)$ via inverse transformation. The most common methods of Laplace Transform inversions are: the Laplace Transform Tables and partial fraction.

Extensive tables of Laplace transform pairs have been developed over the years [Oberhettinger, F.], [Abramowitz, M.]. Here, we present a short table that includes the most common transform pairs and their region of convergence (ROC).

Table 5.3: Table of Laplace Transform Pairs

No.	Time Function $f(t) = \frac{1}{2\pi j}\int_{\sigma-j\infty}^{\sigma+j\infty} F(s)e^{st}\,\mathrm{d}s$	Laplace Transform $F(s) = \int_0^{\infty} f(t)e^{st}\,\mathrm{d}s$	ROC $\mathrm{Re}\, s > \sigma$
1	$\delta(t)$	1	$\mathrm{Re}\, s > 0$
2	$\delta(t-T)$	e^{-sT}	$\mathrm{Re}\, s > 0$
3	$u(t-T)$	$\frac{e^{-sT}}{s}$	$\mathrm{Re}\, s > 0$
4	$e^{-\alpha t}$	$\frac{1}{s+\alpha}$	$\mathrm{Re}\,(s+\alpha) > 0$
5	$\cos\omega t$	$\frac{s}{s^2+\omega^2}$	$\mathrm{Re}\, s > 0$
6	$\sin\omega t$	$\frac{\omega}{s^2+\omega^2}$	$\mathrm{Re}\, s > 0$
7	$t^n, \quad n = 0, 1, \ldots$	$\frac{n!}{s^{n+1}}$	$\mathrm{Re}\, s > 0$
8	$t^n, \quad n = -1, -2, \ldots$	$(-1^n)\frac{\mathrm{d}^n}{\mathrm{d}s^n}F(s) = \left(\frac{1}{s}\right)$	$\mathrm{Re}\, s > 0$
9	$u(t)$	$\frac{1}{s}$	$\mathrm{Re}\, s > 0$
10	$e^{-\alpha t}\cos\omega t$	$\frac{s+\alpha}{(s+\alpha)^2+\omega^2}$	$\mathrm{Re}\,(s+\alpha) > 0$
11	$e^{-\alpha t}\sin\omega t$	$\frac{\omega}{(s+\alpha)^2+\omega^2}$	$\mathrm{Re}\,(s+\alpha) > 0$
12	$\frac{\sin t}{t}$	$\tan^{-1}\frac{1}{s} = \frac{\pi}{2} - \tan^{-1} s$	$\mathrm{Re}\, s > 0$
13	$-t\,f(t)$	$\frac{\mathrm{d}}{\mathrm{d}s}F(s)$	$\mathrm{Re}\, s > \sigma$
14	$\cosh\omega t$	$\frac{s}{s^2-\omega^2}$	$\mathrm{Re}\, s > \omega$
15	$\sinh\omega t$	$\frac{\omega}{s^2-\omega^2}$	$\mathrm{Re}\, s > \omega$
16	$f(t) = f(t+nT)$ $n = 0, 1, 2, \ldots$	$\frac{1}{1-e^{-Ts}}\int_0^T e^{-s\tau} f(\tau)\,\mathrm{d}\tau$	$\left\lvert e^{-Ts}\right\rvert < 1$
17	$\left(\frac{1}{a-b}\right)\left(e^{-t/a} - e^{-t/b}\right)$	$\frac{1}{(1+as)(1+bs)}, a \neq b$	$\mathrm{Re}\, s > \frac{-1}{a}$ and $\frac{-1}{b}$
18	$\frac{1}{a^2}te^{-t/a}$	$\frac{1}{(1+as)^2}$	$\mathrm{Re}\, s > \frac{-1}{a}$
19	$\frac{1}{\omega}e^{-\delta t}\sin\omega t$	$\frac{1}{s^2+2\delta s+(\omega^2+\delta^2)}$	$\mathrm{Re}\,(s+\delta) > 0$
20	$\frac{1}{n!}\left(1-e^{-t/a}\right)^n, a > 0$	$\frac{1}{s}\prod_{i=1}^{i=n}\frac{1}{(as+i)}$	$\mathrm{Re}\left(s+\frac{1}{a}\right) > 0$

(continued)

No.	Time Function $f(t) = \frac{1}{2\pi j}\int\limits_{\sigma-j\infty}^{\sigma+j\infty} F(s)e^{st}\,\mathrm{d}s$	Laplace Transform $F(s) = \int\limits_{0}^{\infty} f(t)e^{st}\,\mathrm{d}s$	ROC $\mathrm{Re}\, s > \sigma$
21	$f(t-T)\,u(t-T)$	$e^{-Ts}\,F(s)$	$\mathrm{Re}\, s > \sigma$
22	$\int\limits_0^t f(\tau)\,\mathrm{d}\tau = f(t) * u(t)$	$\frac{1}{s}F(s) = \left(\frac{1}{s}\right)$	$\mathrm{Re}\, s > \sigma$

5.9.1 Partial Fraction Expansion Method

Consider the function $X(s)$ with isolated singularities. It is a rational function of s and can be written as a ratio of two polynomials of s (factors of the form e^{-sT} represent a delay and are treated as such). Functions of this type are represented as a sum of simple terms and their inverse is easily recognized or obtained from the transform table. Let

$$X(s) = \frac{N(s)}{D(s)} = \frac{b_0 s^m + b_1 s^{m-1} + \cdots + b_m}{s^n + a_1 s^{n-1} + \cdots + a_n} + \sum_{k=0}^{N} c_k s^k \tag{5.18}$$

$$D(s) = (s-s_1)^{r_1}(s-s_2)^{r_2}\cdots(s-s_k)^{r_k} = \prod_{j=1}^{k}(s-s_j)^{r_j}, \sum_{j=1}^{k} r_j = n$$

$D(s)$ is a denominator polynomial with multiple roots $s_1, s_2, \ldots, s_k$.

The partial fraction sum of $X(s)$ is written as

$$X_1(s) = \left[X(s) - \sum_{k=0}^{N} c_k s^k\right] = \sum_{j=1}^{k}\sum_{i=1}^{r_j} \frac{A_{ij}}{(s-s_j)^i} \tag{5.19}$$

Both sides of Eq. 5.19 by $(s-s_j)^{r_j}$ and differentiating $(r_j - i)$ times to yield

$$A_{ij} = \frac{1}{(r_j - i)!}\left[\frac{\mathrm{d}^{r_j-i}}{\mathrm{d}s^{r_j-i}}\left\{(s-s_j)^{r_j}X_1(s)\right\}\right]_{s=s_j} \tag{5.20}$$

Taking the Laplace inverse of Eq. 5.19

$$x(t) = L^{-1}[X(s)] = \sum_{j=1}^{k}\sum_{i=1}^{r_j} A_{ij}L^{-1}\left[\frac{1}{(s-s_j)^i}\right] + \sum_{k=0}^{N} c_k L^{-1}\left[s^k\right]$$

But

$$L^{-1}\left[\frac{1}{(s-s_j)^i}\right] = \frac{t^{i-1}}{(i-1)!}e^{s_j t}$$

$$L^{-1}\left[s^k\right] = \delta^{(k)}(t), \quad k^{\text{th}} \text{ derivative of delta function}$$

Hence, using partial fractions of $X(s)$ in the Eq. 5.18

$$\boxed{\begin{aligned} x(t) &= \sum_{j=1}^{k} e^{s_j t}\left[\sum_{i=1}^{r_j}\frac{t^{i-1}}{(i-1)!(r_j-i)!}\left\{\frac{d^{r_j-i}}{as^{r_j-i}}\left[(s-s_j)^{r_j}X_1(s)\right]\right\}_{s=s_j}\right] \\ &+ \sum_{k=0}^{N} c_k\delta^{(k)}(t) \end{aligned}}$$

In many instances (such as transfer functions) the coefficients $c_k, k = 1, 2, \ldots$ are zero, because physical realizability dictates that the degree of the denominator $D(s)$ should be greater than or equal to the degree of the numerator $N(s)$.

Example 5.7:

Consider the transfer function

$$X(s) = \frac{4s^2 + 16s + 14}{s^3 + 6s^2 + 11s + 6} = \frac{4s^2 + 16s + 14}{(s+1)(s+2)(s+3)}$$

$$n = 3, \qquad r_1 = r_2 = r_3 = 1, \qquad i = 1, j = 1, 2, 3$$

$$X(s) = \frac{A_{11}}{s+1} + \frac{A_{12}}{s+2} + \frac{A_{13}}{s+3}$$

$$A_{11} = (s+1)X(s)|_{s=-1} = \frac{4 - 16 + 14}{(1)(2)} = 1$$

$$A_{12} = (s+2)X(s)|_{s=-2} = \frac{4(-2)^2 + 16(-2) + 14}{(-2+1)(-2+3)} = 2$$

$$A_{13} = (s+3)X(s)|_{s=-3} = \frac{4(-3)^2 + 16(-3) + 14}{(-3+1)(-3+2)} = 1$$

Thus,

$$x(t) = e^{-t} + 2e^{-2t} + e^{-3t}$$

Example 5.8:

For the $X(s)$ below, evaluate $x(t)$

$$X(s) = \frac{s^2 + 3s + 3}{s^3 + 3s^2 + 3s + 1} = \frac{s^2 + 3s + 3}{(s+1)^3}$$

$$n = 3, j = 1, r_1 = 3$$

$$X(s) = \frac{s^2 + 3s + 3}{(s+1)^3} = \frac{A_{11}}{(s+1)} + \frac{A_{21}}{(s+1)^2} + \frac{A_{31}}{(s+1)^3}$$

$$A_{11} = \frac{1}{(3-1)!}\frac{d^2}{ds^2}\left[\frac{s^2 + 3s + 3}{(s+1)^3}(s+1)^3\right]_{s=-1} = 1$$

$$A_{21} = \frac{1}{(3-2)!}\frac{d}{ds}\left[\frac{s^2 + 3s + 3}{(s+1)^3}(s+1)^3\right]_{s=-1} = 1$$

$$A_{31} = \left[\frac{s^2 + 3s + 3}{(s+1)^3}(s+1)^3\right]_{s=-1} = 1$$

From the Laplace Transform pairs Table 5.3

$$x(t) = e^{-t} + te^{-t} + \frac{t^2}{2!}e^{-t}$$

Example 5.9:

Given

$$X(s) = \frac{4s^2 + 25s^3 + 55s^2 + 50s + 14}{(s+1)^3(s+2)(s+3)}$$

$$n = 5, \quad r_1 = 3, r_2 = 1, r_3 = 1$$

$$X(s) = \frac{A_{11}}{(s+1)} + \frac{A_{21}}{(s+1)^2} + \frac{A_{31}}{(s+1)^3} + \frac{A_{12}}{(s+2)} + \frac{A_{13}}{(s+3)}$$

$$A_{11} = \frac{1}{2!}\frac{d^2}{ds^2}\left[(s+1)^3X(s)\right]_{s=-1} = 1$$

$$A_{21} = \frac{d}{ds}\left[(s+1)^3X(s)\right]_{s=-1} = 1$$

$$A_{31} = \left[(s+1)^3X(s)\right]_{s=-1} = -1$$

$$A_{12} = [(s+2)X(s)]_{s=-2} = 2$$

$$A_{13} = [(s+3)X(s)]_{s=-3} = 1$$

Thus,

$$x(t) = e^{-t} + te^{-t} - \frac{t^2}{2}e^{-t} + 2e^{-2t} + e^{-3t}$$

Example 5.10:

Evaluate $x(t)$

$$X(s) = \frac{1 - e^{-sT}}{(s^2 + 2s + 1)(s+2)}$$

A partial fraction is used for a rational function only. Let

$$X(s) = F(s)\left[1 - e^{-sT}\right]$$

$F(s)$ is a rational function, therefore,

$$F(s) = \frac{1}{(s^2 + 2s + 1)(s+2)} = \frac{A_{11}}{(s+1)} + \frac{A_{21}}{(s+1)^2} + \frac{A_{12}}{(s+2)}$$

$$A_{11} = \frac{d}{ds}[(s+1)F(s)]_{s=-1} = -1$$

$$A_{21} = \left[(s+1)^2F(s)\right]_{s=-1} = 1$$

$$A_{12} = [(s+2)F(s)]_{s=-2} = 1$$

Thus

$$x(t) = \left[-e^{-t} + te^{-t} + e^{-2t}\right][u(t) - u(t-T)]$$

where

$$u(t) = \begin{cases} 0 & t < 0 \\ 1 & t \geq 0 \end{cases}$$

$$u(t - T) = \begin{cases} 0 & t < T \\ 1 & t \geq T \end{cases}$$

5.9.2 Laplace Inverse via Contour Integration

Jordan's Lemma and Its Application for the Laplace Inverse

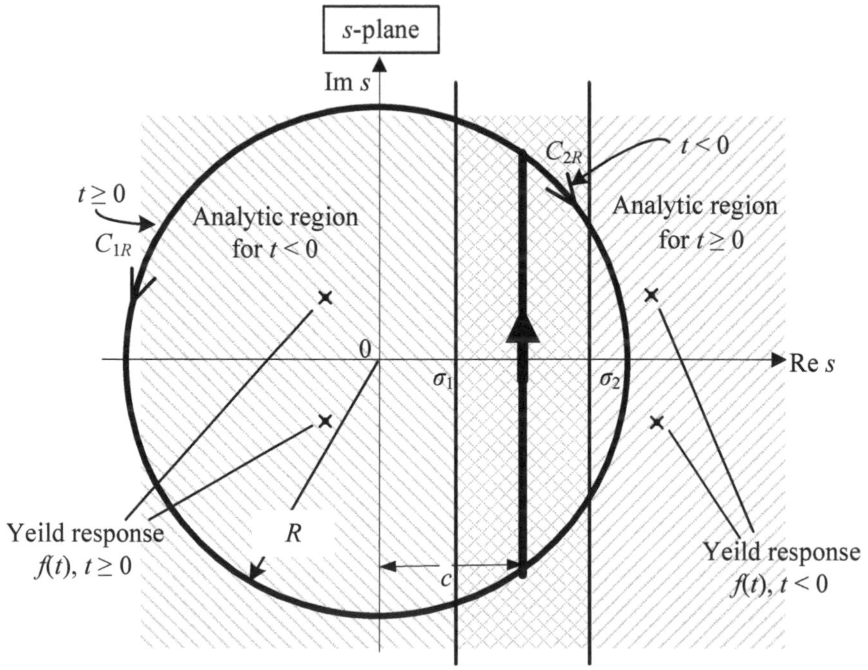

Figure 5.8: Jordan's Lemma for Double-Sided Laplace-Sided Inverse Application

Jordan's Lemma:

Let a function $F(s)$ be analytic in a strip given by $\sigma_1 < \text{Re}\, s < \sigma_2$ as shown in the

Figure 5.8. Outside this strip, $F(s)$ has a finite number of isolated singularities on the LHS and RHS of the s-plane.

Furthermore,

$$\lim_{|s|\to\infty} F(s) = 0$$

Then

$$\lim_{R\to\infty} \int_{C_1R} F(s)e^{st}\,\mathrm{d}s = 0 \quad t \geq 0 \text{ and finite}$$

$$\lim_{R\to\infty} \int_{C_2R} F(s)e^{st}\,\mathrm{d}s = 0 \quad t < 0 \text{ and finite}$$

This lemma was proved in Chapter 3.

We shall use this lemma for the evaluation of Inverse Laplace Transforms, both single-sided and bilateral.

Note:

1. We are using the symbol $F(s)$ for the single-sided Laplace Transform and $F_b(s)$ for the bilateral Laplace Transform.

2. Contour Integration uses Cauchy's Residue theorem and therefore can be only applied to a closed contour.

Single-Sided Inverse Laplace Transform Evaluation

Let us evaluate the following integral:

$$f(t) = L^{-1}[F(s)] = \frac{1}{2\pi j} \int_{c-j\infty}^{c+j\infty} F(s)e^{st}\,\mathrm{d}s, \quad c > \sigma_1 \text{ abcissa of convergence} \tag{5.21}$$

Figure 5.8 shows the contour of integration.

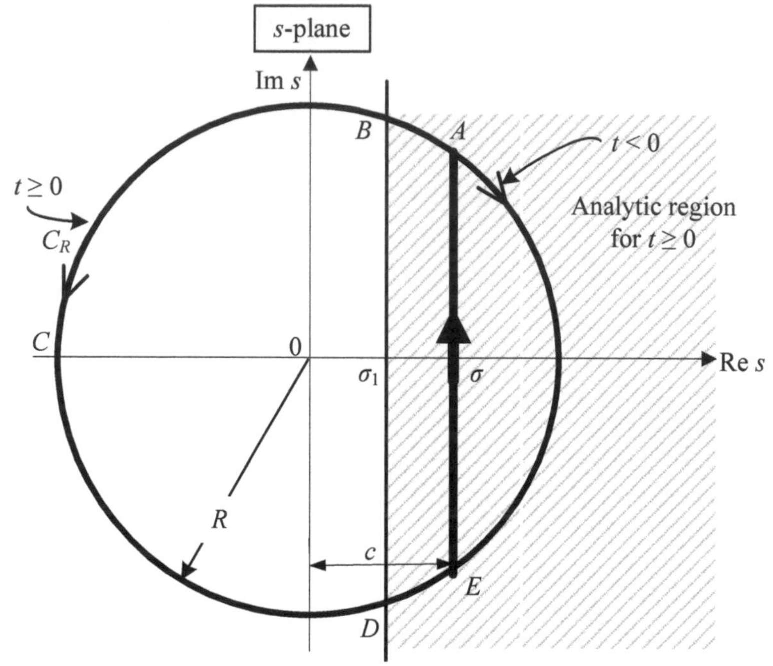

Figure 5.9: Jordan's Lemma for Single-Sided Laplace Inverse Application

Let us name the curve $ABCDE$ as C_R.

The function $F(s)$ is such that:

(a) It is analytic to the right of the line $\text{Re}\, s = \sigma_1$

(b) It has isolated singularities to the left of the line $\text{Re}\, s = c$ given by $s_i, i = 1, 2, \ldots, n$.

(c) $\lim_{|s| \to \infty} F(s) = 0$

Therefore on the basis of Jordan's Lemma, we conclude,

$$\lim_{R \to \infty} \int_{C_R} F(s) e^{st}\, \mathrm{d}s = 0 \qquad \text{for } t \geq 0 \text{ (not for } t < 0)$$

From Eq. 5.21, it follows that

$$f(t) = \lim_{R\to\infty} \frac{1}{2\pi j} \int_{c-jR}^{c+jR} F(s)e^{st} = \lim_{R\to\infty} \frac{1}{2\pi j} \left[\int_{c-jR}^{c+jR} F(s)e^{st}\,\mathrm{d}s + \int_{C_R} F(s)e^{st}\,\mathrm{d}s \right]$$

$$= \lim_{R\to\infty} \frac{1}{2\pi j} \oint F(s)e^{st}\,\mathrm{d}s, \quad t > 0$$

$$= [\text{sum of Residues of } F(s)e^{st} \text{ at the poles of } F(s)$$

$$\text{to the left of the line Re } s = c]$$

Thus,

$$f(t) = \sum_{k=1}^{n} \text{Residue}\left[F(s)e^{st}\right]_{s=s_k}, \quad t > 0 \tag{5.22}$$

Since $F(s)$ is analytic to the right of the line Re $s = \alpha$, and by Jordan's Lemma, the contribution on the closed semicircle vanishes, yielding

$$f(t) = \frac{1}{2\pi j} \oint F(s)e^{st}\,\mathrm{d}s = 0 \quad t < 0$$

If

(1) $F(s) = \dfrac{N(s)}{D(s)} = \dfrac{b_0 s^m + b_1 s^{m-1} + \cdots + b_m}{s^n + a_1 s^{n-1} + \cdots + a_n}, \qquad m < n$

(2) Roots s_k of $D(s)$ are all mutualy distinct, $k = 1, 2, \ldots, n$

Then

$$f(t) = L^{-1}\left[\frac{N(s)}{D(s)}\right] = \sum_{k=1}^{n} \frac{N(s_k)}{D'(s_k)} e^{s_k t}, \qquad D'(s_k) = \frac{\mathrm{d}}{\mathrm{d}s} D(s)\bigg|_{s=s_k}$$

This is known as the Heaviside Formula.

It should be emphasized that since the function $F(s)$ has no singularities on the right hand-side of vertical line Re $s = \alpha_1$, we can move the line Re $s = c$ anywhere to the right of the vertical line Re $s = \alpha_1$ and the value of integral Eq. 5.21 will remain unaltered. This integral Eq. 5.21 is often called the Bromwich Integral and the vertical line Re $s = c$ is the Bromwich line or Bromwich.

5.10 Constant Coefficient Differential Equations

The Laplace transform converts the linear constant coefficient differential equations into algebraic equations, which include terms involving initial conditions. The process is simple. We take the Laplace transform of each term in the equation. (Since we are dealing with causal physical systems, only the single-sided Laplace transform is used in this case). The resulting algebraic equations (instead of tedious differentials equations) are then solved and the inverse Laplace transform is taken to obtain the solution of the original differential equation.

Given:

$$\sum_{i=0}^{n} a_i \frac{\mathrm{d}^{n-i}}{\mathrm{d}t^{n-i}} x(t) = f(t), \qquad \left.\frac{\mathrm{d}^{n-i}}{\mathrm{d}t^{n-i}} x(t)\right|_{t=0} = x^{(n-i)}(0),\ i = 1, \ldots, n \tag{5.23}$$

Taking the Laplace transform on both sides of the above equation

$$\sum_{i=0}^{n} a_i \left[L \frac{\mathrm{d}^{n-i}}{\mathrm{d}t^{n-i}} x(t) \right] = L\left[f(t)\right]$$

$$\sum_{i=0}^{n} a_i \left[s^{n-i} X(s) - \sum_{j=0}^{n-i-1} s^{n-i-1-j} x^{(j)}(0) \right] = F(s)$$

$$X(s) = \frac{F(s)}{\Delta(s)} + \frac{\sum_{i=0}^{n} a_i \sum_{j=0}^{n-i-1} s^{n-i-j-1} x^{(j)}(0)}{\Delta(s)} \tag{5.24}$$

where,

$$\Delta(s) = a_0 s^n + a_1 s^{n-1} + \cdots + a_n$$

Taking the Laplace inverse of Eq. 5.24,

$$x(t) = L^{-1}[X(s)] = L^{-1}\left[\frac{F(s)}{\Delta(s)}\right] + L^{-1}\left[\frac{\sum_{i=0}^{n} a_i \sum_{j=0}^{n-i-1} s^{n-i-j-1} x^{(j)}(0)}{\Delta(s)}\right]$$

$L^{-1}\left[\frac{F(s)}{\Delta(s)}\right]$ is referred to as the Particular Integral (PI) or the forcing term

$L^{-1}\left[\frac{\sum_{i=0}^{n} a_i \sum_{j=0}^{n-i-1} s^{n-i-j-1} x^{(j)}(0)}{\Delta(s)}\right]$ is known as the Transient term due to initial conditions also known as the complimentary function (cf)

Thus, in general the Laplace transform $X(s)$ of the solution $x(t)$ of the differential equation Eq. 5.23 is of the form

$$X(s) = \frac{N(s)}{D(s)} = \frac{b_0 s^m + b_1 s^{m-1} + \cdots + b_m}{a_0 s^n + a_1 s^{n-1} + \cdots + a_n}, \qquad m \le n \tag{5.25}$$

$$D(s) = \prod_{j=1}^{k} (s - s_j)^{r_j} \quad \text{(Multiple poles)}, \quad a_0 = 1$$
$$r_1 + r_2 + \cdots + r_k = n$$

5.11 Computation of $x(t)$ for Causal Processes

1. Use tables, or partial fraction expansion and invert $X(s)$ to obtain $x(t)$.

2. Contour integration involving the residue method can be used as following

$$X(s) = \frac{N(s)}{\prod_{j=1}^{k} (s - s_j)_j^r}, \qquad r_1 + r_2 + \cdots + r_k = n$$

The residue of $X(s)e^{st}$ at $s = s_j$ for the j-th pole at $(s - s_j)$ of order r_j is

$$\text{Res}\left[X(s)e^{st}\right]_{s=s_j} = \frac{1}{(r_j - 1)!}\frac{\mathrm{d}^{r_j-1}}{\mathrm{d}s^{r_j-1}}\left[(s - s_j)^{r_j}X(s)e^{st}\right]_{s=s_j}$$

$$x(t) = L^{-1}[X(s)] = \sum_{j=1}^{k}\frac{1}{(r_j - 1)!}\left[\frac{\mathrm{d}^{r_j-1}}{\mathrm{d}s^{r_j-1}}(s - s_j)^{r_j}X(s)e^{st}\right]_{s=s_j}$$

Note: The partial fraction method is only applicable when the order of the poles of $X(s)$ are not fractional but integer. Contour integration involving residues can be used for both fractional and integer poles.

5.12 Inverse Bilateral Laplace Transform $F_b(s)$

In case the function $f(t)$ is noncausal (such functions appear regularly in stochastic processes involving correlation functions), we use bilateral Laplace transforms $F_b(s)$. In order to recover $f(t)$ from $F_b(s)$, we must know the convergence strip $\sigma_1 < \text{Re}\, s = c < \sigma_2$ and this strip must exist. Using Jordan's Lemma

$$\begin{bmatrix} f(t) \\ t \geq 0 \end{bmatrix} = \frac{1}{2\pi j}\int_{c-j\infty}^{c+j\infty} F_b(s)e^{st}\,\mathrm{d}s = \frac{1}{2\pi j}\oint_{c-j\infty}^{c+j\infty} F_b(s)e^{st}\,\mathrm{d}s$$
$$= [\text{sum of residues of } F_b(s)e^{st} \text{ at the poles of } F_b(s) \text{ to the left of the line } \text{Re}\, s = (\sigma_1 + \epsilon)] \tag{5.26}$$

$$\begin{bmatrix} f(t) \\ t < 0 \end{bmatrix} = \frac{1}{2\pi j}\int_{c-j\infty}^{c+j\infty} F_b(s)e^{st}\,\mathrm{d}s = \frac{1}{2\pi j}\oint_{c-j\infty}^{c+j\infty} F_b(s)e^{st}\,\mathrm{d}s$$
$$= -[\text{sum of residues of } F_b(s)e^{st} \text{ at the poles of } F_b(s) \text{ to the right of the line } \text{Re}\, s = (\sigma_2 - \epsilon)] \tag{5.27}$$

ϵ is a small positive number $\to 0$

Thus, the poles of $F_b(s)$ to the left of $(\sigma_1 + \epsilon)$ yields a time response for positive time and the poles of $F_b(s)$ in the region right of the $(\sigma_2 - \epsilon)$ contribute a response for negative time.

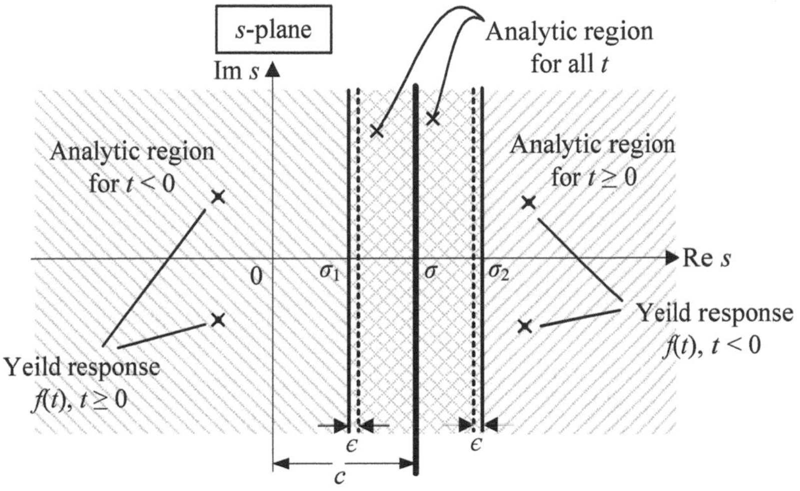

Figure 5.10: Jordan's Lemma

The vertical line $c + j\omega, -\infty < \omega < \infty$ can lie anywhere between the vertical strip formed by $\sigma_1 + j\omega$ and $\sigma_2 + j\omega, -\infty < \omega < \infty$. This is known as the strip of convergence and the function $F(s)$ is analytic (no poles) inside this strip. Without the Strip of Convergence the function $f(t)$ cannot be recovered from $F(s)$ for both $t \geq 0$ and $t < 0$. A general function $f(t), -\infty < t < \infty$ can be written as

$$f(t) = f_1(t)u(t) + f_2(t)u(-t)$$

$$f_1(t) \leq M_1 e^{\sigma_1 t}, \qquad f_2(-t) \leq M_2 e^{-\sigma_2 t}$$

Writing the bilateral Laplace transform,

$$F_b(s) = \int_{-\infty}^{+\infty} f(t)e^{-st}\,\mathrm{d}t = \int_{-\infty}^{0} f_2(t)e^{-st}\,\mathrm{d}t + \int_{0}^{+\infty} f_1(t)e^{-st}\,\mathrm{d}t$$

Substituting $t = -\tau$ and realizing that τ is a dummy variable

$$F_b(s) = \int_0^\infty f_2(-t)e^{st}\,\mathrm{d}t + \int_0^\infty f_1(t)e^{-st}\,\mathrm{d}t = F_2(-s) + F_1(s)$$

where

$$Lf_1(t) = F_1(s) \qquad \mathrm{Re}\, s > \sigma_1$$

$$Lf_2(-t) = F_2(-s) \qquad \mathrm{Re}\, s > \sigma_2$$

$$L_b f(t) = F_b(s) = F_1(s) + F_2(-s) \qquad \sigma_1 < \mathrm{Re}\, s < \sigma_2$$

(strip of convergence)

Note:

1. $F_b(s)$ consists of a sum of two functions $F_1(s)$ and $F_2(-s)$.
2. The vertical strip $\sigma_1 < \mathrm{Re}\, s < \sigma_2$ decides which part of $F_b(s)$ belongs to $F_1(s)$ and which other part belongs to $F_2(-s)$.
3. $F_1(s)$ contains all the poles to the left of the vertical line $c + j\,\mathrm{Im}\, s$ in the s-plane where $\sigma_1 < c < \sigma_2$.
4. $F_2(s)$ contains all the poles to the right of $c + j\,\mathrm{Im}\, s$ where $\sigma_1 < c < \sigma_2$.

Example 5.11:

Given:

$$f(t) = e^{-\alpha t}u(t) + e^{\beta t}u(-t), \qquad \alpha > 0, \beta > 0$$

Find $F_b(s)$.

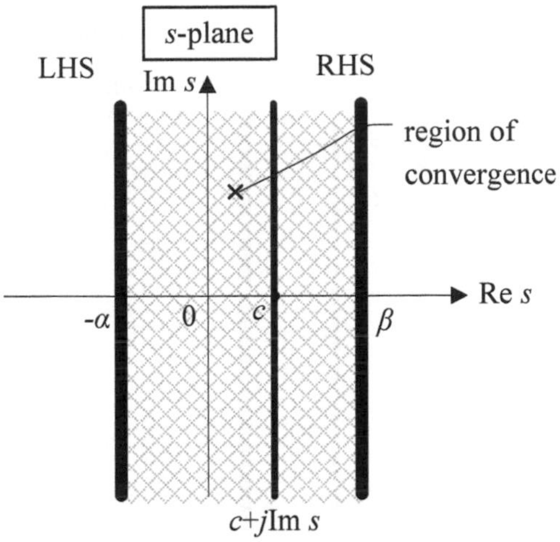

Figure 5.11: Strip of Convergence

Solution:

$$f_1(t) = e^{-\alpha t} \leftrightarrow F_1(s) = \frac{1}{s+\alpha}$$
$$f_2(-t) = e^{\beta t} \leftrightarrow F_2(s) = \frac{1}{s-\beta}$$

$$F_b(s) = F_1(s) + F_2(-s) = \frac{1}{s+\alpha} + \frac{1}{-s-\beta} \qquad -\alpha < \text{Re}\, s < +\beta$$

Example 5.12:

Given:

$$F_b(s) = \frac{1}{(s+1)(s+2)} = \frac{1}{(s+1)} - \frac{1}{(s+2)}, \qquad -2 < \text{Re}\, s < -1$$

Evaluate $f(t)$ Solution:

$$F_1(s) = -\frac{1}{s+2} \leftrightarrow f_1(t) = -e^{-2t}$$
$$F_2(-s) = \frac{1}{s+1} \leftrightarrow f_2(-t) = -e^{-t}$$
$$f(t) = -e^{-2t}u(t) - e^{t}u(-t)$$

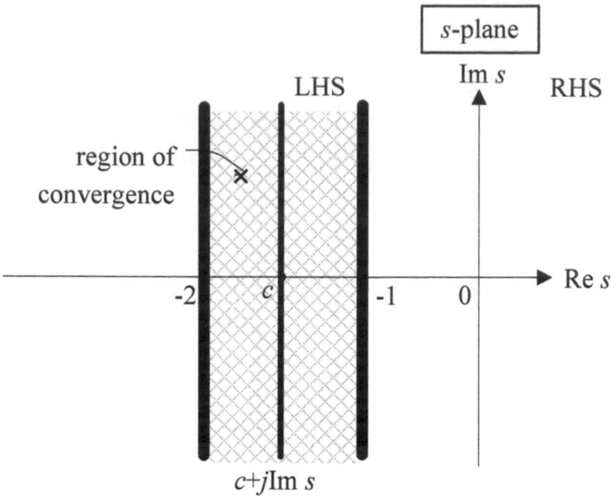

Figure 5.12: Strip of Convergence

Example 5.13:

Given:

$$F_b(s) = \frac{1}{(s+\alpha)^{n+1}} + \frac{1}{(s+\beta)^{n+1}}, \qquad -\alpha < \text{Re}\, s < -\beta$$

Evaluate $f(t)$

Solution:

$$F_1(s) = \frac{1}{(s+\alpha)^{n+1}} \leftrightarrow f_1(t) = \frac{t^n}{n!}e^{-\alpha t}$$
$$F_2(-s) = \frac{1}{(s+\beta)^{n+1}} \leftrightarrow f_2(t) = (-1)^{n+1}\frac{t^n}{n!}e^{+\beta t}$$

Thus,

$$L^{-1}\left[F_b(s)\right] = f(t) = \frac{t^n}{n!}\left[e^{-\alpha t}u(t) + (-1)^{n+1}e^{+\beta t}u(-t)\right]$$

5.13 Transfer Function

Consider a circuit shown in Figure 5.13. There is no initial stored energy in the system and the output variable, and their derivatives are zero at time $t < 0$. The excitation source is applied at $t = 0$.

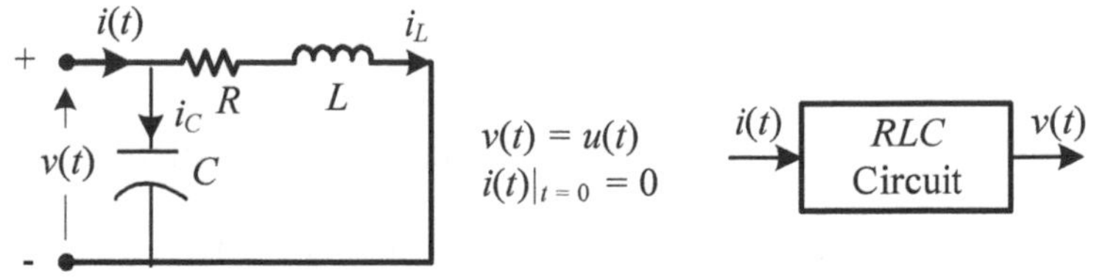

Figure 5.13: Transfer Function Concept

The Kirchoff's current and voltage laws yield

$$\begin{aligned} i(t) &= i_L(t) + i_C(t) \\ v(t) &= Ri_L(t) + L\frac{\mathrm{d}}{\mathrm{d}t}i_L(t) \\ i_C(t) &= C\frac{\mathrm{d}v(t)}{\mathrm{d}t} \end{aligned}$$

Let $i(t)$ be the applied current and $v(t)$ the response voltage. Eliminating $i_L(t)$ and $i_C(t)$, we obtain

$$\frac{\mathrm{d}^2}{\mathrm{d}t^2}v(t) + \frac{R}{L}\frac{\mathrm{d}}{\mathrm{d}t}v(t) + \frac{1}{LC}v(t) = \frac{1}{C}\frac{\mathrm{d}}{\mathrm{d}t}i(t) + \frac{R}{LC}i(t) \tag{5.28}$$

Taking the Laplace transform of both sides of Eq. 5.28 and realizing that all the initial conditions are zero,

$$\frac{V(s)}{I(s)} = \frac{\dfrac{1}{C}s + \dfrac{R}{LC}}{s^2 + \dfrac{R}{L}s + \dfrac{1}{LC}}$$

The function $\dfrac{V(s)}{I(s)}$ is called the "Transfer Impedance" or "Transfer Function" of the circuit. Transfer impedance represents a special characteristic of the circuit in that it depends only upon the element values of the various components and how they are connected. This function does not depend on the initial state of the circuit. In fact, the roots of the denominator $\left(s^2 + \frac{R}{L}s + \frac{1}{LC}\right)$ are the natural frequencies of the circuit.

In general, the input and the output variables of a linear time-invariant dynamic system are related via an nth order differential equation:

$$\sum_{i=0}^{n} a_i \frac{\mathrm{d}^{n-i}}{\mathrm{d}t^{n-i}} y(t) = \sum_{j=0}^{m} b_j \frac{\mathrm{d}^{m-j}}{\mathrm{d}t^{m-j}} f(t) \qquad n \geq m$$

where

$$y(t) = \text{ output variable}$$

$$f(t) = \text{ input variable}$$

$$y^{(n)}(t)\big|_{t=0^-} = 0$$

Taking the Laplace transform of both sides,

$$\text{Transfer Function} = \frac{L[\text{output}]}{L[\text{input}]} \equiv \frac{Y(s)}{F(s)} \equiv H(s) = \frac{\sum_{j=0}^{m} b_j s^{m-j}}{\sum_{i=0}^{n} a_i s^{n-i}}, \qquad n \geq m$$

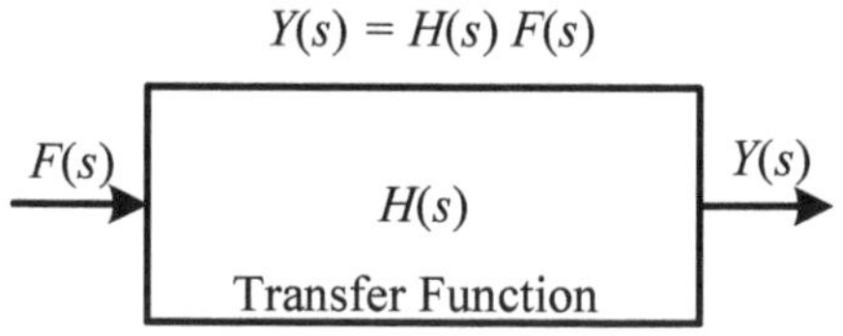

Figure 5.14: Input–Output Relation via Transfer Function

Thus, with initial conditions taken as zero

$$\begin{bmatrix} \text{Laplace transform} \\ \text{of the output} \\ Y(s) \end{bmatrix} = \begin{bmatrix} \text{Transfer} \\ \text{function} \\ H(s) \end{bmatrix} \begin{bmatrix} \text{Laplace transform} \\ \text{of the input} \\ F(s) \end{bmatrix}$$

The transfer function characterizes only the output response of a relaxed system representing no initial energy storage. Total response can only be computed by taking into account initial values of system variables and their derivatives. For many inputs and outputs, the system response matrix $\boldsymbol{Y}(s)$ and its input vector $\boldsymbol{F}(s)$ are related by "**Transfer Function Matrix**" $\boldsymbol{H}(s)$:

$$\boldsymbol{Y}(s) = \boldsymbol{H}(s)\boldsymbol{F}(s) \tag{5.29}$$

$$\begin{aligned} \boldsymbol{H}(s) &= \text{ Transfer Function} && \text{is a } n \times m \text{ matrix} \\ \boldsymbol{Y}(s) &= L[\boldsymbol{y}(t)], && \boldsymbol{y}(t) \text{ is an } n \text{ vector} \\ \boldsymbol{F}(s) &= L[\boldsymbol{f}(t)], && \boldsymbol{f}(t) \text{ is a } m \text{ vector} \end{aligned}$$

Note: $\boldsymbol{H}(s)\boldsymbol{F}(s) \neq \boldsymbol{F}(s)\boldsymbol{H}(s)$, $\boldsymbol{F}(s)\boldsymbol{H}(s)$ may be an invalid expression.

5.14 Impulse Response

An alternate representation of the system by a Transfer Function is the impulse response. Let the input to the system be an impulse. Thus,

$$f(t) = L^{-1}[F(s)] = \delta(t) \quad \text{so that} \quad F(s) = 1$$

Substituting $F(s) = 1$ in Eq. 5.29 yields:

$$Y(s) = H(s) \qquad \text{transform of the response of the system to an impulse input}$$

Taking the Laplace inverse of the above expression

$$y(t) = L^{-1}[Y(s)] = L^{-1}[H(s)] = h(t), \qquad \text{Impulse Response}$$

Hence, $L^{-1}[H(s)]$ is known as the impulse response of the system. It is also sometimes referred to as weighting function of the system for the response to a specified input. Summarizing

$$L^{-1}[H(s)] = h(t) = \text{ Impulse response of the system}$$

$$L[h(t)] = H(s) = \text{ Transfer function of the system}$$

5.15 Convolution for Linear Time-Invariant System

Let us apply an input $\delta(t)$ to the system and measure its response $h(t), 0 \leq t < \infty$. Thus, we assume that $h(t)$ is known.

Question: What is the system response to any general input $f(t)$?

This question has meaning only for linear systems where convolution property holds good. For non-linear systems, there is no way to know the output from the impulse response only.

Answer:

Let us represent $f(t)$ by impulse function as

$$f(t) = \int_{-\infty}^{+\infty} f(\tau)\delta(t-\tau)\,\mathrm{d}\tau = \int_{-\infty}^{+\infty} f(t-\tau)\delta(\tau)\,\mathrm{d}\tau$$

For a time-invariant system

$$\text{Input } \delta(t-\tau) \quad \rightarrow \quad \text{yields output } h(t-\tau)$$

Using superposition (thinking of integration as a summation process in the limit)

$$\text{Input} \int_{-\infty}^{+\infty} f(\tau)\delta(t-\tau)\,d\tau \quad \text{yields} \quad \text{Output } y(t) = \int_{-\infty}^{+\infty} f(\tau)h(t-\tau)\,d\tau \quad (5.30)$$

Response for a Causal System with Causal Input

Let

$$f(t) \equiv 0, \quad t \leq 0 \quad \text{(causal input)} \tag{5.31}$$

$$h(t-\tau) \equiv 0, \quad \tau > t \quad \text{(causal system)} \tag{5.32}$$

Eq. 5.30 can be broken in three terms, yielding

$$y(t) = \int_{-\infty}^{0} f(\tau)h(t-\tau)\,d\tau + \int_{0}^{t} f(\tau)h(t-\tau)\,d\tau + \int_{t}^{\infty} f(\tau)h(t-\tau)\,d\tau$$

The first and the third terms vanish due to Eq. 5.31 and Eq. 5.32. Thus,

$$y(t) = \int_{0}^{t} f(\tau)h(t-\tau)\,d\tau \tag{5.33}$$

It is very easy to see that Eq. 5.33 can also be written as

$$y(t) = \int_{0}^{t} f(t-\tau)h(\tau)\,d\tau \tag{5.34}$$

The integrals Eq. 5.33 and Eq. 5.34 are referred to as the convolution integrals. These convolution integrals can be expressed symbolically as

$$y(t) = f(t) * h(t) = h(t) * f(t) \tag{5.35}$$

where "$*$" is referred to as the convolution. Thus, the output $y(t)$ is a result of the "convolution" of $f(t)$ with $h(t)$ (or vice versa).

Laplace transform viewpoint

$$Y(s) = H(s)F(s) \tag{5.36}$$

$$y(t) = \int_0^t f(\tau)h(t-\tau)\,\mathrm{d}\tau \tag{5.37}$$

$$y(t) = L^{-1}[H(s)F(s)] = \int_0^t f(\tau)h(t-\tau)\,\mathrm{d}\tau = f(t) * h(t) \tag{5.38}$$

Eq. 5.38 can also be validated via the inverse Laplace transform.

Using the definition of the inverse Laplace transform,

$$y(t) = \frac{1}{2\pi j}\int_{c-j\infty}^{c+j\infty} H(s)F(s)e^{st}\,\mathrm{d}s$$

$$H(s) = \int_0^\infty h(t)e^{-st}\,\mathrm{d}t = \int_0^\infty h(\tau)e^{-s\tau}\,\mathrm{d}\tau$$

$$y(t) = \int_0^\infty h(\tau)\left[\frac{1}{2\pi j}\int_{c-j\infty}^{c+j\infty} F(s)e^{s(t-\tau)}\,\mathrm{d}s\right]\mathrm{d}\tau$$

The expression above in brackets is the Laplace inverse that represents $f(t-\tau)$.

$$y(t) = \int_0^\infty h(\tau)f(t-\tau)\,\mathrm{d}\tau$$

$$y(t) = \int_0^t h(\tau)f(t-\tau)\,\mathrm{d}\tau + \int_t^\infty h(\tau)f(t-\tau)\,\mathrm{d}\tau$$

For causal systems $h(t) = 0$ for $t < 0$.

The second term vanishes, yielding

$$y(t) = \int_0^\infty h(\tau)f(t-\tau)\,\mathrm{d}\tau = \int_0^t h(\tau)f(t-\tau)\,\mathrm{d}\tau = L^{-1}[H(s)F(s)]$$

5.16 Frequency Convolution in Laplace Domain

Given two causal functions $f_1(t), f_2(t)$ such that

$$|f_1(t)| \le M_1 e^{\alpha_1 t} \qquad f_1(t) \equiv 0 \quad t < 0$$

$$|f_2(t)| \le M_2 e^{\alpha_2 t} \qquad f_2(t) \equiv 0 \quad t < 0$$

Choose $\alpha_2 > \alpha_1$

Then

$$|f_1(t)f_2(t)| \le |f_1(t)||f_2(t)| = M_1 M_2 e^{(\alpha_1+\alpha_2)t}$$

Thus, the abcissa of convergence of $f_1(t)\, f_2(t)$ is $(\alpha_1 + \alpha_2)$.

Consider the Laplace transform of the product of two time functions,

$$L[f_1(t)f_2(t)] = \int_0^\infty f_1(t)f_2(t)e^{-st}\,\mathrm{d}t \qquad (\alpha_1+\alpha_2) < \sigma = \mathrm{Re}\, s < \infty \tag{5.39}$$

But

$$f_1(t) = \frac{1}{2\pi j}\int_{c-j\infty}^{c+j\infty} F_1(s)e^{st}\,\mathrm{d}s$$

$$= \frac{1}{2\pi j}\int_{c-j\infty}^{c+j\infty} F_1(\lambda)e^{\lambda t}\,\mathrm{d}\lambda, \qquad \alpha_1 < \mathrm{Re}\,\lambda = c < \infty \tag{5.40}$$

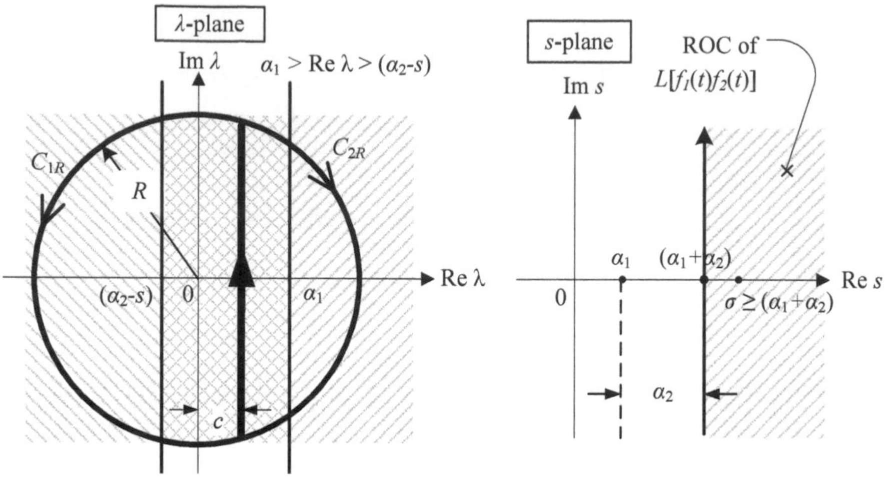

Figure 5.15: Region of Convergence of $f_1(t)f_2(t)$ Transform.

Substituting Eq. 5.40 into Eq. 5.39

$$L[f_1(t)f_2(t)] = \int_0^{\infty}\left[\frac{1}{2\pi j}\int_{c-j\infty}^{c+j\infty} F_1(\lambda)e^{\lambda t}\,\mathrm{d}\lambda\right] f_2(t)e^{-st}\,\mathrm{d}t, \qquad (\alpha_1+\alpha_2) < \mathrm{Re}\, s < \infty$$

Interchanging the order of integration,

$$L[f_1(t)f_2(t)] = \frac{1}{2\pi j}\int_{c-j\infty}^{c+j\infty} F_1(\lambda)\left[\int_0^{\infty} f_2(t)e^{-(s-\lambda)t}\,\mathrm{d}t\right]\mathrm{d}\lambda, \qquad (\alpha_1+\alpha_2) < \mathrm{Re}\, s < \infty$$

Let

$$\int_0^{\infty} f_2(t)e^{-(s-\lambda)t}\,\mathrm{d}t = F(s-\lambda) \qquad \alpha_2 < \mathrm{Re}\,(s-\lambda) < \infty \tag{5.41}$$

$$L[f_1(t)f_2(t)] = \frac{1}{2\pi j}\int_{c-j\infty}^{c+j\infty} F_1(\lambda)F_2(s-\lambda)\,\mathrm{d}\lambda, \quad \begin{cases} \alpha_1 < \mathrm{Re}\,\lambda = c < \infty \\ \alpha_2 < \mathrm{Re}\,(s-\lambda) < \infty \\ \alpha_1+\alpha_2 < \mathrm{Re}\, s < \infty \end{cases} \tag{5.42}$$

If, $\lim_{s\to\infty} |F_1(s)| = \lim_{s\to\infty} |F_2(s)| = 0$, then the closed contour integration in the λ-plane can be accomplished by taking a semicircle to the right or the left of the line $\operatorname{Re}\lambda = c$, as long as Jordan Lemma conditions are satisfied. As a result of these conditions being satisfied, Cauchy's Residue Theorem can be used to compute Eq. 5.42. Figure 5.15 shows the contour of integration to be used. Note that the infinite semicircle can be chosen either to the left or to the right of $c + j\operatorname{Im} s$ yielding the same result. The complex integration expressed by Eq. 5.42 is known as a convolution in the frequency domain and expressed as $F_1(s) * F_2(s)$. For $s = 0, \alpha_2 < \operatorname{Re}\lambda < \alpha_1$. Thus, the line $\operatorname{Re}\lambda = c$ lies between the lines α_1 and α_2. We can close the contour along C_{R1} or C_{R2} and will obtain the same result.

Example 5.14:

Compute the Laplace transform of $f(t) = f_1(t)\, f_2(t)$

$$f_1(t) = e^{-t}u(t), \qquad -1 < \alpha_1 < \infty$$

$$f_2(t) = e^{-3t}u(t), \qquad -3 < \alpha_2 < \infty$$

$$F_1(s) = \frac{1}{s+1}, \quad F_1(\lambda) = \frac{1}{\lambda+1}, \quad F_2(s) = \frac{1}{s+3}, \quad F_2(s-\lambda) = \frac{1}{s+3-\lambda}$$

$$\begin{aligned} L[f_1(t)f_2(t)] &= \frac{1}{2\pi j}\oint_{C_1} \frac{1}{\lambda+1}\left(\frac{-1}{\lambda-s-3}\right) d\lambda \\ &= 2\pi j \text{Residue } [F_1(\lambda)F_2(s-\lambda)]_{\lambda=-1} = \frac{1}{s+4} \end{aligned}$$

or

$$\begin{aligned} L[f_1(t)f_2(t)] &= -\frac{1}{2\pi j}\oint_{C_2} \frac{1}{\lambda+1}\left(\frac{-1}{\lambda-s-3}\right) d\lambda \\ &= -2\pi j \text{ Residue } [F_1(\lambda)F_2(s-\lambda)]_{\lambda=s+3} = \frac{1}{s+4} \end{aligned}$$

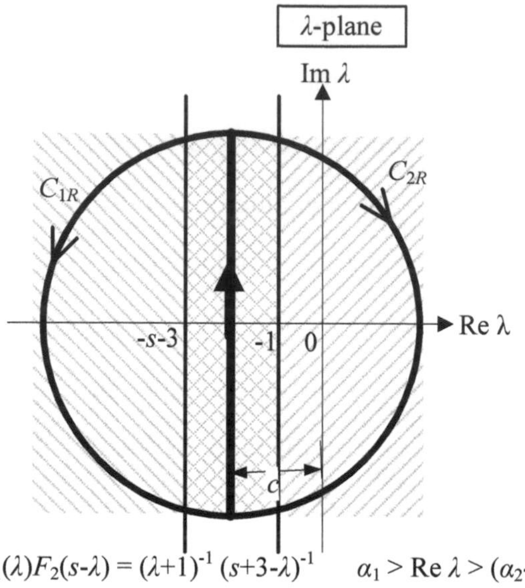

Figure 5.16: Contour Integration

Note that the same result is obtained if either of the contours C_{R1} or C_{R2} are used.

5.17 Parseval's Theorem

Parseval's theorem relates the energy contents of a signal in time domain to the frequency domain. Consider the Eq. 5.42,

$$L[f_1(t)f_2(t)] = \int_0^{\infty} f_1(t)f_2(t)e^{-st}\,\mathrm{d}t = \frac{1}{2\pi j}\int_{c-j\infty}^{c+j\infty} F_1(\lambda)F_2(s-\lambda)\,\mathrm{d}\lambda$$

Substituting $s = 0$ on both sides of the above expression,

$$\int_0^{\infty} f_1(t)f_2(t)\,\mathrm{d}t = \frac{1}{2\pi j}\int_{c-j\infty}^{c+j\infty} F_1(\lambda)F_2(-\lambda)\,\mathrm{d}\lambda = \frac{1}{2\pi j}\int_{c-j\infty}^{c+j\infty} F_1(s)F_2(-s)\,\mathrm{d}s \qquad (5.43)$$

If we choose

$$f_1(t) = f_2(t) = f(t)$$

Then

$$\int_0^{\infty} f^2(t)\,\mathrm{d}t = \int_0^{\infty} |f(t)|^2\,\mathrm{d}t = \frac{1}{2\pi j}\int_{c-j\infty}^{c+j\infty} F(s)F(-s)\,\mathrm{d}s \qquad (5.44)$$

$$s = \sigma + j\omega, \quad c \geq \sigma$$

Parseval's Theorem

If the function is Fourier transformable, we let $s = j\omega, \sigma = 0$ and Eq. 5.44 yields

$$\int_0^{\infty} f^2(t)\,\mathrm{d}t = \frac{1}{2\pi}\int_{-\infty}^{+\infty} F(j\omega)F(-j\omega)\,\mathrm{d}\omega = \int_{-\infty}^{+\infty} \left|\frac{F(j\omega)}{\sqrt{2\pi}}\right|^2 \mathrm{d}\omega \qquad (5.45)$$

Parseval's Theorem

Eq. 5.44 and Eq. 5.45 are known as Parseval's Theorem.

The expression $\int_0^{\infty} f(t)^2\,\mathrm{d}t$ can be considered as the energy expanded in a unit resistor through which a current $f(t)$ is flowing and $\left|F(j\omega)/\sqrt{2\pi}\right|^2$ is referred to as the energy density of the signal $f(t)$.

For a periodic signal $f(t)$, the average power P is related to Fourier coefficients c_k in the following manner

$$P = \frac{1}{T}\int_{-T/2}^{+T/2} f^2(t)\,\mathrm{d}t = \sum_{k=-\infty}^{+\infty} c_k c_k^* = \sum_{k=-\infty}^{+\infty} |c_k|^2$$

Discrete Parseval's Theorem

Interesting enough, the phase contents of $F(j\omega)$ (or c_k) do not play any part in the signal energy (or the power) computation.

5.18 Generation of Orthogonal Signals

Consider a set of signals $f_i(t)$, $i = 1, 2, \ldots, n$ on the time interval $[0, +\infty]$. Let the signals be orthogonal functions namely

$$I_{ij} = \int_0^{+\infty} f_i(t) f_j(t)\, \mathrm{d}t = \begin{cases} 0 & i \neq j \\ K_i = \text{constant} & i = j \end{cases} \tag{5.46}$$

From Eq. 5.44

$$\boxed{I_{ij} = \int_0^{\infty} f_i(t) f_j(t)\, \mathrm{d}t = \frac{1}{2\pi j} \int_{c-j\infty}^{c+j\infty} F_i(s) F_j(-s)\, \mathrm{d}s = 0 \quad i \neq j} \tag{5.47}$$

Orthogonal Signal Generation Algorithm

Eq. 5.47 is useful in generating orthogonal functions in the time interval $[0, +\infty]$. Let

$$F_{ij}(s) = F_i(s) F_j(-s) = \frac{N_{ij}(s)}{D_{ij}(s)} \tag{5.48}$$

where $N_{ij}(s)$ and $D_{ij}(s)$ are rational polynomials in s of degree n_{ij} and d_{ij}, respectively. In order to generate a set of orthogonal functions satisfying Eq. 5.46 we choose a candidate for $F_{ij}(s)$ satisfying the following conditions:

1. $d_{ij} > n_{ij} + 1$.

2. All the roots of $D_{ij}(s)$ are in the left-hand side of the s-plane (LHS) or all of them are in the right-hand side of the s-plane (RHS).

If $F_{ij}(s)$ fulfills the above two requirements, we can show via Jordan's Lemma,

$$I_{ij} = \int_0^\infty f_i(t) f_j(t)\,dt = \frac{1}{2\pi j}\oint F_{ij}(s)\,ds = \frac{1}{2\pi j}\oint F_{ij}(s)\,ds$$

$$= [\text{Sum of Residues of } F_{ij}(s) \text{ at its LHS poles}] \tag{5.49}$$

$$= -[\text{Sum of Residues of } F_{ij}(s) \text{ at its RHS poles}] = 0, i \neq j$$

The integration contour can be closed on either the LHS or the RHS of the s-plane.

Example 5.15:

Kautz Polynomials—Orthogonal Set

Let

$$F_1(s) = \frac{1}{1+s} \qquad f_1(t) = e^{-t}$$

$$F_2(s) = \left(\frac{1-s}{1+s}\right)\left(\frac{1}{2+s}\right) \qquad f_2(t) = 2e^{-t} - 3e^{-2t}$$

$$F_3(s) = \left(\frac{1-s}{1+s}\right)\left(\frac{2-s}{2+s}\right)\left(\frac{1}{3+s}\right) \qquad f_3(t) = 3e^{-t} - 12e^{-2t} + 10e^{-3t}$$

$$\vdots$$

$$F_i(s) = \left[\prod_{k=1}^{i-1}\left(\frac{k-s}{k+s}\right)\right]\left(\frac{1}{i+s}\right), \qquad i = 2, 3, \ldots, n$$

It can be easily shown that for $i > j$

$$F_i(s)F_j(-s) = \left(\frac{1}{j+s}\right)\left[\prod_{k=j+1}^{i-1}\left(\frac{k-s}{k+s}\right)\right]\left(\frac{1}{i+s}\right), \quad i = 2, 3, \ldots, n \tag{5.50}$$

The product $F_i(s)F_j(-s)$ has a denominator of degree two higher than the numerator and all of its poles are in the LHS. Thus,

$$I_{ij} = \int_0^\infty f_i(t) f_j(t)\,dt = \frac{1}{2\pi j}\oint F_i(s)F_j(-s)\,ds = 0, \quad i \neq j \tag{5.51}$$

$$\int_0^\infty f_i^2(t)\,\mathrm{d}t = \frac{1}{2\pi j}\oint \frac{1}{(i+s)(i-s)}\,\mathrm{d}s = \left(\frac{1}{2i}\right), \quad i = j \tag{5.52}$$

The functions $f_i(t)$ are called Kautz Polynomials and are related to the well-known Legendre polynomials via the transformation $x = 1-e^{-t}$, $0 \le x < 1$, $0 < t < \infty$. These functions have an interesting pole and zero pattern. The reader is recommended to study this property further by sketching this pattern.

Example 5.16:

Laguerre Polynomials

Let

$$F_1(s) = \frac{1}{s+1} \qquad f_1(t) = e^{-t}$$

$$F_2(s) = \left(\frac{1}{1+s}\right)\left(\frac{1-s}{1+s}\right) \qquad f_2(t) = 2te^{-t} - e^{-t}$$

$$F_3(s) = \left(\frac{1}{1+s}\right)\left(\frac{1-s}{1+s}\right)\left(\frac{1-s}{1+s}\right) \qquad f_3(t) = 3e^{-t} - te^{-t} + t^2e^{-t}$$

$$\vdots$$

$$F_i(s) = \left(\frac{1}{1+s}\right)\left(\frac{1-s}{1+s}\right)^{i-1}, \qquad i = 1, 2, 3, \ldots, n$$

In general

$$F_{ij}(s) = F_{i+j}(s) = F_i(s)F_j(-s) = \left(\frac{1}{1+s}\right)\left(\frac{1-s}{1+s}\right)^{(i+j-1)}, i > j$$

Once again the conditions of orthogonality are satisfied by $F_{ij}(s)$.

Therefore,

$$I_{ij} = \int_0^\infty f_i(t)f_j(t)\,\mathrm{d}t = \frac{1}{2\pi j}\oint F_i(s)F_j(-s)\,\mathrm{d}s = 0, \quad i \ne j \tag{5.53}$$

These functions are referred to as the **Laguerre Functions**. Here again, the pole zero pattern of various functions is interesting. The study of this pattern suggests various other candidates for $F_{ij}(s)$. To encourage the reader to explore this further, we suggest another set here

$$F_1(s) = \frac{1}{s+1}$$

$$F_2(s) = \left(\frac{1-s}{1+s}\right)\left(\frac{1}{s^2 + 2\xi_1\omega_1 s + \omega_1^2}\right), \qquad \xi_1 > 0$$

$$F_3(s) = \left(\frac{1-s}{1+s}\right)\left(\frac{s^2 - 2\xi_1\omega_1 s + \omega_1^2}{s^2 + 2\xi_1\omega_1 s + \omega_1^2}\right)\left(\frac{1}{s^2 + 2\xi_2\omega_2 s + \omega_2^2}\right), \qquad \xi_2 > 0$$

$$\vdots$$

5.19 The Fourier Transform

This section is devoted to the study of Fourier Transforms and its applications. In section 5.4 the Fourier transform pair relations were established and immediately extended to the Laplace transform pair. The ease with which the Laplace transform can be learned made us postpone the study of the Fourier transform until this section. To recapitulate, the Fourier transform pair is defined as:

$$F(j\omega) = \int_{-\infty}^{+\infty} f(t)e^{-j\omega t}\,\mathrm{d}t, \quad f(t) = \frac{1}{2\pi}\int_{-\infty}^{+\infty} F(j\omega)e^{j\omega t}\,\mathrm{d}\omega$$

Notation: $f(t) \leftrightarrow F(j\omega)$ implies a Transform Pair relationship.
It is also customary that the notation $F(\omega)$ and $F(j\omega)$ is used interchangeably. When it is important to emphasize that $F(\omega)$ is a complex quantity, we shall use $F(j\omega)$ instead.

The function $f(t)$ can be real or complex but we shall consider it as a real function unless stated otherwise. The restriction on Fourier transformability are given in section 5.3. $F(j\omega)$ is a complex quantity. Therefore,

$$F(j\omega) = R(\omega) + jX(\omega) = \sqrt{[R^2(\omega) + X^2(\omega)]}\ \angle \tan^{-1}\frac{X(\omega)}{R(\omega)}$$

where

$$\left.\begin{aligned} R(\omega) &= R(-\omega) \\ X(\omega) &= -X(-\omega) \end{aligned}\right\} \quad \text{only true when } f(t) \text{ is real}$$

Example 5.17:

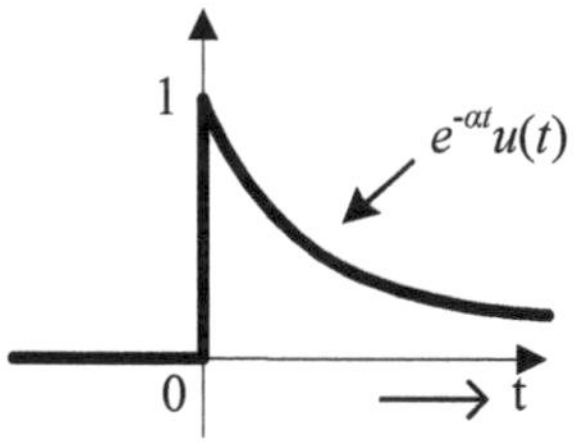

Figure 5.17: Causal Time Exponential

$$f(t) = e^{-\alpha t}u(t), \quad \alpha \geq 0$$

$$F\left[e^{-\alpha t}u(t)\right] = \int_{-\infty}^{+\infty} [e^{-\alpha t}u(t)]\, e^{-j\omega t}\, dt = \int_{0}^{\infty} e^{-(\alpha+j\omega)t}\, dt = \frac{1}{\alpha + j\omega}$$

$$= \frac{\alpha - j\omega}{\alpha^2 + \omega^2} = \frac{\alpha}{\alpha^2 + \omega^2} - \frac{j\omega}{\alpha^2 + \omega^2}$$

As $\alpha \rightarrow 0$ in the limit

$$F[u(t)] = \lim_{\alpha \rightarrow 0}\left[\frac{\alpha}{\alpha^2 + \omega^2} - \frac{j\omega}{\alpha^2 + \omega^2}\right] = \pi\delta(\omega) + \frac{1}{j\omega}$$

Example 5.18:

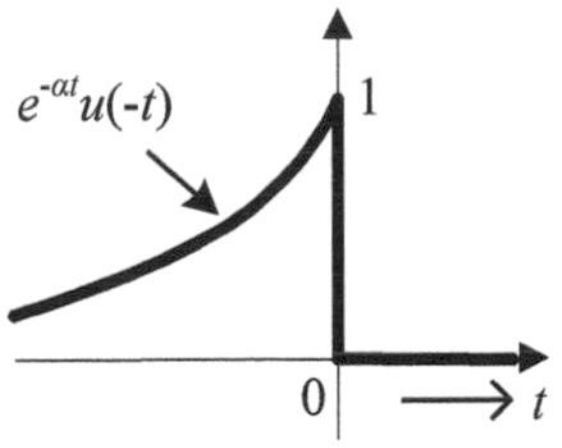

Figure 5.18: Noncausal Time Exponential

$$f(t) = e^{\alpha t}u(-t), \quad \alpha \geq 0$$

$$F\left[e^{\alpha t}u(-t)\right] = \int_{-\infty}^{0} e^{\alpha t}e^{-j\omega t}\,\mathrm{d}t$$

$$t = -\tau,\ \mathrm{d}t = -\,\mathrm{d}\tau,$$

$$F\left[e^{\alpha t}u(-t)\right] = -\int_{\infty}^{0} e^{-\alpha t}e^{j\omega t}\,\mathrm{d}\tau = \int_{0}^{\infty} e^{-\alpha t}e^{j\omega t}\,\mathrm{d}\tau = \frac{1}{\alpha - j\omega}$$

As $\alpha \to 0$, in the limit

$$F\left[u(-t)\right] = \lim_{\alpha\to 0}\left[\frac{1}{\alpha - j\omega}\right] = \lim_{\alpha\to 0}\left[\frac{\alpha}{\alpha^2+\omega^2} + \frac{j\omega}{\alpha^2+\omega^2}\right] = \pi\delta(\omega) - \frac{1}{j\omega}$$

Example 5.19:

Consider $P_T(t)$, a rectangular pulse of unit height and width $2T$, centered at $t = 0$

$$F\left[P_T(t)\right] = \int_{-\infty}^{+\infty} P_T(t)e^{-j\omega t}\,\mathrm{d}t = \int_{-T}^{+T} e^{-j\omega t}\,\mathrm{d}t$$

$$= \left[\frac{e^{-j\omega t}}{-j\omega}\right]_{-T}^{+T} = \frac{2}{\omega}\sin\omega T = 2T\frac{\sin\omega T}{\omega T} = 2T\ \mathrm{sinc}(\omega T)$$

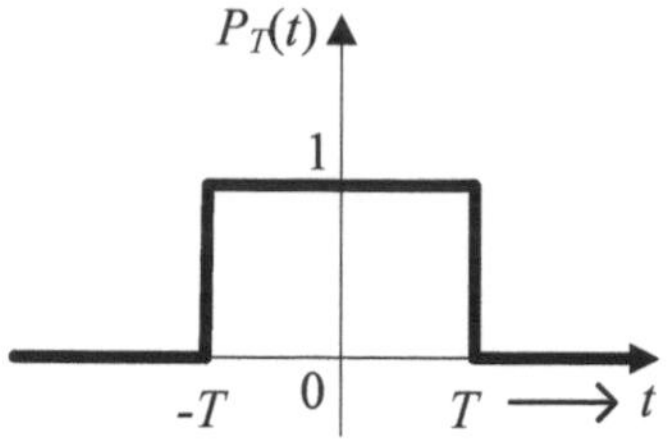

Figure 5.19: Pulse Function

Example 5.20:

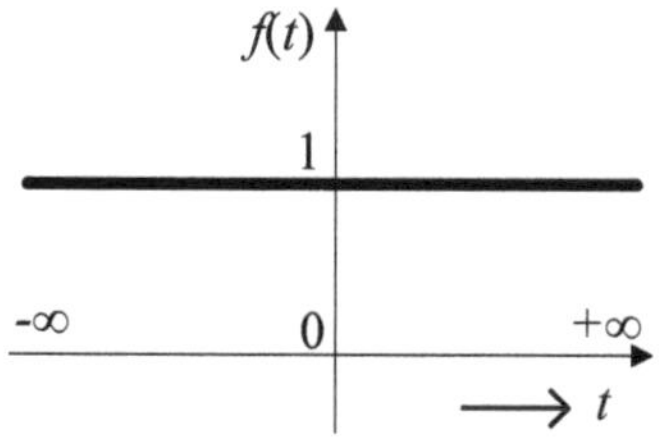

Figure 5.20: Unit Function for All Times

$$f(t) = 1$$

$$F[1] = F[u(t) + u(-t)] = \left(\pi\delta(\omega) + \frac{1}{j\omega}\right) + \left(\pi\delta(\omega) - \frac{1}{j\omega} = 2\pi\delta(\omega)\right)$$

Example 5.21:

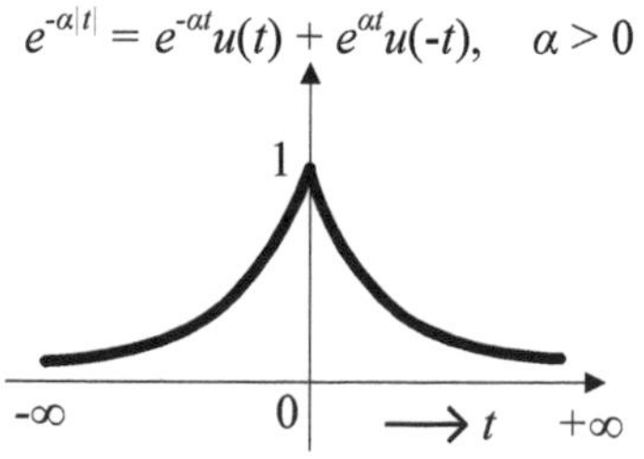

Figure 5.21: Decaying Exponential for Positive and Negative Times

$$
\begin{aligned}
f(t) &= e^{-\alpha|t|} \\
f[f(t)] &= F\left[e^{-\alpha|t|}\right] = F\left[e^{-\alpha t}u(t) + e^{\alpha t}u(-t)\right] \\
&= \frac{1}{\alpha + j\omega} + \frac{1}{\alpha - j\omega} = \frac{2\alpha}{\alpha^2 + \omega^2}
\end{aligned}
$$

Example 5.22:

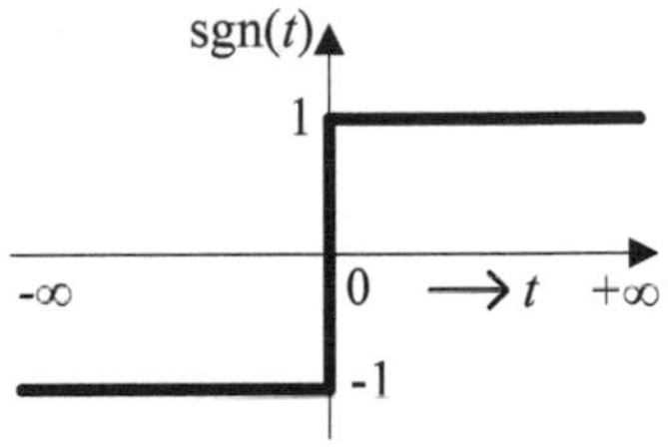

Figure 5.22: Signum Function

$$
\begin{aligned}
f(t) &= \text{sgn}(t) = \frac{|t|}{t} = u(t) - u(-t) \\
F[\text{sgn}(t)] &= F[u(t)] - F[u(-t)] \\
&= \left(\pi\delta(\omega) + \frac{1}{j\omega}\right) + \left(\pi\delta(\omega) - \frac{1}{j\omega}\right) = \frac{2}{j\omega}
\end{aligned}
$$

Example 5.23:

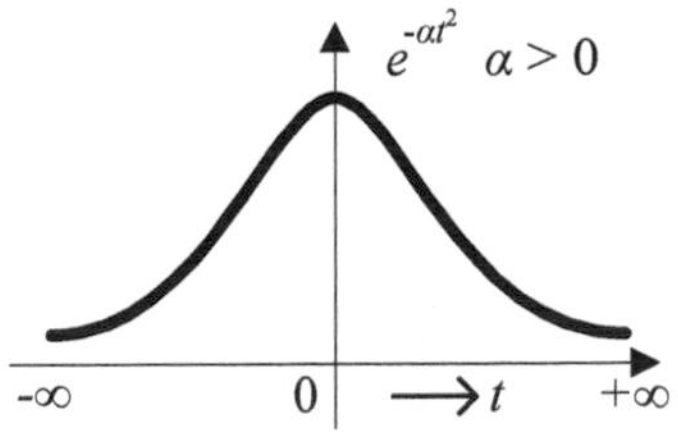

Figure 5.23: Gaussian Function

$$f(t) = e^{-\alpha t^2} \quad \text{Gaussian function}$$

$$F[f(t)] = \int_{-\infty}^{+\infty} e^{-\alpha t^2} e^{-j\omega t}\,\mathrm{d}t \tag{5.54}$$

$$= \int_{-\infty}^{+\infty} e^{-\alpha[t^2 + \frac{j\omega}{\alpha}t]}\,\mathrm{d}t$$

$$= \int_{-\infty}^{+\infty} e^{-\alpha\left[t^2 + \frac{j\omega}{\alpha}t + \left(\frac{j\omega}{2\alpha}\right)^2 - \left(\frac{j\omega}{2\alpha}\right)^2\right]}\,\mathrm{d}t$$

$$= e^{-\frac{\omega^2}{4\alpha}} \int_{-\infty}^{+\infty} e^{-\alpha\left[t + \frac{j\omega}{2\alpha}\right]^2}\,\mathrm{d}t$$

$$= Ie^{-\frac{\omega^2}{4\alpha}} \tag{5.55}$$

where

$$I = \int_{-\infty}^{+\infty} e^{-\alpha\left[t + \frac{j\omega}{2\alpha}\right]^2}\,\mathrm{d}t$$

Let

$$t + \frac{j\omega}{2\alpha} = x$$

$$\mathrm{d}t = \mathrm{d}x$$

$$t \to \pm\infty$$

$$x \to \pm\infty$$

$$\frac{\omega}{2\alpha} = c$$

Thus,

$$I = \int_{jc-\infty}^{jc+\infty} e^{-\alpha^2 x^2}\,\mathrm{d}x = \int_{-\infty}^{+\infty} c^{-\alpha^2 x^2}\,\mathrm{d}x \quad \text{(see Chapter 3)}$$

Computation of I

$$I = \int_{-\infty}^{+\infty} e^{-\alpha x^2}\,\mathrm{d}x = \int_{-\infty}^{+\infty} e^{-\alpha y^2}\,\mathrm{d}y$$

Thus,

$$I^2 = \int_{-\infty}^{+\infty}\int_{-\infty}^{+\infty} e^{-\alpha(x^2+y^2)}\,\mathrm{d}x\,\mathrm{d}y$$

Let

$$x = r\cos\theta, \qquad y = r\sin\theta$$

$$\mathrm{dA} = \mathrm{d}x\,\mathrm{d}y = [\text{Jacobian of } x, y \text{ with respect to } r, \theta] = r\,\mathrm{d}r\,\mathrm{d}\theta$$

Limits of integration are θ from 0 to 2π and r from 0 to ∞. Thus,

$$I^2 = \int_0^{2\pi}\int_0^{\infty} e^{-\alpha r^2} r\,\mathrm{d}r\,\mathrm{d}\theta = 2\pi \int_0^{\infty} e^{-\alpha r^2} r\,\mathrm{d}r$$

Let

$$r^2 = \tau, \qquad 2r\,\mathrm{d}r = \mathrm{d}\tau$$

or

$$I^2 = 2\pi \int_0^{\infty} \frac{1}{2} e^{-\alpha\tau}\,\mathrm{d}\tau = \frac{2\pi}{2\alpha} = \frac{\pi}{\alpha}$$

Thus,

$$I = \sqrt{\frac{\pi}{\alpha}} \qquad (I > 0)$$

Substituting the value of I in Eq. 5.55

$$F\left[e^{-\alpha t^2}\right] = \sqrt{\frac{\pi}{\alpha}} e^{-\frac{\omega^2}{4\alpha}}$$

Example 5.24:

Prove

$$\frac{1}{\pi}\int_{-\infty}^{+\infty}\frac{\sin\omega t}{\omega}\,d\omega = \mathrm{sgn}(t) = \frac{|t|}{t}$$

Proof:

$$f(t) = \frac{1}{\pi}\int_{-\infty}^{+\infty}\frac{\sin\omega t}{\omega}\,d\omega = \frac{1}{\pi}\int_{-\infty}^{+\infty}\frac{e^{j\omega t}-e^{-j\omega t}}{2j\omega}\,d\omega$$
$$= \frac{1}{2\pi}\left[\int_{-\infty}^{+\infty}\frac{e^{j\omega t}}{j\omega}\,d\omega - \int_{-\infty}^{+\infty}\frac{e^{-j\omega t}}{j\omega}\,d\omega\right]$$

For the second term in the bracket let $\omega = -x$ and again realizing that x is a dummy variable

$$-\int_{-\infty}^{+\infty}\frac{e^{-j\omega t}}{j\omega}\,d\omega = +\int_{-\infty}^{+\infty}\frac{e^{jxt}}{jx}\,dx = \int_{-\infty}^{+\infty}\frac{e^{j\omega t}}{j\omega}\,d\omega$$

Thus,

$$f(t) = \frac{1}{2\pi}\int_{-\infty}^{+\infty}\frac{2e^{j\omega t}}{j\omega}\,d\omega = F^{-1}\left[\frac{2}{j\omega}\right] = \mathrm{sgn}(t) = \frac{|t|}{t}$$

Example 5.25:

An infinite train of impulses $S_T(t)$ is defined as

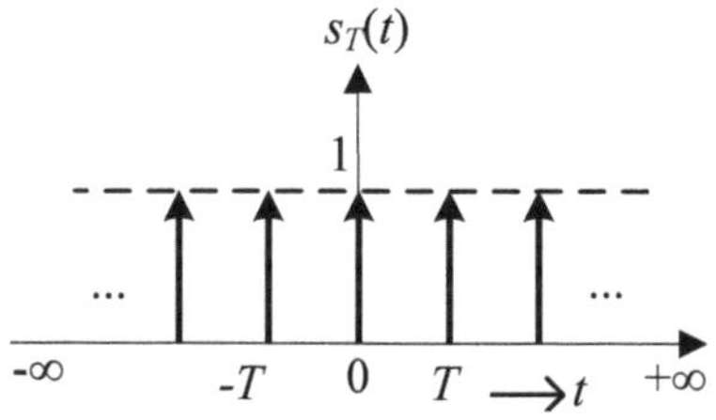

Figure 5.24: Infinite Train of Impulses $S_T(t)$

$$S_T(t) = \sum_{n=-\infty}^{+\infty} \delta(t - nT)$$

Since $S_T(t)$ is a periodic signal with a period T, its Fourier decomposition is

$$S_T(t) = \sum_{n=-\infty}^{+\infty} c_n e^{j\omega_0 nt}, \quad \omega_0 = 2\pi/T$$

$$c_n = \frac{1}{T} \int_{-T/2}^{+T/2} S_T(t) e^{-j\omega_0 nt} = \frac{1}{T}$$

Thus, the Fourier series for the Impulse Train is:

$$S_T(t) = \frac{1}{T} \sum_{n=-\infty}^{+\infty} e^{j\omega_0 nt}$$

The Fourier transform of $S_T(t)$:

$$F\left[S_T(t)\right] = F\left[\frac{1}{T} \sum_{n=-\infty}^{+\infty} e^{j\omega_0 nt}\right] = \frac{2\pi}{T} \sum_{n=-\infty}^{+\infty} \delta(\omega - n\omega_0)$$

5.20 Fourier Transform Properties

Just as in the case of Laplace transforms, there are many properties of the Fourier Transforms that make the transformation of unknown signals an easy extension of the transformation of known signals. We shall make use of some of these properties. From its definition, the Fourier Transform can be looked upon as an expansion of the time function $f(t)$ in terms of an infinite sequence of basis functions. $F(\omega)$ represents the complex (amplitude and phase) contribution of the frequency ω to the signal $f(t)$. Thus, a signal of short duration such as the delta function requires a contribution from all of the frequencies. On the other hand, signals of long duration have a band of relatively smaller frequencies.

Fourier Transform Properties

1. **Linearity**

$$f_i(t) \leftrightarrow F_i(\omega), \qquad a_i \text{ are constants }, \quad i = 1, 2, \ldots, n$$

Then

$$\sum_{i=1}^{n} a_i f_i(t) \leftrightarrow \sum_{i=1}^{n} a_i F_i(\omega)$$

Proof:

The Fourier Transform operator F is linear and hence,

$$F\left[\sum_{i=1}^{n} a_i f_i(t)\right] = \sum_{i=1}^{n} a_i F\left[f_i(t)\right] = \sum_{i=1}^{n} a_i F_i(\omega) \tag{5.56}$$

2. **Symmetry Property**

A look at the Fourier Transform Pair shows a certain symmetry between variables t and $-\omega$. This can be exploited to ease the determination of the transforms of some time functions. This property can be stated as:

If

$$f(t) \leftrightarrow F(\omega)$$

Then

$$F(\omega)\bigg|_{\omega=t} = F(t) \leftrightarrow 2\pi f(-\omega)$$

$$f(-\omega) = f(t)\bigg|_{t=-\omega} \tag{5.57}$$

Proof:

$$F(\omega) = \int_{-\infty}^{+\infty} f(t)e^{-j\omega t}\, \mathrm{d}t = \int_{-\infty}^{+\infty} f(\tau)e^{-j\omega\tau}\, \mathrm{d}\tau, \qquad (t, \tau \text{ are dummy variables})$$

In the above integral let $\omega = t$, then

$$F(t) = \int_{-\infty}^{+\infty} f(\tau)e^{-j\tau t}\,\mathrm{d}\tau$$

Since τ is a dummy variable, let $\tau = -\omega$, then

$$\begin{aligned} F(t) &= -\int_{\infty}^{-\infty} f(-\omega)e^{j\omega t}\,\mathrm{d}\omega = \frac{1}{2\pi}\int_{-\infty}^{+\infty} [2\pi f(-\omega)]\,e^{j\omega t}\,\mathrm{d}\omega \\ &\triangleq F^{-1}[2\pi f(-\omega)] \quad \left(\text{see the definition of } F^{-1}\right) \end{aligned}$$

Thus,

$$f(t) \leftrightarrow F(\omega), \quad F(t) \leftrightarrow 2\pi f(-\omega) \tag{5.58}$$

Example 5.26:

Using symmetry property we can easily show that if

$$p_T(t) \leftrightarrow 2T \sin\frac{\omega T}{\omega T}$$

$$2T\frac{\sin Tt}{Tt} \leftrightarrow 2\pi p_T(-\omega) = 2\pi p_T(\omega) \tag{5.59}$$

Example 5.27:

Derive the Fourier Transform of a triangular pulse $q_T(t)$ and use this transform and Symmetry Property to determine the Fourier Transform $\frac{\sin^2 at}{\pi a t^2}$ Solution:

$$F[q_T(t)] = \int_{-\infty}^{+\infty} q_T(t)e^{-j\omega t}\,\mathrm{d}t = \int_{0}^{T}\left(1 - \frac{t}{T}\right)e^{-j\omega t}\,\mathrm{d}t + \int_{-T}^{0}\left(1 + \frac{t}{T}\right)e^{j\omega t}\,\mathrm{d}t$$

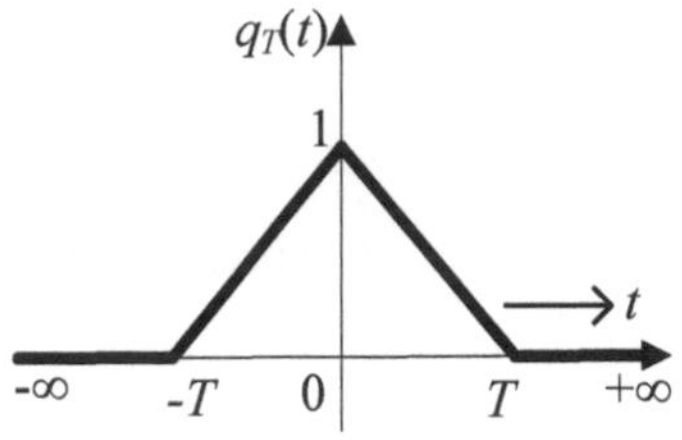

Figure 5.25: Triangular Pulse

Integrating and simplifying the above expression,

$$F[q_T(t)] = F(\omega) = \frac{4}{T}\frac{\sin^2 \omega T/2}{\omega^2}$$

Thus,

$$\frac{4}{T}\frac{\sin^2\left(t\frac{T}{2}\right)}{t^2} \leftrightarrow 2\pi q_T(\omega), \qquad T = 2a$$

Thus,

$$\frac{\sin^2 at}{\pi a t^2} \leftrightarrow q_{2a}(\omega)$$

3. **Scaling**

$$f(t) \leftrightarrow F(\omega)$$

Then

$$f(at) \leftrightarrow \frac{1}{|a|}F\left(\frac{\omega}{a}\right)$$

Proof:

$$F[f(at)] = \int_{-\infty}^{+\infty} f(at)e^{-j\omega t}\,\mathrm{d}t$$

$$\text{Let } at = z$$

$$\mathrm{d}t = \frac{1}{a}\,\mathrm{d}z$$

(i) For $a > 0$

$$F[f(at)] = \frac{1}{a}\int_{-\infty}^{+\infty} f(\tau)e^{-j\frac{\omega}{a}\tau}\,\mathrm{d}\tau = \frac{1}{a}F\left(\frac{\omega}{a}\right)$$

(ii) For $a < 0$

$$F[f(at)] = \frac{1}{a}\int_{\infty}^{-\infty} f(\tau)e^{-j\frac{\omega}{a}\tau}\,\mathrm{d}\tau = -\frac{1}{a}F\left(\frac{\omega}{a}\right)$$

Thus, for any value of a, positive or negative,

$$f(at) \leftrightarrow \frac{1}{|a|}F\left(\frac{\omega}{a}\right) \tag{5.60}$$

This theorem shows that a contraction in time scale represents an expansion in frequency scale and vice versa.

4. **Frequency Shift, Time-Shift, Differentiation, Modulation**

$$\begin{aligned}
f(t) &\leftrightarrow F(\omega)\\
f(t)e^{j\omega_0 t} &\leftrightarrow F(\omega-\omega_0)\\
f(t-t_0) &\leftrightarrow F(\omega)e^{-j\omega t_0}\\
\frac{\mathrm{d}^n}{\mathrm{d}t^n}f(t) &\leftrightarrow (j\omega)^n F(\omega)\\
(-jt)^n f(t) &\leftrightarrow \frac{\mathrm{d}^n}{\mathrm{d}\omega^n}F(\omega)\\
2f(t)\cos\omega_0 t &\leftrightarrow F(\omega+\omega_0)+F(\omega-\omega_0)
\end{aligned} \tag{5.61}$$

5. **Frequency Convolution**

$$f_1(t) \leftrightarrow F_1(\omega)$$

$$f_2(t) \leftrightarrow F_2(\omega)$$

Then

$$f_1(t)f_2(t) \leftrightarrow \frac{1}{2\pi}[F_1(\omega) * F_2(\omega)] = \frac{1}{2\pi}\int_{-\infty}^{+\infty} F_1(\lambda)F_2(\omega-\lambda)\,\mathrm{d}\lambda$$
$$= \frac{1}{2\pi}\int_{-\infty}^{+\infty} F_1(\omega-\lambda)F_2(\omega)\,\mathrm{d}\lambda \tag{5.62}$$

Proof:

$$I = F[f_1(t)f_2(t)] = \int_{-\infty}^{+\infty} f_1(t)f_2(t)e^{-j\omega t}\,\mathrm{d}t$$

This integral can be rewritten as:

$$I = \int_{-\infty}^{+\infty}\left\{\left[\frac{1}{2\pi}\int_{-\infty}^{+\infty} F_1(\lambda)e^{j\lambda t}\,\mathrm{d}\lambda\right] f_2(t)e^{-j\omega t}\,\mathrm{d}t\right\}$$

Interchanging the order of integration and differentiation,

$$I = \frac{1}{2\pi}\int_{-\infty}^{+\infty} F_1(\lambda)\left\{\int_{-\infty}^{+\infty} f_2(t)e^{-j(\omega-\lambda)t}\,\mathrm{d}t\right\}\mathrm{d}\lambda$$

The term in the bracket represents $F_2(\omega-\lambda)$. Thus,

$$I = F[f_1(t)f_2(t)] = \frac{1}{2\pi}\int_{-\infty}^{+\infty} F_1(\lambda)F_2(\omega-\lambda)\,\mathrm{d}\lambda \tag{5.63}$$

The interchange of $f_1(t)$ and $f_2(t)$ yields the second integral in Eq. 5.62.

6. **Time Convolution**

$$F_1(\omega) \leftrightarrow f_1(t)$$

$$F_2(\omega) \leftrightarrow f_2(t)$$

Show

$$F_1(\omega)F_2(\omega) \leftrightarrow f_1(t) * f_2(t) = \int_{-\infty}^{+\infty} f_1(\tau)f_2(t-\tau)\,\mathrm{d}\tau = \int_{-\infty}^{+\infty} f_1(t-\tau)f_2(\tau)\,\mathrm{d}\tau \quad (5.64)$$

Proof:

$$\begin{aligned} F_1(\omega)F_2(\omega) &= F_1(\omega)\left[\frac{1}{2\pi}\int_{-\infty}^{+\infty} f_2(t)e^{-j\omega t}\,\mathrm{d}t\right] = F_1(\omega)\left[\frac{1}{2\pi}\int_{-\infty}^{+\infty} f_2(\tau)e^{-j\omega\tau}\,\mathrm{d}\tau\right] \\ &= \frac{1}{2\pi}\int_{-\infty}^{+\infty} f_2(\tau)\left[F_1(\omega)e^{-j\omega\tau}\right]\mathrm{d}\tau \end{aligned}$$

Taking the Fourier Inverse of both sides

$$\begin{aligned} F^{-1}\left[F_1(\omega)F_2(\omega)\right] &= F^{-1}\left[\frac{1}{2\pi}\int_{-\infty}^{+\infty} f_2(\tau)\left[F_1(\omega)e^{-j\omega T}\right]\mathrm{d}\tau\right] \\ &= \frac{1}{2\pi}\int_{-\infty}^{+\infty} f_2(\tau)\left[F^{-1}F_1(\omega)e^{-j\omega\tau}\right]\mathrm{d}\tau = \frac{1}{2\pi}\int_{-\infty}^{+\infty} f_2(\tau)f_1(t-\tau)\,\mathrm{d}\tau \end{aligned}$$

7. **Parseval's Theorem**

It is very easy to see from Eq. 5.63, that

$$\int_{-\infty}^{+\infty} f^2(t)\,\mathrm{d}t = F\left[f^2(t)\right]_{\omega=0} = \frac{1}{2\pi}\int_{-\infty}^{+\infty} F(\lambda)F(-\lambda)\,\mathrm{d}\lambda = \int_{-\infty}^{+\infty}\left|\frac{F(\omega)^2}{\sqrt{2\pi}}\right|\mathrm{d}\omega \quad (5.65)$$

This expression relates energy in time and frequency domains. Thus, a frequency band of $[\omega_1, \omega_2]$ has an energy content of $\int_{\omega_1}^{\omega_2}\left|\frac{F(j\omega)}{\sqrt{2\pi}}\right|^2 \mathrm{d}\omega$.

Extensive Fourier Transform tables are available in the literature [Oberhettinger, F.], [Abramowitz, M.], and [Bracewell, R.N.]. Table 5.4 presents Fourier Transform Properties for a few important time functions.

Table 5.4: Fourier Transform Properties

1	$f(t)$ – Time-Function	$F(\omega) =$ Fourier Transform of $f(t)$
2	$f(t) = \int_{-\infty}^{+\infty} f(\tau)\,\delta(t-\tau)\,\mathrm{d}\tau,$	$\delta(\omega) = \frac{1}{2\pi}\int_{-\infty}^{+\infty} e^{j\omega t}\,\mathrm{d}t = \delta(-\omega)$
3	$f(t) = \frac{1}{2\pi}\int_{-\infty}^{+\infty} F(\omega)\,e^{j\omega t}\,\mathrm{d}\omega,$	$F(\omega) = \int_{-\infty}^{+\infty} f(t)\,e^{-j\omega t}\,\mathrm{d}t, \quad \int_{-\infty}^{+\infty} \lvert f(t)\rvert\,\mathrm{d}t < \infty$
4	At Discontinuity $f(t) = \frac{1}{2}\,[f(t^+) + f(t^-)]$	
5	$f(t)$	$F(\omega) = R(\omega) + jX(\omega) = A(\omega)\measuredangle\varphi(\omega)$ $F^*(\omega) = R(\omega) - jX(\omega) = A(\omega)\measuredangle - \varphi(\omega)$
6	$f(t) = f_R(t) + jf_I(t)$ $f_R(t) = \frac{1}{2\pi}\int_{-\infty}^{+\infty} (R(\omega)\cos\omega t - X(\omega)\sin\omega t)\,\mathrm{d}\omega$ $f_I(t) = \frac{1}{2\pi}\int_{-\infty}^{+\infty} (R(\omega)\sin\omega t + X(\omega)\cos\omega t)\,\mathrm{d}\omega$	$F(\omega) = R(\omega) + jX(\omega)$ $R(\omega) = \int_{-\infty}^{+\infty} (f_R(t)\cos\omega t + f_I(t)\sin\omega t)\,\mathrm{d}t$ $X(\omega) = \int_{-\infty}^{+\infty} (f_R(t)\sin\omega t - f_I(t)\cos\omega t)\,\mathrm{d}t$
7	$f_1(t)\,f_2(t)$	$\frac{1}{2\pi}\int_{-\infty}^{+\infty} F_1(\lambda)\,F_2(\omega-\lambda)\,\mathrm{d}\lambda = \frac{1}{2\pi}F_1(\omega) * F_2(\omega)$
8	$\int_{-\infty}^{+\infty} f_1(\tau)\,f_2(t-\tau)\,\mathrm{d}\tau$	$F_1(\omega)\,F_2(\omega)$

continued on next page

9	$f(t) = f_R(t)(\text{Real}) \qquad f_I(t) = 0$ $f(t) = \text{Re}\,\frac{1}{2\pi}\int_{-\infty}^{+\infty} F(\omega)\,e^{j\omega t}\,d\omega$ $= \frac{1}{\pi}\int_{0}^{\infty} (R(\omega)\cos\omega t - X(\omega)\sin\omega t)\ dt$	$R(\omega) = \int_{-\infty}^{+\infty} f(t)\cos\omega t\,dt = R(-\omega)$ $X(\omega) = \int_{-\infty}^{+\infty} f(t)\sin\omega t\,dt = -X(-\omega)$ $F(\omega) = F^*(-\omega)$
10	$f(t) = jf_I(t)(\text{Imaginary}) \quad f_R(t) = 0$ $f(t) = \frac{j}{\pi}\int_{0}^{\infty} (R(\omega)\sin\omega t + X(\omega)\cos\omega t)\ d\omega$	$R(\omega) = -R(-\omega) = \int_{-\infty}^{+\infty} f_I(t)\sin\omega t\,dt$ $X(\omega) = X(-\omega) = \int_{-\infty}^{+\infty} f_I(t)\cos\omega t\,dt$ $F(\omega) = -F^*(-\omega)$
11	$f(t) = f(-t) = f_e(t)(\text{even function})$ $f(t) = \frac{1}{\pi}\int_{0}^{\infty} R(\omega)\cos\omega t\,d\omega$	$X(\omega) = 0$ $R(\omega) = 2\int_{0}^{\infty} f(t)\ \cos\omega t\,dt$
12	$f(t) = -f(-t) = f_o(t)(\text{odd function})$ $f(t) = -\frac{1}{\pi}\int_{0}^{\infty} X(\omega)\sin\omega t\,d\omega$	$R(\omega) = 0$ $X(\omega) = -2\int_{0}^{\infty} f(t)\sin\omega t\,dt$
13	$f(t) = f_e(t) + f_o(t)$ $f_e(t) = \frac{1}{2}[f(t) + f(-t)]$ $f_o(t) = \frac{1}{2}[f(t) - f(-t)]$	$f_e(t) \leftrightarrow R(\omega)$ $f_o(t) \leftrightarrow jX(\omega)$

continued on next page

14	$f(t) = \text{Real causal, } f(t) = 0, t < 0$ $f(t) = \frac{2}{\pi}\int_0^\infty R(\omega)\cos\omega t\,\mathrm{d}\omega = -\frac{2}{\pi}\int_0^\infty X(\omega)\sin\omega t\,\mathrm{d}\omega$ $f(0) = \frac{1}{2}f(0^+) = \frac{1}{\pi}\int_0^\infty R(\omega)\,\mathrm{d}\omega$	$X(\omega) = \frac{2}{\pi}\int_0^\infty\int_0^\infty R(y)\sin\omega t\cos yt\,\mathrm{d}y\,\mathrm{d}t$ $R(\omega) = \frac{2}{\pi}\int_0^\infty\int_0^\infty X(y)\sin yt\cos\omega t\,\mathrm{d}y\,\mathrm{d}t$
15	Linearity, $a_1 f_1(t) + a_2 f_2(t)$	$a_1 F_1(\omega) + a_2 f_2(\omega)$
16	Symmetry $f(t)$ $F(t)$	$F(\omega)$ $2\pi f(-\omega)$
17	Time Scale $f(at)$	$\frac{1}{\lvert a\rvert}F\left(\frac{\omega}{a}\right)$
18	Time Shift $f(t - t_0)$	$F(\omega)e^{-j\omega t_0}$
19	Frequency Shift $e^{j\omega_0 t} f(at)$	$\frac{1}{\lvert a\rvert}F\left(\frac{\omega-\omega_0}{\alpha}\right)$
20	Integration $\int_{-\infty}^{t} f(\tau)\,\mathrm{d}\tau$	$\pi F(0)\delta(\omega) + \frac{F(\omega)}{j\omega}$
21	Conjugate $f^*(t)$	$F^*(-\omega)$
22	Moment, $m_n = \int_{-\infty}^{+\infty} t^n f(t)\,\mathrm{d}t$ $(-j)^n m_n = \frac{\mathrm{d}^n}{\mathrm{d}\omega^n} F(0), n = 0, 1, 2, \ldots$	$F(\omega) = \sum_{n=0}^{\infty}(-j)^n m_n \frac{\omega^n}{n!}$

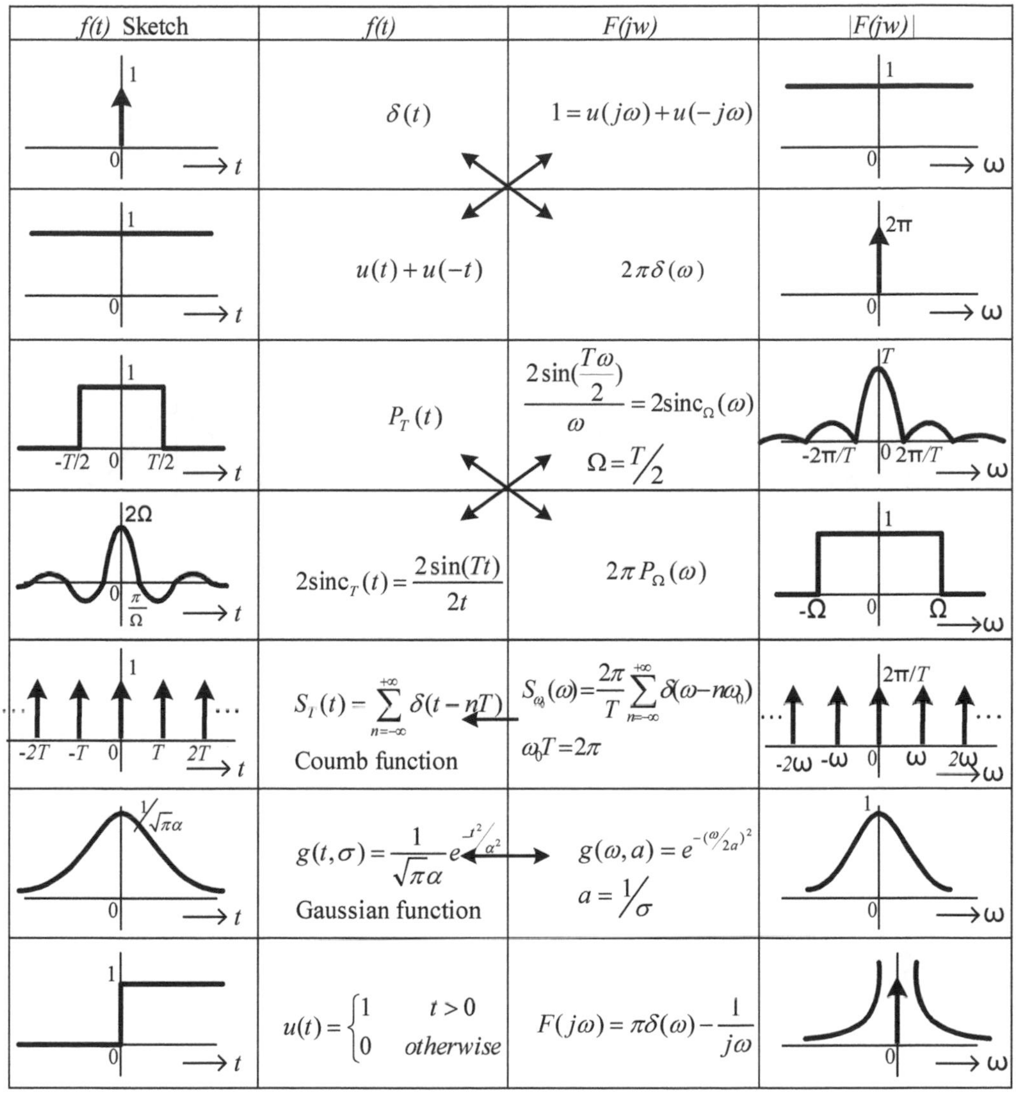

Figure 5.26: Time-Frequency Signals Spectrum

f(t) Sketch	*f(t)*	*F(jw)*	|*F(jw)*|
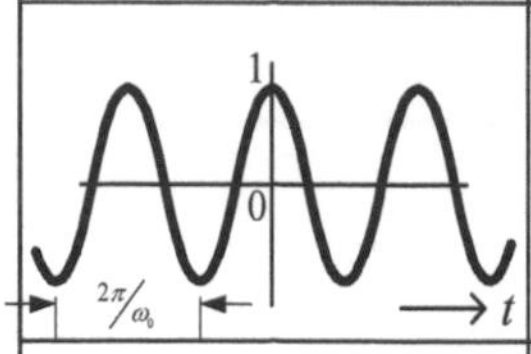	$\cos\omega_0 t$	$\pi[\delta(\omega-\omega_0)+\delta(\omega+\omega_0)]$	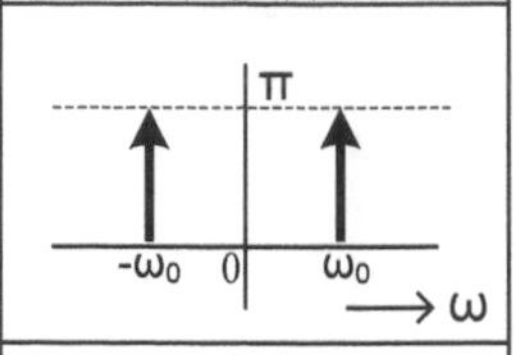
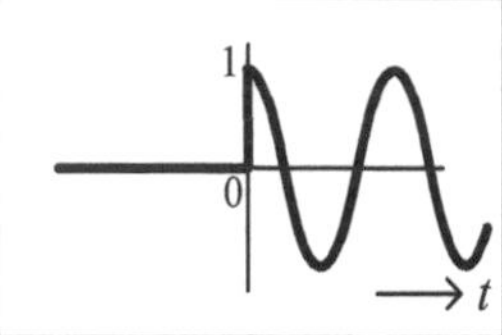	$(\cos\omega_0 t)u(t)$	$\frac{\pi}{2}[\delta(\omega-\omega_0)+\delta(\omega+\omega_0)]$ $-\frac{j\omega}{\omega^2-\omega_0^2}$	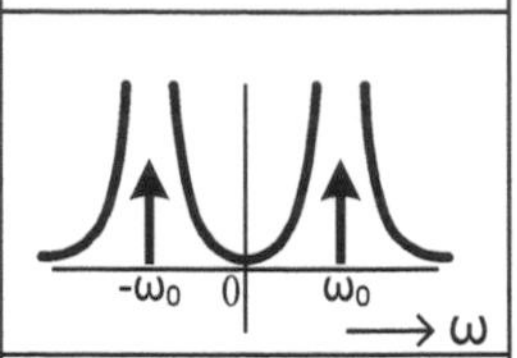
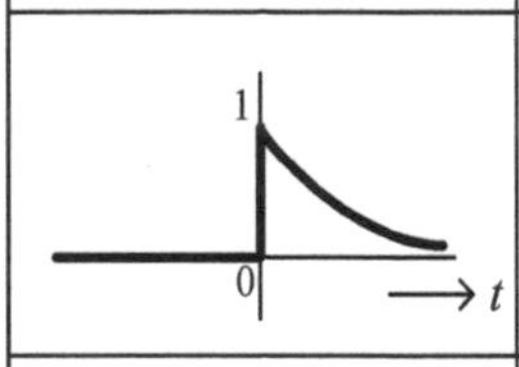	$e^{-\alpha t}u(t)$	$\frac{1}{\alpha+j\omega}$	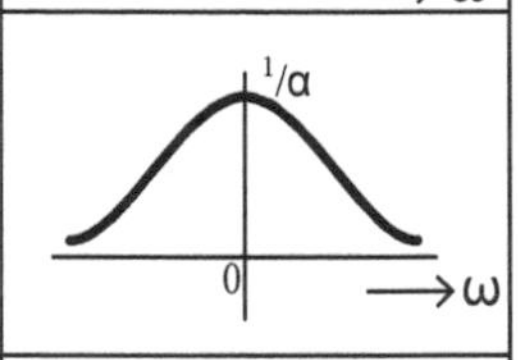
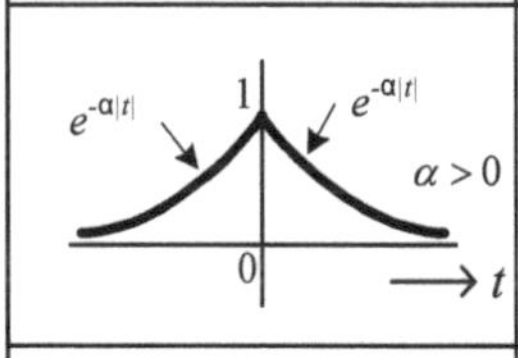	$e^{-\alpha\|t\|}$	$\frac{2\alpha}{\alpha^2+\omega^2}$	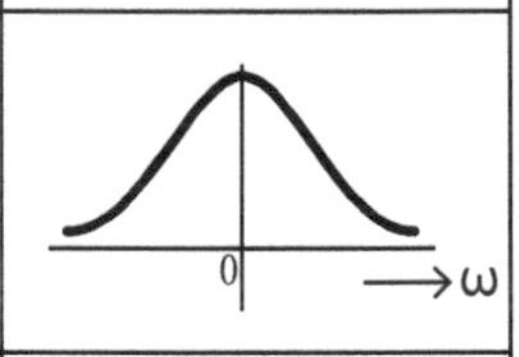
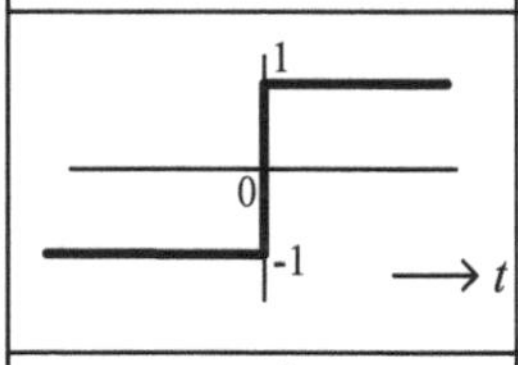	$\text{sqn}(t)$	$\frac{2}{j\omega}$	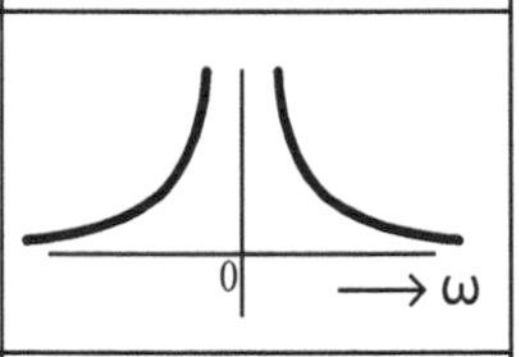
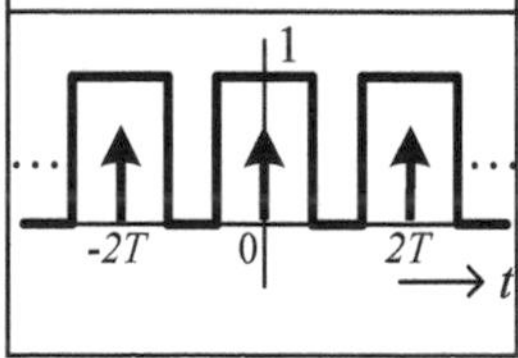	$p_T(t)*\text{Comb}_{2T}(t)$ $\omega_0=\pi/T$	$\sum_{-\infty}^{+\infty}\frac{2}{n}\sin(\frac{n\pi}{2})\delta(\omega-n\omega_0)$	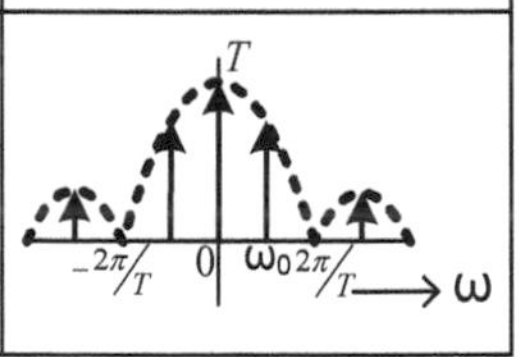

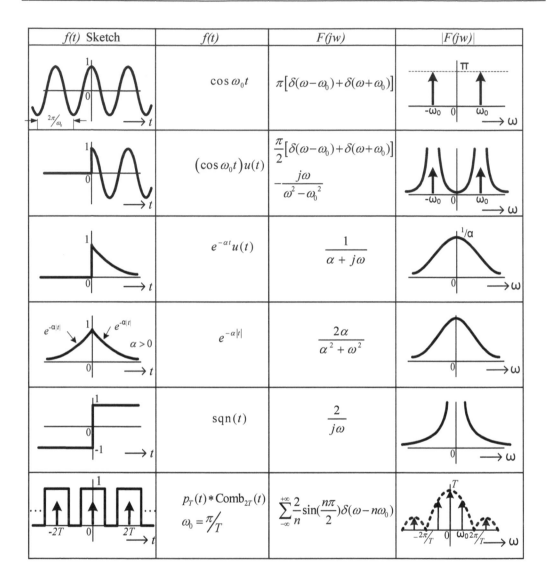

Figure 5.27: Time-Frequency Signals Spectrum Continued

5.21 Fourier Transform Inverse

From the definition of the Fourier Transform Inverse

$$f(t) = F^{-1}[F(\omega)] = f(t) = \frac{1}{2\pi}\int_{-\infty}^{+\infty} F(\omega)e^{j\omega t}\,d\omega \tag{5.66}$$

Let

$$s = j\omega, \qquad d\omega = \frac{1}{j}\,ds$$

Then

$$f(t) = \frac{1}{2\pi j}\int_{-j\infty}^{+j\infty} F\left(\frac{s}{j}\right)e^{st}\,ds$$

Assume

$$|F(\omega)| \to 0 \quad \text{as} \quad \omega \to \infty$$

(i) For $t > 0$

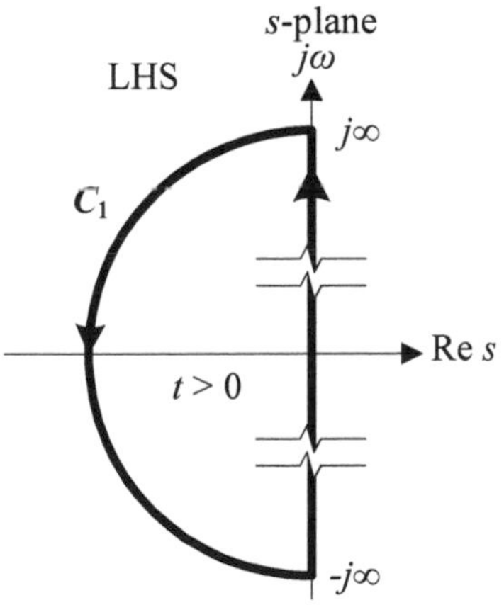

Figure 5.28: Computation of Fourier Transform Inverse for $t > 0$

$\lim_{|s|\to\infty} \left|F(s/j)\,e^{st}\right| \to 0$ when s is in the LHS, the contribution to the contour integral along the infinite semicircle C_1 in the LHS $\to 0$.

From Eq. 5.22

$$f(t) = \frac{2\pi j}{2\pi j}\Bigg[\sum \text{Residues of } \left[F\left(\frac{s}{j}\right)e^{st}\right] \text{ at the poles of } F\left(\frac{s}{j}\right) \text{ in LHS} + \frac{1}{2}\sum \text{Residue of } \left[F\left(\frac{s}{j}\right)e^{st}\right] \text{ at the poles of } F\left(\frac{s}{j}\right) \text{ on } j\omega \text{ axis}\Bigg]$$

(ii) For $t < 0$

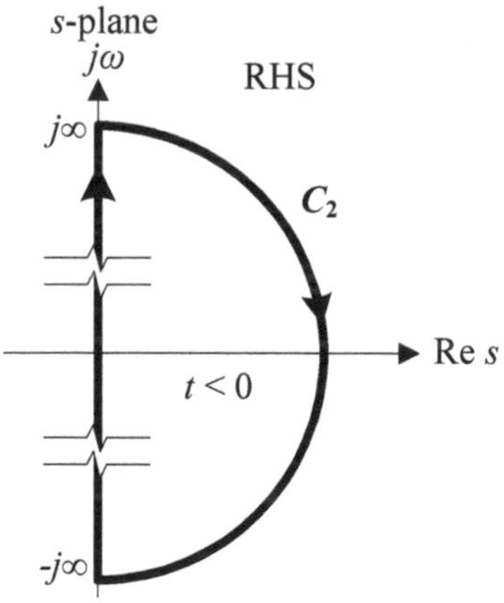

Figure 5.29: Computation of Fourier Transform Inverse for $t < 0$

$\lim_{|s|\to\infty} \left|F\left(\frac{s}{j}\right)e^{st}\right| \to 0$ when s is in the RHS, the contribution to the contour integral along the infinite circle C_2 in the RHS of the s-plane $\to 0$.

$$f(t) = -\frac{2\pi j}{2\pi j}\Bigg[\sum \text{Residues of } \left[F\left(\frac{s}{j}\right)e^{st}\right] \text{ at the poles of } F\left(\frac{s}{j}\right) \text{ in RHS} + \frac{1}{2}\sum \text{Residue of } \left[F\left(\frac{s}{j}\right)e^{st}\right] \text{ at the poles of } F\left(\frac{s}{j}\right) \text{ on } j\omega \text{ axis}\Bigg]$$

Example 5.28: Evaluate the Fourier Inverse of the following function

$$F(\omega) = \frac{2\alpha}{\alpha^2 + \omega^2}, \qquad \alpha > 0$$

Solution:

$$f(t) = \frac{1}{2\pi}\int_{-\infty}^{+\infty} \frac{2\alpha e^{j\omega t}}{\alpha^2+\omega^2}\,\mathrm{d}\omega = \frac{1}{2\pi j}\int_{-j\infty}^{+j\infty} \frac{2\alpha e^{st}}{\alpha^2 - s^2}\,\mathrm{d}s = \frac{1}{2\pi j}\int_{-j\infty}^{+j\infty}\left[\frac{e^{st}}{s+\alpha} - \frac{e^{st}}{s-\alpha}\right]\mathrm{d}s$$

There are no singularities on the $j\omega$ axis. Thus,

For $t > 0$

$$f(t) = \mathrm{Res}\left[\frac{e^{st}}{s+\alpha}\right]_{s=-\alpha} = e^{-\alpha t}$$

For $t < 0$

$$f(t) = -\mathrm{Res}\left[-\frac{e^{st}}{s-\alpha}\right]_{s=\alpha} = e^{\alpha t}$$

Combining the two results,

$$f(t) = F^{-1}\left[\frac{2\alpha}{\alpha^2+\omega^2}\right] = e^{-\alpha t}u(t) + e^{\alpha t}u(-t) = e^{-\alpha|t|}$$

Example 5.29:

Let us consider the application of the Fourier Transform to solve a circuit problem.

The circuit is initially in a relaxed state at $t = 0$. Then a step voltage is applied.

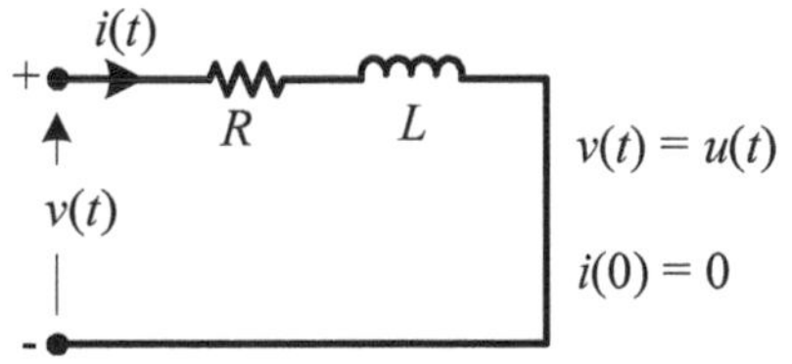

Figure 5.30: Computation of *R–L* Circuit Response via Fourier Transforms

The circuit equation is:

$$Ri + L\frac{\mathrm{d}i}{\mathrm{d}t} = v(t) = u(t) \quad \text{(Step Function)}$$

Taking the Fourier Transform of both sides

$$i(t) \leftrightarrow I(\omega), \qquad \frac{\mathrm{d}}{\mathrm{d}t}i(t) \leftrightarrow j\omega I(\omega)$$

$$(R + j\omega L)I(\omega) = \pi\delta(\omega) + \frac{1}{j\omega}$$

$$I(\omega) = \frac{\pi\delta(\omega)}{R + j\omega L} + \frac{1}{j\omega}(R + j\omega L)$$

Taking the Fourier Inverse of both sides

$$\begin{aligned} i(t) &= F^{-1}\left[\frac{\pi\delta(\omega)}{R + j\omega L}\right] + F^{-1}\left[\frac{1}{j\omega(R + j\omega L)}\right] \\ &= F^{-1}\left[\frac{\pi\delta(\omega)}{R + j\omega L}\right] + \frac{1}{R}F^{-1}\left(\frac{1}{j\omega}\right) - \frac{L}{R}F^{-1}\left(\frac{1}{R + j\omega L}\right) \end{aligned}$$

But

$$\begin{aligned} F^{-1}\left[\frac{1}{j\omega}\right] &= \frac{1}{2\pi}\int_{-\infty}^{+\infty}\frac{e^{j\omega t}}{j\omega}\,\mathrm{d}\omega = \frac{1}{2\pi j}\int_{-j\infty}^{+j\infty}\frac{e^{st}}{s}\,\mathrm{d}s \\ &= \left[\frac{1}{2}\mathrm{Res}\left[\frac{e^{st}}{s}\right]_{s=0}\right] \qquad \text{for } t > 0 \\ &= -\left[\frac{1}{2}\mathrm{Res}\left[\frac{e^{st}}{s}\right]_{s=0}\right] \qquad \text{for } t < 0 \\ &= \frac{1}{2}[u(t) - u(-t)] \end{aligned}$$

$$\begin{aligned} F^{-1}\left[\frac{1}{R + j\omega L}\right] &= \frac{1}{2\pi}\int_{-\infty}^{+\infty}\frac{e^{j\omega t}}{R + j\omega L}\,\mathrm{d}\omega = \frac{1}{2\pi j}\int_{-j\infty}^{+j\infty}\frac{e^{st}}{R + sL}\,\mathrm{d}s \\ &= \left[\mathrm{Res}\,\frac{e^{st}}{R + sL}\right]_{s=-(R/L)} \\ &= \frac{1}{L}e^{-(R/L)t} \qquad \text{for } t > 0 \\ &= 0 \qquad \text{for } t < 0 \end{aligned}$$

Summarizing

$$F^{-1}\left[\frac{\pi\delta(\omega)}{R+j\omega L}\right] = \frac{1}{2\pi}\int_{-\infty}^{+\infty}\frac{\pi\delta(\omega)}{R+j\omega L}\,\mathrm{d}\omega = \frac{1}{2R} = \frac{1}{2R}[u(t)+u(-t)]$$

$$F^{-1}\left[\frac{1}{R+j\omega L}\right] = e^{-(R/L)t}u(t)$$

Hence,

$$i(t) = \frac{1}{2R}[u(t)+u(-t)] + \frac{1}{2R}[u(t)-u(-t)] - \frac{1}{R}e^{-(R/L)t}u(t)$$

or

$$i(t) = \frac{1}{R}\left(1-e^{-(R/L)t}\right)u(t)$$

This problem could have been solved easily by the Laplace Transform.

5.22 Hilbert Transform

The Fourier Transform $F(\omega)$ of a general real time function $f(t)$ is written as:

$$F[f(t)] = F(\omega) = R(\omega) + jX(\omega) \tag{5.67}$$

Normally, $R(\omega)$ and $X(\omega)$ are two independent functions for a general time function $f(t)$. But if the function $f(t)$ is causal, that is $f(t) = 0$ for $t < 0$, then some interesting relationships take place between $R(\omega)$ and $X(\omega)$. In fact, if $R(\omega)$ (or $X(\omega)$) is known for $-\infty < \omega < +\infty$, then its counterpart $X(\omega)$ (or $R(\omega)$) can be calculated via the **Hilbert Transform** relationship.

Hilbert Transforms play an important role in High Frequency Transmitter Engineering where analytic signals are required. Let us derive the Hilbert Transform Pair.

Derivation

A causal function $f(t)$ can be written as

$$0 = f(t)u(-t) \tag{5.68}$$

Taking the Fourier Transform of both sides

$$0 = F[f(t)u(-t)] = F[f(t)] * F[u(-t)] \tag{5.69}$$

$$F[f(t)] = R(\omega) + jX(\omega)$$
$$F(u(-t)] = \pi\delta(\omega) - \frac{1}{j\omega}$$

$$0 = \frac{1}{2\pi}\int_{-\infty}^{+\infty} [R(\lambda) + jX(\lambda)]\left[\pi\delta(\omega - \lambda) - \frac{1}{j(\omega - \lambda)}\right] \mathrm{d}\lambda \tag{5.70}$$

Equating real and imaginary parts

$$0 = \frac{1}{2\pi}\left[\pi R(\omega) - \int_{-\infty}^{+\infty} \frac{X(\lambda)}{\omega - \lambda}\,\mathrm{d}\lambda\right]$$
$$0 = \frac{1}{2\pi}\left[\pi X(\omega) + \int_{-\infty}^{+\infty} \frac{R(\lambda)}{\omega - \lambda}\,\mathrm{d}\lambda\right]$$

$$\left.\begin{aligned} R(\omega) &= \frac{1}{\pi}\int_{-\infty}^{+\infty} \frac{X(\lambda)}{\omega - \lambda}\,\mathrm{d}\lambda = H[X(\lambda)] \\ X(\omega) &= -\frac{1}{\pi}\int_{-\infty}^{+\infty} \frac{R(\lambda)}{\omega - \lambda}\,\mathrm{d}\lambda = H[R(\lambda)] \end{aligned}\right\} \text{Hilbert Transform Pair} \tag{5.71}$$

$$F[f(t)] = R(\omega) + jX(\omega), \quad f(t) \equiv 0 \quad \text{for } t < 0$$

$H[\cdot]$ stands for Hilbert Transform.

Let

$$F(s) = F(\omega)\bigg|_{j\omega=s} = [R(\omega) + jX(\omega)]_{j\omega=s} \tag{5.72}$$

Since $f(t) \equiv 0$ for $t < 0$, all of its singularities are in the LHS of the s-plane. Such a time function is called the Analytic Function (or Regular Function). It plays a very important role in network synthesis.

Consider

$$\omega = t$$

$$R(\omega)\bigg|_{\omega=t} = \hat{g}(t)$$

$$X(\omega)\bigg|_{\omega=t} = g(t)$$

Let us create a complex function $g_a(t)$ such that

$$g_a(t) = g(t) + j\hat{g}(t) \tag{5.73}$$

This function $g_a(t)$ is known as the "Complex analytic time function" whose real and imaginary parts are related via the Hilbert Transform. Namely,

$$g_a(t) = g(t) + j\hat{g}(t)$$

$$\hat{g}(t) = H[g(t)] = \frac{1}{\pi}\int_{-\infty}^{+\infty} \frac{g(\tau)}{(t-\tau)}\, d\tau = g(t) * \frac{1}{\pi t} \tag{5.74}$$

$$g(t) = H^{-1}[\hat{g}(t)] = -\frac{1}{\pi}\int_{-\infty}^{+\infty} \frac{\hat{g}(t)}{(t-\tau)}\, d\tau = -\hat{g}(t) * \frac{1}{\pi t} \tag{5.75}$$

Eqs. 5.74 and 5.75 represent a Hilbert Transform Pair for a Complex Analytic Time Function.

It is easy to show that

$$H[H[g(t)]] = -g(t) \tag{5.76}$$

$$H[\sin \omega t + \varphi] = -\cos(\omega t + \varphi) \tag{5.77}$$

$$H[\cos \omega t + \varphi] = \sin(\omega t + \varphi) \tag{5.78}$$

5.22.1 Hilbert Transform—Inversion of Singular Integrals

Based upon the theory of Hilbert Transform pair, we can state the following theorem [Akhiezer, N.I.]. For any function $F(\omega) \in \boldsymbol{L}^2(-\infty, \infty)$ satisfying the following equation,

$$F(\omega) = \frac{1}{\pi} \int_{-\infty}^{\infty} \frac{G(\lambda)}{\omega - \lambda} \, d\lambda \tag{5.79}$$

there exists a corresponding function $G(\omega) \in \boldsymbol{L}^2(-\infty, \infty)$, which satisfies the following equation

$$G(\omega) = -\frac{1}{\pi} \int_{-\infty}^{\infty} \frac{F(\lambda)}{\omega - \lambda} \, d\lambda \tag{5.80}$$

The above singular integrals are unitary operators in $\boldsymbol{L}^2(-\infty, \infty)$ and satisfy

$$\int_{-\infty}^{\infty} |F(\omega)|^2 \, d\omega = \int_{-\infty}^{\infty} |G(\omega)|^2 \, d\omega \tag{5.81}$$

Note:

(1) If $F(\omega) = F(-\omega)$, then $G(\omega) = -G(-\omega)$

(2) If $F(\omega) = -F(-\omega)$, then $G(\omega) = G(-\omega)$

This leads us to the inversion operator in $\boldsymbol{L}^2(0, \infty)$.

Considering $F(\omega)$ to be an even function of ω,

$$F(\omega) = \frac{2}{\pi} \int_0^\infty \frac{\lambda G(\lambda)}{(\omega^2 - \lambda^2)} \, \mathrm{d}\lambda$$

$$G(\omega) = -\frac{2\omega}{\pi} \int_0^\infty \frac{R(\lambda)}{(\omega^2 - \lambda^2)} \, \mathrm{d}\lambda$$

Furthermore,

$$\int_0^\infty |F(\omega)|^2 \, \mathrm{d}\omega = \int_0^\infty |G(\omega)|^2 \, \mathrm{d}\omega$$

Letting

$$\lambda^2 = t, \quad \omega^2 = s$$

we get

$$F(s) = \frac{1}{\pi} \int_0^\infty \frac{G(t)}{(s - t)} \, \mathrm{d}t$$

$$G(s) = -\frac{1}{\pi} \int_0^\infty \left[\frac{F(t)}{(s - t)} \right] \left[\sqrt{\frac{s}{t}} \right] \mathrm{d}t$$

5.22.2 Physical Realization of Hilbert Tranform of a Function

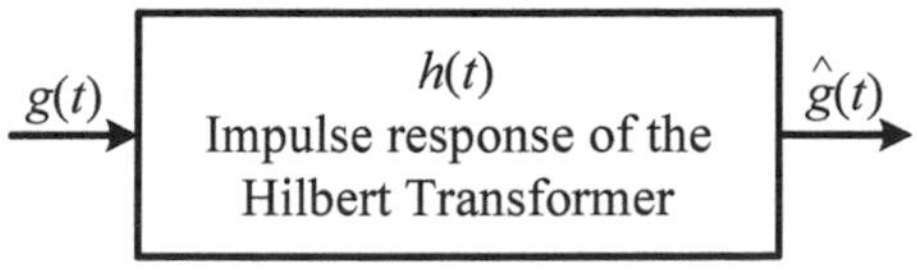

Figure 5.31: Hilbert Transform Realization

Let us consider a filter (or a circuit) whose input is $g(t)$ and its output is its Hilbert

Transform $\hat{g}(t)$. Let $h(t)$ be its impulse response. Then

$$\hat{g}(t) = g(t) * h(t)$$

Comparing the above equation with Eq. 5.75

$$\begin{aligned} h(t) &= \text{Impulse Response of the Hilbert Transformer } = \frac{1}{\pi t} \\ F[h(t)] &= H(\omega) = F\left[\frac{1}{\pi t}\right] = -j(\text{sgn } \omega) \end{aligned} \quad (5.82)$$

The Hilbert Transformer is also called the **Quadrature Filter** since it produces a $-\frac{\pi}{2}$ phase shift for positive frequencies.

$$\begin{aligned} G_a(\omega) &= F[g_a(t)] = F[g(t) + j\hat{g}(t)] \\ &= G(\omega) + j[-j(\text{sgn } \omega)G(\omega)] = \begin{cases} 2G(\omega), & \omega > 0 \\ G(\omega), & \omega = 0 \\ 0, & \omega < 0 \end{cases} \end{aligned} \quad (5.83)$$

Fourier Transform of Analytical Signal

Analytical signals in the time domain result in the suppression of negative frequencies and are very useful in single-side band transmitters.

Example 5.30:

Use the Contour Integration Formula to derive Hilbert Transform Relations

Solution: Let $F(s)$ be an analytic function in the RHS of s-plane as well as on the imaginary axis:

$$F(s)\Big|_{s=j\lambda} = F(\lambda) = R(\lambda) + jX(\lambda)$$

Consider the integral

$$I = \int_{-\infty}^{+\infty} \frac{F(\lambda)}{\omega - \lambda}\, d\lambda = -\int_{-j\infty}^{+j\infty} \frac{F(s/j)}{s - j\omega}\, ds$$

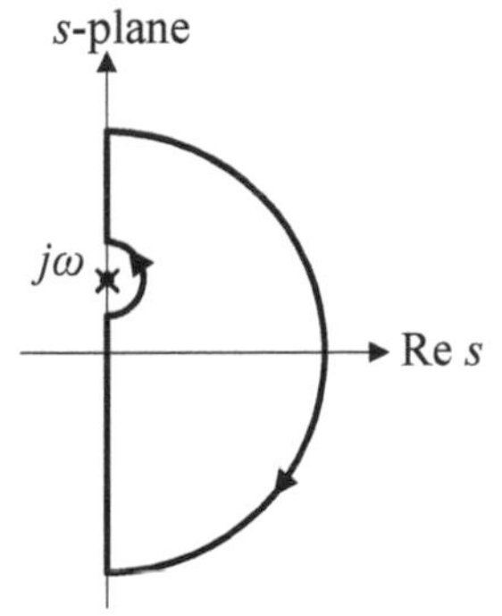

Figure 5.32: Hilbert Transform Derivation via Contour Integration

If $|F(s)| \rightarrow s^{-\alpha}, \alpha > 0$ as $s \rightarrow \infty$, the Jordan's lemma is satisfied, making the contribution from the infinite semicircle to the integral = 0, yielding

$$I = -2\pi j \left[\frac{1}{2} \text{Res} \left[-\frac{F(s/j)}{(s - j\omega)} \right]_{s=j\omega} \right] = +\pi j F(\omega)$$

$$+\pi j F(\omega) = +\pi j [R(\omega) + jX(\omega)] = \int_{-\infty}^{+\infty} \frac{R(\lambda)}{\omega - \lambda}\, d\lambda + j \int_{-\infty}^{+\infty} \frac{X(\lambda)}{\omega - \lambda}\, d\lambda$$

Comparing imaginary and real parts

$$\left. \begin{aligned} R(\omega) &= \frac{1}{\pi} \int_{-\infty}^{+\infty} \frac{X(\lambda)}{\omega - \lambda}\, d\lambda = \text{ even function of } \omega \\ X(\omega) &= -\frac{1}{\pi} \int_{-\infty}^{+\infty} \frac{R(\lambda)}{\omega - \lambda}\, d\lambda = \text{ odd function of } \omega \end{aligned} \right| \qquad (5.84)$$

Recognizing the even and odd character of these functions

$$R(\omega) = \frac{1}{\pi}\int_0^\infty X(\lambda)\left[\frac{1}{\omega-\lambda} - \frac{1}{\omega+\lambda}\right] \mathrm{d}\lambda = \frac{2}{\pi}\int_0^\infty \frac{\lambda X(\lambda)}{\omega^2-\lambda^2}\,\mathrm{d}\lambda$$

$$X(\omega) = -\frac{1}{\pi}\int_0^\infty R(\lambda)\left[\frac{1}{\omega-\lambda} + \frac{1}{\omega+\lambda}\right] \mathrm{d}\lambda = -\frac{2}{\pi}\int_0^\infty \frac{\omega R(\lambda)}{\omega^2-\lambda^2}\,\mathrm{d}\lambda$$

Physically $F(\omega)$ may be a propagation function, a Transfer function or any other familiar network characterization (such as a scattering function).

Example 5.31:

Given

$$R(\omega) = \frac{1}{1+\omega^2}$$

Find $F(\omega) = R(\omega) + jX(\omega)$, analytic in the RHS

Solution:

$$X(\omega) = -\frac{1}{\pi}\int_{-\infty}^{+\infty} \frac{R(\lambda)}{(\omega-\lambda)}\,\mathrm{d}\lambda = -\frac{1}{\pi}\int_{-\infty}^{+\infty} \frac{1}{(\omega-\lambda)(1+\lambda^2)}\,\mathrm{d}\lambda$$

$$= -\frac{1}{\pi}\int_{-j\infty}^{+j\infty} \frac{\mathrm{d}s}{(s-j\omega)(s+1)(s-1)}$$

Enclosing the contour along c_1 in the RHS and using Residue theorem

$$X(\omega) = -\left(\frac{1}{\pi}\right)(-2\pi j)\left\{\frac{1}{2}\mathrm{Res}\left[\frac{1}{s-j\omega} - \frac{1}{s^2-1}\right]_{s=j\omega} + \mathrm{Res}\left[\frac{1}{(s-j\omega)(s+1)(s-1)}\right]_{s=1}\right\}$$

$$= 2j\left[-\frac{1}{2(1+\omega^2)} + \frac{1}{2}\frac{1}{1-j\omega}\right]$$

$$= j\left[-\frac{1}{1+\omega^2} + \frac{1+j\omega}{1+\omega^2}\right]$$

$$= \frac{-\omega}{1+\omega^2}$$

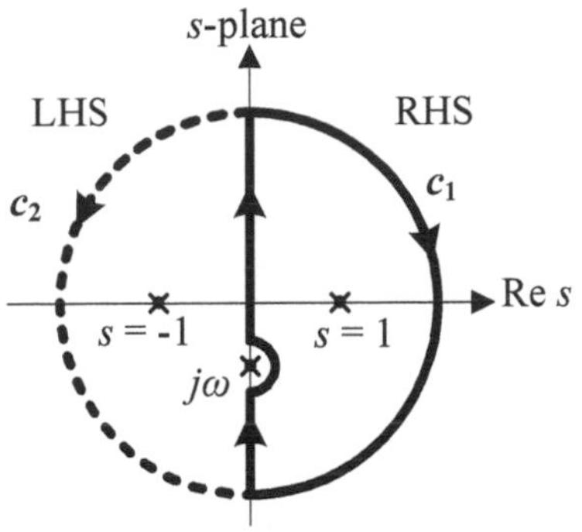

Figure 5.33: Computation of $F(\omega)$ Given $R(\omega)$

Alternatively, we could have closed the contour along c_2 in the LHS along the dotted semicircle yielding

$$\begin{aligned}X(\omega) = -\left(\frac{1}{\pi}\right)(2\pi j)\Big\{\frac{1}{2}\text{Res}\left[\frac{1}{s-j\omega} - \frac{1}{s^2-1}\right]_{s=j\omega} \\ + \text{Res}\left[\frac{1}{(s-j\omega)(s+1)(s-1)}\right]_{s=1}\Big\} \\ = -2j\left[-\frac{1}{2(1+\omega^2)} + \frac{1}{2}\frac{1}{1+j\omega}\right] \\ = -j\left[-\frac{1}{1+\omega^2} + \frac{1-j\omega}{1+\omega^2}\right] \\ = \frac{-\omega}{1+\omega^2}\end{aligned}$$

Thus,

$$F(\omega) = R(\omega) + jX(\omega) = \left(\frac{1-j\omega}{1+\omega^2}\right) = \frac{1}{1+j\omega}$$

5.23 The Variable Parameter Differential Equations

Laplace and Fourier Transforms are useful in dealing with the constant parameter dynamic systems. In this section we shall see how these transformations can be used to solve the differential equations with variable coefficients that are functions of the independent variable.

Consider the differential equation:

$$\boldsymbol{L}_t y = 0, \quad t \text{ is indepedent variable} \tag{5.85}$$

where

$$\boldsymbol{L}_t = \sum_{m=0}^{k} \sum_{n=0}^{r} a_{mn} t^n \boldsymbol{D}^m(t) = \sum_{m=0}^{k} \sum_{n=0}^{r} a_{mn} \boldsymbol{T}_{mn}$$

$$\boldsymbol{D}^m(t) = \frac{\mathrm{d}^m}{\mathrm{d}t^m}$$

$$t^n\, \boldsymbol{D}^m(t) = t^n \frac{\mathrm{d}^m}{\mathrm{d}t^m} = \boldsymbol{T}_{mn}$$

a_{mn} are constants

We seek a solution that is a linear combination of k linearly independent fundamental solutions of the form [Plaschko, P.]

$$y(t) = \int_C \boldsymbol{K}(s,t)\boldsymbol{Y}(s)\,\mathrm{d}s \tag{5.86}$$

where s is a complex variable and C is the path of integration in the s-plane. The variable t can be real or complex and change from the initial value t_1 to the final value t_2 along a curve t_C ($s(A) = t_1$, $s(B) = t_2$).

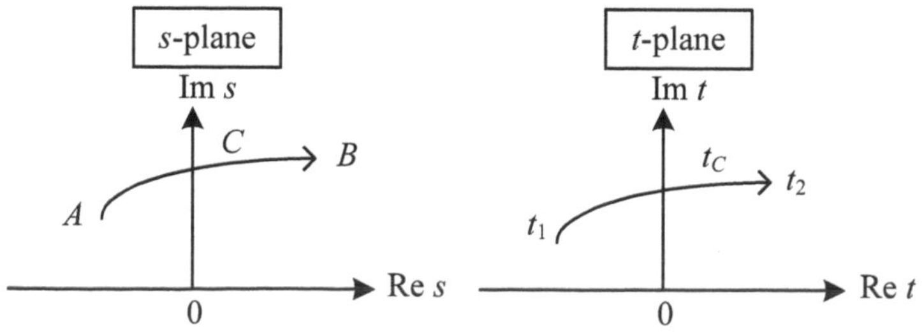

Figure 5.34: The Path of Integration

We shall assume the existence of an analytic function $\boldsymbol{Y}(s)$. The integration is independent of the path of integration joining the initial point A and the terminal point B. The functions $\boldsymbol{K}(s,t)$ are called **Kernel Functions** and may be chosen so as to produce a simple solution of the Eq. 5.85. There are no general rules for the selection of $\boldsymbol{K}(s,t)$. Depending upon the choice of $\boldsymbol{K}(s,t)$ we obtain different Integral Transforms that are suitable for different or special classes of problems. Various important kernel functions are:

$\boldsymbol{K}(s,t) = e^{-st}, e^{-j\omega t}, e^{ast}$	**Generalized Laplace or Fourier Transform-Kernel**
$\boldsymbol{K}(s,t) = (s-t)^a$	**Euler-Kernel**
$\boldsymbol{K}(s,t) = s^{at}$	**Mellin Transform-Kernel**
$\boldsymbol{K}(s,t) = e^{at\sin s}$	**Sommerfeld-Kernel**

Let us apply the **Generalized Laplace-Fourier Transform** to obtain the solution of Eq. 5.85. Take

$$\boldsymbol{K}(s,t) = e^{ast} \qquad (a = -1 \text{ or } \pm j) \tag{5.87}$$

Taking derivatives

$$\begin{aligned} \frac{\mathrm{d}^m}{\mathrm{d}t^m}\boldsymbol{K}(s,t) &= \boldsymbol{D}^m(t)\boldsymbol{K}(s,t) = (as)^m\boldsymbol{K}(s,t) \\ \frac{\mathrm{d}^n}{\mathrm{d}s^n}\boldsymbol{K}(s,t) &= \boldsymbol{D}^n(s)\boldsymbol{K}(s,t) = (at)^n\boldsymbol{K}(s,t) \end{aligned} \tag{5.88}$$

or

$$t^n\boldsymbol{K}(s,t) = a^{-n}\,\boldsymbol{D}^n(s)\boldsymbol{K}(s,t) \tag{5.89}$$

Substituting Eq. 5.86 into Eq. 5.85 we obtain

$$\boldsymbol{T}_{mn}y = t^n\boldsymbol{D}^m(t)y = t^n\int_C [\boldsymbol{D}^m(t)\boldsymbol{K}(s,t)]\,\boldsymbol{Y}(s)\,\mathrm{d}s \tag{5.90}$$

or

$$\boldsymbol{T}_{mn}y = \int_C (t^n)\,(as)^m\, \boldsymbol{K}(s,t)\boldsymbol{Y}(s)\,\mathrm{d}s \tag{5.91}$$

Substituting for $t^n\boldsymbol{K}(s,t)$ from Eq. 5.89, we obtain

$$\boldsymbol{T}_{mn}y = (a^{m-n}) \int_C (s^m\boldsymbol{Y}(s))\,(\boldsymbol{D}^n(s)\boldsymbol{K}(s,t))\,\mathrm{d}s \tag{5.92}$$

The Eq. 5.92 can be integrated by parts n times to solve for $\boldsymbol{T}_{mn}\,y$. Using integration by part once,

$$\begin{aligned}\boldsymbol{T}_{mn}y = (a^{m-n})\Big\{ &\left[(s^m\boldsymbol{Y}(s))\left(\boldsymbol{D}^{n-1}(s)\boldsymbol{K}(s,t)\right)\right]_{s=A}^{s=B} \\ &- \int_C [\boldsymbol{D}(s)\,(s^m\boldsymbol{Y}(s))]\left[\boldsymbol{D}^{n-1}(s)\boldsymbol{K}(s,t)\right]\mathrm{d}s\Big\}\end{aligned} \tag{5.93}$$

The solution is not very tractable for higher order equations. Fortunately, many equations in physics are only of second order and therefore Eq. 5.93 yields interesting results. Let us consider the special case for $n = 1$ and $k = 2$.
Thus, $m = 0, 1, 2$ and $n = 0, 1$

$$\boldsymbol{L}_t = (a_0 + a_1 t)\boldsymbol{D}^2(t) + (b_0 + b_1 t)\boldsymbol{D}(t) + (c_0 + c_1 t) = 0$$

$$a_{20} = a_0 a^2 \qquad a_{10} = b_0 a \qquad a_{00} = c_0$$

$$a_{21} = a_1 a^2 \qquad a_{11} = b_1 a \qquad a_{01} = c_1$$

$$\boldsymbol{L}_t y = \sum_{m=0}^{2}\sum_{n=0}^{1} a_{mn}\boldsymbol{T}_{mn}y = \sum_{m=0}^{2} a_{m0}T_{m0}y + \sum_{m=0}^{2} a_{m1}T_{m1}y = 0 \tag{5.94}$$

$$T_{m0}y = a^m \int_C s^m \boldsymbol{Y}(s)\boldsymbol{K}(s,t)\,\mathrm{d}s \tag{5.95}$$

$$T_{m1}y = a^{m-1}\left[s^m \boldsymbol{Y}(s)\boldsymbol{K}(s,t)\right]_{s=A}^{s=B} - a^{m-1}\int_C \left[\boldsymbol{D}(s)\left(s^m \boldsymbol{Y}(s)\right)\right]\boldsymbol{K}(s,t)\,\mathrm{d}s \tag{5.96}$$

Furthermore,

$$\sum_{m=0}^{2} a_{m0}(a)^m s^m = F_0(s) = a_0 a^2 s^2 + b_0 a s + c_0 \tag{5.97}$$

$$\sum_{m=0}^{2} a_{m1}(a)^{m-1} s^m = F_1(s) = \frac{1}{a}\left(a_1 a^2 s + b_1 a s + c_1\right) \tag{5.98}$$

Substituting Eq. 5.97 and Eq. 5.98 into Eq. 5.94

$$\begin{aligned} \boldsymbol{L}_t y(t) &= \left[F_1(s)\boldsymbol{Y}(s)\boldsymbol{K}(s,t)\right]_{t=A}^{t=B} \\ &+ \int_C \left[F_0(s)\boldsymbol{Y}(s) - \frac{\mathrm{d}}{\mathrm{d}s}\left(F_1(s)\boldsymbol{Y}(s)\right)\right]\boldsymbol{K}(s,t)\,\mathrm{d}s = 0 \end{aligned} \tag{5.99}$$

The solution $y(t)$ is of the form

$$y(t) = \int_C e^{ast}\boldsymbol{Y}(s)\,\mathrm{d}s, \quad \boldsymbol{K}(s,t) = e^{ast}$$

$$\boldsymbol{L}_t y = 0$$

The function $\boldsymbol{Y}(s)$ is determined from Eq. 5.99 as:

$$\frac{\mathrm{d}}{\mathrm{d}s}\left[F_1(s)\boldsymbol{Y}(s)\right] - F_0(s)\boldsymbol{Y}(s) = 0 \tag{5.100}$$

Subject to conditions

$$\left[F_1(s)\boldsymbol{Y}(s)e^{ast}\right]_{s=A}^{s=B} = 0 \tag{5.101}$$

Note: The original equation was a second order differential equation in the time domain while its solution is represented by a first order equation in a Laplace domain.

The solution of Eq. 5.100 along with Eq. 5.101 should provide two linearly independent solutions because the original equation was of the second order. In case we have two sets of endpoints (A_1, B_1) and (A_2, B_2) we choose two paths, C_1 involving (A_1, B_1) and C_2 involving (A_2, B_2), to obtain two independent solutions depending on a parameter λ.

$$y_1(t,\lambda) = \int_{C_1} e^{ast}\mathbf{Y}(s,\lambda)\,\mathrm{d}s$$

$$y_2(t,\lambda) = \int_{C_2} e^{ast}\mathbf{Y}(s,\lambda)\,\mathrm{d}s$$

For a closed curve C ($A = B$) the end condition, Eq. 5.101 is automatically fulfilled. In this case we have two possibilities:

1) Find a different path of integration or use the property of invariance by using a new variable and replace t with $\pm t$ yielding two independent solutions:

$$z = \gamma t, \qquad \gamma = \pm 1$$

Thus, $y(t,\lambda)$ and $y(-t,\lambda)$ are two independent solutions.

This is not always feasible. We shall see that for $\lambda = n$ (a positive integer), we may find that $y(t,\lambda)$ and $y(-t,\lambda)$ are not linearly independent.

2) Find a fundamental solution $y_1(t)$ and seek another independent solution of the form

$$y_2(t) = y_1(t)\int c(t)\,\mathrm{d}t$$

If $y_1(t)$ is a solution to the n-th order differential equation, then $y_2(t)$ is the solution to the $(n-1)$-th reduced order differential equation of the function $c(t)$. This is a general way to reduce degeneracy and produce new independent fundamental solutions. This method of producing other fundamental solutions is called the Reduction Procedure or the method of "variation of parameters." discussed in chapter 3 (Differential Equations). Let us illustrate in the next section, the above method by an example of generalized error function.

5.24 Generalized Error Function

Let us apply the Integral Transform method to the Error Function. Consider the following equation

$$L_t y = \ddot{y} + \alpha t\dot{y} + \beta y = 0 \tag{5.102}$$

$$y(s) = \int_C y(t)e^{ast}\, \mathrm{d}t$$

$$F_0(s) = a^2 s^2 + \beta$$

$$F_1(s) = \frac{1}{a}(\alpha a s) = \alpha s$$

From Eq. 5.100 and Eq. 5.101

$$\frac{\mathrm{d}}{\mathrm{d}s}\left[\alpha s \boldsymbol{Y}(s)\right] = \left(a^2 s + \beta\right)\boldsymbol{Y}(s)$$

$$\left[\alpha s \boldsymbol{Y}(s) e^{ast}\right]_{t=A}^{t=B} = 0$$

Let us choose

$a = 2, \quad \alpha = -2, \quad \beta = 2\lambda, \quad$ where λ is a real parameter, a variable.

Thus,

$$-2\boldsymbol{Y}(s,\lambda) - 2s\frac{\mathrm{d}\boldsymbol{Y}(s)}{\mathrm{d}s} = \left(4s^2 + 2\lambda\right)\boldsymbol{Y}(s,\lambda)$$

or

$$\frac{\mathrm{d}}{\mathrm{d}s}\boldsymbol{Y}(s,\lambda) = -\frac{\left(2s^2 + \lambda + 1\right)}{s}\boldsymbol{Y}(s,\lambda)$$

or

$$\ln \boldsymbol{Y}(s,\lambda) = -\int\left[\frac{2s^2 + \lambda + 1}{s}\right]\mathrm{d}s$$

$$\begin{aligned}\ln \boldsymbol{Y}(s,\lambda) &= -s^2 - (\lambda+1)\ln s + \text{ constant of integration}\\ &= -s^2 + \ln s^{-(\lambda+1)} + \text{ constant of integration}\end{aligned}$$

Simplifying

$$\boldsymbol{Y}(s,\lambda) = \left[s^{-(\lambda+1)}e^{-s^2}\right]K, \qquad K = \text{ constant of integration}$$

$$y(t,\lambda) = K\int_{c} e^{2st-s^2}s^{-(\lambda+1)}\,\mathrm{d}s$$

Eq. 5.101 takes the form

$$\left[e^{-\lambda}e^{(2st-s^2)}\right]_{s=A}^{s=B} = 0$$

Special Cases

(1) For $\lambda = 0,\ z = it$

$$y(t,0)\Big|_{t=\frac{z}{i}} = \operatorname{erf}(z) = \frac{2}{\sqrt{(\pi)}}\int_0^z e^{-s^2}\,\mathrm{d}s$$

(2) For $\lambda = -1$

$$y(t,-1) = \int_C e^{-s^2+2st}\,\mathrm{d}s$$

(3) $\lambda = n$ (an integer) and C, a closed curve encircling the origin in the s-plane. Redefining $y(t, \lambda)$ as $y(t, n)$. From Eqs. 5.86 and 5.87,

$$y(t, n) = K \oint_C e^{2st-s^2} s^{-(n+1)} \, \mathrm{d}s \tag{5.103}$$

The above integral has a pole of order $(n + 1)$ at $s = 0$.

Differentiating the above integral twice,

$$\ddot{y}(t, n) - 2t\dot{y}(t, n) + 2ny(t, n) = 0$$

In order to compute the above integral we use the **Cauchy's Residue Theorem** which states

$$\frac{\mathrm{d}^n F(z)}{\mathrm{d}z^n}|(z = z_0) = F^{(n)}(z_0) = \frac{n!}{2\pi j} \oint_C \frac{F(z)}{(z - z_0)^{n+1}} \, \mathrm{d}z$$

$F(z)$ is analytic inside C, n is an integer.

Since the integrand integral Eq. 5.103 is analytic inside C and satisfies the residue theorem condition, applying it to Eq. 5.103 yields:

$$y(t, n) = \left[\frac{\mathrm{d}^n}{\mathrm{d}s^n} \left(e^{2st-s^2} \right) \right]_{s=0}, \qquad K = \frac{n!}{2\pi j}$$

This set of functions $y(n, t)$ are n-th order **Hermite Polynomials** $\boldsymbol{H_n(t)}$.

$$y(0, t) = \boldsymbol{H}_0(t) \quad = 1$$

$$y(1, t) = \boldsymbol{H}_1(t) \quad = 2t$$

$$y(2, t) = \boldsymbol{H}_2(t) \quad = 4t^2 - 2$$

$\vdots$

$$y(2n,t) = \boldsymbol{H}_{2n}(t) \qquad = \boldsymbol{H}_{2n}(-t) \qquad \text{(even functions of time)}$$

$$y(2n+1,t) = \boldsymbol{H}_{2n+1}(t) \qquad = -\boldsymbol{H}_{2n+1}(-t) \qquad \text{(odd functions of time)}$$

Note: Only one fundamental solution is available due to the odd and even property of these functions. As suggested earlier, the second solution is either obtained by another integration path or by a method of the variation of parameters.

(4) If the parameter λ is a negative fractional number, we can obtain two independent solutions. The first solution takes the form:

$$y_1(t,\lambda) = K\int_C \frac{e^{-s^2+2st}}{s^{\lambda+1}}\,\mathrm{d}s = K\int_C I(s,t,\lambda)\,\mathrm{d}s \tag{5.104}$$

For the fractional negative λ there is a branch point at $s = 0$. Figure 5.35 shows a branch cut along the real axis in s-plane and the path C of integration consisting of C_1, the circle R of radius $\rho(\rho \to 0)$ and C_2. This gives us the first fundamental solution $y_1(t,\lambda)$, λ not an integer.

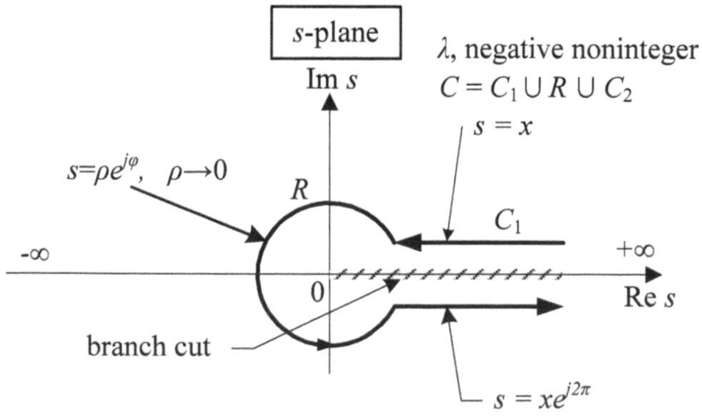

Figure 5.35: Integration Path, C (Fractional Negative λ) for Solution $y_1(t,\lambda)$

Computation of integral Eq. 5.104

- **Integration over R**

$$s = \rho e^{j\varphi}, \qquad \rho \to 0, \qquad 0 \geq \varphi \geq 2\pi$$

It is easy to see that

$$\lim_{\substack{s\to 0\\ \lambda>0}} |I(s,t,\lambda)| \simeq \lim_{\substack{s\to 0\\ \lambda>0}} \left|\frac{1}{s^{\lambda+1}}\right| \to \infty \tag{5.105}$$

Hence, integration over R diverges for $\lambda > 0$. Therefore, we shall deal only with negative noninteger values of λ, $|\lambda| > 1$.

- **Integration over C_1**

$$s = xe^{j\varphi_+} \qquad \varphi_+ = 0, \qquad 0 > x > \infty \tag{5.106}$$

$$s^{-(\lambda+1)} = x^{-(\lambda+1)}$$

- **Integration over C_2**

$$s = xe^{j\varphi_-} \qquad \varphi_- = 2\pi, \qquad 0 > x > \infty$$

$$s^{-(\lambda+1)} = x^{-(\lambda+1)}e^{-j(\lambda+1)(2\pi)} = x^{-(\lambda+1)}e^{-j\lambda 2\pi} \tag{5.107}$$

Thus, the first fundamental solution

$$\begin{aligned} y_1(t,\lambda) &= -K\int_0^\infty \left(e^{(2xt-x^2)}\right)\left(x^{-(\lambda+1)}\right) \mathrm{d}x + K\int_0^\infty e^{(2xt-x^2)}x^{-(\lambda+1)}e^{-j\lambda 2\pi}\,\mathrm{d}x \\ &= -2Ki\sin(\pi\lambda)e^{-j\pi\lambda}\int_0^\infty x^{-(\lambda+1)}e^{(2xt-x^2)}\,\mathrm{d}x, \quad \lambda \notin N_+, |\lambda| > 1 \end{aligned} \tag{5.108}$$

Note: When λ is a negative integer the solution $y_1(t, \lambda) \equiv 0$ For the second solution $y_2(t, \lambda)$, we shall use the mirror image of the path C with respect to the y-axis in the s-plane as shown in Figure 5.36.

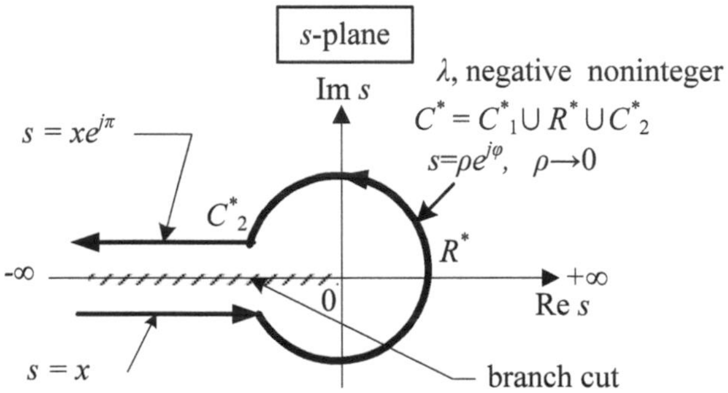

Figure 5.36: Integration Path, C^* for Solution $y_2(t, \lambda)$ of Eq. 5.104

Thus, the second fundamental solution is given by $y_2(t, \lambda) = y_1(-t, \lambda)$

(5) **Approximation for** $|t| \to 0, \lambda \notin N_+$

Let e^{2xt} be approximated as

$$e^{2xt} \simeq \sum_{n=0}^{\infty} (2t)^n x^n \qquad \text{for } |t| \text{ small}$$

The solution $y_1(t, \lambda)$ gets simplified as

$$y_1(t, \lambda) = B_0(\lambda) \sum_{n=0}^{\infty} \frac{(2t)^n}{n!} A_n(\lambda) \qquad |t| \to 0$$

$$B_0(\lambda) = (-2Ki \sin \pi\lambda) e^{-j\pi\lambda}, \qquad A_n(\lambda) = \int_0^{\infty} x^{n-\lambda-1} e^{-x^2} \, dx$$

Bibliography

[Abramowitz, M.] Abramowitz, M., and Stegun, I.A. *Handbook of Mathmatical Functions with Formulae, Graphs and Mathematical Tables,* New York: Dover, 1019–1030, 1972.

[Akhiezer, N.I.] Akhiezer, N.I., and Glazman, I.M. *Theory of Linear Operators in Hilbert Space,* New York: Dover,1993

[Bracewell, R.N.] Bracewell, R.N. *The Fourier Transform and Its Applications,* 3rd ed., Boston: McGraw Hill, 2000.

[Oberhettinger, F.] Oberhettinger, F. *Tables of Laplace Transforms,* New York: Springer-Verlag, 1973.

[Plaschko, P.] Plaschko, P. and Brod, K. *Here Mathematische Methoden fur Ingenier und Physiker,* Berlin Heidelberg: Springer-Verlag, 1989.

Chapter 6

Digital Systems, Z-Transforms, and Applications

6.1 Introduction

Information technology convergence continues due to progress in digital techniques. Music, movies, the Internet, and HDTV are all becoming digital media. The preference to digital is that instead of accessing every magnitude of analog signal, all digital signals have to do is differentiate between 1 and 0 (high and low, or on and off). For every device, the general principles from analog to digital conversion are the same. The process involves sampling and digitization. Analog to digital converters (ADC) do this task of sampling and digitization. The clock frequency at which the signal is sampled is crucial to the accuracy of analog to digital conversion. The reverse process involves digital to analog conversion (DAC). The output of DAC is a series of analog signals that are of the staircase type, involving the signal to be constant between two sampled instances (higher order holds can further smooth this process if necessary).

Sampled or discrete systems are modeled by the difference equations. In

Chapter 2 we showed the analogy between the solutions of differential and difference equations. The same analogy exists between Laplace (or Fourier) Transforms and Z-Transforms involving the discrete systems. We will treat the subject of discrete systems in detail in the following pages.

6.2 Discrete Systems and Difference Equations

A continuous system has continuous inputs as well as continuous outputs. By contrast, a discrete system is defined as one in which both the inputs and the outputs, known as system signals, vary only at discrete equally spaced moments of time, known as sampling instances. We may consider these signals to be constant in between these intervals. The continuous systems are described by the differential equations, whereas the discrete systems are described by the difference equations as demonstrated in Figure 6.1

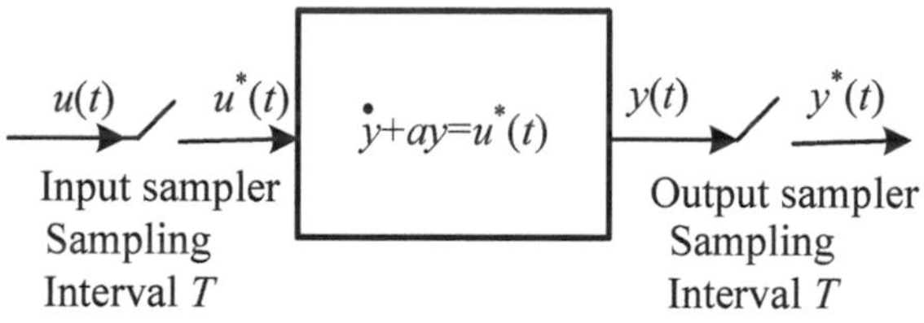

Figure 6.1: Continuous System with Sampler at the Input and the Output

Let the continuous system be described by a simple differential equation:

$$\dot{y} + ay = u^*(t), \qquad y(0) = y_0 \tag{6.1}$$

$$\left.\begin{aligned} u^*(t) &= \text{sampled system input} = \sum_{m=0}^{n-1} u(mT)\delta(t - mT) \\ y^*(t) &= \text{sampled system output} = \sum_{m=0}^{n} y(mT)\delta(t - mT) \end{aligned}\right\} \quad \text{For } t \le nT \tag{6.2}$$

The solution of Eq. 6.1 is

$$y(t) = e^{-at}\left[y(0) + \sum_{m=0}^{n-1} u(mT)\int_{0}^{t} e^{a\tau}\delta(\tau - mT)\,\mathrm{d}\tau\right]$$

or

$$y(t) = e^{-at}\left[y(0) + \sum_{m=0}^{n-1} u(mT)e^{amT}\right] \tag{6.3}$$

$$y(nT) = e^{-anT}\left[y(0) + \sum_{m=0}^{n-1} u(mT)e^{amT}\right] \tag{6.4}$$

$$y((n+1)T) = e^{-a(n+1)T}\left[y(0) + \sum_{m=0}^{n-1} u(mT)e^{amT} + u(nT)e^{anT}\right]$$

$$y((n+1)T) = e^{-aT}\left[e^{-anT}\left(y(0) + \sum_{m=0}^{n-1} u(mT)e^{amT}\right) + u(nT)\right] \tag{6.5}$$

Thus,

$$y((n+1)T) = e^{-aT}y(nT) + e^{-aT}u(nT) \tag{6.6}$$

Thus, a sampler at the input and the output of a continuous system results in a difference equation description. It is often convenient to normalize the time as $\hat{t} = t/T$. Let the sampled value of a general continuous signal $f(\hat{t})$ be represented as $f(n)$ at $t = nT$, $\hat{t} = n$. From here on, without loss of generality, we will consider the time to be normalized with respect to T. Figure 6.2 shows the discrete system.

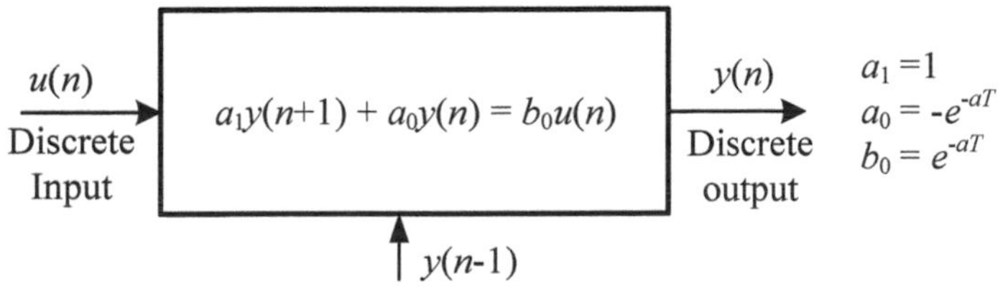

Figure 6.2: Equivalent Discrete System

Important Note:

- $u^*(\hat{t})$ can be considered as a discrete sequence $\{u(n)\}_{n=0}^{\infty}$.
- $u^*(\hat{t})$ can also be considered as a sequence of time impulses $\sum_{n=0}^{\infty} u(\hat{t})\delta(\hat{t} - n)$.
- Both descriptions are useful depending upon the application.

6.2.1 *k*-th Order Difference of a Discrete Function

1. First order difference of a discrete function $f(n)$ is defined as:

$$\Delta f(n) = f(n+1) - f(n)$$

2. Second order difference of a discrete function equals:

$$\Delta^2 f(n) = \Delta f(n+1) - \Delta f(n) = f(n+2) - 2f(n+1) + f(n)$$

3. k-th order difference equals:

$$\Delta^k f(n) = \Delta^{k-1} f(n+1) - \Delta^{k-1} f(n)$$

Using the expression above, it is easy to show that

4.
$$\Delta^k f(n) = \sum_{\nu=0}^{k} (-1)^{\nu} \binom{k}{\nu} f(n+k-\nu) \tag{6.7}$$

where

$$\binom{k}{\nu} = \text{binomial coefficient} = \frac{k!}{\nu!(k-\nu)!}$$

A k-th order LTI Discrete system describes a sequence $\{u(n+i)\}_{i=0}^{k-1}$, $\{y(n+i)\}_{i=0}^{k-1}$, which results in a discrete output $y(n+k)$ is given by

$$\sum_{i=0}^{k} a_i y(n+i) = \sum_{i=0}^{k-1} b_i u(n+i), \quad n = 1, 2, \ldots \tag{6.8}$$

$y(i), \quad i = 0, 1, \ldots, k-1$ are considered as known initial values

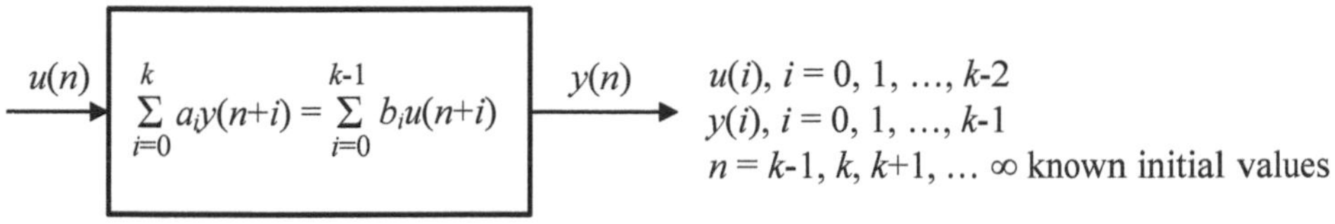

Figure 6.3: k-th order Discrete System Description

Another way to describe this dicrete system is via higher order differences,

$$\sum_{i=0}^{k} c_i \Delta^i y(n) = \sum_{i=0}^{k-1} d_i \Delta^i u(n) \tag{6.9}$$

$\Delta^i y(0), \quad i = 0, 1, \ldots, k-1$ are considered as known initial differences

$$a_{k-i} = \sum_{\nu=0}^{i} c_{k-\nu} (-1)^{i-\nu} \binom{k-\nu}{i-\nu}$$

$$c_{k-i} = \sum_{\nu=0}^{i} a_{k-\nu} \binom{k-\nu}{i-\nu}$$

$$b_{k-1-i} = \sum_{\nu=0}^{i} d_{k-1-\nu} (-1)^{i-\nu} \binom{k-1-\nu}{i-\nu}$$

$$d_{k-1-i} = \sum_{\nu=0}^{i} b_{k-1-\nu} \binom{k-1-\nu}{i-\nu}$$

6.2.2 Building Blocks of the Discrete Systems

In general, there are four building blocks for the discrete systems.

(i) **The Delay Element**

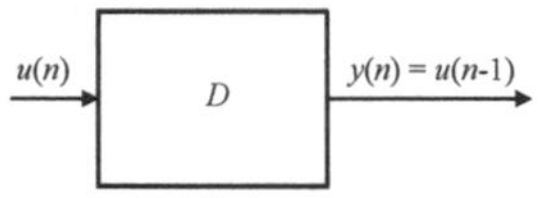

Figure 6.4: Delay Element

(ii) **The Adder Element**

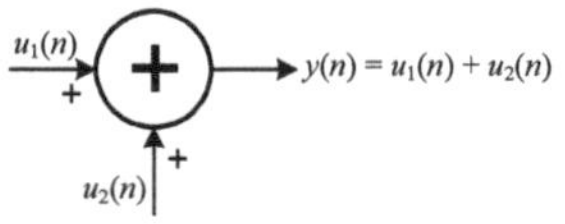

Figure 6.5: Adder Element

(iii) **The Gain Element**

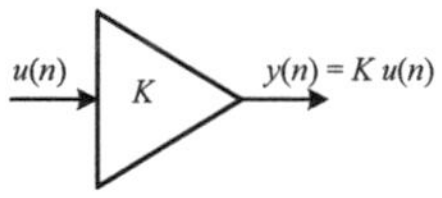

Figure 6.6: Gain Element

(iv) **The Accumulator**

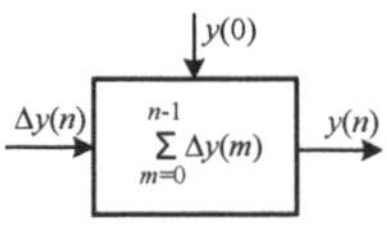

Figure 6.7: Accumulator

6.3 Realization of a General Discrete System

$$u(n) = \sum_{\nu=0}^{k} a_\nu x(n+\nu) = \sum_{i=0}^{k} a_i x(n+i) \tag{6.10}$$

$$y(n) = \sum_{\nu=0}^{k-1} a_\nu x(n+\nu) = \sum_{i=0}^{k-1} b_i x(n+i) \tag{6.11}$$

Proof: From Eq. 6.11

$$\sum_{i=0}^{k} a_i y(n+i) = \sum_{i=0}^{k}\sum_{\nu=0}^{k-1} a_i b_\nu x(n+\nu+i) = \sum_{\nu=0}^{k-1} b_\nu \left[\sum_{i=0}^{k} a_i x(n+i+\nu)\right] \tag{6.12}$$

From Eq. 6.10, $$\sum_{i=0}^{k} a_i x(n+i+\nu) = u(n+\nu) \tag{6.13}$$

From Eqs. 6.12 and 6.13, $$\sum_{i=0}^{k} a_i y(n+i) = \sum_{\nu=0}^{k-1} b_\nu u(n+\nu) = \sum_{i=0}^{k-1} b_i u(n+i)$$

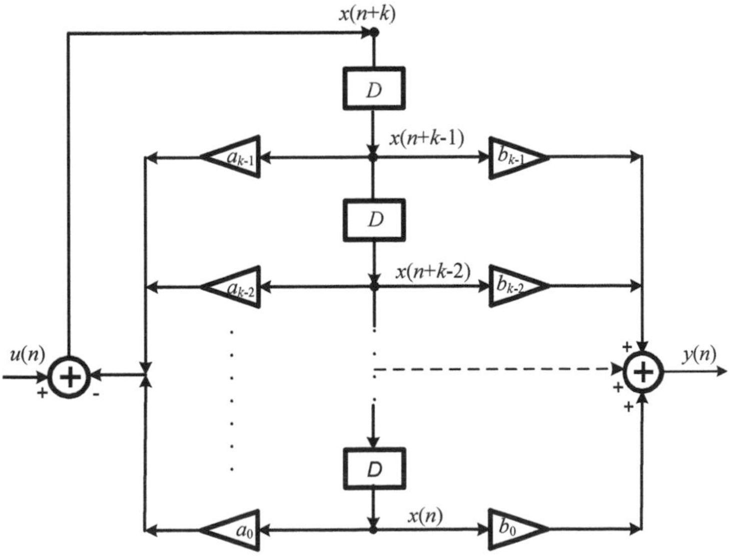

Figure 6.8: Realization of a General Discrete System

Thus Eqs. 6.10, and 6.11 are equivalent to Eq. 6.8. Eqs. 6.10 and 6.11 are realized as a block diagram in Figure 6.8. We assume $a_k = 1$ as a normalizing constant.

6.4 Z-Transform for the Discrete Systems

Just as the Laplace Transform is a convenient tool to study the behavior of the continuous systems, the Z-Transform represents a convenient and organized method for studying the discrete systems.

Definition of Z-Transform

Given a function $f(\hat{t})$, its **Laplace Transform** is defined as

$$L[f(\hat{t})] = \int_0^\infty f(\hat{t})e^{-s\hat{t}}\, d\hat{t}, \qquad \left(\hat{t} \text{ is normalized with respect to } T \text{ i.e., } \hat{t} = \frac{t}{T}\right)$$

($f(t)$ and $f(\hat{t})$ will be used interchangeably.)

Let $f(\hat{t})$ be discretized to yield $f(n)$. The Laplace Transform of this discretized function is:

$$L[f^*(\hat{t})] = \int_0^\infty \sum_{n=0}^{\infty} f(n)\delta(\hat{t} - n)e^{-s\hat{t}}\, d\hat{t} = \sum_{n=0}^{\infty} f(n)e^{-ns}$$

Let $z = e^s$, a new complex variable.

$$\text{One-sided Z-Transform,} \quad F(z) = Z[f(n)] = \sum_{n=0}^{\infty} f(n)z^{-n} \tag{6.14}$$

$$\text{Two-sided Z-Transform,} \quad F_b(z) = Z_b[f(n)] = \sum_{n=-\infty}^{+\infty} f(n)z^{-n} \tag{6.15}$$

At present we are only interested in single-sided Z-Transforms. We shall discuss the double-sided Laplace Transform in the later sections.

For the Eq. 6.14 to be useful it is necessary that the above series should converge. For the reasons of convergence, we shall consider only functions $f(n)$ of the form

$$|f(n)| < MR^n, \qquad n \geq 0, R > 0, M > 0 \tag{6.16}$$

To ensure the convergence of the summation in Eq. 6.14, we shall apply a ratio test that states that,

$$\left\| \left[\frac{z^{-(n+1)} f(n+1)}{z^{-n} f(n)} \right] \right\| < 1, \quad n > 0 \tag{6.17}$$

The inequality Eq. 6.17 implies that the infinite series converges in the region in the z-plane shown in Figure 6.9.

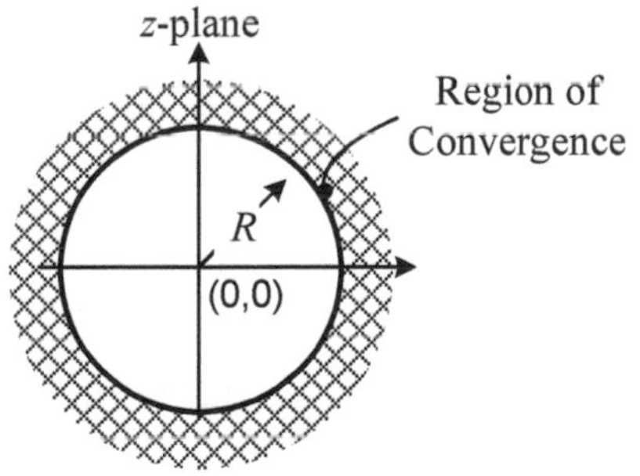

Figure 6.9: Z-Transform Convergence Domain of Exponential Growth Functions

Thus,

$$|R/z| < 1, \quad |z| > R \text{ (radius of convergence)}, \quad |f(n)| \leq MR^n$$

Example 6.1:

Evaluate the Z-Transform of:

$$f(n) = u(n) = \begin{cases} 1 & n \geq 0 \\ 0 & \text{otherwise} \end{cases} \qquad \text{(discrete step function)}$$

Solution:

$$Z[u(n)] \triangleq \sum_{n=0}^{\infty} u(n)z^{-n} = \sum_{n=0}^{\infty} z^{-n} = \frac{1}{1-z^{-1}} = \frac{z}{z-1}$$

Region of convergence is $\quad |z| > R > 1$

Example 6.2:

Evaluate the Z-Transform of:

$$f(n) = e^{-\alpha n}u(n)$$

Solution:

$$Z[f(n)] = \sum_{n=0}^{\infty} e^{-\alpha n}z^{-n} = \sum_{n=0}^{\infty} \left(e^{-\alpha n}z^{-1}\right)^{n} = \frac{1}{1-e^{-\alpha}z^{-1}} = \frac{z}{z-e^{-\alpha}}$$

$$|z| > e^{-\alpha}$$

Important:

Single-sided Transforms are used for causal functions, which are identically equal to zero for negative value of the argument n. The function $f(n) \equiv 0,\ n < 0$ is a causal function.

6.5 Fundamental Properties of Z-Transforms

Table 6.1 represents a collection of Z-Transform properties and the rules which enable Transforms of various discrete functions to be evaluated in a very easy manner. Except for algebraic manipulations, the Transforms of most functions requires summing the series of the form

$$S_N = \sum_{\nu=0}^{N} z^{-\nu}, \qquad |z^{-1}| < 1 \tag{6.18}$$

S_N can be determined as:

$$S_N = 1 + z^{-1} + z^{-2} + \ldots + z^{-N}$$

$$z^{-1}S_N = z^{-1} + z^{-2} + \ldots + z^{-N} + z^{-(N+1)}$$

or

$$\left(1 - z^{-1}\right) S_N = 1 - z^{-(N+1)}$$

$$S_N = \frac{1 - z^{-(N+1)}}{1 - z^{-1}}$$

$$\lim_{\substack{N\to\infty \\ |z^{-1}|<1}} z^{-(N+1)} = 0$$

$$S_\infty = \sum_{n=0}^{\infty} z^{-n} = \frac{1}{1 - z^{-1}} = \frac{z}{z - 1}$$

Another relation that is very useful in various proofs is:

$$\frac{1}{2\pi j} \oint_c z^k \, \mathrm{d}z = \begin{cases} 0 & k = 0, +1, \pm 2, \pm 3, \ldots \\ 1 & k = -1 \end{cases}$$

where c is a circle with the region of convergence surrounding the origin.

Table 6.1: Z-Transform Properties

Property	Function $f(n), n > 0$ $f(n) = \oint\limits_{\substack{c\to R \\ \lvert z\rvert>R}} \frac{F(z)z^{n-1}}{2\pi j} \mathrm{d}z$	Z-Transform of $f(n)$ $F(z) = \sum\limits_{n=0}^{\infty} f(n)z^{-n}$ $\lvert f(n)\rvert < MR^n,\ M > 0$
Linearity	$f_\nu(n), \quad \nu = 1, \cdots, k$ $f(n) = \sum\limits_{\nu=1}^{k} a_\nu f_\nu(n)$	$F_\nu(z), \quad \nu = 1, \cdots, k$ $F(z) = \sum\limits_{\nu=1}^{k} a_\nu F_\nu(z)$
Time Delay	$f(n \pm k)$	$z^{\pm k}[F(z) \mp \sum\limits_{\nu=0}^{k-1} z^{\mp\nu} f(\pm\nu)]$

(continued)

Property	Function $f(n), n > 0$ $f(n) = \oint\limits_{\substack{c \to R \\ \lvert z\rvert > R}} \frac{F(z)z^{n-1}}{2\pi j}\,\mathrm{d}z$	Z-Transform of $f(n)$ $F(z) = \sum\limits_{n=0}^{\infty} f(n)z^{-n}$ $\lvert f(n)\rvert < MR^n,\ M > 0$
Scaling Variable z	$a^{\pm n} f(n)$	$F(a^{\mp 1}z)$
Multiplication by $n!/k!$	$\frac{n!}{k!} f(n)$	$\left(-z\frac{\mathrm{d}}{\mathrm{d}z}\right)^k F(z)$
Transform of Differences	$\Delta f(n) = f(n+1) - f(n)$	$(z-1)F(z) - zf(0)$
	$\Delta^2 f(n) = \Delta f(n+1) - \Delta f(n)$	$(z-1)^2 F(z) - z(z-1)f(0)$ $-z\Delta f(0)$
	$\Delta^k f(n) = \Delta^{k-1} f(n+1)$ $- \Delta^{k-1} f(n)$	$(z-1)^k F(z)$ $-z\left[\sum\limits_{\nu=0}^{k-1} (z-1)^{k-\nu-1} \Delta^\nu f(0)\right]$ $\Delta^0 f(0) = f(0)$
	$F(z) = \frac{z}{z-1} \sum\limits_{\nu=0}^{k-1} \frac{\Delta^\nu f(0)}{(z-1)^\nu} + \frac{1}{(z-1)^k} Z\left[f(n)\right]$	
Inverse Transform	$f(n) = \frac{1}{2\pi j} \oint\limits_{\lvert z\rvert > R} F(z)z^{n-1}dz$	
Infinite Sum	$\sum\limits_{n=0}^{\infty} f(n)$	$\lim\limits_{z \to 1} F(z)$
Finite Sum	$\sum\limits_{m=0}^{n-1} f(m)$	$\frac{F(z)}{z-1}$
Time Convolution	$f_1(n) * f_2(n) = \sum\limits_{m=0}^{n} f_1(m) f_2(n-m)$ $= \sum\limits_{m=0}^{n} f_1(n-m) f_2(m)$	$F_1(z)F_2(z)$
z-Domain Convolution	$f_1(n) f_2(n)$	$\frac{1}{2\pi j} \oint F_1(\lambda) F_2(z/\lambda) \frac{\mathrm{d}\lambda}{\lambda}$
Initial and Final Value Theorem	$f(0) = \lim\limits_{z \to \infty} F(z)$ $f(\infty) = \lim\limits_{z \to 1} (z-1)F(z)$	

(continued)

Property	Function $f(n), n > 0$ $f(n) = \oint\limits_{\substack{c \to R \\ \lvert z\rvert > R}} \frac{F(z)z^{n-1}}{2\pi j} \mathrm{dz}$	Z-Transform of $f(n)$ $F(z) = \sum\limits_{n=0}^{\infty} f(n)z^{-n}$ $\lvert f(n)\rvert < MR^n,\ M > 0$
"z" is known as shift operator $z \to$ Forward Shift $z^{-1} \to$ Backward shift $f(n+k) = k$ step Forward Shifted Function $f(n)$ $f(n-k) = k$ step Backward Shifted Function $f(n)$		

Important: $(z - 1)$ plays a similar role in the z-Transform as s the in Laplace Transform. Most of the properties in Table 6.1 follow directly from the definition except for some algebra. We shall prove a few of these properties that may not be obvious.

Property 1: Linearity

Z-Transform is a linear operator, implying

$$Z[f(n)] = Z\left[\sum_{\nu=1}^{k} a_\nu f_\nu(n)\right] = \sum_{\nu<1}^{k} a_\nu Z[f_\nu(n)] = \sum_{\nu=1}^{k} a_\nu F_\nu(z)$$

Property 2: Time Delay Property of a Discrete Sequence

If

$f(n) \leftrightarrow F(z)$ (reads "Z-Transform of $f(n)$ is $F(z)$")

Then

$$f(n+k) \leftrightarrow z^k \left[F(z) - \sum_{\nu=0}^{k-1} z^{-\nu} f(\nu)\right] \tag{6.19}$$

Proof:

$$Z[f(n+k)] = \sum_{n=0}^{\infty} z^{-n} f(n+k), \qquad (\text{let } n+k=\nu)$$

$$= \sum_{\nu=k}^{\infty} z^{-(\nu-k)} f(\nu) = z^k \left[\sum_{\nu=0}^{\infty} z^{-\nu} f(\nu) - \sum_{\nu=0}^{k-1} z^{-\nu} f(\nu)\right]$$

$$= z^k \left[F(z) - \sum_{\nu=0}^{k-1} z^{-\nu} f(\nu)\right]$$

$$f(n+k) \leftrightarrow z^k \left[F(z) - \sum_{\nu=0}^{k-1} z^{-\nu} f(\nu)\right] \tag{6.20}$$

If, $f(0) = f(1) = \ldots = f(k-1) = 0$

Then, $f(n+k) \leftrightarrow z^k F(z)$ k-step forward shift operator

Similarly,

$$f(n-k) \leftrightarrow z^{-k} \left[F(z) + \sum_{\nu=0}^{k-1} z^{\nu} f(-\nu)\right] \tag{6.21}$$

If, $f(-1) = f(-2) = \ldots = f(-k) = 0$

Then, $f(n-k) \leftrightarrow z^{-k} F(z)$ k-step backward shift operator

Example 6.3: Determine the Z-Transform of $f(n) = u(n-k)$.

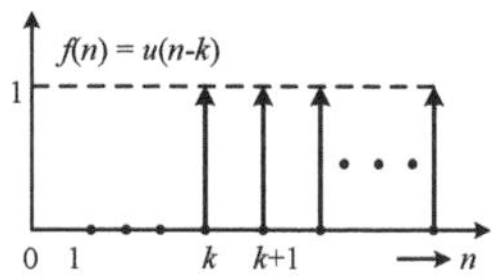

Figure 6.10: Z-Transform of a Delayed Step Function

Solution:

Initial conditions $f(-\nu) = 0, \quad \nu = 1, 2, \ldots, k$

But

$$f(n) = u(n) \leftrightarrow \frac{z}{z-1}$$

Therefore,

$$f(n-k) = u(n-k) \leftrightarrow z^k\left(\frac{z}{z-1}\right)$$

Example 6.4: Determine the Z-Transform of:

$$f(n) = \delta(n-k)$$

Solution:

$$\delta(n) \leftrightarrow 1$$

From Eq. 6.21, $\delta(n-k) \leftrightarrow z^{-k}$

Example 6.5: Compute thc Z-Transform of:

$$f(n) = e^{-\alpha(n-k)}u(n-k)$$

Solution:

$$u(n) \leftrightarrow \frac{1}{1-z^{-1}}$$

From the scaling variable z property, of Table 6.1

$$e^{-\alpha n}u(n) \leftrightarrow \frac{1}{1-(ze^{\alpha})^{-1}} = \frac{z}{z-e^{-\alpha}}$$

From the time delay property of Table 6.1

$$e^{-\alpha(n-k)}u(n-k) \leftrightarrow \frac{z^{-k}z}{z-e^{-\alpha}} = \frac{z^{-k+1}}{z-e^{-\alpha}}$$

Property 5: Transform of Differences of the Discrete Sequence

This is also known as the Euler-Abel Transformation. Via this transformation we are able to accelerate the convergence of the series provided the series is decaying fast for large values of n.

Let

$$\Delta f(n) = f(n+1) - f(n)$$

Taking the Z-Transform of both sides

$$Z[\Delta f(n)] = z[F(z) - f(0)] - F(z) = (z-1)F(z) - zf(0) \tag{6.22}$$

Similarly

$$\Delta^2 f(n) = \Delta f(n+1) - \Delta f(n)$$

From Eq. 6.22 and the time delay property

$$\begin{aligned} Z\left[\Delta^2 f(n)\right] &= z\,[Z[\Delta f(n)] - \Delta f(0)] - z\,[Z[\Delta f(n)]] \\ &= (z-1)\,[Z[\Delta f(n)] - z\Delta f(0)] \\ &= (z-1)\,[(z-1)F(z) - zf(0)] - z\Delta f(0) \end{aligned}$$

$$\Delta^2 f(n) \leftrightarrow (z-1)^2 F(z) - z\,[(z-1)f(0) - \Delta f(0)]$$

Repeating the above process

$$\Delta^k f(n) \leftrightarrow (z-1)^k F(z) - z\sum_{\nu=0}^{k-1}(z-1)^{k-1-\nu}\Delta^\nu f(0) \tag{6.23}$$

In case $f(n)$ is a complicated function, but its k-th order difference is a simple function, then $Z[f(n)]$ is easily computed as

$$F(z) = \frac{1}{(z-1)^k}\left[Z\left[\Delta^k f(n)\right]\right] + \frac{z}{z-1}\sum_{\nu=0}^{k-1}\frac{\Delta^\nu f(0)}{(z-1)^\nu} \tag{6.24}$$

Now we can determine the smallest value of k for which

$$\Delta^k f(n) = 0$$

resulting in

$$F(z) = \frac{z}{z-1} \sum_{\nu=0}^{k-1} \frac{\Delta^\nu f(0)}{(z-1)^\nu} \tag{6.25}$$

In case, all $\Delta^\nu f(0)$ are 0, $\nu = 0, \ldots, k-1$ then

$$Z\left[\Delta^k f(n)\right] = (z-1)^k F(z)$$

Example 6.6: Compute the Z-Transform of $f(n) = n$

Solution:

$$\begin{aligned} & f(n) = n && \Delta^0 f(0) = f(0) = 0 \\ & \Delta f(n) = 1 && \Delta f(0) = 1 \\ & \Delta^k f(n) = 0, \quad k \geq 2 && \Delta^\nu f(0) = 0, \quad \nu > 1 \end{aligned}$$

From Eq. 6.25, $k = 2$

$$F(z) = \left(\frac{z}{z-1}\right)\left[f(0) + \frac{\Delta f(0)}{(z-1)}\right] = \left(\frac{z}{z-1}\right)\left[0 + \frac{1}{z-1}\right] = \frac{z}{(z-1)^2}$$

Example 6.7: Compute the Z-Transform of $f(n) = n^2$

Solution:

$$\begin{aligned} & f(n) = n^2 && \Delta^0 f(0) = f(0) = 0 \\ & \Delta f(n) = (n+1)^2 - n^2 = 2n + 1 && \Delta f(0) = 1 \\ & \Delta^2 f(n) = [2(n+1) + 1] - [2n + 1] = 2 && \Delta^2 f(0) = 2 \\ & \Delta^k f(n) = 0, \quad k > 3 && \Delta^\nu f(0) = 0, \quad \nu > 2 \end{aligned}$$

Using Eq. 6.25, for $k = 3$

$$F(z) = \frac{z}{(z-1)}\left[\Delta f(0) + \frac{\Delta f(0)}{(z-1)} + \frac{\Delta^2 f(0)}{(z-1)^2}\right]$$
$$= \frac{z}{(z-1)^2}\left[0 + 1 + \frac{2}{z-1}\right] = \frac{z(z+1)}{(z-1)^3}$$

Example 6.8: Evaluate

$$F(z) = \sum_{n=0}^{\infty}(n+1)^2 z^{-n}, \qquad |z^{-1}| < 1$$

Solution

$$f(n) = n^2 + 2n + 1$$

$$f(0) = 1, \qquad f(1) = 4, \qquad f(2) = 9$$
$$\Delta f(0) = 3, \qquad \Delta f(1) = 5, \qquad \Delta f(2) = 0$$
$$\Delta^2 f(0) = 2, \qquad \Delta^2 f(1) = 0, \qquad \Delta^2 f(2) = 0$$

$$F(z) = \frac{1}{1-z^{-1}} + \frac{3z^{-1}}{1-z^{-1}} + \frac{2z^{-2}}{(1-z^{-1})^2}$$

It is easy to show that

$$\binom{n}{m} \leftrightarrow \frac{z}{(z-1)^{m+1}} \tag{6.26}$$

Property 6: Z-Transform Inverse Written as $Z^{-1}[F(z)]$

$$Z^{-1}[F(z)] \equiv f(n) = \frac{1}{2\pi j}\oint_{|z|>R} F(z)z^{n-1}\,dz$$

Proof:

$$F(z) = \sum_{m=0}^{\infty} f(m)z^{-m}$$

Multiplying both sides by z^{n-1} and integrating around a circle c with the center at the origin and radius $|z| > R$

$$\oint_{|z|\geq R} F(z)z^{n-1}\,\mathrm{d}z = \sum_{m=0}^{\infty} f(m) \oint_{|z|\geq R} z^{n-m-1}\,\mathrm{d}z$$

But

$$\oint_{|z\geq R|} z^{n-m+1}\,\mathrm{d}z = \begin{cases} 2\pi j, & n = m \\ 0, & n \neq m \end{cases}$$

Thus,

$$\frac{1}{2\pi j}\oint_{|z|\geq R} f(z)z^{n-1}\,\mathrm{d}z = f(n) \tag{6.27}$$

It should be realized that $F(z)$ is analytic for $|z| > R$ and therefore it can be expanded in the Taylor series at points outside $|z| > R$ and the various coefficients of expansion represents the function $f(n)$.

Note: Many books define analytic functions that can be represented by a convergent Taylor series inside the unit circle and therefore an analytic function inside a unit circle is represented by $\sum_{n=0}^{\infty} a_n z^n$. This may result in confusion but can be avoided if we realize that in Z-Transforms the complex variable functions are analytic outside the unit circle while in H_p problems (Hardy spaces) the analytic region is inside the unit circle/disk (Disk Algebra).

Property 7: Transform of the Infinite Sum of the Discrete Sequence, $\sum_{n=0}^{\infty} f(n)$

From the definition

$$F(z) = \sum_{n=0}^{\infty} f(n)z^{-n}$$

Letting $z \to 1$

$$\sum_{n=0}^{\infty} f(n) = \lim_{z \to 1} F(z) \tag{6.28}$$

Property 8: Finite Sum of a Discrete Sequence

$$g(n) = \sum_{m=0}^{n-1} f(m), \quad g(0) = 0$$

$$\Delta g(n) = g(n+1) - g(n) = \sum_{n=0}^{n} f(m) - \sum_{m=0}^{n-1} f(m) = f(n)$$

Taking the Z-Transform of both sides of the equation

$$F(z) = Z\left[\Delta g(n)\right] = (z-1)Z[g(n)] - zg(0)$$

$$Z[g(n)] = Z\left[\sum_{m=0}^{n-1} f(m)\right] = \frac{F(z)}{z-1} \tag{6.29}$$

$$\sum_{m=0}^{n-1} f(n) = Z^{-1}\left[\frac{F(z)}{(z-1)}\right]$$

Hence, Eq. 6.29 is useful in the summing of certain series.

Example 6.9: Evaluate:

$$g(n) = \sum_{m=1}^{n} m = \sum_{m=0}^{n-1} (m+1)$$

Solution:

It is easy to see that

$$g(n) = \frac{n(n+1)}{2} = \binom{n+1}{2}$$

Let us use Eq. 6.29 to find $g(n)$.

The function $f(n)$ is:

$$f(n) = n + 1 = \Delta g(n) = \binom{n+1}{2} - \binom{n}{2}$$

$$F(z) = \frac{z}{(z-1)^2} + \frac{z}{z-1}$$

$$\frac{F(z)}{z-1} = \frac{z}{(z-1)^3} + \frac{z}{(z-1)^2} \leftrightarrow \binom{n}{2} + \binom{n}{1} = \binom{n+1}{2}$$

Example 6.10: Evaluate:

$$g(n) = \sum_{m=0}^{n} m^2 = \sum_{m=0}^{n-1} (m+1)^2$$

Solution:

$$f(n) = (n+1)^2 = \left(n^2 + 2n + 1\right) = \Delta g(n)$$

$$\left(n^2 + 2n + 1\right) \leftrightarrow F(z) = \frac{z(z+1)}{(z-1)^3} + \frac{2z}{(z-1)^2} + \frac{z}{z-1}$$

Thus,

$$g(n) = \sum_{m=0}^{n} m^2 \leftrightarrow \frac{F(z)}{z-1} = \frac{z(z+1)}{(z-1)^4} + \frac{2z}{(z-1)^3} + \frac{z}{(z-1)^2}$$
$$= z\left[\frac{z}{(z-1)^4}\right] + \frac{z}{(z-1)^4} + \frac{2z}{(z-1)^3} + \frac{z}{(z-1)^2}$$

Taking the Z-Transform inverse

$$\sum_{m=0}^{n-1} m^2 = \binom{n+1}{3} + \binom{n}{3} + 2\binom{n}{2} + \binom{n}{1}$$
$$= \frac{(n+1)(n)(n-1)}{6} + \frac{n(n-1)(n-2)}{6} + n(n-1) + n$$
$$= \frac{n(n+1)(2n-1)}{3!}$$

Property 9: Time Convolution of Discrete Causal Sequence

This property has very important application in determing the response of a discrete system to various inputs. We are required to show:

If

$$F_1(z) \leftrightarrow f_1(n), \quad F_2(z) \leftrightarrow f_2(n)$$

Then

$$\begin{aligned} F_1(z)F_2(z) \leftrightarrow f_1(n) * f_2(n) &= \sum_{m=0}^{n} f_1(m) f_2(n-m) \\ &= \sum_{m=0}^{n} f_1(n-m) f_2(m) \end{aligned} \tag{6.30}$$

Proof:

An important point in the proof is to realize that due to causality

$$f_i(n-m) \equiv 0 \quad m > n, \quad i = 1, 2$$

From the definition

$$F_2(z) \triangleq \sum_{m=0}^{\infty} z^{-m} f_2(m) \qquad (m \text{ is a dummy variable})$$

$$F_1(z)F_2(z) = F_1(z) \sum_{m=0}^{\infty} z^{-m} f_2(m) = \sum_{m=0}^{\infty} \left[F_1(z) z^{-m}\right] f_2(m)$$

Now

$$F_1(z) \leftrightarrow f_1(n), \quad F_1(z) z^{-m} \leftrightarrow f_1(n-m)$$

Thus,

$$F_1(z)F_2(z) \leftrightarrow \sum_{m=0}^{\infty} f_1(m)f_2(n-m)$$
$$= \sum_{m=0}^{n} f_1(n-m)f_2(m) + \sum_{m=n+1}^{\infty} f_1(n-m)f_2(m)$$

The second term in the above expression is zero because $f(n-m) \equiv 0$ for $m > n$. Thus,

$$F_1(z)F_2(z) \leftrightarrow \sum_{m=0}^{n} f_1(n-m)f_2(m) \quad (6.31)$$

Interchanging $F_1(z)$ and $F_2(z)$ and following the same arguments,

$$F_2(z)F_1(z) \leftrightarrow \sum_{m=0}^{n} f_1(m)f_2(n-m)$$

Example 6.11: Given

$$F(z) = \frac{z}{(z-\alpha_1)(z-\alpha_2)}, \qquad \operatorname{Re} z > \alpha_1, \alpha_2. \quad \text{Determine } f(n)$$

Solution:

$$F(z) = F_1(z)F_2(z) = \frac{z}{(z-\alpha_1)(z-\alpha_2)}$$

$$F_1(z) = \frac{z}{z-\alpha_1}$$

$$F_2(z) = \frac{1}{z-\alpha_2}$$

$$\frac{z}{z-\alpha_1} \leftrightarrow \alpha_1^n u(n)$$

$$\frac{1}{z-\alpha_2} \leftrightarrow \alpha_2^{n-1} u(n-1)$$

Thus,

$$F(z) \leftrightarrow \sum_{m=0}^{n} [\alpha_2^{m-1} u(m-1)][\alpha_1^{n-m} u(n-m)] = \sum_{m=0}^{n} \alpha_2^{m-1} \alpha_1^{n-m}$$

$$= \alpha_1^{n-1} \sum_{m=0}^{n-1} \left(\frac{\alpha_2}{\alpha_1}\right)^m = \alpha_1^{n-1} \left[\frac{1 - \left(\frac{\alpha_2}{\alpha_1}\right)^n}{1 - \frac{\alpha_2}{\alpha_1}}\right]$$

$$= \begin{cases} \dfrac{\alpha_1^n - \alpha_2^n}{\alpha_1 - \alpha_2} & (\alpha_1 \neq \alpha_2) \\ n\alpha_1^{n-1} & (\alpha_1 = \alpha_2) \end{cases}$$

Property 10: Z-Domain Convolution (or Frequency Domain Convolution)

$$f_1(n) \leftrightarrow F_1(z) \qquad f_1(n) < M_1 R_1, \quad 0 < n < \infty, \quad M_1 > 0$$

$$f_2(n) \leftrightarrow F_2(z) \qquad f_2(n) < M_2 R_2, \quad 0 < n < \infty, \quad M_2 > 0$$

Show:

$$f_1(n) f_2(n) \leftrightarrow \frac{1}{2\pi j} \oint_{\substack{|\lambda|>R_1 \\ |z|>R_1 R_2}} F_1(\lambda) F_2(z/\lambda) \frac{\mathrm{d}\lambda}{\lambda} = \frac{1}{2\pi j} \oint_{\substack{|\lambda|>R_1 \\ |z|>R_1 R_2}} F_1(z/\lambda) F_2(\lambda) \frac{\mathrm{d}\lambda}{\lambda}$$

Proof:

$$f_1(n) = \frac{1}{2\pi j} \oint_{|\lambda|>R_1} F_1(\lambda) \lambda^{n-1} \, \mathrm{d}\lambda$$

Thus,

$$f_1(n) f_2(n) = \frac{1}{2\pi j} \oint_{|\lambda|>R_1} F_1(\lambda) \lambda^{n-1} f_2(n) \, \mathrm{d}\lambda$$

Taking the Z-Transform of both sides

$$Z[f_1(n)f_2(n)] = \sum_{n=0}^{\infty} f_1(n)f_2(n)z^{-n}$$

$$= \frac{1}{2\pi j} \oint_{\substack{|\lambda|>R_1 \\ |z/\lambda|>R_2}} F_1(\lambda)\left[\sum_{n=0}^{\infty} z^{-n}\lambda^{n-1}f_2(n)\right] \mathrm{d}\lambda$$

The term in the bracket corresponds to $\dfrac{F_2(z/\lambda)}{\lambda}$.

Hence,

$$f_1(n)f_2(n) \leftrightarrow \oint_{\substack{|\lambda|>R_1 \\ |z|>R_1R_2}} F_1(\lambda)F_2(z/\lambda)\frac{\mathrm{d}\lambda}{\lambda}$$

The interchange of $f_1(n)$ and $f_2(n)$ yields the second expression.

Example 6.12: Determine the Z-Transform of n^2 via complex convolution.

Solution:

$$n^2 = (n)(n)$$

$$f_1(n) = f_2(n) = n, \qquad R_1 > 1, \quad R_2 > 1$$

$$n \leftrightarrow \frac{z}{(z-1)^2}$$

$$Z\left[n^2\right] = \frac{1}{2\pi j} \oint_{\substack{\lambda>1 \\ |z|>1}} \left[\frac{\lambda}{(\lambda-1)^2}\right]\left[\frac{z/\lambda}{(z/\lambda-1)^2}\right]\frac{\mathrm{d}\lambda}{\lambda}$$

$$= \frac{1}{2\pi j} \oint_{\substack{\lambda>1 \\ |z|>1}} \left[\frac{1}{(\lambda-1)^2}\right]\left[\frac{\lambda z}{(z-\lambda)^2}\right] \mathrm{d}\lambda = \frac{1}{2\pi j} \oint_{\substack{\lambda>1 \\ |z|>1}} F(\lambda)\,\mathrm{d}\lambda$$

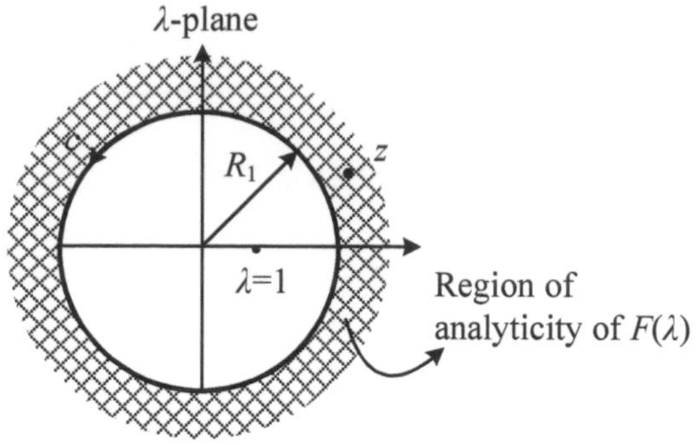

Figure 6.11: Region of Analyticity of $F(\lambda)$ Outside the Circle c, $|\lambda| > R_1$

Thus,

$$Z\left[n^2\right] = \left[\text{Residue of } \frac{\lambda z}{(z-\lambda)^2} \text{ at the double pole } \lambda = 1\right]$$
$$= \frac{d}{d\lambda}\left[\frac{\lambda z}{(z-\lambda)^2}\right]_{\lambda=1} = \frac{z}{(z-1)^2} - \frac{2z}{(z-1)^3} = \frac{z(z+1)}{(z-1)^3}$$

Property 11: Initial and Final Value Theorem

From the definition

$$F(z) \triangleq \sum_{n=0}^{\infty} f(n)z^{-n} = f(0) + \frac{f(1)}{z} + \frac{f(2)}{z^2} + \cdots$$

Initial Value Theorem

$$\lim_{z\to\infty} F(z) = f(0)$$

Furthermore,

$$Z[\Delta f(n)] \equiv \sum_{n=0}^{\infty} \Delta f(n)z^{-n} = (z-1)F(z) - zf(0)$$

In the limit as $z \to 1$

$$\sum_{n=0}^{\infty} \Delta f(n) = \lim_{z\to 1}(z-1)F(z) - f(0)$$

But

$$\sum_{n=0}^{\infty} \Delta f(n) = \sum_{n=0}^{\infty}[f(n+1) - f(n)] = \lim_{n\to\infty} f(n) - f(0)$$

Thus,

Final Value Theorem

$$\lim_{n\to\infty} f(n) = f(\infty) = \lim_{z\to 1}(z-1)F(z)$$

Table 6.2: Table of the Single-Sided Z-Transform Pairs

No.	$f(n) = \frac{1}{2\pi j}\oint F(z)z^{n-1}\,dz$ $\lvert f(n)\rvert < MR^n,\ n > 0$	$F(z) = \sum_{n=0}^{\infty} f(n)z^{-n}$ $\lvert z\rvert > R$	Region of Convergence R
1	$f(n)$	$F(z)$	$\lvert z\rvert > R$
2	$\delta(n)$	1	$\lvert z\rvert > 0$
3	$u(n) = 1$	$z/(z-1)$	$\lvert z\rvert > 1$
4	$n\,u(n) = n$	$z/(z-1)^2$	$\lvert z\rvert > 1$
5	$n^2u(n) = n^2$	$z(z+1)/(z-1)^3$	$\lvert z\rvert > 1$
6	a^n	$z/(z-a)$	$\lvert z\rvert > \lvert a\rvert$
7	$\sin\beta n$	$z\sin\beta/(z^2 - 2z\cos\beta + 1)$	$\lvert z\rvert > 1$
8	$\cos\beta n$	$\frac{z(z-\cos\beta)}{z^2-2z\cos\beta+1}$	$\lvert z\rvert > 1$
9	$\sum_{m=0}^{n-1} m$	$z/(z-1)^3$	$\lvert z\rvert > 1$

(continued)

No.	$f(n) = \frac{1}{2\pi j}\oint F(z)z^{n-1}\,dz$ $\lvert f(n)\rvert < MR^n,\ n > 0$	$F(z) = \sum_{n=0}^{\infty} f(n)z^{-n}$ $\lvert z\rvert > R$	Region of Convergence R
10	$\binom{n}{m} = \frac{n!}{m!(n-k)!}$	$z/(z-1)^{n+1}$	$\lvert z\rvert > 1$
11	$\sum_{m=0}^{n-1} m^2$	$z(z+1)/(z-1)^4$	$\lvert z\rvert > 1$
12	$ne^{-\alpha n}$	$e^{-\alpha}z/(z-e^{-\alpha})^2$	$\lvert z\rvert > \lvert e^{-\alpha}\rvert$
13	$\binom{n+k}{k}a^{n-k}$	$z^{k+1}/(z-a)^{k+1}$	$\lvert z\rvert > \lvert a\rvert$
14	$a^n/n!,\ n =$ odd integer	$\sinh a/z$	$\lvert z\rvert > 0$
15	$a^n/n!,\ n =$ even integer	$\cosh a/z$	$\lvert z\rvert > 0$
16	$a^n/n!$	$e^{a/z}$	$\lvert a/z\rvert < 1$

Note: Interestingly, $(z-1)$ plays the same role in the Z-Transform as s plays in the Laplace Transform, $(z-1) = e^s - 1 \approx 1 + s - 1 = s$ for small $|s|$. It is easy to compute the Z-Transform from the definitions provided we can sum their resulting series. For the sake of convenience, a short table of Z-Transform pairs is provided in Table 6.2, along with the region of convergence where the Z-Transform is an analytic function.

6.6 Evaluation of $f(n)$, Given Its Z-Transform

Given a single-sided Z-Transform $F(z)$, which is analytic outside $|z| > R$ in the z-plane and has all its singularities in a region $|z| < R$. We are required to find the function $f(n)$, such that $f(n) = Z^{-1}[F(z)]$, $|f(n)| < MR^n$ for $n > 0$.

In what follows we present three methods of finding $f(n)$, when $F(z)$ is given

Method #1 Taylor Series Expansion about Infinity ($z \to \infty$)

$$F(z)|_{z=1/\lambda} = F(1/\lambda) = F_1(\lambda), \quad \lambda = \frac{1}{z}$$

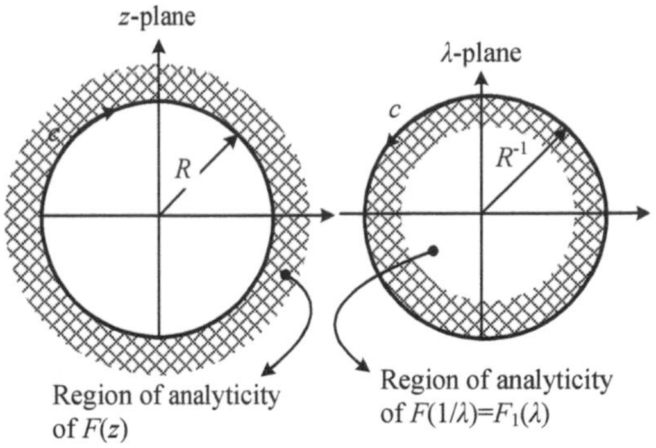

Figure 6.12: Analytic Region for $F(z)$ and $F(1/\lambda)$

Since $F_1(\lambda)$ is analytic inside the region surrounded by c in the λ-plane, which includes the origin. It can be expanded in the Taylor series as

$$F_1(\lambda) = \sum_{n=0}^{\infty} a_n \lambda^n, \quad \text{thus}$$

$$F_1(\lambda) = F(1/\lambda) = \sum_{n=0}^{\infty} a_n \lambda^n = \sum_{n=0}^{\infty} f(n)\lambda^n, \text{ implies } f(n) = a_n, \; n = 0, 1, 2, \ldots$$

This method is useful either when the first few terms of the sequence $f(n)$ are nonzero or the sequence can be recognized as a closed-form function $f(n)$. Usually $F(z)$ is a rational function involving the ratio of two polynomials. Both the numerator and denominator polynomials are arranged in the ascending powers of z^{-1} and a long division is performed yielding the quotient in the ascending power of z^{-1}. In general, this method has only limited uses. Let us show this procedure by a general example.

Example 6.13: Given

$$F(z) = \frac{N(z)}{D(z)} = \frac{\sum_{k=0}^{n} b_k z^{-k}}{\sum_{k=0}^{d} a_k z^{-k}}, \qquad n \geq d, \quad \text{find } f(n)$$

Solution:

Let

$$F(z) = \frac{\sum_{k=0}^{n} b_k z^{-k}}{\sum_{k=0}^{d} a_k z^{-k}} = \left(z^{-(n-d)}\right)\left(\sum_{i=0}^{\infty} c_i z^{-i}\right) \tag{6.32}$$

Equating coefficients of powers of z^{-1} on both sides of Eq. 6.32, we obtain

$$\begin{aligned}
&a_0 c_0 = b_0 \\
&a_0 c_1 + a_1 c_0 = b_1 \\
&a_0 c_2 + a_1 c_1 + a_2 c_0 = b_2 \\
&\vdots \\
&a_0 c_n + a_1 c_{n-1} + \cdots + a_n c_0 = b_n \\
&a_0 c_j + a_1 c_{j-1} + \cdots + a_j c_0 = b_j \\
&a_j = 0 \qquad j > d \\
&b_j = 0 \qquad j > n
\end{aligned} \tag{6.33}$$

Solving these equations (at least the first few terms) is easy. Once the c_k's are known then, $f(n) = Z^{-1}[F(z)]$

Example 6.14: Given

$$F(z) = \frac{z(2z-3)}{z^2 - 3z + 2} = \frac{2 - 3z^{-1}}{1 - 3z^{-1} + 2z^{-2}} \qquad R > 2, \qquad n = d = 2$$

$$a_0 = 1, \quad a_1 = 3, \quad a_2 = 2, \qquad b_0 = 2, \quad b_1 = -3, \quad b_2 = b_3 = \ldots = 0$$

From Eq. 6.33

$$\begin{aligned}
&c_0 = 2 && \to c_0 = 2 \\
&c_1 - 3c_0 = -3 && \to c_1 = 3 \\
&c_2 - 3c_1 + 2c_0 = 0 && \to c_2 = 5 \\
&c_3 - 3c_2 + 2c_1 = 0 && \to c_3 = 9
\end{aligned} \qquad f(n) = c_n$$

$$f(0) = 2, \quad f(1) = 3, \quad f(2) = 5, \quad f(3) = 9$$

The general term of the sequence is not so obvious.

We shall see later that this sequence results in $f(n) = (1)^n + (2)^n$

Method #2 Partial Fraction Expansion

The objective, as in the case of Laplace Transforms is to expand $F(z)$ as a sum of factors that are easily recognizable transforms of the elementary functions of n such as exponentials, sines, cosines, impulses, and their delayed forms. Since the Z-Transform of many elementary functions contain z in the numerator, it is helpful to find partial fractions of $F(z)/z$ and then use Table 6.2 to obtain the sequence $f(n)$.

Example 6.15:

Given

$$F(z) = \frac{3z}{z^2 + z - 2}, \qquad R > 2$$

Evaluate $f(n)$

Solution:

$$F_1(z) = \frac{F(z)}{z} = \frac{3}{z^2 + z - 2} = \frac{3}{(z-1)(z-2)} = \frac{A}{z-1} + \frac{B}{z+2}$$

$$A = \text{Res}\,[F_1(z)]_{z=1} = [F_1(z)(z-1)]_{z=1} = 1$$

$$B = \text{Res}\,[F_1(z)]_{z=-2} = [F_1(z)(z+2)]_{z=-2} = -1$$

$$F(z) = \frac{z}{z-1} - \frac{z}{z+2}$$

$$f(n) = (1)^n - (-2)^n$$

Example 6.16:

Given

$$F(z) = \frac{3z^2}{(z-1)^2(z+2)}, \qquad R > 2, \quad \text{Evaluate } f(n)$$

Solution:

$$F_1(z) = \frac{F(z)}{z} = \frac{3z}{(z-1)^2(z+2)} = \frac{A}{(z-1)^2} + \frac{B}{(z-1)} + \frac{C}{(z+2)}$$

$$\begin{aligned} C &= [F_1(z)(z+2)]_{z=-2} = -2/3 \\ A &= \left[F_1(z)(z-1)^2\right]_{z=1} = 1 \\ B &= \frac{\mathrm{d}}{\mathrm{d}z}\left[F_1(z)(z-1)^2\right]_{z=1} = 2/3 \end{aligned}$$

Thus,

$$F(z) = \frac{z}{(z-1)^2} + \frac{2}{3}\left(\frac{z}{z-1}\right) - \frac{2}{3}\left(\frac{z}{z+2}\right)$$

From Table 6.2

$$f(n) = n + \frac{2}{3} - \frac{2}{3}(-2)^n$$

Method #3 Evaluation of $f(n)$ via Complex Contour Integration

Recalling that

$$f(n) = \frac{1}{2\pi j} \oint_{|z|>R} F(z) z^{n-1}\, \mathrm{d}z$$

where the contour $c : |z| = R$ encloses all the singularities of $F(z)$, and is analytic outside the contour c.

From Cauchy's Residue theorem

$$f(n) = \begin{cases} \left[\text{Sum of residues of the poles of } F(z)z^{n-1} \text{ inside } c \right] & n \geq 0 \\ 0 & n < 0 \end{cases}$$

Example 6.17:

$$F(z) = \frac{N(z)}{\prod_{k=1}^{m}(z - z_k)^{r_k}}, \qquad |z| > z_k, \quad k = 1, 2, \ldots, m$$

where $N(z)$ is a polynomial in z

$F(z)$ has poles at $z = z_k$ of the order r_k, $\quad k = 1, 2, \ldots, m$.

$$f(n) = \sum_{k=0}^{m} \frac{1}{(r_k - 1)!} \frac{\mathrm{d}^{r_k - 1}}{\mathrm{d}z^{r_k - 1}} \left[F(z) z^{n-1} (z - z_k)^{r_k} \right]_{z = z_k}$$

Example 6.18:

Given

$$F(z) = \frac{3z^2}{(z-1)^2(z+2)}, \qquad R > 2$$

Evaluate $f(n)$ using contour integration

Solution:

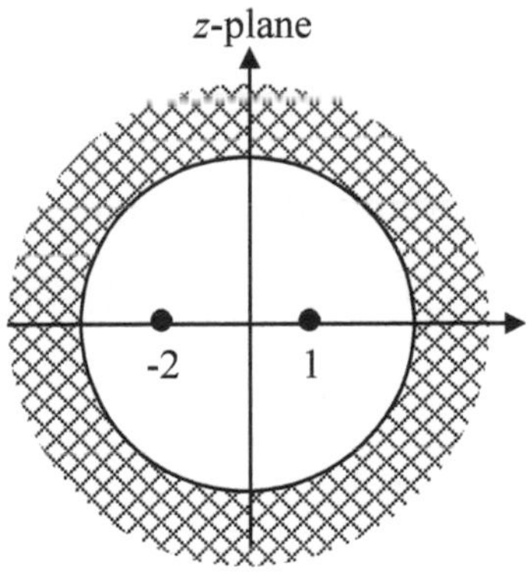

Figure 6.13: Location of Poles of $F(z)$

Two poles are at $z = 1$ and $z = -2$, thus for $|z| > 2$ the function is analytic.

$$\mathrm{Res}\left[F(z)z^{n-1}\right]_{z=1} = \frac{\mathrm{d}}{\mathrm{d}z}\left[\frac{(3z^2)\, z^{n-1}}{(z+2)}\right]_{z=1} = n + 1 - \frac{1}{3} = \left(n + \frac{2}{3}\right)$$

$$\mathrm{Res}\left[F(z)z^{n-1}\right]_{z=-2} = -\frac{2}{3}(-2)^n$$

$$f(n) = \left(n + \frac{2}{3}\right) - \frac{2}{3}(-2)^n$$

6.7 Difference Equations via Z-Transforms

Just as Laplace Transforms are very convenient for the solutions of differential equations, the Z-Transforms are an equally convenient way of studying difference equations. Using the Z-Transform, the linear difference equations are converted into algebraic equations involving Z-Transform of the unknown functions along with all the initial conditions. These equations yield a specific expression for the Z-Transforms of the unknown functions whose Transform inverse is obtained via any of the methods described in the previous section.

Consider a k-th order linear difference equation

$$\sum_{\nu=0}^{k} a_\nu y(n+k-\nu) = \sum_{\nu=0}^{k-1} b_\nu u(n+k-\nu)$$

where $u(n)$ represents the input forcing function and $y(n)$ is the resulting output. The coefficients a_ν, b_ν $(\nu = 0, 1, \ldots, k-1)$ and the initial conditions $y(r) = y_r$, $u(r) = u_r$ $(r = 0, 1, \ldots, k-1)$ are known.

From Property 2 of Table 6.2

$$\left[\sum_{\nu=0}^{k} a_\nu z^{k-\nu}\right] Y(z) = \left[\sum_{\nu=0}^{k-1} b_\nu z^{k-\nu}\right] U(z) + \sum_{\nu=0}^{k-1}\sum_{r=0}^{k-\nu-1} [a_\nu y_\nu - b_\nu u_r] z^{k-\nu-r} \tag{6.34}$$

Let,

$$\sum_{\nu=0}^{k} a_\nu z^{k-\nu} = D(z), \quad \sum_{\nu=0}^{k-1} b_\nu z^{k-\nu} = N(z)$$

$$\sum_{\nu=0}^{k-1}\sum_{r=0}^{k-\nu-1} [a_\nu y_r - b_\nu u_r] z^{k-\nu-r} = N_I(z) \tag{6.35}$$

Eq. 6.34 is reduced to the form

$$Y(z) = \frac{N(z)}{D(z)} U(z) + \frac{N_I(z)}{D(z)} \tag{6.36}$$

Taking the inverse Transform of both sides of Eq. 6.36, $y(n)$ is obtained, which is made up of two parts, $y(n) = y_f(n) + y_I(n)$

$$y_f(n) = Z^{-1}\left[\frac{N(z)}{D(z)}U(z)\right]$$ The contribution to $y(n)$ due to the forcing function $u(n)$. This is analogous to the particular integral term (PI) that appears in the continuous systems.

$$y_I(n) = Z^{-1}\left[\frac{N_I(z)}{D(z)}\right]$$ The contribution to $y(n)$ due to initial conditions. These are referred to as transient terms, because in the stable system they usually vanish relatively fast.

$D(z) = \sum_{\nu=0}^{k} a_\nu z^{k-\nu}$ is known as the characteristic polynomial.

$[y(n) = y_f(n) + y_I(n)]$ is the total response to the input and the initial conditions.

If

$$U(z) = \frac{N_u(z)}{D_u(z)} = \text{(Ratio of two polynomials in } z\text{)}$$

then

$$\begin{aligned} Y(z) &= \frac{N(z)}{D(z)}\frac{N_u(z)}{D_u(z)} + \frac{N_I(z)}{D(z)} = \frac{N(z)N_u(z) + N_I(z)D_u(z)}{D(z)D_u(z)} \\ &= \frac{\hat{N}(z)}{\hat{D}(z)} = \frac{c_0 + c_1 z + \cdots + c_M z^M}{d_0 + d_1 z + \cdots + d_N z^N}, \qquad M \le N \end{aligned} \tag{6.37}$$

We can now take the Inverse Transform of Eq. 6.37 to obtain $y(n)$.

6.7.1 Causal Systems Response $y(n)$ $(Y(z) = \hat{N}(z)/\hat{D}(z))$

1. **$\hat{D}(z)$ has simple roots**

   $\hat{D}(z) = 0$ has only simple roots at z_ν $(\nu = 1, 2, \ldots, N)$

$$Y(z) = \frac{\hat{N}(z)}{\hat{D}(z)} = \sum_{\nu=1}^{N} \frac{C_\nu}{(z - z_\nu)} \tag{6.38}$$

where

$$C_\nu = \left[\frac{\hat{N}(z)}{\hat{D}(z)}(z - z_\nu)\right]_{z=z_\nu} = \left[\frac{\hat{N}(z)}{\frac{\mathrm{d}}{\mathrm{d}z}\hat{D}(z)}\right]_{z=z_\nu} = \frac{\hat{N}(z_\nu)}{\hat{D}'(z_\nu)}$$

$$\nu = 1, 2, \ldots, N$$

Now

$$\frac{1}{z - z_\nu} \leftrightarrow \begin{cases} z_\nu^{(n-1)} & n \geq 0 \\ 0 & n < 0 \end{cases}$$

Hence,

$$y(n) = \sum_{\nu=1}^{N} \frac{\hat{N}(z_\nu)}{\hat{D}'(z_\nu)}(z_\nu)^{n-1}, \qquad n \geq 1 \tag{6.39}$$

2. $\hat{D}(z)$ **has multiple roots**

$\hat{D}(z) = 0$ has multiple roots at z_ν of multiplicity $r_\nu \geq 1$, $\nu = 1, 2, \ldots, p$, $r_1 + r_2 + \ldots + r_p = N$

Then

$$Y(z) = \frac{\hat{N}(z)}{\hat{D}(z)} = \sum_{\nu=1}^{p} \sum_{\mu=0}^{r_\nu - 1} C_{\nu\mu} \frac{z}{(z - z_\nu)^{\mu+1}} \tag{6.40}$$

where

$$C_{\nu\mu} = \frac{1}{(r_\nu - \mu - 1)!} \frac{\mathrm{d}^{r_\nu - \mu - 1}}{\mathrm{d}z^{r_\nu - \mu - 1}} \left[\frac{\hat{N}(z)}{z\hat{D}(z)}(z - z_\nu)^{r_\nu}\right]_{z=z_\nu}$$

Now

$$\frac{z}{(z - z_\nu)^{\mu+1}} \leftrightarrow \binom{n}{\mu}(z_\nu)^{n-\mu} = \left[\frac{n!}{(n-\mu)!\mu!}\right](z_\nu)^{n-\mu}$$

Thus, $y(n)$ is obtained as

$$y(n) = \sum_{\nu=1}^{p} \sum_{\mu=0}^{r_\nu - 1} C_{\nu\mu} \binom{n}{\mu}(z_\nu)^{n-\mu} \tag{6.41}$$

6.7.2 Digital Transfer Function

Transfer Function of Discrete Systems

In case all of the initial conditions are zero, the response of the system Eq. 6.7 is given by the Z-Transform equation

$$Y(z) = \frac{N(z)}{D(z)} U(z)$$

or

$$\frac{Y(z)}{U(z)} = \left\{ \frac{\text{Z-Transform of the output}}{\text{Z-Transform of the input}} \atop \text{all initial conditions zero} \right\} = H(z) = \frac{N(z)}{D(z)} \tag{6.42}$$

$H(z)$ is known as the Transfer Function of the discrete system.

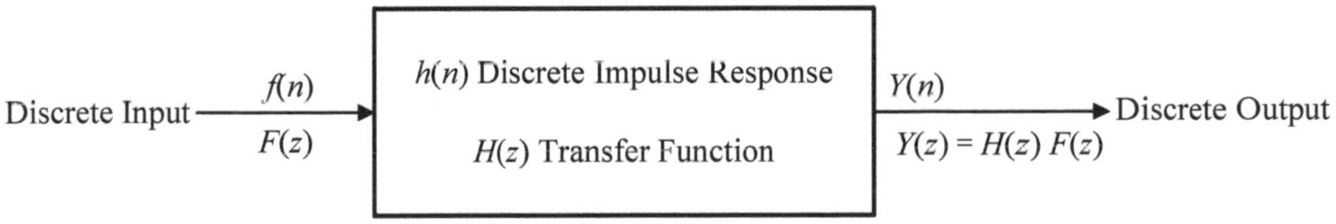

Figure 6.14: Transfer Function

It is obvious that,

$$Z^{-1}[H(z)] = h(n) = \text{ Impulse response of the discrete system.}$$

From the convolution theorem

$$Y(n) = \sum_{m=-\infty}^{\infty} u(m)h(n-m) = \sum_{m=-\infty}^{\infty} u(n-m)h(m) \tag{6.43}$$

For causal systems

$$Y(n) = \sum_{m=0}^{n} u(m)h(n-m) = \sum_{m=0}^{n} u(n-m)h(m) \tag{6.44}$$

The following observations are in order

1. Z-Transfer Function $H(z)$ of a system is the Z-Transform of its impulse response $h(n)$,

$$H(z) = \sum_{n=0}^{\infty} h(n)z^{-n} \tag{6.45}$$

2. Z-Transform $Y(z)$ of the output $y(n)$ is equal to the Discrete Transfer Function of the system multiplied by the Z-Transform $U(z)$ of its input $u(n)$ (all initial conditions zero)

$$Y(z) = H(z)U(z)$$

3. Consider two systems in cascade with Transfer Function $H_1(z)$ and $H_2(z)$ respectively.

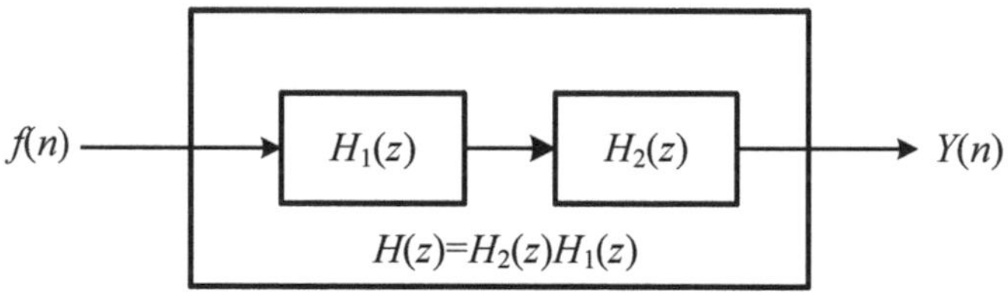

Figure 6.15: Transfer Function of Cascaded Systems

The Transfer Function of the composite system is $H(z) = H_2(z)H_1(z)$ where order of multiplication is important for the matrix transfer functions.

6.7.3 Representation of Digital Transfer Function

$$H(z) = \frac{N(z)}{D(z)} = \frac{Y(z)}{U(z)}$$

$$N(z) = b_0 + b_1 z^{-1} + \ldots + b_k z^{-k}, \quad D(z) = 1 + a_1 z^{-1} + \ldots + a_k z^{-k}$$

Let us introduce a dummy variable $X(z)$ (state variable) such that

$$\frac{X(z)}{U(z)} = \frac{1}{D(z)}, \quad \frac{Y(z)}{X(z)} = N(z)$$

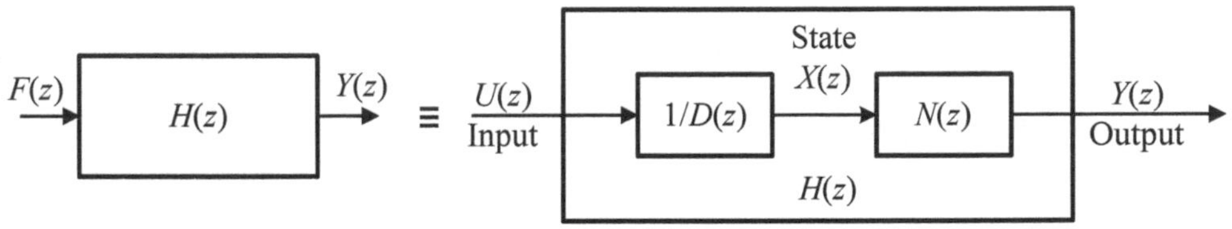

Figure 6.16: State Variable Representation of a Discrete System

$$U(z) = D(z)X(z) = X(z) - \left(\sum_{i=1}^{k} a_i z^{-i}\right) X(z) \tag{6.46}$$

$$Y(z) = N(z)X(z) = \sum_{i=0}^{k} b_i z^{-i} X(z) \tag{6.47}$$

The Transfer Function $H(z)$ represents a Nonrecursive, Transversal, or Finite-Impulse Response (FIR) system, because its output is represented in terms of finite numbers of its most recent state variable values.

Figure 6.17 shows Block Diagram Interconnection.

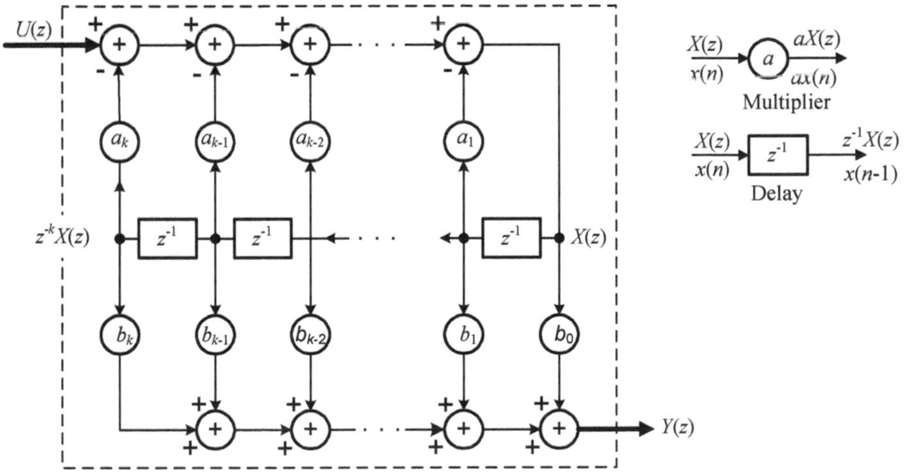

Figure 6.17: Block Diagram for Z-Transfer Function

6.8 Computation for the Sum of the Squares

In most system designs minimization of the sum of squares of the discrete error plays a very vital role. Figure 6.18 shows a general optimal system design problem.

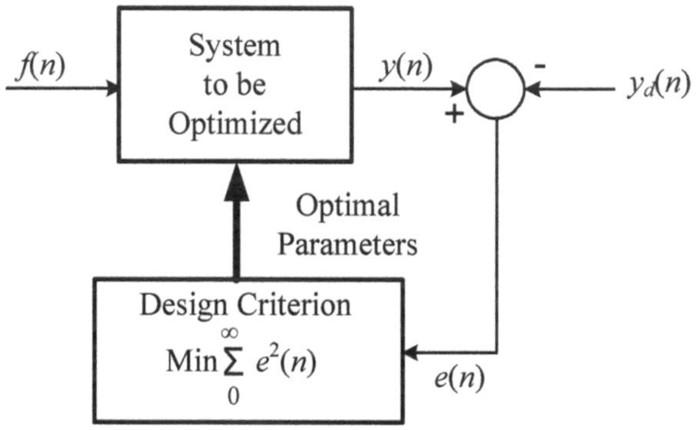

Figure 6.18: Optimal Minimum Sum of Squared Error Design

The explicit computation of $\sum_{n=0}^{\infty} e^2(n)$ in terms of the coefficients of its transform $E(z)$ is a very desirable closed form result and shall be presented here. It is understood that $\lim_{n\to\infty} e(n) = 0$ and therefore is a stable function of discrete time n.

6.8.1 Sum of Squared Sampled Sequence

Method #1 Method of Analytic Expansion of a Meromorphic Function

Let

$$E(z) = \frac{N(z)}{D(z)}$$

This method is useful when $N(z)$ and $D(z)$ are given as the power series. Consider two stable signals $f_1(n), f_2(n)$.

From Property 10 of Table 6.1

$$Z[f_1(n)f_2(n)] \equiv \sum_{n=0}^{\infty} z^{-n} f_1(n)f_2(n) = \frac{1}{2\pi j} \oint_{|\lambda|=1} F_1(\lambda)F_2(z/\lambda)\frac{d\lambda}{\lambda} \tag{6.48}$$

Let

$$f_1(n) = f_2(n) = e(n)$$
$$Z[e(n)] = E(z) = \frac{N(z)}{D(z)} \tag{6.49}$$

For $z = 1$, the Eq. 6.48 yields

$$I = \sum_{n=0}^{\infty} e^2(n) = \frac{1}{2\pi j} \oint_{|\lambda|=1} E(\lambda)E(1/\lambda)\frac{d\lambda}{\lambda} = \frac{1}{2\pi j} \oint_{|z|=1} E(z)E(1/z)\frac{dz}{z} \tag{6.50}$$

Let

$$E(z) = \frac{N(z)}{D(z)} = \frac{\sum\limits_{\nu=0}^{k-1} b_\nu z^\nu}{\sum\limits_{\nu=0}^{k} a_\nu z^\nu} \tag{6.51}$$

or

$$\sum_{\nu=0}^{k} a_\nu z^\nu E(z) = \sum_{\nu=0}^{k-1} b_\nu z^\nu \tag{6.52}$$

Multiplying both sides of the above Eq. 6.52 by $E(1/z)z^{-n-1}$ and integrating counterclockwise around $|z| = 1$

$$\sum_{\nu=0}^{k} a_\nu \left[\frac{1}{2\pi j} \oint_{|z|=1} \frac{E(z)E(1/z)}{z^{n+1-\nu}} dz\right] = \sum_{\nu=0}^{k-1} b_\nu \frac{1}{2\pi j} \oint_{|z|=1} \frac{E(1/z)}{z^{n+1-\nu}} dz \tag{6.53}$$

$$n = 0, 1, \ldots, k$$

Let

$$\frac{1}{2\pi j} \oint \frac{E(z)E(1/z)}{z^{n+1-\nu}} dz = I(n-\nu) \tag{6.54}$$

$$\frac{1}{2\pi j} \oint \frac{E(1/z)}{z^{n+1-\nu}} dz = g(n-\nu) \tag{6.55}$$

It is easy to show that $I(n-\nu) = I(-n+\nu)$ (even function of $(n-\nu)$).

To see this, let

$$\lambda = 1/z, \quad d\lambda = -1/z^2\, dz, \quad \frac{dz}{z} = -\frac{d\lambda}{\lambda}$$

Substituting this into Eq. 6.54

$$I(n-\nu) = \frac{1}{2\pi j}\oint \frac{E(z)E(1/z)}{z^{n+1-\nu}}\, dz = -\frac{1}{2\pi j}\oint \frac{E(\lambda)E(1/\lambda)}{\lambda^{n+1-\nu}}\, d\lambda$$
$$= \frac{1}{2\pi j}\oint \frac{E(z)E(1/z)}{z^{-n+1+\nu}}\, dz = I(-n+\nu)$$

Continuing from Eqs. 6.53, 6.54, and 6.55, we obtain

$$\sum_{\nu=0}^{k} a_\nu I(n-\nu) = \sum_{\nu=0}^{k} b_\nu g(n-\nu) = l(n), \quad n = 0, 1, \ldots, k \qquad (6.56)$$

Eqs. 6.56 are a system of $k+1$ equations which are simultaneously solved to obtain $I(n)$ in terms of $l(n)$'s, $n = 0, 1, \ldots, k$. Clearly

$$I(0) = \sum_{n=0}^{\infty} e^2(n)$$

Eq. 6.56 can be written as:

$$\begin{aligned}
&a_0 I(0) + a_1 I(1) + a_2 I(2) + \ldots + a_k I(k) &&= l(0)\\
&a_1 I(0) + (a_0 + a_2) I(1) + \ldots + a_k I(k-1) + 0 &&= l(1)\\
&a_2 I(0) + (a_1 + a_3) I(1) + (a_0 + a_4) I(2) + \ldots + 0 &&= l(2)\\
&\vdots\\
&a_k I(0) + a_{k-1} I(1) + \ldots + a_1 I(k-1) + a_0 I(k) &&= l(k)
\end{aligned}$$

or

$$I = I(0) = \frac{\Delta_1}{\Delta} = \frac{\begin{vmatrix} l(0) & a_1 & a_2 & \cdots & a_{k-1} & a_k \\ l(1) & (a_0 + a_2) & a_3 & \cdots & a_k & 0 \\ l(2) & (a_1 + a_3) & (a_0 + a_4) & \cdots & 0 & \\ \vdots & & & & & \\ l(k) & a_{k-1} & a_{k-2} & \cdots & a_1 & a_0 \end{vmatrix}}{\begin{vmatrix} a_0 & a_1 & a_2 & \cdots & a_{k-1} & a_k \\ a_1 & (a_0 + a_2) & a_3 & \cdots & a_k & 0 \\ a_2 & (a_1 + a_3) & (a_0 + a_4) & \cdots & 0 & \\ \vdots & & & & & \\ a_k & a_{k-1} & a_{k-2} & \cdots & a_1 & a_0 \end{vmatrix}} \tag{6.57}$$

In order to complete the computation of $I(0)$, all we need is a computation algorithm for $l(n)$, $n = 0, 1, \ldots, k$

The computation algorithm for $l(n)$

$$l(n) = \sum_{\nu=0}^{k} b_\nu g(n - \nu), \quad n = 0, 1, \ldots, k$$

$E(1/z)$ is analytic in $|z| \leq 1$.

Let

$$E(1/z) = \frac{\sum\limits_{\mu=0}^{k} b_\mu z^{k-\mu}}{\sum\limits_{\mu=0}^{k} a_\mu z^{k-\mu}} = \sum_{j=0}^{\infty} c_j z^j \tag{6.58}$$

Then,

$$\begin{aligned} g(n - \nu) &= \frac{1}{2\pi j} \oint_{|z|=1} \frac{E(1/z)}{z^{n-\nu+1}} \mathrm{d}z = \sum_{j=0}^{\infty} \frac{c_j}{2\pi j} \oint_{|z|=1} \frac{z^j}{z^{n-\nu+1}} \mathrm{d}z \\ &= c_{n-\nu}, \quad n - \nu \geq 0 \end{aligned} \tag{6.59}$$

Thus,

$$l(n) = \sum_{\nu=0}^{k} b_\nu c_{n-\nu} \tag{6.60}$$

From Eq. 6.60

$$b_k + b_{k-1}z + \cdots + b_0 z^k = \left(a_k + a_{k-1}z + \cdots + a_0 z^k\right)\left(c_0 + c_1 z + \cdots + c_k z^k\right)$$

Equating the powers of z

$$\begin{aligned} b_k &= a_k c_0 \\ b_{k-1} &= a_{k-1}c_0 + a_k c_1 \\ b_{k-2} &= a_{k-2}c_0 + a_{k-1}c_1 + a_k c_2 \\ &\cdots \\ b_0 &= a_0 c_0 + a_1 c_1 + \cdots + a_k c_k \end{aligned} \tag{6.61}$$

Solution of Eq. 6.61 yields $c_{n-\nu}$, which when substituted in Eq. 6.60 yields $l(n)$. **This completes the computation algorithms for $I = \sum_{n=0}^{\infty} e^2(n)$.** The computation can be performed in Maple to determine the effect of different parameters on the minimum value of I.

Method #2 Method of Residues

This method is useful when $E(z)$ is a simple expression easily represented by its poles, all of which lie within the unit circle.

As shown earlier

$$\begin{aligned} I = \sum_{n=0}^{\infty} e^2(n) &= \frac{1}{2\pi j} \oint_{|z|=1} E(z)E(z^{-1})z^{-1}\,\mathrm{d}z \\ &= \text{Sum of Residues}\left[E(z)E(z^{-1})z^{-1}\right] \text{ at the poles of } E(z) \end{aligned} \tag{6.62}$$

We shall illustrate this method with the following example.

Example 6.19: Given

$$e(n) = e^{\lambda n}, \qquad |e^{\lambda}| < 1$$

Compute

$$I = \sum_{n=0}^{\infty} e^2(n)$$

Solution:

$$E(z) = \frac{z}{(z - e^{\lambda n})}$$

$$\begin{aligned} I &= \frac{1}{2\pi j} \oint \frac{z}{(z - e^{\lambda n})} \frac{z^{-1}}{(z^{-1} - e^{\lambda n})} z^{-1}\,\mathrm{d}z \\ &= \frac{1}{2\pi j} \oint \frac{1}{(z - e^{\lambda n})(1 - ze^{\lambda n})}\,\mathrm{d}z = \left[\frac{1}{1 - ze^{\lambda n}}\right]_{z=e^{\lambda n}} = \frac{1}{1 - e^{2\lambda n}} \end{aligned}$$

Method #3 Multiplier Method

Given

$$E(z) = \frac{\sum_{i=0}^{k-1} b_i z^{k-i}}{\sum_{i=0}^{k} a_i z^{k-i}}, \qquad Z[e(n)] = E(z) \tag{6.63}$$

We are required to compute $I = \sum_{n=0}^{\infty} e^2(n)$.

The computation is performed in two steps

Step #1: The transformation of $E(z)$ in Eq. 6.63 to the Difference Equations with initial conditions. From Eq. 6.63

$$\sum_{i=0}^{k} a_i \left(z^{k-i} E(z)\right) = \sum_{i=0}^{k-1} b_i z^{k-i} \tag{6.64}$$

But

$$z^{k-i} E(z) = Z[e(n + k - i)] + z^{k-i} \sum_{m=0}^{k-i-1} e(m) z^{-m} \tag{6.65}$$

From Eqs. 6.64 and 6.65

$$Z\left[\sum_{i=0}^{k} a_i e(n+k-i)\right] = \left[\sum_{i=0}^{k-1} b_i z^{k-i} - \sum_{i-0}^{k-1} a_i \sum_{m=0}^{k-i-1} e(m) z^{k-i-m}\right] \tag{6.66}$$

Let

$$\left[\sum_{i=0}^{k-1} b_i z^{k-i} - \sum_{i=0}^{k-1} a_i \sum_{m=0}^{k-i-1} e(m) z^{k-i-m}\right] = 0 \tag{6.67}$$

Equating powers of z in Eq. 6.67, we obtain initial conditions $e(0), e(1), \ldots, e(k-1)$. The resulting difference equation is

$$\sum_{i=0}^{k} a_i e(n+k-i) = 0 \tag{6.68}$$

The initial conditions are obtained from Eq. 6.67 as following

$$\begin{bmatrix} a_0 & 0 & 0 & \cdots & 0 & 0 \\ a_1 & a_0 & 0 & & 0 & 0 \\ a_2 & a_1 & a_0 & & 0 & 0 \\ \vdots & \vdots & \vdots & & \vdots & \vdots \\ a_{k-2} & a_{k-3} & a_{k-4} & \cdots & a_0 & 0 \\ a_{k-1} & a_{k-2} & a_{k-3} & \cdots & a_1 & a_0 \end{bmatrix} \begin{bmatrix} e(0) \\ e(1) \\ e(2) \\ \vdots \\ e(k-2) \\ e(k-1) \end{bmatrix} = \begin{bmatrix} b_0 \\ b_1 \\ b_2 \\ \vdots \\ b_{k-2} \\ b_{k-1} \end{bmatrix} \tag{6.69}$$

$$e(i) = \frac{b_i - \sum_{j=0}^{i-1} a_{i-j} e(j)}{a_0}, \qquad e(0) = b_0/a_0, \qquad i = 1, 2, \ldots, k-1$$

Step #2: Consider the resulting Difference Eq. 6.68, with initial conditions from the Eq. 6.69, $\sum_{i=0}^{k} a_i e(n+k-i) = 0$.

Let

$$e(n+j) = x_{j+1}(n), \qquad j = 0, 1, \ldots, k \tag{6.70}$$

Eq. 6.68 can be rewritten as

$$\sum_{i=0}^{k} a_i x_{(k+1-i)}(n) = 0$$

or

$$\boxed{\sum_{i=1}^{k+1} a_{(k+1-i)} x_i(n) = 0} \tag{6.71}$$

Let us multiply the Eq. 6.71 with $x_j(n)$ and sum it for $n = 0$ to ∞ yielding

$$\sum_{i=1}^{k+1} a_{(k+1-i)} \sum_{n=0}^{\infty} x_i(n) x_j(n) = 0$$

Let

$$\sum_{n=0}^{\infty} x_i(n) x_j(n) = I_{ij} = I_{ji} \tag{6.72}$$

It is easy to see that

$$I_{ij} = \sum_{n=1}^{\infty} x_{i-1}(n) x_{j-1}(n) = \sum_{n=1}^{\infty} x_i(n-1) x_j(n-1) = \sum_{n=0}^{\infty} x_i(n) x_j(n)$$

Hence,

$$I_{ij} = I_{i-1,j-1} - x_{i-1}(0) x_{j-1}(0), \quad \begin{array}{l} i = 0, 1, \ldots, k \\ j = 0, 1, \ldots, k \end{array} \tag{6.73}$$

$$\sum_{i=1}^{k+1} a_{k+1-i} I_{ij} = 0 \tag{6.74}$$

Eqs. 6.72, 6.73, and 6.74, yield a matrix equation

$$\boldsymbol{AI} = \boldsymbol{q}(0) \tag{6.75}$$

where

$$A = \begin{bmatrix} a_k & a_{k-1} & a_{k-2} & a_{k-3} & \cdots & a_2 & a_1 & a_0 \\ a_{k-1} & (a_k + a_{k-2}) & a_{k-3} & a_{k-4} & \cdots & a_1 & a_0 & 0 \\ a_{k-2} & (a_{k-1} + a_{k-3}) & (a_k + a_{k-4}) & a_{k-5} & \cdots & a_0 & 0 & 0 \\ \vdots & \vdots & \vdots & \vdots & \vdots & \vdots & \vdots & \vdots \\ a_2 & (a_3 + a_1) & (a_4 + a_0) & a_5 & \cdots & a_k & 0 & 0 \\ a_1 & (a_2 + a_0) & a_3 & a_4 & \cdots & a_{k-1} & a_k & 0 \\ a_0 & a_1 & a_2 & a_3 & \cdots & a_{k-2} & a_{k-1} & a_k \end{bmatrix}$$

$$\boldsymbol{I} = \begin{bmatrix} I_{1,1} \\ I_{1,2} \\ \vdots \\ I_{1,k+1} \end{bmatrix}, \quad \boldsymbol{q}(0) = \begin{bmatrix} 0 \\ \boldsymbol{x}^T(0)\Omega_2\boldsymbol{x}(0) \\ \boldsymbol{x}^T(0)\Omega_3\boldsymbol{x}(0) \\ \vdots \\ \boldsymbol{x}^T(0)\Omega_{k+1}\boldsymbol{x}(0) \end{bmatrix}, \quad \boldsymbol{x}(0) = \begin{bmatrix} x_1(0) \\ x_2(0) \\ \vdots \\ x_k(0) \end{bmatrix} = \begin{bmatrix} x(0) \\ x(1) \\ \vdots \\ x(k-1) \end{bmatrix}$$

$$\Omega_2 = \frac{1}{2}\begin{bmatrix} 2a_{k-1} & a_{k-2} & \cdots & a_0 \\ a_{k-2} & & & \\ \cdots & & 0 & \\ a_0 & & & \end{bmatrix}$$

$$\Omega_3 = \frac{1}{2}\begin{bmatrix} 2a_{k-2} & (a_{k-1} + a_{k-3}) & a_{k-4} & \cdots & a_0 & 0 \\ (a_{k-1} + a_{k-3}) & 2a_{k-2} & a_{k-3} & \cdots & a_1 & a_0 \\ a & a & & & & \\ \vdots & \vdots & & 0 & & \\ a_0 & a_1 & & & & \\ 0 & a_0 & & & & \end{bmatrix}$$

$$\Omega_4 = \frac{1}{2}\begin{bmatrix} 2a_{k-3} & (a_{k-2}+a_{k-4}) & (a_{k-1}+a_{k-5}) & a_{k-6} & a_{k-7} & \cdots & a_0 & 0 & 0 \\ (a_{k-2}+a_{k-4}) & 2a_{k-3} & (a_{k-2}+a_{k-4}) & a_{k-5} & a_{k-6} & \cdots & a_1 & a_0 & 0 \\ (a_{k-1}+a_{k-5}) & (a_{k-2}+a_{k-4}) & 2a_{k-3} & a_{k-4} & a_{k-5} & \cdots & a_2 & a_1 & a_0 \\ a_{k-6} & a_{k-5} & a_{k-4} & & & & & & \\ a_{k-7} & a_{k-6} & a_{k-5} & & & & & & \\ \vdots & \vdots & \vdots & & & 0 & & & \\ a_0 & a_1 & a_2 & & & & & & \\ 0 & a_0 & a_1 & & & & & & \\ 0 & 0 & a_0 & & & & & & \end{bmatrix}$$

$$\Omega_k = \frac{1}{2}\begin{bmatrix} 2a_1 & (a_2+a_0) & a_3 & \cdots & a_{k-2} & a_{k-1} & 0 \\ (a_2+a_0) & 2a_1 & (a_2+a_0) & \cdots & a_{k-3} & a_{k-2} & 0 \\ a_3 & (a_2+a_0) & 2a_1 & \cdots & a_{k-4} & a_{k-3} & 0 \\ \vdots & \vdots & \vdots & \ddots & \vdots & \vdots & \vdots \\ a_{k-2} & a_{k-3} & a_{k-4} & \cdots & 2a_1 & (a_2+a_0) & 0 \\ a_{k-1} & a_{k-2} & a_{k-3} & \cdots & (a_2+a_0) & 2a_1 & a_0 \\ 0 & 0 & 0 & \cdots & 0 & a_0 & 0 \end{bmatrix}$$

$$\Omega_{k+1} = \frac{1}{2}\begin{bmatrix} 2a_0 & a_1 & a_2 & \cdots & a_{k-3} & a_{k-2} & a_{k-1} \\ a_1 & 2a_0 & a_1 & \cdots & a_{k-4} & a_{k-3} & a_{k-2} \\ a_2 & a_1 & 2a_0 & \cdots & a_{k-5} & a_{k-4} & a_{k-3} \\ \vdots & \vdots & \vdots & \ddots & \vdots & \vdots & \vdots \\ a_{k-3} & a_{k-4} & a_{k-5} & \cdots & 2a_0 & a_1 & a_2 \\ a_{k-2} & a_{k-3} & a_{k-4} & \cdots & a_1 & 2a_0 & a_1 \\ a_{k-1} & a_{k-2} & a_{k-3} & \cdots & a_2 & a_1 & 2a_0 \end{bmatrix}$$

$$I_{11} = \sum_{n=0}^{\infty} x_1^2(n) = \sum_{n=0}^{\infty} e^2(n)$$

Example 6.20: Consider a 4th order difference equation:

$$x(n+4) + a_1x(n+3) + a_2x(n+2) + a_3x(n+1) + a_4x(n) = 0 \tag{6.76}$$

I_{11} is obtained from the following equations

$$\begin{bmatrix} a_4 & a_3 & a_2 & a_1 & 1 \\ a_3 & (a_2+a_4) & a_1 & 1 & 0 \\ a_2 & (a_3+a_1) & (a_4+1) & 0 & 0 \\ a_1 & (a_2+1) & a_3 & a_4 & 0 \\ 1 & a_1 & a_2 & a_3 & a_4 \end{bmatrix} \begin{bmatrix} I_{11} \\ I_{12} \\ I_{13} \\ I_{14} \\ I_{15} \end{bmatrix} = \begin{bmatrix} 0 \\ \boldsymbol{x}^T(0)\Omega_2\boldsymbol{x}(0) \\ \boldsymbol{x}^T(0)\Omega_3\boldsymbol{x}(0) \\ \boldsymbol{x}^T(0)\Omega_4\boldsymbol{x}(0) \\ \boldsymbol{x}^T(0)\Omega_5\boldsymbol{x}(0) \end{bmatrix}, \quad \boldsymbol{x}(0) = \begin{bmatrix} x(0) \\ x(1) \\ x(2) \\ x(3) \end{bmatrix}$$

6.9 Bilateral Z-Transform $f(n) \leftrightarrow F_b(z)$

Up to now we have assumed that $f(n) \equiv 0$ for $n < 0$. The Z-Transform for these functions defined by Eq. 6.14, is known as single-sided or unilateral Z-Transform. In certain applications involving noncausal systems, $f(n) \not\equiv 0$ for $n < 0$. In these cases we need **Bilateral or double-sided Z-Transform** as defined in Eq. 6.15. Consider a noncausal function such that

$$|f(n)| < M_1R_a^n \quad n \geq 0, \quad M_1 > 0; \quad |f(n)| < M_2R_b^n \quad n < 0, \quad M_2 > 0$$

$$F_b(z) \triangleq \sum_{n=-\infty}^{+\infty} f(n)z^{-n} \qquad \text{Bilateral Z-Transform} \tag{6.77}$$

The above series has positive and negative powers of z. Applying a ratio test we find that the series in Eq. 6.77 converges if

$$|z| > R_a \quad \text{for} \quad n \geq 0$$

$$|z| < R_b \quad \text{for} \quad n < 0$$

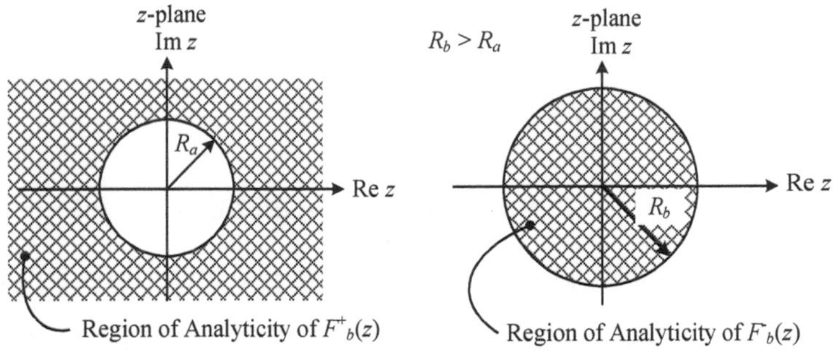

(a) Region of Convergence of $F_b^+(z)$ (b) Region of Convergence of $F_b^-(z)$

Figure 6.19: Region of Convergence $F_b^+(z)$ and $F_b^-(z)$

Thus, the Bilateral Z-Transform exists in a ring in the z-plane defined by $R_a < |z| < R_b$.

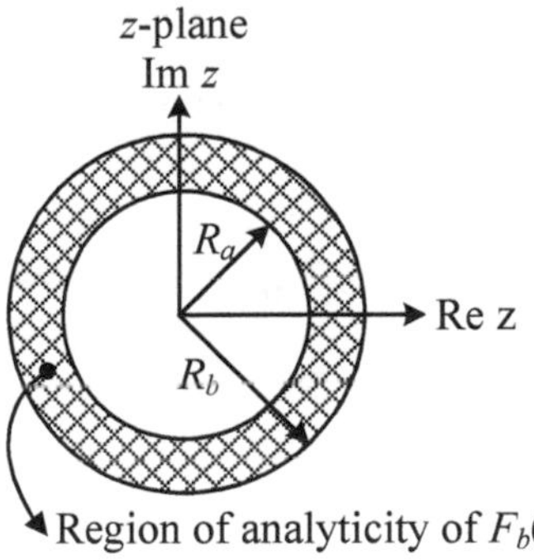

Figure 6.20: Region of Convergence of $F_b(z)$

The inverse bilateral Z-Transform is again obtained by multiplying both sides of Eq. 6.77 with z^{k-1} and realizing that when integrating around a closed curve lying in the region $R_a < |z| < R_b$, only the $1/z$ term yields a nonzero value, yielding

$$f(n) = \frac{1}{2\pi j} \oint_{R_a<|z|<R_b} F_b(z) z^{n-1}\, dz \qquad (6.78)$$

Define

$$\sum_{n=0}^{\infty} f(n)z^{-n} = F^+(z), \quad \sum_{n=-1}^{-\infty} f(n)z^{-n} = F^-(z)$$

$$F_b(z) = F^+(z) + F^-(z)$$

$F^+(z)$ corresponds to $f(n)$ for $n \geq 0$ and all its singularities lie within $|z| < R_a$ and is analytic outside. $F^-(z)$ corresponds to $f(n)$ for $n < 0$ and all its singularities lie in the region $|z| > R_b$ and is analytic inside $|z| < R_b$. The application of the Residue Theorem leads to

$$\begin{aligned} f(n) &= \left[\text{sum of Residues of } F^+(z)z^{n-1} \text{ at the poles of } F^+(z), \quad n \geq 0\right] \\ &= -\left[\text{sum of Residues of } F^-(z)z^{n-1} \text{ at the poles of } F^-(z), \quad n < 0\right] \end{aligned}$$

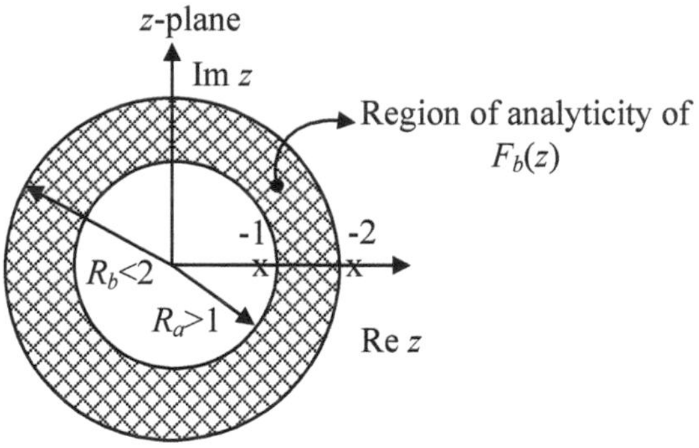

Figure 6.21: Location of the Poles of $F_b(z)$ and Region of Convergence

Example 6.21: Given:

$$F(z) = \frac{2z - 1}{(z-1)^2(z-2)} \quad R_a = 1, \; R_b = 2, \quad \text{Evaluate } f(n)$$

Solution:

$$\frac{F(z)}{z} = \frac{z(2z-1)}{(z-1)^2(z-2)} = \frac{A}{z-1} + \frac{B}{(z-1)^2} + \frac{C}{z-2}, \; A = -4, \; B = -1, \; C = 6$$

$$F^{+}(z) = -\frac{4z}{z-1} - \frac{z}{(z-1)^2}$$
$$F^{-}(z) = \frac{6z}{z-2}$$

$$\text{Res}\left[\frac{z^{n-1}z}{z-1}\right]_{z=1} = z^n|_{z=1} = 1$$
$$\text{Res}\left[\frac{z^{n-1}z}{(z-1)^2}\right]_{z=1} = \frac{\mathrm{d}}{\mathrm{d}z}[z^n]_{z=1} = nz^{n-1}\Big|_{z=1} = n$$
$$\text{Res}\left[\frac{z^{n-1}z}{z-2}\right]_{z=2} = [z^n]_{z=2} = 2^n$$

Thus,

$$f(n) = \begin{cases} (-4-n) & n \geq 0 \\ 6\,(2^n) & n < 0 \end{cases}$$

6.10 Summation of the Series via Z-Transforms

In this section we present some important series which frequently occur in many physical system studies. For many other series, the book by L.B.W. Jolly, "Summation of Series," [Jolly, L.B.W.] is a valuable source. The application of the Z-Transform is very helpful in the summation of many of these series.

Table 6.3: Some Important Series

No.	Series	Closed Form Expression
1	$\sum_{n=n_1}^{n=n_2} z^n$	$z^{n_1}\left(\frac{1-z^{(n_2+1-n_1)}}{1-z}\right)$ $n_2 > n_1, \quad \lvert z\rvert < 1$
2	$\sum_{n=0}^{\infty}(n+1)z^n$	$\left(\frac{1}{1-z^2}\right), \quad \lvert z\rvert < 1$
3	$\sum_{k=1}^{n} k(k+1)$	$\frac{1}{3}(n)(n+1)(n+2)$

(continued)

No.	Series	Closed Form Expression
4	$\sum_{k=1}^{n}(k)(k+1)(k+2)$	$\frac{1}{4}(n)(n+1)(n+2)(n+3)$
5	$\sum_{k=1}^{n} k(k+1)(k+2)(k+3)$	$\frac{1}{5}(n)(n+1)(n+3)(n+4)$
6	$\sum_{k=1}^{n} \frac{k^2 4^k}{(k+1)(k+2)}$	$\left(\frac{1}{3}\right)4^{(n+1)}\left(\frac{n-1}{n+2}\right)+\frac{2}{3}$
7	$\sum_{k=1}^{n} \frac{z^k}{1-(z^k)^2}$	$\frac{1}{1-z}-\frac{1}{1-z^n}, \qquad \lvert z\rvert < 1$
8	$\sum_{k=1}^{\infty} \frac{z^k}{(1-z^k)^2}$	$= \left(\frac{z}{1-z}\right)$ when $\lvert z\rvert^2 < 1$ $= -\frac{1}{(z-1)} \qquad \lvert z\rvert > 1$
9	$\sum_{n=0}^{\infty} \frac{-1/2z^2}{(n^2+z^2)}$	$\frac{\pi}{2z}\left(\frac{1+e^{-2\pi z}}{1-e^{-2\pi z}}\right)$
10	$\sum_{n=0}^{\infty}(-1)\frac{1}{(n^2+z^2)}-\frac{1}{2z^2}$	$\frac{\pi}{z}\left(\frac{1}{e^{\pi z}-e^{-\pi z}}\right)$
11	$\sum_{k=1}^{n} z_1^k\left(z_1^k+z_2^k\right)$	$\frac{z_1^2\left(z_1^{2n}-1\right)}{z_1^2-1}+\frac{z_1z_2\left(z_1^n z_2^n-1\right)}{z_1z_2-1}$ $\lvert z_1\rvert < 1, \quad \lvert z_1z_2\rvert < 1$
12	Zeta Function $\zeta(s,a)=\sum_{n=0}^{\infty}\frac{1}{(a+n)^s}$	$\frac{1}{\Gamma(s)}\int_0^{\infty}\frac{x^{s-1}e^{-ax}}{1-e^{-x}}\,dx$
	where $\frac{1}{(\Gamma(s))}=\frac{1}{2\pi j}\oint_c \frac{e^z}{z^s}\,dz$ When s is an integer n, $\Gamma(s)=\Gamma(n)=(n-1)!, \qquad n>0$	
13	$1+2\sum_{k=1}^{\infty}\frac{1}{1+k^2}$	$\pi\coth\pi$
14	$\sum_{k=0}^{m}(-1)^k\binom{n}{k}$	$(-1)^m\binom{n-1}{m}$
15	$\sum_{k=1}^{n-1}\frac{\sin\pi k}{n}$	$\cot\frac{\pi}{2n}$
16	$\sum_{k=0}^{n}\sin^2 k\theta$	$\frac{n}{2}-\frac{[\cos(n+1)\theta][\sin n\theta]}{2\sin\theta}$
17	$\sum_{k=0}^{n}\cos^2 k\theta$	$\left(\frac{n}{2}+1\right)+\frac{[\cos(n+1)\theta][\sin n\theta]}{2\sin\theta}$
18	$\sum_{k=0}^{n-1}\cot\left(\theta+\frac{k\pi}{n}\right)$	$n\cot n\theta$

(continued)

No.	Series	Closed Form Expression
19	$\sum_{n=0}^{\infty} z^n \cos(2n+1)\theta$	$\left[\frac{(1-z)\cos\theta}{1-2z\cos 2\theta+z^2}\right]$, $\quad\lvert z\rvert < 1$
20	$\sum_{n=0}^{\infty} z^n \sin n\theta$	$\left[\frac{z\sin\theta}{1-2z\cos\theta+z^2}\right]$, $\quad\lvert z\rvert < 1$
21	$\sum_{n=0}^{\infty} z^n \cos n\theta$	$\left[\frac{1-z\cos\theta}{1-2z\cos\theta+z^2}\right]$, $\quad\lvert z\rvert < 1$
22	$\sum_{n=0}^{\infty}(-1)^n \frac{1}{(2n+1)} \cos(2n+1)\theta$	$\left(\frac{\pi}{4}\right)$, $\quad -\frac{\pi}{2} < \theta < \frac{\pi}{2}$
23	$\sum_{n=0}^{\infty} \frac{1}{(2n+1)} \sin(2n+1)\theta$	$= \frac{\pi}{4} \quad 0 < \theta < \pi$ $= 0 \quad \theta = \pi$ $= -\frac{\pi}{4} \quad \pi < \theta < 2\pi$
24	$\sum_{n=0}^{\infty} \frac{1}{1+4n} - \sum_{n=0}^{\infty} \frac{1}{3+4n}$	$\frac{\pi}{4}$
25	$\sum_{n=0}^{\infty} \frac{z^n}{n!}$	e^z, $\quad\lvert z\rvert < 1$
26	$\sum_{n=1}^{\infty} \frac{z^n}{n}$	$\ln\left(\frac{1}{1-z}\right)$, $\quad\lvert z\rvert < 1$
27	$\sum_{n=1}^{\infty} \frac{z^n}{n(n+1)}$	$1-\left(z^{-1}-1\right)\ln\left(\frac{1}{1-z}\right)$, $\quad\lvert z\rvert < 1$
28	$\sum_{k=0}^{n}(2k+1)^2$	$\frac{1}{3}n(2n-1)(2n+1)$
29	$\sum_{k=0}^{n}(2k+1)^3$	$n^2\left(2n^2-1\right)$
30	$\sum_{k=0}^{n}(k+1)^2(2)^{k+1}$	$2^n\left(2n^2-4n+6\right)-6$
31	$\sum_{k=0}^{n}(k+1)(k+2)^2$	$\frac{1}{12}(n)(n+1)(n+2)(3n+5)$
32	$\sum_{n=0}^{\infty} \frac{(z)^{-(2n+1)}}{(2n+1)}$	$\tanh^{-1}\frac{1}{z} = \coth^{-1} z$, $\quad\lvert z\rvert > 1$

6.11 Sampled Signal Reconstruction

6.11.1 Introduction

From the Fourier Transform, we noticed that, a time signal with a constant value had only one frequency component, namely zero. On the other hand, a signal that varies abruptly, such as $\delta(t)$, has all the frequencies in its spectrum. This suggests that there exists a relationship between rate of change of a function $f(t)$ and its frequency contents in its Fourier Transform $F(j\omega)$. This concept leads us to the **Sampling Theorem** (due to Shannon) that states that if a function $f(t)$ is "Band-limited," then knowing the sampled values $f(n\Delta T),\ n = 0, \pm 1, \pm 2, \ldots, \pm\infty$, we can fully recover the original continous function $f(t)$. Obviously the sampling interval ΔT is important. If the sampling interval ΔT is small, the very advantage of the sampling is lost. On the other hand, if the sampling interval ΔT is too large, then the original continous function cannot be recovered with any real accuracy. As an illustration consider the function $\sin \omega t$ as shown in Figure 6.22

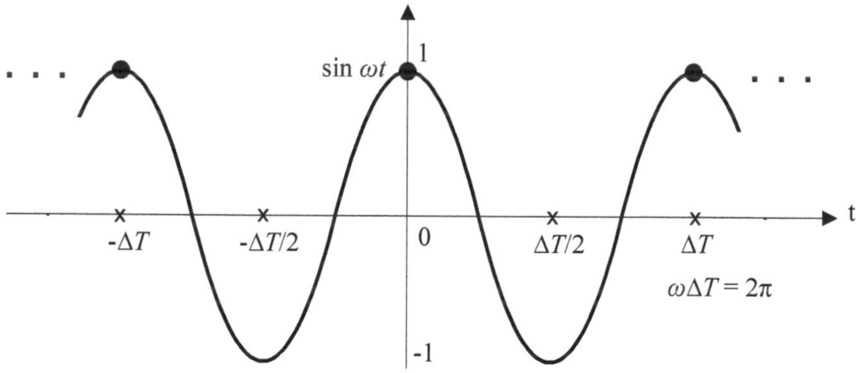

Figure 6.22: $\sin \omega t$ Sampled Every $2\pi/\omega$ Seconds

If the function $f(t) = \sin \omega t$ is sampled every $\Delta T = \dfrac{2\pi}{\omega}$ second, the sampled function will always have the value 1 and there is no way that we can recover the original function from these sampled values. This is why it is important that the value of the sampling interval ΔT should have a correct, small enough value if we want to recover an errorfree original function from the sampled values. In what follows we shall develop an exact relationship between the original function $f(t)$ and its sampled values $f(n\Delta T)$ and guidelines for the sampling interval ΔT such that no information in the continuous system is lost by sampling and the original signal can be fully recovered from these sampled values.

6.11.2 Sampling of a Band-Limited Time-Continuous Signal and Its Exact Reconstruction from Sampled Values

Consider a time-continuous signal $f(t)$. Its Fourier Transform is:

$$F[f(t)] = F(j\omega) \qquad -\infty < \omega < \infty$$

Let us sample this signal at time intervals ΔT and sampling frequency $\omega_s = 2\pi/\Delta T$. The sampled time signal is $f^*(t) = f(t) \sum_{n=-\infty}^{+\infty} \delta(t - n\Delta T) = f(t)S_{\Delta T}(t)$. The function $S_{\Delta T}(t)$ is a periodic function. Its Fourier expansion is:

$$S_{\Delta T}(t) = \frac{1}{\Delta T} \sum_{n=-\infty}^{+\infty} e^{jn\omega_s t} = \frac{1}{\Delta T} \sum_{-\infty}^{+\infty} e^{-jn\omega_s t}$$

$$f^*(t) = \frac{1}{\Delta T} \sum_{n=-\infty}^{+\infty} f(t)e^{-jn\omega_s t}, \qquad \omega_s = 2\pi/\Delta T \tag{6.79}$$

Taking the Fourier Transform of the function $f^*(t)$,

$$F\left[f^*(t)\right] = F^*(j\omega) = \frac{1}{\Delta T} \sum_{n=-\infty}^{+\infty} F\left[f(t)e^{-jn\omega_s t}\right], \quad F^*(j\omega) = \frac{1}{\Delta T} \sum_{n=-\infty}^{+\infty} F(j(\omega + n\omega_s))$$

Keeping in mind that $F(j\omega)$ may contain all the frequencies from $-\infty$ to $+\infty$ (unlimited bandwidth), let us ask the following question.

Given

$$F^*(j\omega) = \frac{1}{\Delta T} \sum_{n=-\infty}^{+\infty} F(j(\omega + n\omega_s))$$

Is it possible for us to filter this signal spectrum $F^*(j\omega)$ such that we recover the signal spectrum $F(j\omega)$? The answer to this question in general is **No**. But this question directs us to the solution of a practical and important problem involving band-limited signals discussed below. Consider a frequency band-limited signal $f(t)$, such that

$$F[f(t)] = F(j\omega) = 0, \qquad 0 \leq |\omega| \leq \omega_b \tag{6.80}$$

$$2\omega_b = \text{ Spectral bandwidth of the signal } f(t)$$

Let us sample this signal $f(t)$ at a sampling frequency $\omega_s = 2\pi/\Delta T$, such that $\omega_s \geq 2\omega_b$ (the bandwidth). This results in $F^*(j(\omega + n\omega_s))$, which are copies of $F(j\omega)$ with the magnitude $1/\Delta T$ such that there is no spectrum overlapping (aliasing) between $F^*(j\omega), F^*(j(\omega \pm 1\omega_s)) \ldots F^*(j(\omega \pm n\omega_s))$ as shown in Figure 6.23(a). It is clear from Figure 6.23(b) that when $\omega_s < 2\omega_b$ there is a spectrum overlapping (aliasing).

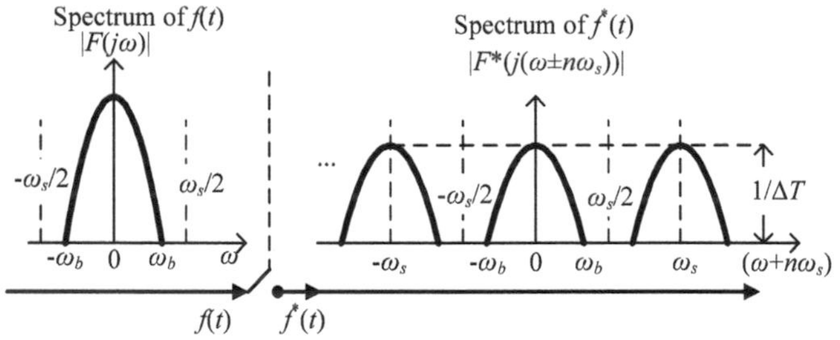

(a) Spectrum of $F(j\omega)$ and $\dfrac{1}{\Delta T}F^*(j(\omega + n\omega_s))$ for $\omega_s > 2\omega_b$.

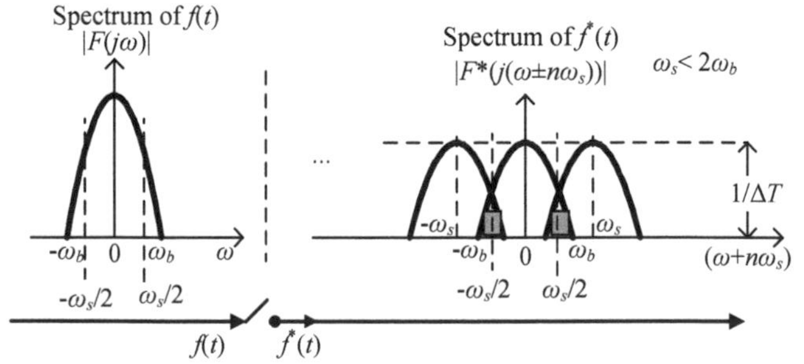

(b) Spectrum of $F(j\omega)$ and $\dfrac{1}{\Delta T}F^*(j(\omega + n\omega_s))$ for $\omega_s < 2\omega_b$.

Figure 6.23: Sampling of the Frequency Band-Limited Signals

It is easy to see that if we pass $f^*(t)$, $\omega_s \geq 2\omega_b$ through a low-pass recovery filter whose Transfer function $H(j\omega)$ is

$$H(j\omega) = \Delta T = \text{ Constant }, \qquad -\omega_b \leq \omega \leq \omega_b \tag{6.81}$$

$$= 0 \quad \text{otherwise}$$

$$h(t) = F^{-1}H(j\omega) = \frac{\sin(\pi t/\Delta T)}{(\pi t/\Delta T)}$$

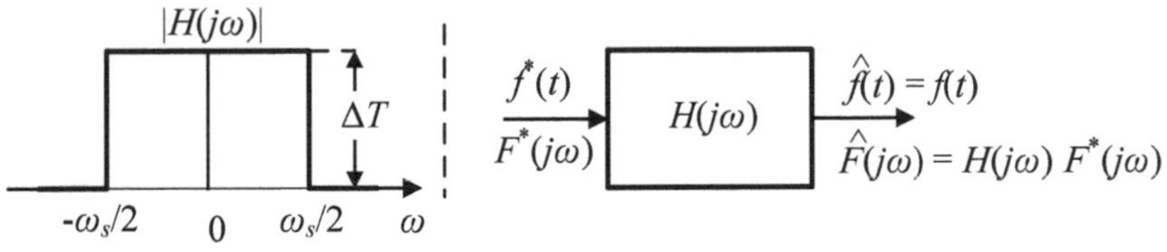

Figure 6.24: Transfer Function $H(j\omega)$ of a Low-Pass Recovery Filter (Reconstruction Filter).

The output of the recovery filter $\hat{F}(j\omega)$ is

$$\hat{F}(j\omega) = F^*(j\omega)H(j\omega) = F\left[f^*(t)\right]H(j\omega)$$

Taking the inverse of both sides

$$\hat{f}(t) = F^{-1}\left[F^*(j\omega)H(j\omega)\right] = f^*(t) * h(t)$$

where

$$f^*(t) = \sum_{n=-\infty}^{+\infty} f(n\Delta T)\delta(t - n\Delta T)$$

$$h(t) = F^{-1}H(j\omega) = \left(\frac{1}{2\pi}\right)\left(\frac{2\pi}{\omega_s}\right)\int_{-\omega_s/2}^{+\omega_s/2} e^{j\omega t}\, d\omega = \frac{\sin(\pi t/\Delta T)}{(\pi t/\Delta T)}$$

Thus,

$$\hat{f}(t) = f^*(t) * h(t) = \sum_{n=-\infty}^{+\infty} f(n\Delta T)\left[\delta(t - n\Delta T) * \frac{\sin(\pi t/\Delta T)}{\pi t/\Delta T}\right]$$

$$= \sum_{n=-\infty}^{+\infty} f(n\Delta T)\frac{\sin(\pi(t - n\Delta T)/\Delta T)}{\pi(t - n\Delta T)/\Delta T} \simeq f(t), \quad \Delta T = \frac{2\pi}{\omega_s}$$

Thus,

Time Sampling Theorem for a Frequency Band-Limited Signal

The signal $f(t)$ and its sampled version $f(n\Delta T)$ are related as

$$f(t) = \sum_{n=-\infty}^{+\infty} f(n\Delta T)\frac{\sin(\pi(t - n\Delta T)/\Delta T)}{(\pi(t - n\Delta T)/\Delta T)}, \qquad \omega_s = \frac{2\pi}{\Delta T} > 2\omega_b$$

$\Delta T = \dfrac{2\pi}{\omega_b}$ is called the **Nyquist Sampling Rate**

This result was first arrived at by Shannon. It essentially states that a frequency band-limited signal can be exactly interpolated from its samples provided the sampling rate $2\pi/\Delta T$ is greater than or equal to twice the largest frequency present in its spectrum. In practice the sampling rates are much higher than the theoretical minimum. Now we can appreciate the transmission of a band-limited signal via digital conversion. Analog signals are subject to all kinds of distortions. We convert the analog signal to a digital signal via ADC and transmit the digital signal, the signal at the receiving end is accurately received without spurious noises. The received signal is now converted at the receiving end via reconstructing filter with the impulse response $\left(\dfrac{\sin \pi t/\Delta T}{\pi t/\Delta T}\right)$ as shown in Figure 6.25. Furthermore, Figure 6.26 shows the time and frequency responses of the reconstruction of a band-limited signal.

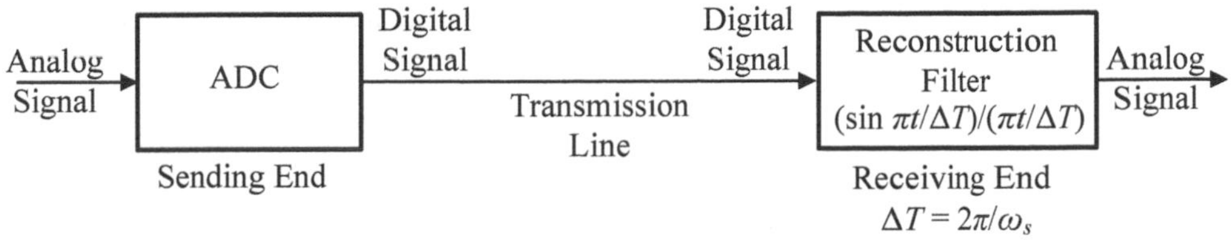

Figure 6.25: Distortionless Transmission

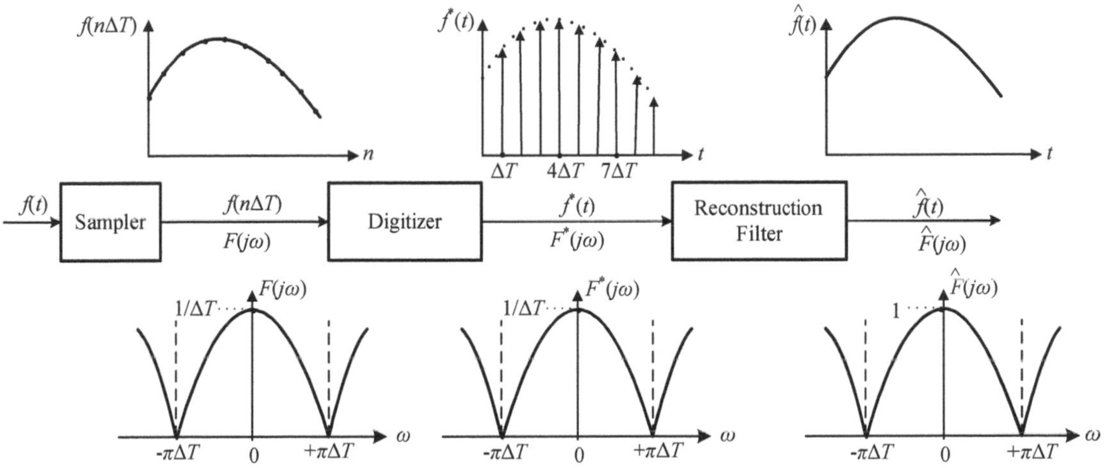

Figure 6.26: Reconstruction of a Band-Limited Signal

6.11.3 Fourier Series Revisited

In Chapter 1, we discussed the periodic signals and their Fourier series expansion from the vector spaces consideration. In this section we shall study the Fourier series from the computational viewpoint.

Fourier Series

Consider a periodic function $f(t)$ with a period T specified in the interval $-T/2 \leq t \leq T/2$, and satisfying the following conditions (**Dirichlet Conditions**):

1. $\displaystyle\int_{-T/2}^{T/2} |f(t)|\, \mathrm{d}t < \infty$

2. For $-T/2 < t < T/2$, the function $f(t)$ has a finite number of maxima and minima and points of discontinuity $t = t_i,\ i = 1, \ldots, n$.

3. At the point of discontinuity $t = t_i$, the function $f(t)$ is defined as:

$$f(t_i) = \frac{1}{2}\left[f\left(t_i^+\right) + f\left(t_i^-\right)\right]$$

Important: Such a function can be written as a Fourier series expansion.

$$f(t) = \sum_{n=-\infty}^{+\infty} c_n e^{jn\omega_0 t}, \quad \omega_0 = \frac{2\pi}{T}$$

$$\begin{aligned} c_n &= \frac{1}{T}\int_{-T/2}^{T/2} f(t)e^{-jn\omega_0 t}\,dt \quad \text{(complex quantity)},\ n = 0, \pm 1, \pm 2, \ldots \\ &= a_n + jb_n \quad (a_n \text{ and } b_n \text{ are real}) \end{aligned}$$

For $f(t)$ = a real function

$$c_{-n} = \overline{c_n} = a_n - jb_n$$

Notation:

$f(t) \leftrightarrow c_n$ implies

$$c_n = \frac{1}{T}\int_{-T/2}^{T/2} f(t)e^{-jn\omega_0 t}\,dt$$

$$f(t) = \sum_{n=-\infty}^{+\infty} c_n e^{jn\omega_0 t}$$

$f(t) \leftrightarrow c_n$ implies the time function $f(t)$ and its Fourier Coefficients c_n.

Table 6.4: Properties of Fourier Series

No.	Time Function $f(t)=\sum_{n=-\infty}^{+\infty} c_n e^{jn\omega_0 t},\ \omega_0=\frac{2\pi}{T}$	Fourier Coefficient $c_n=\frac{1}{T}\int_{-T/2}^{+T/2} f(t)e^{-jn\omega_0 t}\,\mathrm{d}t$
1	Time Delay $f(t-t_0)$	$c_n e^{-jn\omega_0 t_0}$
2	Circular Time Convolution $f_1(t) * f_2(t)=\frac{1}{T}\int_{-T/2}^{+T/2} f_1(\tau)f_2(t-\tau)\,\mathrm{d}\tau$	$c_{1n}c_{2n}$
3	Fourier Coefficient Convolution $f_1(t)f_2(t)$	$c_{1n} * c_{2n}=\sum_{k=-\infty}^{+\infty} c_{1n-k}c_{2k}$
4	Parseval's Theorem $f^2(t)$ $\frac{1}{T}\int_{-T/2}^{T/2} f^2(t)\,\mathrm{d}t$	$\sum_{k=-\infty}^{+\infty} c_{n-k}c_k$ $\sum_{k=-\infty}^{+\infty} c_{-k}c_k$ $=\sum_{k=-\infty}^{+\infty} \lvert c_k\rvert^2$ when $f(t)$ is real
5	$\sum_{n=-\infty}^{+\infty} \delta(t+nT)$	$\frac{1}{T}$
6	Poisson Sum $\sum_{n=-\infty}^{+\infty} f(t+nT)$	$\frac{1}{T}F[f(t)]_{\omega=n\omega_0}=\frac{1}{T}F(jn\omega_0)$
7	z-Transform as Fourier Series $F\left(e^{j\omega T}\right)=\sum_{n=-\infty}^{+\infty} f(n)e^{-jn\omega T}$ $=\sum_{n=-\infty}^{+\infty} f(n)z^{-n}\Big\rvert_{z=e^{j\omega T}}$	$f(n)=\frac{T}{2\pi}\int_{-\pi/T}^{\pi/T} F\left(e^{j\omega T}\right)e^{jn\omega T}\,\mathrm{d}\omega$ $=\frac{1}{2\pi}\int_{-\pi}^{\pi} F\left(e^{j\theta}\right)e^{jn\theta}\,\mathrm{d}\theta$

Most of the properties of the Table 6.4 follow from the definition.

We shall prove a few of these properties

Property 1: Fourier Series of Delayed Time Function $f(t - t_0)$

From Fourier expansion $f(t) = \sum_{n=-\infty}^{\infty} c_n e^{jn\omega_0 t}, \quad \omega_0 = 2\pi/T$

Replacing t with $t - t_0$ on both sides of the above expression,

$$f(t - t_0) = \sum_{n=-\infty}^{\infty} c_n e^{jn\omega_0(t-t_0)} = \sum_{n=-\infty}^{\infty} \left(c_n e^{-jn\omega_0 t_0}\right) e^{jn\omega_0 t}, \text{ yielding}$$

$$f(t) \leftrightarrow c_n$$

$$f(t - t_0) \leftrightarrow c_n e^{-jn\omega_0 t_0} \tag{6.82}$$

Property 2: Time Domain Circular Convolution

$$\text{If} \quad f_1(t) \leftrightarrow c_{1n}, \quad f_2(t) \leftrightarrow c_{2n}, \text{ then}$$

$$f_1(t) * f_2(t) = \left[\frac{1}{T}\int_{-T/2}^{T/2} f_1(\tau) f_2(t-\tau)\, d\tau\right] \leftrightarrow c_{1n} c_{2n}$$

Proof:

$$\frac{1}{T}\int_{-T/2}^{T/2} f_1(\tau) f_2(t-\tau)\, d\tau = \frac{1}{T}\int_{-T/2}^{T/2} \left(\sum_{n=-\infty}^{+\infty} c_{1n} e^{jn\omega_0\tau}\right)\left(\sum_{k=-\infty}^{+\infty} c_{2k} e^{jk\omega_0(t-\tau)}\right) d\tau$$

$$= \sum_{k=-\infty}^{+\infty}\left[\sum_{n=-\infty}^{+\infty} c_{1n} c_{2k}\left(\frac{1}{T}\int_{-T/2}^{T/2} e^{j(n-k)\omega_0\tau}\, d\tau\right)\right] e^{jk\omega_0 t}$$

But,

$$\frac{1}{T}\int_{-T/2}^{T/2} e^{j(n-k)\omega_0\tau}\, d\tau = \delta(n-k) = \begin{cases} 0 & n \neq k \\ 1 & n = k \end{cases}$$

Thus,

$$f_1(t) * f_2(t) = \frac{1}{T}\int_{-T/2}^{T/2} f_1(\tau)f_2(t-\tau)\,\mathrm{d}\tau = \sum_{n=-\infty}^{\infty}(c_{1n}c_{2n})\,e^{jn\omega_0 t} \tag{6.83}$$

or

$$f_1(t) * f_2(t) \leftrightarrow c_{1n}c_{2n}$$

Property 5: Fourier Series of Impulse Train (or Comb Function) $S_T(t)$

Show,

$$S_T(t) = \sum_{n=-\infty}^{+\infty}\delta(t+nT) = \frac{1}{T}\sum_{n=-\infty}^{+\infty}e^{jn\omega_0 t}, \quad \omega_0 = \frac{2\pi}{T}$$

Proof:

Since function $\sum_{n=-\infty}^{\infty}\delta(t+nT)$ is a periodic function, its Fourier series can be written as

$$S_T(t) = \sum_{n=-\infty}^{\infty}\delta(t+nT) = \sum_{n=-\infty}^{\infty}c_n e^{jn\omega_0 t}$$

Multiplying both sides with $(1/T)e^{-jk\omega_0 t}$ and integrating between the interval $-T/2$ to $T/2$

$$\sum_{n=-\infty}^{\infty}\frac{1}{T}\int_{-T/2}^{T/2}\delta(t+nT)e^{-jk\omega_0 t}\,\mathrm{d}t = \sum_{n=-\infty}^{\infty}\frac{c_n}{T}\int_{-T/2}^{T/2}e^{j(n-k)\omega_0 t}\,\mathrm{d}t$$

or

$$\frac{1}{T}e^{jkn\omega_0 T} = \sum_{n=-\infty}^{\infty}\frac{c_n}{T}\int_{-T/2}^{T/2}e^{j(n-k)\omega_0 t}\,\mathrm{d}t$$

But

$$e^{jkn\omega_0 T} = 2\pi jkn = 1$$

$$\int_{-T/2}^{T/2}e^{j(n-k)\omega_0 t}\,\mathrm{d}t = T\delta(n-k)$$

Thus,

$$\frac{1}{T} = \sum_{n=-\infty}^{\infty} \left(\frac{c_n}{T}\right)[T\delta(n-k)] = \sum_{n=-\infty}^{\infty} c_n\delta(n-k)$$

or

$$\frac{1}{T} = c_k = c_n$$

Thus,

$$S_T(t) = \sum_{n=-\infty}^{\infty} \delta(t+nT) \leftrightarrow \frac{1}{T} \tag{6.84}$$

Property 6: Poisson's Sum

If,

$$F(j\omega) = F[f(t)] = \int_{-\infty}^{+\infty} f(t)e^{-j\omega t}\,\mathrm{d}t$$

Then,

$$\sum_{n=-\infty}^{+\infty} f(t+nT) = \frac{1}{T}\sum_{n=-\infty}^{+\infty} F(jn\omega_0)e^{jn\omega_0 t} \qquad \text{(Poisson's Sum)}$$

Proof:

$$\begin{aligned}
\sum_{n=-\infty}^{+\infty} f(t+nT) &= \sum_{n=-\infty}^{+\infty}\left[\int_{-\infty}^{+\infty} f(\tau)\delta(t+nT-\tau)\,\mathrm{d}\tau\right] \\
&= \int_{-\infty}^{+\infty} f(\tau)\left[\sum_{n=-\infty}^{+\infty} \delta(t+nT-\tau)\right]\mathrm{d}\tau \\
&= \int_{-\infty}^{+\infty} f(\tau)S_T(t-\tau)\,\mathrm{d}\tau \\
&= f(t) * S_T(t) \\
&= \frac{1}{T}\sum_{n=-\infty}^{+\infty}\left[\int_{-\infty}^{+\infty} f(\tau)e^{-jn\omega_0\tau}\,\mathrm{d}\tau\right]e^{jn\omega_0 t} \\
&= \frac{1}{T}\sum_{n=-\infty}^{+\infty} F(jn\omega_0)e^{jn\omega_0 t}
\end{aligned}$$

Both sides of the above expression can be considered as the output of a filter whose impulse response is $f(t)$ (or Transfer function $F(s)$) and whose input is either a train of impulses $\sum_{n=-\infty}^{\infty} \delta(t+nT)$ or a summation of exponentials $\sum_{n=-\infty}^{\infty} 1/T e^{jn\omega_0 t}$

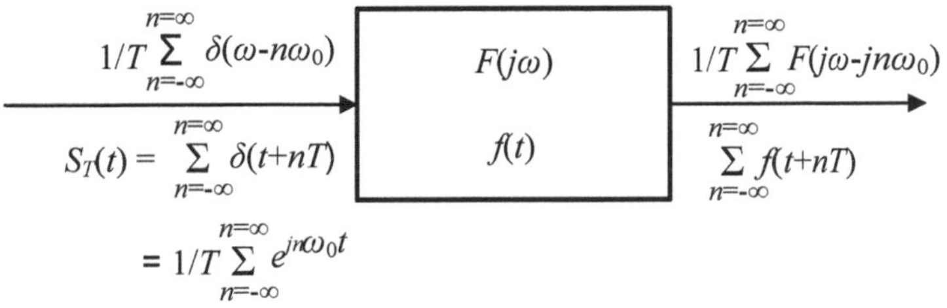

Figure 6.27: Filter with Comb Function $S_T(t)$ Input.

The next table presents a short collection of the Fourier series of some periodic and aperiodic functions.

Table 6.5: Short Collection of the Fourier Series

No.	Fourier Series of Periodic and Aperiodic Signals
1	Triangular Wave $f(t) = -f(-t)$ $f(t) = \frac{8k}{\pi^2}\sum_{n=0}^{\infty}\frac{1}{(2n+1)^2}\sin\left(\frac{(2n+1)\pi}{2}\right)\sin(2n+1)\frac{\omega_0 t}{T}$ 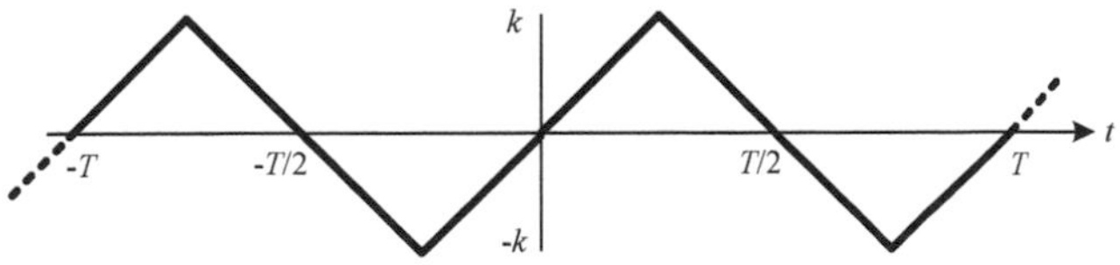

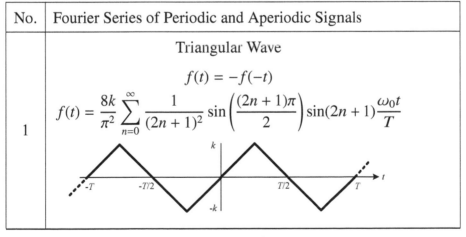

(continued)

No.	Fourier Series of Periodic and Aperiodic Signals
2	Square Wave $f(t) = -f(-t)$ $f(t) = \frac{4k}{\pi} \sum_{n=0}^{\infty} \frac{1}{(2n+1)} \sin(2n+1)\frac{\omega_0 t}{T}$ For $T = 2\pi$ $f(t) = \frac{4k}{\pi} \sum_{n=0}^{\infty} \frac{1}{(2n+1)}$ 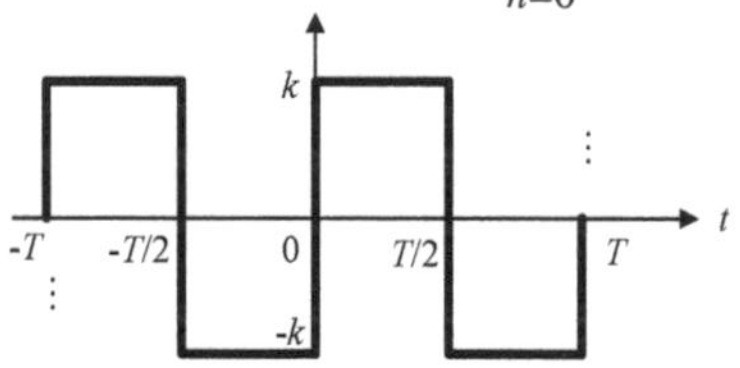
3	Decaying Exponential $f(t) = e^{-t}, \ -1 \le t \le 1$ $f(t) = \sum_{-\infty}^{+\infty} \frac{(-1)^n (1 - jn\pi) \sinh(1) e^{jn\pi t}}{1 + n^2\pi^2}$ 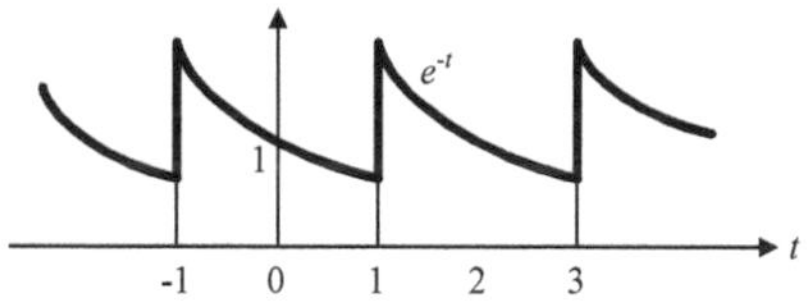
4	Full Sine Wave-Rectified $f(t) = k \sin t, \ 0 \le t \le \pi$ $f(t) = \frac{2k}{\pi} - \frac{2k}{\pi} \sum_{n=1}^{\infty} \frac{\left(e^{jn\omega_0 t/T} + e^{-jn\omega_0 t/T}\right)}{(4n^2+1)}, \quad \omega_0 = \frac{2\pi}{T}$ 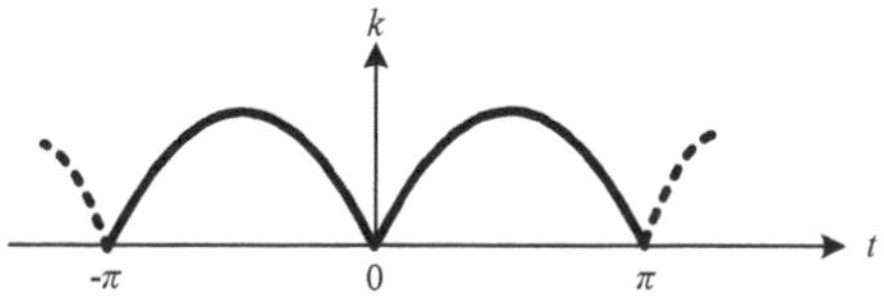

(continued)

No.	Fourier Series of Periodic and Aperiodic Signals
5	Periodic Gate Function $f(t) = k, \ \|t\| \leq d$ $f(t) = 0, \text{for } d < \|t\| \leq T$ $f(t) = \frac{kd}{T} \sum_{n=-\infty}^{+\infty} \text{sinc}\left(\frac{n\pi d}{T}\right) e^{j2\pi nt/T}, \quad \omega_0 = \frac{2\pi}{T}$ 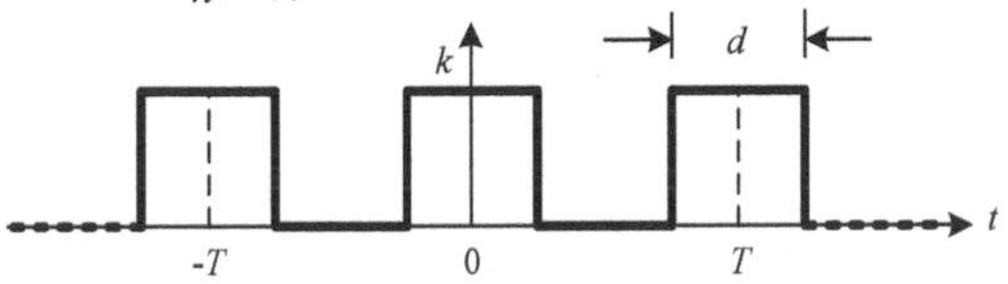
6	Quadratic Function $f(t) = t^2, \ -\pi \leq t \leq \pi$ $f(t) = \frac{\pi^2}{3} + 4\left(\sum_{n=1}^{\infty}(-1)^n \frac{\cos nt}{n^2}\right)$ 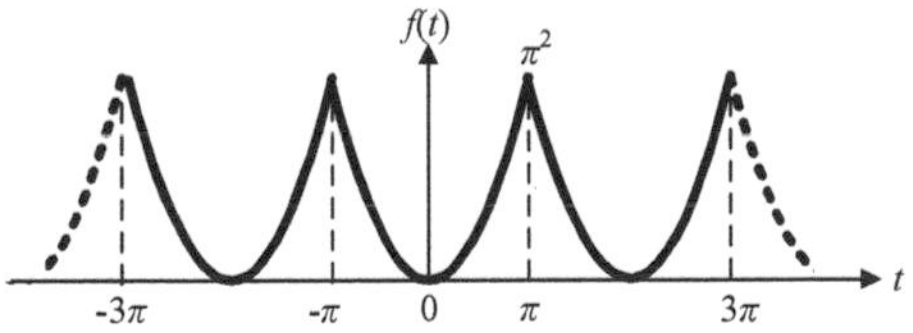
7	$f(t) = t,$ $-\pi < t < \pi$ $f(t) = 2\left(\sum_{n=1}^{\infty}(-1)^{n+1} \frac{\sin t}{n}\right) = t$ 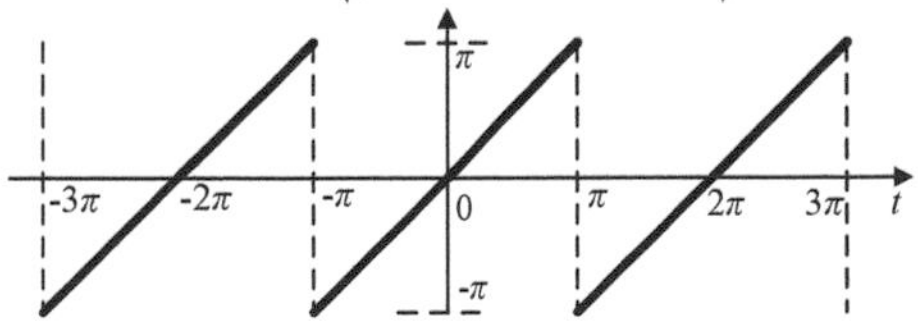
8	$f(t) = At^2 + Bt + c,$ $0 \leq t \leq 2\pi$ $f(t) = \frac{4A}{3}\pi^2 + B\pi + c + 4A \sum_{n=1}^{\infty} \frac{\cos nt}{n^2} + 4\left(\pi A + \frac{B}{2}\right) \sum_{n=1}^{\infty} \frac{\sin nt}{n}$

(continued)

No.	Fourier Series of Periodic and Aperiodic Signals
	Special Case
(i)	$A = C = 0, B = 1/2$ $t = \pi + 2\sum_{n=1}^{\infty} \frac{\sin nt}{n}$
(ii)	$A = \frac{1}{4\pi}, B = -\frac{1}{2}$ $\sum_{n=1}^{\infty} \frac{\cos nt}{n^2} = \frac{3t^2 - 6\pi t + 2\pi^2}{12}$ 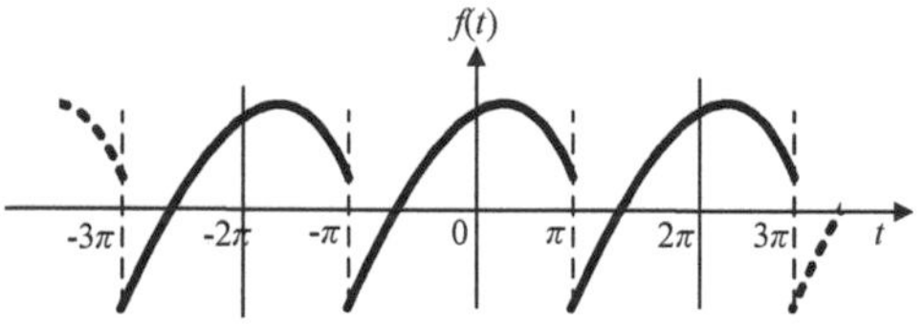
9	(i) From Fourier Series of $f(t) = t^2, \quad -\pi \leq t < \pi$ $\sum_{n=1}^{\infty} (-1)^{n+1} \frac{\cos nx}{n^2} = \frac{\pi^2 - 3t^2}{12}$ (ii) From Fourier Series of $f(t) \begin{cases} = 1 & 0 < x < \pi \\ = -1 & \pi < x < 2\pi \end{cases}$ $\frac{\pi}{4} = \sum_{n=0}^{\infty} \frac{\sin(2n+1)t}{(2n+1)}$ (iii) From (8) $\frac{\pi^2}{6} = \sum_{n=1}^{\infty} \frac{1}{n^2}$ $\frac{\pi^2}{12} = 1 - \sum_{n=1}^{\infty} \frac{1}{(n+1)^2}$ $\frac{\pi}{4} = \sum_{n=0}^{\infty} (-1)^n \frac{1}{(2n+1)}$

6.11.4 Discrete Fourier Transforms or Discrete Fourier Series and Fast Fourier Transform Computation Algorithm

Introduction

Up to now we have looked at the Transform theory only from a theoretical viewpoint. The time function $f(t)$ is viewed as known for $-\infty < t < \infty$. Its Fourier Transform $F(j\omega)$ is supposed to be computed at all frequencies $-\infty < \omega < \infty$. In case the function $f(t)$ is periodic, theoretically one is required to find all its coefficients c_n, $n = 0, \pm 1, \ldots, \pm\infty$. But in most physical problems requiring some sort of physical measurements, the function $f(t)$ is known only at discrete and finite time intervals. Furthermore, the computing considerations dictate that the Transforms be computed at discrete frequencies. This involves two basic steps of replacing infinite sums and infinite integrals with finite sums. Thus, the need for the study of Transforms involving finite sums. For convenience, in this section, sometimes we shall use the symbol $F(f)$ instead of $F(j\omega)$.

Aperiodic Signal Representation by Its Discrete Fourier Series (DFS)

Most of the practical signals are time-duration limited. From the Fourier Transform pair, it is clear that the signal cannot be time-duration limited and frequency Band-Limited simultaneously or vice versa. However, in most situations the "tail" of $F(j\omega)$ can be neglected because it contains a negligible fraction of the signal energy. In what follows, we shall use two approaches to obtain the discrete Fourier series of an Aperiodic signal.

1. Approach #1

   Both the time function $f(t)$ and its Fourier Tranform $F(j\omega)$ are treated as "sample and hold" sequence of N points. The definition of Fourier Transform and its inverse is used to arrive at the discrete Fourier Transform function.

2. Approach #2

   The limited time duration Aperiodic function $f(t)$ is sampled for the time duration and then periodically repeated. The corresponding frequency spectrum is also sampled and periodically repeated. The Fourier series expansion is used to arrive at the discrete Transform pair.

The Fourier Transform pair for a time-limited and a band-limited signal is:

$$F(f) = f_0 \int_0^T f(t)e^{-j2\pi f_0 t}\,\mathrm{d}t, \qquad f(t) = \frac{1}{f_0}\int_0^{f_0} F(f)e^{j2\pi f_0 t}\,\mathrm{d}f$$

Derivation of DFT

Approach #1

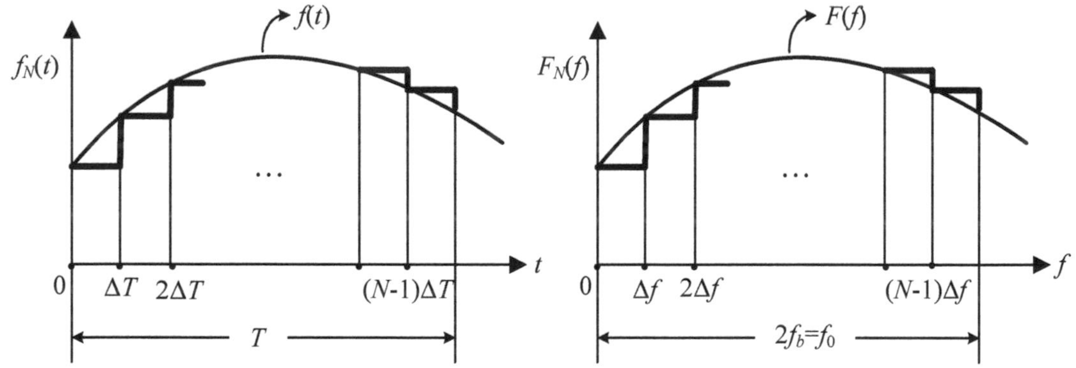

Figure 6.28: N-Point Staircase Representation of $f(t)$ and Its Fourier Spectra $F(f)$

$$f_N(t) = \sum_{k=0}^{N}[\Delta f(k\Delta T)]u(t - k\Delta T), \quad 0 \le t \le N\Delta t$$

$$F_N(t) = \sum_{k=0}^{N}\Delta F(k\Delta f)u(f - k\Delta f), \quad 0 \le f \le N\Delta f$$

$f_0 = N\Delta f$ $\quad$ N frequency samples, $\quad$ $T = N\Delta T$ $\quad$ N time samples

We shall choose, as explained later, $(\Delta T)(\Delta f) = \frac{1}{N}$, $\quad T f_0 = N$

Derivation

From the definition

$$f(t) = \int_0^{\infty} f(\tau)\delta(t-\tau)\,\mathrm{d}\tau = \int_0^{T} f(\tau)\delta(t-\tau)\,\mathrm{d}\tau \tag{6.85}$$

$$f(\tau) = 0 \quad 0 < \tau < T \tag{6.86}$$

$$\delta(t-\tau) = \int_{-\infty}^{\infty} e^{j2\pi(t-\tau)f}\,\mathrm{d}f = \int_0^{f_0} e^{j2\pi(t-\tau)f}\,\mathrm{d}f \tag{6.87}$$

(since the spectrum of interest is from 0 to f_0)

From Eqs. 6.85 and 6.87

$$f(t) = \int_0^{T}\int_0^{f_0} f(\tau)e^{j2\pi(t-\tau)f}\,\mathrm{d}f\,\mathrm{d}\tau = \frac{1}{f_0}\int_0^{f_0}\left[f_0\int_0^{T} f(\tau)e^{-j2\pi\tau f}\,\mathrm{d}\tau\right]e^{j2\pi tf}\,\mathrm{d}f$$

The multiplying factor of f_0 and $1/f_0$ is used to comply with the established definitions but are not really necessary. Let

$$f_0\int_0^{T} f(\tau)e^{-j2\pi\tau f}\,\mathrm{d}\tau = f_0\int_0^{T} f(t)e^{-j2\pi tf}\,\mathrm{d}t = F(f) \tag{6.88}$$

$$f(t) = \frac{1}{f_0}\int_0^{f_0} F(f)e^{j2\pi tf}\,\mathrm{d}f \tag{6.89}$$

Define

$$f_N(k) = f_N(k\Delta T) = f(t)|_{t-k\Delta T}$$

$$F_N(n) = F_N(n\Delta f) = F(f)|_{f=n\Delta f}$$

In Eqs. 6.88 and 6.89, replacing

$$t \text{ with } k\Delta T,\ f \text{ with } n\Delta f,\ f(t) \text{ with } f_N(k),\ F(f) \text{ with } F_N(n)$$

$$\text{Integration with summation} \begin{cases} f(t) \text{ is constant between } k\Delta T \text{ and } (k+1)\Delta T \\ F(f) \text{ is constant between } n\Delta f \text{ and } (n+1)\Delta f \end{cases}$$

we obtain

$$F_N(n) = N\Delta T\Delta f \sum_{k=0}^{N-1} f_N(k) e^{j2\pi nk\Delta T\Delta f}$$

$$f_N(k) = \frac{\Delta f}{N\Delta f} \sum_{n=0}^{N-1} F_N(n) e^{-j2\pi nk\Delta t\Delta f}$$

Let $\Delta T\Delta f = \frac{1}{N}$, $\quad e^{j2\pi/N} = W_N$

We obtain the discrete Fourier Transform pair (DFT)

$$F_N(n) = \sum_{k=0}^{N-1} f_N(k) W_N^{nk} \tag{6.90}$$

$$f_N(k) = \frac{1}{N} \sum_{k=0}^{N-1} F_N(n) W_N^{-nk} \tag{6.91}$$

Assuming negligible aliasing (as long as $n\Delta f \le 1/(2\Delta T)$ or $2n \le N$), the discrete Fourier Transform $F_N(n)$ adequately represents the Fourier Transform $F(f)$ of a time-limited and a frequency band-limited function.

Note : The time duration T of the signal $f(t)$ can be extended to any larger time $T_1 > T$ by defining $f(t) \equiv 0$ for $T < t < T_1$ (zero time padding). Similarly the bandwidth can be extended to f_1 by defining $F(f) \equiv 0$ for $f_0 < f < f_1$

Approach #2—Periodic Representation of Time-Limited Aperiodic Signal

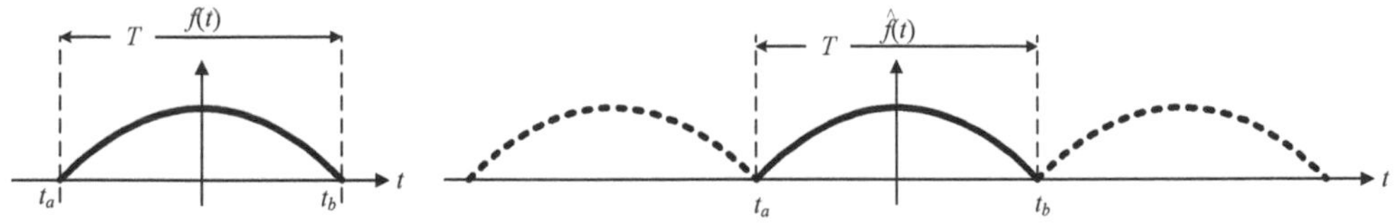

Figure 6.29: Time-Limited Aperiodic Signal and Its Periodic Representation

To derive DFS of the time-limited Aperiodic signal $f(t)$ we construct a periodic signal $\hat{f}(t)$ such that

$$\hat{f}(t) = f(t), \qquad t_a \leq t < t_b, \qquad t_b - t_a = T$$

$$\hat{f}(t) = f(t) = 0 \qquad \text{outside the range } (t_a, t_b)$$

$$\hat{f}(t) = \hat{f}(t + nT)$$

Now $\hat{f}(t)$ is periodic in the range of interest $-\infty < t < \infty$. It is the same signal as $f(t)$ but only in the range (t_a, t_b).

Furthermore, changing the time by translation results in $0 \leq t \leq T$.

Using the Fourier series expansion

$$\hat{f}(t) = f(t) = \sum_{n=-\infty}^{+\infty} c_n e^{jn\omega_0 t}, \qquad \omega_0 = \frac{2\pi}{T}, \qquad n\omega_0 = \omega$$

$$t_a \leq t \leq t_b \tag{6.92}$$

$$t_b - t_a = T$$

$$c_n = \frac{1}{T} \int_0^T f(t) e^{-jn\omega_0 t}\, \mathrm{d}t, \qquad n = 0, \pm 1, \ldots, \pm\infty$$

c_n are known as Fourier coefficients belonging to frequency $\omega = n\omega_0$.

Since our measurements of $f(t)$ are usually at discrete uniform intervals of time, we consider the sampled values of $f(t)$ at $t = k\Delta T,\ k = 0, 1, \ldots, N-1, \Delta T = T/N$.

Let us define

$$f(t)\Big|_{t=k\Delta T} = f_N(k), \quad 0 \le t \le T, \quad \Delta T = T/N \quad (N\text{-point sequence}) \tag{6.93}$$

$$e^{-jn\omega_0 t}\Big|_{t=k\Delta T} = e^{-j(nk)(\omega_0 \Delta T)} = \left(e^{-j2\pi/N}\right)^{nk} = W_N^{-nk} \tag{6.94}$$

$$c_N(n) = n\text{-th Fourier coefficient for } N\text{-point sequence } f_N(k) \begin{cases} k = 0, 1, \ldots, N-1 \\ n = 0, \pm 1, \ldots, \pm\infty \end{cases}$$

$$W_N = e^{j2\pi/N}, \quad N\text{-th root of unity (very important entity)} \tag{6.95}$$

Thus, the Discrete Fourier Series **DFS** is

DFS of a N-point Discrete Sequence

$$\begin{aligned} f_N(k) &= \sum_{n=-\infty}^{+\infty} c_N(n) W_N^{(nk)} \qquad n = 0, \pm 1, \ldots, \pm\infty \\ c_N(n) &= \frac{1}{N} \sum_{k=0}^{N-1} f_N(k) W_N^{-(nk)} \qquad k = 0, 1, \ldots, N-1 \end{aligned} \tag{6.96}$$

The only drawback in the computation of Eq. 6.96 is that there are infinite coefficients $c_N(n)$ to be computed. From Eq. 6.95, we realize an **interesting character of** $c_N(n)$ **and** W_N, having the **cyclic property** with a period N and hence requires only N coefficients $c_N(n)$ to be computed.

Looking at the coefficient $c_N(n + mN), \ m = 0, \pm 1, \pm 2, \ldots, \pm\infty$

$$N c_N(n + mN) = \sum_{k=0}^{N-1} f_N(k) \left(W_N\right)^{-(n+mN)k} = \sum_{k=0}^{N-1} f_N(k) W_N^{-(nk)} W_N^{-(nkN)} \tag{6.97}$$

But

$$W_N^{-nkN} = \left(W_N^N\right)^{-(nk)} = (1)^{-(nk)} = 1 \tag{6.98}$$

Hence,

$$Nc_N(n + mN) = \sum_{k=0}^{N-1} f_N(k) W_N^{-(nk)} = Nc_N(n) \qquad (6.99)$$

Therefore, $c_N(n)$ is a periodic sequence with a period N and we need only compute $c_N(n)$ for $n = 0, 1, \ldots, N-1$ even though the summation in Eq. 6.96 requires n from $-\infty$ to $+\infty$.

Thus, Eq. 6.96 can be rewritten as

$$f_N(k) = \frac{1}{N}\left[\sum_{m=-\infty}^{+\infty} Nc_N(0 + mN) + \sum_{m=-\infty}^{+\infty} Nc_N(1 + mN) W_N^k + \cdots \right.$$
$$\left. \cdots + \sum_{m=-\infty}^{+\infty} Nc_N(N - 1 + mN) W_N^{k(N-1)} \right]$$

or

$$f_N(k) = \frac{1}{N} \sum_{n=0}^{N-1} \left[\sum_{m=-\infty}^{+\infty} Nc_N(n + mN) \right] W_N^{nk} \qquad (6.100)$$

Let us define

$$\sum_{m=-\infty}^{+\infty} Nc_N(n + mN) = F_N(n) \qquad (6.101)$$

Summarizing the above results

$$\begin{aligned} f_N(k) &= \frac{1}{N} \sum_{n=0}^{N-1} F_N(n) W_N^{(nk)} & \qquad f_N(k) &\leftrightarrow F_N(n) \\ F_N(n) &= \sum_{k=0}^{N-1} f_N(k) W_N^{-(nk)} & \qquad F_N(n) &\leftrightarrow f_N(k) \end{aligned} \qquad (6.102)$$
$$W_N = e^{j2\pi/N} = N\text{-th root of unity}$$

It is interesting to note that $f_N(k)$ is an N-point time data of $f(t)$ for $0 \leq t \leq T$ belonging to the sampling time $t_k = k/\Delta T = k\omega_0/2\pi$ and computes its Fourier spectrum. On the other hand, $F_N(n)$ is an N-point Frequency data which computes the N-point time data.

One can easily obtain $F_N(n)$ from $f_N(k)$ by multiplying both sides of the first equation in Eq. 6.102 with W_N^{-pk} and the summation of all the products so formed from $k = 0, 1, \ldots, N-1$ such that

$$N\sum_{k=0}^{N-1} f_N(k)W_N^{-pk} = \sum_{k=0}^{N-1}\sum_{n=0}^{N-1} F_N(n)W_N^{(n-p)k} = \sum_{n=0}^{N-1} F_N(n)\left[\sum_{k=0}^{N-1} W_N^{(n-p)k}\right]$$

Now

$$\sum_{k=0}^{N-1} W_N^{(n-p)k} = \begin{cases} \dfrac{1-W_N^{(p-n)N}}{1-W_N} = 0 & p \neq n, \quad W_N^{(p-n)N} = 1 \\ N & p = n, \quad W_N^0 = 1 \end{cases}$$

Hence,

$$\boxed{F_N(n) = \sum_{k=0}^{N-1} f_N(k)W_N^{-(nk)}}$$

6.11.5 Computation of $F_N(n)$ from $f_N(k)$ and Vice Versa

From Eq. 6.102

$$f_N(0) = \frac{1}{N}\left[F_N(0) + F_N(1) + \cdots + F_N(N-1)\right]$$

$$f_N(1) = \frac{1}{N}\left[F_N(0) + F_N(1)W_N + \cdots + F_N(N-1)W_N^{(N-1)}\right]$$

$$\cdots$$

$$f_N(N-1) = \frac{1}{N}\left[F_N(0) + F_N(1)W_N^{(N-1)} + \cdots + F_N(N-1)\,(W_N)^{(N-1)^2}\right]$$

or

$$\boldsymbol{f}_N = \frac{1}{N}\boldsymbol{W}_N\boldsymbol{F}_N$$

where

$$\boldsymbol{W}_N = \begin{bmatrix} 1 & 1 & \cdots & 1 \\ 1 & W_N & & W_N^{(N-1)} \\ \vdots & \vdots & & \\ 1 & W_N^{(N-1)} & \cdots & W_N^{(N-1)^2} \end{bmatrix}, \quad \boldsymbol{f}_N = \begin{bmatrix} f_N(0) \\ f_N(1) \\ \cdots \\ f_N(N-1) \end{bmatrix}, \quad \boldsymbol{F}_N = \begin{bmatrix} F_N(0) \\ F_N(1) \\ \cdots \\ F_N(N-1) \end{bmatrix}$$

Note:

$$\left.\begin{array}{l} F_N(n) = F_N(n+mN) \\ f_N(k) = f_N(k+mN) \end{array}\right| \begin{array}{l} \text{both are periodic sequences with a period } N \\ k, n = 0, 1, \cdots, N-1 \qquad m = 0, \pm 1, \ldots, \pm\infty \end{array}$$

It is easy to verify that

$$\boldsymbol{W}_N^{-1} = \frac{1}{N}\begin{bmatrix} 1 & 1 & \cdots & 1 \\ 1 & W_N^{-1} & & W_N^{-(N-1)} \\ \vdots & & & \\ 1 & W_N^{-(N-1)} & \cdots & W_N^{-(N-1)^{N-1}} \end{bmatrix}$$

$$\boldsymbol{W}_N^a \boldsymbol{W}_N^b = \boldsymbol{W}_N^{(a+b)}$$

6.11.6 Aliasing Error of Numerical Computation of DFT Due to the Time-Limited Signal Restriction

Once again the reader is reminded that from the Fourier Transform properties it is clear that Time-Limited signals are not Frequency Band-limited and vice versa. This is an inherent source of error that can be reduced by the judicial choice of N. But the error can never be eliminated. The coefficient Nc_n represents the actual $(n\omega_0)$ frequency content while $F_N(n)$ is the computed value. In general Nc_n and

$F_N(n)$ are two different quantities.

The difference $e_N(n) = (F_N(n) - Nc_n)$ is called the Aliasing Error

In fact $F_N(n)$ is sometimes referred to as "aliased Fourier coefficient of $f(t)$." In general if Nc_n are known one can easily find $F_N(n)$ via Eq. 6.101. But there is no way to find Nc_n from given $F_N(n)$ in a general case. However, if the time function is time-limited as well as band-limited, with its largest harmonic $M\omega_0 < (N/2)\omega_0$, then given $F_N(n)$ yields Nc_n. For such a time function $f(t)$,

$$f(t) = \sum_{n=-M}^{M} c_n e^{jn\omega_0 t}, \quad 2M + 1 < N$$
$$c_n \equiv 0 \quad \text{when} \quad |n| > M$$

From Eq. 6.101

$$F_N(n) = c_n$$

implying

$$e_N(n) = 0 \qquad -M < n \leq M$$

For a general case, if $M \geq N/2$ then even though we can find $f_N(k)$ and $F_N(n)$, the original time function $f(t)$ is nonrecoverable.

Table 6.6: Properties of **DFT**

No.	Original Time Function $f_N(k)$, $k = 0, 1, \ldots, N-1$	Discrete Fourier Transform $F_N(n)$, $n = 0, 1, \ldots, N-1$
	$W_N = e^{j2\pi/N}$	
1	Definition $f_N(k) = \frac{1}{N}\sum_{n=0}^{N-1} F_N(n) W_N^{nk}$	$F_N(n) = \sum_{k=0}^{N-1} f_N(k) W^{-nk}$
	$f_N(k) \leftrightarrow F_N(n)$	**DFT**
	$F_N(n) \leftrightarrow f_N(k)$	**DFT** Inverse

(continued)

No.	Original Time Function $f_N(k)$ $k = 0, 1, \dots, N-1$	Discrete Fourier Transform $F_N(n)$ $n = 0, 1, \dots, N-1$
	$W_N = e^{j2\pi/N}$	
2	Reversal Property $f_N(-k)$	$F_N(-n)$
3	Time Shift $f_N(k-m)$	$W_N^{-km} F_N(n)$
4	Frequency Shift $W_N^{km} f_N(k)$	$F_N(n-m)$
5	Cyclic Convolution ⊛ $f_N(1(k)) \circledast f_N(2(k))$	$NF_N(1(n))F_N(2(n))$
6	Product Theorem $f_N(1(k))f_N(2(k))$	$\sum_{m=0}^{N-1} F_N(1(m))F_N(2(n-m))$
7	Sum of Sequence $\sum_{k=0}^{N-1} f_N(k)$	$F_N(0)$
8	Initial Value $f_N(0)$	$N \sum_{n=0}^{N-1} F_N(n)$
9	Parseval's Theorem $\sum_{k=0}^{N-1} \lvert f_N(k)\rvert^2$	$N \sum_{n=0}^{N-1} \lvert F_N(n)\rvert^2$

Example 6.22: Apply 4-point and 8-point **DFT** to:

$$\text{Cosinusoid} \quad f(t) = 2\cos \pi t, \; T = 2$$

Solution: 4-point **DFT**.

$$N = 4, \; W = e^{j2\pi/4} = j, \; \omega_0 = \pi$$

$$f(t) = \sum_{n=-3}^{+3} c_n e^{2\pi jnt}$$

Function $f(t)$ is sampled at $t_k = \dfrac{k}{2}, \quad k = 0, 1, 2, 3$

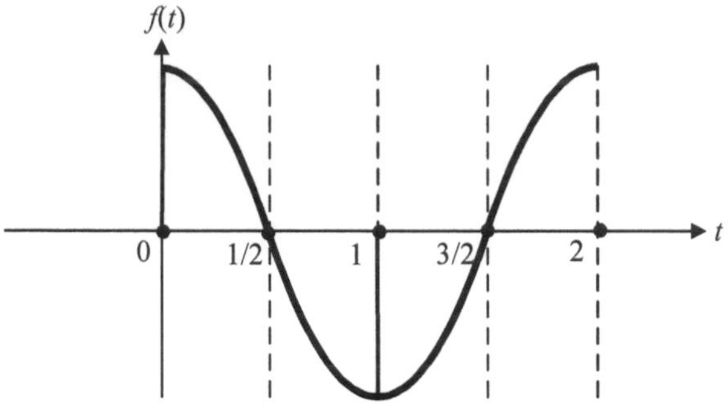

Figure 6.30: Sampled Cosinusoid

Thus,

$$f_4^T = \left[f_4(0)\ f_4(1)\ f_4(2)\ f_4(3)\right] = \left[2\ 0\ -2\ 0\right]$$

$$F_4(n) = \sum_{k=0}^{3} f_4(k) W_4^{-nk} = \sum_{k=0}^{3} (f_4(k))\,(j)^{-nk}, \quad n = 0, 1, 2, 3$$

Thus,

$$F_4^T = \left[F_4(0)\ F_4(1)\ F_4(2)\ F_4(3)\right] = \left[0\ 1\ 0\ 1\right]$$

From the Fourier coefficients viewpoint

$$c_1 = c_{-1} = 1$$

$$c_0 = c_2 = c_{-2} = c_3 = c_{-3} = 0$$

Function $f(t)$ is band-limited and therefore using Eq. 6.101

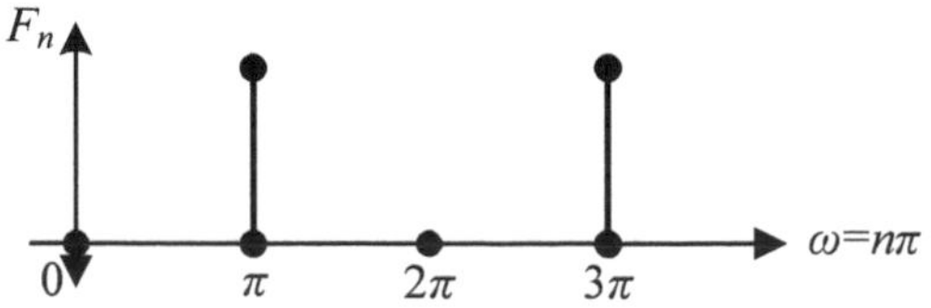

Figure 6.31: Fourier Coefficients for a 4-Point FFT

$$F_4(0) = c_0 = 0$$

$$F_4(1) = c_1 + c_{-3} = 1$$

$$F_4(2) = c_2 + c_{-2} = 0$$

$$F_4(3) = c_3 + c_{-1} = 1$$

Applying 8-point **DFT**

$$f(t) = \sum_{n=-3}^{3} c_n e^{jn\omega_0 t}, \quad c_{-n} = c_n$$

$$N = 8, \quad W_8 = e^{j2\pi/8}, \quad M = 3$$

Thus, $N > 2M + 1$ is satisfied and Eq. 6.101 holds.

Thus,

$$F_8(0) = c_0, \quad F_8(1) = c_1, \quad F_8(2) = c_2, \quad F_8(3) = c_3$$

$$F_8(4) = 0, \quad F_8(5) = c_{-3}, \quad F_8(6) = c_{-2}, \quad F_8(7) = c_{-1}$$

From Eq. 6.101, we can solve for various c's.

Check

$$F_8(4) = \sum_{k=0}^{7} f_8(k) W_8^{4k} = \sum_{k=0}^{7} f_8(k) e^{j\pi k}$$

$$f_8(k) = f(t)\Big|_{t=\frac{2\pi k}{8\omega_0}} = \sum_{n=-3}^{3} c_n e^{j\frac{2\pi nk}{8}} = \sum_{n=-3}^{3} c_n e^{j\frac{\pi nk}{4}}, \quad k = 0, \pm 1, \pm 2, \pm 3$$

Let $x = e^{j\pi/4}$

$$F_8(4) = \sum_{k=0}^{7} \left[\sum_{n=-3}^{3} c_n x^{nk} \right] x^{4k} = \sum_{n=-3}^{+3} c_n x^{n+4} \left[\sum_{k=0}^{7} x^k \right]$$

But

$$\sum_{k=0}^{7} x^k = \frac{1-x^8}{1-x} = \frac{1-e^{(j\pi/4)8}}{1-e^{j\pi/4}} = \frac{1-e^{j2\pi}}{1-e^{j\pi/4}} = 0$$

Thus, $F_8(4) = 0$

Example 6.23:

Given

$$\begin{aligned} F_{16}(n) &= 1 \qquad -4 \le n \le 4 \\ &= 0 \qquad 5 \le n \le 11 \qquad \text{(padding)} \end{aligned}$$

Compute $f_{16}(k), \quad k = 0, \ldots, 15$

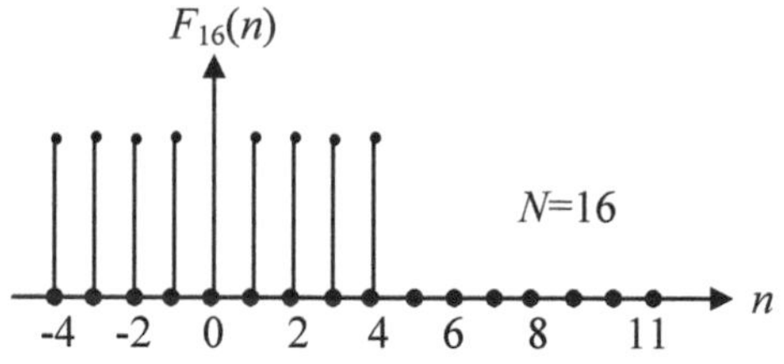

Figure 6.32: Fourier Coefficients for a 16-Point FFT

Solution:

$$f_{16}(k) = \frac{1}{16}\sum_{n=-4}^{4} F_{16}(n)W_{16}^{nk} = W_{16}^{-4k}\left(1 + W_{16}^{k} + W_{16}^{2k} + \ldots + W_{16}^{8k}\right), \quad W_{16} = e^{j2\pi/16}$$

$$= \frac{W_{16}^{-4k}\left(1 - W_{16}^{9k}\right)}{1 - W_{16}^{k}} = \frac{\sin 9\pi k/16}{\sin \pi k/16}$$

Frequency Sampling Theorem for a Time Truncated Signal

The time sampling theorem states that a time signal whose Fourier Transform is 0 outside the interval $[-\omega_b, \omega_b]$ may be sampled with a sampling interval less than or equal to $1/2\omega_b$ and reconstructed via the Sinc interpolation function without any loss of information.

Question:

Can we interchange of roles of time and frequency (via Fourier Transform) and sample the frequency spectrum and thereby obtain the exact Fourier Transform of the time truncated signal? The answer is **Yes**, provided the time truncated signal is processed as following:

(1) The time truncated signal $f(t)\{0 \leq t < T\}$ is sampled at discrete times $\{0, \Delta T, 2\Delta T, \ldots, (N-1)\Delta T;\ N\Delta T = T\}$ and its discrete Fourier Transform $F_N(n)$ is computed.

Where

$$F_N(n) = F(n\Delta f) \quad \text{where } \Delta f = 1/N\Delta T, \quad n = 0, 1, \ldots, n-1$$

(2) Periodically continue the computation of $F(n\Delta f)$ by realizing that

$$F((n + mN)\Delta f) = F(n\Delta f)$$

(3) Use the Sinc interpolating function discussed earlier to obtain

$$F(f) = \sum_{n=-\infty}^{+\infty} F(n\Delta f)\left[\frac{\sin\left(\pi(f - n\Delta f)/\Delta f\right)}{\pi(f - n\Delta f)/\Delta f}\right]$$

6.11.7 The Fast Fourier Transform (FFT)

Most numerical computations involve additions and multiplications. Addition is a relatively negligible time-consuming process compared to multiplication and hence, the "curse of multiplication dimensionality" plays an important role in efficient computing. The evaluation of **DFT** represented by sums in Eq. 6.102, if done in a brute force manner, requires $(N-1)$ multiplications (additions neglected) for every k ranging from 0 to $N - 1$, roughly requiring $(N - 1)^2$ multiplications. For large N, the computing time is prohibitive. Many researchers recognized the periodic nature of Eq. 6.102 and devised means of reducing the number of multiplication operations. The origin for the simplification of the computation of Eq. 6.102 goes back to Gauss in 1805. The so-called Decimation algorithm was proposed by Danielson and Lanzcos. In 1965, Cooley and Tuckey were the first to present a systematic algorithm for the computation of such expressions that dramatically reduces the number of multiplications and hence the computation time. This algorithm is called **Fast Fourier Transform (FFT)** and represents a clever strategy of "divide and conquer." The first step involves dividing the sum of N^2 multiplications in Eq. 6.102 into two sums involving $2[(N/2)^2 + (N/2)]$ multiplications. The process is repeated again for each sum until $N/2^r = 1$ is arrived at (assume N to be even). The number of multiplications via this algorithm approaches $O(N \log N)$. This FFT algorithm is important in different fields where enormous computing is demanded. The FFT algorithm can be applied to any problem involving observational data with a band-limited frequency spectrum. The enormous saving in computation for large N can be graphically seen in Figure 6.33.

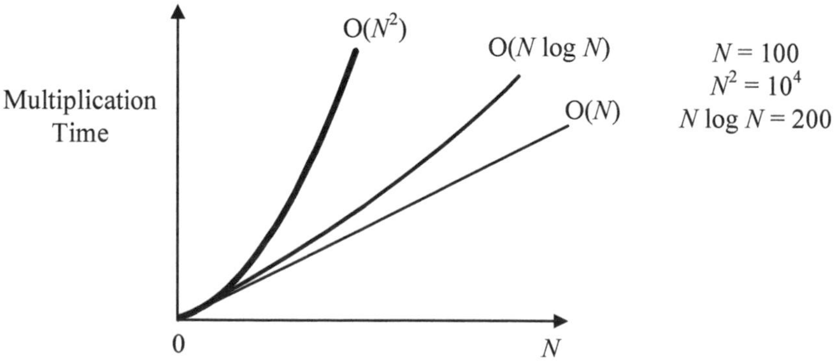

Figure 6.33: Computational Complexity of Multiplications

This concern about the number of multiplications and computation speeds can be seen as following:

Most of the computation problems, involving observations or data, take the form

$$\boldsymbol{A}\boldsymbol{x} = \boldsymbol{b}$$

The matrix $\boldsymbol{A}$ might be sparse (O(N)) or dense (O(N^2)). The solution involves computation of $\boldsymbol{A}^{-1}$, which in general is expensive for dense matrices $\boldsymbol{A}$ with large N. On the other hand, when $\boldsymbol{A}$ is sparse, any scheme that is fast and requires lesser multiplications and storage is greatly appreciated. "Fast Fourier Transform (FFT)" is one of these algorithms for computing $\boldsymbol{A}^{-1}$ economically.

6.11.8 Numerical Computation of DFT via Fast Fourier Transform—FFT

The philosophy of FFT can be summed up as following: A computationally intensive recursive problem can be divided into two halves and each half can further be divided into two halves and so on, thereby reducing the complexity. The FFT algorithm, along with fast digital computers, have been responsible for an explosion in digital signal processing, communications, control, and micromedia.

For the sake of convenience and without loss of generality let us define the N-point discrete data $f_N(k)$ and its Discrete Fourier Transform $F_N(n)$ as following:

$$\begin{aligned} f_N(k) &= f(k, N) = f(k), && k = 0, 1, \ldots, N-1 \\ F_N(n) &= F(n, N) = F(n), && n = 0, 1, \ldots, N-1 \\ W_N &= e^{j2\pi/N}, && N\text{-th root of unity} \end{aligned}$$

Interesting to note:

$$1 + W_N + W_N^2 + \ldots + W_N^{N-1} = 0$$

$$W_N^{N/2} = -1$$

$$W_N^{(N/2)+i} = W_N^{N/2} W_N^i = -W_N^i \quad i = 0, 1, \ldots, (N/2) - 1$$

$$W_N^a W_N^b = W_N^{a+b}$$

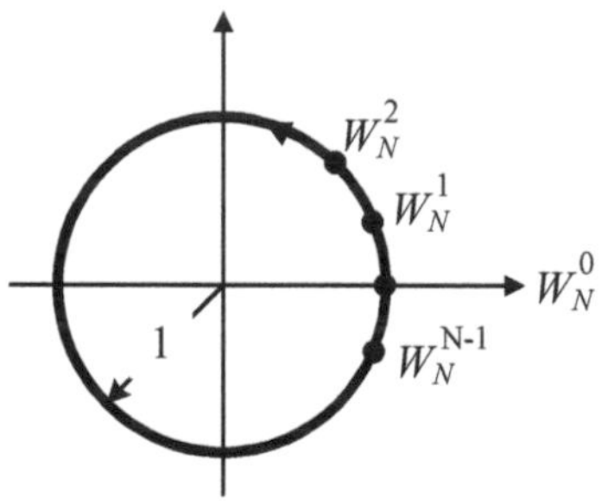

Figure 6.34: N-th Roots of Unity

The **DFT** pair is

$$f(k) = f(k, N) = \frac{1}{N} \sum_{n=0}^{N-1} F(n) W_N^{(nk)}, \quad N\text{-point Computed values of } f(k) \quad (6.103)$$

$$F(n) = F(n, N) = \sum_{k=0}^{N-1} f(k) W_N^{-(nk)}, \quad N\text{-point computed values of } F(n) \quad (6.104)$$

The above two equations show that the same amount of computation is required whether $\{f(k)\}_{k=0}^{(N-1)}$ is the data and $\{F(n)\}_{n=0}^{(N-1)}$ is to be computed or vice versa. In what follows, we shall compute $\{F(n)\}_0^{N-1}$ given $\{f(k)\}_0^{N-1}$ as the data. We choose

$$N = 2^r, \qquad r = 1, 2, \ldots, \quad \text{an integer} \qquad (6.105)$$

In case $N \neq 2^r$, the sequence $\{f(k)\}_0^{N-1}$ can be packed with some zeroes either at the front or the rear end of the sequence (known as zero padding). While there are several ways to compute FFT, the two most frequently used algorithms are:

1. **Decimation-in-Time–algorithm**

   The N-point data is divided into even and odd halves

   $\{f(0), f(2), \ldots, f(N/2-1)\}$ and $\{f(1), f(3), \ldots, f(N-1)\}$

2. **Decimation-in-Frequency–algorithm**

   The N-point data is divided into two halves

   $\{f(0), f(1), \ldots, f(N/2-1)\}$ and $\{f(N/2), f(N/2+1), \ldots, f(N-1)\}$

The computation time required is the same for both these schemes. In what follows, we shall use the Decimation-in-Time algorithm.

Task: Compute $F(n), n = 0, \ldots, n-1$, given $f(k), k = 0, \ldots, n-1$:

$$f(k) = f(k, N) = \frac{1}{N} \sum_{n=0}^{N-1} F(n)\, W_N^{(nk)} \tag{6.106}$$

Utilizing the even and odd terms summations

$$Nf(k,N) = \underbrace{\left[\sum_{n=0}^{N/2-1} F(2n)\, (W_N)^{(2nk)}\right]}_{\textbf{DFT} \text{ of even points}} + \underbrace{\left[\sum_{n=0}^{N/2-1} F(2n+1)\, (W_N)^{(2n+1)k)}\right]}_{\textbf{DFT} \text{ of odd points}} \tag{6.107}$$

But

$$W_N^2 = e^{(j(2\pi/N)\times 2)} = e^{\left(j\frac{2\pi}{N/2}\right)} = W_{N/2}$$

Thus,

$$Nf(k,N) = \left[\sum_{n=0}^{N/2-1} F(2n)\, (W_{N/2})^{(nk)}\right] + W_N^k \left[\sum_{n=0}^{N/2-1} F(2n+1)\, (W_{N/2})^{(nk)}\right] \tag{6.108}$$

Furthermore,

$$Nf\,(k+N/2, N) = \left[\sum_{n=0}^{N/2-1} F(2n)\,(W_{N/2})^{(nk)}\right] - W_N^k \left[\sum_{n=0}^{N/2-1} F(2n+1)\,(W_{N/2})^{(nk)}\right]$$

Let us define

$$\sum_{n=0}^{N/2-1} F(2n)\,(W_{N/2})^{(nk)} = Nf(k, N/2, 0)$$
$$\sum_{n=0}^{N/2-1} F(2n+1)\,(W_{N/2})^{(nk)} = Nf(k, N/2, 1)$$

Thus,

$$\left[\sum_{n=0}^{N/2-1} F(2n)\,(W_{N/2})^{(nk)}\right] = Nf\,(k, N/2, 0) \quad \text{FFT of even terms}$$
$$\left[\sum_{n=0}^{N/2-1} F(2n+1)\,(W_{N/2})^{(nk)}\right] = Nf\,(k, N/2, 1) \quad \text{FFT of odd terms} \tag{6.109}$$

$$f(k, N) = f(k, N/2, 0) + W_N^k f(k, N/2, 1)$$
$$f(k + N/2, N) = f(k, N/2, 0) - W_N^k f(k, N/2, 1) \tag{6.110}$$

and

$$\frac{1}{2}\left[f(k, N) + f\,(k + N/2, N)\right] = f\,(k, N/2, 0)$$
$$\frac{1}{2}\,(W_N)^{-k}\left[f(k, N) - f\,(k + N/2, N)\right] = f\,(k, N/2, 1) \tag{6.111}$$

The symbol "0" and "1" stand for even and odd, respectively. This binary notation is a key to bookkeeping of the recursive calculations. The computational algorithm is known as the "butterfly."Let us use the Eq. 6.111 as the computation algorithm. $f(k, N), f(k + N/2, N)$ $(k = 0, 1, \ldots, (N/2) - 1)$ is our input data and $f(k, N/2, 0), f(k, N/2, 1)$ $(k = 0, 1, \ldots (N/2) - 1)$ is our output from the "butterfly."

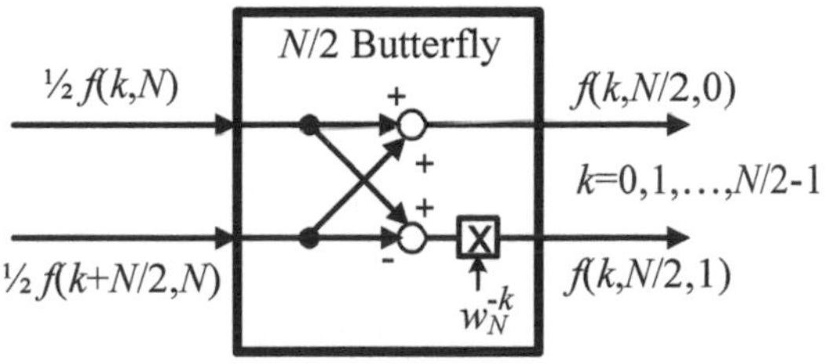

Figure 6.35: $(N/2)$ Point Transforms for the k-th Time Function—Remove N!

Each of the series in Eq. 6.109 can be again split into two halves as following:

$$Nf(k, N/2, 0) = \left[\sum_{n=0}^{N/4-1} F(4n)\,(W_{N/4})^{(nk)}\right] + (W_{N/4})^{(2k)}\left[\sum_{n=0}^{N/4-1} F(2(2n+1))\,(W_{N/4})^{(nk)}\right]$$

$$Nf(k+N/4, N/2, 0) = \left[\sum_{n=0}^{N/4-1} F(4n)\,(W_{N/4})^{(nk)}\right] - (W_{N/4})^{(2k)}\left[\sum_{n=0}^{N/4-1} F(2(2n+1))\,(W_{N/4})^{(nk)}\right]$$

$$Nf(k, N/2, 1) = \left[\sum_{n=0}^{N/4-1} F(4n+1)\,(W_{N/4})^{(nk)}\right] + (W_{N/4})^{(2k)}\left[\sum_{n=0}^{N/4-1} F(4n+3)\,(W_{N/4})^{(nk)}\right]$$

$$Nf(k+N/4, N/2, 1) = \left[\sum_{n=0}^{N/4-1} F(4n+1)\,(W_{N/4})^{(nk)}\right] - (W_{N/4})^{(2k)}\left[\sum_{n=0}^{N/4-1} F(4n+3)\,(W_{N/4})^{(nk)}\right],\quad k = 0,\ldots,\frac{N}{4}-1$$

or

$$f(k, N/2, 0) = f(k, N/4, 0, 0) + (W_{N/4})^{(2k)}\,f(k, N/4, 0, 1)$$

$$f(k+N/4, N/2, 0) = f(k, N/4, 0, 0) - (W_{N/4})^{(2k)}\,f(k, N/4, 0, 1)$$

$$f(k, N/2, 1) = f(k, N/4, 1, 0) + (W_{N/4})^{(2k)}\,f(k, N/4, 1, 1)$$

$$f(k+N/4, N/2, 1) = f(k, N/4, 1, 0) - (W_{N/4})^{(2k)}\,f(k, N/4, 1, 1)$$

Thus,

$$
\begin{aligned}
f(k, N/4, 0, 0) &= \frac{1}{2}\,[f(k, N/2, 0) + f(k + N/4, N/2, 0)] \\
f(k, N/4, 0, 1) &= \frac{1}{2}\,(W_{N/4})^{-(2k)}\,[f(k, N/2, 0) - f(k + N/4, N/2, 0)] \\
f(k, N/4, 1, 0) &= \frac{1}{2}\,[f(k, N/2, 1) + f(k + N/4, N/2, 1)] \\
f(k, N/4, 1, 1) &= \frac{1}{2}\,(W_{N/4})^{-(2k)}\,[f(k, N/2, 1) - f(k + N/4, N/2, 1)]
\end{aligned}
$$

$k = 0, 1, \ldots, (N/4) - 1$. We terminate this process after r steps, where $N = 2^r$.

Computation Algorithm for FFT

Define $f(k, N) = f(k, N)$

$$
k = 0, 1, \ldots, (N/2^p) - 1, \quad p = 0, 1, 2, \ldots, \ p \le r, \quad N = 2^r
$$

Given

A sequence of data, $f(k, N/2^p, i_0, i_1, \ldots, i_p)$, $\quad (i_p = 0 \text{ or } 1 : \text{binary number})$

Compute: A new updated sequence of numbers

$$
\begin{aligned}
f(k, N/2^{p+1}, i_0, i_1, \ldots, i_p, 0) = \frac{1}{2}\Big[& f(k, N/2^p, i_0, i_1, \ldots, i_p) \\
& + f(k + N/2^{(p+1)}, N/2^p, i_0, i_1, \ldots, i_p)\Big]
\end{aligned}
$$

$$
\begin{aligned}
f(k, N/2^{p+1}, i_0, i_1, \ldots, i_p, 1) = \frac{1}{2} W_{N/2^p}^{-k}\Big[& f(k, N/2^p, i_0, i_1, \ldots, i_p) \\
& - f(k + N/2^{(p+1)}, N/2^p, i_0, i_1, \ldots, i_p)\Big]
\end{aligned}
$$

For a given r, computing stops when $p = r$.

The final outputs $F(n, N)$, $n = 0, \ldots, N - 1$ are determined as following:

$$
Nf(0, 1, i_0, i_1, \ldots, i_r) = F(n, N)
$$

The index n is converted from binary to decimal as

$$n = \sum_{j=1}^{r} i_j 2^j = i_0 + i_1 2 + i_2 2^2 + \ldots + i_r 2^r$$

This concludes our explanation of the **FFT** computational algorithm.

4-point FFT: 4-point **FFT** is summarized as following:

$$\left.\begin{aligned} f(0) = f(0,4) &= \tfrac{1}{4}\left[F(0) + F(1) + F(2) + F(3)\right] \\ f(1) = f(1,4) &= \tfrac{1}{4}\left[F(0) + F(1)W_4 + F(2)W_4^2 + F(3)W_4^3\right] \\ f(2) = f(2,4) &= \tfrac{1}{4}\left[F(0) + F(1)W_4^2 + F(2)W_4^4 + F(3)W_4^6\right] \\ f(3) = f(3,4) &= \tfrac{1}{4}\left[F(0) + F(1)W_4^3 + F(2)W_4^6 + F(3)W_4^9\right] \end{aligned}\right\} \tag{6.112}$$

where

$$\left.\begin{aligned} f(0,2,0) &= \tfrac{1}{4}\left[F(0) + F(2)\right] \\ f(0,2,1) &= \tfrac{1}{4}\left[F(1) + F(3)\right] \\ f(1,2,0) &= \tfrac{1}{4}\left[F(0) + F(2)W_4^2\right] \\ f(1,2,1) &= \tfrac{1}{4}\left[F(1) + F(3)W_4^2\right] \end{aligned}\right\} \tag{6.113}$$

$$\begin{aligned} f(0,1,0,0) &= \frac{1}{4}[F(0)] = \frac{1}{2}[f(0,2,0) + f(1,2,0)] \\ f(0,1,1,0) &= \frac{1}{4}[F(1)] = \frac{1}{2}[f(0,2,1) + f(1,2,1)] \\ f(0,1,0,1) &= \frac{1}{4}[F(2)] = \frac{1}{2}[f(0,2,0) - f(1,2,0)] \\ f(0,1,1,1) &= \frac{1}{4}[F(3)] = \frac{1}{2}[f(0,2,1) - f(1,2,1)] \end{aligned}$$

8-Point FFT: Figures 6.36 and 6.37 show the $N = 8$-point FFT and its binary branches.

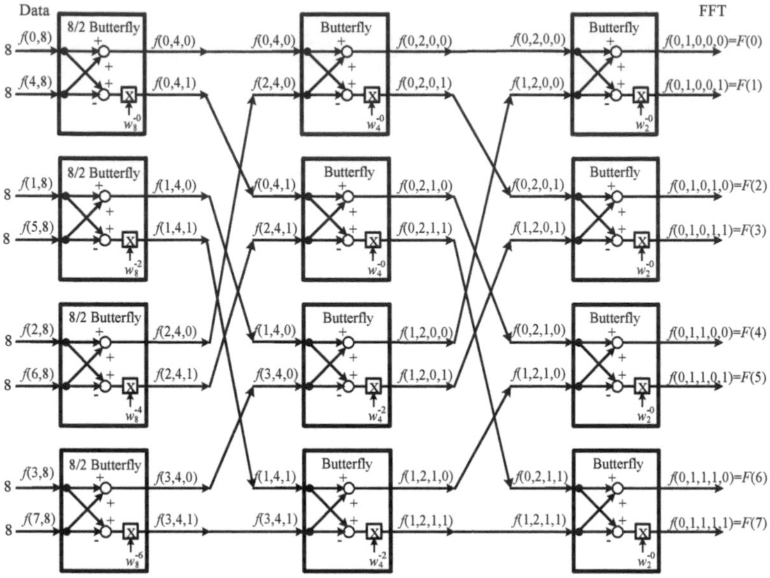

Figure 6.36: FFT for the 8-Point Data

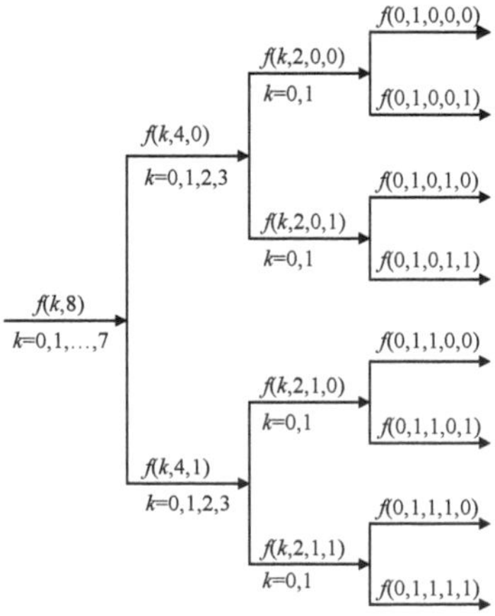

Figure 6.37: Binary Tree for a Sequence of an 8-Point FFT

This form of FFT is referred to as the Cooley-Tuckey radix 2 index. The following Table 6.7 shows the "unscrambling" and bit reversal operation of the indices. The last three digits of $f(\cdot,\cdot,\cdot,\cdot,\cdot)$ represents its bit index, referred to as "Gray" code.

Table 6.7: Bit Reversal

Data $f(k,N)$	Binary Form	Bit Reversal	Final Index $F(n,N)$, FFT
0	000	000	0
4	100	001	1
2	010	010	2
6	110	011	3
1	001	100	4
5	101	101	5
3	011	110	6
7	111	111	7

6.11.9 FFT in Two Dimensions

Given: $f(k_1,k_2), \quad 0 \le k \le N_1 - 1, \quad 0 \le k_2 \le N_2 - 1$

The two-dimensional Discrete Fourier Transform (DFS) can be defined as:

$$F(n_1,n_2) = \sum_{k_2=0}^{N_2-1} \sum_{k_1}^{N_1-1} f(k_1,k_2)e^{-j2\pi(n_1k_1+n_2k_2)}$$

It is obvious that the two-dimensional FFT can be computed by first taking the index-1 constant and computing the one-dimensional FFT on the index-2 of the original function and then again computing the one-dimensional FFT on the remaining index-2.

Thus,

$$F(n_1,n_2) = \text{FFT-on-}k_1\ (\text{FFT-on-}k_2(f(k_1,k_2)))$$

The order of indices can be reversed. The procedure can be extended to higher dimensions.

6.11.10 Appendix: Accelerating Power Series Convergence

This method is called the **Euler-Abel Transformation** see [Demidovich, B.P.]. Consider the convergent series

$$F(z) = \sum_{n=0}^{\infty} f(n)z^{-n}, \qquad |z| > 1$$

Let

$$F(z) = f(0) + Z^{-1}\sum_{n=1}^{\infty} f(n)z^{-n+1} = f(0) + Z^{-1}F_1(z)$$

It is easy to see that

$$F_1(z) = \sum_{n=0}^{\infty} f(n+1)z^{-n} = \frac{f(0)}{1-z^{-1}} + \frac{1}{1-z^{-1}}\sum_{n=0}^{\infty} \Delta f(n)z^{-n}$$

or

$$F(z) = f(0) + Z^{-1}F_1(z) = \frac{f(0)}{1-z^{-1}} + \frac{z^{-1}}{1-z^{-1}}\sum_{n=0}^{\infty} \Delta f(n)z^{-n} \tag{6.114}$$

Euler-Abel Transformation

Repeating this Transformation

$$F(z) = \sum_{n=0}^{\infty} f(n)z^{-n} = \frac{f(0)}{1-z^{-1}} + \frac{z^{-1}}{1-z^{-1}}\left(\frac{\Delta f(0)}{1-z^{-1}} + \frac{z^{-1}}{1-z^{-1}}\sum_{n=0}^{\infty} \Delta^2 f(n)z^{-n}\right)$$

Repeating this process k times.

$$F(z) = \sum_{n=0}^{\infty} f(n)z^{-n} = \left[\sum_{r=0}^{k-1} [\Delta^r f(0)]\frac{z^{-r}}{(1-z^{-1})^{r+1}}\right] + \left(\frac{z^{-1}}{1-z^{-1}}\right)^k \sum_{n=0}^{\infty} [\Delta^r f(n)]\,z^{-n}$$

This formula is advantageous only when higher order $\Delta^r f(n)$ decay fast for $n \to$ large. In particular if

$$f(n) = P(n), \qquad \text{an integral polynomial in } n \text{ of degree } p$$

Then

$$F(z) = \sum_{n=0}^{\infty} P(n)z^{-1} = \sum_{r=0}^{p} [\Delta^r P(0)]\left(\frac{(z^{-1})^r}{(1-z^{-1})^{r+1}}\right)$$

Example 6.24:

Evaluate

$$F(z) = \sum_{n=0}^{\infty}(n+1)^2 z^{-n}, \qquad |z^{-1}| < 1$$

Solution

$$P(n) = n^2 + 2n + 1$$

$$\begin{array}{lll} P(0) = 1, & P(1) = 4, & p(2) = 9 \\ \Delta P(0) = 3, & \Delta P(1) = 5, & \Delta P(2) = 0 \\ \Delta^2 P(0) = 2, & \Delta^2 P(1) = 0, & \Delta^2 P(2) = 0 \end{array}$$

$$F(z) = \frac{1}{1-z^{-1}} + \frac{3z^{-1}}{1-z^{-1}} + \frac{2z^{-2}}{(1-z^{-1})^2}$$

Bibliography

[Brigham, E.O.] Brigham, E.O., and Morrow, R.E. The Fast Fourier Transform, *IEEE Spectrum* 4, 63–70, December 1967.

[Cooley, J.W.] Cooley, J.W., Lewis, P.A.W., and Welch, P.D. Historical Notes on the Fast Fourier Transform, *IEEE Transactions Audio Electroacoustics* AU-15; 76–79, June 1967. (Also published in *Proceedings of IEEE* 55(10), 1675–1677, June 1967.)

[Cooley, J.W.] Cooley, J.W., and Tuckey, J.W. An Algorithm for the Machine Calculation of Complex Fourier Series, *Math Computation* 19, 297–305, 1965.

[Demidovich, B.P.] Demidovich, B.P., and Maron. A. *Computational Mathematics,* Moscow: Mir Publishers, (Translated from Russian by Yankovsky G.), 1970.

[Gabel, R.A.] Gabel, R.A., and Roberts, R.A. *Signals and Systems,* New York: John Wiley and Sons, Inc., 1980.

[Gustavson, F.G.] Gustavson, F.G. High Performance Linear Algebra Algorithms Using New Generalized Data Structures for Matrices, *IBM Journal of Research and Development,* 47(1), January 1997.

[Jolly, L.B.W.] Jolly, L.B.W. *Summation of Series,* New York: Dover Publications, Inc., 1961.

[Lathi, B.P.] Lathi, B.P. *Linear Systems And Signals,* Berkeley: Cambridge Press, 1992.

[Papoulis, A.] Papoulis, A. *The Fourier Integral and Its Applications,* 2nd Ed., New York: McGraw-Hill, 1984.

[Ragazzini, J.R.] Ragazzini, J.R., and Franklin, G.F. *Sampled-Data Control Systems,* New York: McGraw-Hill Book Company Inc., 1958.

[Salekhov, G.] Salekhov, G. *On the Theory of the Calculation of Series,* (Text in Russian.), UMN 4:4(32), 50-Ű82, 1949.

[Schwartz, M.] Schwartz, M. and Shaw L. *Signal Processing: Discrete Spectral Analysis, Detection and Estimation,* New York: McGraw-Hill, 1975.

[Stroud, A.H.] Stroud, A.H., and Secrest, D. *Gaussian Quadrature Formulas,* Englewood Cliffs: Prentice-Hall, 1966.

[Tolstov, G.P.] Tolstov, G.P. *Fourier Series,* New York: Dover Publications, Inc., (Translated from Russian by Richard Silverman.)

Chapter 7

State Space Description of Dynamic Systems

7.1 Introduction

Most of the material covered in the text up to now deal with the complex variable transform description of dynamic systems without any emphasis on the system structure or any regard to the system synthesis or design. In this chapter, the state variable description of the system is presented, which helps us look at the **internal structure** of the system along with the input-output system performance. The **State Variable** representation highlights the role played by energy storage elements such as capacitors, reactors, springs and masses, etc. Instead of just the input-output model, we are able to examine the effect of various driving inputs on the different internal components and their performance (controllability concept) and the influence of the different sensing and measuring components on the outputs (observability concept). In general, the state space description yields much more information about the system than the transfer function and often leads to methodical synthesis and design algorithms. The state variable single order differential equation model for the n-th order system is ideally suitable for digital simulation. Besides offering a concise and compact notation, the well-developed apparatus of linear vector spaces and matrix algebra yields rich dividends. In what follows we present methods for deriving state space equations and methods for solving them using matrix algebra.

7.2 State Space Formulation

7.2.1 Definition of the State of a System

The state of a dynamical system at some time $t = t_0$ is defined by any complete independent set of system variables whose values at $t = t_0$, along with the knowledge of input functions for $t \geq t_0$, completely determines the future system behavior. In general, the variables associated with each of the energy storage elements can be taken as state variables.

Example 7.1:

Consider a simple circuit represented by a resistor R, inductor L, and a capacitor C. The variables associated with both L and C are current and voltage and can be taken as state variables.

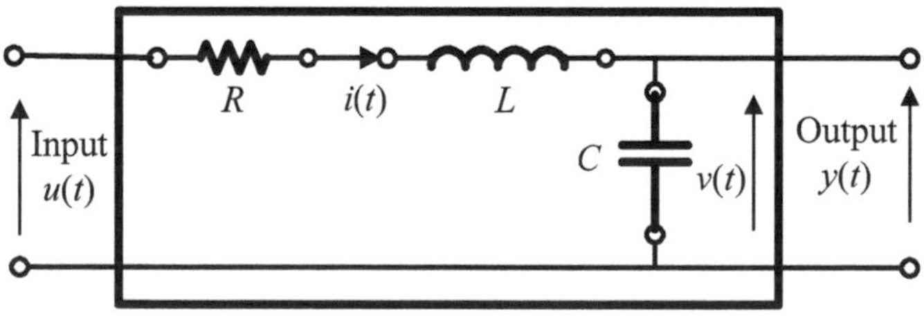

Figure 7.1: R, L, C Network Illustrating State Variables

$$u(t) = \text{Input Voltage}$$

$$y(t) = \text{Output Voltage}$$

The network equations are

$$\left.\begin{aligned} R\,i(t) + L\frac{\mathrm{d}i}{\mathrm{d}t} + v(t) = u(t) \\ \frac{1}{c}\int_0^t i(\tau)\,\mathrm{d}\tau = v(t) \end{aligned}\right\} \quad \text{(state equations)}$$

$$y(t) = v(t) \qquad \text{(output equations)}$$

Let

$$q(t) = \int_0^t i(\tau)\,\mathrm{d}\tau = x_1(t)$$

$$i(t) = x_2(t)$$

The network equations are:

$$\begin{aligned} \dot{x}_1 &= x_2 \\ \dot{x}_2 &= -\frac{1}{LC}x_1 - \frac{R}{L}x_2 + u(t) \\ y(t) &= \frac{1}{c}x_1(t) \end{aligned}$$

Knowledge of the variables $x_1(t)$ and $x_2(t)$ at any given time $t = t_0$ along with $u(t)$ for $t \geq t_0$ is enough to completely compute the future values of the variables $x_1(t), x_2(t)$, and the output $y(t)$. It is interesting to note that $x_1(t)$ and $x_2(t)$ are the outputs of the integrators. Thus, it seems reasonable to choose the output of the integrators as the "State Variables" of the system. Later on we shall see that the minimum number of integrators required to simulate a system has great significance (McMillan Degree). It should be realized that the choice of state variables is not unique because a different set of state variables may fully describe the same physical system.

7.2.2 State Variable Formulation—n-th Order System

Consider a scalar differential equation

$$x^{(n)}(t) + a_1 x^{(n-1)}(t) + \cdots + a_n x(t) = b_0\, u(t) \tag{7.1}$$

where $u(t)$ is the input. The output equation is given by:

$$y = x(t) + d\,u(t)$$

Let

$$\begin{aligned} x_1(t) &= x(t) \\ x_2(t) &= \dot{x}(t) \\ &\vdots \\ x_n(t) &= x^{(n-1)}(t) \end{aligned} \tag{7.2}$$

Eq. 7.1 can be rewritten in the form

$$x^{(n)}(t) = \dot{x}_n = -a_n x_1(t) - a_{n-1} x_2(t) + \cdots - a_1\, x_n(t) + b_0\, u(t) \tag{7.3}$$

Combining Eq. 7.2 and Eq. 7.3:

$$\begin{bmatrix} \dot{x}_1 \\ \dot{x}_2 \\ \vdots \\ \\ \dot{x}_n \end{bmatrix} = \begin{bmatrix} 0 & 1 & \cdots & & 0 \\ 0 & 0 & 1 & \cdots & 0 \\ \vdots & & & & \\ 0 & 0 & \cdots & 0 & 1 \\ -a_n & -a_{n-1} & \cdots & & -a_1 \end{bmatrix} \begin{bmatrix} x_1(t) \\ x_2(t) \\ \vdots \\ \\ x_n(t) \end{bmatrix} + \begin{bmatrix} 0 \\ 0 \\ \vdots \\ 0 \\ b_0 \end{bmatrix} u(t)$$

$$y(t) = x_1(t) = \begin{bmatrix} 1 \; 0 \cdots \; 0 \end{bmatrix} \begin{bmatrix} x_1(t) \\ x_2(t) \\ \vdots \\ x_n(t) \end{bmatrix} + du(t) \tag{7.4}$$

In the matrix form:

$$\begin{aligned} \dot{\boldsymbol{x}} &= \boldsymbol{A}\,\boldsymbol{x} + \boldsymbol{b}u \\ y &= \boldsymbol{c}^T \boldsymbol{x} + d\,u \end{aligned} \tag{7.5}$$

where $\boldsymbol{A}$, $\boldsymbol{b}$, $\boldsymbol{c}$ and d are obvious from Eq. 7.4.

7.2.3 State Variable Formulation of a General System

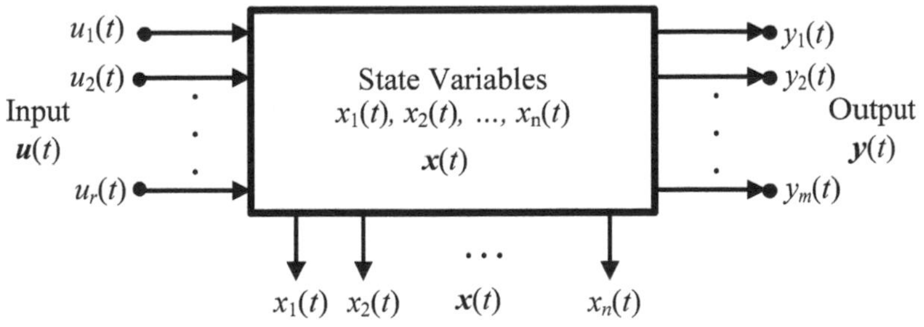

Figure 7.2: Schematic n-State Variable Description

A schematic state variable representation for an n-dimensional dynamic system is shown in Figure 7.2. Consider a general n-th order dynamical system with r inputs $u_1(t), u_2(t), \ldots, u_r(t)$ and m outputs $y_1(t), y_2(t), \ldots, y_m(t)$. The general system equations are given as:

$$\frac{\mathrm{d}x_i}{\mathrm{d}t} = f_i\left(x_1, x_2, \ldots, x_n; u_1, u_2, \ldots, u_r, t\right), \quad i = 1, 2, \ldots, n$$

$$y_k(t) = g_k\left(x_1, x_2, \ldots, x_n; u_1, u_2, \ldots, u_r, t\right), \quad k = 1, 2, \ldots, m$$

For a linear time-invariant system, these equations take a simpler form:

$$\dot{x}_i = \sum_{j=1}^{n} a_{ij}x_j + \sum_{l=1}^{r} b_{il}u_l, \qquad i = 1, 2, \ldots, n$$

$$y_k = \sum_{p=1}^{n} c_{ip}x_p + \sum_{q=1}^{r} d_{iq}u_q, \qquad k = 1, 2, \ldots, n$$

In the matrix form:

$$\dot{\boldsymbol{x}} = \boldsymbol{A}\,\boldsymbol{x} + \boldsymbol{B}\,\boldsymbol{u} \tag{7.6}$$

$$\boldsymbol{y} = \boldsymbol{C}\,\boldsymbol{x} + \boldsymbol{D}\,\boldsymbol{u} \tag{7.7}$$

Eqs. 7.6 and 7.7 represent a Linear time-invariant (LTI) System where:

$\boldsymbol{x}(t)$ = State variable (n-vector) with components $x_1(t), x_2(t), \ldots, x_n(t)$

$\boldsymbol{y}(t)$ = Output variables (m-vector) with components $y_1(t), y_2(t), \ldots, y_m(t)$

$\boldsymbol{u}(t)$ = Input variables (r-vector) with components $u_1(t), u_2(t), \ldots, u_r(t)$

$\boldsymbol{A}$ = $(n \times n)$ System matrix with coefficients a_{ij} $(i, j = 1, 2, \ldots, n)$

$\boldsymbol{B}$ = $(n \times r)$ Excitation matrix with coefficients b_{ij} $(i = 1, \ldots, n; j = 1, \ldots, r)$

$\boldsymbol{C}$ = $(m \times n)$ Output matrix with coefficients c_{ij} $(i = 1, \ldots, m; j = 1, \ldots, n)$

$\boldsymbol{D}$ = $(n \times r)$ Matrix with coefficients d_{ij} $(i = 1, \ldots, m; j = 1, \ldots, n)$

When A, B, C, and D are functions of time then the system is Linear Time Varying (LTV). In general, $\boldsymbol{D}$ matrix has little dynamical influence on the system output and without any loss of generality it can be taken as zero. We shall very often state "given a dynamical system $\boldsymbol{A}, \boldsymbol{B}, \boldsymbol{C}$, and $\boldsymbol{D}$"; by that, we imply a dynamical system described by Eqs. 7.6 and 7.7. If we consider $x_i(t)$ $(i = 1, 2, \ldots, n)$ as the coordinates of a vector $\boldsymbol{x}(t)$ referred to the i-th basis in an n-dimension Euclidean space, then the state of the system at any time t can be represented by a point $\boldsymbol{x}(t)$ in this n-dimensional state space. The solution of Eq. 7.6 represents a trajectory in this space on which $\boldsymbol{x}(t)$ moves with time. This is why Eq. 7.6 and Eq. 7.7 are called the **State Space Representation** of a dynamical system. Figure 7.3 represents the block diagram for the state space representation.

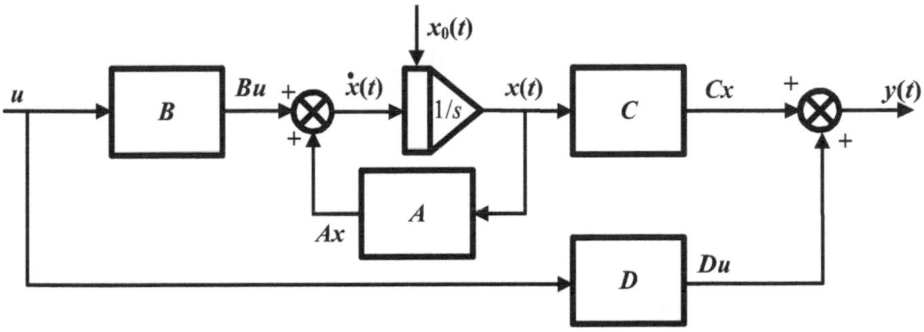

Figure 7.3: Block Diagram for State Space Representation

Note: Throughout the rest of this chapter, single-input single-output system will be designated as SISO whereas the multi-input multi-output system will be referred to as MIMO.

7.3 State Variables Selection

There is no cut and dry rule for the selection of the state variables. Physical laws dictating each system determine the proper state variables. In rigid body dynamics involving Newtonian mechanics or Lagrangian dynamics, the **position** and **velocity** of a particular element (or stored potential and kinetic energy) may be chosen as state variables. For the circuit analysis the **voltage** across a set of **independent capacitors** and **current** through a set of **independent inductors** yields a proper set of **state variables**. If a set of capacitors or voltage sources form a complete circuit, then their voltage (or charges) are dependent. Similarly, if a set of inductors or current sources are the only elements connected to a node, then their currents (or fluxes) are dependent. All those capacitors and inductors that avoid the above situation have their voltages and currents as independent variables and as such can be chosen as **independent state variables**.

The circuit system itself is an interconnection of resistors, capacitors, inductors, springs, masses, etc. In the following section we present some general methods for deriving the state space equations for a general dynamical system.

7.4 Methods of Deriving State Variable Equations

7.4.1 Lagrangian Set of Equations of Motion

All the quantities below are referred to as "Generalized Variables" or "Parameters."

$$x_i(t) = i\text{-th position coordinates}$$

$$\dot{x}_i(t) = i\text{-th velocity coordinates}$$

$$f_i(t) = \text{External forces influencing the } i\text{-th coordinates}$$

$$m_{ij} = m_{ji} = \text{Kinetic energy storage elements}$$

$$k_{ij} = k_{ji} = \text{Potential energy storage elements}$$

$$d_{ij} = d_{ji} = \text{Energy dissipative elements}$$

$$i, j = 1, 2, \ldots, n$$

The above quantities are written in matrix notations as $\boldsymbol{x}, \dot{\boldsymbol{x}}, \boldsymbol{f}, \boldsymbol{M}, \boldsymbol{K},$ and $\boldsymbol{D}$ respectively. The kinetic, potential, and dissipative energy as well as the input power to the system can be expressed as follows:

$$T = \text{Kinetic energy} = \frac{1}{2}\sum_{i=1}^{n}\sum_{j=1}^{n} m_{ij}\dot{x}_i\,\dot{x}_j = \frac{1}{2}\dot{\boldsymbol{x}}^T\boldsymbol{M}\,\dot{\boldsymbol{x}} \tag{7.8}$$

$$V = \text{Potential energy} = \frac{1}{2}\sum_{i=1}^{n}\sum_{j=1}^{n} k_{ij}x_i\,x_j = \frac{1}{2}\boldsymbol{x}^T\boldsymbol{K}\,\boldsymbol{x} \tag{7.9}$$

$$F = \text{Dissipative power} = \frac{1}{2}\sum_{i=1}^{n}\sum_{j=1}^{n} d_{ij}\dot{x}_i\,\dot{x}_j = \frac{1}{2}\dot{\boldsymbol{x}}^T\boldsymbol{D}\,\dot{\boldsymbol{x}} \tag{7.10}$$

$$I = \text{Input power} = \sum_{i=1}^{n} f_i\,\dot{x}_i = \dot{\boldsymbol{x}}^T\boldsymbol{f} \tag{7.11}$$

$$L = \text{Lagrangian} = T - V$$

$$T, V, F, I, \text{ and } L \text{ are all scalar quantities} \tag{7.12}$$

The modified Lagrangian equations for each set of coordinates are (see Chapter 8):

$$\frac{\mathrm{d}}{\mathrm{d}t}\left(\frac{\partial L}{\partial \dot{x}_i}\right) - \frac{\partial L}{\partial x_i} + \frac{\partial}{\partial \dot{x}_i}(F - I) = 0 \qquad i = 1, 2, \ldots, n \tag{7.13}$$

In matrix form,

$$\frac{\mathrm{d}}{\mathrm{d}t}\,(\nabla_{\dot{x}})\,L - \nabla_x L + \nabla_{\dot{x}}(F - I) = 0 \tag{7.14}$$

Example 7.2:

Consider a mechanical system of Figure 7.4.

Let the velocity of mass m_2 be the output.

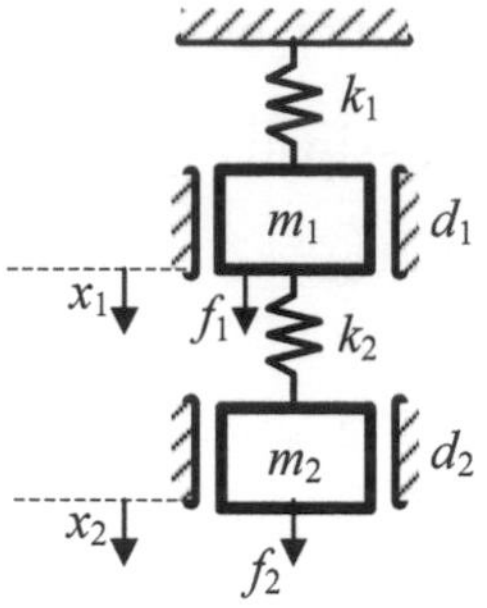

Figure 7.4: State Variable Equations for a Mechanical System

$$T = \frac{1}{2}m_1\dot{x}_1^2 + \frac{1}{2}m_2\dot{x}_2^2, \quad V = \frac{1}{2}k_1x_1^2 + k_2x_2^2$$

$$F = \frac{1}{2}d_1\dot{x}_1^2 + \frac{1}{2}d_2\dot{x}_2^2, \quad I = (f_1 + m_1g)\,\dot{x}_1 + (f_2 + m_2g)\,\dot{x}_2$$

$$g = \text{ acceleration due to gravity}$$

From Eq. 7.13

$$m_1\ddot{x}_1 + d_1\dot{x}_1 + k_1x_1 - k_2x_2 = f_1 + m_1g$$

$$m_2\ddot{x}_2 + d_2\dot{x}_2 + k_2x_2 - k_1x_1 = f_2 + m_2g$$

$$y = \dot{x}_2$$

$$\dot{x}_1 = x_3, \qquad \dot{x}_2 = x_4$$

The state variable equations are

$$\frac{d}{dt}\begin{bmatrix} x_1 \\ x_2 \\ x_3 \\ x_4 \end{bmatrix} = \begin{bmatrix} 0 & 0 & 1 & 0 \\ 0 & 0 & 0 & 1 \\ -k_1/m_1 & k_2/m_1 & -d_1/m_1 & 0 \\ k_1/m_2 & -k_2/m_2 & 0 & -d_2/m_2 \end{bmatrix}\begin{bmatrix} x_1 \\ x_2 \\ x_3 \\ x_4 \end{bmatrix} + \begin{bmatrix} 0 & 0 \\ 0 & 0 \\ 1/m_1 & 0 \\ 0 & 1/m_2 \end{bmatrix}\begin{bmatrix} f_1 + m_1g \\ f_2 + m_2g \end{bmatrix}$$

$$y = \begin{bmatrix} 0 & 0 & 0 & 1 \end{bmatrix} \begin{bmatrix} x_1 \\ x_2 \\ x_3 \\ x_4 \end{bmatrix}$$

The corresponding $\boldsymbol{A}, \boldsymbol{B}, \boldsymbol{C}$, and $\boldsymbol{D}$ matrices as well as $\boldsymbol{x}, \boldsymbol{u}$, and $\boldsymbol{y}$ variables are obvious.

Example 7.3:

Derive the state variable equations for the following electric circuit

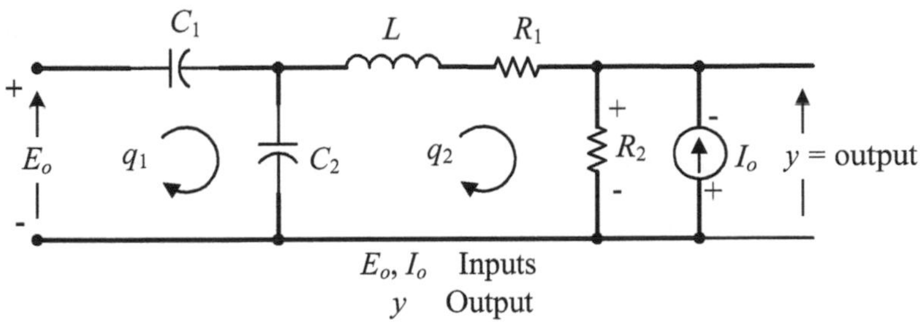

Figure 7.5: State Variable Equations for the Electric Circuit

In this example

$$\begin{aligned}
\text{Charges } q_1, q_2 &\equiv \text{ generalized coordinates} \\
\text{Currents } \dot{q}_1, \dot{q}_2 &\equiv \text{ generalized velocities} \\
\text{Inductances } &\equiv \text{ masses } m_{ij} \\
\text{Capacitors } &\equiv \text{ spring constants } K_{ij} \\
\text{Resistors } &\equiv \text{ frictional constants } d_{ij}
\end{aligned}$$

$$T = \frac{1}{2}L\dot{q}_2^2, \quad V = \frac{1}{2}C_1 q_1^2 + \frac{1}{2}C_2 (q_1 - q_2)^2$$

$$F = \frac{1}{2}(R_1 + R_2)\dot{q}_2^2, \quad I = E_0\dot{q}_1 - I_0 R_2 (\dot{q}_2 + I_0)$$

From Eq. 7.13, the state variable equations are:

$$0 + C_1 q_1 + C_2 (q_1 + q_2) - E_0 = 0$$
$$L \ddot{q}_2 - C_2 (q_1 - q_2) + (R_1 + R_2) \dot{q}_2 + I_0 R_2 = 0$$

$$y = R_2 \dot{q}_2 + I_0 R_2$$

Normally one would expect three single order state variable equations. But the first equation is only algebraic in nature and can be used to eliminate q_1 from the second equation, yielding only one second order equation (or two single order state variable equations). This is simply due to the fact that C_1, C_2, and E_0 form a capacitors-only loop and in this situation the voltages across C_1 and C_2 do not form independent variables. We shall further discuss this phenomenon in the next section. The resultant second order differential equation is:

$$L \ddot{q}_2 + (R_1 + R_2)\, \dot{q}_2 + \frac{C_1 C_2}{C_1 + C_2} q_2 = \frac{C_2}{C_1 + C_2} E_0 - I_0 R_2$$

The above equations can be written in matrix form as a state equation. A second method for deriving the state variables description of the networks involves graph theory and has been presented in many excellent books and literature [Desoer, C.A.]. In what follows, we briefly outline a general procedure for the derivation of state variable equations of a general network.

7.4.2 Formulation of the State Variable Equations of an Electric Network Using Linear Graph Theory

This section is written with a view that the state variable equations can be generated for a general network via computer algorithms if the topology of the network, element values as well as the current and voltage sources, are given. The linear graph and its associated network have the same topological information and are considered equivalent for the purpose of deriving the state variable equations.

(A) **Simple Circuits Approach #1**

In simpler networks, the replacement of the capacitors and the inductors by their fixed voltage and current sources respectively and the application of superposition theorem and Kirchoff's current and voltage laws suffice to generate the required state variable equations and needs only simple graph theory concepts. This is clear from the following example

Example 7.4:

Consider a simple circuit shown in Figure 7.6a. The capacitors and inductors are replaced by their respective voltage and current sources and redrawn in the Figure 7.6b.

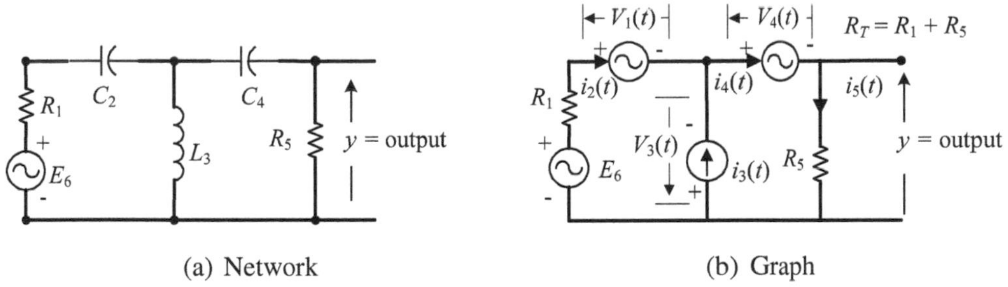

Figure 7.6: A Simple Network, Capacitors, and Inductors Replaced by Their Respective Voltage and Current Sources

Three required state variable equations represent the current through the inductor and voltages across the capacitors:

$$\dot{v}_2 = \frac{1}{C_2} i_2, \qquad \dot{v}_4 = \frac{1}{C_4} i_4$$

$$\dot{i}_3 = \frac{1}{L_3} v_3, \qquad y = R_5 i_5$$

Variables i_2, i_4, v_3, and i_5 are obtained via the superposition of fixed sources v_2, v_4, i_3, and input source E_6. The following matrix shows the relationship of each source to the required variables. **A current source equal to zero implies an open circuit and a voltage source equal to zero implies a short circuit.** From Kirchoff's current and voltage laws:

Variable \ Source	v_2	v_4	i_3	E_6
i_2	$-1/R_T$	$-1/R_T$	$-R_5/R_T$	$1/R_T$
i_4	$-1/R_T$	$-1/R_T$	$-R_1/R_T$	$1/R_T$
v_3	R_5/R_T	R_1/R_T	R_1R_5/R_T	$-R_5/R_T$
i_5	$-1/R_T$	$-1/R_T$	R_1/R_T	$1/R_T$

$R_T = R_1 + R_5$

Thus, the state variable equations in matrix form are:

$$\frac{\mathrm{d}}{\mathrm{d}t}\begin{bmatrix} v_2 \\ v_4 \\ i_3 \end{bmatrix} = \begin{bmatrix} -1/C_2R_T & -1/C_2R_T & -R_5/C_2R_T \\ -1/C_4R_T & -1/C_4R_T & R_1/C_2R_T \\ R_5/L_3R_T & R_1/L_3R_T & R_1R_5/L_3R_T \end{bmatrix}\begin{bmatrix} v_2 \\ v_4 \\ i_3 \end{bmatrix} + \begin{bmatrix} 1/C_2R_T \\ 1/C_4R_T \\ -R_5/L_3R_T \end{bmatrix} E_6$$

$$y = \begin{bmatrix} -R_5/R_T & -R_5/R_T & -R_1R_5/R_T \end{bmatrix}\begin{bmatrix} v_2 \\ v_4 \\ i_3 \end{bmatrix} + \begin{bmatrix} R_5/R_T \end{bmatrix} E_6$$

(B) **Complex Circuits Approach #2**

For circuits with more elements it is advantageous to use graph theory for the systematic generation of state variable network equations by computer algorithms when the network topology and its element values along with the voltage and the current sources are known. Figure 7.7 shows an electric circuit and its corresponding linear graph. Each circuit element is replaced by a line segment, maintaining the connection features of the original circuit. The line segments and the nodes are labeled as in the original network.

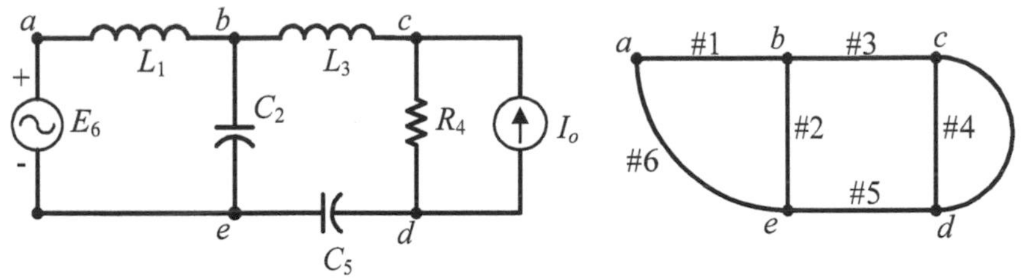

Figure 7.7: An Electric Network and Its Corresponding Linear Graph

The Following steps are taken to derive the state equations:

1. **Select a State Tree (also called a Normal Tree)**

   - A tree of linear graph is any such graph that contains all the **nodes** (or **vertices**) but no **circuits** (or **loops**).
   - The line segments (or elements) connecting any two nodes are called **branches**. The rest of the graph elements not in the tree are called **links** (or **chords**).
   - A minimal set of tree branches whose removal results in the network to be disconnected such that there are no loops, is called a **cutset** of the connected network (or the graph).
   - Every **link** when restored to the original network forms an independent loop while every tree branch with some other other branches and some links form an independent cutset.
   - A network with n nodes and e line segments has $(n - 1)$ branches and $(e - n + 1)$ links. Two segments joining the same two nodes are counted as one.

- A **Normal** (or **State**) **Tree** is selected as following:
  (i) Select all voltage sources as branch segments of the tree as a first priority.
  (ii) Select the maximum number of permissible (independent) capacitors as branch segments of the tree as the second choice. In case there are capacitors-only loops, split these capacitors so that the maximum number of them can be included in the tree. The remainder are excess capacitors and may be used as links.
  (iii) Next include the resistors.
  (iv) If necessary, include a minimum number of inductors. This situation only arises due to the existence of the inductors-only cutsets.
  (v) Current sources are the last choice for the completion of a tree. Normally all current sources should appear as links if possible.

  A tree so selected contains a maximum set of independent capacitors as its branches and whose links contain the maximum set of the independent inductors. Thus, a **Normal Tree** is selected.

Further steps are:

2. Assign to each of the capacitors C_k in the tree a state variable voltage $v_k(t)$ and to each of the inductors L_j in the link a state variable current $i_j(t)$. The capacitors in the links and the inductors in the tree branches are not assigned any state variables. Replace each of the rest of the branch segments (resistors) as a fixed voltage source and each of the rest of the link segments (resistors) as a fixed current source.
3. Write all the fundamental cutset and loop equations using Kirchoff's current and the Kirchoff's voltage laws. Cutset equations for the capacitor tree branches and the loop equations for the inductive links form the basis for the set of state variables for a given network. The rest of the

fundamental cutset and loop equations are used to eliminate the unwanted variables, such as resistor voltages and resistor currents.

4. To obtain the state variable set, write the independent link inductor voltages and independent branch capacitor currents as

$$v_j(t) = L_j \frac{d}{dt} i_j(t), \qquad i_k = C_k \frac{d}{dt} v_k(t)$$

5. A Similar procedure is used to express the output variables as a combination of the input variables.

Example 7.5:

Let us consider the circuit in Figure 7.8 and illustrate the various steps discussed above. Figure 7.8 shows the selected Normal Tree, redrawn in heavy lines.

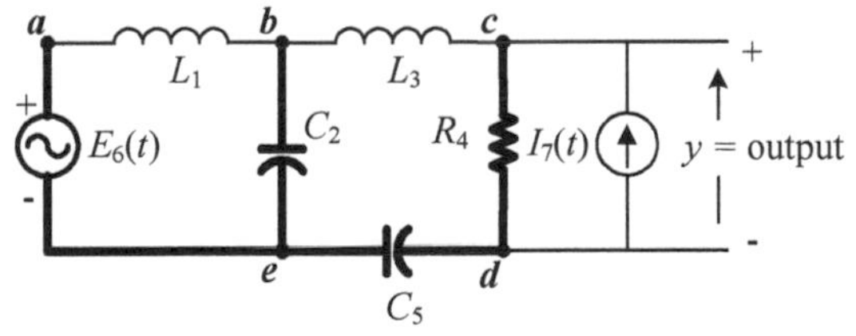

Figure 7.8: Normal State Tree (Heavy Lines) for the Circuit 7.7

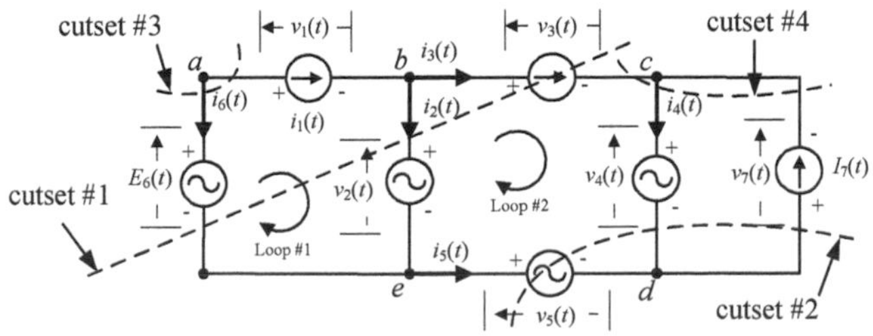

Figure 7.9: State Variables from Cutsets and Loops.

The state variable equations are:

(1) Two loop equations involving voltages

(2) Two cutset equations involving currents (see Figure 7.9)

Independent Loop #1	$v_1(t) = -v_2(t) + E_6(t)$	$v_1(t) = L_1 \frac{di_1(t)}{dt}$ $v_3(t) = L_3 \frac{di_3(t)}{dt}$
Independent Loop #2	$v_3(t) = v_2(t) + v_5(t) - v_4(t)$	$v_4(t) = R_4 i_4(t)$
Independent Cutset #1	$i_2(t) = -i_3(t) - i_6(t)$	$i_2(t) = C_2 \frac{dv_2(t)}{dt}$
Independent Cutset #2	$i_5(t) = -i_4(t) + I_7(t)$	$i_5(t) = C_5 \frac{dv_5(t)}{dt}$

Thus, the resultant single order state variable equations are:

$$
\begin{aligned}
\frac{d}{dt} i_1(t) &= -\frac{1}{L_1} v_2(t) + \frac{1}{L_1} E_6(t) \\
\frac{d}{dt} i_3(t) &= +\frac{1}{L_3} v_2(t) + \frac{1}{L_3} v_5(t) - \frac{R_4}{L_3} i_4(t) \\
\frac{d}{dt} v_2(t) &= -\frac{1}{C_3} i_3(t) - \frac{1}{C_3} i_6(t) \\
\frac{d}{dt} v_5(t) &= -\frac{1}{C_5} i_4(t) - \frac{1}{C_5} I_7(t)
\end{aligned}
$$

The other two cutset equations are used to eliminate $i_4(t)$ and $i_6(t)$. These are:

(1) Cutset through links #1 and #6, $i_6(t) = -i_1(t)$ (Equation #3)

(2) Cutset through links #3, #4, and #7, $i_4 = i_3(t) + i_7(t)$ (Equation #4)

Finally, the output voltage across R_4 is:

$$y(t) = v_4(t) = R_4\, i_3(t) + R_4\, i_7(t)$$

The resulting state variable equations are:

$$\frac{\mathrm{d}}{\mathrm{d}t}\begin{bmatrix} i_1(t) \\ i_3(t) \\ v_2(t) \\ v_5(t) \end{bmatrix} = \begin{bmatrix} 0 & 0 & -1/L_1 & 0 \\ 0 & -R_4/L_3 & 1/L_3 & 1/L_3 \\ 1/C_3 & -1/C_3 & 0 & 0 \\ 0 & -1/C_5 & 0 & 0 \end{bmatrix}\begin{bmatrix} i_1(t) \\ i_3(t) \\ v_2(t) \\ v_5(t) \end{bmatrix} + \begin{bmatrix} 1/L_1 & 0 \\ 0 & -R_4/L_3 \\ 0 & 0 \\ 0 & 0 \end{bmatrix}\begin{bmatrix} E_6(t) \\ I_7(t) \end{bmatrix}$$

$$y = \begin{bmatrix} 0 \; R_4 \; 0 \; 0 \end{bmatrix}\begin{bmatrix} i_1(t) \\ i_3(t) \\ v_2(t) \\ v_5(t) \end{bmatrix} + \begin{bmatrix} 0 \; R_4 \end{bmatrix}\begin{bmatrix} E_6(t) \\ I_7(t) \end{bmatrix}$$

The number of independent state variables is referred to as the order of the system.

7.5 State Space Concepts

7.5.1 State Space Similarity Transformations

Consider a n-th order linear time-invariant system

$$\dot{\boldsymbol{x}} = \boldsymbol{A}\,\boldsymbol{x} + \boldsymbol{B}\,\boldsymbol{u} \tag{7.15}$$

$$\boldsymbol{y} = \boldsymbol{C}\,\boldsymbol{x} + \boldsymbol{D}\,\boldsymbol{u} \tag{7.16}$$

We can find a new state variable description of the same physical system where the new state variables are a linear combination of the previous state variables. Introducing a new state variables $\hat{x}$ via nonsingular time-invariant matrix $\boldsymbol{T}$:

$$\hat{x} = \boldsymbol{T}\,\boldsymbol{x} = \sum_{j=1}^{n} t_{ij}\,x_j, \qquad i = 1, 2, \ldots, n \tag{7.17}$$

Eq. 7.15 and Eq. 7.16 can be transformed as the following:

$$\begin{aligned} T\dot{x} &= TAIx + TBu &&= TAT^{-1}Tx + TBu \\ y &= CIx + Du &&= CT^{-1}Tx + Du \end{aligned} \qquad (I = \text{Identity Matrix})$$

Introducing

$$TAT^{-1} = \hat{A}, \quad TB = \hat{B}, \quad CT^{-1} = \hat{C}, \quad D = \hat{D}$$

the new state variable equations are:

$$\dot{\hat{x}} = \hat{A}\hat{x} + \hat{B}u \tag{7.18}$$

$$y = \hat{C}\hat{x} + \hat{D}u \tag{7.19}$$

Systems $\{\hat{A}, \hat{B}, \hat{C}, \hat{D}\}$ and $\{A, B, C, D\}$ are considered as equivalent or similar systems. The input-output characteristics of both systems remain the same. This can be seen from the fact that the transfer function of both systems is the same as demonstrated in the next section. T is referred as **Similarity Transformation**.

7.5.2 Transfer Function Matrix from State Space Equations

Taking the Laplace Transform of Eqs. 7.15 and 7.16 (neglecting initial conditions)

$$sx(s) = A\,x(s) + Bu(s)$$

$$y(s) = C\,x(s) + Du(s)$$

or

$$x(s) = (sI - A)^{-1}\,Bu(s) \tag{7.20}$$

$$y(s) = \left[C\,(sI - A)^{-1}\,B\right]u(s) \tag{7.21}$$

Defining:

$$L[\text{ Output vector }] = [\text{ Transfer function matrix }]\ L[\text{ Input vector }]$$

The Transfer function matrix $\boldsymbol{G}(s)$ of system Eqs. 7.15 and 7.16 is

$$\boldsymbol{G}(s) = \boldsymbol{C}\,(s\boldsymbol{I} - \boldsymbol{A})^{-1}\,\boldsymbol{B} + \boldsymbol{D} \tag{7.22}$$

Similarly for the system Eqs. 7.18 and 7.19

$$\hat{\boldsymbol{G}}(s) = \hat{\boldsymbol{C}}\left(s\boldsymbol{I} - \hat{\boldsymbol{A}}\right)^{-1}\hat{\boldsymbol{B}} + \hat{\boldsymbol{D}} \tag{7.23}$$

It is easy to show that $\hat{\boldsymbol{G}}(s) = \boldsymbol{G}(s)$. This can be written as the following:

$$\begin{aligned}
\hat{\boldsymbol{G}}(s) &= \hat{\boldsymbol{C}}\left(s\boldsymbol{I} - \hat{\boldsymbol{A}}\right)^{-1}\hat{\boldsymbol{B}} + \hat{\boldsymbol{D}} = \boldsymbol{C}\boldsymbol{T}^{-1}\left(s\boldsymbol{I} - \boldsymbol{T}\boldsymbol{A}\boldsymbol{T}^{-1}\right)^{-1}\boldsymbol{T}\boldsymbol{B} + \boldsymbol{D} \\
&= \boldsymbol{C}\boldsymbol{T}^{-1}\left(s\boldsymbol{T}\boldsymbol{T}^{-1} - \boldsymbol{T}\boldsymbol{A}\boldsymbol{T}^{-1}\right)^{-1}\boldsymbol{T}\boldsymbol{B} + \boldsymbol{D} \qquad \left(\text{Note } \boldsymbol{I} = \boldsymbol{T}\boldsymbol{T}^{-1}\right) \\
&= \boldsymbol{C}\boldsymbol{T}^{-1}\left[\boldsymbol{T}(s\boldsymbol{I} - \boldsymbol{A})\boldsymbol{T}^{-1}\right]^{-1}\boldsymbol{T}\boldsymbol{B} + \boldsymbol{D} \qquad \left(\text{Note } (\boldsymbol{E}\boldsymbol{F}\boldsymbol{G})^{-1} = \boldsymbol{G}^{-1}\boldsymbol{F}^{-1}\boldsymbol{E}^{-1}\right) \\
&= \boldsymbol{C}\boldsymbol{T}^{-1}\left[\boldsymbol{T}(s\boldsymbol{I} - \boldsymbol{A})^{-1}\boldsymbol{T}^{-1}\right]\boldsymbol{T}\boldsymbol{B} + \boldsymbol{D} \\
&= \boldsymbol{C}\,(s\boldsymbol{I} - \boldsymbol{A})^{-1}\,\boldsymbol{B} + \boldsymbol{D} = \boldsymbol{G}(s)
\end{aligned} \tag{7.24}$$

This shows that the Transfer function matrix that represents the input-output characteristics of a system is invariant under the state variable similarity transformation. The two matrices $\hat{\boldsymbol{A}}$ and $\boldsymbol{A}$ are referred to as **Similar Matrices.** If we define:

$$\boldsymbol{A}(s) = (s\boldsymbol{I} - \boldsymbol{A}), \qquad s \text{ is any parameter}$$

$$\hat{\boldsymbol{A}}(s) = \left(s\boldsymbol{I} - \hat{\boldsymbol{A}}\right) = \left(s\boldsymbol{T}\boldsymbol{T}^{-1} - \boldsymbol{T}\boldsymbol{A}\boldsymbol{T}^{-1}\right) = \boldsymbol{T}(s\boldsymbol{I} - \boldsymbol{A})\boldsymbol{T}^{-1}$$

Then $\boldsymbol{A}(s)$ and $\hat{\boldsymbol{A}}(s)$ are also similar matrices.

Furthermore,

$$\det \boldsymbol{A} = \det \hat{A}$$

$$\det \boldsymbol{A}(s) = \det \hat{A}(s)$$

$$\det(\boldsymbol{A}(s)) = \det[s\boldsymbol{I} - \boldsymbol{A}] = \Delta_A(s) = P_n(s)$$

The quantity $\Delta_A(s)$ is an n-th order polynomial in s and is known as the **characteristic polynomial of the matrix A**. It is clear that two similar matrices have identical characteristic polynomials.

Important Facts:

1. Two different Transfer functions represent two different systems.

2. Two different state space representations that are related via Similarity Transformations yield the same Transfer function.

7.5.3 Canonical Realizations of a Given Transfer Function

Single-Input Single-Output (SISO) Systems

State space equations provide much more information about the internal structure of the system than the transfer function. A system may have state variables over which the overall system has no control, even though the output of the system may be perfectly controllable. Similarly, there may be some hidden oscillations in the system that may not be observed in the output and may make the system nonlinear and hence may change its behavior in a detrimental fashion, even though the mathematically modeled system from the input-output viewpoint may look perfectly acceptable. We shall precisely define and explain these concepts later on. As discussed earlier, the state equations of the system can be written in different canonical forms from the viewpoint of "Controllability" and "Observability." (These terms are explained in later pages.)

(i) **Four Canonical Realizations**

Control vector $\boldsymbol{u}$ effects the state $\boldsymbol{x}$ and the output $\boldsymbol{y}$ through the matrix $\boldsymbol{B}$ (see Figure 7.3). Thus, for studying controllability and designing the control function $\boldsymbol{u}$ a simpler form of $\boldsymbol{B}$ is useful. We present here two forms referred to as **controller** and **controllability** canonical forms. Similarly there are two other realizations known as **observer** and **observability** forms.

Given the transfer function:

$$G(s) = \frac{Y(s)}{U(s)} = \frac{N(s)}{D(s)} + d = \frac{Y_1(s)}{U(s)} + d \tag{7.25}$$

where

$$N(s) = b_1 s^{m-1} + b_2 s^{m-2} + \cdots + b_m$$
$$D(s) = s^n + a_1 s^{n-1} + a_2 s^{n-2} + \cdots + a_n$$

For $G(s)$ to represent a transfer function the degree m of the numerator $N(s)$ is less than or equal to the degree n of the denominator $D(s)$.

For $m = n$, the Transfer function is Proper.

For $m < n$, it is referred to as Strictly Proper ($d \equiv 0$).

Let

$$\frac{Y_1(s)}{U(s)} = \left[\frac{X(s)}{U(s)}\right]\left[\frac{Y_1(s)}{X(s)}\right] = \left[\frac{1}{D(s)}\right][N(s)] \tag{7.26}$$

and redefine

$$\frac{X(s)}{U(s)} = \frac{1}{D(s)}$$
$$Y_1(s) = N(s)X(s)$$
$$Y(s) = Y_1(s) + dU(s)$$

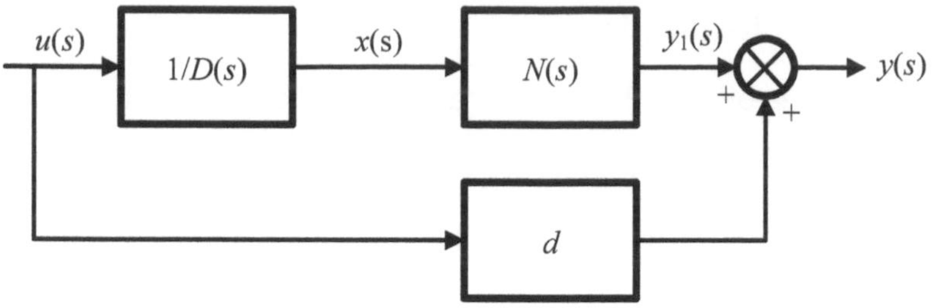

Figure 7.10: Transfer Function to be Realized

$$x^{(n)} + a_1 x^{(n-1)} + \cdots + a_n x^{(0)} = u \tag{7.27}$$

$$b_1 x^{(m-1)} + b_2 x^{(m-2)} + \cdots + b_m x^{(0)} + d\,u = y, \quad x^{(0)} = x \tag{7.28}$$

These equations can be simulated as four different realizations referred to as controller, controllability, observer, and observability.

(a) **Controller Canonical Form—Realization #1**

Eq. 7.27 can be represented by n single order equations

$$\left.\begin{array}{ll} x & = x_1 \\ \cdots & \\ \dot{x}_{n-1} = & \qquad\qquad x_n \\ \dot{x}_n & = -a_n x_1 - a_{n-1} x_2 + \cdots - a_1 x_n + u \\ y & = b_m x_1 + b_{m-1} x_2 + \cdots + b_1 x_m + d\,u \end{array}\right\} \tag{7.29}$$

Thus,

$$\boldsymbol{A} = \begin{bmatrix} 0 & 1 & 0 \cdots & 0 & 0 \\ 0 & 0 & 1 & 0 & 0 \\ & & \vdots & & \\ 0 & 0 & 0 & 0 & 1 \\ -a_n & -a_{n-1} & \cdot & \cdot & \cdot\ -a_1 \end{bmatrix}, \quad \boldsymbol{b} = \begin{bmatrix} 0 \\ 0 \\ \vdots \\ 0 \\ 1 \end{bmatrix}, \quad \boldsymbol{c} = \begin{bmatrix} b_m \\ b_{m-1} \\ \vdots \\ b_2 \\ b_1 \end{bmatrix}, \quad \boldsymbol{d} = d$$

If we neglect d, the control variable appears in one place only in the vector $\boldsymbol{b}$, giving it a very simple form.

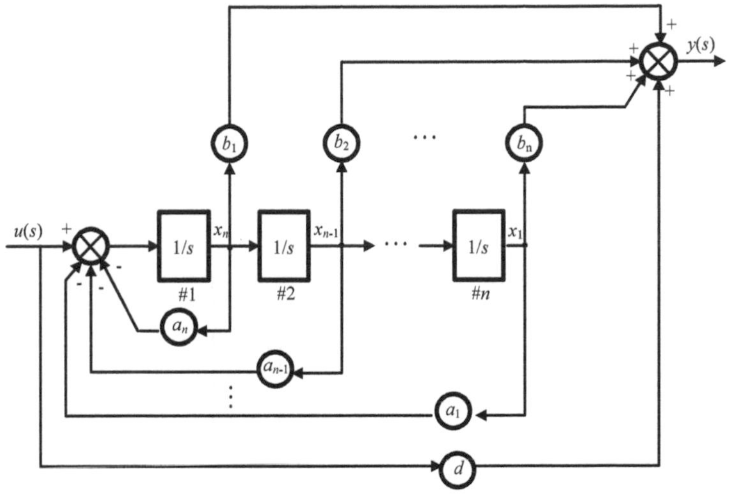

Figure 7.11: Controller Realization of a Transfer Function

(b) **Controllability Canonical Form—Realization #2**

Consider the same differential equation representing $G(s)$ in Eq. 7.25

$$x^{(n)} + a_1 x^{(n-1)} + \cdots + a_{n-1}\dot{x} + a_n x = b_0 u^{(n)} + b_1 u^{(n-1)} + \ldots + b_n u \quad (7.30)$$

$$y = x$$

Let

$$\left.\begin{aligned} & x = x_n + b_0\, u \\ & \dot{x} + a_1 x = x_{n-1} + b_1\, u + b_0\, \dot{u} \\ & \ddot{x} + a_1\, \dot{x} + a_2 x = x_{n-2} + b_2 u + b_1 \dot{u} + b_0 \ddot{u} \\ & \vdots \\ & x^{(n-1)} + a_1 x^{(n-2)} + \cdots + a_{n-1} x = x_1 + b_{n-1} u + \cdots + b_0 u^{(n-1)} \end{aligned}\right\} \quad (7.31)$$

Differentiating Eq. 7.31 and utilizing Eq. 7.30,

$$\begin{aligned}
\dot{x}_1 &= -a_n x_n && +b_n u \\
\dot{x}_2 &= -a_{n-1} x_n + x_1 && +b_{n-1} u \\
&\vdots \\
\dot{x}_n &= -a_1 x_n + x_{n-1} && +b_1 u \\
y &= x_n && +b_0 u
\end{aligned} \tag{7.32}$$

$$\boldsymbol{A} = \begin{bmatrix} 0\,0\,0 \cdots 0 & -a_n \\ 1\,0\,0 \cdots 0 & -a_{n-1} \\ 0\,1\,0 \cdots 0 & -a_{n-2} \\ & \vdots \\ 0\,0\,0 \cdots 1 & -a_n \end{bmatrix}, \quad \boldsymbol{b} = \begin{bmatrix} b_n \\ b_{n-1} \\ b_{n-2} \\ \vdots \\ b_1 \end{bmatrix}, \quad \boldsymbol{c} = \begin{bmatrix} 0 \\ 0 \\ 0 \\ \vdots \\ 1 \end{bmatrix}, \quad \boldsymbol{d} = b_0$$

In this form all the coordinates $x_1, x_2, \ldots, x_n$ are driven by the control variable u, unless some of the b coefficients are zero. Vector $\boldsymbol{c}$ is of a very simple form with all zeros except unity at one place. There are n integrators and the output of each integrator is taken as a state variable.

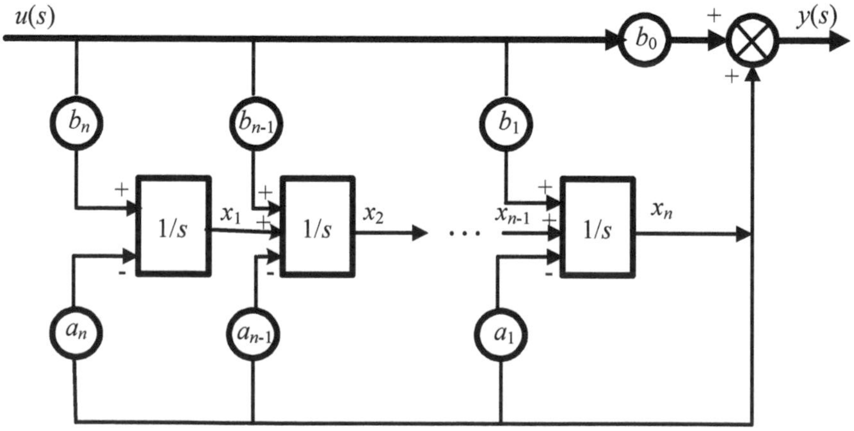

Figure 7.12: Controllability Canonical Form—Realization #2

(c) **Observer Canonical Form—Realization #3**

Eq. 7.25 can be rewritten as:

$$D(s)Y_1(s) - N(s)U(s) = 0, \qquad Y(s) = Y_1(s) + dU(s)$$

$$Y_1(s) = -\left[\frac{1}{s}(a_1Y_1(s) - b_1U(s)) + \frac{1}{s^2}(a_2Y_1(s) - b_2U(s)) + \ldots + \frac{1}{s^n}(a_nY_1(s) - b_nU(s))\right]$$

$$Y(s) = Y_1(s) + dU(s)$$

Again, the output of each integrator represents a state variable.

The state variable equations in observer form are:

$$\begin{aligned}
y_1 &= x_1 \\
\dot{x}_1 &= -(a_1x_1 - b_1u) + x_2 \\
\dot{x}_2 &= -(a_2x_1 - b_2u) + x_3 \\
&\vdots \\
\dot{x}_{n-1} &= -(a_{n-1}x_1 - b_{n-1}u) + x_n \\
\dot{x}_n &= -(a_nx_1 - b_nu) \\
y &= y_1 + du
\end{aligned}$$

$$\boldsymbol{A} = \begin{bmatrix} -a_1 & 1 & 0 & \cdots & 0 \\ -a_2 & 0 & 1 & \cdots & 0 \\ \vdots & \vdots & \vdots & & \\ -a_{n-1} & 0 & 0 & \cdots & 1 \\ -a_n & 0 & 0 & \cdots & 0 \end{bmatrix} \quad \boldsymbol{b} = \begin{bmatrix} b_1 \\ b_2 \\ \vdots \\ b_{n-1} \\ b_n \end{bmatrix} \quad \boldsymbol{c} = \begin{bmatrix} 1 \\ 0 \\ \vdots \\ 0 \\ 0 \end{bmatrix} \quad \boldsymbol{d} = d$$

The above equations are simulated using n integrators as follows:

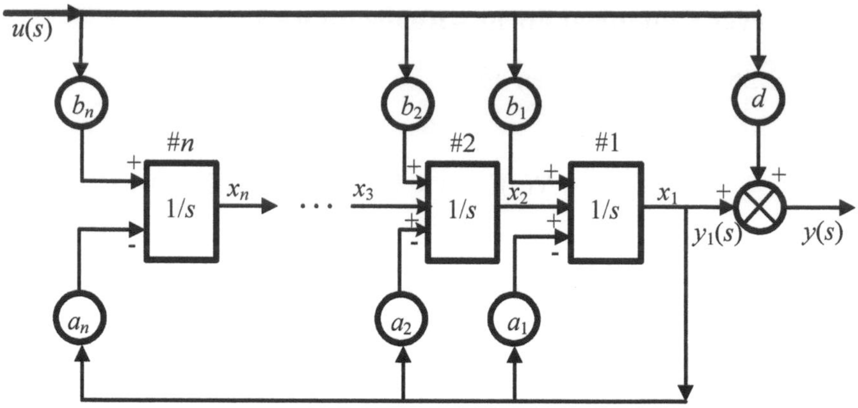

Figure 7.13: Observer Canonical Form—Realization #3

(d) **Observability Canonical Form—Realization #4**

Given:

$$G(s) = \frac{Y(s)}{U(s)} = \frac{N(s)}{D(s)}, \quad d = 0$$

This is rewritten as:

$$D(s)\,Y(s) = N(s)\,U(s)$$

or

$$y^{(n)} + \sum_{i=0}^{n-1} a_{n-i} y^{(i)} = \sum_{i=0}^{n-1} b_{n-i} u^{(i)}, \quad y^{(i)} = \frac{\mathrm{d}^i}{\mathrm{d}t^i} y, \quad u^{(i)} = \frac{\mathrm{d}^i}{\mathrm{d}t^i} u \tag{7.33}$$

Let us choose a set of variables

$$\left.\begin{aligned} y &= x_1 \\ \dot{x}_1 &= x_2 + \beta_1 u \\ \dot{x}_2 &= x_3 + \beta_2 u \\ &\vdots \\ \dot{x}_{n-1} &= x_n + \beta_{n-1} u \\ \dot{x}_n &= -\sum_{i=0}^{n-1} a_{n-i} x_{i+1} + \beta_n u \end{aligned}\right\} \tag{7.34}$$

These equations can be rewritten as

$$\left.\begin{aligned}
y &= x_1 \\
\dot{y} &= x_2 + \beta_1 u \\
&\vdots \\
y^{(n-1)} &= x_n + \beta_1 u^{(n-2)} + \beta_2 u^{(n-3)} + \cdots + \beta_{n-1} u \\
y^{(n)} &= \dot{x}_n + \beta_1 u^{(n-1)} + \beta_2 u^{(n-2)} + \cdots + \beta_{n-1}\dot{u} + \beta_n u
\end{aligned}\right\} \tag{7.35}$$

$$\begin{aligned}
y^{(i)} &= x_{i+1} + \sum_{k=1}^{i} \beta_k n^{(i-k)} \qquad i = 0, \ldots, n-1 \\
y^{(n)} &= -\sum_{i=0}^{n-1} a_{n-i} x_{i+1} + \beta_n u
\end{aligned} \tag{7.36}$$

Substituting Eq. 7.35 into Eq. 7.33, we obtain

$$\sum_{i=0}^{n-1} a_{n-i} \sum_{k=1}^{i} \beta_k u^{(i-k)} + \beta_n u = \sum_{i=0}^{n-1} b_{n-i} u^{(i)} \tag{7.37}$$

Rearranging indices, we obtain

$$\sum_{i=0}^{n} \left[\sum_{k=1}^{i} a_{i-k} \beta_k \right] u^{(n-i)} = \sum_{i=1}^{n} b_i u^{(n-i)} \tag{7.38}$$

Combining Eqs. 7.34, 7.35, and 7.38, we obtain

$$\begin{aligned}
\dot{x}_j &= x_{j+1} + \beta_j u \qquad j = 1, 2, \ldots, n-1 \\
\dot{x}_n &= -\sum_{i=1}^{n-1} a_{n-i} x_{i+1} + \beta_n u \\
b_i &= \sum_{k=1}^{i} a_{i-k} \beta_k \qquad i = 1, 2, \ldots, n \quad a_0 = 1 \\
y &= x_1
\end{aligned} \tag{7.39}$$

or

$$\dot{x} = Ax + \beta u$$
$$y = c^\top x + d$$

where

$$A = \begin{bmatrix} 0 & 1 & & 0 & 0 \\ 0 & 0 & & 0 & 0 \\ & & \vdots & & \\ 0 & 0 & & 0 & 1 \\ -a_n & -a_{n-1} & \cdots & -a_2 & -a_1 \end{bmatrix}, \; \beta = \begin{bmatrix} \beta_1 \\ \beta_2 \\ \vdots \\ \beta_{n-1} \\ \beta_n \end{bmatrix} = \begin{bmatrix} 1 & 0 & \cdots & 0 \\ a_1 & 1 & \ldots & 0 \\ \vdots & & \vdots & \\ a_{n-2} & a_{n-3} & \ldots & 0 \\ a_{n-1} & a_{n-2} & & 1 \end{bmatrix}^{-1} \begin{bmatrix} b_1 \\ b_2 \\ \vdots \\ b_{n-1} \\ b_n \end{bmatrix}$$

$$c^T = \begin{bmatrix} 1 & 0 & 0 & \cdots & 0 & 0 \end{bmatrix}, \quad d = 0$$

As evident, the vector c is simple here, involving only one entry.

Note: The reader should verify these equations for $n = 3$.

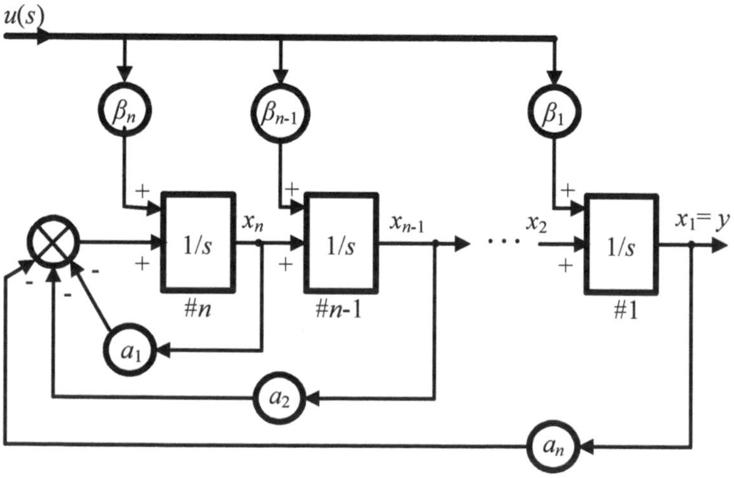

Figure 7.14: Observability Canonical Form—Realization #4

Figure 7.14 shows the output of every integrator as a state variable.

Similarity Transformation to Obtain a Companion or Normal Matrix

In our treatment of various Controller forms, we always considered the system matrix $\boldsymbol{A}$ to have the companion (normal) matrix form because it is easy to compute its characteristic polynomial, eigenvalues, and the corresponding eigenvectors. In reality, a general matrix $\boldsymbol{A}$ may have all its elements $a_{11}, a_{12}, \cdots$ etc. It is very useful to be able to compute a similarity transformation that transforms a general matrix $\boldsymbol{A}$ into a Companion matrix $\boldsymbol{A}_E$. Let us seek a matrix $\boldsymbol{T}$ such that

$$\boldsymbol{T}\boldsymbol{A}\boldsymbol{T}^{-1} = \boldsymbol{A}_E \tag{7.40}$$

where

$$\boldsymbol{T} = \begin{bmatrix} \boldsymbol{t}_1^T \\ \boldsymbol{t}_2^T \\ \vdots \\ \boldsymbol{t}_n^T \end{bmatrix}, \qquad \boldsymbol{A} = \begin{bmatrix} a_{11} & a_{12} & \cdots & a_{1n} \\ a_{21} & a_{22} & \cdots & a_{2n} \\ \vdots & & & \\ a_{n1} & a_{n2} & \cdots & a_{nn} \end{bmatrix} = \begin{bmatrix} \boldsymbol{a}_1^T \\ \boldsymbol{a}_2^T \\ \vdots \\ \boldsymbol{a}_n^T \end{bmatrix}$$

and

$$\boldsymbol{A}_E = \begin{bmatrix} 0 & 1 & 0 & \cdots & 0 \\ 0 & 0 & 1 & \cdots & 0 \\ \vdots & & & & \\ 0 & 0 & 0 & \cdots & 1 \\ -a_n & -a_{n-1} & -a_{n-2} & \cdots & -a_1 \end{bmatrix}$$

Thus,

$$\boldsymbol{T}\boldsymbol{A} = \boldsymbol{A}_E\boldsymbol{T} \tag{7.41}$$

We can always arbitrarily choose one of the rows of $\boldsymbol{T}$.

Let

$$\boldsymbol{t}_1 = \begin{bmatrix} 1 \\ 0 \\ \vdots \\ 0 \end{bmatrix} \tag{7.42}$$

Then

$$\boldsymbol{TA} = \begin{bmatrix} \boldsymbol{t}_1^T \\ \boldsymbol{t}_2^T \\ \vdots \\ \boldsymbol{t}_n^T \end{bmatrix} \boldsymbol{A} = \begin{bmatrix} \boldsymbol{t}_1^T \boldsymbol{A} \\ \boldsymbol{t}_2^T \boldsymbol{A} \\ \vdots \\ \boldsymbol{t}_n^T \boldsymbol{A} \end{bmatrix} \tag{7.43}$$

Also

$$\boldsymbol{A}_E \boldsymbol{T} = \begin{bmatrix} 0 & 1 & \cdots & 0 \\ 0 & 0 & \cdots & 0 \\ \vdots & & & \\ -a_n & -a_{n-1} & \cdots & -a_1 \end{bmatrix} \begin{bmatrix} \boldsymbol{t}_1^T \\ \boldsymbol{t}_2^T \\ \vdots \\ \boldsymbol{t}_n^T \end{bmatrix} = \begin{bmatrix} \boldsymbol{t}_2^T \\ \boldsymbol{t}_3^T \\ \vdots \\ -\sum_{i=1}^n a_i \boldsymbol{t}_{n+1-i}^T \end{bmatrix} \tag{7.44}$$

From Eqs. 7.41, 7.42, 7.43, and 7.44

$$\left.\begin{aligned} &\boldsymbol{t}_1^T = \begin{bmatrix} 1 \; 0 \; 0 \cdots 0 \end{bmatrix} \\ &\boldsymbol{t}_2^T = \boldsymbol{t}_1^T \boldsymbol{A} = \begin{bmatrix} \boldsymbol{a}_1^T \end{bmatrix} \\ &\boldsymbol{t}_3^T = \boldsymbol{t}_2^T \boldsymbol{A} = \begin{bmatrix} \boldsymbol{t}_2^T \boldsymbol{a}_1 \; \boldsymbol{t}_2^T \boldsymbol{a}_2 \cdots \boldsymbol{t}_2^T \boldsymbol{a}_n \end{bmatrix} = \boldsymbol{t}_1^T \boldsymbol{A}^2 \\ &\vdots \\ &\boldsymbol{t}_n^T = \boldsymbol{t}_{n-1}^T \boldsymbol{A} = \boldsymbol{t}_1^T \boldsymbol{A}^{n-1} \end{aligned}\right\} \tag{7.45}$$

The last equation yields

$$\boldsymbol{t}_n^T \boldsymbol{A} = -\sum_{i=1}^{n} a_i \boldsymbol{t}_{n+1-i}^T \tag{7.46}$$

From Eqs. 7.45 and 7.46, we obtain

$$\boldsymbol{t}_1^T \left[\boldsymbol{A}^n + a_1\boldsymbol{A}^{n-1} + \cdots + a_n\boldsymbol{I}\right] = \boldsymbol{0}$$

implying,

$$\boldsymbol{A}^n + a_1\boldsymbol{A}^{n-1} + \cdots + a_n\boldsymbol{I} = \boldsymbol{0} \quad \text{Cayley-Hamilton Theorem.}$$

Furthermore,

$$\boldsymbol{T} = \begin{bmatrix} \boldsymbol{t}_1^T \\ \boldsymbol{t}_2^T \\ \vdots \\ \boldsymbol{t}_n^T \end{bmatrix} = \begin{bmatrix} \boldsymbol{t}_1^T \\ \boldsymbol{t}_1^T\boldsymbol{A} \\ \vdots \\ \boldsymbol{t}_1^T\boldsymbol{A}^{n-1} \end{bmatrix}$$

This transformation enables us to transform any matrix to its Companion form.

Important Observations

1. **Controller Canonical Form**

   Control variable u appears as an input to one integrator only. This control signal passes through n integrators and thus gets integrated n times. This form is suitable for control variables implementation.

2. **Controllability Canonical Form**

   Control variable $u(t)$ appears as a direct input to each of the integrators and as such influences each of the variables directly and can make the controllability test easier.

3. **Observer Canonical Form**

   Only the observed variable x_1 appears directly in the output (besides control variable u). This makes the implementation of variable x_1 as a feedback signal in the controller easy.

4. **Observability Canonical Form**

   In this form (for $u = 0$) the output y and all its derivatives can be measured as the output of the integrators, making the observability test simple.

5. The properties of control and observation of a system are the inherent consequences of the system structure and will be studied in detail via matrix algebra (explained later).

6. In fact, the controllable canonical form and observable canonical form can be viewed as dualities of each other in the sense that if the input u and output y are interchanged and the direction of the signals flow represented by the arrows are reversed, we can obtain observer realization from controller realization and vice versa. The same is true of the controllability and the observability realizations.

7. **Why Integration and not Differentiation**

   - Dynamical systems are usually low frequency devices. For simulation purposes we prefer integration and usually try to avoid the differentiation for the simple reason that a nonideal world contains unwanted "noise" consisting of high frequency components. Integration is a smoothing process that eliminates these high frequency components. On the other hand, differentiating enhances these noisy high frequency components. This can be seen follows: Defining the noise as $n(t) = A\, e^{j\omega t},\; A \approx 0.1,\; \omega \approx 10^6$ integrating

   $$I(n(t)) = \int n(t)\, \mathrm{d}t = \frac{1}{j}\left(\frac{A}{\omega}\right)e^{j\omega t} \quad \frac{A}{\omega} = 10^{-7} \text{ (a very small number)}$$

   For practical purposes $I(n(t)) \; = \; 0$, showing that the integration process suppresses high frequency noise. Similarly the effect of differentiation can be seen as:

$$D(n(t)) = \frac{\mathrm{d}}{\mathrm{d}t}n(t) = j(A\omega)e^{j\omega t}, \quad A\omega = 10^5 \quad \text{(large value)}$$

Thus, the detrimental effect of $n(t)$ is accentuated by differentiation.

7.5.4 Controllability and Observability Concepts

What does the Controllability and Observability concepts mean with regard to understanding and design of control systems? Before we state the precise definitions of Controllability and Observability, it is appropriate to understand the importance of these concepts. We shall illustrate these with some simple examples. Consider the following system:

$$\begin{bmatrix} \dot{x}_1 \\ \dot{x}_2 \end{bmatrix} = \begin{bmatrix} \lambda_1 & 0 \\ 0 & \lambda_2 \end{bmatrix} \begin{bmatrix} x_1 \\ x_2 \end{bmatrix} + \begin{bmatrix} 0 \\ 1 \end{bmatrix} u$$
$$y = \begin{bmatrix} 1 & 0 \end{bmatrix} \begin{bmatrix} x_1 \\ x_2 \end{bmatrix} \tag{7.47}$$

The system has two coordinates or "modes." Can we influence both these modes via the control signal u? The answer is **No**. The coordinate x_1 is driven only by its initial energy or initial condition and control u has no influence on the system. We conclude that the coordinate x_1 is not **Controllable**. Now, let us look at the output y. Can we obtain any information about the variable x_2 by measuring the output y? The answer is again **No**. The output u cannot be used to "observe" the variable x_2. We conclude that the "**controllability**" and the "**observability**" concepts are associated with the structure of the matrices $\boldsymbol{A}, \boldsymbol{B}, \boldsymbol{c}$, of the system. These concepts are essential in designing controllers that have the desired effect on the system outputs. We shall explain these concepts from a geometric viewpoint by projecting the state space $\boldsymbol{x}$ into the **Controllable** and the **Observable Subspaces**. Let us first give the precise definition for both of these attributes and derive the corresponding criteria.

7.5.5 Controllability Definition and Criterion

System: $\dot{\boldsymbol{x}} = \boldsymbol{A}\boldsymbol{x} + \boldsymbol{b}u$ or the pair $(\boldsymbol{A}, \boldsymbol{b})$ is Controllable from any initial state $\boldsymbol{x}(0)$ to any final state $\boldsymbol{x}(T)$ if there exists a control $\boldsymbol{u}(t)$ that steers the system from $\boldsymbol{x}(0)$ to $\boldsymbol{x}(T)$ in a finite time, $0 < t \leq T$, otherwise the system is uncontrollable. If $\boldsymbol{A}$ is diagonal, it is obvious that if any $b_i = 0$, then that mode cannot be controlled. Since we are only interested in the influence of the control variable $u(t)$ on the state $\boldsymbol{x}(t)$, the control influenced part of $\boldsymbol{x}(t)$ can be written as:

$$\left(e^{-\boldsymbol{A}T}\boldsymbol{x}(T) - \boldsymbol{x}(0)\right) = \boldsymbol{x}_1(T) = \int_0^T e^{-\boldsymbol{A}\tau}\boldsymbol{b}u(\tau)\,\mathrm{d}\tau$$

Expanding $e^{-\boldsymbol{A}\tau}$ via Cayley-Hamilton theorem:

$$\boldsymbol{x}_1(T) = \int_0^T \left(\sum_{i=0}^{n-1} \alpha_i(\tau)\boldsymbol{A}^i\right)\boldsymbol{b}u(\tau)\,\mathrm{d}\tau = \sum_{i=0}^{n-1} \boldsymbol{A}^i\boldsymbol{b}\left(\int_0^T \alpha_i(\tau)u(\tau)\,\mathrm{d}\tau\right)$$

Let

$$\text{Let} \quad \int_0^T \alpha_i(\tau)u(\tau)\,\mathrm{d}\tau = v_i, \qquad \boldsymbol{v}^T = \begin{bmatrix} v_0 \; v_2 \; \dots \; v_{n-1} \end{bmatrix}$$

Thus,

$$\boldsymbol{x}_1(t) = \begin{bmatrix} \boldsymbol{b} \; \boldsymbol{A}\boldsymbol{b} \; \dots \; \boldsymbol{A}^{n-1}\boldsymbol{b} \end{bmatrix}\begin{bmatrix} \boldsymbol{v} \end{bmatrix} = \boldsymbol{C}(\boldsymbol{A}, \boldsymbol{b})\boldsymbol{v}$$

In order that $u(t)$ (control) dependent vector $\boldsymbol{v}$ can be computed for a given $\boldsymbol{x}_1(T)$, it is necessary and sufficient that the matrix $\boldsymbol{C}(\boldsymbol{A}, \boldsymbol{b})$ is invertible. This is the controllability criterion.

7.5.6 Observability Definition and Criterion

Consider the system:

$$\dot{\boldsymbol{x}} = \boldsymbol{A}\,\boldsymbol{x} + \boldsymbol{b}\,u$$
$$y = \boldsymbol{c}^T\,\boldsymbol{x}$$

or the pair $(\boldsymbol{A}, \boldsymbol{c})$ is observable if for any initial state $\boldsymbol{x}(0)$ and given control $u(t)$, there exists a finite time $T > 0$, such that knowledge of $y(t)$ over the interval 0 to T sufficies to determine the initial state $\boldsymbol{x}(0)$ uniquely, otherwise the system is not observable. From the knowledge of $\boldsymbol{x}(0)$ it is easy to compute $\boldsymbol{x}(t)$ as follows:

$$\boldsymbol{x}(t) = \boldsymbol{e}^{\boldsymbol{A}t}\boldsymbol{x}(0) + \int_0^t \boldsymbol{e}^{\boldsymbol{A}(t-\tau)}\boldsymbol{b}u(\tau)\,\mathrm{d}\tau$$
$$y(t) = \boldsymbol{c}^T\boldsymbol{x}(t)$$

Since the influence of the known control $u(t)$ on $y(t)$ can be easily computed, we are only interested in the contribution of the intial condition $\boldsymbol{x}(0)$ to the output variable. This is computed as:

$$\left[y(t) - \boldsymbol{c}^T \int_0^t \boldsymbol{e}^{\boldsymbol{A}(t-\tau)}\boldsymbol{b}u(\tau)\,\mathrm{d}\tau\right] = y_1(t) = \boldsymbol{c}^T\boldsymbol{e}^{\boldsymbol{A}t}\boldsymbol{x}(0), \quad 0 \le t \le T$$

Let us observe $y(t)$ and hence, $y_1(t)$ at $t = t_1, t_2, \ldots, t_n$. Expanding $\boldsymbol{e}^{\boldsymbol{A}t}$ in the powers of $\boldsymbol{A}$ via the Cayley-Hamilton Theorem, the above equation takes the form:

$$\begin{bmatrix} y_1(t_1) \\ y_1(t_2) \\ \vdots \\ y_1(t_n) \end{bmatrix} = \begin{bmatrix} \alpha_0(t_1) & \alpha_1(t_1) & \cdots & \alpha_{n-1}(t_1) \\ \alpha_0(t_2) & \alpha_1(t_2) & \cdots & \alpha_{n-1}(t_2) \\ & & \vdots & \\ \alpha_0(t_n) & \alpha_1(t_n) & \cdots & \alpha_{n-1}(t_n) \end{bmatrix} \begin{bmatrix} \boldsymbol{c}^T \\ \boldsymbol{c}^T\boldsymbol{A} \\ \vdots \\ \boldsymbol{c}^T\boldsymbol{A}^{n-1} \end{bmatrix} \boldsymbol{x}(0)$$

For the set of observation $y_1(t_1), y_1(t_2), \ldots, y_1(t_n)$, the vector $\boldsymbol{x}(0)$ can only be determined if and only if the observability matrix

$$O(C, A) = \left[\boldsymbol{c} \; \boldsymbol{A}^T \boldsymbol{c} \cdots \left(\boldsymbol{A}^{n-1} \right)^T \boldsymbol{c} \right] \quad \text{is invertible}$$

7.5.7 Controllability–Observability Geometric Interpretation

Consider a single-input single-output system:

$$\dot{\boldsymbol{x}} = \boldsymbol{A}\boldsymbol{x} + \boldsymbol{b}u, \qquad y = \boldsymbol{c}^T \boldsymbol{x}, \qquad \boldsymbol{x}(t_0) = \boldsymbol{x}_0 \tag{7.48}$$

1. **Case 1: Matrix A with Distinct Eigenvalues**

   For simplicity let A have distinct eigenvalues $\lambda_1, \lambda_2, \ldots, \lambda_n$ and associated with each of the eigenvalues are n-independent eigenvectors $\boldsymbol{p}_1, \boldsymbol{p}_2, \ldots, \boldsymbol{p}_n$, such that

$$\boldsymbol{A}\, \boldsymbol{p}_i = \lambda_i\, \boldsymbol{p}_i, \qquad i = 1, 2, \ldots, n \tag{7.49}$$

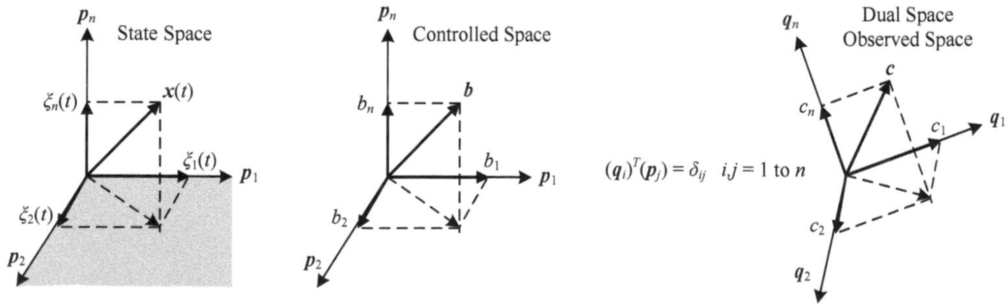

(a) State modes (b) Controlled coefficients (c) Observed space coefficients

Figure 7.15: Vectors $\boldsymbol{x}(t)$, $\boldsymbol{b}$, and $\boldsymbol{c}$ in State, Dual Space-Geometric Representation

This simply implies that any vector $\boldsymbol{x}(t)$ in n-dimensional state space can be projected onto the n-dimensional space spanned by eigenvectors $\boldsymbol{p}_1, \boldsymbol{p}_2, \ldots, \boldsymbol{p}_n$.

Thus,

$$x(t) = \sum_{i=1}^{n} \xi_i(t)\, p_i$$

It is a most descriptive way of describing the motion of $x(t)$ with time. The components $\xi_i(t)$ are referred to as the modes of the system. In general, the particular "axis" p_i and p_j may not be orthogonal. The easiest way to compute $\xi_i(t)$ from $x(t)$ and p_i is to realize that corresponding to the set $p_i \quad (i = 1, 2, \ldots, n)$ there is a set of n-independent dual vectors $q_i, \quad (i = 1, 2, \ldots n)$ such that:

$$q_i^T\, p_i = \delta_{ij} \begin{cases} 1 & i = j \\ 0 & i \neq j \end{cases}$$

$$A^T\, q_i = \lambda_i\, q_i \qquad q_i \text{ is a eigenvector of } A^T$$

Proceeding with the solution of Eq. 7.48 we obtain.

$$x(t) = e^{At} x(0) + \int_0^t e^{A(t-\tau)} b\, u(\tau)\, d\tau \tag{7.50}$$

$$y(t) = c^T e^{At}\, x(0) + \int_0^t c^\top\, e^{A(t-\tau)} b\, u(\tau)\, d\tau \tag{7.51}$$

Decomposing vectors $x(t), x(0)$, and b along p_i and c along q_i

$$\left.\begin{aligned} x(0) &= \sum_{i=1}^{n} \xi_i(0)\, p_i \\ x(t) &= \sum_{i=1}^{n} \xi_i(t)\, p_i \\ b &= \sum_{i=1}^{n} b_i\, p_i \\ A\, p_i &= \lambda_i\, p_i \\ e^{At}\, p_i &= e^{\lambda_i t}\, p_i \\ c^\top &= \sum_{j=1}^{n} c_j\, q_j^T \end{aligned}\right\} \tag{7.52}$$

From Eqs. 7.50, 7.51, and 7.52

$$x(t) = \sum_{i=0}^{n} \left[e^{\lambda_i(t)} \left(\xi_i(0) + b_i \int_0^t e^{-\lambda_i \tau} u(\tau)\, d\tau \right) \right] p_i = \sum_{i=1}^{n} \xi_i(t)\, p_i \quad (7.53)$$

$$y(t) = \sum_{j=1}^{n} c_j q_j^T \sum_{i=1}^{n} \xi_i(t)\, p_i = \sum_{i=1}^{n} c_i \xi_i(t) \quad (7.54)$$

where

$$\xi_i(t) = e^{\lambda_i t} \left(\xi_i(0) + b_i \int_0^t e^{-\lambda_i \tau} u(\tau)\, d\tau \right)$$

$$\xi(0) = [p_1, p_2, \ldots, p_n]^{-1} x(0)$$

any $b_i = 0$ implies that the i-th mode is uncontrollable

any $c_i = 0$ implies that the i-th mode is not observable

Controllability–Observability criteria for Matrix A with Distinct Eigenvalues

2. **Case 2: Matrix A with Multiple Eigenvalues**

For distinct eigenvalues, we obtain n-independent eigenvectors, one independent eigenvector for each distinct eigenvalue and therefore the invariant subspaces of $\boldsymbol{A}$ are all one-dimensional vectors. For multiple eigenvalues, the dimension of the invariant subspace of $\boldsymbol{A}$ associated with each eigenvalue is equal to its dimension. If λ_i has multiplicity r_i, the dimension of its invariant subspace is r_i. Without loss of generality let $\lambda_i = \lambda$ be of multiplicity n.

The generalized eigenvectors $p_1, p_2, \ldots, p_n$ in this case are:

$$\begin{aligned} A\, p_1 &= \lambda\, p_1 \\ A\, p_i &= \lambda\, p_i + p_{i-1} \qquad i = 2, 3, \ldots, n \end{aligned} \quad (7.55)$$

Note: These generalized eigenvectors are independent and can be computed via Eq. 7.55 for a given $\boldsymbol{A}$.

Let us use Eq. 7.55 to decompose the system.

$$\begin{aligned} \dot{\boldsymbol{x}} &= \boldsymbol{A}\,\boldsymbol{x} + \boldsymbol{b}\,u, \qquad \boldsymbol{x}(0) = \boldsymbol{x}_0 \\ y &= \boldsymbol{c}^T \boldsymbol{x} \end{aligned}$$

Using Eq. 7.52

$$\begin{aligned} \boldsymbol{x}(t) &= \sum_{i=1}^{n} \xi_i(t)\,\boldsymbol{p}_i \\ \dot{\boldsymbol{x}}(t) &= \sum_{i=1}^{n} \dot{\xi}_i(t)\,\boldsymbol{p}_i \\ \boldsymbol{x}_0 &= \sum_{i=1}^{n} \xi_i(0)\,\boldsymbol{p}_i \\ \boldsymbol{A}\,\boldsymbol{p}_i &= \lambda\,\boldsymbol{p}_i + \boldsymbol{p}_{i-1} \qquad i = 2, 3, \ldots, n \\ \boldsymbol{c}^T &= \sum_{j=1}^{n} c_j\,\boldsymbol{q}_j^T \\ \boldsymbol{b} &= \sum_{i=1}^{n} b_i\,\boldsymbol{p}_i \end{aligned}$$

Substitution of Eq. 7.55 into the above equations:

$$\begin{aligned} \sum_{i=1}^{n} \dot{\xi}_i\,\boldsymbol{p}_i &= \sum_{i=1}^{n} \xi_i(t)\,(\lambda\,\boldsymbol{p}_i + \boldsymbol{p}_{i-1}) + \sum_{i=1}^{n} b_i\,\boldsymbol{p}_l \\ y(t) &= \sum_{i=1}^{n} c_i\,\xi_i(t) \end{aligned}$$

Equating coefficients of $\boldsymbol{p}_i$, yields

$$\begin{aligned} \dot{\xi}_i(t) &= \lambda_i\,\xi_i(t) + \xi_{i-1}(t) + b_i\,u(t) \qquad i = 1, 2, \ldots, n-1 \\ \dot{\xi}_n(t) &= \lambda_n\,\xi_n(t) + b_n\,u(t) \\ \boldsymbol{\xi}(0) &= [\boldsymbol{p}_1, \boldsymbol{p}_2, \ldots, \boldsymbol{p}_n]^{-1}\,\boldsymbol{x}(0) \\ y(t) &= \sum_{i=1}^{n} c_i\,\xi_i(t) \end{aligned} \tag{7.56}$$

Matrix A with Multiple Eigenvalues

- $\xi_i(t)$ are the modes of $\boldsymbol{x}(t)$ representing the system state variables.
- c_i represents the contribution by the mode $\xi_i(t)$ to the output $y(t)$.
- First term of $\xi_i(t)$ depends upon $\xi_i(0)$, which is due to initial condition $\boldsymbol{x}(0)$ only. The second term shows the influence of $\xi_{i-1}(t)$ on the i-th mode, while the third term of $\xi_i(t)$ involving the coefficients b_i determines how much influence the control $u(t)$ has on the particular mode $\xi_i(t)$.

We are now in a position to give a precise interpretation of Controllability and Observability concepts from a Geometrical viewpoint.

7.5.8 Geometric Controllability–Observability Criterion

(a) Controllability criterion for the pair $(\boldsymbol{A}, \boldsymbol{b})$—distinct eigenvalues of $\boldsymbol{A}$

Let

$$\left.\begin{aligned} \boldsymbol{b} &= \sum_{i=1}^{n} b_i\, \boldsymbol{p}_i \\ \boldsymbol{A}\,\boldsymbol{b} &= \sum_{i=1}^{n} b_i\, \lambda_i\, \boldsymbol{p}_i \\ \boldsymbol{A}^2\,\boldsymbol{b} &= \sum_{i=1}^{n} b_i\, \lambda_i^2\, \boldsymbol{p}_i \\ &\vdots \\ \boldsymbol{A}^{n-1}\,\boldsymbol{b} &= \sum_{i=1}^{n} b_i\, \lambda_i^n\, \boldsymbol{p}_i \end{aligned}\right\} \tag{7.57}$$

These equations can be written in a $n \times n$ matrix form, known as the controllability matrix:

$$\boldsymbol{C}(\boldsymbol{A}, \boldsymbol{b}) = \begin{bmatrix} \boldsymbol{b} & \cdots & \boldsymbol{A}^{n-1}\boldsymbol{b} \end{bmatrix} = \begin{bmatrix} \boldsymbol{p}_1 & \boldsymbol{p}_2 & \cdots & \boldsymbol{p}_n \end{bmatrix} \begin{bmatrix} b_1 & \lambda_1\, b_1 & \cdots & \lambda_1^{n-1} b_1 \\ b_2 & \lambda_2\, b_2 & \cdots & \lambda_2^{n-2}\, b_2 \\ \vdots & \vdots & & \\ b_n & \lambda_n\, b_n & \cdots & \lambda_n^{n-1}\, b_n \end{bmatrix} \tag{7.58}$$

$\lambda_1, \lambda_2, \ldots, \lambda_n$ are all distinct.

If any $b_i = 0$, matrix $\boldsymbol{C}(\boldsymbol{A}, \boldsymbol{b})$ has a rank less than n implying that the system $(\boldsymbol{A}, \boldsymbol{b})$ is not completely controllable. For the system $(\boldsymbol{A}, \boldsymbol{b})$ to be completely controllable,

(1) $\boldsymbol{C}(\boldsymbol{A}, \boldsymbol{b})$ is Invertible

(2) $\boldsymbol{C}(\boldsymbol{A}, \boldsymbol{b})$ has the rank n

(3) $\boldsymbol{C}(\boldsymbol{A}, \boldsymbol{b})$ has n-linearly independent rows or columns.

(4) None of the elements $b_j = 0, \; j = 1, 2, \ldots, n$

It is also easy to prove that the same test is true for multiple eigenvalues.

(b) Observability criterion for the pair $(\boldsymbol{A}, \boldsymbol{c})$: If any $c_j = 0$, then the j-th mode is absent from the output y and hence, not observable. If all $c_j \neq 0$, then the system is completely observable and the observability conditions can be derived as following:

$$\boldsymbol{c}^T \boldsymbol{A}^k = \sum_{j=1}^{n} \lambda_j^k c_j \boldsymbol{q}_j^T, \quad c_j \neq 0, \quad k = 0, 1, 2, \ldots, n-1$$

Thus, the observability matrix $\boldsymbol{O}(\boldsymbol{A}, \boldsymbol{c})$ is:

$$\boldsymbol{O}(\boldsymbol{A}, \boldsymbol{c}) = \begin{bmatrix} \boldsymbol{c}^T \\ \boldsymbol{c}^T \boldsymbol{A} \\ \boldsymbol{c}^T \boldsymbol{A}^2 \\ \vdots \\ \boldsymbol{c}^T \boldsymbol{A}^{n-1} \end{bmatrix} = \begin{bmatrix} c_1 & c_2 & \cdots & c_n \\ \lambda_1 c_1 & \lambda_2 c_2 & \cdots & \lambda_n c_n \\ \vdots & & & \\ \lambda_1^{n-1} c_1 & \lambda_2^{n-1} c_2 & \cdots & \lambda_n^{n-1} c_n \end{bmatrix} \begin{bmatrix} \boldsymbol{q}_1^T \\ \boldsymbol{q}_2^T \\ \vdots \\ \boldsymbol{q}_n^T \end{bmatrix} \tag{7.59}$$

If any $c_j = 0$, matrix $\boldsymbol{O}(\boldsymbol{A}, \boldsymbol{c})$ has a rank less than n. Thus, for a system $(\boldsymbol{A}, \boldsymbol{c})$ to be completely observable:

(1) $\boldsymbol{O}(\boldsymbol{A}, \boldsymbol{c})$ is Invertible

(2) $\boldsymbol{O}(\boldsymbol{A}, \boldsymbol{c})$ has the rank n

(3) $\boldsymbol{O}(\boldsymbol{A}, \boldsymbol{c})$ has n-linearly independent rows or columns.

Note: If the terminal time T is finite then a limited region in the state space is reachable and observable. Furthermore, if $T \to \infty$, then the whole state space can be reached and observed provided the above criterion is satisfied.

Example 7.6:

Given an $(\boldsymbol{A}, \boldsymbol{b}, \boldsymbol{c})$ system

$$\boldsymbol{x}(k+1) = \boldsymbol{A}\boldsymbol{x}(k) + \boldsymbol{b}u(k), \quad \boldsymbol{x}(k) \text{ is an } n \text{ vector}, \quad u(k) \text{ is 1 vector}$$
$$y(k) = \boldsymbol{c}^T\boldsymbol{x}(k)$$

Determine:

(i) The initial state $\boldsymbol{x}(0)$ from the observed output $y(0), y(1), \cdots, y(n-1)$.

(ii) The controls $u(0), \cdots, u(n-1)$, steering the system from $\boldsymbol{x}(0)$ to the origin.

Solution:

(i) Let us choose the controls $u(k)$ to be zero at all instances.

Then

$$\boldsymbol{x}(k+1) = \boldsymbol{A}\boldsymbol{x}(k)$$

Thus,

$$\begin{aligned} y(0) &= \boldsymbol{c}^T\boldsymbol{x}(0) \\ y(1) &= \boldsymbol{c}^T\boldsymbol{x}(1) = \boldsymbol{c}^T\boldsymbol{A}\boldsymbol{x}(0) \\ y(2) &= \boldsymbol{c}^T\boldsymbol{x}(2) = \boldsymbol{c}^T\boldsymbol{A}^2\boldsymbol{x}(0) \\ &\vdots \\ y(n-1) &= \boldsymbol{c}^T\boldsymbol{x}(n-2) = \boldsymbol{c}^T\boldsymbol{A}^{n-1}\boldsymbol{x}(0) \end{aligned}$$

Hence

$$\begin{bmatrix} y(0) \\ y(1) \\ \vdots \\ y(n-1) \end{bmatrix} = \begin{bmatrix} \boldsymbol{c}^T \\ \boldsymbol{c}^T \boldsymbol{A} \\ \vdots \\ \boldsymbol{c}^T \boldsymbol{A}^{n-1} \end{bmatrix} \boldsymbol{x}(0) = \boldsymbol{O}(\boldsymbol{A}, \boldsymbol{c})\boldsymbol{x}(0)$$

Inverting the Observability matrix $\boldsymbol{O}(\boldsymbol{A}, \boldsymbol{c})$, we determine $\boldsymbol{x}(0)$.

(ii) Initial state $\boldsymbol{x}(0)$ is computed from (i), we now apply controls $u(0), u(1), \cdots, u(n-1)$, yielding

$$\begin{aligned} \boldsymbol{x}(1) &= \boldsymbol{A}\boldsymbol{x}(0) + \boldsymbol{b}u(0) \\ \boldsymbol{x}(2) &= \boldsymbol{A}\boldsymbol{x}(1) + \boldsymbol{b}u(1) = \boldsymbol{A}^2\boldsymbol{x}(0) + bu(1) + \boldsymbol{A}\boldsymbol{b}u(0) \\ &\vdots \\ \boldsymbol{x}(n) &= \boldsymbol{0} = \boldsymbol{A}^n\boldsymbol{x}(0) + \boldsymbol{b}u(n-1) + \boldsymbol{A}\boldsymbol{b}u(n-2) + \cdots + \boldsymbol{A}^{(n-1)}\boldsymbol{b}u(0) \end{aligned}$$

or

$$-\boldsymbol{A}^n\boldsymbol{x}(0) = \begin{bmatrix} \boldsymbol{b} & \boldsymbol{A}\boldsymbol{b} & \cdots & \boldsymbol{A}^{(n-1)}\boldsymbol{b} \end{bmatrix} \begin{bmatrix} u(0) \\ u(1) \\ \vdots \\ u(n-1) \end{bmatrix} = \boldsymbol{C}(\boldsymbol{A}, \boldsymbol{b}) \begin{bmatrix} u(0) \\ u(1) \\ \vdots \\ u(n-1) \end{bmatrix}$$

Inverting the Controllability matrix $\boldsymbol{C}(\boldsymbol{A}, \boldsymbol{b})$ yields the controls $u(0), u(1), \cdots, u(n-1)$. The solution to (i) and (ii) requires that both the observability and the controllability matrices should be of the rank n.

7.5.9 MIMO Systems Observability–Controllability Criterion

We shall make use of the Cayley-Hamilton Theorem to compute $\boldsymbol{e}^{\boldsymbol{A}t}$ and derive the Controllability and Observability criterion for multi-input, multi-output systems.

System Equations

$$\left.\begin{aligned} \dot{\boldsymbol{x}} &= \boldsymbol{A}\,\boldsymbol{x} + \boldsymbol{B}\,\boldsymbol{u}\ , \qquad && \boldsymbol{x}\ n\text{-vector},\ \boldsymbol{u}\ r\text{-vector} \\ \boldsymbol{y} &= \boldsymbol{C}\,\boldsymbol{x} \qquad , && \boldsymbol{y}\ m\text{-vector} \end{aligned}\right\} \tag{7.60}$$

The characteristic polynomial of matrix $\boldsymbol{A}$ is,

$$P(\lambda) = \Delta_A(\lambda) = \lambda^n + a_1\,\lambda^{n-1} + a_2\,\lambda^{n-2} + \ldots + a_n$$

From the Cayley-Hamilton theorem

$$\Delta_A(\boldsymbol{A}) = \boldsymbol{A}^n + a_1\,\boldsymbol{A}^{n-1} + a_2\,\boldsymbol{A}^{n-2} + \ldots + a_n\,\boldsymbol{I} = \boldsymbol{0} \tag{7.61}$$

From the definition of $\boldsymbol{e}^{\boldsymbol{A}\,t}$

$$\boldsymbol{e}^{\boldsymbol{A}\,t} = \boldsymbol{I} + \boldsymbol{A}\,t + \frac{\boldsymbol{A}^2 t^2}{2!} + \ldots + \frac{\boldsymbol{A}^n t^n}{n!}, \qquad n \to \infty \tag{7.62}$$

Eq. 7.61 can be used to express $\boldsymbol{A}^n$ and all the higher powers of $\boldsymbol{A}$ above n in terms of $\boldsymbol{A}^{(n-1)}$ and lower powers yielding:

$$\boldsymbol{e}^{\boldsymbol{A}\,t} = \alpha_0(t)\boldsymbol{I} + \alpha_1(t)\boldsymbol{A} + \alpha_2\boldsymbol{A}^2 + \ldots + \alpha_{n-1}(t)\boldsymbol{A}^{n-1} \tag{7.63}$$

where

$$\alpha_0(0) = 1$$

$$\alpha_i(0) = 0, \quad i = 1, 2, \ldots, (n-1)$$

The easiest way to compute $\alpha_i(t) \quad (i = 0, 1, \ldots, n-1)$ is to realize that,

$$\frac{\mathrm{d}}{\mathrm{d}t}\left(\boldsymbol{e}^{\boldsymbol{A}t}\right) = \boldsymbol{A}\boldsymbol{e}^{\boldsymbol{A}t}$$

From Eqs. 7.63 and 7.61

$$\sum_{i=0}^{n-1} \dot{\alpha}_i(t)\,\boldsymbol{A}^i = \sum_{i=0}^{n-1} \alpha_i(t)\,\boldsymbol{A}^{i+1} = \sum_{i=0}^{n-1} \alpha_{i+1}(t)\,\boldsymbol{A}^i + \alpha_1(t)(-1)\left(\sum_{i=0}^{n-1} a_i \boldsymbol{A}^i\right) \tag{7.64}$$

Equating the powers of $\boldsymbol{A}$ on both sides of Eq. 7.64, one obtains

$$\begin{aligned}
\dot{\alpha}_{n-1}(t) &= -a_1\,\alpha_{n-1}(t) + \alpha_{n-2}(t) & \alpha_{n-1}(0) &= 0\\
\dot{\alpha}_{n-2}(t) &= -a_2\,\alpha_{n-2}(t) \qquad + \alpha_{n-3}(t) & \alpha_{n-2}(0) &= 0\\
&\vdots & &\vdots\\
\dot{\alpha}_1(t) &= -a_1\,\alpha_1(t) \qquad\qquad + \alpha_2(t) & \alpha_1(0) &= 0\\
\dot{\alpha}_0(t) &= -a_n\,\alpha_0(t) & \alpha_0(0) &= 1
\end{aligned} \tag{7.65}$$

The solution to the above equation yields $\alpha_i(t)$, $\quad i = 0, \ldots, n-1$ and can be used to obtain the controllability and observability criterion.

Multi-Input Controllability—Given the Pair $(\boldsymbol{A}, \boldsymbol{B})$ and the Initial State $\boldsymbol{x}(0)$.

$$\dot{\boldsymbol{x}} = \boldsymbol{A}\,\boldsymbol{x} + \boldsymbol{B}\,\boldsymbol{u} \quad , \quad \boldsymbol{x}(0) = \boldsymbol{x}_0$$

$$\left(\boldsymbol{e}^{-\boldsymbol{A}\,t}\,\boldsymbol{x}(t) - \boldsymbol{x}(0)\right) = \hat{\boldsymbol{x}}(t) = \int_0^t \boldsymbol{e}^{-\boldsymbol{A}\,\tau}\boldsymbol{B}\,u(\tau)\,\mathrm{d}\tau$$

From the Eq. 7.63,

$$\hat{\boldsymbol{x}}(t) = \int_0^t \sum_{i=0}^{n-1} \alpha_i(-\tau)\,\boldsymbol{A}^i\,\boldsymbol{B}\,\boldsymbol{u}(\tau)\,\mathrm{d}\tau \tag{7.66}$$

$$\text{Let,} \quad \int_0^t \alpha_i(-\tau)\,\boldsymbol{u}(\tau)\,\mathrm{d}\tau = \boldsymbol{v}_i(t) \quad \text{an } r\text{-vector} \tag{7.67}$$

Thus, the Eq. 7.66 can be expressed as,

$$\hat{\boldsymbol{x}}(t) = \sum_{i=0}^{n-1} \boldsymbol{A}^{n-i}\,\boldsymbol{B}\,\boldsymbol{v}_i(t) \tag{7.68}$$

$$\hat{x}(t) = \begin{bmatrix} B \; AB \; \cdots \; A^{n-1}B \end{bmatrix} \begin{bmatrix} v_0(t) \\ v_1(t) \\ \vdots \\ v_{n-1}(t) \end{bmatrix} \tag{7.69}$$

$$\hat{x}(t) = C(A, B)v(t) \tag{7.70}$$

$$C(A, B) = \begin{bmatrix} B \; AB \; \cdots \; A^{n-1}B \end{bmatrix}$$

Matrix $C(A, B)$ is the controllability matrix. It has n rows and $n \times r$ columns. $v(t)$ is $n \times r$ vector. In order to solve for unique controls $u(t)$, we need only have n-linearly independent rows or columns for matrix $C(A, B)$. Hence, the pair (A, B) is controllable if and only if the controllability matrix $C(A, B)$ has the rank n.

Multi-Output Observability Test

Given the pair, (C, A): $\dot{x}(t) = Ax(t)$, $\quad y(t) = C\,x(t)$, $\qquad C$ is $m \times n$ matrix

Observe the vector $y(t)$ and all its derivatives at $t = 0$. The solution of $x(t)$ yields,

$$\begin{aligned} y(t) &= C\,e^{A\,t}\,x(t) \\ \dot{y}(t) &= C\,A\,e^{A\,t}\,x(t) \\ &\vdots \\ y^{(n-1)}(t) &= C\,A^{n-1}\,x(t) \end{aligned}$$

Letting $t = 0$ we obtain

$$\begin{bmatrix} y(0) \\ \dot{y}(0) \\ \vdots \\ y^{n-1}(0) \end{bmatrix} = \begin{bmatrix} C \\ CA \\ \vdots \\ CA^{n-1} \end{bmatrix} x(0) = O(A, C)x(0) \tag{7.71}$$

In order to compute $x(0)$ it is necessary and sufficient that the matrix $O(A, C)$ has the rank n. If we know $\boldsymbol{x}(0)$ we can compute $\boldsymbol{x}(t)$ for all times as

$$\boldsymbol{x}(t) = \boldsymbol{e}^{\boldsymbol{A}t}\boldsymbol{x}(0)$$

As a practical matter, it is not always possible to compute all the derivatives of the output $\boldsymbol{y}(t)$. Another alternate proof of observability is given below.

Alternate Simple Proof of Observability Theorem for the MIMO System

The state vector $\boldsymbol{x}(t)$ and output vector $\boldsymbol{y}(t)$ can be written as:

$$\boldsymbol{x}(t) = e^{\boldsymbol{A}t}\boldsymbol{x}(0) \tag{7.72}$$

$$\boldsymbol{y}(t) = \boldsymbol{C}e^{\boldsymbol{A}t}\boldsymbol{x}(0) \tag{7.73}$$

By observing the output vector $\boldsymbol{y}(t)$ for $0 < t \leq T$, we determine the initial state vector $\boldsymbol{x}(0)$ and hence $\boldsymbol{x}(t)$ for all times.

Proof:

From the Cayley-Hamilton Theorem

$$\boldsymbol{y}(t)|_{t=t_j} = \boldsymbol{y}(t_j) = \left[\boldsymbol{C}\sum_{i=0}^{n-1}\alpha_i(t_j)\boldsymbol{A}^i\right]\boldsymbol{x}(0), \quad j = 1, 2, \ldots, k, \quad mk \geq n$$

or

$$\begin{bmatrix} \boldsymbol{y}(t_1) \\ \boldsymbol{y}(t_2) \\ \vdots \\ \boldsymbol{y}(t_k) \end{bmatrix} = \begin{bmatrix} \alpha_0(t_1)\boldsymbol{I}_m & \alpha_1(t_1)\boldsymbol{I}_m & \cdots & \alpha_{n-1}(t_1)\boldsymbol{I}_m \\ \alpha_0(t_2)\boldsymbol{I}_m & \alpha_1(t_2)\boldsymbol{I}_m & \cdots & \alpha_{n-1}(t_2)\boldsymbol{I}_m \\ & & \vdots & \\ \alpha_0(t_k)\boldsymbol{I}_m & \alpha_1(t_k)\boldsymbol{I}_m & \cdots & \alpha_{n-1}(t_k)\boldsymbol{I}_m \end{bmatrix} \begin{bmatrix} \boldsymbol{C} \\ \boldsymbol{CA} \\ \vdots \\ \boldsymbol{CA}^{n-1} \end{bmatrix} \boldsymbol{x}(0), \quad \boldsymbol{I}_m \text{ is a } m\text{-column unity matrix}$$

Defining Σ by the coefficients $\alpha_i(t_f)$ and $\boldsymbol{y}$ by $\boldsymbol{y}(t_j)$, $i = 0, \ldots, n-1$, $j = 1, 2, \ldots, k$,

$$\boldsymbol{y} = \boldsymbol{\Sigma}\boldsymbol{O}(\boldsymbol{C}, \boldsymbol{A})\boldsymbol{x}(0), \quad \boldsymbol{\Sigma} \text{ has } m \times k \text{ rows and } n \text{ columns.} \tag{7.74}$$

The vector $\boldsymbol{x}(0)$ can be determined from Eq. 7.74 **if and only if** we can find n-linearly independent rows of the Observability matrix $O(\boldsymbol{C}, \boldsymbol{A})$. Thus the system Eq. 7.72 represented by pair $(\boldsymbol{A}, \boldsymbol{C})$ is completely observable **if and only if** the Observability matrix

$$O(\boldsymbol{C}, \boldsymbol{A}) = \begin{bmatrix} \boldsymbol{C} \\ \boldsymbol{CA} \\ \vdots \\ \boldsymbol{CA}^{n-1} \end{bmatrix} \quad \text{has the rank } n.$$

7.5.10 Canonical Controllable–Observable Decomposition

Let us illustrate the difference between the state variable and Transfer function representation via the following example.

Example 7.7:

Consider the following system:

$$G(s) = \frac{s^2 + 3s + 2}{s^3 + 6s^2 + 11s + 6} = \frac{Y(s)}{U(s)}$$

If we were to simulate it in the given form, it represents a third order system,

$$\dddot{x} + 6\ddot{x} + 11\dot{x} + 6x = u$$

$$y = \ddot{x} + 3\dot{x} + 2x$$

Yielding the state variable representation,

$$\boldsymbol{A} = \begin{bmatrix} 0 & 1 & 0 \\ 0 & 0 & 1 \\ -6 & -11 & -6 \end{bmatrix}, \quad \boldsymbol{b} = \begin{bmatrix} 0 \\ 0 \\ 1 \end{bmatrix}, \quad \boldsymbol{c} = \begin{bmatrix} 2 \\ 3 \\ 1 \end{bmatrix}$$

In reality the system is only a first order (due to poles-zeros cancellation):

$$G(s) = \frac{s^2 + 3s + 2}{s^3 + 6s^2 + 11s + 6} = \frac{(s+1)(s+2)}{(s+1)(s+2)(s+3)} = \frac{1}{s+3}$$

yielding:

$$\boldsymbol{A} = [-3], \quad \boldsymbol{b} = [1], \quad \boldsymbol{c} = [1]$$

Obviously, the two poles have been canceled with two zeros. So the two state space representations are entirely different. This is due to the fact that the real system has some states that are not influenced by the control and also there are some state or "modes" that cannot be observed in the output. For simulation and design purposes, we need an algorithm that yields a canonical decomposition algorithm, which yields only the controllable and the observable states and allows us to recognize the uncontrollable and unobservable states. Some of the unobserved states may be unstable, resulting in internal unstability while the output states are stable.
Let us consider the following algorithms:

(a) **Controllable Decomposition**

Given an $\boldsymbol{A}, \boldsymbol{B}, \boldsymbol{C}$ system:

$$\dot{\boldsymbol{x}} = \boldsymbol{A}\boldsymbol{x} + \boldsymbol{B}\boldsymbol{u}, \quad \boldsymbol{y} = \boldsymbol{C}\boldsymbol{x}, \quad n\text{-th order system}$$

$$\text{Rank}[C(\boldsymbol{A}, \boldsymbol{B})] = n_1 < n \quad \text{not completely controllable.}$$

Transformation: $\hat{\boldsymbol{x}} = \boldsymbol{T}_c\boldsymbol{x}$, yields:

$$\dot{\hat{\boldsymbol{x}}} = \hat{\boldsymbol{A}}\boldsymbol{x} + \hat{\boldsymbol{B}}\boldsymbol{u}, \quad \hat{\boldsymbol{A}} = \boldsymbol{T}_c\boldsymbol{A}\boldsymbol{T}_c^{-1}, \quad \hat{\boldsymbol{B}} = \boldsymbol{T}_c\boldsymbol{B}$$

$$\hat{\boldsymbol{y}} = \hat{\boldsymbol{C}}\boldsymbol{x}, \quad \hat{\boldsymbol{C}} = \boldsymbol{C}\boldsymbol{T}^{-1}, \quad \text{where}$$

$$\boldsymbol{T}^{-1} = \left[\hat{\boldsymbol{q}}_1 \ \hat{\boldsymbol{q}}_2 \cdots \hat{\boldsymbol{q}}_{n_1} \mid \hat{\boldsymbol{q}}_{n_1+1} \ \hat{\boldsymbol{q}}_{n_1+2} \cdots \hat{\boldsymbol{q}}_n\right] = \left[\boldsymbol{Q}_1 \, \boldsymbol{Q}_2\right]$$

Where,

$$\{\hat{\boldsymbol{q}}_k\}_1^{n_1} \begin{cases} \text{are } n_1\text{-linearly independent vectors} \\ \text{of the controllability matrix } C(\boldsymbol{A}, \boldsymbol{B}) \end{cases}$$

$$\{\hat{\boldsymbol{q}}_l\}_{l=n_1+1}^{n} \begin{cases} \text{are } (n - n_1) \text{ linearly independent vectors} \\ \text{chosen at will to make } \boldsymbol{T}_c \text{ invertible with the full rank.} \end{cases}$$

This transformation makes the pole/zero canceled pairs visible. Transformed system takes the form:

$$\begin{bmatrix} \dot{\boldsymbol{x}}_c \\ \dot{\boldsymbol{x}}_{uc} \end{bmatrix} = \begin{bmatrix} \hat{\boldsymbol{A}}_c & \hat{\boldsymbol{A}}_{12} \\ \boldsymbol{0} & \hat{\boldsymbol{A}}_{uc} \end{bmatrix} \begin{bmatrix} \hat{\boldsymbol{x}}_c \\ \hat{\boldsymbol{x}}_{uc} \end{bmatrix} + \begin{bmatrix} \hat{\boldsymbol{B}}_c \\ \boldsymbol{0} \end{bmatrix} \boldsymbol{u}, \quad \text{Pair } (\hat{\boldsymbol{A}}_c, \hat{\boldsymbol{B}}_c) \text{ is controllable.}$$

$$\begin{bmatrix} \hat{y} \end{bmatrix} = \begin{bmatrix} \hat{\boldsymbol{C}}_c & \hat{\boldsymbol{C}}_{uc} \end{bmatrix} \begin{bmatrix} \hat{\boldsymbol{x}}_c \\ \hat{\boldsymbol{x}}_{uc} \end{bmatrix}, \quad \hat{\boldsymbol{A}}_c \text{ is a } n_1 \times n_1 \text{ matrix, (controllable part of } \boldsymbol{A})$$

The transformed System Transfer function is:

$$\hat{G}_c(s) = \hat{\boldsymbol{C}}_c(s\boldsymbol{I} - \hat{\boldsymbol{A}}_c)\hat{\boldsymbol{B}}_C = \boldsymbol{C}(s\boldsymbol{I} - \boldsymbol{A})^{-1}\boldsymbol{B}$$

Proof and Algorithm:

1. Select $\{\hat{\boldsymbol{q}}_k\}_{k=1}^{n_1}$ from $C(\boldsymbol{AB})$ to form a set of n_1-linearly independent vectors.
2. $\{\boldsymbol{A}\hat{\boldsymbol{q}}_k\}_{k=1}^{n_1}$ also form a set of n_1-linearly independent vectors and hence, can be expressed as a linear combination of $\{\hat{\boldsymbol{q}}_k\}_{k=1}^{n_1}$. Similarly $\{\boldsymbol{A}\hat{\boldsymbol{q}}_k\}_{k=n_1+1}^{n}$ can also be expressed as a linear combination of the linearly independent set $\{\hat{\boldsymbol{q}}_k\}_{k=1}^{n}$
3. Consider the sequence $\{\boldsymbol{A}\hat{\boldsymbol{q}}_k\}_{k=1}^{n_1}$ and the corresponding representation:

$$\begin{bmatrix} \boldsymbol{A}\hat{\boldsymbol{q}}_1 & \cdots & \boldsymbol{A}\hat{\boldsymbol{q}}_{n_1} \end{bmatrix} = \begin{bmatrix} \hat{\boldsymbol{q}}_1 & \hat{\boldsymbol{q}}_2 & \cdots & \hat{\boldsymbol{q}}_{n_1} \end{bmatrix} \begin{bmatrix} s_1 & s_2 & \cdots & s_{n_1} \end{bmatrix} = \begin{bmatrix} \hat{\boldsymbol{q}}_1 & \hat{\boldsymbol{q}}_2 & \cdots & \hat{\boldsymbol{q}}_{n_1} \end{bmatrix} \begin{bmatrix} \hat{\boldsymbol{A}}_c \end{bmatrix}$$

$$\{s_i\}_{i=1}^{n_1} \text{ are } n \times 1 \text{ vectors}$$

Augmenting the above equation with linearly independent vector $\{\hat{\boldsymbol{q}}_k\}_{k=n_1+1}^{n}$

$$\left[\boldsymbol{A}\hat{\boldsymbol{q}}_1 \cdots \boldsymbol{A}\hat{\boldsymbol{q}}_{n_1}\right] = \left[\hat{\boldsymbol{q}}_1\ \hat{\boldsymbol{q}}_2 \cdots \hat{\boldsymbol{q}}_{n_1} \mid \hat{\boldsymbol{q}}_{n_1+1} \cdots \hat{\boldsymbol{q}}_n\right]\begin{bmatrix}\hat{\boldsymbol{A}}_c \\ \boldsymbol{0}\end{bmatrix} = \boldsymbol{T}_c^{-1}\begin{bmatrix}\hat{\boldsymbol{A}}_c \\ \boldsymbol{0}\end{bmatrix} \tag{7.75}$$

4. Similarly the rest of the vectors $\{\boldsymbol{A}\hat{\boldsymbol{q}}_k\}_{k=n_1+1}^{n}$ can be written as:

$$\left[\boldsymbol{A}\hat{\boldsymbol{q}}_{n_1+1} \cdots \boldsymbol{A}\hat{\boldsymbol{q}}_{n}\right] = \left[\hat{\boldsymbol{q}}_1 \cdots \hat{\boldsymbol{q}}_{n_1} \mid \hat{\boldsymbol{q}}_{n_1+1} \cdots \hat{\boldsymbol{q}}_n\right]\begin{bmatrix}\hat{\boldsymbol{A}}_{12} \\ \hat{\boldsymbol{A}}_{uc}\end{bmatrix} = \boldsymbol{T}_c^{-1}\begin{bmatrix}\hat{\boldsymbol{A}}_{12} \\ \hat{\boldsymbol{A}}_{uc}\end{bmatrix} \tag{7.76}$$

Combining the two equations, Eqs. 7.75 and 7.76, yields

$$\left[\boldsymbol{A}\hat{\boldsymbol{q}}_1\ \boldsymbol{A}\hat{\boldsymbol{q}}_2 \cdots \boldsymbol{A}\hat{\boldsymbol{q}}_n\right] = \left[\hat{\boldsymbol{q}}_1\ \hat{\boldsymbol{q}}_2 \cdots \hat{\boldsymbol{q}}_n\right]\begin{bmatrix}\hat{\boldsymbol{A}}_c & \hat{\boldsymbol{A}}_{12} \\ \boldsymbol{0} & \hat{\boldsymbol{A}}_{uc}\end{bmatrix} = \boldsymbol{T}_c^{-1}\begin{bmatrix}\hat{\boldsymbol{A}}_c & \hat{\boldsymbol{A}}_{12} \\ \boldsymbol{0} & \hat{\boldsymbol{A}}_{uc}\end{bmatrix}$$

Thus,

$$\boldsymbol{A}\boldsymbol{T}_c^{-1} = \boldsymbol{T}_c^{-1}\begin{bmatrix}\hat{\boldsymbol{A}}_c & \hat{\boldsymbol{A}}_{12} \\ \boldsymbol{0} & \hat{\boldsymbol{A}}_{uc}\end{bmatrix}, \quad \boldsymbol{T}_c^{-1} = \left[\hat{\boldsymbol{q}}_1\ \hat{\boldsymbol{q}}_2 \cdots \hat{\boldsymbol{q}}_n\right]$$

or

$$\hat{\boldsymbol{A}} = \boldsymbol{T}_c\boldsymbol{A}\boldsymbol{T}_c^{-1} = \begin{bmatrix}\hat{\boldsymbol{A}}_c & \hat{\boldsymbol{A}}_{12} \\ \boldsymbol{0} & \hat{\boldsymbol{A}}_{uc}\end{bmatrix}$$

5. Expressing the columns of $\boldsymbol{B}$ in terms of $\{\hat{\boldsymbol{q}}_i\}_{n_1}^{i=1}$

$$\boldsymbol{B} = \left[\boldsymbol{b}_1 \cdots \boldsymbol{b}_r\right] = \left[\hat{\boldsymbol{q}}_1 \cdots \hat{\boldsymbol{q}}_{n_1}\right]\left[\boldsymbol{\beta}_1 \cdots \boldsymbol{\beta}_r\right] = \left[\hat{\boldsymbol{q}}_1 \cdots \hat{\boldsymbol{q}}_{n_1}\right]\left[\hat{\boldsymbol{B}}_c\right] \tag{7.77}$$

where $\{\boldsymbol{\beta}_i\}_{i=1}^{r}$ are $n_1 \times 1$ vectors.. Augmenting the above Eq. 7.77 with the rest of the linearly independent vectors $\{\hat{\boldsymbol{q}}_k\}_{n}^{k=n_1+1}$,

$$\boldsymbol{B} = \left[\hat{\boldsymbol{q}}_1\ \hat{\boldsymbol{q}}_2 \cdots \hat{\boldsymbol{q}}_{n_1} \mid \hat{\boldsymbol{q}}_{n_1+1} \cdots \hat{\boldsymbol{q}}_n\right]\begin{bmatrix}\hat{\boldsymbol{B}}_c \\ \boldsymbol{0}\end{bmatrix}$$

$$T_c B = \begin{bmatrix} \hat{B}_c \\ 0 \end{bmatrix}$$

6. Using the same argument, $CT_c^{-1} = \begin{bmatrix} \hat{C}_c & \hat{C}_{uc} \end{bmatrix}$, the controllable subsystem is:

$$\dot{\hat{x}}_c = \hat{A}_c \hat{x}_c + \hat{B}_c u, \quad \hat{y}_c = \hat{C}_c \hat{x}_c$$

$$\hat{G}_c(s) = \hat{C}_c \left(sI - \hat{A}_c \right) \hat{B}_c$$

It should be noted that, $\dot{\hat{x}}_{uc} = \hat{A}_{uc} \hat{x}_{uc}$ represents the uncontrollable and unstables modes even though the Transfer function looks stable.

(b) **Observable Decomposition**

This Transformation makes the unobservable states visible and we can eliminate them. Transformed system takes the form:

$$\begin{bmatrix} \dot{\hat{x}}_o \\ \dot{\hat{x}}_{uo} \end{bmatrix} = \begin{bmatrix} \hat{A}_o & 0 \\ \hat{A}_{21} & \hat{A}_{uo} \end{bmatrix} \begin{bmatrix} \hat{x}_o \\ \hat{x}_{uo} \end{bmatrix} + \begin{bmatrix} \hat{B}_o \\ \hat{B}_{uo} \end{bmatrix} u$$

$$\begin{bmatrix} y \end{bmatrix} = \begin{bmatrix} \hat{C}_o & 0 \end{bmatrix} \begin{bmatrix} \hat{x}_o \\ \hat{x}_{uo} \end{bmatrix}$$

$$\text{Rank} \begin{bmatrix} O(A, C) \end{bmatrix} = n_2 < n$$

Proof and Algorithm:

(1) Choose n_2-linearly independent rows from $O(A, C)$. Designate them as $t_1^T, t_2^T, \cdots, t_{n_2}^T$.

(2) Choose $(n_1 - n_2)$ linearly independent rows $\{t_i\}_n^{i=n_2+1}$, such that the matrix formed by $\{t_k^T\}_n^{k=1}$ has a full rank n.

(3) Consider the sequence $\{t_k^T A\}_{n_2}^{k=1}$ and the corresponding representation.

$$\begin{bmatrix} t_1^T A \\ t_2^T A \\ \vdots \\ t_{n_2}^T A \end{bmatrix} = \begin{bmatrix} t_1^T \\ t_2^T \\ \vdots \\ t_{n_2}^T \end{bmatrix} A = \begin{bmatrix} \hat{A}_0 \end{bmatrix} \begin{bmatrix} T_0 \end{bmatrix} = \begin{bmatrix} \hat{A}_0 \mid \mathbf{0} \end{bmatrix} \begin{bmatrix} T_0 \\ \hline t_{n_2+1}^T \\ \vdots \\ t_n^T \end{bmatrix}$$

Where

$$\hat{A}_0 = \begin{bmatrix} r_1^T \\ r_2^T \\ \vdots \\ r_{n_2}^T \end{bmatrix}, \quad n_2 \times n_2 \text{ matrix}, \quad T_0 = \begin{bmatrix} t_1^T \\ t_2^T \\ \vdots \\ t_{n_2}^T \end{bmatrix}$$

Similarly

$$\begin{bmatrix} t_{n_2+1}^T \\ \cdots \\ t_n^T \end{bmatrix} A = \begin{bmatrix} r_{n_2+1}^T \\ \cdots \\ r_n^T \end{bmatrix} \begin{bmatrix} T_0 \\ \hline t_{n_2+1}^T \\ \vdots \\ t_n^T \end{bmatrix} = \begin{bmatrix} \hat{A}_{21} & \hat{A}_{u0} \end{bmatrix} \begin{bmatrix} T_0 \\ \hline t_{n_2+1}^T \\ \vdots \\ t_n^T \end{bmatrix}$$

Thus,

$$\begin{bmatrix} t_1^T \\ t_2^T \\ \vdots \\ t_{n_2}^T \\ \hline t_{n_2+1}^T \\ \vdots \\ t_n^T \end{bmatrix} A = \begin{bmatrix} r_1^T \\ r_2^T \\ \vdots \\ r_{n_2}^T \\ \hline r_{n_2+1}^T \\ \vdots \\ r_n^T \end{bmatrix} \begin{bmatrix} t_1^T \\ t_2^T \\ \vdots \\ t_{n_2}^T \\ \hline t_{n_2+1}^T \\ \vdots \\ t_n^T \end{bmatrix} = \begin{bmatrix} \hat{A}_0 & \mathbf{0} \\ \hat{A}_{21} & \hat{A}_{u0} \end{bmatrix} \begin{bmatrix} t_1^T \\ t_2^T \\ \vdots \\ t_{n_2}^T \\ \hline t_{n_2+1}^T \\ \vdots \\ t_n^T \end{bmatrix}$$

Thus,

$$T_0 A = \hat{A}_0 T_0$$

or

$$\hat{A}_0 = T_0 A T_0^{-1}$$

Similarly

$$C_0 = C T_0^{-1} = \begin{bmatrix} \hat{C}_0 & \mathbf{0} \end{bmatrix}, B_0 = \begin{bmatrix} \hat{B}_0 \\ \hat{B}_{u0} \end{bmatrix}$$

Observable system equations are:

$$\begin{aligned} \dot{\hat{x}}_0 &= \hat{A}_0 \hat{x}_0 + \hat{B}_0 u \\ \hat{y}_0 &= \hat{C}_0 \hat{x}_0 \\ \hat{G}_0(s) &= \hat{C}_0 \left(sI - \hat{A}_0 \right) \hat{B}_0 \end{aligned}$$

7.5.11 Kalman Decomposition for SISO–Geometric Viewpoint

Given:

$$\begin{aligned} &\dot{x} = Ax + bu \\ &y = c^T x \\ &b^T = \begin{bmatrix} b_1 & b_2 & \cdots & b_n \end{bmatrix} \\ &c^T = \begin{bmatrix} c_1 & c_2 & \cdots & c_n \end{bmatrix} \end{aligned}$$

When both the controllable and observable Similarity Transformations are applied together, we arrive at four subsystems, known as the “Kalman Decomposition.” For controller design one is interested only in the subspace in which system modes are both **Controllable** and **Observable**. The state variable x and the output y can be projected on this subspace and hence the system dimension can be reduced. Let

$$\left.\begin{array}{ll} b_i \neq 0 & i = 1, 2, \ldots, k \\ b_i = 0 & i = k+1, \ldots, n \\ \text{Similarly} & \\ c_j \neq 0 & j = m, m+1, \ldots, l \\ c_j = 0 & j = 1, 2, \ldots, m-1 \\ c_j = 0 & j = l+1, l+2, \ldots, n \\ m < k < l & \end{array}\right\} \tag{7.78}$$

The total state space of the matrix $\boldsymbol{A}$ is spanned by eigenvectors $\boldsymbol{p}_1, \boldsymbol{p}_2, \ldots, \boldsymbol{p}_n$ and its dual space $\boldsymbol{q}_1, \boldsymbol{q}_2, \ldots, \boldsymbol{q}_n$.

$$\boldsymbol{q}_i^T \boldsymbol{p}_j = \delta_{ij} \begin{cases} 1 & i = j \\ 0 & i \neq j \end{cases}$$

From Figure 7.16, we arrive at the following:

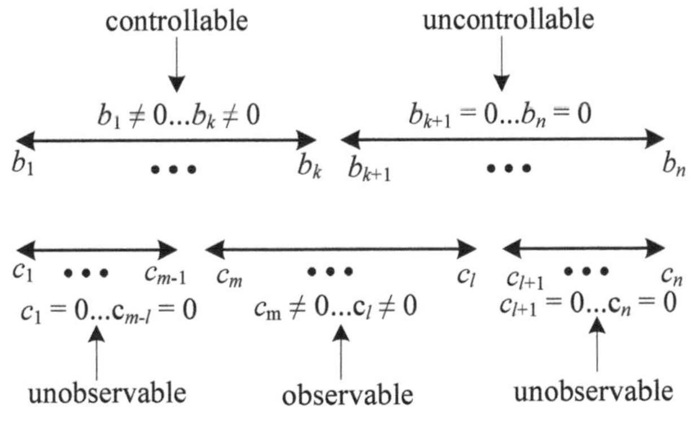

Figure 7.16: Decomposition—Observable and Controllable Subsystem

(1) Subspace spanned by $\boldsymbol{p}_1, \boldsymbol{p}_2, \ldots, \boldsymbol{p}_{m-1}$ is controllable but not observable.

(2) Subspace spanned by $\boldsymbol{p}_m, \boldsymbol{p}_{m+1}, \ldots, \boldsymbol{p}_k$ is both controllable and observable.

(3) Subspace spanned by $\boldsymbol{p}_{k+1}, \boldsymbol{p}_{k+2}, \ldots, \boldsymbol{p}_l$ is not controllable but observable.

(4) Subspace spanned by $\boldsymbol{p}_{l+1}, \boldsymbol{p}_{l+2}, \ldots, \boldsymbol{p}_n$ is neither controllable nor observable.

$$q_i^T p_j = \delta_{ij}, \quad i, j = m, m+1, \cdots, k$$

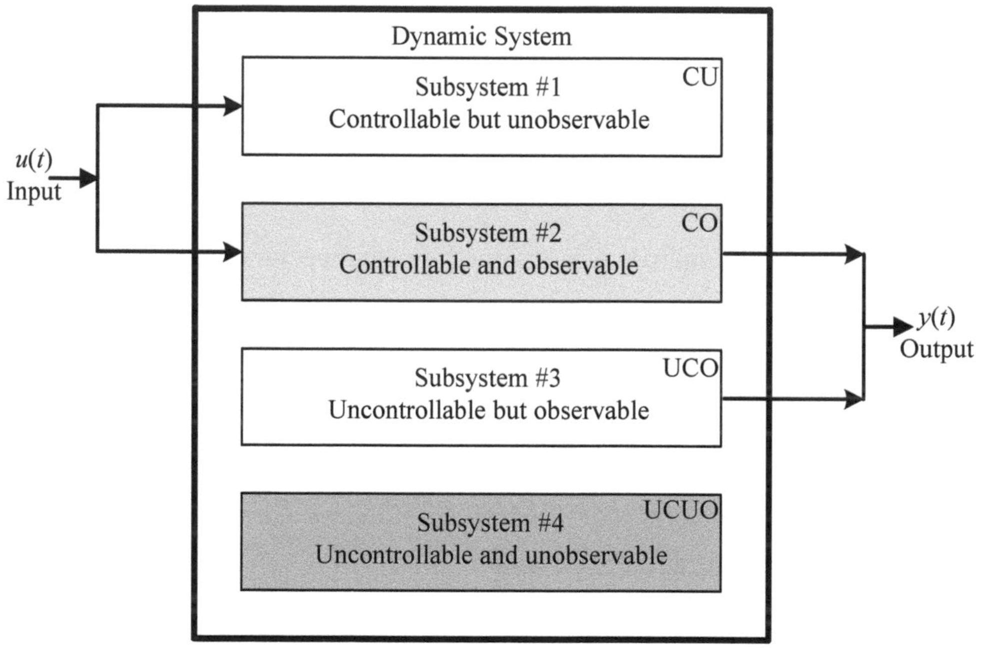

Figure 7.17: Dynamic System Kalman Decomposition

Therefore only the subsystem #2 is both controllable and observable.

We shall project the system motion onto this both controllable and observable subspace. Hence, we seek a projection matrix $\boldsymbol{P}$ whose rows are given by vectors $\boldsymbol{q}_m^T, \boldsymbol{q}_{m+1}^T, \ldots, \boldsymbol{q}_k^T$, and column vectors $\boldsymbol{p}_1, \boldsymbol{p}_2, \cdots, \boldsymbol{p}_n$ such that

$$\boldsymbol{x}_{co} = \boldsymbol{P}\boldsymbol{x} = \left(\begin{bmatrix} \boldsymbol{q}_m^T \\ \boldsymbol{q}_{m+1}^T \\ \vdots \\ \boldsymbol{q}_k^T \end{bmatrix} \sum_{i=m}^{k} x_i \boldsymbol{p}_i \right) = \begin{bmatrix} x_m \\ x_{m+1} \\ \vdots \\ x_k \end{bmatrix} = \begin{array}{l} \text{Controllable and} \\ \text{Observable Coordinates} \end{array}$$

The projection matrix $\boldsymbol{P}$ isolates both the controllable and observable coordinates. Extending the above argument, for a MIMO system, the reduced system state equations are:

$$\dot{\boldsymbol{x}}_{co} = \hat{\boldsymbol{A}}_{co}\,\boldsymbol{x}_{co} + \hat{\boldsymbol{B}}_{co}\,u(t), \quad \hat{\boldsymbol{A}} = \boldsymbol{P}\,\boldsymbol{A}, \quad \hat{\boldsymbol{B}}_{co} = \boldsymbol{P}\,\boldsymbol{B}, \quad \hat{\boldsymbol{C}}_{co} = \boldsymbol{P}\boldsymbol{C} \tag{7.79}$$

$$\boldsymbol{y}_{co} = \boldsymbol{P}\boldsymbol{y} = \boldsymbol{C}_{co}\boldsymbol{x}_{co} \tag{7.80}$$

$$\hat{\boldsymbol{A}} = \left[\hat{\boldsymbol{A}}_{co}\ \hat{\boldsymbol{A}}_{12}\right] \tag{7.81}$$

$$\hat{G}_{co}(s) = \hat{\boldsymbol{C}}_{co}\left(s\boldsymbol{I} - \hat{\boldsymbol{A}}_{co}\right)\hat{\boldsymbol{B}}_{co} \tag{7.82}$$

Note: Subscript *co* stands for the controllable and observable, which in reality is the most important part of the system.

Kalman Decomposition Form

$$\begin{bmatrix} \dot{\boldsymbol{x}}_{co} \\ \dot{\boldsymbol{x}}_{cuo} \\ \dot{\boldsymbol{x}}_{uco} \\ \dot{\boldsymbol{x}}_{ucuo} \end{bmatrix} = \begin{bmatrix} \hat{\boldsymbol{A}}_{co} & & \hat{\boldsymbol{A}}_{13} & 0 \\ \hat{\boldsymbol{A}}_{21} & \hat{\boldsymbol{A}}_{cuo} & \hat{\boldsymbol{A}}_{23} & \hat{\boldsymbol{A}}_{24} \\ 0 & 0 & \hat{\boldsymbol{A}}_{uco} & 0 \\ 0 & 0 & \hat{\boldsymbol{A}}_{43} & \hat{\boldsymbol{A}}_{ucuo} \end{bmatrix} \begin{bmatrix} \boldsymbol{x}_{co} \\ \boldsymbol{x}_{cuo} \\ \boldsymbol{x}_{uco} \\ \boldsymbol{x}_{ucuo} \end{bmatrix} + \begin{bmatrix} \boldsymbol{B}_{co} \\ \boldsymbol{B}_{cuo} \\ 0 \\ 0 \end{bmatrix} \boldsymbol{u}(t)$$

$$\left[y\right] = \left[\hat{\boldsymbol{C}}_{co}\ 0\ \hat{\boldsymbol{C}}_{uco}\ 0\right] \begin{bmatrix} \hat{\boldsymbol{x}}_{co} \\ \hat{\boldsymbol{x}}_{cuo} \\ \hat{\boldsymbol{x}}_{uco} \\ \hat{\boldsymbol{x}}_{ucuo} \end{bmatrix}$$

The controllable and observable subsystem is called **Minimal Realization**, which is equivalent to the Transfer function. Furthermore, for operational reasons, we like to be sure that uncontrollable and unobservable modes are not unstable. That may require further scrutiny and control system responsibility.

7.5.12 Controllability and Observability Grammian

In the earlier sections we discussed the controllability and observability test for

$$\dot{\boldsymbol{x}} = \boldsymbol{A}\boldsymbol{x} + \boldsymbol{B}\boldsymbol{u} \quad , \quad \boldsymbol{x}(0) = \boldsymbol{x}_0 \quad , \quad 0 \leq t \leq t_f$$
$$y = \boldsymbol{C}\boldsymbol{x}$$

We concluded that for a completely controllable and completely observable system, the controllability matrix $\boldsymbol{C}(\boldsymbol{A},\boldsymbol{B})$ and the observability matrix $\boldsymbol{O}(\boldsymbol{A},\boldsymbol{C})$ must have full rank n, where

$$\boldsymbol{C}(\boldsymbol{A},\boldsymbol{B}) = \begin{bmatrix} \boldsymbol{B} \ \boldsymbol{A}\boldsymbol{B} \ \ldots \ \boldsymbol{A}^{n-1}\boldsymbol{B} \end{bmatrix} \qquad \text{(has full rank } n\text{)}$$

$$\boldsymbol{O}(\boldsymbol{A},\boldsymbol{C}) = \begin{bmatrix} \boldsymbol{C} \\ \boldsymbol{C}\boldsymbol{A} \\ \vdots \\ \boldsymbol{C}\boldsymbol{A}^{n-1} \end{bmatrix} \qquad \text{(has full rank } n\text{)}$$

Study of these concepts from a quadratic function point of view is as follows.

Mean Squared Quadratic Computation

$$\dot{\boldsymbol{x}} = \boldsymbol{A}\boldsymbol{x}, \quad \boldsymbol{x}(0) = \boldsymbol{x}_0 \tag{7.83}$$

$\boldsymbol{A}$ is Hurwitz stable (eigenvalues of $\boldsymbol{A}$ have real part positive). Let us compute,

$$I(t) = \int_t^{\infty} \boldsymbol{x}^T \boldsymbol{Q} \boldsymbol{x}\, \mathrm{d}t, \qquad \boldsymbol{Q} \geq 0 \text{ (positive semidefinite)} \tag{7.84}$$

We seek a solution of the form $I(t) = \boldsymbol{x}^T \boldsymbol{S} \boldsymbol{x},\ \boldsymbol{S} > 0$ (positive definite)

$$I(\infty) = 0, \quad \boldsymbol{x}(\infty) = 0, \text{ System being stable}$$

Differentiating both sides of Eq. 7.84

$$\dot{I} = \dot{\boldsymbol{x}}^T \boldsymbol{S}\boldsymbol{x} + \boldsymbol{x}^T \boldsymbol{S}\dot{\boldsymbol{x}} = -\boldsymbol{x}^T \boldsymbol{Q}\boldsymbol{x}$$

From the Eq. 7.83

$$\boldsymbol{A}^T \boldsymbol{S} + \boldsymbol{S}\boldsymbol{A} = -\boldsymbol{Q} \qquad \text{Liapunov equation} \tag{7.85}$$

This is the well-known Liapunov equation where $\boldsymbol{S}$ is a positive definite matrix and $\boldsymbol{Q}$ can be positive definite or positive semidefinite. This equation appears regularly in control systems as well as the study of differential equations (see Chapter 3). A positive definite matrix $\boldsymbol{S}$ is denoted as $\boldsymbol{S} > 0$. Associated with it is a positive definite quadratic scalar function

$$V = \boldsymbol{x}^T \boldsymbol{S}\boldsymbol{x}, \qquad V > 0$$

Controllability Grammian:

Consider the $\boldsymbol{A}, \boldsymbol{B}$ system,

$$\dot{\boldsymbol{x}} = \boldsymbol{A}\boldsymbol{x} + \boldsymbol{B}\boldsymbol{u}, \quad \boldsymbol{x}(0) = \boldsymbol{x}_0, \qquad 0 \le t \le t_f$$

The solution to the above equation at time $t = t_f$ is:

$$\hat{\boldsymbol{x}}(t_f) = \left(\boldsymbol{x}(t_f) - \boldsymbol{e}^{At_f}\boldsymbol{x}(0)\right) = \int_0^{t_f} \boldsymbol{e}^{A(t_f-\tau)}\boldsymbol{B}u(\tau)\,\mathrm{d}\tau \tag{7.86}$$

where

$$\boldsymbol{e}^{A(t_f)} = \Phi(t_f) \quad , \boldsymbol{e}^{A(t_f-\tau)} = \boldsymbol{\Phi}(t_f - \tau) = \boldsymbol{\Phi}(t_f)\boldsymbol{\Phi}(-\tau) \qquad \boldsymbol{\Phi}(0) = \boldsymbol{I}$$

We seek a control $\boldsymbol{u}(t)$, $(0 < t < t_f)$, such that Eq. 7.86 is satisfied. This is an inverse solution problem. It is clear that the matrix $\boldsymbol{\Phi}(t_f - \tau)\boldsymbol{B}$ is not invertible. So we choose,

$$u(t) = \boldsymbol{B}^T \boldsymbol{e}^{A^T(t_f-\tau)} \boldsymbol{S}_c^{-1}(t_f)\hat{\boldsymbol{x}}(t_f) \qquad 0 < t \leq t_f \tag{7.87}$$

Substituting Eq. 7.87 into Eq. 7.86, we obtain

$$\hat{\boldsymbol{x}}(\tau_f) = \left[\int_0^{t_f} \boldsymbol{e}^{A(t_f-\tau)} \boldsymbol{B}\boldsymbol{B}^T \boldsymbol{e}^{A^T(t_f-\tau)}\, \mathrm{d}\tau\right] \boldsymbol{S}_c^{-1}(t_f)\hat{\boldsymbol{x}}(t_f) \tag{7.88}$$

Implying

$$S_c(t_f) = \int_0^{t_f} \boldsymbol{e}^{A(t_f-\tau)} \boldsymbol{B}\boldsymbol{B}^T \boldsymbol{e}^{A^T(t_f-\tau)}\, \mathrm{d}\tau \tag{7.89}$$

Thus, there exists a control $\boldsymbol{u}(t)$, $(0 < t \leq t_f)$, which can drive the system from any given initial state $\boldsymbol{x}(0)$ to any final state $\boldsymbol{x}(t_f)$ in the interval $[0, t_f]$ provided the matrix $\boldsymbol{S}_c(t_f)$ in Eq. 7.89 is positive definite (nonsingular) and hence invertible. The matrix $\boldsymbol{S}_c(t_f)$ is called the Controllability Grammian. The Requirement of nonsingularity of the Grammian $\boldsymbol{S}_c(t_f)$ is an alternate controllability criterion for the pair $(\boldsymbol{A}, \boldsymbol{B})$. Since t_f is any general value of t, $0 \leq t \leq t_f$, we shall define the general controllability Grammian to be,

$$\boldsymbol{S}_c(t) = \int_0^{t} \boldsymbol{e}^{A(t-\tau)} \boldsymbol{B}\boldsymbol{B}^T \boldsymbol{e}^{A^T(t-\tau)}\, \mathrm{d}\tau \qquad 0 \leq t \leq t_f \tag{7.90}$$

It is easy to see via change of variables, that

$$\boldsymbol{S}_c(t) = \int_0^{t} \boldsymbol{e}^{A(t-\tau)} \boldsymbol{B}\boldsymbol{B}^T \boldsymbol{e}^{A^T(t-\tau)}\, \mathrm{d}\tau = \int_0^{t} \boldsymbol{e}^{A\tau} \boldsymbol{B}\boldsymbol{B}^T \boldsymbol{e}^{(A^T\tau)}\, \mathrm{d}\tau$$

Important Properties of the Controllability Grammian $S_c(t_f)$

1. **Pair $(A, \boldsymbol{B})$ is completely controllable if and only if $S_c(t_f)$ is nonsingular**

   We give another proof of the above statement via contradiction.

   Let us assume that $\boldsymbol{S}_c(t_f)$ is singular for a pair $(\boldsymbol{A}, \boldsymbol{B})$ that is controllable.

   Consider an arbitrary n-vector $\boldsymbol{v}(t_f) \neq 0$ such that

$$\boldsymbol{v}^T(t_f)\boldsymbol{S}_c(t_f)\boldsymbol{v}(t_f) = 0 \tag{7.91}$$

   Eq. 7.91 implies

$$\boldsymbol{v}^T(t_f)\boldsymbol{S}_c(t_f)\boldsymbol{v}(t_f) = \int_0^{t_f} \boldsymbol{v}^T(t_f)\boldsymbol{e}^{A(t_f-\tau)}\boldsymbol{B}\boldsymbol{B}^T\boldsymbol{e}^{A^T(t_f-\tau)}\boldsymbol{v}(t_f)\,\mathrm{d}\tau = 0$$

   The above equation yields:

$$\boxed{\boldsymbol{v}^T(t_f)\boldsymbol{e}^{A(t_f-\tau)}\boldsymbol{B}\boldsymbol{B}^T\boldsymbol{e}^{A^T(t_f-\tau)}\boldsymbol{v}(t_f) = 0 \qquad 0 \leq \tau \leq t_f} \tag{7.92}$$

   Since $\boldsymbol{v}(t_f)$ is any arbitrary vector, let us choose

$$\boldsymbol{v}(t_f) = \boldsymbol{e}^{-A^T(t_f-\tau)}\boldsymbol{x}(0), \qquad \text{yielding}$$

$$\boldsymbol{x}^T(0)\boldsymbol{B}^T\boldsymbol{B}\boldsymbol{x}(0) = 0$$

   which is impossible for a general $\boldsymbol{x}(0)$ unless $\boldsymbol{B} \equiv 0$, resulting in a contradiction. **Thus, $S_c(t_f)$ must be nonsingular and so also $S_c(t)$.**

2. If the pair $(\boldsymbol{A}, \boldsymbol{B})$ is completely controllable and $\boldsymbol{A}$ is Hurwitz (stable, with all eigenvalues of $\boldsymbol{A}$ with a negative real part), then as $t_f \to \infty$, and $x(t_f) \to 0$

$$\boldsymbol{S}_c(\infty) = \boldsymbol{S}_c = \int_0^{\infty} \boldsymbol{e}^{A\tau}\boldsymbol{B}\boldsymbol{B}^T\boldsymbol{e}^{(A^T\tau)}\,\mathrm{d}\tau \tag{7.93}$$

Proof:

$$e^{-At}x(t) = x(0) + \int_{0}^{t} e^{-A\tau} Bu(\tau)\,\mathrm{d}\tau$$

For a Hurwitz matrix, A, as $t \to \infty$, $x(\infty) \to 0$ yielding:

$$0 = x(0) + \int_{0}^{\infty} e^{-A\tau} B^T u(\tau)\,\mathrm{d}\tau$$

Let us choose:

$$u(\tau) = -Be^{-(A^T\tau)} S_c^{-1}(\infty)x(0) \tag{7.94}$$

Thus,

$$S_c = S_c(\infty) = -\int_{0}^{\infty} e^{-A\tau} BB^T e^{-(A^T\tau)}\,\mathrm{d}\tau = \int_{0}^{\infty} e^{A\tau} BB^T e^{(A^T\tau)}\,\mathrm{d}\tau \tag{7.95}$$

Premultiplying S_c with A and postmultiplying S_c with A^T and adding

$$AS_c + S_c A^T = \int_{0}^{\infty} \frac{\mathrm{d}}{\mathrm{d}\tau}\left(e^{+A\tau} BB^T e^{(A^T\tau)}\right)\mathrm{d}\tau$$

Since the integrand goes to zero at $\tau = \infty$, we obtain

$$AS_c + S_c A^T = -BB^T \qquad B \neq 0 \tag{7.96}$$

Furthermore, it is obvious from Eq. 7.93 that S_c is positive definite.

3. If the pair (A, B) is completely controllable in the interval $[t_0, t_f]$, then for any $t_0 \leq t \leq t_f$.

$$S_c(t_0, t_f) = e^{A(t_f - t)} S_c(t_0, t) e^{A^T(t_f - t)} + S_c(t, t_f), \;\; S_c(t_0, t) \text{ is positive definite.}$$

4. The controllability matrix $\boldsymbol{C}(\boldsymbol{A},\boldsymbol{B})$ has the full rank n, and

$$\boldsymbol{C}(\boldsymbol{A},\boldsymbol{B})\,\boldsymbol{C}^T(\boldsymbol{A},\boldsymbol{B}) > 0 \quad \text{(positive definite)} \tag{7.97}$$

5. The system is completely controllable if and only if the augmented matrix

$$\boldsymbol{A}(\lambda,\boldsymbol{B}) = [(\lambda I - \boldsymbol{A}), \boldsymbol{B}] \qquad (n \times (n+r)) \text{ matrix}$$

has full rank n for every eigenvalue λ of $\boldsymbol{A}$. This implies that the rows of $(\lambda I - \boldsymbol{A})^{-1}\boldsymbol{B}$ are linearly independent functions of every eigenvalue λ of $\boldsymbol{A}$.

Observability Grammian:

Given the $(\boldsymbol{A},\boldsymbol{C})$ system

$$\dot{\boldsymbol{x}} = \boldsymbol{A}\boldsymbol{x}, \quad \boldsymbol{y} = \boldsymbol{C}\boldsymbol{x}, \quad \boldsymbol{x}(0) = \boldsymbol{x}_0, \quad 0 < t \le t_f \tag{7.98}$$

The solution is

$$\boldsymbol{y}(t) = \boldsymbol{e}^{\boldsymbol{A}(t)}\boldsymbol{C}\boldsymbol{x}(0)$$

The output $\boldsymbol{y}(t)$ is observed for interval $[0, t_f]$ and we are required to determine the initial state, $\boldsymbol{x}(0)$ and hence $\boldsymbol{x}(t) = \boldsymbol{e}^{\boldsymbol{A}(t)}\boldsymbol{x}(0)$. Let us construct a performance index

$$I(t_f) = \int_0^{t_f} \left[\boldsymbol{y}(\tau) - \boldsymbol{e}^{\boldsymbol{A}\tau}\boldsymbol{C}\boldsymbol{x}(0)\right]^T \left[\boldsymbol{y}(\tau) - \boldsymbol{e}^{\boldsymbol{A}\tau}\boldsymbol{C}\boldsymbol{x}(0)\right] \mathrm{d}\tau$$

$\boldsymbol{x}(0)$ is so chosen as to minimize $I(t_f)$. The minimum of $I(t_f)$ is given by

$$\boldsymbol{x}(0) = \boldsymbol{S}_o^{-1}(t_f) \int_0^{t_f} \boldsymbol{e}^{(\boldsymbol{A}^T\tau)}\boldsymbol{C}^T\boldsymbol{y}(\tau)\,\mathrm{d}\tau \tag{7.99}$$

where

$$S_o(t_f) = \int_0^{t_f} e^{(A^T\tau)}C^T C e^{A\tau}\,\mathrm{d}\tau > 0 \quad \text{(positive definite matrix)} \tag{7.100}$$

and

$$S_o(\infty) = S_o = \int_0^{\infty} e^{(A^T\tau)}C^T C e^{A\tau}\,\mathrm{d}\tau$$

$$S_o A + A^T S_o = -C^T C$$

A is Hurwitz

The matrices $S_o(t_f), S_o$ are known as **Observability Grammians.**

Important Properties of Observability Grammian

1. For a completely Observable system A, C

   a. $S_o(t_f) > 0$ (positive definite)

   b. $S_o(t_f)$ is nonsingular (full rank n)

   c. $O^T(A, C) = \left[C^T \; A^T C^T \; \ldots \; \left(A^{n-1}\right)^T C^T \right]$ is full rank n

2. If all the eigenvalues of A have real parts negative (Hurwitz) and the system is completely observable, then the observability Grammian can be expressed as

$$S_o(\infty) = S_o = \int_0^{\infty} e^{(A^T\tau)}C^T C e^{A\tau}\,\mathrm{d}\tau > 0$$

   and

$$A^T S_o + S_o A = -C^T C$$

3. In order to compute $x(0)$ (and hence, $x(t) = e^{At}x(0)$) the observed value of the

output vector, $\boldsymbol{y}(t)$ at only one time t is not enough. It has to be observed for a period of time $[0, t_f]$ and then via Eq. 7.99, $\boldsymbol{x}(0)$ is computed. Of course, we can observe $\boldsymbol{y}(t)$ and its n derivatives at one time instead.

4. The system Eq. 7.98 is completely observable if and only if the augmented matrix

$$\boldsymbol{A}(\lambda, \boldsymbol{C}) = \begin{bmatrix} (\lambda \boldsymbol{I} - \boldsymbol{A}) \\ \boldsymbol{C} \end{bmatrix}$$

has a full rank n for every eigenvalue of $\boldsymbol{A}$. This implies that the colums of $\boldsymbol{C}\,(\lambda \boldsymbol{I} - \boldsymbol{A})^{-1}$ are linearly independent for every eigenvalue λ of $\boldsymbol{A}$.

Dual (Adjoint) Systems

Let us compare both the controllability and observability Grammian

$$\boldsymbol{S}_c(\infty) = \int_0^\infty \boldsymbol{e}^{-A\tau} \boldsymbol{B}\boldsymbol{B}^T \boldsymbol{e}^{-(A^T\tau)}\, d\tau = \boldsymbol{S}_c > 0$$

$$\boldsymbol{S}_o(\infty) = \int_0^\infty \boldsymbol{e}^{(A^T\tau)} \boldsymbol{C}^T \boldsymbol{C} \boldsymbol{e}^{A\tau}\, d\tau = \boldsymbol{S}_o > 0$$

If we replace $\boldsymbol{A}$ with $-\boldsymbol{A}^T$ and $\boldsymbol{B}$ with $\boldsymbol{C}^T$, then $\boldsymbol{S}_c$ implies $\boldsymbol{S}_o$ and $\boldsymbol{S}_o$ implies $\boldsymbol{S}_c$ This gives us the duality theorem of controllability and observability.

Duality Theorem

Given system (Sy) and its adjoint (ADSy):

$$\left.\begin{aligned} \dot{\boldsymbol{x}} &= \boldsymbol{A}\boldsymbol{x} + \boldsymbol{B}\boldsymbol{u} \\ \boldsymbol{y} &= \boldsymbol{C}\boldsymbol{x} \end{aligned}\right| \quad \text{(Sy)} \quad n\text{-th order system}$$

$$\left.\begin{aligned} \dot{\boldsymbol{\lambda}} &= -\boldsymbol{A}^T\boldsymbol{\lambda} + \boldsymbol{C}^T\boldsymbol{u} \\ \boldsymbol{y}_\lambda &= \boldsymbol{B}^T\boldsymbol{\lambda} \end{aligned}\right| \quad \text{(ADSy)} \quad n\text{-th order system}$$

If the system (Sy) with pair $(\boldsymbol{A}, \boldsymbol{B})$ is controllable, then the system (ADSy) with pair $(-\boldsymbol{A}^T, \boldsymbol{B}^T)$ is observable. Furthermore, if the system (Sy) with pair $(\boldsymbol{A}, \boldsymbol{C})$ is observable, then the adjoint system (ADSy) with pair $(-\boldsymbol{A}^T, \boldsymbol{C}^T)$ is controllable.

Balancing Similarity Transformations

We shall seek a Similarity Transformation such that the Grammians of the Transformed systems are **diagonal matrices** and these matrices are equal. Such Transformations are referred to as Balancing Similarity Transformations. There are other balancing transformations where these diagonal matrices may be different. We are aided in the above quest by the fact that positive definite matrices appearing in the Grammians can be diagonalized via **unitary matrices**.

- **System**:

$$\dot{\boldsymbol{x}} = \boldsymbol{A}\boldsymbol{x} + \boldsymbol{B}\boldsymbol{u}$$
$$\boldsymbol{y} = \boldsymbol{C}\boldsymbol{x}$$

- **Original Grammians**:

$$\begin{aligned} &\boldsymbol{A}\boldsymbol{S}_c + \boldsymbol{S}_c\boldsymbol{A}^T + \boldsymbol{B}\boldsymbol{B}^T = \boldsymbol{0} \quad \text{(Controllability)} \\ &\boldsymbol{S}_o\boldsymbol{A} + \boldsymbol{A}^T\boldsymbol{S}_o + \boldsymbol{C}^T\boldsymbol{C} = \boldsymbol{0} \quad \text{(Observability)} \end{aligned} \tag{7.101}$$

- **Transformed System**:

$$\begin{aligned} &\hat{\boldsymbol{x}} = \boldsymbol{T}\boldsymbol{x}, \quad \boldsymbol{x} = \boldsymbol{T}^{-1}\hat{\boldsymbol{x}} \\ &\dot{\hat{\boldsymbol{x}}} = \hat{\boldsymbol{A}}\hat{\boldsymbol{x}} + \hat{\boldsymbol{B}}\boldsymbol{u}, \quad \hat{\boldsymbol{A}} = \boldsymbol{T}\boldsymbol{A}\boldsymbol{T}^{-1}, \quad \hat{\boldsymbol{B}} = \boldsymbol{T}\boldsymbol{B} \\ &\boldsymbol{y} = \hat{\boldsymbol{C}}\hat{\boldsymbol{x}}, \quad \hat{\boldsymbol{C}} = \boldsymbol{C}\boldsymbol{T}^{-1} \end{aligned} \tag{7.102}$$

- **Transformed Grammians**:

$$\hat{A}\hat{S}_c + \hat{S}_c\hat{A}^T + \hat{B}\hat{B}^T = 0$$
$$\hat{S}_o\hat{A} + \hat{A}^T\hat{S}_o + \hat{C}^T\hat{C} = 0 \tag{7.103}$$

We seek a Transformation T_b such that

$$\hat{S}_c = \hat{S}_o = \Sigma, \quad \text{(a diagonal matrix often referred as diag } \Sigma) \tag{7.104}$$

Solution:

From Eqs. 7.102 and 7.103

$$T_bAT_b^{-1}\hat{S}_c + \hat{S}_c\left(T_b^T\right)^{-1}A^T\left(T_b^T\right) + T_bBB^T\left(T_b^T\right) = 0$$
$$\hat{S}_oT_bAT_b^{-1} + \left(T_b^T\right)^{-1}A^T\left(T_b^T\right)\hat{S}_o + \left(T_b^T\right)^{-1}C^TCT_b^{-1} = 0$$

or

$$\left.\begin{array}{l} T_b\left[A\left(\left(T_b^T\right)\hat{S}_c\left(T_b^T\right)^{-1}\right) + \left(T_b^{-1}\hat{S}_c\left(T_b^T\right)^{-1}\right)A^T + BB^T\right]\left(T_b^T\right) = 0 \\ \left(T_b^T\right)^{-1}\left[\left(\left(T_b^T\right)\hat{S}_oT_b\right)A + A^T\left(\left(T_b^T\right)\hat{S}_oT_b\right) + C^TC\right]T_b^{-1} = 0 \end{array}\right| \tag{7.105}$$

Comparing Eqs. 7.101, 7.104, and 7.105

$$T_b^{-1}\hat{S}_c\left(T_b^T\right)^{-1} = S_c, \quad \hat{S}_c = T_bS_c\left(T_b^T\right) = \Sigma$$
$$\left(T_b^T\right)\hat{S}_oT_b = S_o, \quad \hat{S}_o = \left(T_b^T\right)^{-1}S_oT_b^{-1} = \Sigma$$

Furthermore,

$$T_b^{-1}\hat{S}_c\hat{S}_oT_b = S_cS_o$$

or

$$\hat{S}_c\hat{S}_o = T_bS_cS_oT_b^{-1}$$

Define:

$$\boldsymbol{\Lambda} = \boldsymbol{\Sigma}^2 = \boldsymbol{T}_b\,[\boldsymbol{S}_c\boldsymbol{S}_o]\,\boldsymbol{T}_b^{-1}, \quad \text{a diagonal matrix} \tag{7.106}$$

The above expression is achievable because the matrix $\boldsymbol{S}_c\boldsymbol{S}_o$ is positive definite and hence diagonalizable. Moreover, the eigenvectors of $(\boldsymbol{S}_c\boldsymbol{S}_o)$ represent the columns of the matrix $\boldsymbol{T}_b$. Thus,

$$\boldsymbol{S}_c\boldsymbol{S}_o = \boldsymbol{T}_b^{-1}\boldsymbol{\Lambda}\boldsymbol{T}_b, \quad \boldsymbol{\Lambda} = \text{Diag}\left(\hat{\lambda}_1, \hat{\lambda}_2, \cdots, \hat{\lambda}_n\right)$$

It should be noted, that the Transformation $\boldsymbol{T}$ is not unique. In order to obtain another Transformation, let $\boldsymbol{S}_c = \boldsymbol{L}_c^T\boldsymbol{L}_c$ (Cholesky decomposition, Chapter 2) From Eq. 7.106:

$$\boldsymbol{\Sigma} = \boldsymbol{T}_b\boldsymbol{S}_c\left(\boldsymbol{T}_b^T\right) = \boldsymbol{T}_b\boldsymbol{L}_c^T\boldsymbol{L}_c\left(\boldsymbol{T}_b^T\right)$$

Using unitary matrices, we can rewrite the above equations as:

$$\boldsymbol{\Sigma} = \boldsymbol{\Sigma}^{1/2}\boldsymbol{U}^T\left(\boldsymbol{L}_c^T\right)^{-1}\boldsymbol{L}_c^T\boldsymbol{L}_c\boldsymbol{L}_c^{-1}\boldsymbol{U}\boldsymbol{\Sigma}^{1/2} \qquad \boldsymbol{U}^T = \boldsymbol{U}^{-1} \quad (\boldsymbol{U}\text{ being Unitary})$$

Thus,

$$\boldsymbol{T}_b = \boldsymbol{\Sigma}^{1/2}\boldsymbol{U}^T\left(\boldsymbol{L}_c^T\right)^{-1}, \quad \boldsymbol{T}_b^{-1} = \boldsymbol{L}_c^T\boldsymbol{U}\boldsymbol{\Sigma}^{-1/2} \tag{7.107}$$

Substituting Eq. 7.107 in Eq. 7.106, we obtain

$$\hat{\boldsymbol{S}}_o = \boldsymbol{\Sigma} = \boldsymbol{\Sigma}^{-1/2}\boldsymbol{U}^T\boldsymbol{S}_o\boldsymbol{L}_c^T\boldsymbol{U}\boldsymbol{\Sigma}^{-1/2} \tag{7.108}$$

which implies

$$\boldsymbol{U}^T\boldsymbol{L}_c\boldsymbol{S}_o\boldsymbol{L}_c^T\boldsymbol{U} = \boldsymbol{\Sigma}^2$$

or

$$\boldsymbol{L}_c\boldsymbol{S}_o\boldsymbol{L}_c^T = \left(\boldsymbol{U}^T\right)^{-1}\boldsymbol{\Sigma}^2\boldsymbol{U}^{-1} = \boldsymbol{U}\boldsymbol{\Sigma}^2\boldsymbol{U}^{-1} = \boldsymbol{U}\boldsymbol{\Lambda}\boldsymbol{U}^{-1}$$

Remark about Grammian vs. Matrix Controllability and Observability Test

Both the Grammian and the matrix tests provide necessary and sufficient conditions for controllability and observability. The matrix test involves only the parameter conditions, while the Grammian represents the integral conditions involving the time history of the system. There is no basic difference if the system is linear time-invariant. But for the time varying systems they may lead to a different implementation. Another important reason for studying the Grammian is the **model reduction** problem that allows an approximation to the original system with a lower order system. As quadratic functions, the Grammians represent a measure of the energy contents of the system. In fact the controllability Grammian $\boldsymbol{S}_c$ is a measure of the energy of the dominant controllable states, whereas the observability Grammian $\boldsymbol{S}_o$ is a measure of the energy in the dominant observable states. Hence:

$$\{\hat{\lambda}_i\}_{i=1}^n = \{\sigma_i^2\}_{i=1}^n \quad \text{are eigenvalues of the matrix} \quad \boldsymbol{L}_c \boldsymbol{S}_o \boldsymbol{L}_c^T$$

The Balancing Transformation is, $\boldsymbol{T}_b = \boldsymbol{L}_c^T \boldsymbol{U} \boldsymbol{\Sigma}^{-1/2}$

We shall use balancing Similarity Transformation to obtain "desired" Grammians for a transformed system, more suitable for order reduction. Such transformations are known as model reduction "Balancing Similarity Transformations."

Order Reduction via Balanced Transformations

Let us arrange the diagonal elements of $\boldsymbol{\Sigma}$ in the decreasing order as: $\sigma_1 > \sigma_2 > \cdots > \sigma_n$ and partition $\boldsymbol{\Sigma}$ into submatrices:

$$\boldsymbol{\Sigma} = \begin{bmatrix} \Sigma_1 & 0 \\ 0 & \Sigma_2 \end{bmatrix}$$

$\boldsymbol{\Sigma_1}$ is $k \times k$ diagonal matrix with elements $\sigma_1, \sigma_2 \cdots \sigma_k$

$\boldsymbol{\Sigma_2}$ is $(n-k) \times (n-k)$ diagonal matrix with elements $\sigma_{k+1}, \sigma_{k+2} \cdots \sigma_n$

The new transformed balanced system $(\boldsymbol{A}_b, \boldsymbol{B}_b, \boldsymbol{C}_b)$:

$$\boldsymbol{A}_b = \boldsymbol{T}_b\boldsymbol{A}\boldsymbol{T}_b^{-1} = \begin{bmatrix} \boldsymbol{A}_{11} & \boldsymbol{A}_{12} \\ \boldsymbol{A}_{21} & \boldsymbol{A}_{22} \end{bmatrix}, \boldsymbol{B}_b = \boldsymbol{T}_b\boldsymbol{B} = \begin{bmatrix} \boldsymbol{B}_1 \\ \boldsymbol{B}_2 \end{bmatrix}$$

$$\boldsymbol{C}_b = \boldsymbol{C}\boldsymbol{T}_b^{-1} = \begin{bmatrix} \boldsymbol{C}_1 & \boldsymbol{C}_2 \end{bmatrix}$$

Yielding:

$G(s) = \boldsymbol{C}^T (s\boldsymbol{I} - \boldsymbol{A})\, \boldsymbol{B}$ Original system Transfer Function

$G_r(s) = \boldsymbol{C}_1^T (s\boldsymbol{I} - \boldsymbol{A}_{11})\, \boldsymbol{B}_1$ Reduced order Transfer Function

The reduced order model is stable and has a $\boldsymbol{H}^\infty$ error bound:

$$||G(s) - G_r(s)||^\infty \leq 2 \sum_{i=k+1}^{n} \sigma_i$$

Example 7.8:

The controllable canonical realization of a SISO system has the form:

$$\dot{\boldsymbol{x}} = \boldsymbol{A}\boldsymbol{x} + \boldsymbol{b}u, \quad y = \boldsymbol{c}^T\boldsymbol{x}$$

$$\boldsymbol{A} = \begin{bmatrix} 0 & 1 & 0 \\ 0 & 0 & 1 \\ -a_3 & -a_2 & -a_1 \end{bmatrix}, \quad \boldsymbol{b} = \begin{bmatrix} 0 \\ 0 \\ 1 \end{bmatrix}, \quad \boldsymbol{c}^T = \begin{bmatrix} c_2 & c_1 & c_0 \end{bmatrix}$$

We shall reduce the 3rd order system to 2nd order via Balanced Transformation.

- Compute $\boldsymbol{S}_c$ from Eq. 7.101

$$\begin{bmatrix} 0 & 1 & 0 \\ 0 & 0 & 1 \\ -a_3 & -a_2 & -a_1 \end{bmatrix} \begin{bmatrix} s_{c11} & s_{c12} & s_{c13} \\ s_{c12} & s_{c22} & s_{c23} \\ s_{c13} & s_{c23} & s_{c33} \end{bmatrix} + \begin{bmatrix} s_{c11} & s_{c12} & s_{c13} \\ s_{c12} & s_{c22} & s_{c23} \\ s_{c13} & s_{c23} & s_{c33} \end{bmatrix} \begin{bmatrix} 0 & 0 & -a_3 \\ 1 & 0 & -a_2 \\ 0 & 1 & -a_1 \end{bmatrix} + \boldsymbol{b}\boldsymbol{b}^T = \begin{bmatrix} 0 & 0 & 0 \\ 0 & 0 & 0 \\ 0 & 0 & 0 \end{bmatrix}$$

Yielding:

$$
\begin{aligned}
d_{11} &= 2s_{c12} = 0 \\
d_{12} &= s_{c22} + s_{c13} = 0 \\
d_{13} &= s_{c23} - (a_3 s_{c11} + a_2 s_{c12} + a_1 s_{c13}) = 0 \\
d_{22} &= 2s_{c23} = 0 \\
d_{23} &= s_{c33} - (a_3 s_{c12} + a_2 s_{c22} + a_1 s_{c23}) \\
d_{33} &= 1 - 2\,(a_3 s_{c13} + a_2 s_{c23} + a_1 s_{c33}) = 0
\end{aligned}
$$

Solving the above $n(n+1)/2$ equations:

$$
\boldsymbol{S}_c = \begin{bmatrix} \frac{a_1}{2a_3\,(a_1 a_2 - a_3)} & 0 & \frac{-1}{2\,(a_1 a_2 - a_3)} \\ 0 & \frac{1}{2\,(a_1 a_2 - a_3)} & 0 \\ \frac{-1}{2\,(a_1 a_2 - a_3)} & 0 & \frac{a_2}{2a_3\,(a_1 a_2 - a_3)} \end{bmatrix}
$$

- Similarly, we compute $\boldsymbol{S}_o$ from Eq. 7.101

$$
\begin{bmatrix} s_{o11} & s_{o12} & s_{o13} \\ s_{o12} & s_{o22} & s_{o23} \\ s_{o13} & s_{o23} & s_{o33} \end{bmatrix} \begin{bmatrix} 0 & 1 & 0 \\ 0 & 0 & 1 \\ -a_3 & -a_2 & -a_1 \end{bmatrix} + \begin{bmatrix} 0 & 0 & -a_3 \\ 1 & 0 & -a_2 \\ 0 & 1 & -a_1 \end{bmatrix} \begin{bmatrix} s_{o11} & s_{o12} & s_{o13} \\ s_{o12} & s_{o22} & s_{o23} \\ s_{o13} & s_{o23} & s_{o33} \end{bmatrix} + \boldsymbol{c}\boldsymbol{c}^T = \begin{bmatrix} 0 & 0 & 0 \\ 0 & 0 & 0 \\ 0 & 0 & 0 \end{bmatrix}
$$

Solving the above equations,

$$
s_{o11} = \frac{2c_1 c_2 a_3 a_2 a_1 - 2c_1 c_2 a_3^2 - a_2^2 c_2^2 a_1 + a_2 c_2^2 a_3 - a_3 c_2^2 a_1^2 + 2a_3^2 c_2 c_0 a_1 - c_0^2 a_3^3}{2a_3\,(-a_2 a_1 + a_3)}
$$

$$
s_{o12} = \frac{-2a_3 c_2 c_0 a_2 a_1 + c_2^2 a_1^2 a_2 + c_0^2 a_3^2 a_2 + a_3^2 c_1^3}{2a_3\,(a_2 a_1 - a_3)}
$$

$$
s_{o13} = \frac{c_2^2}{2a_3}
$$

$$s_{o22} = \frac{\begin{bmatrix} -2a_3c_1c_0a_2a_1 + 2c_1c_0a_3^2 + a_3c_1^2a_2 - 2a_3c_2c_0a_2 + a_1^3c_2^2 \\ +a_1^2a_3c_1^2 - 2a_1^2a_3c_2c_0 + a_1c_0^2a_3^2 + a_2^2c_0^2a_3 + c_2^2a_3 \end{bmatrix}}{2a_3\left(a_2a_1 - a_3\right)}$$

$$s_{o23} = \frac{-2a_3c_2c_0a_1 + c_2^2a_1^2 + a_3c_1^2a_1 + a_3^2c_0^2}{2a_3\left(a_2a_1 - a_3\right)}$$

$$s_{o33} = \frac{c_0^2a_3a_2 + c_2^2a_1^2 + a_3c_1^2 - 2a_3c_2c_0}{2a_3\left(a_2a_1 - a_3\right)}$$

- Solve for $\boldsymbol{L}_c$ via: $\boldsymbol{L}_c^T\boldsymbol{L} = \boldsymbol{S}_c$

- Determine Hankel singular values,

$$\det\left|\lambda\boldsymbol{I} - \boldsymbol{L}_c\boldsymbol{S}_o\boldsymbol{L}_c^T\right| = \left(\lambda - \sigma_1^2\right)\left(\lambda - \sigma_2^2\right)\left(\lambda - \sigma_3^2\right) \text{ such that } \sigma_1 > \sigma_2 > \sigma_3$$

- Choose

$$\boldsymbol{\Sigma}_r = \begin{bmatrix} \sigma_1 & 0 \\ 0 & \sigma_2 \end{bmatrix}$$

Balancing Similarity Transformation matrix, $\boldsymbol{T}_b$ is, $\boldsymbol{T}_b^{-1} = \boldsymbol{L}_c^T\boldsymbol{U}_r\boldsymbol{\Sigma}_r^{-1/2}$

$\boldsymbol{U}_r$ is a 3×2 orthogonal matrix whose columns are eigenvectors corresponding to eigenvalues σ_1^2 and σ_2^2. The symbol r stands for the reduced model. Cholesky decompositions of $\boldsymbol{S}_c$ gives $\boldsymbol{L}_c$ as:

$$\boldsymbol{S}_c = \boldsymbol{L}_c^T\boldsymbol{L}_c \quad \text{(We can use } \boldsymbol{S}_o = \boldsymbol{L}_o^T\boldsymbol{L}_o\text{)}$$

- The third order system has been reduced into the following second order system $(\boldsymbol{A}_r, \boldsymbol{b}_r, \boldsymbol{c}_r)$:

$$\boldsymbol{A}_r = \boldsymbol{T}_b\boldsymbol{A}\boldsymbol{T}_b^{-1}$$

$$\boldsymbol{b}_r = \boldsymbol{T}_b\boldsymbol{b}, \quad \boldsymbol{c}_r^T = \boldsymbol{c}_r^T\boldsymbol{T}_b^{-1}$$

7.5.13 State Variable Feedback Control via State Observers

Earlier stable feedback control systems are based upon the premise that we are able to measure all the state variables, true or not. This involves a complete knowledge of the state space, which is not necessarily the case. Hence, the need to find control strategies that do not involve measurements of all the state variables.

Consider the typical control system:

$$\dot{\boldsymbol{x}}(t) = \boldsymbol{A}\boldsymbol{x}(t) + \boldsymbol{B}\boldsymbol{u}(t)$$

$$y(t) = \boldsymbol{C}\boldsymbol{x}(t)$$

Using a quadratic performance optimization criterion, the control algorithm is:

$$\boldsymbol{u}(t) = -\boldsymbol{K}(t)\boldsymbol{x}(t)$$

Note: $\boldsymbol{y}(t)$ is ignored in the control system synthesis here. Instead all the components of the state vector $\boldsymbol{x}$ are required for computation of $u(t)$.

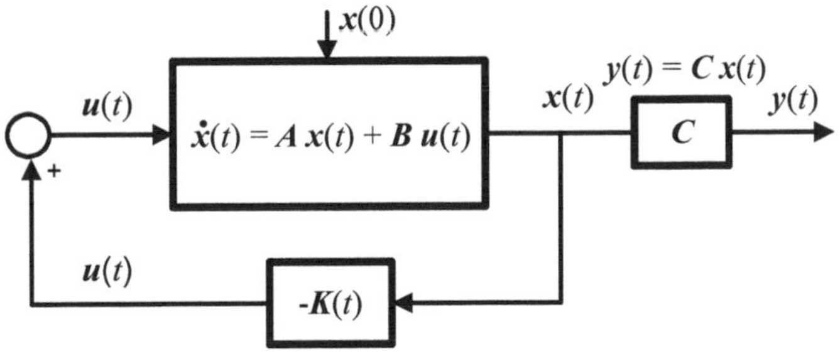

Figure 7.18: State Variable Feedback Controller.

In reality, the only measurable variables are $\boldsymbol{y}(t)$, and the state variable vector $\boldsymbol{x}(t)$ is probably not measurable in its totality. We shall try to overcome this deficiency via the so-called "state observer or state estimator." using information from $\boldsymbol{y}(t)$

Observer, or State Estimator Algorithm:

As stated earlier, if the initial condition state vector $\boldsymbol{x}(0)$ is known the problem of state estimation is trivial (see Eq. 7.72). If $\hat{\boldsymbol{x}}(t)$ is an asymptotic approximation of $x(t)$, we can generate $\hat{x}(t)$ via computer modeling as follows:

$$\left.\begin{aligned} &\dot{\hat{\boldsymbol{x}}}(t) = \boldsymbol{A}\hat{\boldsymbol{x}}(t) + \boldsymbol{B}\boldsymbol{u}(t), \quad \hat{\boldsymbol{x}}(t_0) = \boldsymbol{x}(0) \\ &\boldsymbol{u}(t) = -\boldsymbol{K}(t)\hat{\boldsymbol{x}}(t) \\ &\hat{\boldsymbol{y}}(t) = \boldsymbol{C}\hat{\boldsymbol{x}}(t) \end{aligned}\right\} \text{Computer Simulation}$$

But the proposed state observer scheme is impractical for the following reasons:

(i) Such an observer results in an open-loop control, not useful in most situations.

(ii) The initial state $\boldsymbol{x}(0)$ is not available or is contaminated with "noise." In fact, the whole observability problem stems from ignorance of the initial state $\boldsymbol{x}(0)$. This leads us to the conclusion that we have to make use of the output $\boldsymbol{y}(t)$ along with $\boldsymbol{A}, \boldsymbol{B}$, and $\boldsymbol{C}$ matrices to design a dynamic observer yielding $\hat{\boldsymbol{x}}(t)$. Furthermore, it is important to realize that just like $u(t)$, $\boldsymbol{x}(0)$ is an input to the system that drives the system and yields together with $u(t)$, the output $\boldsymbol{y}(t)$. This leads us to the following observer design shown in Figure 7.19.

The observer equations are:

$$\dot{\hat{\boldsymbol{x}}} = \boldsymbol{A}\hat{\boldsymbol{x}} + \boldsymbol{B}\boldsymbol{u} + E(\boldsymbol{y} - \hat{\boldsymbol{y}}), \quad \hat{\boldsymbol{y}} = \boldsymbol{C}\hat{\boldsymbol{x}} \tag{7.109}$$

$\hat{\boldsymbol{x}}$ = The observer output, $\hat{\boldsymbol{y}}$ = The estimate of the system output

Eq. 7.109 can be rewritten as:

$$\dot{\hat{\boldsymbol{x}}} = (\boldsymbol{A} - \boldsymbol{EC})\hat{\boldsymbol{x}} + \boldsymbol{B}\boldsymbol{u} + \boldsymbol{E}\boldsymbol{y}, \quad \hat{\boldsymbol{y}} = \boldsymbol{C}\hat{\boldsymbol{x}}, \quad \boldsymbol{u} = -\boldsymbol{K}\hat{\boldsymbol{x}}$$

$$\dot{\hat{\boldsymbol{x}}} = (\boldsymbol{A} - \boldsymbol{EC} - \boldsymbol{BK})\hat{\boldsymbol{x}} + \boldsymbol{E}\boldsymbol{y} \tag{7.110}$$

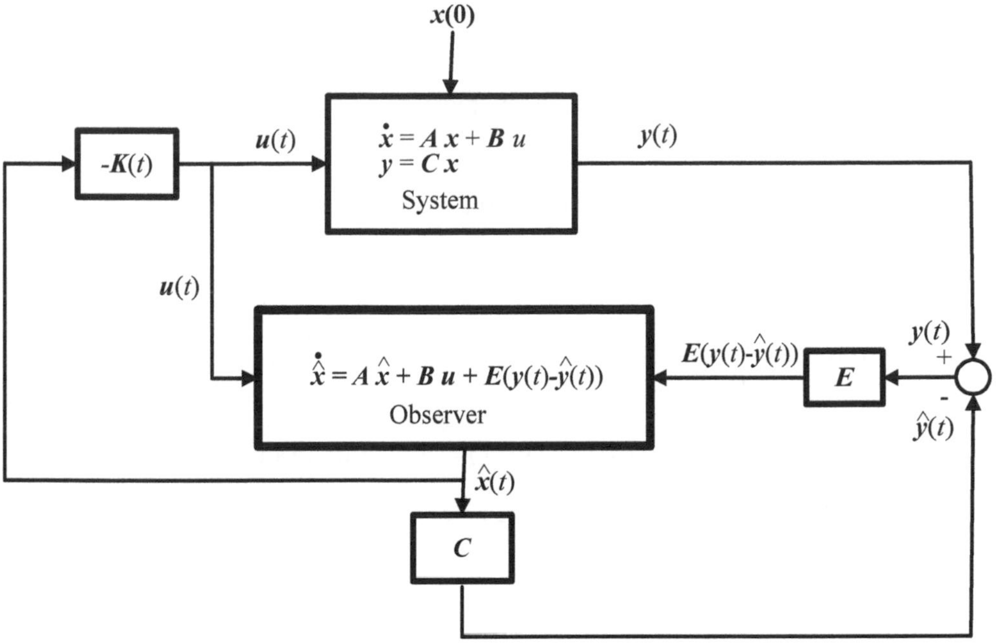

Figure 7.19: Observer Design.

This shows that the choice of $\boldsymbol{E}$ matrix is critical for convergence of $\hat{\boldsymbol{x}}(t)$ to $\boldsymbol{x}(t)$ in the shortest possible time. Furthermore, simplification can be achieved by using the error variable and $u(t)$ as follows:

$$e = x - \hat{\boldsymbol{x}}, \quad \boldsymbol{u} = -\boldsymbol{K}\hat{\boldsymbol{x}}$$

Thus,

$$\dot{\boldsymbol{x}} = (\boldsymbol{A} - \boldsymbol{BK})\boldsymbol{x} - \boldsymbol{Ke}$$

$$\dot{\boldsymbol{e}} = (\boldsymbol{A} - \boldsymbol{EC})\boldsymbol{e}$$

When $\boldsymbol{E}$ is properly chosen so that the eigenvalues of $(\boldsymbol{A} - \boldsymbol{BK})$ and $(\boldsymbol{A} - \boldsymbol{EC})$ have negative real parts located at optimal chosen locations, we have robust control and a good observer system, known as the **Luenberger Observer.**

7.5.14 Controllability–Observability Time-Varying Systems

Consider a Linear Time Varying (LTV) System:

$$\begin{aligned} \dot{\boldsymbol{x}} &= \boldsymbol{A}(t)\boldsymbol{x} + \boldsymbol{B}(t)\boldsymbol{u}\,, \quad \boldsymbol{x}(t_0) = \boldsymbol{x}_0 \\ y &= \boldsymbol{c}(t)\boldsymbol{x} \end{aligned} \tag{7.111}$$

The nonsingular state transition matrix equation is given by:

$$\begin{aligned} &\dot{\boldsymbol{\varphi}}(t,\tau) = \boldsymbol{A}(t)\boldsymbol{\varphi}(t,\tau)\,, \quad \boldsymbol{\varphi}(0,0) = \boldsymbol{\varphi}(\tau,\tau) = \boldsymbol{I}\,, \quad \text{the identity matrix} \\ &\boldsymbol{\varphi}(t_2,t_1) = \boldsymbol{\varphi}(t_2,\tau)\boldsymbol{\varphi}(\tau,t_1) \qquad t_1 \leq \tau \leq t_2 \end{aligned} \tag{7.112}$$

We shall first use the Grammian approach to study the properties of Controllability and Observability for the time varying systems. Except for some simple systems, the determination of $\boldsymbol{\varphi}(t,\tau)$ is rather a tedious task, if not impossible.

Definition of Controllability for LTV Systems

Given a specified time t_0 and initial condition $\boldsymbol{x}(0)$, if there exists a control $u(t), t_0 \leq t \leq t_1$, which transfers the initial state from $\boldsymbol{x}(0)$ to $\boldsymbol{x}(t_1)$, then the system is controllable otherwise the system is uncontrollable. Specified times t_0 and t_1 are of the essence.

Definition of Observability for LTV Systems

Given a specified time t_0 and the input $\boldsymbol{u}(t)$ and the output $y(t)$, for all t, $t_0 \leq t \leq t_1$, if it is possible to uniquely determine the initial state, $\boldsymbol{x}(t_0) = \boldsymbol{x}(0)$, then all the system states are defined as observable otherwise some or all of the states may be unobservable.

Controllability Criterion #1 for LTV Systems

Consider the Controllability positive definite Grammian:

$$S_c(t_1,t_0) = \int_{t_0}^{t_1} \boldsymbol{\varphi}(t_1,\tau)\boldsymbol{B}(\tau)\boldsymbol{B}^T(\tau)\boldsymbol{\varphi}^T(t_1,\tau)\,\mathrm{d}\tau > 0 \tag{7.113}$$

The pair $(\boldsymbol{A}(t), \boldsymbol{B}(t))$ representing Eq. 7.111 is controllable at $t = t_0$ if and only if the Grammian Eq. 7.113 exists for any final time, $t_1 > t_0$.

Proof: Controllability Grammian

The solution to Eq. 7.111 is similar as in the time-invariant case, except $\boldsymbol{e}^{\boldsymbol{A}(t-\tau)}$ is replaced with $\boldsymbol{\varphi}(t, \tau)$, yielding:

$$\boldsymbol{x}(t) = \boldsymbol{\varphi}(t, t_0)\boldsymbol{x}(t_0) + \int_{t_0}^{t} \varphi(t, \tau)\boldsymbol{B}(\tau)u(\tau)\,\mathrm{d}\tau \tag{7.114}$$

Following the earlier reasoning, let

$$\begin{aligned} x(t_1) - \boldsymbol{\varphi}(t_1, t_0)\boldsymbol{x}(t_0) &= \hat{\boldsymbol{x}}(t_1, t_0) \\ \boldsymbol{u}(t) &= +\boldsymbol{B}^T(t)\boldsymbol{\varphi}^T(t_1, \tau)S_c^{-1}(t_1, t_0)\hat{\boldsymbol{x}}(t_1, t_0) \end{aligned} \tag{7.115}$$

From Eq. 7.115, the Eq. 7.114 can be rewritten as:

$$\hat{x}(t_1, t_0) = \left[\int_{t_0}^{t_1} \varphi(t_1, \tau)\boldsymbol{B}(\tau)\boldsymbol{B}^T(\tau)\boldsymbol{\varphi}^T(t_1, \tau)\,\mathrm{d}\tau\right]\boldsymbol{S}_c^{-1}(t_1, t_0)\hat{\boldsymbol{x}}(t_1, t_0), \text{ yielding}$$

$$\boldsymbol{S}_c(t_1, t_0) = \int_{t_0}^{t_1} \varphi(t_1, \tau)\boldsymbol{B}(\tau)\boldsymbol{B}^T(\tau)\boldsymbol{\varphi}^T(t_1, \tau)\,\mathrm{d}\tau, \quad \text{(nonsingular Grammian)} \tag{7.116}$$

Existence of a nonsingular Grammian, $\boldsymbol{S}_c(t_1, t_0)$ assures us that for a given t_0 we can arrive at the final state, $\boldsymbol{x}(t_1)$ from any initial state, $\boldsymbol{x}(t_0)$.

Observability Criterion #1 for LTV Systems

- Since $\boldsymbol{u}(t)$ has no effect on observability, we let $\boldsymbol{u}(t) = 0$.
- We shall observe the output vector $y(t)$ for all $t, t_0 \leq t \leq t_1$ and use these measurements to compute the state $\boldsymbol{x}(t_0)$ and the associated criterion.

Proof: Observability Grammian

From Eqs. 7.111 and 7.112,

$$\boldsymbol{y}(t) = \boldsymbol{C}(t)\boldsymbol{\varphi}(t, t_0)\boldsymbol{x}(t_0), \quad \boldsymbol{x}(t_0) \text{ being an initial state vector.} \tag{7.117}$$

Right multiply Eq. 7.117 with $\boldsymbol{\varphi}^T(t, t_0)\boldsymbol{C}^T(t)$ and integrating from t_0 to t_1,

$$\int_{t_0}^{t_1} \boldsymbol{\varphi}^T(\tau, t_0)\boldsymbol{C}^T(\tau)\boldsymbol{y}(\tau)\,\mathrm{d}\tau = \left(\int_{t_0}^{t_1} \boldsymbol{\varphi}^T(\tau, t_0)\boldsymbol{C}^T(\tau)\boldsymbol{C}(\tau)\boldsymbol{\varphi}(\tau, t_0)\,\mathrm{d}\tau\right)\boldsymbol{x}(t_0)$$

$$\boldsymbol{x}(t_0) = \boldsymbol{S}_o^{-1}(t_1, t_0)\int_{t_0}^{t} \boldsymbol{\varphi}^T(\tau, t_0)\boldsymbol{C}^T(\tau)\boldsymbol{y}(\tau)\,\mathrm{d}\tau \tag{7.118}$$

$$\boldsymbol{S}_o(t_1, t_0) = \left(\int_{t_0}^{t_1} \boldsymbol{\varphi}^T(\tau, t_0)\boldsymbol{C}^T(\tau)\boldsymbol{C}(\tau)\boldsymbol{\varphi}(\tau, t_0)\,\mathrm{d}\tau\right) > 0 \tag{7.119}$$

Thus, the criterion of observability is that the Grammian $\boldsymbol{S}_o(t_1, t_o)$ is a nonsingular.

Controllability Criterion #2 for LTV System

An easier way to derive the controllability criterion is through the duality principle discussed below. Consider a dual to the system described in Eq. 7.111:

$$\begin{aligned} \dot{\boldsymbol{\lambda}}(t) &= -\boldsymbol{A}^T(t)\boldsymbol{\lambda}(t) \\ \boldsymbol{y}_\lambda(t) &= \boldsymbol{B}^T(t)\boldsymbol{\lambda}(t) \end{aligned} \tag{7.120}$$

The Hamiltonian $H(t)$ and its derivative can be written as:

$$H(t) = \boldsymbol{\lambda}^T(t)\boldsymbol{x}(t)$$

$$\dot{H}(t) = \dot{\boldsymbol{\lambda}}^T(t)\boldsymbol{x}(t) + \boldsymbol{\lambda}^T(t)\dot{\boldsymbol{x}}(t) = -\boldsymbol{\lambda}^T(t)\boldsymbol{A}(t)\boldsymbol{x}(t) + \boldsymbol{\lambda}^T(t)\boldsymbol{A}(t)\boldsymbol{x}(t) = 0$$

Thus, $H(t)$ is stationary and constant, implying, $\boldsymbol{\lambda}^T(t_0)\boldsymbol{x}(t_0) = \boldsymbol{\lambda}^T(t_1)\boldsymbol{x}(t_1)$

The system Eq. 7.111 is being steered from the initial state $x(t_0)$ at $t = t_0$ to the final state, $\boldsymbol{x}(t_1)$ at $t = t_1$ via the control $u(t)$, while the system Eq. 7.120 is being observed for the duration, $t_0 \le t \le t_1$. Invoking the duality principle, the observability condition for Eq. 7.120 are the same as the controllability conditions for system Eq. 7.111. So let us derive observability conditions for the system Eq. 7.120. Let

$$\boldsymbol{B}(t) = \boldsymbol{B}_0(t)$$

Thus,

$$\boldsymbol{y}_\lambda(t) = \boldsymbol{B}_o^T(t)\boldsymbol{\lambda}(t)$$

Taking the derivative

$$\dot{\boldsymbol{y}}_\lambda(t) = \left(\dot{\boldsymbol{B}}_o^T(t) - \boldsymbol{B}_o^T(t)\boldsymbol{A}^T(t)\right)\boldsymbol{\lambda}(t) = \boldsymbol{B}_1^T(t)\boldsymbol{\lambda}(t)$$

Let

$$\boldsymbol{B}_1^T(t) = +\dot{\boldsymbol{B}}_o^T(t) - \boldsymbol{B}_o^T(t)\boldsymbol{A}^T(t)$$

Following the same reasoning:

$$\boldsymbol{y}_\lambda^{(k)}(t) = \boldsymbol{B}_k^T(t)\boldsymbol{\lambda}(t)$$

where

$$\boldsymbol{B}_k^T(t) = \dot{\boldsymbol{B}}_{k-1}^T(t) - \boldsymbol{B}_{k-1}^T(t)\boldsymbol{A}^T(t) \quad k = 1, 2, \cdots, n-1$$

Thus,

$$\begin{bmatrix} \boldsymbol{y}_\lambda^{(0)}(t) \\ \boldsymbol{y}_\lambda^{(1)}(t) \\ \vdots \\ \boldsymbol{y}_\lambda^{(n-1)}(t) \end{bmatrix} = \begin{bmatrix} \boldsymbol{B}_0^T(t) \\ \boldsymbol{B}_1^T(t) \\ \vdots \\ \boldsymbol{B}_{n-1}^T(t) \end{bmatrix} \boldsymbol{\lambda}(t), \qquad t_0 \le t \le t_1 \tag{7.121}$$

This implies that the system Eq. 7.111 is controllable if and only if

$$C(\boldsymbol{A}(t),\boldsymbol{B}(t)) = \begin{bmatrix}\boldsymbol{B}_0(t) & \boldsymbol{B}_1(t) & \cdots & \boldsymbol{B}_{n-1}(t)\end{bmatrix}^T \quad \text{has a rank } n. \tag{7.122}$$

Observability Criterion #2

$$\dot{\boldsymbol{x}}(t) = \boldsymbol{A}(t)\boldsymbol{x}(t), \quad \boldsymbol{x}(t) = \boldsymbol{\varphi}(t,t_0)\boldsymbol{x}(t_0)$$

$$\boldsymbol{y}(t) = \boldsymbol{C}(t)\boldsymbol{x}(t), \quad \boldsymbol{y}(t) = \boldsymbol{C}(t)\boldsymbol{\varphi}(t,t_0)\boldsymbol{x}(t_0)$$

Let

$$\boldsymbol{C}_o(t) = \boldsymbol{C}(t)$$

Taking the derivative of $\boldsymbol{y}(t)$

$$\dot{\boldsymbol{y}}(t) = \dot{\boldsymbol{C}}(t)\boldsymbol{x}(t) + \boldsymbol{C}(t)\dot{\boldsymbol{x}}(t) = \left(\dot{\boldsymbol{C}}(t) + \boldsymbol{C}(t)\boldsymbol{A}(t)\right)\boldsymbol{x}(t) = \boldsymbol{C}_1(t)\boldsymbol{x}(t)$$

where

$$\dot{\boldsymbol{C}}(t) + \boldsymbol{C}(t)\boldsymbol{A}(t) = \dot{\boldsymbol{C}}_o(t) + \boldsymbol{C}_o(t)\boldsymbol{A}(t) = \boldsymbol{C}_1(t)$$

Similarly

$$y^{(k)}(t) = \boldsymbol{C}_k(t)\boldsymbol{x}(t)$$

$$\boldsymbol{C}_k(t) = \dot{\boldsymbol{C}}_{k-1}(t) + \boldsymbol{C}_{k-1}(t)\boldsymbol{A}(t), \quad k = 1,\cdots,n-1$$

Hence,

$$\begin{bmatrix}\boldsymbol{y}(t)\\ \dot{\boldsymbol{y}}(t)\\ \vdots \\ \boldsymbol{y}^{(n-1)}(t)\end{bmatrix} = \begin{bmatrix}\boldsymbol{C}_0(t)\\ \boldsymbol{C}_1(t)\\ \vdots \\ \boldsymbol{C}_{n-1}(t)\end{bmatrix}\boldsymbol{x}(t) \tag{7.123}$$

This implies

$$O(\boldsymbol{A}(t), \boldsymbol{C}(t)) = \begin{bmatrix} \boldsymbol{C}_0(t) \\ \boldsymbol{C}_1(t) \\ \vdots \\ \boldsymbol{C}_{n-1}(t) \end{bmatrix} \quad \text{has a rank } n., \quad t_0 \le t \le t_1$$

This concludes discussion of controllability–observability of LTV systems.

7.5.15 SISO Controller Design–Closed-Loop Poles Placement

Consider a controllable system:

$$\begin{aligned} \dot{\boldsymbol{x}} &= \boldsymbol{A}\boldsymbol{x} + \boldsymbol{b}u \\ u &= -\boldsymbol{k}^T\boldsymbol{x} \end{aligned}, \quad \boldsymbol{A} = \begin{bmatrix} a_{11} & a_{12} & \cdots & a_{1n} \\ a_{21} & a_{22} & \cdots & a_{2n} \\ \vdots \\ a_{n1} & a_{n2} & \cdots & a_{nn} \end{bmatrix}, \quad \boldsymbol{b} = \begin{bmatrix} b_1 \\ b_2 \\ \vdots \\ b_n \end{bmatrix}$$

The closed-loop characteristic polynomial of the above control system is chosen as:

$$p_c(s) = s^n + p_1 s^{n-1} + p_2 s^{n-2} + \cdots + p_n = \prod_{i=1}^{n} (s - \lambda_i) \qquad (7.124)$$

where λ_i are the prescribed closed-loop poles. From the Cayley-Hamilton theorem,

$$P_c(A) = \boldsymbol{A}^n + p_1 \boldsymbol{A}^{n-1} + \cdots + p_n \boldsymbol{I} \neq \boldsymbol{0} \quad \text{(closed-loop characteristic polynomial)}$$

$$P(A) = \boldsymbol{A}^n + a_1 \boldsymbol{A}^{n-1} + \cdots + a_n \boldsymbol{I} = \boldsymbol{0} \quad \text{(system characteristic polynomial, } u = 0)$$

We are required to determine the gain vector $\boldsymbol{k}$ such that the closed-loop poles coincide with the prescribed closed-loop poles. Let us transform the system to the companion form via similarity transformation:

$$\boldsymbol{T}\hat{\boldsymbol{x}} = \hat{\boldsymbol{x}}, \quad \boldsymbol{T}\boldsymbol{A}\boldsymbol{T}^{-1} = \hat{\boldsymbol{A}}, \quad \boldsymbol{T}\boldsymbol{b} = \hat{\boldsymbol{b}}, \quad \boldsymbol{k}^T\boldsymbol{T}^{-1} = \hat{\boldsymbol{k}}^T$$

$$\hat{\boldsymbol{A}} = \begin{bmatrix} 0 & 1 & 0 & \cdots & 0 \\ 0 & 0 & 1 & \cdots & 0 \\ \vdots & & & & \\ 0 & 0 & 0 & \cdots & 1 \\ -a_n & -a_{n-1} & -a_{n-2} & \cdots & -a_1 \end{bmatrix}, \quad \hat{\boldsymbol{b}} = \begin{bmatrix} 0 \\ 0 \\ \vdots \\ 0 \\ 1 \end{bmatrix}, \quad \hat{\boldsymbol{k}} = \begin{bmatrix} k_1 \\ k_2 \\ \vdots \\ k_n \end{bmatrix}$$

The transformation $\boldsymbol{T}$ is:

$$\boldsymbol{T} = \begin{bmatrix} \boldsymbol{t}_1^T \\ \boldsymbol{t}_1^T\boldsymbol{A} \\ \vdots \\ \boldsymbol{t}_1^T\boldsymbol{A}^{n-1} \end{bmatrix}$$

But $\boldsymbol{t}_1$ is not a free vector. It is computed as

$$\boldsymbol{T}\boldsymbol{b} = \begin{bmatrix} \boldsymbol{t}_1^T\boldsymbol{b} \\ \boldsymbol{t}_1^T\boldsymbol{A}\boldsymbol{b} \\ \vdots \\ \boldsymbol{t}_1^T\boldsymbol{A}^{n-1}\boldsymbol{b} \end{bmatrix} = \begin{bmatrix} 0 \\ 0 \\ \vdots \\ 1 \end{bmatrix}$$

$$\boldsymbol{t}_1 = \begin{bmatrix} b^T \\ b^T\boldsymbol{A}^T \\ \vdots \\ b^T\left(\boldsymbol{A}^T\right)^{n-1} \end{bmatrix}^{-1} \begin{bmatrix} 0 \\ 0 \\ \vdots \\ 1 \end{bmatrix}$$

The closed-loop system takes the form:

$$\dot{\hat{\boldsymbol{x}}} = \left(\hat{\boldsymbol{A}} - \hat{\boldsymbol{b}}\hat{\boldsymbol{k}}^T\right)\hat{\boldsymbol{x}}$$

where

$$\left(\hat{A} - \hat{b}\hat{k}^T\right) = \begin{bmatrix} 0 & 1 & 0 \cdots & 0 \\ 0 & 0 & 1 \cdots & 0 \\ \vdots & & & \\ 0 & 0 & 0 \cdots & 1 \\ -\left(a_n + \hat{k}_1\right) & -\left(a_{n-1} + \hat{k}_2\right) & \cdots & -\left(a_1 + \hat{k}_n\right) \end{bmatrix}$$

Thus, the closed-loop characteristic polynomial takes the form:

$$p_c(s) = s^n + \left(a_1 + \hat{k}_1\right) s^{n-1} + \left(a_2 + \hat{k}_2\right) s^{n-2} + \cdots + \left(a_n + \hat{k}_n\right) \tag{7.125}$$

Comparing Eqs. 7.124 and 7.125

$$\hat{\boldsymbol{k}} = \begin{bmatrix} \hat{k}_1 \\ \hat{k}_2 \\ \vdots \\ \hat{k}_n \end{bmatrix} = \begin{bmatrix} (p_n - a_n) \\ (p_{n-1} - a_{n-1}) \\ \vdots \\ (p_1 - a_1) \end{bmatrix}$$

$$\boldsymbol{k}^T = \hat{\boldsymbol{k}}^T \boldsymbol{T} = (p_n - a_n)\, \boldsymbol{t}_1^T + \boldsymbol{t}_1^T (p_{n-1} - a_{n-1})\, \boldsymbol{A} + \cdots + \boldsymbol{t}_1^T (p_1 - a_1)\, \boldsymbol{A}^{n-1}$$

or

$$\boldsymbol{k}^T = \boldsymbol{t}_1^T \left(\sum_{i=1}^{n} p_i \boldsymbol{A}^{n-i} - \sum_{i=1}^{n} a_i \boldsymbol{A}^{n-i} \right)$$

From the Cayley-Hamilton Theorem

$$\sum_{i=1}^{n} a_i \boldsymbol{A}^{n-i} = -\boldsymbol{A}^n$$

or

$$\boldsymbol{k}^T = \boldsymbol{t}_1^T \left[\boldsymbol{A}^n + \sum_{i=1}^{n} p_i \boldsymbol{A}^{n-i} \right] = \boldsymbol{t}_1^T \boldsymbol{P}_c(\boldsymbol{A})$$

Hence, the controller gain vector

$$k = P_c\left(A^T\right)\begin{bmatrix} b^T \\ b^T A^T \\ \vdots \\ b^T\left(A^T\right)^{n-1} \end{bmatrix}^{-1}\begin{bmatrix} 0 \\ 0 \\ \vdots \\ 1 \end{bmatrix}$$

7.5.16 Minimal Realization of Time-Invariant Linear Systems

There are two problems in Control Systems Analysis as well as Synthesis:

Problem #1 Determination of the Controllable and the Observable parts of the given system (A, B, C) and its simulation. This problem has been discussed in the last section via the Kalman decomposition. We determined a Similarity transformation T that allowed us to determine the most important component of the system which is completely controllable and completely observable. This part of the total control system is called the "Minimal Realization." This realization yields a system that can be simulated with the least number of first order differential equations. This number is also called the **Schmidt-McMillan degree**.

- Schmidt Transformation

  Original System: $\dot{x} = Ax + Bu, \quad y = Cx$

- Transformed System via Similarity Transformation T,

$$\dot{\hat{x}} = \hat{A}\hat{x} + \hat{B}u, \quad y = \hat{C}x$$

$$\hat{x} = Tx, \quad \hat{A} = TAT^{-1}, \quad \hat{B} = TB, \quad \hat{C} = CT^{-1}$$

$$CA^iB = \hat{C}\hat{A}^iB$$

$$\hat{\boldsymbol{A}} = \begin{bmatrix} \hat{\boldsymbol{A}}_{co} & 0 & * & 0 \\ * & * & * & * \\ 0 & 0 & * & 0 \\ 0 & 0 & * & * \end{bmatrix}, \quad \hat{\boldsymbol{B}} = \begin{bmatrix} \hat{\boldsymbol{B}}_{co} \\ * \\ 0 \\ 0 \end{bmatrix}, \quad \hat{\boldsymbol{C}} = \begin{bmatrix} \hat{\boldsymbol{C}}_{co} & 0 & * & * \end{bmatrix}$$

"$*$" stand for irrelevant matrices.

The system $\left(\hat{\boldsymbol{A}}_{co}, \hat{\boldsymbol{B}}_{co}, \hat{\boldsymbol{C}}_{co}\right)$ is the minimal realization system and can be realized as a Transfer function.

$$\hat{H}(s) = \hat{\boldsymbol{C}}_{co}\left(s\boldsymbol{I} - \hat{\boldsymbol{A}}_{co}\right)\hat{\boldsymbol{B}}_{co}$$

$\hat{H}(s)$ can be simulated with a minimum number of integrators. In the rest of this section the Triple $\left(\hat{\boldsymbol{A}}_{co}, \hat{\boldsymbol{B}}_{co}, \hat{\boldsymbol{C}}_{co}\right)$ will be designated as $(\boldsymbol{A}, \boldsymbol{B}, \boldsymbol{C})$.

Problem #2 Minimal Realization Algorithm from Input and Output Data

$$\dot{\boldsymbol{x}} = \boldsymbol{A}\boldsymbol{x} + \boldsymbol{B}\boldsymbol{u}, \quad \boldsymbol{y} = \boldsymbol{C}\boldsymbol{x}$$

$\boldsymbol{A}$ is $n \times n$ matrix, $\boldsymbol{B}$ is $n \times r$ matrix and $\boldsymbol{C}$ is $m \times n$ matrix.

Assumption: We shall assume that degree n is known and the system is controllable and observable. When n is unknown, some modifications are required in the resulting algorithm. Taking the Laplace Transform:

$$\boldsymbol{y}(s) = \boldsymbol{C}\left(s\boldsymbol{I} - \boldsymbol{A}\right)^{-1}\boldsymbol{B}\boldsymbol{u}(s)$$

It's Laplace Inverse, $$\boldsymbol{y}(t) = \int_0^{\infty} \boldsymbol{C}e^{\boldsymbol{A}(t-\tau)}\boldsymbol{B}\boldsymbol{u}(\tau)\,\mathrm{d}\tau$$

$$\boldsymbol{H}(t) = \boldsymbol{C}e^{\boldsymbol{A}t}\boldsymbol{B}$$

Taking the series expansion of e^{At} and using Cayley-Hamilton theorem:

$$H(t) = \sum_{i=0}^{n-1} C\left(\alpha_i(t)A^i\right)B = \sum_{i=0}^{n-1} \alpha_i(t)CA^iB$$

From the knowledge of $\alpha_i(t)$ and CA^iB, $i = 0, 1, \cdots, n$ one can compute $H(t)$. The coefficients $\alpha_i(t)$ can be computed as discussed in Chapter 2.

Define

$$CA^iB = H_i, \quad i = 0, 1, \cdots, n \quad \textbf{Markov Parameter} \text{ matrices}$$

The realization problem can be restated as the following:

1. Compute the Markov parameters from the input output data.
2. From the knowledge of the Markov parameters, and the degree n of the system compute the matrices A, B, C

SISO Minimal Realization for Single-Input Single-Output Systems

To explain the Realization algorithm, consider the SISO system.

$$\dot{x} = Ax + bu, \quad y = c^T x, \quad H(s) = c^T (sI - A)^{-1} b = \sum_{i=1}^{\infty} c^T A^i b \frac{1}{s^i}$$

$$A = \begin{bmatrix} 0 & 0 \cdots & 0 & -a_1 \\ 1 & 0 & 0 & -a_2 \\ & & & \vdots \\ 0 & 0 & 1 & -a_{n-1} \\ 0 & 0 & 0 & -a_n \end{bmatrix}, \quad b = \begin{bmatrix} 1 \\ 0 \\ \vdots \\ 0 \\ 0 \end{bmatrix}, \quad c^T = \begin{bmatrix} c_1 & c_2 & \cdots & c_n \end{bmatrix}$$

$$h_i = c^T A^i b, \quad i = 0, 1, 2, \cdots$$

The degree n and coefficients a_i, $i = 1, 2, \cdots, n$ are unknown.

Step #1 Determination of the dimension of the matrix $\boldsymbol{A}$, degree n

(i) Let us construct a Hankel matrix for some arbitrarily large n:

$$\boldsymbol{H}(n) = \begin{bmatrix} h_0 & h_1 & h_2 & \cdots & h_{n-1} \\ h_1 & h_2 & h_3 & \cdots & h_n \\ h_2 & h_3 & h_4 & \cdots & h_{n+1} \\ \vdots & & & & \\ h_{n-1} & h_n & h_{n+1} & \cdots & h_{2n-2} \end{bmatrix}$$

(ii) Compute the determinants

$$\Delta_1 = \begin{bmatrix} h_0 \end{bmatrix}, \quad \Delta_2 = \begin{bmatrix} h_0 & h_1 \\ h_1 & h_2 \end{bmatrix}, \quad \Delta_3 = \begin{bmatrix} h_0 & h_1 & h_2 \\ h_1 & h_2 & h_3 \\ h_2 & h_3 & h_4 \end{bmatrix}, \quad \cdots$$

If $\Delta_1, \Delta_2, \cdots, \Delta_{k-1}$ are nonzero, but $\Delta_k = 0$, then $n = (k-1)$ is the dimension of the matrix $\boldsymbol{A}$.

Step #2: Determination of $a_1, a_2, \cdots, a_n$ and hence the system matrix, $\boldsymbol{A}$

From the Cayley-Hamilton Theorem:

$$\boldsymbol{A}^n = -\sum_{i=1}^{n} a_i \boldsymbol{A}^{n-i}$$

or

$$\boldsymbol{c}^T \boldsymbol{A}^n \boldsymbol{b} = -\sum_{i=1}^{n} a_i \boldsymbol{c} \boldsymbol{A}^{n-i} \boldsymbol{b} = -\sum_{i=1}^{n} a_i h_{n-i}, \quad h_i = \boldsymbol{c}^T \boldsymbol{A}^i \boldsymbol{b}$$

or

$$h_n = -\sum_{i=1}^{n} a_i h_{n-i}$$

In general, h_i is $m \times r$ matrix (in this case, $m = 1, r = 1$, thus a scalar.)

Hence,

$$\begin{bmatrix} h_n \\ h_{n+1} \\ \vdots \\ h_{2n-1} \end{bmatrix} = \begin{bmatrix} h_0 & h_1 & \cdots & h_{n-1} \\ h_1 & h_2 & \cdots & h_n \\ \vdots & \vdots & & \vdots \\ h_{n-1} & h_n & \cdots & h_{2n-2} \end{bmatrix} \begin{bmatrix} -a_n \\ -a_{n-1} \\ \vdots \\ -a_1 \end{bmatrix} \tag{7.126}$$

Let

$$\boldsymbol{a}^T = \begin{bmatrix} a_n & a_{n-1} & \cdots & a_1 \end{bmatrix}^T$$

Thus,

$$\boldsymbol{h}(n) = -\boldsymbol{H}(n)\boldsymbol{a}$$

$$\boldsymbol{a} = -\left(\boldsymbol{H}(n)\right)^{-1}\boldsymbol{h}(n)$$

where $\boldsymbol{H}(n)$ is a Hankel matrix as explained earlier. There are some interesting algorithms to compute $\boldsymbol{H}^{-1}(n)$ because of its structure as a bordered matrix. When the system is controllable and observable, the Hankel matrix, $\boldsymbol{H}(n)$ is nonsingular. This completes the determination of the matrix $\boldsymbol{A}$ and its dimension n. As an observation,

$$\begin{bmatrix} h_1 & h_2 & \cdots & h_n \\ h_2 & h_3 & \cdots & h_{n+1} \\ \vdots & & & \vdots \\ h_n & h_{n+1} & \cdots & h_{2n-1} \end{bmatrix} = \begin{bmatrix} h_0 & h_1 & \cdots & h_{n-1} \\ h_1 & h_2 & \cdots & h_n \\ \vdots & & & \vdots \\ h_{n-1} & h_{n-2} & \cdots & h_{2n-2} \end{bmatrix} \begin{bmatrix} 0 & 0 & \cdots & 0 & -a_n \\ 1 & 0 & \cdots & 0 & -a_{n-1} \\ \vdots & \vdots & \cdots & \vdots & \vdots \\ 0 & 0 & \cdots & 1 & -a_1 \end{bmatrix} \tag{7.127}$$

or

$$\hat{\boldsymbol{H}}(n) = \boldsymbol{H}(n)\boldsymbol{A}$$

or

$$\boldsymbol{A} = \boldsymbol{H}^{-1}(n)\hat{\boldsymbol{H}}(n) \tag{7.128}$$

Step #3 Recursive algorithm for inversion of matrix, $\boldsymbol{H}^{-1}(n+1)$ for large n.

Define

$$\boldsymbol{H}(n+1) = \begin{bmatrix} h_0 & h_1 & \cdots & h_n \\ h_1 & h_2 & \cdots & h_{n+1} \\ \vdots & & & \vdots \\ h_n & h_{n+1} & \cdots & h_{2n} \end{bmatrix} = \left[\begin{array}{c|c} \boldsymbol{H}(n) & \boldsymbol{h}(n) \\ \hline \boldsymbol{h}^T(n) & h(2n) \end{array}\right]$$

The reader can verify that

$$\boldsymbol{H}^{-1}(n+1) = \left[\begin{array}{c|c} \boldsymbol{A}(n) & \boldsymbol{b}(n) \\ \hline \boldsymbol{b}^T(n) & \alpha(n) \end{array}\right]$$

where

$$\boldsymbol{A}(n) = \left[\boldsymbol{H}(n) - \frac{1}{h_{2n}}\boldsymbol{h}(n)\boldsymbol{h}^T(n)\right]^{-1}$$

$$\boldsymbol{b}(n) = -\frac{-1}{h_{2n}}\left[\boldsymbol{H} - \frac{1}{h_{2n}}\boldsymbol{h}(n)\boldsymbol{h}^T(n)\right]^{-1} h(n)$$

$$\alpha(n) = \frac{1}{h_{2n}} + \frac{1}{(h_{2n})^2}\boldsymbol{h}^T(n)\left[\boldsymbol{H}(n) - \frac{1}{h_{2n}}\boldsymbol{h}(n)\boldsymbol{h}^T(n)\right]^{-1}\boldsymbol{h}(n)$$

Step #4 Determination of the matrix $\boldsymbol{C}$

In general $\boldsymbol{C}$ is a $m \times n$ matrix. Here we consider the case of $m = 1$.

Looking at the Markov parameters:

$$h_i = \boldsymbol{c}^T\boldsymbol{A}^i\boldsymbol{b}, \quad i = 0, 1, 2, \cdots$$

Rewriting

$$h_i = \left(A^i b\right)^T c, \quad i = 0, 1, \cdots$$

$$b = \begin{bmatrix} 1 \\ 0 \\ 0 \\ \vdots \\ 0 \end{bmatrix}, \quad Ab = \begin{bmatrix} 0 & 0 & \cdots & 0 & -a_n \\ 1 & 0 & \cdots & 0 & -a_{n-1} \\ 0 & 1 & \cdots & 0 & -a_{n-2} \\ \vdots & \vdots & \cdots & \vdots & \vdots \\ 0 & 0 & \cdots & 1 & -a_1 \end{bmatrix} = \begin{bmatrix} 0 \\ 1 \\ 0 \\ \vdots \\ 0 \end{bmatrix}$$

In general

$$A^i b = \begin{bmatrix} 0 \\ \vdots \\ 1 \\ \vdots \\ 0 \end{bmatrix} \quad \leftarrow \quad (i+1)\text{-th row}$$

Hence,

$$\begin{bmatrix} h_0 \\ h_1 \\ \vdots \\ h_{n-1} \end{bmatrix} = \begin{bmatrix} 1 & 0 & \cdots & 0 \\ 0 & 1 & \cdots & 0 \\ \vdots & \vdots & & \vdots \\ 0 & 0 & \cdots & 1 \end{bmatrix} \begin{bmatrix} c_1 \\ c_2 \\ \vdots \\ c_n \end{bmatrix} = \begin{bmatrix} c_1 \\ c_2 \\ \vdots \\ c_n \end{bmatrix}$$

This completes the computation of A, c, and the degree n.

Minimal Realization for Multiple-Input Multiple-Output (MIMO) Systems

Minimal realization problem for MIMO systems is similar to SISO systems.

Step #1 MIMO System Description

$$\dot{x} = Ax + Bu, \quad y = Cx, \quad H(t) = Ce^{At}B = Y(t) \quad \text{(Impulse Response)}$$

$$
\boldsymbol{A} = \begin{bmatrix} 0 & 0 & \cdots & 0 & -a_n \\ 1 & 0 & \cdots & 0 & -a_{n-1} \\ 0 & 1 & \cdots & 0 & -a_{n-2} \\ \vdots & \vdots & & \vdots & \vdots \\ 0 & 0 & \cdots & 1 & -a_1 \end{bmatrix} = \begin{bmatrix} \boldsymbol{e}_2 & \boldsymbol{e}_3 & \cdots & \boldsymbol{e}_{(n-1)} & -\boldsymbol{a} \end{bmatrix}
$$

$$(n \times n)$$

$$
\boldsymbol{B} = \begin{bmatrix} 1 & 0 & \cdots & 0 \\ 0 & 1 & & 0 \\ 0 & 0 & & 0 \\ \vdots & \vdots & & \\ 0 & 0 & & 1 \end{bmatrix} = \begin{bmatrix} \boldsymbol{e}_1 & \boldsymbol{e}_2 & \cdots & \boldsymbol{e}_r \end{bmatrix}
$$

$$(n \times r)$$

$$
\boldsymbol{C} = \begin{bmatrix} c_{11} & c_{12} & \cdots & c_{1n} \\ \vdots & & & \\ c_{m1} & c_{m2} & \cdots & c_{mn} \end{bmatrix} = \begin{bmatrix} \boldsymbol{c}_1^T \\ \vdots \\ \boldsymbol{c}_m^T \end{bmatrix}
$$

$$(m \times n)$$

Markov Parameters matrices are known from impulse response and represent,

$$
\begin{aligned}
&\boldsymbol{Y}(0) = \boldsymbol{CB} \\
&\dot{\boldsymbol{Y}}(0) = \boldsymbol{CAB} \\
&\vdots \\
&\boldsymbol{Y}^{(2n-1)}(0) = \boldsymbol{CA}^{(2n-1)}\boldsymbol{B}
\end{aligned}
\quad , \quad
\begin{aligned}
&\boldsymbol{Y}^{(k)}(0) = \begin{bmatrix} \boldsymbol{y}_1^{(k)}(0) & \cdots & \boldsymbol{y}_r^{(k)}(0) \end{bmatrix} \\
&k = 0, 1, \cdots, (2n-1)
\end{aligned}
$$

$\boldsymbol{Y}^{(n)}(0)$ can be written via the Cayley-Hamilton theorem as:

$$
\boldsymbol{Y}^{(n)}(0) = \boldsymbol{CA}^n\boldsymbol{B} = \boldsymbol{C}\begin{bmatrix} -a_1\boldsymbol{A}^{n-1} & -a_2\boldsymbol{A}^{n-2} & \cdots & -a_n\boldsymbol{I}_n \end{bmatrix}\boldsymbol{B}
$$

$$\begin{bmatrix} \boldsymbol{Y}^{(n)}(0) \\ \boldsymbol{Y}^{(n-1)}(0) \\ \vdots \\ \boldsymbol{Y}^{(2n-1)}(0) \end{bmatrix} = \begin{bmatrix} \boldsymbol{Y}^{(0)}(0) & \boldsymbol{Y}^{(1)}(0) & \cdots & \boldsymbol{Y}^{(n-1)}(0) \\ \boldsymbol{Y}^{(1)}(0) & \boldsymbol{Y}^{(2)}(0) & \cdots & \boldsymbol{Y}^{(n-2)}(0) \\ \vdots & & & \\ \boldsymbol{Y}^{(n-1)}(0) & \boldsymbol{Y}^{(n-2)}(0) & \cdots & \boldsymbol{Y}^{(2n-2)}(0) \end{bmatrix} \begin{bmatrix} -\boldsymbol{I}_r a_n \\ -\boldsymbol{I}_r a_{n-1} \\ \vdots \\ -\boldsymbol{I}_r a_1 \end{bmatrix} \tag{7.129}$$

$\boldsymbol{I}_n$ is $n \times n$ identity matrix and $\boldsymbol{I}_r$ is $r \times r$ identity matrix. $\boldsymbol{Y}^{(0)}(0), \boldsymbol{Y}^{(1)}(0), \cdots \boldsymbol{Y}^{(2n-2)}(0)$ are $m \times r$ matrices. From each of these matrices, select an element from the i-th row and the j-th column. Call this element

$$y_{ij}^{(k)}(0), \quad k = 0, 1, \cdots, 2n-1, \; i = 1, 2, \cdots, m, \; j = 1, 2, \cdots, r$$

Eq. 7.129 yields:

$$\begin{bmatrix} y_{ij}^{(n)}(0) \\ y_{ij}^{(n-1)}(0) \\ \vdots \\ y_{ij}^{(2n-1)}(0) \end{bmatrix} = \begin{bmatrix} y_{ij}^{(0)}(0) & y_{ij}^{(1)}(0) & \cdots & y_{ij}^{(n-1)}(0) \\ y_{ij}^{(1)}(0) & y_{ij}^{(2)}(0) & \cdots & y_{ij}^{(n-2)}(0) \\ \vdots & & & \\ y_{ij}^{(n-1)}(0) & y_{ij}^{(n-2)}(0) & \cdots & y_{ij}^{(2n-2)}(0) \end{bmatrix} \begin{bmatrix} -a_n \\ -a_{n-1} \\ \vdots \\ -a_1 \end{bmatrix} \tag{7.130}$$

The above matrix is a $n \times n$ Hankel matrix of the rank n. Inverting it yields the parameters of the matrix $\boldsymbol{A}$. The solution is valid for any one selection i, j. But in practice, we may have to use greater than one i or j to arrive at some statistically optimum result. Thus in essence, we have determined the matrix $\boldsymbol{A}$ for the MIMO system.

Step #2 Determination of the matrix $\boldsymbol{C}$

Let $\boldsymbol{b}_k$ be a column vector $\boldsymbol{e}_k$. Furthermore,

$$\boldsymbol{A}^i \boldsymbol{b}_k = \boldsymbol{b}_{i+k}$$

$$\boldsymbol{e}_i = \boldsymbol{b}_i$$

$$C = \begin{bmatrix} c_1^T \\ c_2^T \\ \vdots \\ c_m^T \end{bmatrix} = \begin{bmatrix} c_1^T \\ c_2^T \\ \vdots \\ c_m^T \end{bmatrix} \begin{bmatrix} b_1 \; b_2 \; \cdots \; b_m \end{bmatrix} = \begin{bmatrix} c_1^T \\ c_2^T \\ \vdots \\ c_m^T \end{bmatrix} \begin{bmatrix} b_1 \; Ab_1 \; \cdots \; A^{m-1}b_1 \end{bmatrix} \tag{7.131}$$

Furthermore,

$$\begin{bmatrix} c_1^T \\ c_2^T \\ \vdots \\ c_m^T \end{bmatrix} \begin{bmatrix} b_1 \; Ab_1 \; \cdots \; A^{m-1}b_1 \end{bmatrix} = \begin{bmatrix} y_{11}^{(0)}(0) \; y_{12}^{(1)}(0) \; \cdots \; y_{1m}^{(m-1)}(0) \\ y_{21}^{(0)}(0) \; y_{22}^{(1)}(0) \; \cdots \; y_{2m}^{(m-1)}(0) \\ \vdots \\ y_{m1}^{(0)}(0) \; y_{m2}^{(1)}(0) \; \cdots \; y_{mm}^{(m-1)}(0) \end{bmatrix}$$

$$C = \begin{bmatrix} y_{11}^{(0)}(0) \; y_{12}^{(1)}(0) \; \cdots \; y_{1m}^{(m-1)}(0) \\ y_{21}^{(0)}(0) \; y_{22}^{(1)}(0) \; \cdots \; y_{2m}^{(m-1)}(0) \\ \vdots \\ y_{m1}^{(0)}(0) \; y_{m2}^{(1)}(0) \; \cdots \; y_{mm}^{(m-1)}(0) \end{bmatrix}$$

This completes the minimal realization problem.

MIMO Transfer Function Realization in the State Variable Form

Consider a $m \times r$ transfer function matrix $G(s)$ representing a system with r inputs, m outputs, and n states written in the form

$$G(s) = \left(\frac{1}{D(s)}\right) N(s) \tag{7.132}$$

where $D(s)$ is a Least Common Multiple (LCM) polynomial of all the elements of $G(s)$ such that

$$G(s) = \begin{bmatrix} G_{11}(s) & \cdots & G_{1r}(s) \\ \vdots & & \\ G_{m1}(s) & \cdots & G_{mr}(s) \end{bmatrix} = \frac{1}{D(s)} \begin{bmatrix} N_{11}(s) & \cdots & N_{1r}(s) \\ \vdots & & \\ N_{m1}(s) & \cdots & N_{mr}(s) \end{bmatrix} \tag{7.133}$$

$$D(s) = s^n + a_1 s^{n-1} + \cdots + a_{n-1} s + a_n$$

$$N_{ij}(s) = b_{ij}(1)s^{n-1} + b_{ij}(2)s^{n-2} + \cdots + b_{ij}(n)$$

Thus,

$$G_{ij}(s) = \frac{\left[\sum_{k=1}^{n} b_{ij}(k)s^{n-k}\right]}{\left[\sum_{l=0}^{n} a_l s^{n-l}\right]}, \quad a_0 = 1 \tag{7.134}$$

The transfer function matrix $\boldsymbol{G}(s)$ described in the state variable form as:

$$\boldsymbol{G}(s) = \boldsymbol{C}\,(s\boldsymbol{I} - \boldsymbol{A})^{-1}\,\boldsymbol{B}, \quad \boldsymbol{G}_{ij}(s) = \boldsymbol{c}_i^T\,(s\boldsymbol{I} - \boldsymbol{A})^{-1}\,\boldsymbol{b}_j \tag{7.135}$$

We are required to determine the matrices $\boldsymbol{A}, \boldsymbol{B}, \boldsymbol{C}$, given $\boldsymbol{G}(s)$

Step #1 Determination of matrix A

Since the LCM (least common multiple) polynomial denominator $D(s)$ is known, we choose $\boldsymbol{A}$ in the companion matrix form representing the n state variables as:

$$\boldsymbol{A} = \begin{bmatrix} 0 & 1 & 0 & \cdots & 0 \\ 0 & 0 & 1 & \cdots & 0 \\ \vdots & & & & \\ 0 & 0 & 0 & \cdots & 1 \\ -a_n & -a_{n-1} & -a_{n-2} & \cdots & -a_1 \end{bmatrix}$$

Note: As discussed earlier, any square matrix can be converted into the companion form via the Similarity Transformation.

Step #2 Markov Parameter representation of $G_{ij}(s)$

Let

$$(s\boldsymbol{I} - \boldsymbol{A})^{-1} = \sum_{k=0}^{\infty} (1/s)^{k+1}\,\boldsymbol{A}^k$$

Thus,

$$G(s) = C\,(sI - A)^{-1}\,B = \sum_{k=0}^{\infty} (1/s)^{k+1}\,CA^k B$$

$$G_{ij}(s) = \sum_{k=0}^{\infty} (1/s)^{k+1}\,c_i^T A^k b_j \tag{7.136}$$

$$c_i^T A^k b_j = \gamma_{ij}(k) \quad \text{(scaler).}$$

$$G_{ij}(s) = \sum_{k=0}^{\infty} (1/s)^{k+1}\,\gamma_{ij}(k) \tag{7.137}$$

Step #3 Determination of Markov Parameters

From Eqs. 7.134, and 7.137

$$\sum_{k=0}^{\infty} (1/s)^{k+1}\,\gamma_{ij}(k) = \left[\sum_{k=1}^{n-1} \left(b_{ij}(k)\right) s^{n-1-k}\right]\left[\sum_{l=0}^{n} a_l s^{n-l}\right] \tag{7.138}$$

Equating powers of s on each side of Eq. 7.138, we can compute $\gamma_{ij}(k)$ in terms of known parameters $b_{ij}(k)$ and a_l.

We need only compute $\gamma_{ij}(k), k = 0, 1, \cdots, n-1,\ i = 1, 2, \cdots, m,\ j = 1, 2, \cdots, r$. These parameters $\gamma_{ij}(k)$ are referred to as Markov parameters.

Step #4 Determination of $c_i, (1, 2, \cdots, n)$

We shall use only the first column of $G(s)$, namely $G_{i1}(s)$, to determine the vectors c_i. This computation is dependent on our choice of b_1. Let $b_1 = e_1$

The corresponding Markov parameters $\gamma_{i1}(k)$ yield:

$$\begin{bmatrix} b_1^T \\ b_1^T A^T \\ \vdots \\ b_1^T (A^{n-1})^T \end{bmatrix} \begin{bmatrix} \\ c_i \\ \\ \end{bmatrix} = \begin{bmatrix} \gamma_{i1}(0) \\ \gamma_{i1}(1) \\ \vdots \\ \gamma_{i1}(n-1) \end{bmatrix}, \quad i = 1, 2, \cdots, m$$

or

$$c_i = \begin{bmatrix} b_1^T \\ b_1^T A^T \\ \vdots \\ b_1^T (A^{n-1})^T \end{bmatrix}^{-1} \begin{bmatrix} \gamma_{i1}(0) \\ \gamma_{i1}(1) \\ \vdots \\ \gamma_{i1}(n-1) \end{bmatrix}, \quad i = 1, 2, \cdots, m$$

Step #5 Determination of $b_j (j = 2, 3, \cdots, r)$

Having computed $c_i, (i = 1, 2, \cdots, m)$ from Step 4, we shall use the rest of the columns of $G(s)$, namely $G_{ij}(s), i = 1, 2, \cdots, m; j = 2, 3, \cdots, r$, to determine the rest of b_j,

$$\begin{bmatrix} c_i^T \\ c_i^T A \\ \vdots \\ c_i^T A^{n-1} \end{bmatrix} \begin{bmatrix} b_j \end{bmatrix} = \begin{bmatrix} \gamma_{ij}(0) \\ \gamma_{ij}(1) \\ \vdots \\ \gamma_{ij}(n-1) \end{bmatrix}, \quad i = 1, 2, \cdots, m \quad j = 2, 3, \cdots, r$$

or

$$\begin{bmatrix} b_j \end{bmatrix} = \begin{bmatrix} c_i^T \\ c_i^T A \\ \vdots \\ c_i^T A^{n-1} \end{bmatrix}^{-1} \begin{bmatrix} \gamma_{ij}(0) \\ \gamma_{ij}(1) \\ \vdots \\ \gamma_{ij}(n-1) \end{bmatrix}, \quad i = 1, 2, \cdots, m \quad j = 2, 3, \cdots, r$$

This completes the algorithm for transfer function realization.

Bibliography

[Anderson, B.D.] Anderson, B.D.O. and Lin, Y. Controller Reduction: Concepts and Approaches, *IEEE Transactions on Automatic Control*, 34(8), 802–812, 1989.

[Antoulas, A.C.] Antoulas, A.C., Santag E.D. and Yamamato Y. Controllability and Observability, *Wiley Encyclopedia of Electrical and Electronics Engineering,* Edited by J.G. Webster, Vol. 4, 264–281, 2006.

[Bragam, W.L.] Bragam, W.L. *Modern Control Theory,* New York: Prentice Hall Inc., 1991.

[Desoer, C.A.] Desoer, C.A. and Vidyasagar, M. *Feedback Systems: Input-Output Properties,* New York: Academic Press, 1975.

[Desoer, C.A.] Desoer, C.A. *Notes for a Second Course on Linear Systems* New York: Van Nostrand Reinhold Company, 1970.

[Desoer, C.A.] Desoer, C.A. and Kuh, E.E. *Basic Circuit Theory,* New York: McGraw-Hill Book Co., 1966.

[Doyak, J.C.] Doyak, J.C., Francis, B. and Tannenbaum, A. *Feedback Control Theory,* New York: Dover Publications, 2009.

[Kalman, R.E.] Kalman, R.E. Contributions to the Theory of Optimal Control, *Bol. Soc. Mathem.*, 5, 102–119, 1960.

[Foellinger, O.] Foellinger, O. *Regelungs Technik,* Heidelberg: Dr. Alfred Huetig Verlag, ISBN 3-7785-1137-8.

[Moore, B.C.] Moore, B.C. Principle Component Analysis in Linear Systems: Controllability, Observability and Model Reduction, *IEEE Transactions on Automatic Control* AC–26(1), 17–32, 1981.

[Silverman, L.M.] Silverman, L.M. and Anderson, B.D.O Controllability, Observability and Stability of Linear Systems, *Siam J. Control,* 6, 121–129, 1968.

[Sontag, E.D.] Sontag, E.D. *Mathematical Control Theory,* New York: Springer Verlag, 1998.

[VidyaSagar, M.] VidyaSagar, M. *Control Systems Synthesis: A Factorization Approach,* Cambridge, MA: MIT Press, 1985.

Chapter 8

Calculus of Variations

8.1 Introduction

The calculus of variations plays a very important role in the design of optimal control systems that can be stated as an extremal value problem. The purpose of this chapter is to make the reader familiar with the principles of variational calculus and thus prepare him for the synthesis of optimal control systems. In the next section, we introduce preliminaries involving the calculus of maxima and minima and then derives the various results in the calculus of variations.

8.2 Maxima, Minima, and Stationary Points

1. **Extrema of a function of a single variable**

   Given: A scalar function $V = f(y)$ of a single variable y.

   The extrema points of a function are defined as those where its slope vanishes:

$$\frac{\mathrm{d}V}{\mathrm{d}y} = \frac{\mathrm{d}}{\mathrm{d}y} f(y) = 0$$

Let $y = y^*$ be one of the extrema points. Assuming all the necessary derivatives exist and are continous, we can expand $f(y)$ in the Taylor series about y^*:

$$\Delta V = \Delta f(y) = f(y^* + \Delta y) - f(y^*) = \left.\frac{\mathrm{d}f}{\mathrm{d}y}\right|_{y=y^*} \Delta y + \frac{1}{2}\left.\frac{\mathrm{d}^2 f}{\mathrm{d}y^2}\right|_{y=y^*} \Delta y^2 + \text{H.O.T.} \quad (8.1)$$

Neglecting higher order terms

$$\Delta V = \Delta f(y) \approx \left.\frac{\mathrm{d}f}{\mathrm{d}y}\right|_{y=y^*} \Delta y + \frac{1}{2}\left.\frac{\mathrm{d}^2 f}{\mathrm{d}y^2}\right|_{y=y^*} \Delta y^2$$

The first term is called the "**First Variation**" δf or the "variation" and the second term is called the **Second Variation**. At the extrema point $y = y^*$ it is necessary that the first variation, $\delta f = \frac{\mathrm{d}f}{\mathrm{d}y}\Delta y$, vanishes for an arbitrarily small Δy. Thus,

$$\frac{\mathrm{d}f}{\mathrm{d}y} = 0 \quad \text{at} \quad y = y^* \qquad \textbf{Necessary condition for an extrema}$$

Change in $f(y)$ in the extremal point neighborhood is approximated by the $(\Delta y)^2$ term. The classification of the extremals is given by the following:

$$\left.\frac{\mathrm{d}f}{\mathrm{d}y}\right|_{y=y^*} = 0 \quad \text{and} \quad \left.\frac{\mathrm{d}^2 f}{\mathrm{d}y^2}\right|_{y=y^*} \begin{cases} > 0, & \text{then } f(y) \text{ has a local minimum} \\ < 0, & \text{then f(y) has a local maximum} \\ = 0, & \text{then } f(y) \text{ has a "saddle" point} \end{cases}$$

Example 8.1:

Find the extremal and its classification for

$$f(y) = \tan^{-1} y - \tan^{-1} ky \quad ,0 < k < 1$$

Taking the derivative,

$$\frac{\mathrm{d}f}{\mathrm{d}y} = \frac{\mathrm{d}}{\mathrm{d}y}\tan^{-1} y - \frac{\mathrm{d}}{\mathrm{d}y}\tan^{-1} ky = \frac{1}{1+y^2} - \frac{k}{1+(ky)^2} = 0$$

or

$$(1 + (ky)^2) - k(1 + y^2) = 0$$

or

$$y = y^* = \sqrt{\frac{1}{k}}$$

Taking the second derivative at $y = y^*$

$$\frac{\mathrm{d}^2 f}{\mathrm{d}y^2} = \left[\frac{-2y}{(1+y^2)^2} + \frac{2k^3 y}{(1+k^2y^2)^2}\right]\Bigg|_{x=\sqrt{\frac{1}{k}}} = \frac{-2k^{3/2}(1-k)}{(1+k)^2}$$

For the given k,

$$\frac{\mathrm{d}^2 f}{\mathrm{d}y^2} < 0 \text{ yielding a maximum at } y^* = \sqrt{\frac{1}{k}}$$

2. **Extrema of a function of several variables.**

Given a scalar function of several variables $y_1, y_2, \ldots, y_l$. Let

$$V = f(\boldsymbol{y}) = f(y_1, y_2, \ldots, y_l)$$

$$[\, y_1 \; y_2 \; \cdots \; y_l \,]^T \qquad l \times 1 \text{ vector}$$

Let us assume that this function $f(\boldsymbol{y})$ has an extremal value at $\boldsymbol{y} = \boldsymbol{y}^*$.

This extremal value is obtained by perturbing $\boldsymbol{y} = \boldsymbol{y}^*$ to $\boldsymbol{y} = \boldsymbol{y}^* + \triangle\boldsymbol{y}$ and obtaining a variational equation,

$$\delta V = \left(\boldsymbol{\nabla}_{\boldsymbol{y}} f\right)^T \cdot \triangle\boldsymbol{y} = 0 \tag{8.2}$$

$\left(\nabla_y f\right)$ is defined as the gradient of the scaler function, f with respect to $\boldsymbol{y}$.

$$\left(\nabla_y f\right) = \left[\frac{\partial f}{\partial y_1}\ \frac{\partial f}{\partial y_2} \cdots \frac{\partial f}{\partial y_l}\right]^T \quad \text{(Column vector)} \tag{8.3}$$

"T" denotes the transpose of the row vector.

Since $\triangle \boldsymbol{y}$ is arbitrary, let us choose $\triangle \boldsymbol{y}$ in the same direction as the vector $(\nabla_y f)$ yielding

$$\delta V = \left|\nabla_y f\right|^T \cdot |\triangle \boldsymbol{y}| = 0$$

The only way this can be true for any arbitrary value of $\triangle \boldsymbol{y}$ is when

$$\left(\nabla_y f\right)\Big|_{\boldsymbol{y}=\boldsymbol{y}^*} = \boldsymbol{0} \tag{8.4}$$

Thus, the necessary conditions for an extremal are

$$\frac{\partial}{\partial y_i} f\Big|_{y_i=y_i^*} = 0, \qquad i = 1, 2, \ldots, l \tag{8.5}$$

The nature of the extremal (minimum, maximum, or a point of inflection [saddle point]) at $\boldsymbol{y} = \boldsymbol{y}^*$ can be determined by looking at the second partial derivatives

$$\frac{\partial^2}{\partial y_i\, \partial y_j} V = V_{y_i,y_j} \quad \text{at the point} \quad \boldsymbol{y} = \boldsymbol{y}^*$$

involving the following partial derivative matrix:

$$V_{\boldsymbol{yy}} = \boldsymbol{F} = \begin{bmatrix} V_{y_1,y_1} & V_{y_1,y_2} & \cdots & V_{y_1,y_l} \\ V_{y_2,y_1} & V_{y_2,y_2} & \cdots & V_{y_2,y_l} \\ \vdots & \vdots & & \vdots \\ V_{y_l,y_1} & V_{y_l,y_2} & \cdots & V_{y_l,y_l} \end{bmatrix} \tag{8.6}$$

The above matrix is called **Hessian**.

When computed at $\boldsymbol{y} = \boldsymbol{y}^*$, $\quad \boldsymbol{F} = \boldsymbol{F}^* = V^*_{\boldsymbol{yy}}$

(a) is positive definite when $\boldsymbol{y}^*$ is a minimum.

(b) is negative definite when $\boldsymbol{y}^*$ is a maximum.

(c) is neither positive definite nor negative definite when $\boldsymbol{y}^*$ is a point of inflection implying no maxima or a minima.

(d) when $\boldsymbol{y}^*$ is a stationary point then any of the determinants of the diagonal minors of $\boldsymbol{F}$ are equal to zero.

$\boldsymbol{F}^*$ is positive definite if all the following determinants computed at $\boldsymbol{y} = \boldsymbol{y}^*$ are positive.

$$\frac{\partial^2 V}{\partial y_1^2} = V_{y_1 y_1} > 0$$

$$\begin{vmatrix} V_{y_1 y_1} & V_{y_1 y_2} \\ V_{y_2 y_1} & V_{y_2 y_2} \end{vmatrix} > 0$$

$$\vdots$$

$$\begin{vmatrix} V_{y_1 y_1} & V_{y_1 y_2} & \cdots & V_{y_1 y_l} \\ V_{y_2 y_1} & V_{y_2 y_2} & \cdots & V_{y_2 y_l} \\ \vdots & & & \\ V_{y_l y_1} & V_{y_l y_2} & \cdots & V_{y_l y_l} \end{vmatrix} > 0 \tag{8.7}$$

If any of the above determinants computed at $\boldsymbol{y} = \boldsymbol{y}^*$ vanish, we have a semidefinite matrix. If the above determinants computed at $\boldsymbol{y} = \boldsymbol{y}^*$ in Eq. 8.7 are all negative, we have a negative definite matrix. If some determinants are positive and the others are negative then we have an indefinite matrix.

Example 8.2:

Determine the extremal of the following function of two variables:

$$V = f(y) = \frac{1}{(y_1 - 1)^2 + (y_2 - 1)^2 + k^2}$$

Solution:

$$f_{y_1} = \frac{\partial f}{\partial y_1} = \frac{(-2)(y_1 - 1)}{[(y_1 - 1)^2 + (y_2 - 1)^2 + k]^2} = 0$$

$$f_{y_2} = \frac{\partial f}{\partial y_2} = \frac{(-2)(y_2 - 1)}{(y_1 - 1)^2 + (y_2 - 1)^2 + k^2} = 0$$

Solving the above equations simultaneously, the extremal occurs at:

$$y_1^* = 1, \quad y_2^* = 1$$

Furthermore,

$$f_{y_1y_1} = \left.\frac{\partial^2 f}{\partial y_1 \partial y_1}\right|_{\substack{y_1^*=1 \\ y_2^*=1}} = \frac{-2}{k^4}$$

$$f_{y_1y_2} = \left.\frac{\partial^2 f}{\partial y_1 \partial y_2}\right|_{\substack{y_1^*=1 \\ y_2^*=1}} = 0$$

$$f_{y_2y_2} = \left.\frac{\partial^2 f}{\partial y_2 \partial y_2}\right|_{\substack{y_1^*=1 \\ y_2^*=1}} = \frac{-2}{k^2}$$

From the second derivatives we conclude that $y_1^* = y_2^* = 1$ yields a maximum.

Note: Variational equations are always computed at the extremal value $\boldsymbol{y} = \boldsymbol{y}^*$.

8.2.1 Extremal of a Function Subject to Single Constraint

Consider the extremal of a function $V_1 = f(\boldsymbol{y})$

$$\text{subject to constraint,} \quad V_2 = g(\boldsymbol{y}) = 0 \tag{8.8}$$

V_1 and V_2 are both scalar. We form a new augmented scalar function V,

$$V = V_1 + V_2 = f(\boldsymbol{y}) + \lambda g(\boldsymbol{y}) \tag{8.9}$$

where λ is an unknown variable (often referred to as the **Lagrange multiplier**), to be determined along with the variables $\boldsymbol{y}$. The extremal value of the augmented function V is given by the variational equation

$$\delta V = \left(\nabla_y (f + \lambda g)\right)^T \cdot \Delta \boldsymbol{y} = 0 \tag{8.10}$$

The $l + 1$ conditions for the extremal are

$$\frac{\partial}{\partial y_i}(f + \lambda g) = \frac{\partial}{\partial y_i} f + \lambda \frac{\partial g}{\partial y_i} = 0 \quad \text{at } y_i = y_i^* \quad i = 1, 2, \ldots, l \tag{8.11}$$

and

$$\text{Constraint equation } g(\boldsymbol{y}) = 0 \tag{8.12}$$

The resultant $(l + 1)$ equations are solved simultaneously to obtain the extremal coordinates $y_1^*, y_2^*, \ldots, y_l^*$ and λ.It should be emphasized that we have considered λ to be a independent Lagrange multiplier variable. In the later sections involving optimal control these Lagrange multiplier variables may be given an interesting interpretation of steering control functions. In many cases the proper formulation of an extremal problem automatically leads us to the correct solution.

Example 8.3:

Planimetric Problem of Johannes Kepler

Find the maximal area of a rectangle whose vertices lie on a unit circle.

Solution:

Let y_1, y_2 be the coordinates of the vertex of the rectangle in the first quadrant. It is easy to see that if all the vertices (y_1, y_2), $(y_1, -y_2)$, $(-y_1, y_2)$, and $(-y_1, -y_2)$ lie on the circumference of the unit circle, the area of the rectangle is $4y_1y_2$.

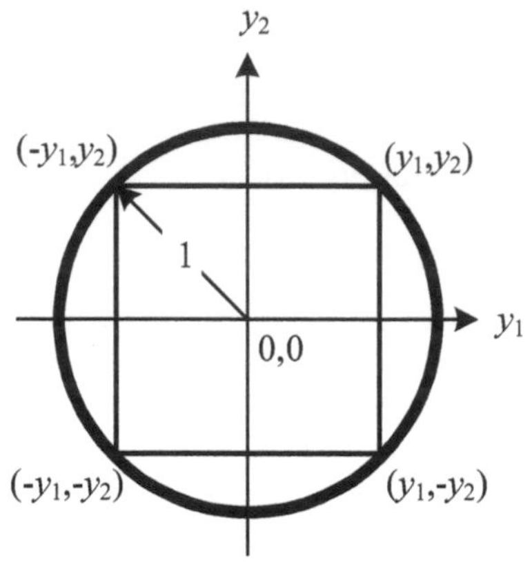

Figure 8.1: Maximum Area Rectangle inside a Circle—Kepler's Problem

Thus, we are looking at the extremal of a scalar function.

$$V_1(\mathbf{y}) = 4y_1y_2, \qquad (y_1 \geq 0, y_2 \geq 0, \text{ obvious constraint})$$

subject to the constraint $g(\mathbf{y}) = y_1^2 + y_2^2 - 1 = 0$

$$V(\mathbf{y}) = 4y_1y_2 + \lambda(y_1^2 + y_2^2 - 1)$$

The variational equations are:

$$\frac{\partial V}{\partial y_1} = 4y_2 + 2\lambda y_1 = 0$$
$$\frac{\partial V}{\partial y_2} = 4y_1 + 2\lambda y_2 = 0$$
$$\frac{\partial V}{\partial \lambda} = y_1^2 + y_2^2 - 1 = 0$$

The constrained extremal solution is,

$$y_1^* = y_2^* = \frac{1}{\sqrt{2}}, \qquad \lambda^* = 2$$

The reader can test the second derivatives and verify that indeed this is a maximal solution. The above problem can be generalized as follows:

Find the extremal value of n-variables with positive values,

$$V_1(\mathbf{y}) = k\prod_{i=1}^{n} y_i$$

$$\text{subject to the condition,}\quad g(\mathbf{y}) = \mathbf{y}^T\mathbf{y} - c^2 = \sum_{i=1}^{n} y_i^2 - c^2 = 0$$

Example 8.4:

For a given surface area k^2, inscribe a cylinder of maximum volume.

Let

$r = y_1 =$ radius of the cylinder, $\quad y_1 > 0$

$l = y_2 =$ length of the cylinder, $\quad y_2 > 0$

$V_1(\mathbf{y}) =$ Volume of the cylinder $= \pi r^2 l = \pi y_1^2 y_2$

Constraint: Surface area of the cylinder $= 2(\pi r^2) + 2\pi r l = 2\pi y_1^2 + 2\pi y_1 y_2 = k^2$

$$\text{Optimizing function}\qquad V_1(\mathbf{y}) = \pi y_1^2 y_2$$

$$\text{Constraint}\qquad g(\mathbf{y}) = (2\pi y_1^2 + 2\pi y_1 y_2 - k^2) = 0$$

$$V(\mathbf{y}) = \pi y_1^2 y_2 + \lambda(2\pi y_1^2 + 2\pi y_1 y_2 - k^2)$$

The variational equations are

$$\frac{\partial V}{\partial y_1} = 2\pi y_1 y_2 + \lambda(4\pi y_1 + 2\pi y_2) = 0$$
$$\frac{\partial V}{\partial y_2} = \pi y_1^2 + \lambda(2\pi y_1) = 0$$
$$\frac{\partial V}{\partial \lambda} = 2\pi y_1^2 + 2\pi y_1 y_2 - k^2 = 0$$

The extremal solution yields

$$y_1 = \frac{k}{\sqrt{6\pi}}, \quad y_2 = k\sqrt{\frac{2}{3\pi}}, \quad \lambda = \frac{k}{2\sqrt{6\pi}}$$

8.2.2 Extremal of a Function Subject to Multiple Constraints

Consider the extremal for a scalar function

$$V_1(\boldsymbol{y}) = f(\boldsymbol{y}) \tag{8.13}$$

subject to multiple constraints

$$\boldsymbol{g}(\boldsymbol{y}) = \boldsymbol{0} \tag{8.14}$$

where $\boldsymbol{g} = \boldsymbol{g}(\boldsymbol{y})$ is a m-dimensional vector with components $g_1, g_2, \ldots, g_m$.

Let us form the augmented scalar function

$$V = f + \boldsymbol{\lambda}^T \cdot \boldsymbol{g} \tag{8.15}$$

where $\boldsymbol{\lambda}$ is a m-dimensional vector and $\boldsymbol{y}$ is l-dimensional.

The first variational equation of V is

$$\delta V = \left(\nabla_y \left(f + \boldsymbol{\lambda}^T \cdot \boldsymbol{g}\right)\right)^T \cdot \Delta \boldsymbol{y} = 0 \tag{8.16}$$

From Eq. 8.14 and Eq. 8.16 we have

$$\begin{aligned} &g_i(\boldsymbol{y}) = 0 && i = 1, 2, \ldots, m \\ &\frac{\partial}{\partial y_j} f + \boldsymbol{\lambda}^T \cdot \frac{\partial}{\partial y_j} \boldsymbol{g} = 0 && j = 1, 2, \ldots, l \end{aligned} \tag{8.17}$$

Eq. 8.17 represents $(l + m)$ equations to be solved simultaneously.

Example 8.5:

Minimum Norm Solution. Let $\boldsymbol{A}$ be $m \times n$ with rank $\boldsymbol{A} = m < n$.

Find the minimum norm solution of the equation $\boldsymbol{Ay} = \boldsymbol{b}$.

Solution: This problem can be formulated as a minimization problem:

Minimize

$$V_1(\boldsymbol{y}) = \frac{1}{2}\boldsymbol{y}^T\boldsymbol{y}$$

Subject to the constraints, $\boldsymbol{g}(\boldsymbol{y}) = (\boldsymbol{Ay} - \boldsymbol{b}) = \boldsymbol{0}$

Form the augmented function, $V(\boldsymbol{y}) = \frac{1}{2}\boldsymbol{y}^T\boldsymbol{y} + \boldsymbol{\lambda}^T(\boldsymbol{Ay} - \boldsymbol{b})$

Extremal Conditions yield:

$$\begin{aligned} \boldsymbol{y} &= -\boldsymbol{A}^T\boldsymbol{\lambda} \\ \boldsymbol{\lambda}^* &= -(\boldsymbol{AA}^T)^{-1}\boldsymbol{b} \\ \boldsymbol{y}^* &= \boldsymbol{A}^T(\boldsymbol{AA}^T)^{-1}\boldsymbol{b} \end{aligned}$$

Note: In what follows the variable $\boldsymbol{y}$ may be function of another independent variable t (time, distance etc). We shall use $\delta\boldsymbol{y}$ as the first variation of $\boldsymbol{y}$ ignoring it as a variable of t and $\mathrm{d}\boldsymbol{y}$ as thc total variation of $\boldsymbol{y}$ considering it's time dependence.

8.3 Definite Integral Extremal (Functional)—Euler-Lagrange, Variable Endpoints

Let us find the extremal of an integral (called a **Functional**)

$$J = \int_{t_1}^{t_2} f_0(t, \boldsymbol{y}, \dot{\boldsymbol{y}})\ \mathrm{d}t \qquad t_1 \le t \le t_2 \tag{8.18}$$

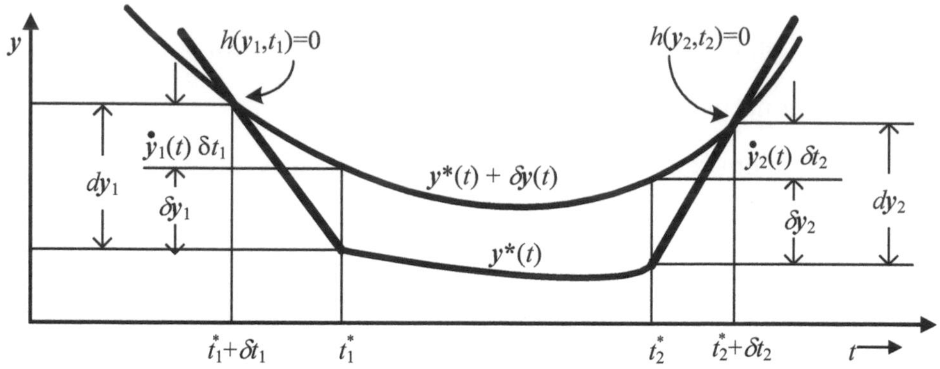

Figure 8.2: Extremal Curve for a Definite Integral

$$\left.\begin{aligned} h_1(\mathbf{y}_1, t_1) &= 0 \\ h_2(\mathbf{y}_2, t_2) &= 0 \end{aligned}\right] \quad \text{Boundary points}$$

subject to the dynamics

$$\dot{\mathbf{y}} = \frac{\mathrm{d}}{\mathrm{d}t}\mathbf{y}\ t, \text{ is an independent variable.} \tag{8.19}$$

We are supposed to find $\mathbf{y}$ as a function of t so that J takes extremal values. The symbol δ represents the first variation of quantities such as $J, \mathbf{y}$, and t. This first variation is assumed infinitesimal. Thus if J is replaced with $J^* = J + \delta J$, then $\left|\frac{\delta J}{J}\right| << 1$

Let

$$\left.\begin{aligned} t^* &= t + \delta t \\ t_1^* &= t_1 + \delta t_1 \\ t_2^* &= t_2 + \delta t_2 \\ \mathbf{y}^* &= \mathbf{y} + \delta \mathbf{y} \\ \dot{\mathbf{y}}^* &= \frac{\mathrm{d}}{\mathrm{d}t}(\mathbf{y} + \delta \mathbf{y}) = \dot{\mathbf{y}} + \delta \dot{\mathbf{y}} \end{aligned}\right\} \tag{8.20}$$

The quantities t_1 and t_2 may be fixed or allow variation.

"$*$" is referred to as extremal value.

The first variational expression of Eq. 8.18 is

$$\delta J = \int_{t_1^*}^{t_2^*} f_0\,(t, \boldsymbol{y}^*, \dot{\boldsymbol{y}}^*)\ \mathrm{d}t - \int_{t_1}^{t_2} f_0\,(t, \boldsymbol{y}, \dot{\boldsymbol{y}})\ \mathrm{d}t \tag{8.21}$$

The endpoints $(\boldsymbol{y}_1, t_1)$ and $(\boldsymbol{y}_2, t_2)$ lie on the extremal curves

$$\begin{aligned} h_1(\boldsymbol{y}_1, t_1) &= 0 \\ h_2(\boldsymbol{y}_2, t_2) &= 0 \end{aligned} \qquad \text{as shown in Figure 8.2}$$

It should be noted that in Eq. 8.21, t represents a dummy variable.

Now

$$\begin{aligned} \int_{t_1^*}^{t_2^*} f_0\,(t, \boldsymbol{y}^*, \dot{\boldsymbol{y}}^*)\ \mathrm{d}t &= \int_{t_1+\delta t_1}^{t_2+\delta t_2} f_0\,(t, \boldsymbol{y}^*, \dot{\boldsymbol{y}}^*)\ \mathrm{d}t \\ &= \int_{t_1}^{t_2} f_0\,(t, \boldsymbol{y}^*, \dot{\boldsymbol{y}}^*)\ \mathrm{d}t + \int_{t_2}^{t_2+\delta t_2} f_0\,(t, \boldsymbol{y}^*, \dot{\boldsymbol{y}}^*)\ \mathrm{d}t - \int_{t_1}^{t_1+\delta t_1} f_0\,(t, \boldsymbol{y}^*, \dot{\boldsymbol{y}}^*)\ \mathrm{d}t \end{aligned}$$

If δt_1 and δt_2 are considered small, we can write the first order approximation as

$$\int_{t_2}^{t_2+\delta t_2} f_0\,(t, \boldsymbol{y}^*, \dot{\boldsymbol{y}}^*)\ \mathrm{d}t - \int_{t_1}^{t_1+\delta t_1} f_0\,(t, \boldsymbol{y}^*, \dot{\boldsymbol{y}}^*)\ \mathrm{d}t \approx [f_0\,(t, \boldsymbol{y}^*, \dot{\boldsymbol{y}}^*)]\,[\delta t_2 - \delta t_1] \tag{8.22}$$

Thus,

$$\delta J = \int_{t_1}^{t_2} [f_0\,(t, \boldsymbol{y}^*, \dot{\boldsymbol{y}}^*) - f_0\,(t, \boldsymbol{y}, \dot{\boldsymbol{y}})]\ \mathrm{d}t + [f_0^*\ \delta t]\Big|_1^2$$

or

$$\delta J = \int_{t_1}^{t_2} \delta f_0^*\ \mathrm{d}t + [f_0^*\ \delta t]\Big|_1^2 \tag{8.23}$$

where

$$\delta f_0 = f(t, \mathbf{y}^*, \dot{\mathbf{y}}^*) - f(t, \mathbf{y}, \dot{\mathbf{y}})$$

From the chain rule

$$\delta f_0 = \left(\nabla_{\mathbf{y}} f_0\right)^T \cdot \delta \mathbf{y} + \left(\nabla_{\dot{\mathbf{y}}}\right)^T \cdot \delta \dot{\mathbf{y}} \tag{8.24}$$

From Eq. 8.23 and Eq. 8.24

$$\delta J = \int_{t_1}^{t_2} \left[\left(\nabla_{\mathbf{y}}\right)^T \cdot \delta \mathbf{y} + \left(\nabla_{\dot{\mathbf{y}}} f_0\right)^T \cdot \delta \dot{\mathbf{y}}\right]\Bigg|_{\substack{\mathbf{y}=\mathbf{y}^* \\ \dot{\mathbf{y}}=\dot{\mathbf{y}}^*}} \mathrm{d}t + \left[f_0^* \, \delta t\right]\Bigg|_1^2 \tag{8.25}$$

Furthermore,

$$\left(\nabla_{\dot{\mathbf{y}}} f_0\right)^T \cdot \delta \dot{\mathbf{y}} = \frac{\mathrm{d}}{\mathrm{d}t}\left[\left(\nabla_{\dot{\mathbf{y}}} f_0\right)^T \cdot \delta \mathbf{y}\right] - \frac{\mathrm{d}}{\mathrm{d}t}\left[\left(\nabla_{\dot{\mathbf{y}}} f_0\right)^T\right] \cdot \delta \mathbf{y} \tag{8.26}$$

The derivative $\delta \dot{\mathbf{y}}$ can be eliminated from Eq. 8.26 to yield

$$\delta J = \int_{t_1}^{t_2} \left\{\left[\frac{\mathrm{d}}{\mathrm{d}t}\left(\nabla_{\dot{\mathbf{y}}} f_0\right)^T - \left(\nabla_{\mathbf{y}} f_0\right)^T\right] \cdot \delta \mathbf{y}\right\} \mathrm{d}t + \left[f_0^* \, \delta t + \left(\nabla_{\dot{\mathbf{y}}} f_0\right)^T \cdot \delta \mathbf{y}\right]\Bigg|_1^2 \tag{8.27}$$

The above expression is computed at $\mathbf{y} = \mathbf{y}^*, \dot{\mathbf{y}} = \dot{\mathbf{y}}^*$.

From here onwards without any possibility of confusion we shall assume that all the derivatives are computed at the extremal values $\mathbf{y} = \mathbf{y}^*, \dot{\mathbf{y}} = \dot{\mathbf{y}}^*$.

For J to be extremum, it is evident that for any arbitrary variation in $\delta \mathbf{y}$, the variation $\delta J \to 0$ when

$$\frac{\mathrm{d}}{\mathrm{d}t}\left(\nabla_{\dot{\mathbf{y}}} f_0\right) - \left(\nabla_{\mathbf{y}} f_0\right) = \mathbf{0} \tag{8.28}$$

and

$$\left[f_0 \, \delta t + \left(\nabla_{\dot{\mathbf{y}}} f_0\right)^T \cdot \delta \mathbf{y}\right]\Bigg|_1^2 = 0 \tag{8.29}$$

If the endpoints are variable then the total variation due to time and the trajectory variation is $\mathrm{d}\boldsymbol{y}$ of $\boldsymbol{y}$, where

$$\mathrm{d}\boldsymbol{y} = \boldsymbol{y}^*(t^*) - \boldsymbol{y}(t) \tag{8.30}$$

$$\boldsymbol{y}^*(t) = \boldsymbol{y}(t) + \delta\boldsymbol{y} \qquad t^* = t + \delta t \qquad (\mathrm{d}t \equiv \delta t) \tag{8.31}$$

Thus,

$$\mathrm{d}\boldsymbol{y} = \boldsymbol{y}^*(t + \delta t) - \boldsymbol{y}(t) \approx \boldsymbol{y}^*(t) + \dot{\boldsymbol{y}}^*\delta t - \boldsymbol{y}(t) \approx \delta\boldsymbol{y} + \dot{\boldsymbol{y}}\delta t \tag{8.32}$$

Hence, from Eqs. 8.30, 8.31, and 8.32, Eq. 8.29 can be rewritten in the final form

$$\left[\left(f_0 - \left(\nabla_{\dot{y}} f_0\right)^T \cdot \dot{\boldsymbol{y}}\right)\delta t + \left(\nabla_{\dot{y}} f_0\right)^T \cdot \mathrm{d}\boldsymbol{y}\right]\Bigg|_1^2 = 0 \tag{8.33}$$

Equations 8.28 are the famous **Euler-Lagrange equations** and represent the necessary conditions for the extremum. Equations 8.29 or 8.33 are known as transversality conditions and are automatically satisfied if f_0 is fixed at the boundaries. Summarizing the Euler-Lagrange equations are

$$\left.\begin{aligned} &\frac{\mathrm{d}}{\mathrm{d}t}\left(\nabla_{\dot{y}} f_0\right) - \left(\nabla_{y} f_0\right) = \mathbf{0} \\ &\frac{\mathrm{d}}{\mathrm{d}t}\left(\frac{\partial f_0}{\partial \dot{y}_i}\right) - \frac{\partial}{\partial y_i} f_0 = 0 \end{aligned}\right\} \quad \text{Euler-Lagrange} \tag{8.34}$$

and

$$\left.\begin{aligned} &\left[f_0\,\delta t + \left(\nabla_{\dot{y}} f_0\right)^T \cdot \delta\boldsymbol{y}\right]\Bigg|_1^2 = 0 \\ &\left[f_0\,\delta t + \sum_{i=1}^{l} \delta y_i \frac{\partial}{\partial \dot{y}_i} f_0\right]\Bigg|_1^2 = 0 \end{aligned}\right\} \quad \begin{gathered}\text{Transversality}\\ \text{conditions}\\ \text{at the boundary}\end{gathered} \tag{8.35}$$

$$\mathrm{d}\boldsymbol{y} = \delta\boldsymbol{y} + \dot{\boldsymbol{y}}\delta t$$

or

$$\mathrm{d}y_i = \delta y_i + \dot{y}_i\delta t \qquad i = 1, 2, \ldots, l$$

Alternate Form of the Euler-Lagrange Equations

Using the chain-rule for differentiation

$$\frac{\mathrm{d}}{\mathrm{d}t}f_0 - \frac{\partial}{\partial t}f_0 = \left(\nabla_y f_0\right)^T \cdot \dot{y} + \left(\nabla_{\dot{y}} f_0\right)^T \cdot \ddot{y} \tag{8.36}$$

$$\frac{\mathrm{d}}{\mathrm{d}t}\left[\left(\nabla_{\dot{y}} f_0\right)^T \cdot \dot{y}\right] = \frac{\mathrm{d}}{\mathrm{d}t}\left(\nabla_{\dot{y}} f_0\right)^T \cdot \dot{y} + \left(\nabla_{\dot{y}} f_0\right)^T \cdot \ddot{y} \tag{8.37}$$

Eq. 8.37 can be combined with Eq. 8.28 to give

$$\frac{\mathrm{d}}{\mathrm{d}t}\left[\left(\nabla_{\dot{y}} f_0\right)^T \cdot \dot{y}\right] = \left(\nabla_y f_0\right)^T \cdot \dot{y} + \left(\nabla_{\dot{y}} f_0\right)^T \cdot \ddot{y} \tag{8.38}$$

Comparing Eq. 8.36 and Eq. 8.38, we obtain

$$\frac{\mathrm{d}}{\mathrm{d}t}f_0 - \frac{\partial}{\partial t}f_0 = \frac{\mathrm{d}}{\mathrm{d}t}\left[\left(\nabla_{\dot{y}} f_0\right)^T \cdot \dot{y}\right] \tag{8.39}$$

Hence, an alternate form for the Euler-Lagrange equation is

$$\frac{\mathrm{d}}{\mathrm{d}t}\left(f_0 - \left(\nabla_{\dot{y}} f_0\right)^T \cdot \dot{y}\right) = \frac{\partial}{\partial t}f_0 \tag{8.40}$$

Example 8.6:

Johann Bernoulli's Brachistochrone Problem (1696)

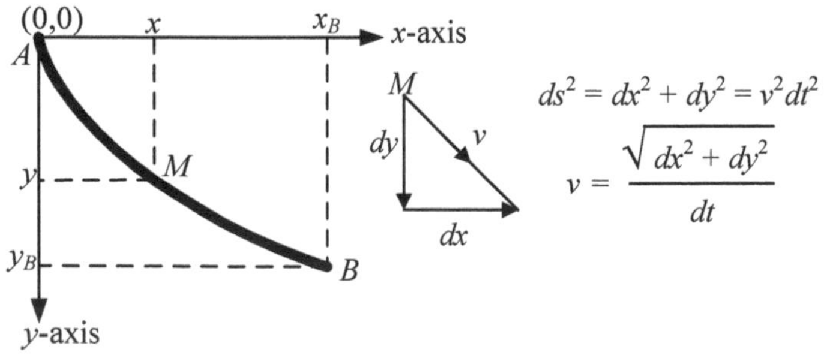

Figure 8.3: Brachistochrone Problem

Find the curve joining the two given points in a vertical plane such that a particle travels from the point A to the point B in the shortest possible time under gravity (assuming no friction).

Solution:

From Figure 8.3

$$\text{Kinetic energy of the particle } T = \frac{1}{2}mv^2$$
$$\text{Potential energy of the particle } V = mgy$$

As the particle moves to lower heights, the potential energy gets converted into the kinetic energy, yielding

$$T = V$$

or

$$\frac{1}{2}mv^2 = mgy$$

Thus,

$$v = \sqrt{2gy} = \frac{\mathrm{d}s}{\mathrm{d}t}$$

Furthermore,

$$\mathrm{d}^2s = \mathrm{d}^2x + \mathrm{d}^2y$$

Thus,

$$\mathrm{d}t = \frac{\mathrm{d}s}{v} = \frac{\sqrt{\mathrm{d}x^2 + \mathrm{d}y^2}}{\sqrt{2gy}} = \frac{\sqrt{1+\dot{y}^2}}{\sqrt{2gy}}, \quad \dot{y} = \frac{\mathrm{d}y}{\mathrm{d}x}$$

Thus, the problem is to find a curve $y(x)$ so as to achieve the minimum of

$$T = J(y, \dot{y}) = \int_0^T \mathrm{d}t = \int_0^{x_B} \left(\frac{\sqrt{1+\dot{y}^2}}{\sqrt{2gy}} \right) \mathrm{d}x \tag{8.41}$$

$$y(0) = 0, \quad y(x_B) = y_B$$

Note: This problem is analogous to the trajectory of a light rays passing between two points in a medium in the shortest possible time.

The alternate Euler-Lagrange equation yields:

$$\frac{\mathrm{d}}{\mathrm{d}t}\left[\frac{\sqrt{(1+\dot{y}^2)}}{\sqrt{2gy}} - \frac{\dot{y}^2}{\sqrt{2gy}\sqrt{1+\dot{y}^2}}\right] = 0, \quad \dot{y} = \frac{\mathrm{d}y}{\mathrm{d}x}$$

Simplifying the above equation:

$$\frac{\mathrm{d}}{\mathrm{d}t}\left[\frac{1}{\left(\sqrt{2gy}\right)\left(\sqrt{1+\dot{y}^2}\right)}\right] = 0$$

Integrating

$$\left(\sqrt{2gy}\right)\left(\sqrt{1+\dot{y}^2}\right) = c^2$$

Thus,

$$\dot{y} = \frac{\mathrm{d}y}{\mathrm{d}x} = \sqrt{\frac{k-y}{y}}$$

This is the differential equation of a cycloid. With some algebraic manipulations,

$$x = \frac{k}{2}(t - \sin t) + k_1, \quad y = \frac{k}{2}(1 - \cos t)$$

where t is a paramter in this cycloid equation.

Example 8.7:

$$J = \int_a^b f(y, \dot{y}, \ddot{y}, \ldots, y^{(n)}, t)\,\mathrm{d}t \quad \text{(Initial and terminal points } y(a), y(b) \text{ are fixed)}$$

The necessary conditions for the minimum is:

$$\frac{\delta f}{\delta y} - \frac{\mathrm{d}}{\mathrm{d}t}\frac{\delta f}{\delta \dot{y}} + \frac{\mathrm{d}^2}{\mathrm{d}t^2}\frac{\delta f}{\delta \ddot{y}} - \frac{\mathrm{d}^3}{\mathrm{d}t^3}\frac{\delta f}{\delta \dddot{y}} + \ldots = 0$$

or

$$\sum_{i=0}^{n}(-1)^i\frac{d^i}{dt^i}\left(\frac{\partial}{\partial y^{(i)}}f\right)=0 \qquad \text{General Euler-Lagrange Equation}$$

Example 8.8:

Determine the minimum of

$$J=\int_{t_1}^{t_2}L(y,\dot{y},t)\,dt$$

where

$$L = \text{Lagrangian} = \frac{1}{2}p(t)\dot{y}^2+\frac{1}{2}q(t)y^2+f(t)y$$

From Euler-Lagrange equations,

$$\frac{d}{dt}(p(t)\dot{y})-qy-f=0$$

Example 8.9:

Poisson's equation

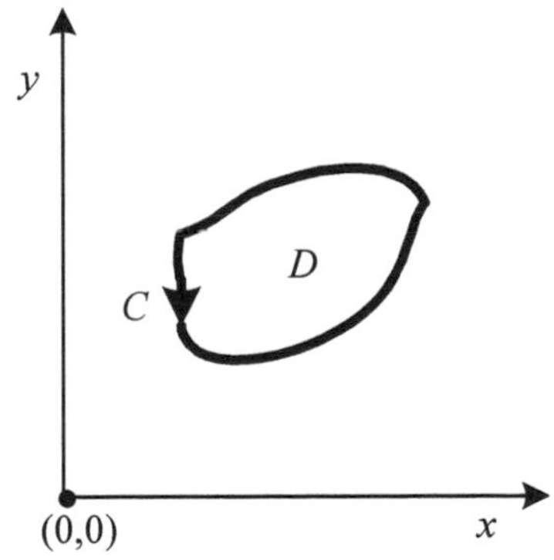

Figure 8.4: Poisson's Equation

Find the minimum of

$$J(u)=\iint_D\left[\frac{1}{2}(u_x^2+u_y^2)+f(x,y)u\right]dx\,dy$$

$u(c)$ is specified on the closed boundary with D as its interior region.

$$L(u, x, y) = \frac{1}{2}\left(u_x^2 + u_y^2\right) + f(x, y)u$$

The resultant Euler-Lagrange equation is:

$$\frac{\partial L}{\partial u} - \frac{\partial}{\partial x}\left(\frac{\partial L}{\partial u_x}\right) - \frac{\partial}{\partial y}\left(\frac{\partial L}{\partial u_y}\right) = 0$$

The necessary condition for a minimum is

$$\nabla^2 u = u_{xx} + u_{yy} = \frac{\partial^2 u}{\partial x^2} + \frac{\partial^2 u}{\partial y^2} = f(x, y)$$

The extension to three dimensions independent variables x, y, and z is obvious.

Example 8.10:

Hamiltonian Principle of the "Path of least action."

Physicists define the action as

$$S = \int_{t_i}^{t_f} L(\boldsymbol{q}, \dot{\boldsymbol{q}}, t)\,\mathrm{d}t$$

$$T = \text{Kinetic Energy}$$

$$V = \text{Potential Energy}$$

$$L = T - V = \text{Lagrangian}$$

The variables $\boldsymbol{q}$ represent "generalized coordinates."

The variational equation for the least path action is $\delta S = 0$

The resultant Euler-Lagrange equations are:

$$\frac{\mathrm{d}}{\mathrm{d}t}\left(\nabla_{\dot{q}} L(\boldsymbol{q}, \dot{\boldsymbol{q}})\right) - \nabla_{q} L(\boldsymbol{q}, \dot{\boldsymbol{q}}) = \boldsymbol{0} \tag{8.42}$$

Important: In the Eq. 8.42, the internal forces are considered as conservative implying

$$\frac{\mathrm{d}}{\mathrm{d}t}\left(\nabla_{\dot{q}}L\right) = \nabla_q V$$

If this is not the case, the above equation can be modified.

From chain-rule

$$\frac{\mathrm{d}L}{\mathrm{d}t} - \frac{\partial L}{\partial t} = (\nabla_q L)^T \cdot \dot{\boldsymbol{q}} + (\nabla_{\dot{q}} L)^T \frac{\mathrm{d}}{\mathrm{d}t}\dot{\boldsymbol{q}} \tag{8.43}$$

From Eq. 8.42 and Eq. 8.43

$$\frac{\mathrm{d}L}{\mathrm{d}t} - \frac{\partial L}{\partial t} = \frac{\mathrm{d}}{\mathrm{d}t}(\nabla_{\dot{q}} L)^T \cdot \dot{\boldsymbol{q}} + (\nabla_{\dot{q}} L)^T \frac{\mathrm{d}}{\mathrm{d}t}\dot{\boldsymbol{q}}$$

$$-\frac{\partial L}{\partial t} = -\frac{\mathrm{d}L}{\mathrm{d}t} + \frac{\mathrm{d}}{\mathrm{d}t}\left((\nabla_{\dot{q}} L)^T \cdot \dot{\boldsymbol{q}}\right)$$

$$-\frac{\partial L}{\partial t} = \frac{\mathrm{d}}{\mathrm{d}t}\left(-L + (\nabla_{\dot{q}} L)^T \dot{\boldsymbol{q}}\right)$$

Let us define

$\boldsymbol{p} = \nabla_{\dot{q}} L$	generalized momentum
$H = -L + (\nabla_{\dot{q}} L)^T \dot{\boldsymbol{q}}$	Hamiltonian

Then

$$H = \boldsymbol{p}^T \dot{\boldsymbol{q}} - L$$

Utilizing the Hamiltonian instead of the Lagrangian, the Euler-Lagrange equations can be replaced by Hamilton-Jacobi Equations (H-J equations) as following:

$$\left.\begin{aligned} H &= \boldsymbol{p}^T \dot{\boldsymbol{q}} - L \\ \dot{\boldsymbol{q}} &= \nabla_p H \\ \dot{\boldsymbol{p}} &= -\nabla_q H \\ \frac{\partial L}{\partial t} &= -\frac{\partial H}{\partial t} \end{aligned}\right\} \text{Hamilton-Jacobi Equations (H-J)}$$

In the conservative systems only:

$$\boldsymbol{p}^T \cdot \dot{\boldsymbol{q}} = (\boldsymbol{M}\dot{\boldsymbol{q}})^T \dot{\boldsymbol{q}} = \dot{\boldsymbol{q}}^T \boldsymbol{M}\boldsymbol{q} = 2T \qquad T \text{ is kinetic energy}$$

$$H = \boldsymbol{p}^T \cdot \dot{\boldsymbol{q}} - L = 2T - (T - V) = T + V$$

$\boldsymbol{M}$ = Generalized mass matrix (very often diagonal). As an example, inductance matrix for electrical circuits.

Evolution Equation and Poisson's Bracket:

Let

$$A = A(\boldsymbol{p}, \boldsymbol{q}, t) \quad \text{where} \quad \boldsymbol{p}, \boldsymbol{q} \text{ satisfy H-J equations}$$

It is easy to see,

$$\frac{\mathrm{d}A}{\mathrm{d}t} = \frac{\partial A}{\partial t} + \left[(\nabla_q A)^T \cdot (\nabla_p H) - (\nabla_p A)^T \cdot (\nabla_q H)\right] = \frac{\partial A}{\partial t} + [A, H] \quad (8.44)$$

where $[A, H]$ is called **Poisson's Bracket** and the Eq. 8.44 is called **Evolution Equation**.

8.4 Integral Extremal with Multiple Constraints

We wish to minimize

$$J = \int_{t_1}^{t_2} f_0(t, \boldsymbol{y}, \dot{\boldsymbol{y}}) \, \mathrm{d}t \quad (8.45)$$

Subject to conditions

$$\boldsymbol{g}(t, \boldsymbol{y}, \dot{\boldsymbol{y}}) = \boldsymbol{0} \quad (8.46)$$

where $\boldsymbol{y}$ is an l-dimensional vector with components $y_1, y_2, \ldots, y_l$ and $\boldsymbol{g}$ is a n-dimensional vector with components $g_1, g_2, \ldots, g_n$.

The augmented scalar function is

$$V = \left(f_0 + \boldsymbol{\lambda}^T \boldsymbol{g}\right) = V(t, \boldsymbol{y}, \dot{\boldsymbol{y}}, \boldsymbol{\lambda}) \tag{8.47}$$

We will now wish to find the extremal of J_1:

$$J_1 = \int_{t_1}^{t_2} V(t, \boldsymbol{y}, \dot{\boldsymbol{y}}, \boldsymbol{\lambda})\, \mathrm{d}t \tag{8.48}$$

The necessary extremal Euler-Lagrange equations are:

$$\begin{aligned} &\left(\frac{\mathrm{d}}{\mathrm{d}t}\left(\nabla_{\dot{y}} V\right) - \nabla_{y} V\right) = \boldsymbol{0} \\ &\left(\frac{\mathrm{d}}{\mathrm{d}t}\left(\nabla_{\dot{\lambda}} V\right) - \nabla_{\lambda} V\right) = \boldsymbol{g}(t, \boldsymbol{y}, \dot{\boldsymbol{y}}) = \boldsymbol{0} \quad \text{because} \quad \nabla_{\dot{\lambda}} V = \boldsymbol{0} \end{aligned} \tag{8.49}$$

The transversality conditions are

$$\left[f_0 \delta t + \left(\nabla_{\dot{y}} V\right)^T \cdot \delta \boldsymbol{y}\right]\Big|_1^2 = 0 \tag{8.50}$$

and they are augmented by the auxiliarly specified end conditions

$$\varphi_k(t, y_1, y_2) = 0 \qquad k = 1, 2, \ldots, p \tag{8.51}$$

8.5 Mayer's Form

We wish to minimize

$$J = h(t, \boldsymbol{y}_1, \boldsymbol{y}_2) \qquad \text{where } \boldsymbol{y} = \boldsymbol{y}_1, \boldsymbol{y}_2 \text{ at times } t_1 \text{ and } t_2 \text{ respectively} \tag{8.52}$$

Subject to constraints

$$\boldsymbol{g}(t, \boldsymbol{y}, \dot{\boldsymbol{y}}) = \boldsymbol{0} \tag{8.53}$$

We can again form a function

$$J_1 = \int_{t_1}^{t_2} V \,\mathrm{d}t + h(t, \boldsymbol{y}_1, \boldsymbol{y}_2) \tag{8.54}$$

$$V = \boldsymbol{\lambda}^T \cdot \boldsymbol{g} \tag{8.55}$$

as shown in Eq. 8.27

$$\mathrm{d}J_1 = \int_{t_1}^{t_2} \left[\frac{\mathrm{d}}{\mathrm{d}t}\left(\nabla_{\dot{y}} V\right)^T - \left(\nabla_y V\right)^T \right] \cdot [\delta \boldsymbol{y}] \,\mathrm{d}t + \left[V\delta t + \left(\nabla_{\dot{y}} V\right)^T \cdot \delta \boldsymbol{y} \right]\Bigg|_1^2 + \delta h \tag{8.56}$$

The conditions for extremal are

$$\frac{\mathrm{d}}{\mathrm{d}t}\left(\nabla_{\dot{y}} V\right) - \left(\nabla_y V\right) = 0$$

$$\nabla_\lambda V = \boldsymbol{g}(t, \boldsymbol{y}, \dot{\boldsymbol{y}}) = 0 \qquad \text{(because } \nabla_{\dot{\lambda}} = 0\text{)} \tag{8.57}$$

$$\left[V\delta t + \left(\nabla_{\dot{y}} V\right)^T \cdot \delta \boldsymbol{y} \right]\Bigg|_1^2 + \delta h = 0 \tag{8.58}$$

Realizing that

$$\mathrm{d}\boldsymbol{y} = \delta \boldsymbol{y} + \dot{\boldsymbol{y}}\delta t \quad , \quad \delta h \approx \mathrm{d}h$$

We obtain the transversality conditions:

$$\left[\left(V - \left(\nabla_{\dot{y}} V\right)^T \cdot \dot{\boldsymbol{y}} \right) \mathrm{d}t + \left(\nabla_{\dot{y}} V\right)^T \cdot \mathrm{d}\boldsymbol{y} \right]\Bigg|_1^2 + \mathrm{d}h = 0 \tag{8.59}$$

along with end constraints $\varphi_k(t, \boldsymbol{y}_1, \boldsymbol{y}_2) = 0$.

It is possible that sometimes the optimum solutions $\boldsymbol{y}$ may be made of piece-wise continuous arcs as shown in Figure 8.5.

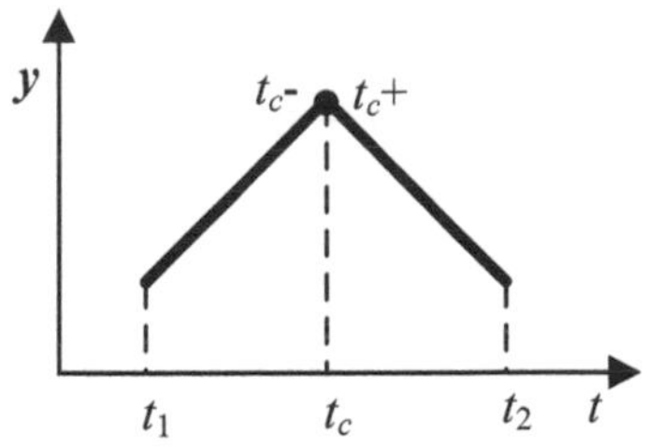

Figure 8.5: Corner Point at $t = t_c$

In that case the transversality condition can be broken down into parts. Thus, for example the coordinates $\mathbf{y}$ may have discontinuities. In order that the relationships of transversality conditions, due to the discontinuities are preserved, we should have, from Eq. 8.59

$$\left[\left(\nabla_{\dot{y}}V\right)^T \cdot \dot{\mathbf{y}}\right]^+ = \left[\left(\nabla_{\dot{y}}V\right)^T \cdot \dot{\mathbf{y}}\right]^- \tag{8.60}$$

and

$$\left[\nabla_{\dot{y}}V\right]^+ = \left[\nabla_{\dot{y}}V\right]^- \tag{8.61}$$

where - and + stand for the quantities being evaluated just before and just after the discontinuity at t_c. These are called **Erdmann-Weierstrass Corner Conditions**.

8.6 Bolza's Form

Determine the extremum of:

$$J = \int_{t_1}^{t_2} f_0(t, \mathbf{y}, \dot{\mathbf{y}})\,\mathrm{d}t + h(t, \mathbf{y}_1, \mathbf{y}_2) \tag{8.62}$$

subject to constraints

$$\mathbf{g}(t, \mathbf{y}, \dot{\mathbf{y}}) = \mathbf{0} \tag{8.63}$$

Let us again form a function V such that

$$J_1 = \int_{t_1}^{t_2} V\,\mathrm{d}t + h(t, y_1, y_2) \tag{8.64}$$

such that

$$V = \left(f_0 + \boldsymbol{\lambda}^T \cdot \boldsymbol{g}\right) \tag{8.65}$$

The conditions for extremal are as given in Eq. 8.57 and Eq. 8.58 and repeated here as

$$\frac{\mathrm{d}}{\mathrm{d}t}\left(\nabla_{\dot{y}} V\right) - \left(\nabla_{y} V\right) = \mathbf{0} \tag{8.66}$$

$$\nabla_{\lambda} V = \boldsymbol{g}(t, \boldsymbol{y}, \dot{\boldsymbol{y}}) = \mathbf{0} \tag{8.67}$$

$$\left[V\delta t + \left(\nabla_{\dot{y}} V\right)^T \cdot \delta \boldsymbol{y}\right]\Bigg|_1^2 + \delta h = 0 \tag{8.68}$$

8.7 Variational Principles and Optimal Control

In a typical optimization problem only few variables are the "control variables" while the other variables are controlled or "steered" by these control variables. A typical example is a rocket engine that produces a thrust profile involving the magnitude and the angle of the thruster to achieve some required flight trajectory. In this case,

Let

$$\boldsymbol{y} = \begin{bmatrix} \boldsymbol{x} \\ \boldsymbol{u} \end{bmatrix}$$

where $\boldsymbol{x}$ is an n vector and $\boldsymbol{u}$ is an r vector, $n + r = l$

$$V = V(\boldsymbol{y}, \dot{\boldsymbol{y}}, t) = V(\boldsymbol{x}, \dot{\boldsymbol{x}}, \boldsymbol{u}, \dot{\boldsymbol{u}}, t) \qquad \text{(Optimization function)}$$

$$\boldsymbol{g} = \boldsymbol{g}(\boldsymbol{y}, \dot{\boldsymbol{y}}, t) = \boldsymbol{g}(\boldsymbol{x}, \dot{\boldsymbol{x}}, \boldsymbol{u}, \dot{\boldsymbol{u}}, t) = 0 \qquad \text{(Constraints)}$$

As before

$$J = \int_{t_1}^{t_2} V \, \mathrm{d}t + h \tag{8.69}$$

Such that

$$V = \left[f_0 + (\boldsymbol{\lambda})^T \cdot \boldsymbol{g} \right] \tag{8.70}$$

The necessary conditions of extremum are

$$\frac{\mathrm{d}}{\mathrm{d}t}(\nabla_{\dot{\boldsymbol{x}}} V) - (\nabla_{\boldsymbol{x}} V) = \boldsymbol{0} \tag{8.71}$$

$$\frac{\mathrm{d}}{\mathrm{d}t}(\nabla_{\dot{\boldsymbol{u}}} V) - (\nabla_{\boldsymbol{u}} V) = \boldsymbol{0} \tag{8.72}$$

$$(\nabla_{\lambda} V) = \boldsymbol{g}(\boldsymbol{y}, \dot{\boldsymbol{y}}, t) = \boldsymbol{0} \qquad (\text{because } \nabla_{\dot{\lambda}} V \equiv \boldsymbol{0}) \tag{8.73}$$

$$\left[V\delta t + (\nabla_{\dot{\boldsymbol{x}}} V)^T \cdot \boldsymbol{\delta x} + (\nabla_{\dot{\boldsymbol{u}}} V)^T \cdot \boldsymbol{\delta u} \right] \Bigg|_1^2 + \delta h = 0 \tag{8.74}$$

Eqs. 8.71 to 8.74, along with the other prescribed boundary conditions, result in solutions of the $(2n + r)$ variables $\boldsymbol{x}, \boldsymbol{u}$, and $\boldsymbol{\lambda}$. Let us simplify these equations for the following control system:

$$\dot{\boldsymbol{x}} = \boldsymbol{f}(\boldsymbol{x}, \boldsymbol{u}), \qquad \boldsymbol{x}(0) = \boldsymbol{x}_0, \quad \boldsymbol{x}(t_f) = \boldsymbol{x}_{t_f} \tag{8.75}$$

Find the control law that minimizes, $J(\boldsymbol{u}) = \int_0^{t_f} f_0(\boldsymbol{x}(t), \boldsymbol{u}(t)) \, \mathrm{d}t + h(\boldsymbol{x}(t_f))$ (8.76)

The contraint equations can be written as:

$$\boldsymbol{g}(\boldsymbol{x}, \dot{\boldsymbol{x}}, \boldsymbol{u}) = \dot{\boldsymbol{x}} - f(\boldsymbol{x}, \boldsymbol{u}) = \boldsymbol{0}$$

As before

$$J(\boldsymbol{u}) = \int_0^{t_f} V \, \mathrm{d}t + h(x_{t_f})$$

$$V = f_0(\boldsymbol{x},\boldsymbol{u}) - \boldsymbol{\lambda}^T \cdot (\dot{\boldsymbol{x}} - \boldsymbol{f}(\boldsymbol{x},\boldsymbol{u}))$$

The necessary extremum conditions are

$$\begin{aligned}
&\frac{\mathrm{d}}{\mathrm{d}t}\boldsymbol{\nabla}_{\dot{\lambda}} - \boldsymbol{\nabla}_{\lambda} = \boldsymbol{0} && \Rightarrow \dot{\boldsymbol{x}} = \boldsymbol{f}(\boldsymbol{x},\boldsymbol{u}) \\
&\frac{\mathrm{d}}{\mathrm{d}t}\boldsymbol{\nabla}_{\dot{x}} - \boldsymbol{\nabla}_{x} = \boldsymbol{0} && \Rightarrow -\dot{\boldsymbol{\lambda}} = +(\nabla_x f^T)\boldsymbol{\lambda} + \nabla_x f_0(\boldsymbol{x},\boldsymbol{u}) \\
&\frac{\mathrm{d}}{\mathrm{d}t}\boldsymbol{\nabla}_{\dot{u}} - \boldsymbol{\nabla}_{u} = \boldsymbol{0} && \Rightarrow \boldsymbol{0} = +(\nabla_u f^T)\boldsymbol{\lambda} + \nabla_u f_0(\boldsymbol{x},\boldsymbol{u})
\end{aligned} \tag{8.77}$$

$$\left[V\delta t + \boldsymbol{\lambda}^T\delta \boldsymbol{x}\right]\Big|_0^{t_f} + \partial h = 0 \Rightarrow \left[V\delta t + \boldsymbol{\lambda}^T\delta x\right]\Big|_0^{t_f} + \partial h = 0$$

8.8 Hamilton-Jacobi—Euler-Lagrange Equations

The concept of Hamiltonian and the "least action principle" provides a compact notation for the optimal control problems. Introducing the Hamiltonian H:

$$H = H(\boldsymbol{x},\boldsymbol{\lambda},\boldsymbol{u}) = \boldsymbol{\lambda}^T \cdot \boldsymbol{f}(\boldsymbol{x},\boldsymbol{u}) + f_0(\boldsymbol{x},\boldsymbol{u})$$

The Euler-Lagrange equations can be rewritten as Hamilton-Jacobi equations as:

$$\begin{aligned}
\dot{\boldsymbol{x}} &= \boldsymbol{\nabla}_{\lambda} H \\
\dot{\boldsymbol{\lambda}} &= -\boldsymbol{\nabla}_x H \qquad \text{(often called \textbf{Adjoint Equations})} \\
\boldsymbol{0} &= \boldsymbol{\nabla}_u H
\end{aligned}$$

Optimal control $\boldsymbol{u}$ is such that:

$$\min_{u=u^*}[H(\boldsymbol{x},\boldsymbol{\lambda},u)] = \min_{|\delta u|\to 0}[H(\boldsymbol{x},\boldsymbol{\lambda},\boldsymbol{u}^* + \delta u) - H(\boldsymbol{x},\boldsymbol{\lambda},u^*)] = 0$$

In most physical situations the control $\boldsymbol{u}$ is an "actuator" whose output is bounded, namely $|\boldsymbol{u}| \leq |\boldsymbol{u}|_{\max}$. In this situation the extremum may occur at the boundary of the **control region**. The resultant variational equations for the control at the boundary:

$$\nabla_u H = 0 \quad \text{are not valid.}$$

However, extremal or the "least action principle" of the Hamiltonian is still valid and was rigorously proved by academician L.S. Pontryagin [Pontryagin, L.].

8.9 Pontryagin's Extremum Principle

Given:

$$\dot{\boldsymbol{x}} = \boldsymbol{f}(\boldsymbol{x}, \boldsymbol{u})$$

Find the control law that minimizes

$$J(\boldsymbol{u}) = \int_0^{t_f} f_0(\boldsymbol{x}, \boldsymbol{u}, t)\, \mathrm{d}t + h(\boldsymbol{x}(t_f))$$

Define

$$H(\boldsymbol{x}, \boldsymbol{\lambda}, \boldsymbol{u}) = f_0(\boldsymbol{x}, \boldsymbol{u}, t) + \boldsymbol{\lambda}^T \cdot \boldsymbol{f}(\boldsymbol{x}, \boldsymbol{u})$$
$$H^* = H[\boldsymbol{x}^*, \boldsymbol{\lambda}^*, \boldsymbol{u}^*, t] = \min_u [t(\boldsymbol{x}^*, \boldsymbol{\lambda}^*, \boldsymbol{u}, t)]$$

The Euler-Lagrange-Pontryagin equations are

$$\dot{\boldsymbol{x}} = \nabla_\lambda H$$
$$\dot{\boldsymbol{\lambda}} = -\nabla_x H$$

$$H^* = H[\boldsymbol{x}^*, \boldsymbol{\lambda}^*, \boldsymbol{u}^*, t] \leq H(\boldsymbol{x}, \boldsymbol{u}, \boldsymbol{\lambda}, t)$$

We shall see later that the transversality conditions discussed earlier result in the two point boundary value problems where $\boldsymbol{x}(0)$ and $\boldsymbol{\lambda}(t_f)$ are known, but $\boldsymbol{\lambda}(0)$ and $\boldsymbol{x}(t_f)$ are unknown.

Example 8.11:

Minimum time planer, vehicle transfer problem via Pontryagin's principle.

Given

$$\begin{aligned}\dot{x}_1 &= x_2 \\ \dot{x}_2 &= u\end{aligned} \qquad \text{or} \qquad \dot{\boldsymbol{x}} = \boldsymbol{A}\boldsymbol{x} + \boldsymbol{b}u,$$

$$A = \begin{bmatrix} 0 & 1 \\ 0 & 0 \end{bmatrix}, \qquad \boldsymbol{b} = \begin{bmatrix} 0 \\ 1 \end{bmatrix}$$

Initial conditions: $x_1(0)$, $x_2(0)$ represents any points on $x_1 - x_2$ plane at $t = 0$

Final conditions: $x_1(t_f) = 0,\ x_2(t_f) = 0$ (origin in $x_1 - x_2$ plane at $t = t_f$)

Optimality functional is

$$J = \int_{t_o}^{t_f} (1)\,\mathrm{d}t = t_f - t_o$$

The Hamiltonian is:

$$H(\boldsymbol{x}, \boldsymbol{\lambda}, \boldsymbol{u}) = 1 + \lambda_1 x_2 + \lambda_2 u$$

The Euler-Lagrange equations are:

$$\begin{aligned}\dot{x}_1 &= x_2 \\ \dot{x}_2 &= u \\ \dot{\lambda}_1 &= 0 \\ \dot{\lambda}_2 &= -\lambda_1\end{aligned} \qquad \begin{aligned}&(\dot{\boldsymbol{x}} = \boldsymbol{A}\boldsymbol{x} + \boldsymbol{b}u) \\ &(\dot{\boldsymbol{\lambda}} = -\boldsymbol{A}^T\boldsymbol{\lambda})\end{aligned} \tag{8.78}$$

The control that minimizes the Hamiltonian for $|u| \leq 1$ is:

$$\boxed{u = -\text{Sign}\,(\boldsymbol{\lambda}^T\boldsymbol{b}) = -\text{Sign}\,\lambda_2, \quad |u| = \pm 1 \text{ or } 0}$$

The Solution of adjoint equations yield:

$$\lambda_1 = c_1$$

$$\lambda_2 = c_2 - c_1 t$$

c_1 and c_2 depend upon $x_1(0)$ and $x_2(0)$. This kind of control is called the **bang-bang control** and u takes only three values $+1$, -1, and 0. The terminal conditions require that at time $t = t_f$ the trajectory should be at the origin in the x_1-x_2 plane and control should take the value 0 thereafter.

The control law takes the form

$$u = -\text{Sign } \lambda_2 = -\text{Sign } (c_2 - c_1 t)$$

Control sign is changed at the instant when

$$c_2 - c_1 t = 0 \qquad \text{or} \qquad t_s = \text{ switching instant } = \frac{c_2}{c_1}$$

Furthermore, at the instant $t = t_f$, control is taken off and therefore $u(t) \equiv 0, \quad t \geq t_f$

Note: In general, it is not easy to find the switching instances. In this problem there is only one switching instant and the control profile is either (a) or (b) as shown in Figure 8.6.

Switching Curves

Consider those points $x_1(0), x_2(0)$ in the $x_1(t)$-$x_2(t)$ plane, which can be steered to the origin by applying the control $u = +1$ and then zero or $u = -1$ and then zero. When the system is at the origin, the control is removed and takes the value zero. The trajectories terminating at the origin are called the **switching curves**. We shall see that "switching" will take place on these curves.

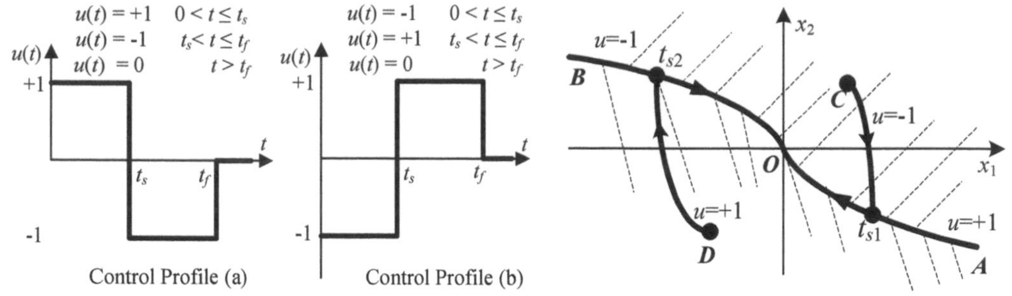

(a) Switching Curve for the bang-bang control (b) Optimal Switching Profile

Figure 8.6: Switching Curves

These switching curves are determined as follows:

1) $u(t) = +1, \qquad x_1(t_f) = 0, \qquad x_2(t_f) = 0$

Solutions of the differential Eqs. 8.78 are

$$x_2(t_f) = 0 = t_f + x_2(0)$$
$$x_1(t_f) = 0 = \frac{1}{2}t_f^2 + x_2(0)t_f + x_1(0)$$

Eliminating $t_f, x_1(0) = \frac{1}{2}x_2^2(0), \quad x_2(0) \leq 0, \quad u = +1.$

This is the equation of a parabola represented by the curve AO in Figure 8.6(b)

2) $u(t) = -1, \qquad x_1(t_f) = 0, \qquad x_2(t_f) = 0$

The Resultant solution is

$$x_2(t_f) = 0 = -t_f + x_2(0)$$
$$x_1(t_f) = 0 = -\frac{1}{2}t_f^2 + x_2(0)t_f + x_1(0)$$

Eliminating $t_f, -x_1(0) = \frac{1}{2}x_2^2(0), \quad x_2(0) > 0.$

This is also the equation of a parabola represented by the curve BO in Figure 8.6(b)

Any point C representing $x_1(0), x_2(0)$ above the switching curve AOB is driven by $u = -1$ upto the point t_{s_1} and then $u = +1$ and so steered to the desired point (namely the origin) in the minimum time. Control is removed at this point. Similarly any point D representing $x_1(0), x_2(0)$ below the switching curve AOB is driven by $u = +1$ upto point t_{s_2} and then $u = -1$ to the origin and then control takes zero value. Clearly due to discontinuity at t_{s_1} and t_{s_2}, the calculus of variations becomes invalid but Weierstrasse's discontinuity equations still apply.

Example 8.12:

Space Navigation Optimal Trajectory (TPBV problem).

A simple and practical method for solving two-point boundary value problem is set forth in [Hyde, P.]. Here we show how it can be applied to the optimum rocket launch problem. To simplify the computations, we will restrict ourselves to the two-dimensional case.

Object: To launch a rocket into an orbit with given insertion altitude and velocity, using as little fuel as possible. The control is performed by steering the orientation of the rocket and fuel consumption rate.

Development of equations governing the motion of the rocket [Lawden, D.].

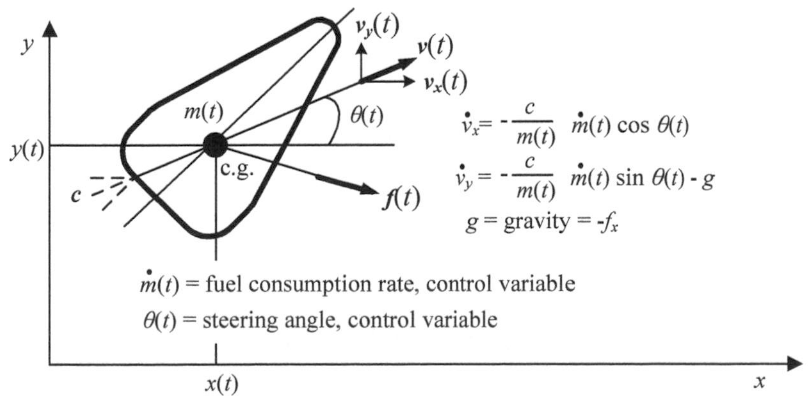

Figure 8.7: Simple Modeling of the Rocket Launch

Let:

$m(t) =$ Total mass of the rocket involving rocket proper and propellant.

$\boldsymbol{f}(t) =$ Total external forces on the rocket

$\boldsymbol{v}(t) =$ The velocity of the c.g. of the rocket

$\delta t =$ The duration of the time in which the rocket mass is decreased by $\delta m(t)$

$\boldsymbol{p} =$ The total linear momentum of the rocket w.r.t its c.g.

$\delta \boldsymbol{p} = \begin{cases} \text{The forward momentum of the rocket proper plus the momentum} \\ \text{due to outgoing exhaust particles} \end{cases}$

$\boldsymbol{c} =$ Exhaust velocity, in the opposite direction as the rocket velocity $\boldsymbol{v}$, $\|\boldsymbol{c}\| = c$

Conservation of Momemtum Equation

The momemtum equation is given by

$$\boldsymbol{p} = m(t)(\boldsymbol{v}(t) - \boldsymbol{c})$$

Let, $\delta \boldsymbol{v}(t)$ be the change in velocity during δt and $\delta m(t)$ be the decrease in the rocket mass ($\delta m(t) >= 0$). Neglecting the second-order term we obtain

$$\delta \boldsymbol{p} = m\delta \boldsymbol{v} + (-\delta m)(-\boldsymbol{c}) = m\delta \boldsymbol{v} + \delta m\boldsymbol{c}$$

furthermore, from conservation of momentum

$$\delta \boldsymbol{p} = \boldsymbol{f}(t)\delta t, \quad \boldsymbol{f}(t)\delta t = m(t)\delta \boldsymbol{v}(t) + \delta m(t)\boldsymbol{c}$$

Dividing both sides of the above equation with δt taking the limit, we get

$$m(t)\dot{\boldsymbol{v}} = \boldsymbol{f}(t) - \dot{m}(t)\boldsymbol{c}, \quad \boldsymbol{f}(t) \text{ represents the external forces}$$

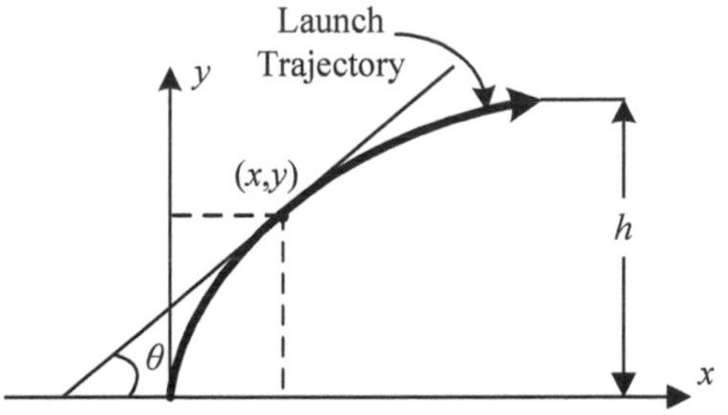

Figure 8.8: Rocket Launch Trajectory

In what follows, $\boldsymbol{f}(t) \equiv 0$ and include the gravity term for the final set of equations. To set up our equations of motion, we define a planar (x, y) rectangular coordinate system centered at the launch site. Let (x, y) be the rocket position, $(\dot{x}, \dot{y})$ its velocity components in x and y directions, and $\theta(t)$ its centre of gravity (c.g.) position angle with respect to the horizontal axis x . Converting the above equations in (x, y) coordinates and including the effect of gravity, we get

$$\ddot{x} = -\frac{c}{m(t)}\dot{m}(t)\cos\theta(t)$$

$$\ddot{y} = -\frac{c}{m(t)}\dot{m}(t)\sin\theta(t) - g, \qquad g = \text{ gravitational acceleration}$$

Rocket motion is controlled by only two variables, namely the steering of the angle $\theta(t)$ and the propellant consumption $\dot{m}(t)$. In state variable form:

$$x = x(t) = x_1 \qquad \text{(horizontal position)}$$

$$v_x = \dot{x}(t) = x_2 \qquad \text{(horizontal velocity)}$$

$$y = y(t) = x_3 \qquad \text{(vertical position)}$$

$$v_y = \dot{y}(t) = x_4 \qquad \text{(vertical velocity)}$$

$$m(t) = x_5 \quad \text{variable fuel mass}$$

Furthermore,

$$\dot{m}(t) = u_1(t), \qquad u_1(t) \text{ is the first control variable}$$

and

$$\theta(t) = u_2(t), \qquad u_2(t) \text{ is the second control variable}$$

These equations in the state variable form are:

$$\begin{aligned}
\dot{x}_1 &= x_2 = v_x \\
\dot{x}_2 &= -\frac{c}{x_5} u_1(t) \cos u_2(t) = \dot{v}_x \\
\dot{x}_3 &= x_4 = v_y \\
\dot{x}_4 &= -\frac{c}{x_5} u_1(t) \sin u_2(t) - g = \dot{v}_y \\
\dot{x}_5 &= u_1(t) = \dot{m}(t) \quad \text{(fuel consumption equation.)} \\
\sqrt{v_x^2 + v_y^2} &= \|v\|
\end{aligned}$$

The initial state variable conditions:

$$x_1(0) = x_2(0) = x_3(0) = x_4(0) = 0$$

$$x_5(0) = m_0$$

The desired terminal conditions at the final time t_f are:

$$x_3(t_f) = h \qquad \text{insertion height}$$

$$x_2(t_f) = v_x(t_f) \qquad \text{horizontal insertion velocity}$$

$$x_4(t_f) = 0$$

We desired to choose optimal control profiles $u_1^*(t)$ and $u_2^*(t)$ so as to use the minimum $x_5(t)$ fuel. Thus, the optimal performance index is:

$$I = \int_0^{t_f} \dot{x}_5(t)\,\mathrm{d}t = m_0 - x_5(t_f) = m_0 - m(t_f)$$

This problem is exactly the Bolza's form Eq. 8.62, namely

$$I = \int_{t_1}^{t_2} f_0(t, \boldsymbol{y}, \dot{\boldsymbol{y}})\,\mathrm{d}t + h_0(t, \boldsymbol{y}_1, \boldsymbol{y}_2)$$

$$\text{subject to constraints,} \quad g(t, \boldsymbol{y}, \dot{\boldsymbol{y}}) = 0$$

This problem can be restated as following:

$$\text{Minimize,} \quad I = \int_0^{t_f} g(\boldsymbol{x}, \boldsymbol{u}, t)\,\mathrm{d}t + \phi(\boldsymbol{x}(t_f), t_f)$$

$$\begin{aligned} \text{subject to constraints,} \quad \dot{\boldsymbol{x}} &= \boldsymbol{f}(\boldsymbol{x}, \boldsymbol{u}, t) && n\text{-differential equations.} \\ \boldsymbol{x}(0) &= \boldsymbol{x}_0 && n\text{-initial conditions.} \\ \boldsymbol{\psi}(\boldsymbol{x}(t_f), t_f) &= \boldsymbol{0} && m\text{-terminal conditions.} \end{aligned}$$

$$\text{Hamiltonian of the system,} \quad H = g + \boldsymbol{\lambda}^T(t)\boldsymbol{f} + \frac{\mathrm{d}}{\mathrm{d}t}\left[\phi + \boldsymbol{\mu}^T\boldsymbol{\psi}\right]$$

where $\boldsymbol{\lambda}(t)$ is n-dimensional adjoint state vector Lagrange multiplier and $\boldsymbol{\mu}$ is an m-dimensional Lagrange multiplier. The necessary conditions for optimal solutions:

$$\boldsymbol{\nabla}_u g - \boldsymbol{\nabla}_u \boldsymbol{f}^T \boldsymbol{\lambda}(t) = \boldsymbol{0}$$

$$\dot{\boldsymbol{\lambda}}(t) = -\boldsymbol{\nabla}_x g - \boldsymbol{\nabla}_x \boldsymbol{f}^T \boldsymbol{\lambda}^T(t) - \frac{\mathrm{d}}{\mathrm{d}t}\left(\boldsymbol{\nabla}_x \phi + \boldsymbol{\nabla}_x \boldsymbol{\psi}^T \boldsymbol{\mu}\right), \qquad \boldsymbol{\lambda}(t_f) = 0$$

$$\dot{\boldsymbol{x}} = \boldsymbol{f}(\boldsymbol{x}, \boldsymbol{u}, t)$$

$$\boldsymbol{\psi}(\boldsymbol{x}(t_f), t_f) = 0$$

For the problem at hand

$$\begin{aligned}
& g \equiv 0 \\
& \phi(\boldsymbol{x}(t_f), t_f) = m_0 - m(t_f) = m_0 - x_5(t_f) \\
& \boldsymbol{u} = [\, u_1(t) \quad u_2(t) \,]^T = [\, \dot{m}(t) \quad \theta(t) \,]^T \\
& \boldsymbol{x} = [\, x \quad y \quad \dot{x} \quad \dot{y} \quad m(t) \,]^T \\
& \boldsymbol{x}(0) = [\, 0 \quad 0 \quad 0 \quad 0 \quad m_0 \,]^T \\
& \boldsymbol{\psi}(t_f) = [\, y(t_f) = h, \;\; \dot{x}(t_f) = v_x(t_f), \;\; \dot{y}(t_f) = 0 \,]^T
\end{aligned}$$

The resultant equations for launching the rocket into the prescribed orbit are:

$$\left.\begin{aligned}
& \lambda_1(t)\left(\frac{c}{m(t)}\right)\cos\theta(t) + \lambda_2(t)\left(\frac{c}{m(t)}\right)\sin\theta(t) = 0 \\
& -\lambda_1(t)\left(\frac{c}{m(t)}\right)u_1(t)\sin\theta(t) + \lambda_2(t)\left(\frac{c}{m(t)}\right)\cos\theta(t) - \lambda_5(t) = 0
\end{aligned}\right\} \tag{8.79}$$

$\theta(t)$ represents the steering angle of the center of gravity of the rocket.

The adjoint variables differential equations are:

$$\left.\begin{aligned}
& \dot{\lambda}_1(t) = -\lambda_3(t) \\
& \dot{\lambda}_2(t) = -\lambda_4(t) \\
& \dot{\lambda}_3(t) = 0 \\
& \dot{\lambda}_4(t) = 0 \\
& \dot{\lambda}_5(t) = -\left(\frac{c}{m^2(t)}\right)(u_1(t)\cos\theta(t))\lambda_1(t) - \left(\frac{c}{m^2(t)}\right)(u_1(t)\sin\theta(t))\lambda_2(t)
\end{aligned}\right\} \tag{8.80}$$

$$\lambda_3(t_f) = \lambda_5(t_f) \tag{8.81}$$

$$\left.\begin{aligned} x(0) &= 0 \\ y(0) &= 0 \\ v_x(0) &= 0 \\ v_y(0) &= 0 \\ m(0) &= m_0 \end{aligned}\right\} \tag{8.82}$$

$$\left.\begin{aligned} y(t_f) &= h, \\ u(t_f) &= v_x(t_f), \\ v_y(t_f) &= 0 \end{aligned}\right\} \tag{8.83}$$

The system variable differential equations are:

$$\left.\begin{aligned} \dot{v}_x(t) &= -\frac{c}{m(t)}u_1(t)\cos\theta(t) \\ \dot{v}_y(t) &= -\frac{c}{m(t)}u_1(t)\sin\theta(t) \\ \dot{x}(t) &= v_x(t) \\ \dot{y}(t) &= v_y(t) \\ \dot{m}(t) &= u_1(t) \end{aligned}\right\} \tag{8.84}$$

Before the final form of the equations are arrived at, we must represent the control variables as functions of Lagrange multipliers $\lambda_1(t)$ and $\lambda_2(t)$. This is achieved from Eqs. 8.79 as:

$$\cos\theta(t) = \frac{\lambda_2(t)}{\sqrt{\lambda_1^2(t) + \lambda_2^2(t)}}, \qquad \sin\theta(t) = \frac{\lambda_1(t)}{\sqrt{\lambda_1^2(t) + \lambda_2^2(t)}}$$

$$u_1(t) = \frac{(\lambda_5(t)m(t))\left(\sqrt{\lambda_1^2(t) + \lambda_2^2(t)}\right)}{c\left(\lambda_2^2(t) - \lambda_1^2(t)\right)}$$

$$u_2(t) = \theta(t)$$

Summarizing the final Two-Point Boundary Value Problem (TPBVP) (nonlinear) equations:

$$
\begin{aligned}
\dot{x}_1 &= \dot{v}_x(t) = -\frac{(\lambda_2(t)\lambda_5(t))}{\left(\lambda_2^2(t) - \lambda_1^2(t)\right)} \\
\dot{x}_2 &= \dot{v}_y(t) = -\frac{\lambda_1(t)\lambda_5(t)}{\left(\lambda_2^2(t) - \lambda_1^2(t)\right)} \\
\dot{x}_3 &= \dot{x}(t) = v_x(t) \\
\dot{x}_4 &= \dot{y}(t) = v_y(t) \\
\dot{x}_5 &= u_1 = \dot{m}(t) \\
\dot{\lambda}_1(t) &= -\lambda_3(t) \\
\dot{\lambda}_2(t) &= -\lambda_4(t) \\
\dot{\lambda}_3(t) &= 0 \\
\dot{\lambda}_4(t) &= 0 \\
\dot{\lambda}_5(t) &= -2\frac{\lambda_1(t)\lambda_2(t)\lambda_5(t)}{\left(\lambda_2^2(t) - \lambda_1^2(t)\right)} \\
u_1 &= \dot{m}(t) = \frac{\lambda_5(t)m(t)\left(\sqrt{\lambda_1^2(t) + \lambda_2^2(t)}\right)}{c\left(\lambda_2^2(t) - \lambda_1^2(t)\right)} \\
u_2 &= \theta(t) = \frac{\lambda_1(t)}{\lambda_2(t)}
\end{aligned}
$$

Initial Conditions

$$x_1(0) = x_2(0) = x_3(0) = x_4(0) = 0, \qquad m(0) = m_0$$

Final Conditions

$$
\begin{aligned}
\lambda_3(t_f) = \lambda_5(t_f) = 0, &\quad x_3(t_f) = v_x(t_f) \\
x_4(t_f) = y(t_f) = h, &\quad x_5(t_f) = m(t_f)
\end{aligned}
$$

We solve the above system equations using the method described in [Hyde, P.] and [Lawden, D.].

To obtain the starting solutions and hence the initial estimate on the adjoint variables $\lambda_i(t)$, we linearize the above equations about the initial values $(u(0), v(0), x(0), y(0), \lambda_1(0), \lambda_2(0), \lambda_3(0), \lambda_4(0), \lambda_5(0))$. Some initial intelligent estimates about $\lambda_i(0)$ are required.

8.10 Dynamic Programming

We now turn to the subject of dynamic programming for optimal control policies. Dynamic programming was a pioneering work started by Professor Bellman and his colleagues. This approach provides an alternate route to optimality as compared with the calculus of variations (notice the similarity between these two approaches). The two main principles involved are the **Bellman Principles**.

#1. **Embedding Principle or Principle of Invariance**

Given a control system with a fixed initial state and a fixed time of operation, we embed the optimal control policy into a general formulation involving variable initial states and a variable time of operation. This results in a whole set of problems where the required solution is embedded inside the general solution and thus provides the solution to the specific problem. This is a very computationally intensive procedure and the "curse of dimensionality" creeps in.

#2. **Principle of Optimality**

Let $x(t)$, as shown in Figure 8.9, be an optimal trajectory under the optimal decision making policy.

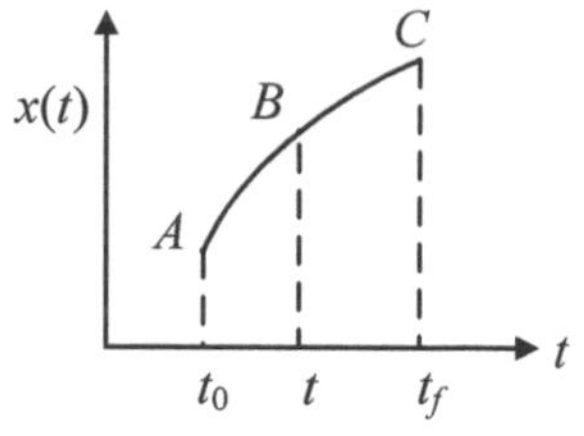

Figure 8.9: Optimal Trajectory

This optimal trajectory has the following fundamental property.

Whatever the initial state and the initial control (or the decision), all the subsequent controls (or the decisions) must be optimal for the trajectory to be optimal (of course, the initial decision must also be optimal). Thus, if AC represents an optimal trajectory for the interval (t_0, t_f), then regardless of how we arrive from point A at time t_0 to point B at time t), the end part BC of this trajectory with the interval (t, t_f); $(t_0 \leq t \leq t_f)$ must be optimal. This is the famous Optimality Principle of Professor Bellman. The ensuing results are the same as from Pontryagin's maximum principle.

Let us apply this optimality principle to the process

$$\dot{\boldsymbol{x}} = \boldsymbol{f}(\boldsymbol{x}, \boldsymbol{u}), \quad \text{where } \boldsymbol{u} \text{ is the control.}$$

and find an optimal control law that minimizes

$$J(\boldsymbol{u}) = h(\boldsymbol{x}(t_f)) + \int_{t_0}^{t_f} f_0(\boldsymbol{x}, \boldsymbol{u})\, \mathrm{d}t \qquad \text{final state } \boldsymbol{x}(t_f) \text{ is fixed}$$

Let us introduce

$$\hat{V}(\boldsymbol{x}, t) = \min_{\boldsymbol{u}(t)} \left[h(\boldsymbol{x}(t_f)) + \int_{t}^{t_f} f_0(\boldsymbol{x}, \boldsymbol{u})\, \mathrm{d}t \right], \qquad t_0 \leq t \leq t_f$$

Applying the principle of optimality

$$\hat{V}(\boldsymbol{x},t) = \min_{\boldsymbol{u}(t)} \left[h(\boldsymbol{x}(t_f)) + \int_{t+\mathrm{d}t}^{t_f} f_0(\boldsymbol{x},\boldsymbol{u})\,\mathrm{d}t + \int_{t}^{t+\mathrm{d}t} f_0(\boldsymbol{x},\boldsymbol{u})\,\mathrm{d}t \right], \qquad t_0 \le t \le t_f$$

The first two terms in the above expression represent $\hat{V}(\boldsymbol{x} + \mathrm{d}\boldsymbol{x}, t + \mathrm{d}t)$.

Thus,

$$\hat{V}(\boldsymbol{x},t) = \min_{\boldsymbol{u}(t)} \left[\hat{V}(\boldsymbol{x} + \mathrm{d}\boldsymbol{x}, t + \mathrm{d}t) + f_0(\boldsymbol{x},\boldsymbol{u})\,\mathrm{d}t \right]$$

Using the first variation

$$\hat{V}(\boldsymbol{x},t) = \min_{\boldsymbol{u}(t)} \left[\hat{V}(\boldsymbol{x},t) + \frac{\partial \hat{V}}{\partial t}\,\mathrm{d}t + \left[(\nabla_x \hat{V})^T \dot{\boldsymbol{x}} \right] \mathrm{d}t + f_0(\boldsymbol{x},\boldsymbol{u})\,\mathrm{d}t \right]$$

The resultant equations are:

$$\frac{\partial \hat{V}}{\partial t} + \min_{\boldsymbol{u}(t)} \left[f_0(\boldsymbol{x},\boldsymbol{u}) + (\boldsymbol{\nabla}_x \hat{V})^T \boldsymbol{f}(\boldsymbol{x},\boldsymbol{u}) \right] = 0$$
$$\hat{V}(\boldsymbol{x},t_f) = h(\boldsymbol{x}(t_f)) \qquad (8.85)$$

The above equations are known as Hamilton-Jacobi-Bellman (HJB) equations. The expression inside the bracket should be same as the Hamiltonian described in section 8.8 for the optimal control $\boldsymbol{u}$, therefore,

$$\left(\boldsymbol{\nabla}_x \hat{V} \right) = \boldsymbol{\lambda} \qquad (8.86)$$

Example 8.13:

Consider the system

$$\dot{\boldsymbol{x}} = \boldsymbol{A}\boldsymbol{x} + \boldsymbol{B}\boldsymbol{u}, \qquad \boldsymbol{x}_0 = \boldsymbol{x}(0)$$

Determine optimal $\boldsymbol{u}(t)$ such that:

$$J(\boldsymbol{u}) = \frac{1}{2} \int_0^{t_f} \left[\boldsymbol{x}^T \boldsymbol{Q} \boldsymbol{x} + \boldsymbol{u}^T \boldsymbol{R} \boldsymbol{u} \right] \mathrm{d}t \quad \text{is minimized}$$

$\boldsymbol{x}$ is n vector, $\quad \boldsymbol{u}$ is r vector

$\boldsymbol{A}$ is $n \times n$ matrix, $\quad \boldsymbol{B}$ is $n \times r$ matrix

$\boldsymbol{Q}$ is positive semidefinite constant matrix

$\boldsymbol{R}$ is positive definite constant matrix

Two cases involving terminal time will be considered.

Case 1: Finite terminal time t_f

$$\begin{aligned} H &= H(\boldsymbol{x}, \boldsymbol{\lambda}, \boldsymbol{u}, t) = \frac{1}{2}\boldsymbol{x}^T\boldsymbol{Q}\boldsymbol{x} + \frac{1}{2}\boldsymbol{u}^T\boldsymbol{R}\boldsymbol{u} + \boldsymbol{\lambda}^T(\boldsymbol{A}\boldsymbol{x} + \boldsymbol{B}\boldsymbol{u}) \\ \nabla_u H &= \boldsymbol{R}\boldsymbol{u} + \boldsymbol{B}^T\boldsymbol{\lambda} = \boldsymbol{0} \end{aligned}$$

or

$$\boldsymbol{u} = -\boldsymbol{R}^{-1}\boldsymbol{B}^T\boldsymbol{\lambda}$$

Thus,

$$\min H(\boldsymbol{x}, \boldsymbol{\lambda}, \boldsymbol{u}, t) = \frac{1}{2}\boldsymbol{x}^T\boldsymbol{Q}\boldsymbol{x} - \frac{1}{2}\boldsymbol{\lambda}^T\boldsymbol{B}\boldsymbol{R}^{-1}\boldsymbol{B}^T\boldsymbol{\lambda} + \boldsymbol{\lambda}^T\boldsymbol{A}\boldsymbol{x}$$

The resultant Hamilton-Jacobi-Bellman equations are:

$$\boxed{\begin{aligned} &\frac{\partial \hat{V}}{\partial t} + \frac{1}{2}\boldsymbol{x}^T\boldsymbol{Q}\boldsymbol{x} - \frac{1}{2}\boldsymbol{\lambda}^T\boldsymbol{B}\boldsymbol{R}^{-1}\boldsymbol{B}^T\boldsymbol{\lambda} + \boldsymbol{\lambda}^T\boldsymbol{A}\boldsymbol{x} = 0 \\ &\nabla_x\hat{V} = \boldsymbol{\lambda}, \quad \left(\nabla_x\hat{V}\right)_{t=t_f} = \boldsymbol{\lambda}(t_f) = \nabla_{x(t_f)}h(\boldsymbol{x}(t_f)) \end{aligned}} \tag{8.87}$$

We seek a quadratic solution for $\hat{V}(\boldsymbol{x}, t)$ of the following form:

$$\hat{V}(\boldsymbol{x}, t) = \frac{1}{2}\boldsymbol{x}^T\boldsymbol{P}(t)\boldsymbol{x}, \text{ where } \boldsymbol{P}(t) \text{ is a symmetric matrix}$$

$$\lambda = \left(\nabla_x \hat{V}\right) = P(t)x$$
$$\frac{\partial \hat{V}}{\partial t} = \frac{1}{2} x^T \dot{P}(t) x$$
$$\lambda^T A x = x^T P^T(t) A x = \frac{1}{2}\left[x^T (P(t)A + A^T P(t))x\right] \quad \text{(Symmetric Form)}$$

Eq. 8.87 can now be written (Riccati equation) as:

$$\frac{1}{2} x^T \left[\dot{P}(t) + Q + P(t)A + A^T P(t) - P(t) B R^{-1} B^T P(t)\right] x = 0, \quad P(t_f) = 0$$

Since $x(t)$ is not always zero, the expression inside the bracket must be zero. Summarizing, the optimal control $u(t)$ is given by

$$u(t) = -R^{-1} B^T P(t) x(t) = -K(t) x(t)$$
$$\dot{P}(t) + Q + P(t)A + A^T P(t) - P(t) B R^{-1} B^T P(t) = \mathbf{0} \tag{8.88}$$
$$P(t_f) = 0$$

The $n(n+1)/2$ differential equations involving elements of the matrix $P(t)$ are called **Riccati Differential Equations** and u is the optimal feedback control.

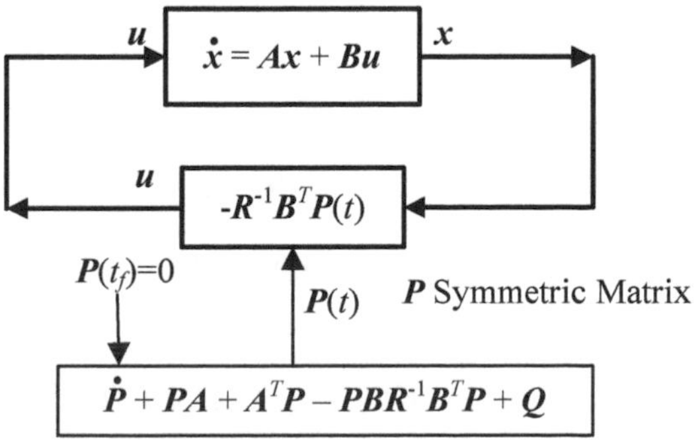

Figure 8.10: Riccati Computer

Case 2. **Infinite time interval,** $t_f \to \infty$

Since the process involves the infinite terminal time, $\hat{V}(\boldsymbol{x}, t)$ is the same for any finite t and depends only on the initial state $\boldsymbol{x}(t_0)$, yielding

$$\frac{\partial \hat{V}(\boldsymbol{x}, t)}{\partial t} = 0$$

Thus,

$$\boldsymbol{P}(t) = \boldsymbol{P}, \quad n \times n \text{ symmetric matrix with constant entries}$$
$$\dot{\boldsymbol{P}} = 0$$

The control law takes the form:

$$\boldsymbol{u}(t) = -\boldsymbol{K}\boldsymbol{x} = -\boldsymbol{R}^{-1}\boldsymbol{B}^T\boldsymbol{P}\boldsymbol{x}$$

where

$$\boldsymbol{P}\boldsymbol{A} + \boldsymbol{A}^T\boldsymbol{P} - \boldsymbol{P}\boldsymbol{B}\boldsymbol{R}^{-1}\boldsymbol{B}^T\boldsymbol{P} + \boldsymbol{Q} = \boldsymbol{0} \qquad \text{Algebraic Riccati-Equation}$$

Notice that the gain matrix $\boldsymbol{K}$ is a constant instead of a function of time.

Instead of using dynamic programming, let us solve this problem via an alternate approach involving Euler-Lagrange equations using the Hamiltonian.

$$H = \boldsymbol{\lambda}^T(\boldsymbol{A}\boldsymbol{x} + \boldsymbol{B}\boldsymbol{u}) + \frac{1}{2}[\boldsymbol{x}^T\boldsymbol{Q}\boldsymbol{x} + \boldsymbol{u}^T\boldsymbol{R}\boldsymbol{u}]$$

$$\dot{\boldsymbol{x}} = \boldsymbol{\nabla}_\lambda H = \boldsymbol{A}\boldsymbol{x} + \boldsymbol{B}\boldsymbol{u}, \qquad \boldsymbol{x}_0 = \boldsymbol{x}(0)$$
$$\dot{\boldsymbol{\lambda}} = -\boldsymbol{\nabla}_x H = -\boldsymbol{A}^T\boldsymbol{\lambda} + \boldsymbol{Q}\boldsymbol{x}$$

$$\mathbf{0} = \nabla_u H = \boldsymbol{B}^T \boldsymbol{\lambda} + \boldsymbol{R}\boldsymbol{u}, \quad \text{or} \quad \boldsymbol{u} = -\boldsymbol{R}^{-1}\boldsymbol{B}^T \boldsymbol{\lambda}$$

$$\boldsymbol{\lambda}(t_f) = \mathbf{0}$$

These are the same equations that were obtained via Pontryagin's maximum principle as well as dynamic programming and now through Euler-Lagrange equations involving calculus of variations. Thus, there exists a similarity between the calculus of variations, the dynamic programming and the maximum principle. In general, the above system of equations, represents $2n$ differential equations with n known initial conditions $\boldsymbol{x}(0)$ and n known terminal conditions $\boldsymbol{\lambda}(t_f) = \mathbf{0}$, yielding a Two-Point Boundary Value Problem (TPBVP). As discussed earlier, these equations in general are nonlinear and hard to solve. For this particular simpler case these equations can be solved as follows:

$$\dot{\boldsymbol{y}} = \begin{bmatrix} \dot{\boldsymbol{x}} \\ \dot{\boldsymbol{\lambda}} \end{bmatrix} = \begin{bmatrix} \boldsymbol{A} & -\boldsymbol{R}^{-1}\boldsymbol{B}^T \\ \boldsymbol{Q} & -\boldsymbol{A}^T \end{bmatrix} \begin{bmatrix} \boldsymbol{x} \\ \boldsymbol{\lambda} \end{bmatrix} \qquad 2n \times 2n \text{ matrix}$$

Let

$$\boldsymbol{M} = \begin{bmatrix} \boldsymbol{A} & -\boldsymbol{R}^{-1}\boldsymbol{B}^T \\ \boldsymbol{Q} & -\boldsymbol{A}^T \end{bmatrix}$$

Thus,

$$\dot{\boldsymbol{y}} = \boldsymbol{M}\boldsymbol{y}$$

The resultant solution is

$$\boldsymbol{y}(t) = \boldsymbol{e}^{\boldsymbol{M}t}\boldsymbol{y}(0)$$

where

$$\boldsymbol{e}^{\boldsymbol{M}t} = \begin{bmatrix} \boldsymbol{\Phi}_{xx}(t) & \boldsymbol{\Phi}_{x\lambda}(t) \\ \boldsymbol{\Phi}_{\lambda x}(t) & \boldsymbol{\Phi}_{\lambda\lambda}(t) \end{bmatrix} \qquad 2n \times 2n \quad \text{Transition Matrix}$$

Therefore,

$$\boldsymbol{x}(t) = \boldsymbol{\Phi}_{xx}(t)\boldsymbol{x}(0) + \boldsymbol{\Phi}_{x\lambda}(t)\boldsymbol{\lambda}(0)$$
$$\boldsymbol{\lambda}(t) = \boldsymbol{\Phi}_{\lambda x}(t)\boldsymbol{x}(0) + \boldsymbol{\Phi}_{\lambda\lambda}(t)\boldsymbol{\lambda}(0)$$

$\boldsymbol{x}(0)$ is known and $\boldsymbol{\lambda}(0)$ is unknown. But we can find $\boldsymbol{\lambda}(0)$ in terms of $\boldsymbol{x}(0)$ as

$$\boldsymbol{\lambda}(t_f) = \boldsymbol{0} = \Phi_{\lambda x}(t_f)\boldsymbol{x}(0) + \Phi_{\lambda\lambda}(t_f)\boldsymbol{\lambda}(0)$$

yielding

$$\boldsymbol{\lambda}(0) = -\boldsymbol{\Phi}_{\lambda\lambda}^{-1}(t_f)\boldsymbol{\Phi}_{\lambda x}(t_f)\boldsymbol{x}(0) = -\boldsymbol{\Phi}_{\lambda\lambda}(-t_f)\boldsymbol{\Phi}_{\lambda x}(t_f)\boldsymbol{x}(0)$$

The resultant equation for $\boldsymbol{\lambda}(t)$ in terms of the known $\boldsymbol{x}(0)$ is:

$$\boldsymbol{\lambda}(t) = \left[\boldsymbol{\Phi}_{\lambda x}(t) - \boldsymbol{\Phi}_{\lambda\lambda}(t)\boldsymbol{\Phi}_{\lambda\lambda}(-t_f)\boldsymbol{\Phi}_{\lambda x}(t_f)\right]\boldsymbol{x}(0) = \boldsymbol{G}(t)\boldsymbol{x}(0)$$

This control law $\boldsymbol{u}(t)$ is an open-loop control and is given by

$$\boldsymbol{u}(t) = -\boldsymbol{R}^{-1}\boldsymbol{B}^T\boldsymbol{G}(t)\boldsymbol{x}(0) \tag{8.89}$$

Notice: The above control law Eq. 8.89 is an open-loop control. It depends on the initial conditions $x(0)$ but has the drawback that it does not compensate for errors in computation of $x(0)$ as is usually the case in a closed-loop control law. We can convert the above equation into closed-loop control by computing

$$\boldsymbol{x}(0) = \left[\boldsymbol{\Phi}_{xx}(t) - \boldsymbol{\Phi}_{x\lambda}(t)\boldsymbol{\Phi}_{\lambda\lambda}(-t_f)\boldsymbol{\Phi}_{\lambda x}(t_f)\right]^{-1}\boldsymbol{x}(t)$$

The final closed-loop control law is:

$$\boldsymbol{u}(t) = -\boldsymbol{K}(t)\boldsymbol{x}(t)$$

where

$$\boldsymbol{K}(t) = \boldsymbol{R}^{-1}\boldsymbol{B}^T\left[\boldsymbol{\Phi}_{xx}(t) - \boldsymbol{\Phi}_{x\lambda}(t)\Phi_{\lambda\lambda}(-t_f)\boldsymbol{\Phi}_{\lambda x}(t_f)\right]^{-1}$$

Example 8.14:

N-Stage Dynamic Programming Process.

Consider an N stage decision process. The process starts at x_0 and at every i-th stage, $i = 0, 1, \ldots, N-1$ some "control" or decision u_i are taken to obtain an optimal return function.

The process is described as a scalar equation:

$$x_{n+1} = ax_n + u_n, \qquad x_0 = c \tag{8.90}$$

The return or cost function is: $$\hat{V}_N = \min_{\{u_n\}_0^{N-1}} \frac{1}{2}\left[\sum_{n=0}^{N-1}\left(x_n^2 + u_n^2\right) + x_N^2\right] \tag{8.91}$$

From our continous case knowledge, the function $\hat{V}(x, t)$ is only a function of x_0, so in this discrete case let us choose a candidate for the function $\hat{V}_N$ to be:

$$\hat{V}_N(x_0) = \frac{1}{2}K_N x_0^2 \tag{8.92}$$

From Principle of Optimality, the $(N-i)$ stage process starting at state x_i:

$$\hat{V}_{N-i}(x_i) = \min_{\{u_k\}_{k=i}^{N-i-1}} \frac{1}{2}\left[\sum_{k=i}^{N-i-1}\left(x_k^2 + u_k^2\right) + x_{N-i}^2\right] = \frac{1}{2}K_{N-i}x_i^2 \tag{8.93}$$

or

$$\hat{V}_{N-i}(x_i) = \min_{u_i} \left[x_i^2 + u_i^2 + \hat{V}_{N-i-1}(x_{i+1}) \right] \tag{8.94}$$

or

$$\hat{V}_{N-i}(x_i) = \min_{u_i} \frac{1}{2} \left[x_i^2 + u_i^2 + K_{N-i-1} x_{i+1}^2 \right] = \min_{u_i} \frac{1}{2} \left[x_i^2 + u_i^2 + K_{N-i-1}(a x_i + u_i)^2 \right] \tag{8.95}$$

The minimum is obtained at

$$\frac{\partial}{\partial u_i} \hat{V}_{N-i} = 0$$

or

$$u_i = -a K_{N-i-1} \left(1 + K_{N-i-1}\right)^{-1} x_i, \quad i = 0, \ldots, N-1 \tag{8.96}$$

$$= G_{N-i} x_i \tag{8.97}$$

Plugging the expression for u_i into Eq. 8.95 we obtain $\hat{V}_{N-i}(x_i)$ as an explicit function of x_i^2 and hence, K_{N-i} as:

$$K_{N-i} = \left[1 + \left(1 + K_{N-i-1}^{-1} \right) G_{N-i} \right] \boldsymbol{x}_i^2$$

The above example can be extended to the multivariable case when x_n is k-dimensional.

Important Facts:

1. Total return function or the cost function is finite as $N \to \infty$.
2. Optimal control is linear.
3. The gain G_{N-i} becomes constant as $N \to \infty$.
4. Method can be easily extended to the stochastic systems.

Bibliography

[Caratheodory, C.] Caratheodory, C. *Calculus of Variations and Partial Differential Equations of the First Order.* Vols. 1 and 2. (Translated from German by R.B. Dean and J.J. Brandstatter). San Francisco: Holden-Day Inc., 1965, 1967.

[Gelfand, I.] Gelfand, I.M. and Fomin, S.V. *Calculus of Variations,* Englewood Cliff, NJ: Prentice Hall Inc., 1963.

[Hyde, P.] Hyde, P. A Simple Method for the Numerical Solution of Two-Point Boundary-Value Problems, M.S. Thesis, University of Pennsylvania, 1960.

[Kailath, T.] Kailath, T. *Linear Systems,* Englewood Cliff, NJ: Prentice Hall Inc., 1980.

[Kalman, R.] Kalman, R.E. Contributions to the Theory of Optimal Control. *Bol. Soc. Mat. Mexicana*, 1960.

[Lawden, D.] Lawden, D.F. *Optimal Trajectories for Space Navigation,* London: Butterworths Mathematical Texts, 1963.

[Petrov, I.] Petrov, I.P. *Variational Methods in Optimal Control Theory,* (Translated from Russian by M.D. Friedman.) New York: Academic Press Inc., 1968.

[Pontryagin, L.] Pontryagin, L.S., Boltyanskii, V.G., Gamkrelidze, R.V., and Mishenko, E.F. *The Mathematical Theory of Optimal Processes.* (Translated by D.E. Brown.) New York: Macmillan, 1964.

Chapter 9

Stochastic Processes and Their Linear Systems Response

Information signals, such as television, wireless, voice and data, which are handled by communication systems are by nature probabilistic. The same is true of automatic control systems, large radar with antennae, airplanes flying in gusts of wind, or ships sailing in stormy seas. Each situation like this is characterized by a "stochastic model." In this chapter, we present the probability definitions, random variables, probability distributions and density functions, and other concepts such as the mean and variance of a random variable. The Central Limit Theorem and various other inequalitites are derived. We shall discuss the representation of Wiener or Brownian Motion via random walk. The geometric representation of random variables and stochastic processes in Hilbert space is also presented. Various types of random processes such as stationary, ergodic, etc., are described along with the ideas of auto- and cross-correlation of system inputs and system outputs. This is followed by the series representation of stochastic processes via orthogonal functions and Karhunen-Loeve expansion. Wiener and Kalman filters minimizing the effect of additive noise in signals are also derived. Since our audience is engineers and physical scientists, we have tried to find a good balance between mathematical rigor and understandable proofs.

9.1 Preliminaries

9.1.1 Probability Concepts and Definitions

We define an event as the outcome of an "experiment." A collection of events is represented by what is known as a **set**. We shall not distinguish between the events or the set to which they belong unless it is pertinent. Let events $A, B, \ldots$ etc.. form a complete set F such that

(i) $A, B \in F$: $A \cup B$ implies A **OR** B (union), $A \cap B$ implies A **AND** B (intersection)

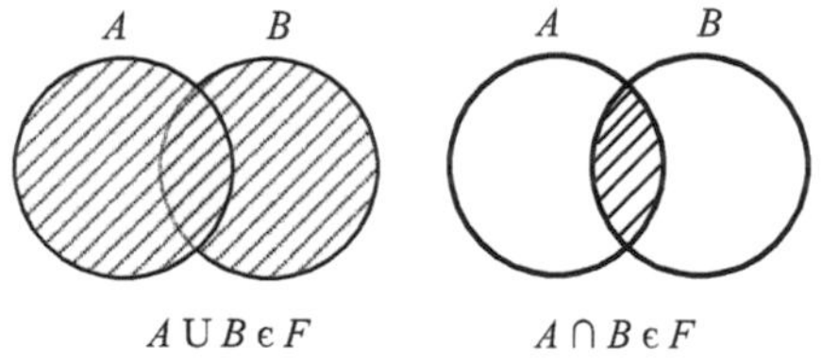

Figure 9.1: Union and Intersection

Furthermore, $A^c, B^c \in F$, where "c" stands for the compliment, $A \cup A^c \equiv F$

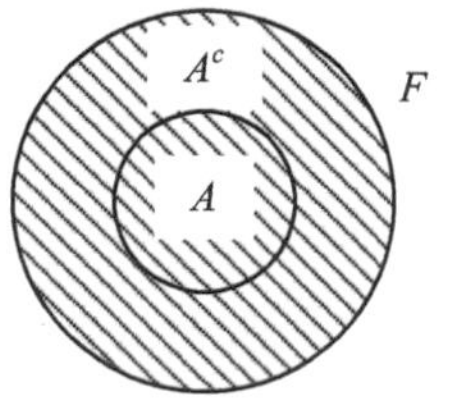

Figure 9.2: $A \cup A^c \equiv F$

Set $(A - B)$ represents $A - (A \cap B)$ implying $(A - B) \equiv (A - (A \cap B))$.

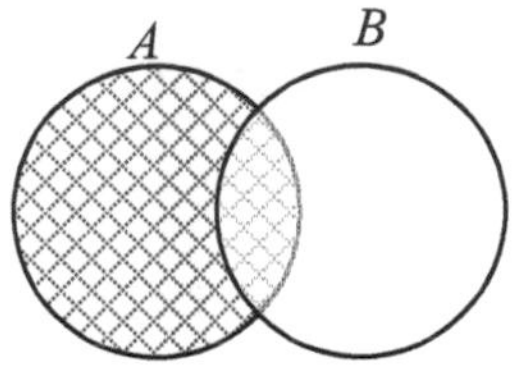

Figure 9.3: $A - B \equiv A - (A \cap B)$

(ii) $A_1, A_2, \ldots \in F$ implies $\cup_{i=1}^{\infty} A_i \in F$, $\cap_{i=1}^{\infty} A_i \in F$

(iii) $\Pr(A)$ is called the **Probability of the Event** A, or the set A :

$$\Pr(A) \geq 0 \tag{9.1}$$

$$\Pr\left[\cup_{i=1}^{\infty} A_i\right] = \sum_{i=1}^{\infty} \Pr(A_i) = 1 \qquad \text{(pairwise disjoint events)} \tag{9.2}$$

$$\Pr\left(A_i \cap A_j\right) = 0 \quad i \neq j \qquad \text{(pairwise disjoint events)} \tag{9.3}$$

The set ϕ is called an **Empty Set**

$$\Pr(\phi) \equiv 0$$

$$\Pr(A^c) = 1 - \Pr(A)$$

(iv) If event A has no effect on event B then they are considered "independent." If event A precludes event B or vice versa then they are "mutually exclusive." **Two independent events cannot be mutually exclusive**.

(v) $\Pr(A \cup B) = \Pr(A) + \Pr(B) - \Pr(A \cap B)$ (Inclusion-Exclusion principle). This simply states that the probability of union of events A and B can be computed by adding the probabilities of events A and B independently happening and subtracting the probability of their intersection (A or B).

$$\Pr(A - B) = \Pr(A) - \Pr(A \cap B) \tag{9.4}$$

$$\Pr(B - A) = \Pr(B) - \Pr(A \cap B) \tag{9.5}$$

Sometimes $A \cap B$ is written as AB and

$$\Pr(A \cap B) = \Pr(AB)$$

$$\frac{\Pr(A \cup B)}{\Pr(B)} = \Pr(A/B) \quad \text{Bayes Rule} \tag{9.6}$$

Example 9.1:

Let $\{B_i\}_{i=1}^N$ be a full set of mutually exclusive events.

Show that $\Pr(A)$ can be computed from $\{\Pr(A \cap B_i)\}_{i=1}^N$.

Proof:

$$\cup_{i=1}^N B_i = I, \quad \text{where } I \text{ is the whole sample space}$$

Then

$$A = AI = \cup_{i=1}^N AB_i = \cup_{i=1}^N A \cap B_i$$

Thus

$$\Pr(A) = \sum_{i=1}^N \Pr(A \cup B_i) = \sum_{i=1}^N \left[\frac{\Pr(A \cup B_i)}{\Pr(B_i)}\right] \Pr(B_i)$$

Using Bayes rule, the above expression gets simplified as

$$\boxed{\Pr(A) = \sum_{i=1}^N \Pr(A/B_i)\Pr(B_i)}$$

Exercise 9.1:

Using the inclusion-exclusion principle of two events, show that for the mutually exclusive events A, B, and C:

$$\Pr(A \cup B \cup C) = [\Pr(A) + \Pr(B) + \Pr(C)] - [\Pr(A \cap B) + \Pr(B \cap C) + \Pr(C \cap A)] + [\Pr(A \cap B \cap C)]$$

(vi) $\Pr(A/B) = \dfrac{\Pr(A \cap B)}{\Pr(B)} = \dfrac{\Pr(B/A) \cdot \Pr(A)}{\Pr(B)}$ **(Bayes Rule)**

(vii) For $A \subseteq B$ implying that A is a subset of B or is "contained" in B, implying.

$$\Pr(A) \leq \Pr(B)$$

The probability of B occurring when A has still not occurred is:

$$\Pr(B \backslash A) = \Pr(B) - \Pr(A)$$

(viii) if A and B are independent

$$\Pr(A \cap B) = \Pr(A) \cdot \Pr(B) \tag{9.7}$$

$$\Pr(A/B) = \Pr(A) \tag{9.8}$$

$$\Pr(A \cup B) = \Pr(A) + \Pr(B) - \Pr(A) \cdot \Pr(B) \tag{9.9}$$

(ix) **Basic combinatorics and counting rules**

Given a set of n distinct elements, a **Permutation** is defined as a set of elements selected from the given set with any possible ordering or arrangement of the elements. A **Combination**, on the other hand, is defined as a set of elements selected from the given set without regard to order or arrangement of the elements. For a selected set of given size the number of permutations are much larger than the number of combinations. Consider a set of n

elements from which all the elements are to be selected at a time. The number of possible permutations are given by $n!$ On the other hand, the number of combinations is only 1.

- In general, the number of ways of obtaining an ordered subset or permutations of k elements from a set of n elements is given by

$$P(n,k) = \frac{n!}{(n-k)!} = \prod_{i=1}^{k}(n+1-i)$$

The number of combinations of an unordered subset of k elements from a set of n elements is given by

$$C(n,k) = \frac{n!}{(n-k)!\,k!} = P(n,k)/k! = C(n,n-k)$$

- Let n be the number of elements in the given set, n_i be the number of identical elements in the i-th group from the set, $i = 1, \ldots, N$

$$\sum_{i=1}^{N} n_i = n$$

The number of permutations of these n objects is given as:

$$P(n; n_1, n_2, \cdots, n_N) = \prod_{i=1}^{N}\left[\frac{\left(n - \sum_{j=0}^{i-1} n_j\right)!}{n_i!\left(n - \sum_{j=0}^{i} n_j\right)!}\right] = \frac{n!}{\prod_{i=1}^{N} n_i!}, \qquad n_0 = 0$$

The number of combinations of these n objects is given as:

$$C(n; n_1, n_2, \cdots, n_N) = \prod_{i=1}^{N} C\left(\left(n - \sum_{k=1}^{i} n_k\right), n_{i+1}\right) = P(n; n_1, n_2, \cdots, n_k)$$

Let n_i be the number of ways in which a sequence of events E_i can occur, $i = 1, 2, \ldots, N$. We are interested in how many ways the various events can occur. Following two counting rules are helpful.

1 **Multiplication Rule:** Independent events occur in $\prod_{i=1}^{N} n_i$ ways.

2 **Addition Rule:** Mutually Exclusive events occur in $\sum_{i=1}^{N} n_i$ ways.

9.1.2 Random Variables

Random variables are described by their probability distribution functions called "PDF". Associated with the random variable X is a **Distribution Function** $F_X(x)$,

$$F_X(x) = \Pr(X \leq x) \tag{9.10}$$

This $F_X(\cdot)$ is called as the **Probability Distribution Function** or Cumulative Distributive Function (CDF) of the random variable X. The function $F_X(\cdot)$ has the following properties

(i) $F_X(\cdot)$ is monotonically increasing from 0 at $x = -\infty$ to 1 at $x = +\infty$,

$$\lim_{x \to -\infty} F_X(x) = 0 \qquad \lim_{x \to +\infty} F_X(x) = 1$$

$G_X(x) = (1 - F_X(x))$ is the **complementary Distribution Function**

For any $a, b \in R,$) (a real set) and $a \leq b$

$$\Pr(X \in (a, b])) = \Pr(a < X \leq b) = \Pr((a, b]) = F_X(b) - F_X(a) \tag{9.11}$$

Very often, $f_X(x)$ and $F_X(x)$ will be interchangeably used for $f(x)$ and $F(x)$ when pertinent.

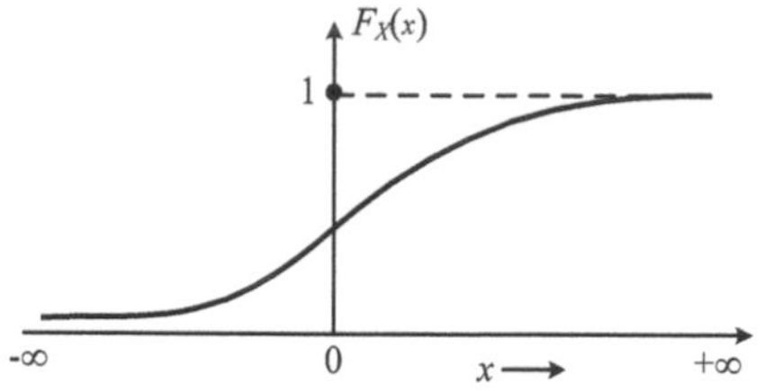

Figure 9.4: Cumulative Distribution Function, CDF

9.2 Continous RV and Probability Density Function

Let the random variable X be continous. Then there exists a probability density function (pdf), defined as $f(x)$ or $f_X(x)$ (different from "PDF") such that

$$\Pr(a < X \leq b) = \int_a^b f(x)\,\mathrm{d}x, \quad \text{for } a = -\infty, b = x, \text{ we define}$$

$$\Pr(X \leq x) = \Pr((-\infty, x]) = F(x) = \int_{-\infty}^{x} f(y)\,\mathrm{d}y$$

$$f(x) = \frac{\mathrm{d}}{\mathrm{d}x} F(x) \geq 0 \qquad \forall\ x$$

$$\Pr(X \in I) = \int_I f(y)\,\mathrm{d}y \quad (I \text{ is open, closed or half-closed interval})$$

$$\Pr(X = x) = \int_x^x f(y)\,\mathrm{d}y = 0, \quad \Pr(x < X \leq x + \mathrm{d}x) = f(x)\,\mathrm{d}x$$

$$\frac{f(a)}{f(b)} = \lim_{\epsilon \to 0^+} \left[\frac{F(a < X \leq a + \epsilon)}{F(b < X \leq b + \epsilon)} \right]$$

$$\int_{-\infty}^{+\infty} f(x)\,\mathrm{d}x = 1, \quad f(x) \text{ is akin to the "mass density."}$$

9.2.1 Expected Value, Variance, and Standard Deviation of RV

(a) **Expected value** $E[X]$

Expected value is equivalent to the average value if we look at a large ensemble of the random variable X and is denoted by $E[X]$ or μ_x.

$$\mu_x = E[X] = \int_{-\infty}^{+\infty} x f(x)\,\mathrm{d}x \tag{9.12}$$

In general, $E[g(x)] \neq g(E[X])$. For example, $E[X^2] \neq (E[X])^2$

Sometimes the mean value μ_x is also called the **Mathematical Expectation**. If X_1 and X_2 are two independent random variables,

$$E[X_1 X_2] = E[X_1]E[X_2] = \mu_{x_1}\mu_{x_2}$$

(b) **Variance** Var [X]

It is defined as the expected value of the square of the difference between the random variable X and $E[X]$ and is represented by σ_x^2.

$$\sigma_x^2 = E[(X - \mu_x)^2] = \int_{-\infty}^{+\infty} (x - \mu_x)^2 f(x)\,\mathrm{d}x \tag{9.13}$$

$$\sigma_x = \text{ standard deviation (from the mean) } = \sqrt{\text{Var }[X]}$$

(c) ***n*-th order statistical moment**

$$m_{nx} = \int_{-\infty}^{+\infty} x^n f(x)\,\mathrm{d}x$$

In order to determine the pdf of a smooth function $f(x)$, all its statistical moments are required (Taylor series expansion).

9.2.2 Discrete Random Variable

Let X be a discrete random variable with an ordered range $r = \{k_i\}_{i=1}^{N}$ of discrete values. Define the counterpart to pdf and PDF for the discrete case as:

$$f(k) = \Pr(X = k) = \Pr(k) \qquad \{k \in r\} \tag{9.14}$$

$$F_X(n) = \sum_{k=k_1}^{n} f(k) \qquad \{n \in r\} \tag{9.15}$$

$$F_X(n_1 \le k \le n_2) = \sum_{k=n_1}^{n_2} f(k) \tag{9.16}$$

Example 9.2:

The probability of an item being defective is q. Given N items, what is the probability that at the most k of the items are defective.

Solution:

$$\Pr(0 \le k \le N) = \sum_{k=0}^{N} \binom{N}{k} (q)^k (1-q)^{N-k}$$

($(1-q)$ is the probability that the item is not defective)

9.2.3 n-Dimensional Random Variables Distribution

Let $\{X_i\}_{i=1}^{n}$ be random variables for $n \in N$. The n-dimensional **joint distribution function** is defined as

$$F(x_1, x_2, \cdots, x_n) = \Pr(X_1 \le x_1, X_2 \le x_2, \cdots, X_n \le x_n) \tag{9.17}$$

The **joint probability density** function is given by:

$$f(x_1, x_2, \cdots, x_n) = \frac{\partial^n}{\prod_{i=1}^{n} \partial x_i} (F(x_1, x_2, \cdots, x_n)) \tag{9.18}$$

The **marginal probability density** of the random variable X_i is:

$$f_{X_i}(x_i) = f(x_i)$$

$$f(x_i) = \underbrace{\int_{-\infty}^{+\infty}\int_{-\infty}^{+\infty}\cdots\int_{-\infty}^{+\infty}}_{((n\text{-}1)\text{ integrals})} f(x_1,\cdots,x_{i-1},x_i,x_{i+1},\cdots,x_n)\,\mathrm{d}x_n,\cdots,\ \mathrm{d}x_{i+1},\ \mathrm{d}x_{i-1}\cdots\mathrm{d}x_1$$

Note: The variable x_i is not integrated.

The marginal probability distribution function of X_i is:

$$F_{X_i}(x_i) = \int_{-\infty}^{x_i} f_{X_i}(x)\,\mathrm{d}x = \int_{-\infty}^{x_i} f(x)\,\mathrm{d}x \tag{9.19}$$

9.2.4 Two-Dimensional Random Variables (Bivariate)

Let X, Y be a pair of continuous random variables in some region R_z of X, Y space. (X, Y) is represented by a point (x, y) in the x, y plane. Region R_z may include the entire x, y plane, $(-\infty < x < \infty,\ -\infty < y < \infty)$ or some isolated regions. Then

$$F(x, y \in R_z) = \iint_{x,y\in R_z} f(x, y)\ \mathrm{d}y\ \mathrm{d}x$$

where

$$f(x, y) = \lim_{\substack{\mathrm{d}x\to 0\\ \mathrm{d}y\to 0}} \frac{F(x < X \le x + \mathrm{d}x, y < Y \le y + \mathrm{d}y)}{\mathrm{d}x\ \mathrm{d}y} = \frac{\partial^2}{\partial x \partial y} F(x, y)$$

If X and Y are **independent variables**, then knowing the value of X does not effect the value of Y, and therefore,

$$f(x, y) = f(x) f(y) \qquad \text{(independent variables)}$$

$f(x, y)$ is often referred to as bivariate joint probability density function. The independence is designated as $X \perp Y$ (orthogonal).Independence and orthogonality of random variables is used interchangeably.

- **Conditional Expectation with Random Variables**

  Let

$$\Pr[(X = x) \cap (Y = y)] = \rho(x, y)$$

$$\Pr[(X = x)/Y = y] = \frac{\rho(x, y)}{\Pr(Y = y)} = \frac{\rho(x, y)}{\int_{-\infty}^{+\infty} \rho(x, y)\, \mathrm{d}x}$$

$$E[X/Y = y] = \frac{\int_{-\infty}^{+\infty} x\rho(x, y)\, \mathrm{d}x}{\int_{-\infty}^{+\infty} \rho(x, y)\, \mathrm{d}x}$$

  It can be easily shown that

$$E[E[X/Y]] = E[X]$$

$$E[YE[X/Y]] = E[YX] = \int_{-\infty}^{+\infty}\int_{-\infty}^{+\infty} xy\rho(x, y)\, \mathrm{d}x\, \mathrm{d}y$$

9.2.5 Bivariate Expectation, Covariance

(a) **Bivariate Expectation**

Consider a random variable Z, which is a function of two random variables,

$$Z = g(X, Y)$$

R_z is the region in the x, y plane where $g(x, y) \leq z$ implying

$$\{(X, Y) \in R_z\} = \{Z \leq z\} = \{g(X, Y) \leq z\}$$

Then the statistics of Z are:

$$F_Z(z) = \Pr[Z \le z] = \Pr\{(X, Y) \in R_z\} = \iint_{R_z} f(x, y)\, dx\, dy$$

and

$$f(z)\, dz = \Pr[z < Z \le (z + dz)] = \iint_{\Delta R_z} f(x, y)\, dx\, dy$$

$$E[g(x, y)] = \iint_{x,y \in R_z} g(x, y) f(x, y)\, dy\, dx \tag{9.20}$$

(i) If $g(x, y) = g(x)$, then $E[g(x, y)] = \iint_{x,y \in R_z} g(x) f(x, y)\, dy\, dx$

(ii) If $g(x, y) = g(y)$, then $E[g(x, y)] = \iint_{x,y \in R_z} g(y) f(x, y)\, dy\, dx$

(iii) If $g(x, y) = (x + y)$, then $E[g(x, y)] = \iint_{x,y \in R_z} (x + y) f(x, y)\, dy\, dx$

or

$$E[X + Y] = E[X] + E[Y]$$

Examples of the statistics of various functions of (X, Y)

Following two examples, involve integration in X, Y-plane.

1. $Z = g(X, Y) = X + Y$

   Region R_z for $(x + y) \le z$ is to the left of the line $x + y = z$ (line in the x, y plane with a slope -1 and an intercept equal to z on the x-axis).

$$F_Z(z) = \int_{-\infty}^{+\infty} \int_{-\infty}^{z-y} f(x, y)\, dx\, dy = \int_{-\infty}^{z-x} \int_{-\infty}^{+\infty} f(x, y)\, dx\, dy$$

$$[f_Z(z)]\, dz = \left[\int_{-\infty}^{+\infty} \int_{-\infty}^{z-y} f(z - y, y)\, dy \right] dz = \left[\int_{-\infty}^{z-x} \int_{-\infty}^{+\infty} f(x, z - x)\, dx \right] dz$$

When X and Y are independent variables,

$$f_Z(z) = \int_{-\infty}^{+\infty} f_X(z-y) f_Y(y)\,\mathrm{d}y = \int_{-\infty}^{+\infty} f_X(x) f_Y(z-x)\,\mathrm{d}x \qquad \text{(Convolution)}$$

2. $Z = \left(X^2 + Y^2\right)^{1/2}, x^2 + y^2 = r^2, r \le z$

   R_z is inside of the circle $x^2 + y^2 \le z^2, \quad z > 0$

$$x = r\cos\theta, \qquad y = r\sin\theta, \qquad \mathrm{d}x\,\mathrm{d}r = r\,\mathrm{d}r\,\mathrm{d}\theta$$

$$f(x, y) = f(\sqrt{x^2 + y^2}) = g(r)$$

Thus

$$F_Z(z) = \int\int_{R_z} \left(x^2 + y^2\right)^{1/2} f(x, y)\,\mathrm{d}x\,\mathrm{d}y = \int_0^{2\pi}\int_0^{Z} rg(r)\,\mathrm{d}r\,\mathrm{d}\theta$$

$$= 2\pi \int_0^{Z} rg(r)\,\mathrm{d}r$$

(b) **Bivariate Covariance**

$$\mathrm{Cov}\,[X, Y] = \sigma_{xy} = E[(X - \mu_x)(Y - \mu_y)] \qquad \text{(Scalar)}$$

Thus,

$$\sigma_{xx} = \sigma_x^2 = \mathrm{Cov}\,[X, X] = \mathrm{Var}\,[X] = E[(X - \mu_x)^2]$$

$$\sigma_{yy} = \sigma_y^2 = \mathrm{Cov}\,[Y, Y] = \mathrm{Var}\,[Y] = E[(Y - \mu_y)^2]$$

$$\sigma_{xy} = \mathrm{Cov}\,[X, Y] = E[(X - \mu_x)(Y - \mu_y)] = E[XY] - E[X]E[Y]$$

If we let

$$X = \begin{bmatrix} X_1 \\ \vdots \\ X_n \end{bmatrix} \qquad \text{(Euclidian vector)}$$

Then the covariance matrix is given as

$$\text{Cov}\,[X] = E[(X - \mu_x)(X - \mu_x)^T] = \Sigma_X \qquad \text{(Matrix)} \tag{9.21}$$

$$E[X] = \mu_x \qquad \text{(Vector)}$$

$$\Sigma_X = \begin{bmatrix} \sigma^2_{x_1} & \sigma_{x_1x_2} & \cdots & \sigma_{x_1x_n} \\ \sigma_{x_1x_2} & \sigma^2_{x_2} & \cdots & \sigma_{x_2x_n} \\ \vdots & & & \\ \sigma_{x_1x_n} & \sigma_{x_2x_n} & \cdots & \sigma^2_{x_n} \end{bmatrix}, \quad \Sigma_X \text{ is a } n \times n \text{ nonnegative definite matrix.} \tag{9.22}$$

If all random variables are linearly independent then Σ_X is a positive definite diagonal matrix. The quantities $\sigma_{x_ix_j}$ is the covariance coefficient between X_i and X_j and can be represented as

$$\sigma_{x_ix_j} = \rho_{ij}\sigma_{x_i}\sigma_{x_j}, \quad |\rho_{ij}| \leq 1, \quad \rho_{ij} = 1, \text{ for } i = j$$

ρ_{ij} is correlation between x_i and x_j. For independent variables ρ_{ij} is zero.

If all the variances are the same, $\sigma^2_{x_i} = \sigma^2$, $\sigma_{x_ix_j} = \sigma^2\rho_{|i-j|} \quad i, j = 1, \ldots, N$

$$\Sigma_X = \sigma^2 \begin{bmatrix} 1 & \rho_1 & \cdots & \rho_{(N-1)} \\ \rho_1 & 1 & \cdots & \rho_{(N-2)} \\ \vdots & & & \rho_1 \\ \rho_{(N-1)} & & \rho_1 & 1 \end{bmatrix}$$

It is easy to notice that $\Sigma_X = E[XX^T] - E[X]E[X]^T$

9.2.6 Lindeberg–Feller Central Limit Theorem

Gaussian Stochastic process is one of the few processes for which it is possible to determine the joint probability density function (pdf) from a set of independent identically distributed (iid) random multivariate representing the process. For a Gaussian stochastic process, we need to know only its mean and variance, namely the first two moments. Let X_i, $i = 1, \ldots, N$ be independent, identically distributed (iid) random variables, μ_{x_i} and $\sigma^2_{x_i}$ be the mean and variance of X_i. Define:

$$X = \frac{1}{N}\sum_{i=1}^{N} X_i, \quad N \to \infty \tag{9.23}$$

$$\mu_x = E\left[\frac{1}{N}\sum_{i=1}^{N} X_i\right], \quad N \to \infty$$

$$\hat{\sigma}^2_x = E\left[\frac{1}{N^2}\left(\sum_{i=1}^{N} (X_i - \mu_{x_i})^2\right)\right], \quad N \to \infty$$

Central Limit Theorem

The Central Limit Theorem states that a set of independent and identically distributed random variables $X_1, X_2, \ldots, X_N$ (as $N \to \infty$) with mean μ_x and variance σ^2_x, approaches a **Gaussian** (or **Normal**) distribution with a pdf:

$$f(x) = \left(\frac{1}{\sqrt{2\pi\sigma^2_x}}\right) e^{-\left[(x-\mu_x)^2/2\sigma^2_x\right]}, \quad \sigma_x = \hat{\sigma}_x/\sqrt{N} \tag{9.24}$$

Note:“Gaussian” and “Normal” are used interchangeably.

Preliminaries

The following two results are used in proving the Central Limit theorem.

(1) For $p << 1$ and $N \to \infty$

$$(1 + p)^N = e^{N\ln(1+p)} \tag{9.25}$$

where

$$\ln(1 + p) = \sum_{n=1}^{\infty}(-1)^{n+1}\frac{p^n}{n} = p - \frac{1}{2}p^2 + \text{Higher Order Terms (HOT)} \quad (9.26)$$

This can be easily seen by taking the natural log of both sides of Eq. 9.26,

$$\ln(1 + p)^N = N \ln(1 + p)$$

(2) Higher than second order statistical moments for X_i for $i = 1, 2, \ldots, N$ are of order $O\ (N^{-k}, k > 2)$, and as such can be neglected as N becomes very large.

We shall prove the Central Limit Theorem as follows:

Proof

The joint pdf of independent stochastic variables $X_i, \quad i = 1, 2, \ldots, N$ is:

$$f(x_1, x_2, \ldots, x_N) = \prod_{i=1}^{N} f(x_i), \quad f(x_i) \text{ being the pdf of } X_i \quad (9.27)$$

For indentically distributed random variables, let us designate $f(x_i) = f(x)\ \forall i$.

Taking the Fourier transform of the pdf $f(x_1, x_2, \cdots, x_n)$

$$F[f(x_1, x_2, \ldots, x_n)] = F(j\omega) = \int_{-\infty}^{+\infty} e^{-j\omega x} f(x)\, \mathrm{d}x \quad (9.28)$$

where

$$F(j\omega) = \int_{-\infty}^{+\infty} \cdots \int_{-\infty}^{+\infty} (e^{-j\omega/N \sum_{i=1}^{N} x_i})(\prod_{i=1}^{N} f(x_i))\, \mathrm{d}x_1\, \mathrm{d}x_2 \ldots \mathrm{d}x_n$$

or

$$F(j\omega) = \prod_{i=1}^{N} \left[\int_{-\infty}^{+\infty} e^{-j\omega x_i/N} f(x_i)\, \mathrm{d}x_i \right] \quad (9.29)$$

Since all the variables, X_i, have identical statistical properties

$$F(j\omega) = \left[\int_{-\infty}^{+\infty} e^{-j\omega x/N} f(x)\,\mathrm{d}x\right]^N \tag{9.30}$$

But

$$e^{-j\omega x/N} = \sum_{n=0}^{\infty}\left(1 + \left(\frac{-j\omega}{N}\right)x + \frac{(j\omega)^2}{2N^2}x^2 + \text{H.O.T}\right)$$

H.O.T are the higher order terms of $O(N^{-3})$

$$F(j\omega) = \left[\int_{-\infty}^{+\infty}\left(1 + \left(\frac{-j\omega}{N}\right)x + \frac{(j\omega)^2}{2N^2}x^2 + O(N^{-3})\right) f(x)\,\mathrm{d}x\right]^N$$

Neglecting H.O.T., $F(j\omega) = \left[1 + \left(\frac{-j\omega}{N}\right)\mu_x + \frac{1}{2}\left(\frac{-j\omega}{N}\right)^2\left(\hat{\sigma}_x^2 + \mu_x^2\right)\right]^N$ (9.31)

Let

$$\left(\frac{-j\omega}{N}\right)\mu_x + \frac{1}{2}\left(\frac{-j\omega}{N}\right)^2\left(\hat{\sigma}_x^2 + \mu_x^2\right) = p(\omega)$$

$$F(j\omega) = (1 + p(\omega))^N$$

From Eq. 9.26

$$F(j\omega) = e^{N\ln(1+p(\omega))} \approx e^{N[p(\omega)-(1/2)p(\omega)^2]} \tag{9.32}$$

Now

$$\begin{aligned} p(\omega) - \frac{p(\omega)^2}{2} &= \left[\left(\frac{-j\omega}{N}\right)\mu_x + \frac{1}{2}\left(\frac{-j\omega}{N}\right)^2\left(\hat{\sigma}_x^2 + \mu_x^2\right)\right] \\ &\quad -\frac{1}{2}\left[\left(\frac{-j\omega}{N}\right)\mu_x + \frac{1}{2}\left(\frac{-j\omega}{N}\right)^2\left(\hat{\sigma}_x^2 + \mu_x^2\right)\right]^2 \\ &\approx \left(\frac{-j\omega}{N}\right)\mu_x + \frac{1}{2}\left(\frac{-j\omega}{N}\right)^2\left(\hat{\sigma}_x^2 + \mu_x^2\right) - \frac{1}{2}\left(\frac{j\omega}{N}\right)^2\mu_x^2 \\ &\approx \left(\frac{-j\omega}{N}\right)\mu_x + \frac{1}{2}\left(\frac{j\omega}{N}\right)^2\hat{\sigma}_x^2 \end{aligned} \tag{9.33}$$

From the Eqs. 9.32 and 9.33

$$F(j\omega) = \int_{-\infty}^{+\infty} e^{-j\omega x} f(x)\,\mathrm{d}x = e^{\left(-j\omega\mu_x + (j\omega)^2\hat{\sigma}_x^2/2N\right)} \tag{9.34}$$

Taking the inverse Fourier Transform of Eq. 9.34

$$f(x) = \frac{1}{2\pi}\int_{-\infty}^{+\infty} e^{j\omega x}\cdot e^{-j\omega\mu_x}\cdot e^{-\omega^2\hat{\sigma}_x^2/2N}\,\mathrm{d}\omega$$

$$f(x) = \frac{1}{2\pi}\int_{-\infty}^{+\infty} e^{j\omega(x-\mu_x)-\omega^2\hat{\sigma}_x^2/2N}\,\mathrm{d}\omega \tag{9.35}$$

Let

$$x - \mu_x = a$$

$$\hat{\sigma}_x^2/2N = b = \sigma_x^2/2$$

Then the Eq. 9.35 takes the form,

$$f(x) = \frac{1}{2\pi}\left[\int_{-\infty}^{+\infty} e^{ja\omega - b\omega^2}\,\mathrm{d}\omega\right]$$

We have encountered this integral in Chapter 4 on Fourier Transform and yields

$$f(x) = \frac{1}{2\pi}\left[\sqrt{\frac{\pi}{b}}e^{-a^2/4b}\right]$$

Hence,

$$f(x) = \frac{1}{\sqrt{2\pi\sigma_x^2}}e^{-(x-\mu_x)^2/2\sigma_x^2} \qquad \text{(Gaussian Density Function)}$$

In case there are N-Gaussian distributed random variables (not iid), the joint probability density function can be written as:

$$f(\boldsymbol{x}) = f(x_1, x_2, \cdots, x_n) = \left(\frac{1}{(2\pi)^{N/2}\left(\Delta_{\Sigma_x}\right)^{1/2}}\right) e^{-1/2(x-\mu_x)^T \Sigma_x^{-1}(x-\mu_x)}$$

$$\Delta_{\Sigma_x} = \det(\boldsymbol{\Sigma}_x), \quad (\boldsymbol{\Sigma}_x) \text{ is a covariance matrix}$$

$$\boldsymbol{x}^T = [x_1, x_2, \cdots, x_n]$$

Note: Exponential expressions appear all over the place in electrical signals and stochastic processes. The Gaussian distributed random variables distribution process is the most studied subject in engineering.

Two Useful Formulae

(1) Let us define Laplace's probability integral

$$\Phi(x) = \sqrt{2/\pi} \int_0^x e^{-t^2/2}\,dt = -\Phi(-x)$$

Then

$$\Pr(x_1 < X < x_2) = 1/2\left[\Phi\left(\frac{x_2 - \mu_x}{\sigma_x}\right) - \Phi\left(\frac{x_1 - \mu_x}{\sigma_x}\right)\right]$$

(2) Probability of an event A is:

$$\Pr(A) = \int_{-\infty}^{+\infty} f(x)\Pr(A/x)\,dx$$

Similarly

$$f(x/A) = \frac{f(x)\Pr(A/x)}{\int_{-\infty}^{+\infty} f(x)\Pr(A/x)\,dx} \quad \text{(General Bayes Formula)}$$

9.3 Random Walk, Brownian, and Wiener Process

In physics, the **Brownian Motion** is described as a phenomena in which particles are suspended in a liquid undergoe random molecule collisions. This motion is characterized by the discrete **Random Walk** stochastic variable $W(n)$ represented by $w(n)$. As $n \to \infty$, we obtain a continuous homogenous motion also characterized as the "Wiener Process" described below. Consider a symmetric random walk on the real line, t, where a particle moves from its present position $w(0) = 0$ one step up or one step down a distance $\sqrt{(\mathrm{d}t)}$, with the equal probability of 1/2. The random variables representing this process at each step are independent, **identically distributed** and are known as Bernoulli random variables. This process is described by the equation:

$$w(n) = w(n-1) \pm \sqrt{(\mathrm{d}t)}, \quad n = 1, 2, \ldots$$

where

$$\begin{aligned}
w(n) &= \text{particle position above the origin, after } n \text{ steps, starting at origin} \\
\sqrt{(\mathrm{d}t)} &= \text{step size} \\
n &= \text{total number of steps} \\
k &= \text{number of steps above the origin} \\
(n-k) &= \text{number of steps below the origin}
\end{aligned}$$

$$w(n) = k\sqrt{(\mathrm{d}t)} - (n-k)\sqrt{(\mathrm{d}t)} = (2k-n)\sqrt{(\mathrm{d}t)}$$

The random variable $W(n)$ represents the position $w(n)$ and is written as:

$$W(n) = W_1 + W_2 + \cdots + W_n$$

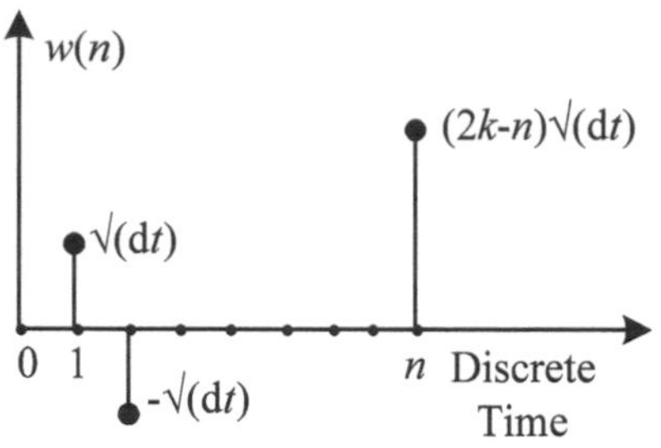

Figure 9.5: Random Walk—Wiener Process

$$\Pr\left[W(n) = w(n) = (2k-n)\sqrt{(\,\mathrm{d}t)}\right] = \binom{n}{k}\left(\frac{1}{2^k}\right)\left(\frac{1}{2^{n-k}}\right) = \binom{n}{k}\left(\frac{1}{2^n}\right)$$

The probability density function (pdf) of the process is:

$$f(w(n)) = \frac{1}{2^n}\sum_{k=0}^{n}\binom{n}{k}\delta(w(n) - (2k-n)\sqrt{(\,\mathrm{d}t)})$$

The variables W_i are identically distributed and independent (iid) and take the value $\sqrt{(\,\mathrm{d}t)}$ or $-\sqrt{(\,\mathrm{d}t)}$ with probability 1/2, yielding

$$E\,[W_i] = 0, \quad \mathrm{Var}\;[W_i] = \sqrt{(\,\mathrm{d}t)}^2 = \mathrm{d}t$$

Notation: iid represents Gaussian RV independent and identically distributed.

$$W(n) = W_1 + W_2 + \cdots + W_n = \sum_{i=1}^{n} W_i$$

$$E\,[W(n)] = \sum_{i=1}^{n} E\,[W_i] = 0$$

$$\mathrm{Var}\;[W(n)] = \sum_{i=1}^{n} \mathrm{Var}\left[W_i^2\right] = n\,\mathrm{d}t$$

Using the Central Limit Theorem, for large n

$$\Pr\left[W(n) = (2k-n)\sqrt{(\,\mathrm{d}t)}\right] \approx \frac{1}{\sqrt{2\pi n\sqrt{(\,\mathrm{d}t)}}}e^{-((2k-n)\sqrt{(\,\mathrm{d}t)}^2/2n\,\mathrm{d}t}$$

Viewed as a function of discrete time n, the random variable $W(n)$ gives the instantaneous position $w(n)$ of a random walker. This property is widely known as discrete **Brownian motion** or the **Wiener process** $W(t)$ for large values of n. Let

$$n\,\mathrm{d}t = t$$
$$(2k-n)\sqrt{(\,\mathrm{d}t)} = w(t)$$

Note: In what follows, we shall use $w(t)$ and $W(t)$ as well as $x(t)$ and $X(t)$ interchangeably. Then the probability density function of the continuous **Wiener process** $W(t)$ with instantaneous value $w(t)$ is given by:

$$f(w(t)) = \frac{1}{\sqrt{2\pi t}}e^{-w^2(t)/2t} \tag{9.36}$$

- Incremental variables $\left[W(n_i) - W(n_j)\right]$, $\;i > j$ depends only upon $\left(n_i - n_j\right)$ steps and therefore represent statistically independent variables.
- Incremental variables $\left(W(t_i) - W(t_j)\right)$, $\;i > j$, depends only upon $\left(t_i - t_j\right)$ and represent statistically independent variables.
- A continous path process $\bar{W}(t)$ can be obtained via linear interpolation as:

$$\bar{W}(t) = W(n) + (W(n+1) - W(n))(t - n\,\mathrm{d}t) \qquad n\,\mathrm{d}t \le t \le (n+1)\,\mathrm{d}t$$

with a mean and variance

$$E[\bar{W}(t)] = 0$$

$$E[\bar{W}(t)^2] = t$$

The incremental variable $\bar{W}(t+s) - \bar{W}(t),\ s > 0$ depends only upon s and is therefore independent of the past variable t. Furthermore,

$$\Pr[\bar{W}(t+s) - \bar{W}(t) \geq x] = \frac{1}{\sqrt{(2\pi s)}} \int_{x}^{+\infty} e^{-y^2/2s}\,\mathrm{d}y$$

Important Properties of the Wiener Processes

1. A Wiener process $W(t)$ represents a Gaussian variable with zero mean and variance t and can also be designated as $W(0, t)$.

2. Wiener increments are defined as $W(t_0), W(t_1) - W(t_0), \ldots, W(t_k) - W(t_{k-1})$ $t_0 < t_1 < \cdots < t_k$. These Wiener increments are Gaussian independent random variables.

3. The process $W(t) - W(s)$ can also be defined as:

$$W(t) - W(s) \approx W(t-s) = W(0, t-s), \quad s \leq t$$

4. Often, it is helpful to think of the Wiener process as an "incremental process."

Statistics of the Wiener Process

- $E[(W(t))^2] = t$
- $E\left[(W(t) - W(s))^2\right] = |t - s|$
- $E[W(t)W(s)] = \min(t, s)$

- $W(s)$ and $W(t-s)$ are uncorrelated implying $E[W(s)W(t-s)] = 0, \quad s \le t$

- $W(t)$ and $W(t-s)$ are not uncorrelated, $E[W(t)W(t-s)] = (t-s), \ \ s \le t$

- Cross-correlation between $w(s)$ and $w(t)$ is:

$$\frac{E[W(s)W(t)]}{\left[E[W^2(s)]E[W^2(t)]\right]^{1/2}} = \frac{\sqrt{s}}{\sqrt{t}} \quad \text{for} \quad t \ge s$$

- From Eq. 9.36, it is obvious that $W(\lambda t) = \sqrt{\lambda} W(t)$

  Hence, it suffices to study the Wiener process for the 0 to 1 time interval and compute its statistical behavior for all times.

- Consider a Wiener process $W(t)$ at time t represented by the variable $w(t)$ having the initial value $w(0)$ at $t = 0$. The pdf of this process is given by the conditional probability density function:

$$f(w(t)/w(0)) = \frac{1}{\sqrt{2\pi t}} e^{-((w(t)-w(0))^2)/2t}, \qquad t > 0$$

- Consider the set of random Wiener process variablcs, $(W(t_1), W(t_2)\ldots, W(t_n))$ for $0 < t_1 < t_2 < \ldots < t_n$. The joint probability density of the set is given by:

$$f\,(w(t_1),\ldots,w(t_n)) = f\,(w(t_n)/w(t_{n-1}))\,f\,(w(t_{n-1})/w(t_{n-2}))\ldots f\,(w(t_1)/w(0))$$

  It is easy to see that:

$$f\,(w(t+s)/w(s)) = \int_R f\,(w(t)/x(t))\,f\,(x(s)/w(s))\;\mathrm{d}x \qquad 0 < s < t$$

  This is the famous Chapman–Kolmogorav equation.

- Wiener process is Markov in the sense that past history has no influence on the future of the process if its present value is specified.

  Thus

$$\Pr\,[W(t_n) \leq w_n / W(t), \quad t \leq t_{n-1}] = \Pr\,[W(t_n) \leq w_n / W(t_n - 1)]$$

  Thus for $t_1 < t_2 < \ldots < t_n$ it follows that:

$$\Pr\,[W(t_n) \leq w_n / W(t_{n-1}), \ldots, W(t_1)] = \Pr\,[W(t_n) \leq w_n / W(t_n - 1)]$$

- For a deterministic dynamical system described by a variable $x(t)$ and its derivative $\frac{\mathrm{d}x}{\mathrm{d}t}$, the variable $x(t)$ as a function of time is of importance. On the other hand, the stochastical dynamic systems is represented by a stochastic process, $X(t)$, whose mean and variance are of importance and not $x(t)$ as a function of time. $X(t)$ being a stochastic variable does not have a time "derivative," its dynamics are described by the differential $\mathrm{d}X$. In order to compute a functional of the stochastic variable $X(t)$, we encounter an extra term not present in the computation of the functional of the deterministic variable $x(t)$. This is arrived at as a result of the Taylor series expansion, retaining only first and second order terms due to the fact that the mean and variance are the only parameters of importance.

  The Wiener process is not differentiable in the usual sense and therefore it is not feasible to talk of $\frac{\mathrm{d}}{\mathrm{d}t}W(t)$ unless we think of **differentials** as "**increments**." Thus, a deterministic differential equation corresponds to a stochastic differential equation as follows:

$$\text{Deterministic DE}: \quad \frac{\mathrm{d}}{\mathrm{d}t}x(t) = Ax(t) + \frac{\mathrm{d}}{\mathrm{d}t}w(t)$$

$$\text{Stochastic DE}: \quad \mathrm{d}X(t) = AX(t)\,\mathrm{d}t + \mathrm{d}W(t)$$

- Incremental Wiener process relations are:

$$W(t+\Delta t) = W(t) + \sqrt{\Delta t}X(0,1)$$

where $X(0,1)$ is a Gaussian variable with zero mean and unity variance

$$W(t+\Delta t) - W(t) = \Delta W(t)$$

$$E\left[\Delta W(t)\right] = E\left[W(t+\Delta t)\right] - E\left[W(t)\right] = 0 - 0 = 0$$

$$E\left[\Delta W^2(t)\right] = E\left[W^2(t+\Delta t) - W^2(t)\right] = (t+\Delta t) - t = \Delta t$$

- Differential Rules

For a Gaussian stochastic process, we are mostly interested in the expected mean value and variance. Taking into consideration the stochastic nature of the variables involving mean and variance, the following rules for multiplication of the differentials are:

$$\left.\begin{aligned} \mathrm{d}t\,\mathrm{d}t &= (\mathrm{d}t)^2 = 0 \\ \mathrm{d}W(t)\,\mathrm{d}t &= \mathrm{d}t\,\mathrm{d}W(t) = 0 \\ \mathrm{d}W(t)\,\mathrm{d}W(t) &= (\mathrm{d}W(t))^2 = \mathrm{d}t \end{aligned}\right\} \tag{9.37}$$

We shall prove these rules and use them to solve stochastic differential equations.

- Taylor series expansion for the increments.

Deterministic variable function $F(x(t),t)$

$$\begin{aligned} \mathrm{d}F &= F(x(t) + \mathrm{d}x(t), t + \mathrm{d}t) - F(x(t),t) \\ &= \frac{\partial F}{\partial t}\mathrm{d}t + \frac{\partial F}{\partial x}\mathrm{d}x + \frac{1}{2}\left[\frac{\partial^2 F}{\mathrm{d}t^2}(\mathrm{d}t)^2 + 2\frac{\partial^2 F}{\partial t \partial x}\mathrm{d}x\,\mathrm{d}t + \frac{\partial^2 F}{\partial x^2}(\mathrm{d}x)^2\right] \\ &\quad + \mathrm{H.O.T} \end{aligned} \tag{9.38}$$

where H.O.T are $O((\mathrm{d}t)^3, (\mathrm{d}x)^3)$.

When $\mathrm{d}x$ and $\mathrm{d}t$ tend to zero, second and higher order terms vanish

$$\mathrm{d}F = \frac{\partial F}{\partial t}\,\mathrm{d}t + \frac{\partial F}{\partial x}\,\mathrm{d}x \tag{9.39}$$

Stochastic variable function

For a general stochastic process $x(t)$, not all the second order terms shall vanish but for a Gaussian process this is true. For a Gaussian random process $x(t)$, the differential DF takes the form:

$$\mathrm{d}F = \frac{\partial F}{\partial t}\,\mathrm{d}t + \frac{\partial F}{\partial X}\,\mathrm{d}X(t) + \frac{1}{2}\frac{\partial^2 F}{\partial X^2}\,(\mathrm{d}X(t))^2 \tag{9.40}$$

where $\frac{\partial F}{\partial x}$ is computed for $X(t) = x(t)$.

This is often referred to as the **Ito differential rule**.

Example 9.3:

Given:

$$F(t) = F(X(t), t), \quad X(t) \text{ is Gaussian}$$

$$\mathrm{d}X(t) = A\,\mathrm{d}t + B\,\mathrm{d}W(t)$$

where $W(t)$ is the Wiener process.

We obtain:

$$\mathrm{d}F = \frac{\partial F}{\partial t}\,\mathrm{d}t + \frac{\partial F}{\partial X}\,\mathrm{d}X + \frac{1}{2}\frac{\partial^2 F}{\partial X^2}\,(\mathrm{d}X)^2$$

or

$$\begin{aligned}\mathrm{d}F = {} & \frac{\partial F}{\partial t}\,\mathrm{d}t + \frac{\partial F}{\partial X}\,[A\,\mathrm{d}t + B\,\mathrm{d}W(t)] \\ & + \frac{1}{2}\frac{\partial^2 F}{\partial X^2}\left[A^2(\mathrm{d}t)^2 + 2AB\,\mathrm{d}W(t)\,\mathrm{d}t + B^2(\mathrm{d}W(t))^2\right]\end{aligned}$$

Since

$$(\mathrm{d}t)^2 = 0, \quad \mathrm{d}W(t)\,\mathrm{d}t = 0, \quad (\mathrm{d}W(t))^2 = \mathrm{d}t$$

We arrive at

$$\mathrm{d}F = \left[\left(\frac{\partial F}{\partial t} + A\frac{\partial F}{\partial X}\right) + \frac{1}{2}B^2\frac{\partial^2 F}{\partial X^2}\right]\mathrm{d}t + B\frac{\partial F}{\partial X}\,\mathrm{d}W(t) \tag{9.41}$$

Due to Ito's Differential rule, only four terms remain.

The difference between the Reimann integral and the Ito integral:

$$\int \frac{\partial F}{\partial x}\,\mathrm{d}x = \left[F(x(t), t) - \frac{\partial f}{\partial t}\,\mathrm{d}t\right] \qquad \text{(Riemann Integral)}$$

$$\left[\int \frac{\partial F}{\partial x}\,\mathrm{d}x\right]_{x=X(t)} = \left[F(x,t) - \frac{\partial F}{\partial t}\,\mathrm{d}t - \frac{1}{2}\int \frac{\partial^2 F}{\partial x^2}\,\mathrm{d}x^2\right]_{x(t)-X(t)} \qquad \text{(Ito Integral)}$$

Different Ways of Constructing Wiener Process

Recall that while deriving the Brownian motion, we looked at it as a weighted summation of independent identically distributed Gaussian variables. We shall use this as an underlying idea for various simulation schemes.

Method #1 "Incremental Formulation with Zero Order Hold"

Let $\{X_i\}_{i=1}^{n}$ represent Gaussian independent and identically distributed variables(Bernoulli random variables) with zero mean and unity variance. Construct the following recursive stochastic process:

$$\hat{W}(t_0) = 0$$

$$\hat{W}(t_k) = \hat{W}(t_{k-1}) + (t_k - t_{k-1})^{1/2} X_k \qquad k = 1, 2, \ldots, n$$

$$\hat{W}(t) = \hat{W}(t_{k-1}), \qquad t_{k-1} \le t \le t_k$$

This piecewise continous process $\hat{W}(t)$ converges to $W(t)$ as $n \to \infty$ and $(t_k - t_{k-1} \to 0)$.

Method #2 "Incremental Formulation with First Order Hold"

The above recursive scheme is modified as:

$$\hat{W}(t) = \hat{W}(t_k) + \left(\frac{t - t_k}{t_k - t_{k-1}}\right)(W(t_k) - W(t_{k-1})), \qquad t_{k-1} \leq t < t_k$$

Method #3 "Karhunen-Loeve Expansion"

We shall discuss this topic in later pages in detail. Suffices to say that we can simulate $W(t)$ as

$$\hat{W}(t) = \sum_{k=0}^{n} X_k \phi_k(t)$$

where X_k are random independent variables and $\phi_k(t)$ are a complete set of orthonormal functions for $0 \leq t \leq T$. As $n \to \infty$, the process $\hat{W}(t)$ converges to Wiener process.

This is essentially approximately the Wiener process through a stochastic Fourier series.

Method #4 "Integration of White Noise"

The white noise process can be written as

$$\hat{N}(t) = \sum X_k \delta(t - k\Delta t), \qquad (n-1)\Delta t < t \leq n\Delta t$$

where X_k are iid RVs that take values $+\Delta t$ and $-\Delta t$ with equal probability. As $n \to \infty$, the process approaches white noise $N(t)$, such that

$$E\left[N(t)N(t-\tau)\right] = \delta(t-\tau)$$

This process $\hat{N}(t)$ is now passed through an integrator yielding $\hat{W}(t)$.

Figure 9.6: Wiener Process as an Integration of White Noise

Thus,

$$\hat{W}(t) = \sum_{k=1}^{n} X_k U(t - k\Delta t), \qquad (n-1)\Delta t < t \leq n\Delta t$$

$$\hat{N}(t) = \frac{\mathrm{d}}{\mathrm{d}t}\hat{W}(t)$$

For large values of n, $\Delta t \to 0$

$$\hat{N}(t) \to \text{Gaussian white noise process} N(t)$$

$$\hat{W}(t) \to \text{Wiener process} W(t)$$

9.3.1 Stochastic Differential and Integral Equations (SDE)

Consider the behavior of a deterministic dynamical system that is described by the deterministic differential equation:

$$\frac{\mathrm{d}x}{\mathrm{d}t} = f(x(t), t), \qquad x(t_0) = x_0 \tag{9.42}$$

We can write the above equation as:

$$\mathrm{d}x(t) = f(x(t), t)\,\mathrm{d}t \tag{9.43}$$

where $\mathrm{d}x(t)$ represents an increment in $x(t)$ for time increment $\mathrm{d}t$.

The stochastic system cannot be described in the form of Eq. 9.42 because $X(t)$ is not differentiable almost everywhere. Eq. 9.43 can be used to describe a Gaussian stochastic process $X(t)$ driven by another Gaussian white noise process $N(t)$ as:

$$\mathrm{d}X(t) = f(X(t), t)\,\mathrm{d}t + g(X(t), t)N(t)\,\mathrm{d}t, \quad N(t)\,\mathrm{d}t = \mathrm{d}W(t) \tag{9.44}$$

$$X(t_0) = X_0 \tag{9.45}$$

We can solve this SDE via the Ito differential rule as follows: Consider:

$$F(X(t)) = \ln X(t)$$

According to the **Ito differential rule**

$$\mathrm{d}F = \frac{\partial F}{\partial t}\,\mathrm{d}t + \frac{\partial F}{\partial X(t)}\,\mathrm{d}X(t) + \frac{1}{2}\frac{\partial^2 F}{\partial X^2(t)}(\,\mathrm{d}X(t))^2 \tag{9.46}$$

But

$$\frac{\partial F}{\partial t} = 0, \quad \frac{\partial F}{\partial X(t)} = \frac{1}{X(t)}, \quad \frac{\partial^2 F}{\partial X^2(t)} = -\frac{1}{X^2(t)}$$

Hence,

$$\mathrm{d}F = \frac{\mathrm{d}X(t)}{X(t)} - \frac{1}{2}\left(\frac{\mathrm{d}X(t)}{X(t)}\right)^2$$

From Eq. 9.44

$$\frac{\mathrm{d}X(t)}{X(t)} = \frac{f(X(t)}{X(t)}\,\mathrm{d}t + \frac{g(X(t))}{X(t)}\,\mathrm{d}W(t)$$

$$\left(\frac{\mathrm{d}X(t)}{X(t)}\right)^2 = \left[\frac{f(X(t))}{X(t)}\right]^2 \mathrm{d}^2t + \left[\frac{g(X(t))}{X(t)}\right]^2 \mathrm{d}^2W(t) + 2\left[\frac{f(X(t))}{X(t)}\right]\left[\frac{g(X(t))}{X(t)}\right] \mathrm{d}t\,\mathrm{d}W(t)$$

Using the Ito differential multiplication rules:

$$(\,\mathrm{d}t)^2 = 0, \quad (\,\mathrm{d}t)(\,\mathrm{d}W(t)) = 0, \quad (\,\mathrm{d}W(t))^2 = \left(\sqrt{\mathrm{d}t}\right)^2 = \mathrm{d}t$$

Thus,

$$\mathrm{d}F = \left[\frac{f(X(t))}{X(t)}\right]\mathrm{d}t + \left(\frac{g(X(t))}{X(t)}\right)\mathrm{d}W(t) - \frac{1}{2}\left(\frac{g(X(t))}{X(t)}\right)^2 \mathrm{d}t$$

Integrating both sides,

$$[\ln X(t)]_{t=t_0}^{t=t} = \int_{t_0}^{t} \left[\left(\frac{f(X(t))}{X(t)}\right) - \frac{1}{2}\left(\frac{g(X(t))}{X(t)}\right)^2\right]\mathrm{d}t + \int_{t_0}^{t} \left(\frac{g(X(t))}{X(t)}\right)\mathrm{d}W(t)$$

Indeed it is important that $F(X(t))$ be a well-behaved function having the above partial derivatives.

Example 9.4:

Solve the following SDE (Stochastic Differential Equation)

$$\mathrm{d}X(t) = \mu X(t)\,\mathrm{d}t + \sigma X(t)\,\mathrm{d}W(t), \quad X(0) = 1$$

From Eq. 9.44

$$f(X(t), t) = \mu X(t), \quad g(X(t), t) = \sigma X(t), \quad F(X(t)) = \ln X(t)$$

Therefore,

$$\mathrm{d}F = \mu\,\mathrm{d}t - \frac{1}{2}\sigma^2\,\mathrm{d}t + \sigma W(t)$$

Integrating both sides

$$\ln X(t) - \ln X_0 = \ln X(t) = \left[\left(\mu - \frac{1}{2}\sigma^2\right)t + \sigma W(t)\right], \quad \ln X_0 = \ln 1 = 0$$

or

$$X(t) = e^{[(\mu - 0.5\sigma^2)t + \sigma W(t)]}$$

The above expression is an essential part of the Black-Scholes pricing options.

Example 9.5:

To illustrate the difference between the Riemann and the Ito integration, consider the following stochastic integral:

$$F(t) = \int_0^t W(t)\,dW(t), \quad W(0) = 0$$

Solution:

$$W(t) = X(t), \quad X(0) = 0, \quad (F(X(t)) = \frac{1}{2}X^2(t), \text{ from Riemann integral})$$

From Stochastic Differential Equation:

$$dX(t) = dW(t), \quad \frac{\partial F(X(t))}{\partial t} = 0, \quad \frac{\partial F(X(t))}{\partial X(t)} = X(t), \quad \frac{\partial^2 F(X(t))}{\partial X^2(t)} = 1$$

From the Ito differential rule Eq. 9.46, $dF(X) = X(t)\,dX(t) + 1/2\,dt$

Integrating both sides

$$[F(X(t))]_0^t = \frac{1}{2}\left(W^2(t) - W^2(0)\right) = \int_0^t W(t)\,dW(t) + \frac{1}{2}t, \quad W(0) = 0$$

$$2\int_0^t W(t)\,dW(t) = W^2(t) - t$$

This is a strange result from the ordinary rules of the Reimann integration. But the probabilistic results are correct, since:

$$E\left[\int_{t_0}^t W(t)\,dW(t)\right] = E\left[\frac{1}{2}(W^2(t) - t)\right]_{t_0}^t = \frac{1}{2}\left[t - t\right]_{t_0}^t = 0$$

9.3.2 Simplified Ito's Theorem and Ito's Differential Rules

If a stochastic function $F(x(t), t)$ is driven by a Wiener process and has continous derivatives $\frac{\partial F}{\partial t}$, $\frac{\partial F}{\partial x}$ and $\frac{\partial^2 F}{\partial x^2}$, then it can be represented as:

$$\Big[F(x(t), t)\Big]_{x(0)=W(0)}^{x(t)=W(t)} = \left[\int_0^t \frac{\partial F}{\partial t}\,\mathrm{d}t + \int_0^t \frac{\partial F}{\partial x}\,\mathrm{d}x(t) + \frac{1}{2}\int_0^t \frac{\partial^2 F}{\partial x^2}\,\mathrm{d}t\right]_{x(t)=W(t)}$$

Interestingly, as pointed out earlier, there is an extra term on the RHS, when compared to the ordinary integrals. From Ito calculus the new rules of differential multiplication are:

$$\mathrm{d}t \cdot \mathrm{d}t = 0, \qquad \mathrm{d}t \cdot \mathrm{d}W(t) = 0, \qquad \mathrm{d}W(t) \cdot \mathrm{d}W(t) = \mathrm{d}t$$

Consider a product of the form, $I = (a_1\,\mathrm{d}t + a_2\,\mathrm{d}W(t)) \cdot (b_1\,\mathrm{d}t + b_2\,\mathrm{d}W(t))$, the only surviving term in the above product is $\mathrm{d}W(t) \cdot \mathrm{d}W(t)$ yielding

$$I = a_2 b_2\,\mathrm{d}W(t) \cdot \mathrm{d}W(t) = a_2 b_2\,\mathrm{d}t$$

If we replace $W(t)$, with a process $X(t)$ driven by $W(t)$, then:

$$\Big[\mathrm{d}F(x(t), t)\Big]_{x(t)=X(t)} = \left[\frac{\partial F}{\partial t}\,\mathrm{d}t + \frac{\partial F}{\partial x}\,\mathrm{d}x(t) + \frac{1}{2}\frac{\partial^2 F}{\partial x^2}\,\mathrm{d}x(t) \cdot \mathrm{d}x(t)\right]_{x(t)=X(t)} \qquad (9.47)$$

This calculus is often referred to as "Box Calculus."

Example 9.6:

Given a SDE:

$$\mathrm{d}X(t) = \left(\mu X^2(t) + \sigma^2 X^3(t)\right)\,\mathrm{d}t + \sigma X^2(t)\,\mathrm{d}W(t), \quad X(0) = 1 \qquad (9.48)$$

Determine $X(t)$

Solution: Let us solve the problem in two different ways.

Method 1: Ito's Theorem

$$F(X) = -\frac{1}{X}, \quad \frac{\partial F}{\partial X} = \frac{d}{dX}\left(\frac{-1}{X}\right) = \frac{1}{X^2}, \quad \frac{\partial^2 F}{\partial X^2} = -\frac{2}{X^3}$$

We apply Ito's theorem to the function $(-1/X)$. Thus

$$\left[d\left(\frac{-1}{X}\right)\right]_{X=X(t)} = \left[\left(\frac{1}{X^2}\right)\left(\mu X^2 + \sigma^2 X^3\right) dt + \frac{1}{2}\left(\frac{-2}{X^3}\right)\left(\sigma X^2\right)^2 dt + \left(\frac{1}{X^2}\right)\left(\sigma X^2\right) dW(t)\right]$$

Simplifying and Integrating both sides,

$$\frac{-1}{X(t)} + \frac{1}{X(0)} = \mu t + \sigma W(t), \qquad X(0) = 1$$

The resulting solution is:

$$X(t) = (1 - \mu t - \sigma W(t))^{-1}$$

Method 2: Let us solve the same problem via Ito's differential rules (Box Calculus)

$$F(x) = \frac{-1}{x}, \quad \frac{\partial F}{\partial t} = 0, \quad \frac{\partial F}{\partial x} = \frac{1}{x^2}, \quad \frac{\partial^2 F}{\partial x^2} = \frac{-2}{x^3}$$

$$dt \cdot dt = 0, \;\; dt \cdot dW = 0, \;\; dW \cdot dW = dt$$

From Eq. 9.48,

$$dx \cdot dx = \left[\left(\mu x^2 + \sigma^2 x^3\right) dt + \sigma x^2 dW\right] \cdot \left[\left(\mu x^2 + \sigma^2 x^3\right) dt + \sigma x^2 dW\right]$$

Simplifying,

$$dx \cdot dx = \sigma^2 x^4 dt$$

From Eq. 9.47,

$$\left[\mathrm{d}\left(\frac{-1}{x}\right)\right]_{x=X(t)} = \left[\left(\frac{1}{x^2}\right)\left(\mu x^2 + \sigma^2 x^3\right)\,\mathrm{d}t + \left(\frac{1}{2}\right)\left(\frac{-2}{x^3}\right)\left(\sigma x^2\right)^2\,\mathrm{d}t + \sigma\,\mathrm{d}W(t)\right]_{x=X(t)}$$

yielding,

$$\left(\frac{-1}{X(t)} + 1\right) = \mu t + \sigma W(t)$$

$$X(t) = (1 - \mu t - \sigma W(t))^{-1}$$

Example 9.7:

Determine (i) $E[e^{\sigma W(t)}]$, (ii) $E[W^k(t)]$, $k > 1$, $W(t)$ is a Wiener process, $W(0) = 0$

Solution:

(i) Let $F(X(t), t) = e^{X(t)}$, $\quad X(0) = 0, \quad F(0) = 1, \quad X(t) = \sigma W(t)$

$$\mathrm{d}X(t) = \sigma\,\mathrm{d}W(t)$$

$$\begin{aligned}\mathrm{d}F(X(t), t) = {} & \frac{\partial}{\partial t}F(X(t), t)\,\mathrm{d}t + \frac{\partial}{\partial X(t)}F(X(t), t)\,\mathrm{d}X(t) \\ & + \frac{1}{2}\frac{\partial^2}{\partial X^2(t)}F(X(t), t)(\,\mathrm{d}X(t))^2\end{aligned}$$

$$\mathrm{d}F(X(t), t) = 0 + F(X(t), t)\sigma\,\mathrm{d}W(t) + \frac{1}{2}F(X(t), t)\sigma^2(\,\mathrm{d}W(t))^2$$

$$E\left[\frac{\mathrm{d}F(X(t))}{F(X(t))}\right] = E\left[\sigma\,\mathrm{d}W(t) + \frac{1}{2}\sigma^2(\,\mathrm{d}W(t))^2\right] = \frac{1}{2}\sigma^2\,\mathrm{d}t$$

Integrating

$$E\,[\ln F(X(t))] = \frac{1}{2}\sigma^2 t$$

Hence

$$E[F(X(t))] = e^{(1/2)\sigma^2 t}, \quad E\left[e^{\sigma W(t)}\right] = e^{(1/2)\sigma^2 t}$$

(ii) Using the power series expansion for both sides of the above equation

$$E\left[\sum_{k=0}^{\infty} \sigma^k \frac{W^k(t)}{k!}\right] = E\left[\sum_{m=0}^{\infty} \left(\frac{1}{2}\right)^m \sigma^{2m} \frac{t^m}{m!}\right]$$

Equating terms with same powers of σ

$$E\left[W^k(t)\right] = 0, \quad k = 1, 3, 5, \ldots$$
$$E\left[\frac{W^k(t)}{k!}\right] = \left(\frac{1}{2}\right)^{k/2} \frac{t^{k/2}}{(k/2)!}, \quad k = 2, 4, 6, \ldots$$

Example 9.8:

This example demonstrates the power of the Ito's Lemma and a new as well as a simple way of deriving the celebrated Black-Scholes equation.

Let

$x(t)$ = Price of the security per share

$W(t)$ = Brownian motion driving the security market

$I = I(x(t), t) = I(x, t)$ = Investment

$F = F(x(t), t) = F(x, t)$ = The hedging financial instrument(put or call options)

$n(t)$ = Number of shares of the security

$V = (n(t))x$

The dynamics of the financial markets are:

$$dx = \mu x \, dt + \sigma x \, dW(t)$$

$$dI = rI \, dt$$

$$dF = dV - dI \quad \text{where} \quad F = V - I$$

μ, σ and r represent "drift," "volatility," and interest rate respectively.

Thus

$$\mathrm{d}x = \mu x\,\mathrm{d}t + \sigma x\,\mathrm{d}W \tag{9.49}$$

$$\mathrm{d}F = rF\,\mathrm{d}t - rV\,\mathrm{d}t + \mathrm{d}V \tag{9.50}$$

$$\mathrm{d}V = \frac{\partial V}{\partial t}\,\mathrm{d}t + \frac{\partial V}{\partial x}\,\mathrm{d}x \tag{9.51}$$

From Ito's Lemma

$$(\mathrm{d}t)^2 = 0, \quad \mathrm{d}t\,\mathrm{d}x = 0, \quad \mathrm{d}t\,\mathrm{d}W = 0, \quad \text{and} \quad (\mathrm{d}W(t))^2 = \mathrm{d}t$$

Derivation of Black-Scholes Equation

Step 1:

Multiply Eq. 9.49 with itself:

$$(\mathrm{d}x)^2 = (\mu x\,\mathrm{d}t)^2 + 2(\mu x)(\sigma x)\,\mathrm{d}W(t)\,\mathrm{d}t + \sigma^2 x^2 (\mathrm{d}W(t))^2$$

Applying Ito's Lemma

$$(\mathrm{d}x)^2 = \sigma^2 x^2\,\mathrm{d}t \tag{9.52}$$

Step 2:

From Taylor series expansion (realizing that $(\mathrm{d}x)^2 \neq 0$)

$$\mathrm{d}F = \mathrm{d}F(x(t), t) = \frac{\partial F}{\partial t}\,\mathrm{d}t + \frac{\partial F}{\partial x}\,\mathrm{d}x + \frac{1}{2}\frac{\partial^2 F}{\partial x^2}(\mathrm{d}x)^2 \tag{9.53}$$

Combining Eqs. 9.52 and 9.53:

$$\mathrm{d}F = \frac{\partial F}{\partial t}\,\mathrm{d}t + \frac{\partial F}{\partial x}\,\mathrm{d}x + \frac{1}{2}\sigma^2 x^2 \frac{\partial^2 F}{\partial x^2}\,\mathrm{d}t \tag{9.54}$$

Furthermore, from Eq. 9.50

$$\mathrm{d}F = rF\,\mathrm{d}t - rV\,\mathrm{d}t + \mathrm{d}V$$

Step 3:

(i) Multiply Eq. 9.54 with $\mathrm{d}x$

$$\mathrm{d}F\,\mathrm{d}x = \frac{\partial F}{\partial x}(\mathrm{d}x)^2 \tag{9.55}$$

(ii) Multiple Eq. 9.50 with $\mathrm{d}x$

$$\mathrm{d}F\,\mathrm{d}x = \mathrm{d}V\,\mathrm{d}x = \left(\frac{\partial V}{\partial t}\mathrm{d}t + \frac{\partial V}{\partial x}\mathrm{d}x\right)\mathrm{d}x = \frac{\partial V}{\partial x}(\mathrm{d}x)^2 \tag{9.56}$$

Comparing Eqs. 9.55 and 9.56:

$$\frac{\partial F}{\partial x}(\mathrm{d}x)^2 = \frac{\partial V}{\partial x}(\mathrm{d}x)^2$$

or

$$\boxed{\frac{\partial F}{\partial x} = \frac{\partial V}{\partial x} = n(t), \quad V = (n(t))x = \frac{\partial F}{\partial x}x} \tag{9.57}$$

Step 4:

From Eqs. 9.51, 9.54, and 9.57, we obtain the Black-Scholes equation

$$\frac{\partial F}{\partial t} + \frac{1}{2}\sigma^2 x^2 \frac{\partial^2 F}{\partial x^2} + rx\frac{\partial F}{\partial x} - rF = 0$$

9.3.3 Optimal Control of the Stochastic Process

Consider a stochastic dynamic system driven by a combination of deterministic control $\boldsymbol{u}(t)$ and stochastic Wiener process $W(t)$:

$$\mathrm{d}\boldsymbol{X}(t) = \boldsymbol{f}(\boldsymbol{X}(t), \boldsymbol{u}(t))\,\mathrm{d}t + \sigma \boldsymbol{b} W(t), \quad \boldsymbol{X}(t_0) = \boldsymbol{X}_0, \boldsymbol{X}(t_f) = \boldsymbol{X}_f \tag{9.58}$$

Find an optimal control law which minimizes:

$$J(\boldsymbol{u}) = E[\int_{t_0}^{t_f} f_0(\boldsymbol{X}(t), \boldsymbol{u}(t))\,\mathrm{d}t] + h(\boldsymbol{X}(t_f)))$$

We shall minimize the above expression w.r.t $\boldsymbol{u}(t)$ using dynamic programming algorithm of optimality and use of Ito's differential rules. Let us introduce

$$\hat{V}(\boldsymbol{X}(t), t) = \min_{\boldsymbol{u}(t)} \left[h(\boldsymbol{X}(t_f)) + \int_{t}^{t_f} f_0(\boldsymbol{X}(t), \boldsymbol{u}(t))\,\mathrm{d}t \right], \qquad t_0 \le t \le t_f$$

Applying the principle of optimality

$$\hat{V}(\boldsymbol{X}(t), t) = \min_{\boldsymbol{u}(t)} \left[h(\boldsymbol{X}(t_f)) + \int_{t+\mathrm{d}t}^{t_f} f_0(\boldsymbol{X}(t), \boldsymbol{u}(t))\,\mathrm{d}t + \int_{t}^{t+\mathrm{d}t} f_0(\boldsymbol{X}(t), \boldsymbol{u}(t))\,\mathrm{d}t \right], \qquad t_0 \le t \le t_f$$

The first two terms in the above expression represent $\hat{V}(\boldsymbol{X}(t) + \mathrm{d}\boldsymbol{X}(t), t + \mathrm{d}t)$. Thus

$$\hat{V}(\boldsymbol{X}(t), t) = \min_{\boldsymbol{u}(t)} \left[\hat{V}(\boldsymbol{X}(t) + \mathrm{d}\boldsymbol{X}(t), t + \mathrm{d}t) + f_0(\boldsymbol{X}(t), \boldsymbol{u}(t))\,\mathrm{d}t \right]$$

$$\hat{V}(\boldsymbol{X}(t), t) = \min_{\boldsymbol{u}(t)} \left[\hat{V}(\boldsymbol{X}(t), t) + \frac{\partial \hat{V}(\boldsymbol{X}(t), t)}{\partial t}\,\mathrm{d}t + \left[\left(\nabla_x \hat{V}(\boldsymbol{X}(t), t) \right)^T \cdot \mathrm{d}\boldsymbol{X}(t) \right] \mathrm{d}t \right.$$
$$\left. + \frac{1}{2} \left[\mathrm{d}\boldsymbol{X}^T(t) \cdot \nabla_{xx} \hat{V}(\boldsymbol{X}(t), t) \cdot \mathrm{d}\boldsymbol{X}(t) \right] \mathrm{d}t + f_0(\boldsymbol{X}(t), \boldsymbol{u}(t))\,\mathrm{d}t \right]$$

yielding

$$\min_{u(t)}\left[\frac{\partial}{\partial t}\hat{V}(\boldsymbol{X}(t),t)\,\mathrm{d}t+\left[\left(\nabla_x\hat{V}(\boldsymbol{X}(t),t)\right)^T\cdot\,\mathrm{d}\boldsymbol{X}(t)\right]\mathrm{d}t\right.$$
$$\left.+\frac{1}{2}\left[\mathrm{d}\boldsymbol{X}^T(t)\cdot\nabla_{xx}\hat{V}(\boldsymbol{X}(t),t)\cdot\,\mathrm{d}\boldsymbol{X}(t)\right]\mathrm{d}t+f_0(\boldsymbol{X}(t),\boldsymbol{u}(t))\,\mathrm{d}t\right]=0$$

Since our performance index is quadratic, we seek a solution of the form:

$$\hat{V}(\boldsymbol{X}(t),t)=\frac{1}{2}\boldsymbol{X}^T(t)\boldsymbol{P}(t)\boldsymbol{X}(t)+c(t),\quad \boldsymbol{P}^T(t)=\boldsymbol{P}(t)\tag{9.59}$$

The resultant optimality equation are:

$$\min_{u(t)}\left[\boldsymbol{X}^T(t)\dot{\boldsymbol{P}}(t)\boldsymbol{X}(t)\,\mathrm{d}t+\dot{c}\,\mathrm{d}t+\left[\boldsymbol{X}^T(t)\cdot\boldsymbol{P}(t)\cdot\,\mathrm{d}\boldsymbol{X}(t)+\,\mathrm{d}\boldsymbol{X}^T(t)\cdot\boldsymbol{P}(t)\cdot\boldsymbol{X}(t)\right]\mathrm{d}t\right.$$
$$\left.+\left[\mathrm{d}\boldsymbol{X}^T(t)\boldsymbol{P}(t)\,\mathrm{d}\boldsymbol{X}(t)\right]\mathrm{d}t+f_0(\boldsymbol{X}(t),\boldsymbol{u}(t))\,\mathrm{d}t\right]=0\tag{9.60}$$

Substituting for $\mathrm{d}\boldsymbol{X}(t)$ from Eq. 9.59 and differentiating wrt $\boldsymbol{u}(t)$, we obtain:

$$\nabla_u f_0(\boldsymbol{X}(t),\boldsymbol{u}(t))+\nabla_u(\boldsymbol{f}^T(\boldsymbol{X}(t),\boldsymbol{u}(t)))\boldsymbol{P}(t)\boldsymbol{X}(t)=\boldsymbol{0}\tag{9.61}$$

Eq. 9.61 yields the control $\boldsymbol{u}(t)$ as a function of $\boldsymbol{X}(t)$ and in combination with Eq.9.61 resulting in the Riccati equation for the matrix $\boldsymbol{P}(t)$. Details of the optimal algorithm are illustrated by the following example.

Example 9.9:

Stochastic Quadratic Performance Index Regulator

$$\mathrm{d}\boldsymbol{X}(t)=(\boldsymbol{A}(t)\boldsymbol{X}(t)+\boldsymbol{B}(t)\boldsymbol{u}(t))\,\mathrm{d}t+\boldsymbol{\sigma}\,\mathrm{d}\boldsymbol{W}(t)$$
$$J(\boldsymbol{u})=E\left[\int_{t_0}^{t_f}\frac{1}{2}\left(\boldsymbol{X}^T(t)\boldsymbol{Q}\boldsymbol{X}^T(t)+\boldsymbol{u}^T(t)\boldsymbol{R}\boldsymbol{u}(t)\right)\mathrm{d}t\right]+\frac{1}{2}\boldsymbol{X}^T(t_f)\boldsymbol{F}\boldsymbol{X}(t_f)$$

Optimal Control, from Eq. 9.61 is:

$$\boldsymbol{R}\boldsymbol{u}(t) + \boldsymbol{B}^T(t)\boldsymbol{P}(t)\boldsymbol{X}(t) = 0$$

$$\boldsymbol{u}(t) = -\boldsymbol{R}^{-1}\boldsymbol{B}(t)\boldsymbol{P}(t)\boldsymbol{X}(t) \tag{9.62}$$

From Eqs. 9.59 and 9.62

$$\frac{1}{2}\boldsymbol{X}^T(t)\left[\dot{\boldsymbol{P}}(t) + \boldsymbol{P}(t)\boldsymbol{A}(t) + \boldsymbol{A}^T(t)\boldsymbol{P}(t) + \boldsymbol{Q} - \boldsymbol{P}(t)\boldsymbol{B}^T(t)\boldsymbol{R}^{-1}\boldsymbol{B}(t)\boldsymbol{P}(t)\right]\boldsymbol{X}(t)\,\mathrm{d}t$$
$$+\left[\dot{c}(t) + \boldsymbol{\sigma}^T\boldsymbol{P}(t)\boldsymbol{\sigma}\right]\mathrm{d}t = 0 \tag{9.63}$$

The solution to Eq. 9.63 yields

$$\dot{\boldsymbol{P}}(t) + \boldsymbol{P}(t)\boldsymbol{A}(t) + \boldsymbol{A}^T(t)\boldsymbol{P}(t) + \boldsymbol{Q} - \boldsymbol{P}(t)\boldsymbol{B}^T(t)\boldsymbol{R}^{-1}\boldsymbol{B}(t)\boldsymbol{P}(t) = 0$$
$$c(t) = \int_t^{t_f}(\boldsymbol{\sigma}^T\boldsymbol{P}(t)\boldsymbol{\sigma})\,\mathrm{d}t = \int_t^{t_f}\mathrm{Trace}(\boldsymbol{\sigma}^T\boldsymbol{\sigma}\boldsymbol{P}(t))\,\mathrm{d}t$$
$$\boldsymbol{P}(t_f) = \boldsymbol{F}, \quad c(t_f) = 0$$

But this controller has a drawback that it has no effect on suppressing the noise variance. Let us consider an alternate approach based upon minimizing the Hamiltonian. Details of this algorithm are explained via the following example.

$$\mathrm{d}X(t) = (aX(t) + bu(t))\,\mathrm{d}t + \sigma\,\mathrm{d}W(t)$$
$$J(u) = E[\int_0^{t_f}\frac{1}{2}(qX^2(t) + ru^2(t))\,\mathrm{d}t] + \int_0^{t_f}\dot{c}(t)\,\mathrm{d}t, \qquad c(t_f) = 0$$

$$\mathrm{d}H(t, X(t), \lambda(t), u(t)) = \lambda(t)[(aX(t) + bu(t))\,\mathrm{d}t + \sigma\,\mathrm{d}W(t)]$$
$$+\frac{1}{2}(qX^2(t) + ru^2(t))\,\mathrm{d}t + \dot{c}(t)\,\mathrm{d}t$$

$$\min_{u(t)}[\,dH] = \min_{u(t)}\left[\frac{\partial H}{\partial t}\,dt + \frac{\partial H}{\partial \lambda(t)}\,d\lambda(t) + \frac{\partial H}{\partial X(t)}\,dX(t) + \frac{\partial H}{\partial u(t)}\,du(t) + \frac{1}{2}\frac{\partial^2 H}{\partial X^2(t)}(\,dX(t))^2\right] = 0$$

From Ito's rules:

$$(\,dX(t))^2 = \sigma^2\,dt, \quad dW(t)\,d(t) = 0, \quad dW(t)\,dW(t) = \sigma^2, \quad dt\,dt = 0$$

Hence,

$$\min_{u(t)}[\,dH] = \min_{u(t)}\left[(aX(t) + bu(t))\,d\lambda(t)\,dt + (a\lambda(t) + qX(t))\,dX(t)\,dt + (b\lambda(t) + ru(t) + q\sigma^2 + \dot{c})\,dt\right] = 0$$

$$\min_{u(t)}\left[(aX(t) + bu(t))\,d\lambda(t)\,dt + (a\lambda(t) + qX(t))(aX(t) + bu(t))\,dt + (b\lambda(t) + ru(t) + q\sigma^2 + \dot{c})\,dt\right] = 0$$

$$d\lambda(t) = -[a\lambda(t) + qX(t)]$$

$$b\lambda(t) + ru(t) = 0, \quad c(t) = \int_t^{t_f} q\sigma^2\,dt$$

$$u(t) = -\left(\frac{b}{r}\right)\lambda(t) \qquad \text{(Control Law)}$$

Feedback controller takes the form, $\lambda(t) = k(t)X(t)$, $k(t)$ being the feedback gain with no variance compensation for the Wiener Process input.

Resulting Riccati equation is $\dot{k} - 2ak + (b^2/r)k^2 + q = 0$

9.3.4 General Random Walk

A stochastic process $X(n)$ is called a **General Random Walk** if it can be presented:

$$X(n) = \alpha X(n-1) + c + N(n), \quad |\alpha| < 1, \; n = 1, 2, \ldots$$

c is a constant, known as the drift

$N(n)$ is white noise with zero mean and variance σ^2

$X(0)$ is the initial value of the process

The solution of the above single order difference equation is:

$$X(n) = \alpha^n X(0) + c\frac{(1-\alpha^n)}{1-\alpha} + \sum_{i=0}^{n-1} \alpha^i N(i)$$

As $n \to \infty$ (large), $\alpha^n \to 0$, we obtain

$$X(n) = \frac{c}{1-\alpha} + \sum_{i=0}^{n-1} \alpha^i N(i), \quad \text{yielding} \quad \mu_x = \frac{c}{1-\alpha}$$

$$\begin{aligned}\text{Cov}\,(X(n), X(m)) &= \text{Cov}\left(\sum_{i=1}^{n-1} \alpha^i N(i), \sum_{j=1}^{m-1} \alpha^j N(j)\right), \quad n, m \to \infty \\ &= \sum_{j=1}^{m-n-1} \sum_{i=1}^{n-1} \text{Cov}\left(\alpha^i N(i), \alpha^j N(j)\right) = \sum_{j=1}^{m-n-1} \alpha^{2j}\sigma^2 = \frac{\alpha^{2(m-n)}}{1-\alpha^2}\sigma^2\end{aligned}$$

Let $m - n - 1 = \tau, \quad n = t$, then

$$\text{Cov}\,(X(t), X(t-\tau)) = \left(\frac{\sigma^2}{1-\alpha^2}\right)\alpha^{2\tau}$$

Filteration (as applied to Financial instruments)

Given a set of random variables $\{X_t : 0 \leq t < \infty\}$, its filteration process $\{F(t)\}_{t>0}$ defines the concept where,

(i) $E[X(t)/F(t)]$ is measurable

(ii) $[F(s)]_{s>0} \subseteq F(t),\ s \leq t$

(iii) Information about $X(t)$ increases with respect to $f(t)$ as t increases

9.3.5 Martingale's Stochastic Process

Let $X(t)$ be a stochastic process denoted by it's realization $x(t)$. This process is Martingale if and only if,

$$E\left[X(t)\right] = \mu_x(t) < \infty, \quad \text{first moment}$$

$$E\left[X(t+\Delta t)/\left(E[X(t)] = \mu_x(t)\right)\right] = \mu_x(t) \quad \text{for all } t \geq 0 \text{ and all } \Delta t > 0$$

Furthermore, if $F(t)$ is a filteration of $X(t)$, then

$$E[X(t+\Delta t)/F(t)] = X(t), \quad \text{for } t \geq 0, \Delta t > 0$$

Value of all future moments = Present value of the moments

In a nutshell, the Martingale's stochastic process is characterized by the fact that the expected values of all the future values of the process are the same as at the present time and have zero "drift" in the expected value.

The essence of the Martingale is the assumption that the "Game" is fair and efficient. If you depart from fairness it is an unjust game. Information contained in the past variable values is fully reflected in the current values. This is a fundamental assumption in the financial engineering which assumes that if there are no arbitrage games, then the market prices reflect risk neutral pricing. This is one of the flaw in the reasoning that financial engineering tools may yield anomalous results.

Example 9.10:

Show that $X(t) = \left(W^2(t) - t\right)$ is a Martingale.

Solution:

The following two conditions should be satisfied:

(i) $E[X(t)] = 0$

(ii) $E[X(t)/X(s)] = X(s)$ for $s \leq t$

It is clear that:

$$E[X(t)] = E[W^2(t) - t] = (t - t) = 0$$

To prove the second condition, we proceed as follows:

$$\begin{aligned} X(t) = \left(W^2(t) - t\right) &= (W(t) - W(s) + W(s))^2 - t \\ &= (W(t) - W(s))^2 + W^2(s) + 2W(s)(W(t) - W(s)) - t \end{aligned}$$

But

$$W(t) - W(s) = W(t - s)$$

$$E[W(s)W(t - s)] = 0$$

$$E[(W(t) - W(s))^2] = E[W^2(t - s)] = t - s$$

$$\begin{aligned} E[X(t)/X(s)] &= E[\left(W^2(t - s) + W^2(s) - t\right)/X(s)] \\ &= (t - s) + W^2(s) - t = W^2(s) - s = X(s) \end{aligned}$$

Thus, $X(t)$ is a Martingale.

Difference between Martingale Process and Markov Process

Every Martingale is a Markov process but every Markov process is not a Martingale. In fact:

$$E[f(X_{n+1})] = g(X_n) \quad \text{is a Markov process} \quad n = 1, 2, \dots$$

$$E[X_{n+1}] = E[X_n/X_{n-1} \cdots X_{n-k}] = X_n, \quad k > 0 \quad \text{is a Martingale.} \quad n = 1, 2, \dots$$

Martingale has no tendency to "drift," rise, or fall.

$$Z_n = (X_n - X_{n-1}) \qquad \text{is also a Martingale.}$$

$$Z_1, Z_2, \dots, Z_k \qquad \text{is an orthogonal set.}$$

9.4 Markov Chains and the Law of Large Numbers

9.4.1 Markov Chains

The Markov chain is a random process where the knowledge of the past experiments do not influence the outcome of the future experiments but depends upon the present experimental results, namely the present probabilities.

Markov Chain Definition

Let X be a discrete random process described by a sequence $X = \{X_i\}_{i=1}^{N}$ of random variables described by a countable set $S = \{S_i\}_{i=1}^{N}$ called the **State Space** S. The chain is made up of moves considered as steps. The process X represents a **Markov Chain** if it satisfies the Markov conditions.

$$P_r\{X_n = s_n/X_0 = s_0, \cdots, X_{n-1} = s_{n-1}\} = P_r\{X_n = s_n/X_{n-1} = s_{n-1}\}$$

The chain is presently in the state s_i at an instant k and moves to state s_j at the instant $(k+1)$ with a probability p_{ij}, known as transition probabilities. In general we should talk about $p_{ij}(k) = P_r\{X_{k+1} = s_j/X_k = s_i\}$

For the moment, we shall only consider the case where $p_{ij}(k)$ are independent of k and hence represented by the symbol p_{ij}. These probabilities are referred to as **Transition Probabilities** and the corresponding matrix

$$\boldsymbol{P} = \{p_{ij}\}, \quad i, j = 1, 2, \cdots, N$$

Markov Chain is symbolized by the **State Probability vector** $\boldsymbol{\pi}(k) = \{\pi_i(k)\}$.

To start the chain, we define an initial state probability vector,

$$\boldsymbol{\pi}(0) = \{\pi_i(0)\}\, i = 1, 2, \cdots, N$$

which defines the initial probability distributions of each state s_i of the state space S. The symbol $\pi_i(k)$ defines the probability the system will occupy the state i after k steps or "transitions." If the initial state, $\boldsymbol{\pi}(\mathbf{0})$, at $k = 0$ is known, then all the entries in $\boldsymbol{\pi}(0)$ are zero except for the entry in the initial state, which is one. In general, $\boldsymbol{\pi}(0)$ is assumed to be known and the sum of its entries is one.

The process driving the **Markov Chains** is a first order matrix difference equation and is defined as:

$$\boldsymbol{\pi}(k+1) = \boldsymbol{P}^T \boldsymbol{\pi}(k), \quad \boldsymbol{\pi}(0) \text{ vector representing the initial probability.} \tag{9.64}$$

It is easy to see

$$\boldsymbol{\pi}(k) = \left(\boldsymbol{P}^T\right)^k \boldsymbol{\pi}(0), \qquad \boldsymbol{P}^{(0)} = \boldsymbol{I}$$

The (i, j)-th entry of $\boldsymbol{P}^k$ is defined as $p_{ij}(k)$ and can be computed as

$$p_{ij}(k) = \sum_{m=1}^{N} p_{im}(k-1) p_{mj} \tag{9.65}$$

The entry $p_{ij}(k)$ gives us the probability that the Markov Chain starts in the state s_i at $k = 0$ and will attain the state s_j after k steps. At each step it is important to realize that

$$\sum_{i=1}^{N} \pi_i(k) = 1 \tag{9.66}$$

This is only possible if and only if

(a) $p_{ij} \geq 0$ for all i, j

(b) $\sum_{i=1}^{N} p_{ij} = 1$ for all j (The column sum is one.)

Such matrices are known as **Stochastic Matrices**.

Eq. 9.64 is a simple (depending upon $\boldsymbol{P}$) first order matrix difference equation, lending itself to the Z-transform yielding

$$\left(z\boldsymbol{I} - \boldsymbol{P}^T\right)\boldsymbol{\pi}(z) = \boldsymbol{\pi}(0)$$

$$\boldsymbol{\pi}(z) = Z[\boldsymbol{\pi}(k)] = \sum_{k=0}^{\infty} z^k \pi(k)$$

or

$$\boldsymbol{\pi}(z) = \left(z\boldsymbol{I} - \boldsymbol{P}^T\right)^{-1} \boldsymbol{\pi}(0)$$

Taking the inverse Z-transform,

$$\pi(k) = \left[\frac{1}{2\pi j} \oint_c (z\boldsymbol{I} - \boldsymbol{P}^T)^{-1} z^{(k-1)}\, \mathrm{d}z\right] \boldsymbol{\pi}(0) \tag{9.67}$$

Since the probabilities are less than one, all the eigenvalues of $\boldsymbol{P}$ lie within the unit circle C. As discussed in Chapter 4,

$$f(P) = \left(z\boldsymbol{I} - \boldsymbol{P}^T\right)^{-1} = \left(\frac{1}{p(z)}\right)\left(\sum_{m=0}^{n-1} \boldsymbol{B}_m(\boldsymbol{P}^T) z^{n-m-1}\right) \tag{9.68}$$

where

$$p(z) = \det\left(z\boldsymbol{I} - \boldsymbol{P}^T\right) = z^n + a_1 z^{n-1} + \cdots + a_n$$

$\boldsymbol{B}_m(\boldsymbol{P}^T)$ matrices are computed as follows.From Eqs. 9.67 and 9.68

$$\boldsymbol{\pi}(k) = \left[\sum_{m=0}^{n-1}\left[\frac{1}{2\pi j}\oint_c\left(\frac{f(z)}{p(z)}\right)z^{n+k-m-2}\,\mathrm{d}z\right]\boldsymbol{B}_m(\boldsymbol{P}^T)\right]\boldsymbol{\pi}(0)$$

or

$$\boldsymbol{\pi}(k) = \left[\sum_{m=0}^{n-1} C_{n-m}(k)\boldsymbol{B}_m(\boldsymbol{P}^T)\right]\boldsymbol{\pi}(0)$$

where, using the residue theorem

$$C_{n-m}(k) = \frac{1}{2\pi j}\oint_c \frac{f(z)}{p(z)} z^{n+k-m-2}\,\mathrm{d}z = \sum_{i=1}^{m}\left(\frac{1}{(r_i-1)!}\right)\frac{\mathrm{d}^{r_i-1}}{\mathrm{d}z^{r_i-1}}\left[f(z)(z-z_i)^{r_i} z^{n+k-m-2}\right]_{z=z_i}$$

$$p(z) = \prod_{i=1}^{m}(z-z_i)^{r_i}, \quad r_1 + r_2 + \cdots + r_m = n, \quad r_i \in N$$

The matrices $\boldsymbol{B}_m(\boldsymbol{P}^T)$ are computed as:

$$\boldsymbol{B}_0(\boldsymbol{P}^T) = \boldsymbol{I}$$

$$\boldsymbol{B}_1(\boldsymbol{P}^T) = \boldsymbol{P}^T + a_1\boldsymbol{I}$$

$$\boldsymbol{B}_2(\boldsymbol{P}^T) = (\boldsymbol{P}^T)^2 + a_1\boldsymbol{P}^T + a_2\boldsymbol{I}$$

$$\boldsymbol{B}_3(\boldsymbol{P}^T) = (\boldsymbol{P}^T)^3 + a_1(\boldsymbol{P}^T)^2 + a_2(\boldsymbol{P}^T) + a_3\boldsymbol{I}$$

$$\vdots$$

$$\boldsymbol{B}_{n-1}(\boldsymbol{P}^T) = (\boldsymbol{P}^T)^{n-1} + a_1(\boldsymbol{P}^T)^{n-2} + a_2(\boldsymbol{P}^T)^{n-3} + \cdots + a_{n-1}\boldsymbol{I}$$

$$\boldsymbol{0} = (\boldsymbol{P}^T)_n + a_1(\boldsymbol{P}^T)^{n-2} + a_2(\boldsymbol{P}^T)^{n-3} + \cdots + a_{n-1}(\boldsymbol{P}^T) + a_n\boldsymbol{I}$$

The difference Equation 9.64 can also be easily solved via the Cayley-Hamilton theorem discussed in Chapter 2 on Matrix Algebra.

9.4.2 Markov's Inequality

For any $k > 0$ and a **nonnegative** random variable X with mean μ_x

$$\Pr\,(X \geq k) \leq \frac{\mu_x}{k}$$

Proof:

$$E[X] = \int_0^{+\infty} xf(x)\,\mathrm{d}x = \int_0^{k} xf(x)\,\mathrm{d}x + \int_k^{\infty} xf(x)\,\mathrm{d}x$$

But

$$\int_0^{k} xf(x)\,\mathrm{d}x \geq 0$$

Hence,

$$E[X] \geq \int_k^{\infty} xf(x)\,\mathrm{d}x \geq k\int_k^{\infty} f(x)\,\mathrm{d}x$$

But

$$\int_k^{\infty} f(x)\,\mathrm{d}x = \Pr\,(X \geq k)$$

Hence,

$$\Pr\,(X \geq k) \leq \frac{\mu_x}{k}$$

9.4.3 Tchebychev's Inequality

The probability density function $f(x)$ takes a maximum value near the expected value μ_x of a random variable. If the variance σ_x is small, then the value of the random variable is close to μ_x. Tchebychev's inequality gives us bounds on $\Pr\,(|X - \mu_x| \geq k)$. Let $f(x)$ be the pdf of a random variable X with

$$E[X] = \mu_x, \quad E\left[(X - \mu_x)^2\right] = \sigma_x^2$$

Tchebychev's Inequality Theorem:

$$\Pr\left(|X - \mu_x| \geq k\right) \leq \sigma_x^2/k^2, \qquad \text{For any } k > 0. \tag{9.69}$$

Proof:

$$\sigma_x^2 = \int_{-\infty}^{+\infty} (x - \mu_x)^2 f(x)\,\mathrm{d}x = \int_{|x-\mu_x|\geq k} (x - \mu_x)^2 f(x)\,\mathrm{d}x + \int_{\mu_x-k}^{\mu_x+k} (x - \mu_x)^2 f(x)\,\mathrm{d}x$$

Since the second integral has positive value

$$\sigma_x^2 \geq \int_{|x-\mu_x|\geq k} (x - \mu_x)^2 f(x)\,\mathrm{d}x \tag{9.70}$$

Furthermore, $(x - \mu_x)^2 \geq k^2$ for $|x - \mu_x| \geq k$.Therefore, Eq. 9.70 takes the form

$$\sigma_x^2 \geq k^2 \int_{|x-\mu_x|\geq k} f(x)\,\mathrm{d}x \tag{9.71}$$

$$\int_{|x-\mu_x|\geq k} f(x)\,\mathrm{d}x = \Pr\,|(X - \mu_x) \geq k| \tag{9.72}$$

$$\Pr\left(|X - \mu_x| \geq k\right) \leq \sigma_x^2/k^2$$

If (σ_x/k) is small then the variable X takes values close to the mean value μ_x. It is important to realize that the exact form of the pdf is not necessary to know and the inequality holds for any pdf.

Note: Tchebychev's inequality can be easily proven from Markov's inequality by replacing X with $|X - \mu_x|^2$ and k with k^2.

$$\Pr\left(|X - \mu_x|^2 \geq k^2\right) \leq E\left[|X - \mu_x|^2\right]/k^2, \quad \text{implying } \Pr\left(|X - \mu_x| \geq k\right) \leq \sigma_x^2/k^2$$

9.4.4 Law of Large Numbers

Let

$$S_n = \sum_{i=1}^{n} X_i, \quad \{X_i\}_{i=1}^{n} \text{ be a sequence of iid random variables with a finite mean.}$$

Theorem #1 Weak Law of Large Numbers

$$\lim_{n\to\infty}\left[\Pr\left(\left|\frac{S_n}{n} - E\left[X_i\right]\right| \geq \epsilon\right)\right] = 0 \qquad \text{for all } i = 1,2,\ldots,n,\ \epsilon > 0$$

Theorem #2 Strong Law of Large Numbers

$$\lim_{n\to\infty}\left[\Pr\left(\lim_{n\to\infty}\left|\frac{S_n}{n} - E\left[X_i\right]\right| \geq \epsilon\right)\right] = 0 \qquad \text{for all } i = 1,2,\ldots,n,\ \epsilon > 0$$

The proof follows from the direct application of Tchebychev's inequality.

9.4.5 Sterling's Formula (Approximation)

This formula is of importance in probabilistic studies and its proof is instructive. Sterling's formula is an approximation for $n!$ in terms of powers of n for large n.

Sterling's Formula
$n! = \left(\frac{n}{e}\right)^n \sqrt{2\pi n} \qquad \text{for large } n$

Proof:

It is easy to see from integration by parts:

$$\Gamma(n+1) = \int_0^\infty t^n e^{-t}\,\mathrm{d}t = n\,\Gamma(n), \qquad n > -1$$

For $n \in$ integer

$$\Gamma(n+1) = n!, \qquad \Gamma(1) = 1$$

Let us rewrite $\Gamma(n+1)$ as an integral in terms of a function $\gamma_n(t) = t^n e^{-t}$:

$$\Gamma(n+1) = \int_0^\infty t^n e^{-t}\,\mathrm{d}t = \int_0^\infty \gamma_n(t)\,\mathrm{d}t = n!, \quad \text{where} \quad \gamma_n(t) = t^n e^{-t}$$

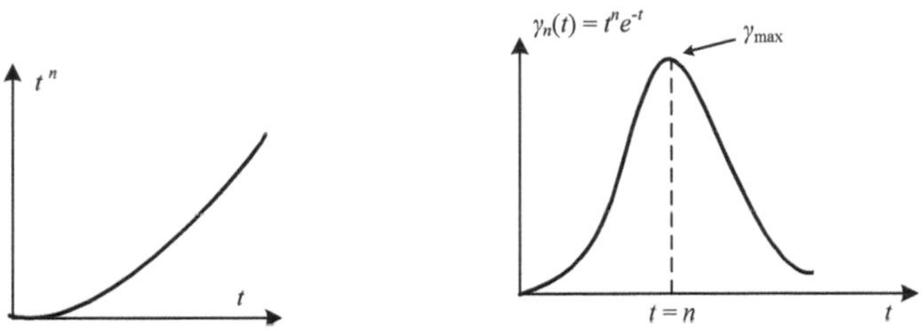

(a) Plot of t^n (b) Plot of $\gamma_n(t)$ vs. t

Figure 9.7: Transformation of t^n via Kernel e^{-t}

The function $\gamma_n(t)$ can be viewed as a transformation of t^n via kernel e^{-t} as shown in Figure 9.7. Notice that for small values of t as n increases, $\Gamma_n(t)$ increases very sharply due to t^n being dominant but eventually e^{-t} prevails and the function $\gamma_n(t)$ is brought down to zero. $\Gamma(n+1)$ can also be viewed as an area under the curve $\gamma_n(t)$ as shown in Figure 9.7. The maximum value $\gamma_{\max}$ of $\gamma_n(t)$ can be obtained by taking its time derivative and setting it equal to zero:

$$\dot{\gamma}_n = t^{n-1} e^{-t}(n-t) = 0$$

yielding

$$\gamma_n(t) = \gamma_{\max} = \gamma_m, \qquad \text{at } t = n$$

$$[\gamma_n(t)]_{\max} = n^n e^{-n} = \left(\frac{n}{e}\right)^n = \gamma_{\max} \tag{9.73}$$

As n increases, the maximum gets sharper and narrower and looks more like a Gaussian-shaped delta function with most of the area concentrated around $t = n$. Let us now look for an approximation between $\gamma_n(t)$ and $\gamma_m e^{-(t-n)^2/\sigma^2}$.

For large values of n, set

$$\gamma_m e^{-(t-n)^2/\sigma^2} = t^n e^{-t} \tag{9.74}$$

where σ is to be determined

Let

$$t - n = x$$

For large values of n, as t varies from 0 to ∞, x varies between $-\infty$ to $+\infty$.

From Eqs. 9.73 and 9.74

$$(n)^n(e)^{-n}e^{-x^2/\sigma^2} = (n+x)^n e^{-(n+x)}$$

or

$$e^{x-x^2/\sigma^2} = \left(1 + \frac{x}{n}\right)^n$$

Taking the natural log, ln of both sides.

$$x - \frac{x^2}{\sigma^2} = n \ln\left(1 + \frac{x}{n}\right)$$

But

$$\ln(1+p) = p - \frac{p^2}{2} + \frac{p^3}{3} - \cdots \approx p - p^2/2 \quad \text{for small values of } p.$$

Let $p = x/n$ where p has a small value for large n

Thus,

$$x - \frac{x^2}{\sigma^2} = x - \frac{x^2}{2n}$$

yielding $\sigma^2 = 2n$ or $\sigma = \sqrt{2n}$. Hence

$$n! = \Gamma(n+1) = \gamma_m \int_{-\infty}^{+\infty} e^{-\frac{x^2}{\sigma^2}}\,\mathrm{d}x = \gamma_m \sqrt{(\sigma\pi)} = \left(\frac{n}{e}\right)^n \sqrt{(2n\pi)}$$

This yields the Sterling's approximation

$$n! = \left(\frac{n}{e}\right)^n \sqrt{2n\pi} \qquad \text{when } n \text{ is a large integer.}$$

9.4.6 Some Important Probability Density Functions

(1) **Uniform Distribution**

$$f(x) = \frac{1}{x_2 - x_1} \qquad x_1 \le x \le x_2 \tag{9.75}$$

(2) **Exponential Distribution**

$$f(x) = \alpha e^{-\alpha x} \qquad 0 \le x < \infty \tag{9.76}$$

(3) **Gaussian or Normal Distribution**

$$f(x) = \frac{1}{\sqrt{2\pi}\sigma_x} e^{-(x-\mu_x)^2/2\sigma_x^2} \qquad -\infty < x < \infty \tag{9.77}$$

(4) **Raleigh Distribution**

$$f(x) = \frac{x}{\sigma_x^2} e^{-x^2/2\sigma_x^2} \tag{9.78}$$

(5) **Weibull Density** (useful in reliability studies)

$$f(x) = \alpha x^{\beta-1} e^{-\alpha x^\beta/\beta} \qquad 0 \le x < \infty \tag{9.79}$$

(6) **Rician Distribution**

$$f_z(r,\theta) = \frac{1}{2\pi\sigma^2}\int_0^{2\pi} e^{(-r^2+\mu^2-2r\mu\cos(\alpha-\theta))/2\sigma^2}\,d\theta \tag{9.80}$$

This distribution is useful in the optimum detection of the envelope of a modulation process. With ω_0 as the carrier frequency, the modulation function $M(t)$ is given as:

$$M(t) = X(t)\cos\omega_0 t - Y(t)\sin\omega_0 t$$

where

$$X(t) = Z(t)\cos\alpha$$

$$Y(t) = Z(t)\sin\alpha$$

where $X(t)$, $Y(t)$ be two independent Gaussian Stochastic functions, with mean values and variance

$$\mu_x = \mu\cos\alpha \qquad \sigma_x = \sigma$$

$$\mu_y = \mu\sin\alpha \qquad \sigma_y = \sigma$$

Let the new stochastic process $Z(t)$ be:

$$Z(t) = \sqrt{X^2(t) + Y^2(t)}$$

We are required to determine the pdf for $Z(t)$ in the region of radius R in the x-y plane

$$f(x,y) = f(x)f(y) = \left(\frac{1}{\sqrt{2\pi\sigma}}\right)^2 e^{-[(x-\mu\cos\alpha)^2+y-(u\sin\alpha)^2]/2\sigma^2}$$

Let

$$x = r\cos\theta, \quad y = r\sin\theta$$

Then

$$f(r,\theta) = \frac{1}{2\pi\sigma^2} e^{-\left[r^2+\mu^2-2r\mu\cos(\alpha-\theta)\right]/2\sigma^2}$$

The **distribution functions** PDF F_z and pdf f_z are:

$$F_z(r,\alpha,\sigma,\mu) = \int_0^{2\pi}\int_0^{r} f(\rho,\theta)\rho\,\mathrm{d}\rho\,\mathrm{d}\theta = \frac{1}{2\pi\sigma^2}\int_0^{2\pi}\int_0^{r} e^{-\left[\rho^2+\mu^2-2\rho\mu\cos(\alpha-\theta)\right]/2\sigma^2}\rho\,\mathrm{d}\rho\,\mathrm{d}\theta$$

$$f_z(r,\alpha,\sigma,\mu) = \frac{\mathrm{d}}{\mathrm{d}r}F_z(r) = \frac{1}{2\pi\sigma^2}\int_0^{2\pi} re^{-\left[r^2+\mu^2-2r\mu\cos(\alpha-\theta)\right]/2\sigma^2}\,\mathrm{d}\theta$$

The expression for $f_z(r,\alpha)$ can be further simplified as:

$$f_z(r,\alpha,\mu,\sigma) = \left[\frac{1}{2\pi}\int_0^{2\pi} e^{(\mu r/\sigma^2)\cos(\alpha-\theta)}\,\mathrm{d}\theta\right]\left[\frac{r}{\sigma^2}e^{-(r^2+\mu^2)/2\sigma^2}\right]$$

The first factor represents the modified Bessel Equations of zero order.

9.5 Stochastic Hilbert Space

9.5.1 Vector Space of Random Variables

Let us construct a Banach or Hilbert space of real valued stochastic variables. The space consists of random variable vectors $\boldsymbol{X}$, $\boldsymbol{Y}$ having a finite mean-squared value. The key result that concerns us is the Karhunen-Loeve expansion, which states that a random process can be represented by a sequence of uncorrelated random variables. Consider two random vectors belonging to the Hilbert Space. For two Eucledian random vectors $\boldsymbol{X}$ and $\boldsymbol{Y}$, the inner product is:

$$E[(\boldsymbol{X},\boldsymbol{Y})] = E[\sum_{i=1}^{N}(X_i,Y_i)] = \sum_{i=1}^{N}E\left[(X_i,Y_i)\right] = \sum_{i=1}^{N}E[|X_i||Y_i|\cos\angle(X_i,Y_i)]$$

Random variable vector $\boldsymbol{X}$ is represented by vector $\boldsymbol{x}$ (a real quantity).

The geometric interpretation of **Cosine inequality** implies,

$$\frac{E[(\boldsymbol{X}, \boldsymbol{Y})]}{E[(\boldsymbol{X}, \boldsymbol{X})]E[(\boldsymbol{Y}, \boldsymbol{Y})]} \leq 1 \tag{9.81}$$

For vectors $\boldsymbol{X}$ and $\boldsymbol{Y}$ in continous space, $E[(\boldsymbol{X}, \boldsymbol{Y})]$ is defined as:

$$E[(\boldsymbol{X}, \boldsymbol{Y})] = \int_{-\infty}^{+\infty}\int_{-\infty}^{+\infty} (\boldsymbol{x}, \boldsymbol{y}) f(\boldsymbol{x}, \boldsymbol{y})\, \mathrm{d}\boldsymbol{y}\, \mathrm{d}\boldsymbol{x}$$

$$E[g(\boldsymbol{X})] = \int_{-\infty}^{+\infty} g(\boldsymbol{x}) f(\boldsymbol{x})\, \mathrm{d}\boldsymbol{x}$$

$$E[(\boldsymbol{X}, \boldsymbol{X})] = \int_{-\infty}^{+\infty} (\boldsymbol{x}, \boldsymbol{x}) f(\boldsymbol{x})\, \mathrm{d}\boldsymbol{x}$$

The distance between two random variables (or vectors) X and Y is:

$$\mathrm{d}(X, Y) = \sqrt{E\,[(X - Y, X - Y)]} = \sqrt{E\,[|X - Y|^2]}$$

Furthermore,

$$X = Y \text{ implies that } \Pr(X = Y) = 1$$

Example 9.11:

If $\boldsymbol{X}$ and $\boldsymbol{Y}$ are N-column matrices (vectors) and $\boldsymbol{K}$ is an $N \times N$ positive definite matrix, then

$$E\,[(\boldsymbol{X}, \boldsymbol{Y})] = E\left[\boldsymbol{X}^T\boldsymbol{Y}\right] = \sum_{i=1}^{N} E\,[X_iY_i]$$

$$E\,[(\boldsymbol{X}, \boldsymbol{KY})] = E\left[\boldsymbol{X}^T\boldsymbol{KY}\right] = \sum_{i=1}^{N}\sum_{j=1}^{N} k_{ij}E[X_iY_j], \quad k_{ij} = k_{ji}$$

The following important random variable properties in the Stochastic Hilbert space can be easily verified using geometric arguments:

- **Linearity**

$$E[\alpha \boldsymbol{X} + \beta \boldsymbol{Y}] = \alpha E[\boldsymbol{X}] + \beta E[\boldsymbol{Y}] \tag{9.82}$$

- **Orthogonality**

$$E[(\boldsymbol{X}, \boldsymbol{Y})] = 0 \rightarrow \boldsymbol{X} \text{ and } \boldsymbol{Y} \text{ are orthogonal and denoted by } \boldsymbol{X} \perp \boldsymbol{Y} \tag{9.83}$$

- **Correlation**

$$\rho = \frac{\sigma_{xy}}{\sigma_x \sigma_y} = \frac{E[(\boldsymbol{X} - \boldsymbol{\mu}_x, \boldsymbol{Y} - \boldsymbol{\mu}_y)]}{\left(\sqrt{E[((\boldsymbol{X} - \boldsymbol{\mu}_x, \boldsymbol{X} - \boldsymbol{\mu}_x))]}\right)\left(\sqrt{E[((\boldsymbol{Y} - \boldsymbol{\mu}_y, \boldsymbol{Y} - \boldsymbol{\mu}_y))]}\right)} \tag{9.84}$$

$$\sigma_{xy} = \rho \, \sigma_x \sigma_y$$

From Cosine inequality

$$|\rho| \leq 1, \quad \left|\sigma_{xy}\right| \leq \sigma_x \sigma_y, \quad \sigma_x^2 \sigma_y^2 \geq \sigma_{xy}^2$$

$\rho = 0$ implies that the vectors are uncorrelated.

Thus for uncorrelated vectors:

$$E[(\boldsymbol{X}, \boldsymbol{Y})] = (E[\boldsymbol{X}], E[\boldsymbol{Y}]) = (\boldsymbol{\mu}_x, \boldsymbol{\mu}_y)$$

$$\boldsymbol{f}(\boldsymbol{x}, \boldsymbol{y}) = \begin{bmatrix} f(x_1) f(y_1) \\ \vdots \\ f(x_n) f(y_n) \end{bmatrix}$$

Uncorrelated random vectors are also called Statistically Independent Vectors. Two random variables can be uncorrelated even though one random variable may be an explicit function of another random variable. On the other hand, if a random variable is a linear function of the other random variable, then they are correlated and their correlation coefficient is unity.

- **Sum of Two Random Variables**

  If

$$\boldsymbol{Z} = \boldsymbol{X} + \boldsymbol{Y}$$

Then

$$\sigma_z^2 = \sigma_x^2 + 2\rho\sigma_x\sigma_y + \sigma_y^2$$

For uncorrelated random variables $\boldsymbol{X}$ and $\boldsymbol{Y}$, $(\rho = 0)$,

$$\sigma_z^2 = \sigma_x^2 + \sigma_y^2$$

For linearly independent RVs, $\rho = 1$ yielding

$$\sigma_z^2 = (\sigma_x + \sigma_y)^2$$

9.5.2 Moment Generating Function or Characteristic Function

Given a random variable X with the probability density function $f(x)$, we define:
Characteristic Function of X as:

$$\Phi_X(j\omega) = \int_{-\infty}^{+\infty} f(x)e^{j\omega x}\,\mathrm{d}x = E\left[e^{j\omega X}\right]$$

(Note that $\Phi_x(-\omega)$ is the Fourier Transform of $f(x)$).

Another way to define the characteristic function is through the Laplace variable s and may be called the **Moment Generating Function**.

Moment Generating Function of X

$$\Phi_X(s) = \int_{-\infty}^{+\infty} f(x)e^{sx}\,\mathrm{d}x = E\left[e^{sX}\right] = \Phi_x(s),\;\; \Phi_x(j\omega) = \Phi_x(s)|_{s=j\omega}$$

The difference between the two is the same as between the Fourier Transform and the Laplace Transform of a function. This transformation leads to the following results:

- $Y = \alpha X + \beta,\;\; \Phi_Y(j\omega) = e^{j\beta\omega}\Phi_X(\alpha j\omega)$ (β, phase shift, α, frequency amplification)
- $\Phi_X(0) = 1$
- X, Y are independent random variables, $\Phi_{X+Y}(j\omega) = \Phi_X(j\omega)\Phi_Y(j\omega)$
- **Inversion Formula for pdf**

$$f(x) = \frac{1}{2\pi}\int_{-\infty}^{+\infty} \Phi_X(j\omega)e^{j\omega x}\,\mathrm{d}\omega,\;\; \text{(Fourier Inverse of the characteristic function)}$$

- **Moment Generation**

$$E\left[X^n e^{sX}\right] = \frac{\mathrm{d}^n}{\mathrm{d}s^n}\Phi_X(s) = \Phi_X^{(n)}(s)$$

$$E\left[X^n\right] = \Phi_X^{(n)}(0) = m_{nx}$$

$$m_{1x} = \mu_x = \Phi_X^{(1)}(0)$$

$$m_{2x} = \sigma_x^2 + \mu_x^2 = \Phi_X^{(2)}(0)$$

$$\Phi_X(s) = \sum_{n=0}^{\infty}\left(\frac{m_{nx}}{n!}\right)s^n$$

For the normal distribution process, only two moments m_1 and m_2 are needed.

- **Discrete Type Random Variables**

If the random variable X takes values $x_n = n$ with probability p_n, then the characteristic function is defined as:

$$\phi_X(\omega) = \sum_{n=1}^{\infty} p_n e^{j\omega x_n} = \sum_{n=1}^{\infty} p_n e^{j\omega n}$$

$$\phi_X(z) = \sum_{n=1}^{\infty} p_n z^{x_n} = \sum_{n=1}^{\infty} p_n z^n \qquad z = e^{j\omega}$$

Example 9.12:

Let X be a random variable with **Poisson distribution**.

This means that X takes a value $x = n$ with a probability p_n, given as

$$p_n = \frac{\lambda^n}{n!} e^{-\lambda}, \qquad n = 0, 1, 2, \ldots, \qquad \lambda > 0$$

$$\Phi_X(s) = \sum_{n=0}^{\infty} \frac{\lambda^n}{n!} e^{-\lambda} \cdot e^{sn} = e^{-\lambda} \sum_{n=0}^{\infty} \frac{(\lambda e^s)^n}{n!}$$

It is easy to verify that

$$\Phi_X(0) = 1$$

$$\frac{\mathrm{d}}{\mathrm{d}s} \Phi_X(s)|_{s=0} = \Phi_X^{(1)}(0) = \lambda$$

$$\frac{\mathrm{d}^2}{\mathrm{d}s^2} \Phi_X(s)|_{s=0} = \Phi_X^{(2)}(0) = \lambda^2 + \lambda$$

Exercise 9.2:

$$f(x) = (1/2)e^{-|x|}, \quad -\infty < x < \infty, \quad \text{Show } E(\omega) = E[e^{j\omega X}] = 1/(1 + \omega^2)$$

$$E(\omega) = 1/(1 + \omega^2), \quad \text{Show } f(x) = (1/2)e^{-|x|}, \quad -\infty < x < \infty$$

9.6 Random or Stochastic Processes

Practical Definition of Stochastic Processes:

A family $(X(t), \xi)$ of two variables t and ξ such that:

- For fixed t, $(X(t), \xi)$ is a random variable X
- For fixed ξ, $(X(t), \xi)$ is a time function $X(t)$

In what follows, we represent this process with $X(t)$ and supress the variable ξ.

Stochastic process $X(t)$ essentially describes the probabilistic behavior of some observable phenomenon that is time dependent. **It is a collection of random variables indexed by time.** Consider a member function $x(t)$ that is one of an infinite number of time functions called "**Ensemble**." This particular member $x(t)$ from the ensemble is called **Sample Function.** At a given time $t = t_1$, the value $x(t_1)$ is a random variable, $X(t_1)$ with some pdf.

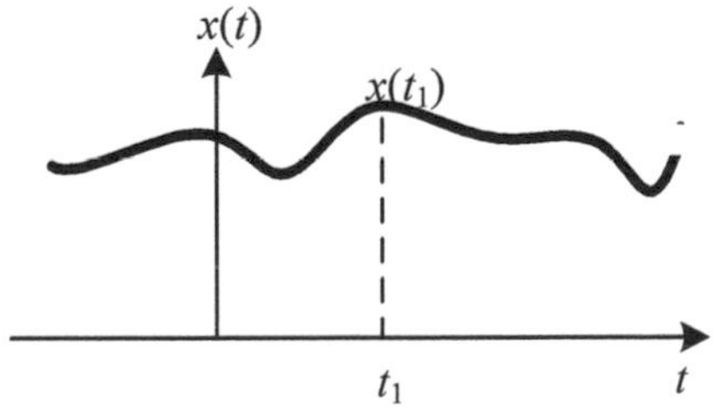

Figure 9.8: Random Function of Time from Ensemble of Random Process

The following are typical examples of the stochastic processes:

- Price and volume of a security.
- Amplitude and phase of a wireless signal.
- Weather forecast.
- Deterministic stochastic process with random parameters.

For example, consider an AC voltage, when A and ϕ are random variables.

$$\Pr(A = 0) = \alpha, \quad \Pr(A = 1) = 1 - \alpha = \beta$$

The variable ϕ has a constant pdf in the interval $(0, \pi/2)$ and 0 elsewhere.

Such a process $X(t)$ representing the AC voltage is:

$$X(t) = A\sin(\omega t + \phi)$$

where the domain of t is a set R of real numbers. If R is a set of integers then the process $X(t)$ is discrete. In reality the stochastic or random process is defined by the ensemble, $X(t, \xi)$, the variable ξ representing a realization from the ensemble of the resulting stochastic process $X(t)$.

Stochastic Process in Hilbert Space, Karhunen-Loeve Expansion

Consider the theory of stochastic processes in the setting of a normed Hilbert or Banach vector space. For each $t, X(t)$ is an element of the Hilbert space of random variables defined earlier and $X(t)$ is a trajectory. This trajectory is continous in the quadratic mean if

$$E\left[|X(t) - \mu_x(t)|\right]^2 \to 0 \qquad \text{for all } t$$

As in the deterministic setting discussed previously, the linear stochastic vector space of importance involving energy and power is $L^2(T_1, T_2)$. In this space we define the inner product of $X(t)$ and $Y(t)$ as:

$$E\left[(X(t), Y(t))\right] = E\left[\int_{T_1}^{T_2} X(t)Y(t)\,\mathrm{d}t\right]$$

Another interesting property of the normed space stochastic processes is that they can be represented via a complete set of linearly independent **nonrandom basis functions,**$\{\phi_i(t)\}_{i=1}^{\infty}$. Just as a deterministic process can be represented as a linear combination of independent basis functions, $\phi_i(t), i = 1, 2, \ldots$, similarly a stochastic process can be represented as a Karhunen-Loeve expansion with **stochastic parameters** as follows:

$$X(t) = \sum_{i=1}^{\infty} X_i \phi_i(t)$$

The set $\{X_i\}_{i=1}^{n}$ is a sequence of random variables.

If $\{\phi_i(t)\}_{i=1}^{\infty}$ is a complete orthogonal set, then:

$$X_i = (X(t), \phi_i(t)) = \int_{T_1}^{T_2} X(t)\phi_i(t)\,\mathrm{d}t$$

Note: Often the stochastic process $X(t),\ -T \le t \le T$ is designated as $(X_t)_{t \in T}$.

9.6.1 Stochastic Process PDF and pdf

- **First Order Distribution**

$$\text{PDF, } F(x, t_1) = P(X(t_1) \le x) \qquad \text{pdf, } f(x, t_1) = \frac{\partial}{\partial x} F(x, t_1)$$

- **Second Order Distribution**

$$\text{Joint PDF} \qquad F(x_1, x_2, t_1, t_2) = P(X(t_1) \le x_1, X(t_2) \le x_2)$$

$$\text{Joint pdf} \qquad f(x_1, x_2, t_1, t_2) = \frac{\partial^2}{\partial x_1 \partial x_2} F(x_1, x_2, t_1, t_2)$$

- ***n*-th Order Distribution**

$$\text{Joint PDF} \qquad F(x_1, \ldots, x_n, t_1, \ldots, t_n) = P(X(t_1) \le x_1, \cdots, X(t_n) \le x_n)$$

$$\text{Joint pdf} \qquad f(x_1, \ldots, x_n, t_1, \ldots, t_n) = \frac{\partial^n}{\partial x_1 \cdots \partial x_n} F(x_1, \cdots, x_n, t_1, \cdots, t_n)$$

9.6.2 Mean, Correlation Functions, and Spectra

In general, for random signals, we are interested in their energy contents and how this energy is modified by the passage of these signals through a linear system. The important statistical parameters for this study are the determination of the first moment (mean) and the second moment (autocorrelation, cross-correlation) of the output of these linear systems.

Let $X(t)$ be a set or "ensemble" of random functions. The process is specified at $t = t_i$ by a set of random variables, $\{X(t_i)\}_{i=1}^{n}, \quad i = 1, 2, \ldots, n.$
Important statistical parameters are:

1. **Mean or Expected Value**

$$E[X(t)] = \mu_x(t)$$

2. **Autocorrelation Function**

$$E[(X(t_1), X(t_2))] = R_{xx}(t_1, t_2)$$

3. **Cross-correlation Function**

$$E[(X(t_1), Y(t_2))] = R_{xy}(t_1, t_2)$$

4. **Auto-covariance** $\sigma_{xx}(t_1, t_1)$, **Cross-covariance** $\sigma_{xy}(t_1, t_2)$

$$\sigma_{xx}(t_1, t_2) = E\left[((X(t_1) - \mu_x(t_1)), (X(t_2) - \mu_x(t_2)))\right]$$

$$\sigma_{xy}(t_1, t_2) = E\left[((X(t_1) - \mu_x(t_1)), \left(Y(t_2) - \mu_y(t_2)\right))\right]$$

$$\mu_x(t_1) = \mu_x(t_2) = u_x(t)$$

$$\sigma_{xx}(t_1, t_2) = R_{xx}(t_1, t_2) - \mu_x^2(t)$$

5. **Average Power of $X(t)$, ($X(t)$ May Be Real or Complex Stochastic Variable)**

   For $t_1 = t_2 = t$, the autocorrelation $R_{xx}(t_1, t_2)$ becomes

$$R_{xx}(t, t) = E\left[|X(t)|^2\right] \qquad \text{Average Power}$$

   In fact

$$E\left[\left|\sum_{i=1}^{n} a_i X(t_i)\right|^2\right] = \sum_{i=1}^{n}\sum_{j=1}^{n} a_i a_j R(t_i, t_j)$$

   In what follows we shall assume that the processes and variables are real, unless and otherwise stated.

9.6.3 Types of Random Processes

1. **Continuous and Discrete Processes**

   Continous Random Process: $X(t)$ is continous in which the random variables $X(t_1)$ and $X(t_2)$ can take any random values within the range of the process values implying that the probability distribution function, PDF, and the probability density function, pdf, is continous.

   Discrete Random Process: Random variable $X(n)$ takes a value $x(n)$ at only certain specified instances such as 0, 1, etc. The pdf is represented by an impulse function $x(k)\delta(n - k)$.

2. **Deterministic Random Process**

   A deterministic random process $X(t)$ can be represented as $X(f(t), \theta)$ where $f(t)$ is deterministic and the parameter θ is a random variable with some specified pdf. Wireless signals often fall in this category. A typical example is specified earlier involving AC voltage with a probabilistic amplitude and phase.

3. **Stationary versus Nonstationary Random Processes**

   If the random process $X(t)$ is such that its mean value and all the higher order moments are independent of time, then the process is known as **stationary**. In this case all probability distribution functions do not depend upon the choice of the time origin. For the stationary case, the joint probability distribution of the n-selected variables satisfies the following relationship.

$$\begin{aligned}\Pr(X_1, \ldots, X_n, t_1, \ldots, t_n) &= \Pr(X(t_1) \le x_1, \ldots, X(t_n) \le x_n) \\ &= \Pr(X(t_1 + t_0) \le x_1, \ldots, X(t_n + t_0) \le x_n)\end{aligned}$$

(Stationary process for every $t_0 \in T$)

In this case, we shall see that the mean and the higher moments are time independent. If the above joint probability density function is time origin t_0 dependent, then the process is considered **nonstationary** and mean values and higher moments of the random process are **time dependent**.

4. **Ergodic versus Nonergodic Random Process**

Given a random process $X(t)$, let us form the two integrals

$$I_n(T) = \frac{1}{2T}\int_{-T}^{+T} x^n(t)\,\mathrm{d}t =< x^n(t) >_T \quad \text{(Temporal average)} \tag{9.85}$$

$$E_n = E\left[X^n(t)\right] = \int_{-\infty}^{+\infty} x^n(t) f(x)\,\mathrm{d}x \quad \text{(Ensemble average)} \tag{9.86}$$

If $E_n = \lim_{T\to\infty} I_n(T)$, then the time averages of these processes are equal to the ensemble averages. Such processes are known as **Ergodic**. Ergodic random processes have a very special property that, one typical member of the "ensemble set" exhibits the statistical behavior of the whole ensemble. For these processes the statistical averages of the ensemble can be computed from the time averages of a single sample provided the ergodic process is stationary.

In general, every stationary process is not ergodic but every ergodic process is stationary. It is generally difficult to prove that a process is ergodic. Following theorem on ergodic stationarity is due to G. Birkhoff.

Birkoffs Ergodic Theorem

A stationary stochastic process with a finite mean is **mean-Ergodic** if the expected value of the process and its temporal averages are equal for almost all samples. To establish the ergodicity of a process we proceed as following:

Let $X(t)$ be stationary process with finite mean μ_x.

We define a new random variable, M_T as:

$$M_T = \frac{1}{2T} \int_{-T}^{+T} X(t)\, \mathrm{d}t \tag{9.87}$$

$$E[M_T] = \frac{1}{2T} \int_{-T}^{+T} E[X(t)]\, \mathrm{d}t = m_T \tag{9.88}$$

$$E\left[(M_T - m_T)^2\right] = \sigma_T^2 \tag{9.89}$$

If $m_T \to \mu_x$ and $\sigma_T^2 \to 0$ as $T \to \infty$, we claim that $X(t)$ is ergodic and of course stationary. So the main advantage of ergodicity in real life, is that from only one sample function, we can compute its statistical mean and variance.

5. **Strict-Sense Stationary Process**

A stochastic process is called Strict-Sense Stationary if all its statistical properties are invariant to a shift to the time origin. This can be written as

$$X(t+\tau) \overset{\text{st}}{=} X(t)$$

where "st" stands for statistically equivalent.

Furthermore, if the process is ergodic then statistical averages can be computed by time averages. For most stochastic processes it is difficult to prove stationarity. However, a weaker property known as wide-sense stationarity can be assumed for most investigations.

6. **Wide-Sense Stationary (WSS) Process**

   A stochastic process $X(t), 0 \leq t \leq T$ is called **auto covariance or wide-sense stationary** if there exists two numbers μ_x and σ_x for every $0 \leq t \leq T$, such that

$$E[X(t)] = \mu_x \qquad \text{Mean}$$

$$\text{Var}[X(t)] = E[((X(t) - \mu_x), (X(t) - \mu_x))] = \sigma_x^2 \qquad \text{Variance}$$

   Furthermore, the autocorrelation function is:

$$R_{xx}(t_1, t_2) = R_{xx}(t_1 - t_2) = R_{xx}(\tau) = E[(X(t_1), X(t_2))], \quad \tau = |(t_1 - t_2)|$$

   In case the process is ergodic as well as wide-sense stationary,

$$R_{xx}(\tau) = E[(X(t), X(t - \tau))] = \lim_{T \to \infty} \frac{1}{2T} \int_{-T}^{+T} x(t)x(t + \tau)\,\mathrm{d}t$$

7. **White Noise Stochastic Process**

   Given a stochastic process such that $R(t_1, t_2) = 0 \qquad t_1 \neq t_2$

$$R(t_1, t_2) = I(t_1)\delta(t_1 - t_2)$$

   Such a process is called **White Noise** because of its "white spectrum." Let

$$t_1 - t_2 = \tau, \quad t = t_1, \quad R(t_1, t_2) = R(t, \tau) = I(t)\delta(\tau)$$

   For a stationary white noise, $\quad R(t_1, t_2) = R(t, \tau) = R(\tau) = I\delta(\tau)$

9.6.4 Autocorrelation Properties of an Ergodic Process

1. Mean-Squared Value

$$R_{xx}(0) = E\left[(X(t), X(t))\right] = E\left[|X(t)|^2\right] = \lim_{T\to\infty} \frac{1}{2T} \int_{-T}^{+T} |X(t)|^2 \, dt$$

2. The autocorrelation function $R_{xx}(\tau)$ is an even function of τ

$$R_{xx}(\tau) = E\left[(X(t), X(t+\tau))\right] = E\left[X(t)X(t-\tau)\right] = R_{xx}(-\tau)$$

3. Largest value of $R_{xx}(\tau)$ occurs at $\tau = 0$.

   This is easily seen from

$$E\left[((X(t) - X(t+\tau)), (X(t) - X(t+\tau)))\right] \geq 0$$

$$E\left[(X(t), X(t)) + (X(t+\tau), X(t+\tau))\right] \geq 2E\left[|X(t)X(t+\tau)|\right]$$

$$2R_{xx}(0) \geq 2R_{xx}(\tau)$$

$$E\left[|X(t+\tau) \pm X(t)|\right]^2 = 2\left[R_{xx}(0) \pm R_{xx}(\tau)\right] \geq 0$$

4. If the autocorrelation function $R(t_1, t_2)$ is zero for $|t_1 - t_2| > a$, then

$$R(t, \tau) = 0 \qquad |\tau| > a$$

   Often this process (for small a) can be replaced by a **white noise** of average intensity

$$I(t) = \int_{-a}^{+a} R(t, \tau)\, d\tau, \quad R(t, \tau) = I(t)\delta(\tau) \quad \text{(white noise)}$$

5. If the autocorrelation function is estimated via measurement of $X(t)$, then it can be computed as

$$R_{xx}(\tau) = \left(\frac{1}{T-\tau}\right)\int_0^{T-\tau} X(t)X(t+\tau)\,\mathrm{d}t \qquad \tau < T$$

or

$$R_{xx}(n\Delta t) = \frac{1}{N-n}\sum_{k=0}^{N-n} X_k X_{k+n} \quad n = 0, 1, \ldots, M, \quad M < N,\ T = N\Delta t,\ \tau = n\Delta t$$

9.6.5 Cross-correlation Functions of Stationary Ergodic Process

Consider sample functions $x(t)$, $x(t+\tau)$ and $y(t)$, $y(t+\tau)$ from two different stationary stochastic processes. Let

$$x(t_1) = X_1, \quad y(t_1) = Y_1, \quad y(t_1+\tau) = Y_2, \quad x(t_1+\tau) = X_2$$

The cross-correlation function is

$$R_{xy}(\tau) = E\left[(X_1, Y_2)\right] = \int_{-\infty}^{+\infty}\left(\int_{-\infty}^{+\infty} x_1 y_2 f(x_1, y_2)\,\mathrm{d}y_2\right)\mathrm{d}y_1 = \frac{1}{2T}\int_{-T}^{+T} x(t)y(t+\tau)\,\mathrm{d}t$$

$$R_{yx}(\tau) = E\left[(Y_1, X_2)\right] = \int_{-\infty}^{+\infty}\left(\int_{-\infty}^{+\infty} y_1 x_2 f(y_1, x_2)\,\mathrm{d}x_2\right)\mathrm{d}y_2 = \frac{1}{2T}\int_{-T}^{+T} y(t)x(t+\tau)\,\mathrm{d}t$$

It is easy to see that

- $R_{xy}(\tau) = R_{yx}(-\tau), \quad \left|R_{xy}(\tau)\right| \leq \left(R_{xx}(0)R_{yy}(0)\right)^{1/2}$
- If $z = x(t) + y(t)$, then $R_{zz}(\tau) = R_{xx}(\tau) + R_{yy}(\tau) + R_{xy}(\tau) + R_{yx}(\tau)$
- If $x(t)$ and $y(t)$ are statistically independent, then $R_{xy}(\tau) = R_{yx}(\tau) = \mu_x\mu_y$

Example 9.13:

Consider a stationary random process $X(t)$ with the autocorrelation function

$$R_{xx}(\tau) = e^{-\alpha|\tau|}$$

Determine the autocorrelation function of a linear combination $Y(t)$

$$Y(t) = aX(t) + bX(t - t_0), \qquad t_0 \geq 0$$

Solution:

$$\begin{aligned} R_{yy}(\tau) &= E\left[(aX(t) + bX(t - t_0))\,(aX(t + \tau) + bX(t - t_0 + \tau))\right] \\ &= \left(a^2 + b^2\right) R_{xx}(\tau) + ab\,(R_{xx}(\tau + t_0) + R_{xx}(\tau - t_0)) \end{aligned}$$

Its maximum value occurs at $\tau = 0$, yielding

$$R_{yy}(0) = \left(a^2 + b^2\right) + 2abe^{-t_0}$$

Example 9.14:

Let us consider a stationary random signal $X(t)$ contaminated with an uncorrelated noise $N(t)$ such that, $Y(t) = X(t) + N(t)$

Given:

$$X(t) = X_0 \cos(\omega t + \theta), \qquad (\theta \text{ being a random variable with zero mean})$$

$$R_{nn}(\tau) = N_0^2 e^{-c|\tau|}, \qquad (\text{autocorrelation function of } N(t))$$

Determine the autocorrelation function $R_{yy}(\tau)$

Solution:

$$R_{xn}(\tau) = R_{nx}(\tau) = 0$$

Thus,

$$R_{yy}(\tau) = R_{xx}(\tau) + R_{nn}(\tau) = \frac{1}{2}X_0^2\cos(\omega\tau) + N_0^2 e^{-c|\tau|}$$

Example 9.15:

Determine the autocorrelation and cross-correlation function matrices when $\boldsymbol{X}(t)$ and $\boldsymbol{Y}(t)$ are real stationary stochastic vectors.

Solution: Let

$$\boldsymbol{X}(t) = \begin{bmatrix} X_1(t) \\ \vdots \\ X_n(t) \end{bmatrix}, \qquad \boldsymbol{Y}(t) = \begin{bmatrix} Y_1(t) \\ \vdots \\ Y_n(t) \end{bmatrix}$$

$$\boldsymbol{R}_{xx}(\tau) = E\left[\boldsymbol{X}(t_1)\boldsymbol{X}^T(t_1+\tau)\right], \qquad \boldsymbol{R}_{xy}(\tau) = E\left[\boldsymbol{X}(t_1)\boldsymbol{Y}^T(t_1+\tau)\right] \quad \text{(Matrix)}$$

$$\boldsymbol{R}_{xy}(\tau) = E\left[\boldsymbol{X}(t_1)\boldsymbol{Y}^T(t_1+\tau)\right], \qquad \boldsymbol{R}_{yx}(\tau) = E\left[\boldsymbol{Y}(t_1)\boldsymbol{X}^T(t_1+\tau)\right] \quad \text{(Matrix)}$$

where for $i, j = 1, 2, \ldots, n$

$$\boldsymbol{R}_{xx}(\tau) = (\boldsymbol{R}_{xx}(\tau))_{ij} = \left(R_{x_ix_j}(\tau)\right) = \frac{1}{2T}\int_{-T}^{+T} X_i(t)X_j(t+\tau)\,\mathrm{d}t$$

$$\boldsymbol{R}_{yy}(\tau) = \left(\boldsymbol{R}_{yy}(\tau)\right)_{ij} = \left(R_{y_iy_j}(\tau)\right) = \frac{1}{2T}\int_{-T}^{+T} Y_i(t)Y_j(t+\tau)\,\mathrm{d}t$$

$$\boldsymbol{R}_{xy}(\tau) = \left(\boldsymbol{R}_{xy}(\tau)\right)_{ij} = \left(R_{x_iy_j}(\tau)\right) = \frac{1}{2T}\int_{-T}^{+T} X_i(t)Y_j(t+\tau)\,\mathrm{d}t$$

$$\boldsymbol{R}_{yx}(\tau) = \left(\boldsymbol{R}_{yx}(\tau)\right)_{ij} = \left(R_{y_ix_j}(\tau)\right) = \frac{1}{2T}\int_{-T}^{+T} Y_i(t)X_j(t+\tau)\,\mathrm{d}t$$

Note: Keep in mind that while here we are using $\boldsymbol{X}$ for a stochastic vector, in linear algebra a vector is represented by a lower case $\boldsymbol{x}$ and $\boldsymbol{X}$ implies a matrix.

Example 9.16:

Consider a stationary stochastic process $X(t)$, which is sampled at t_i, $i = 1, 2, \ldots, N$. Resulting time function vector is defined as:

$$X(t) = \begin{bmatrix} X(t_1) \\ \vdots \\ X(t_n) \end{bmatrix}$$

Determine its autocorrelation and covariance matrix.

Solution:

$$\boldsymbol{R}_{xx} = E \begin{bmatrix} \left[X^2(t_1)\right] & \cdots & [X(t_1)X(t_n)] \\ \vdots & & \\ [X(t_n)X(t_1)] & \cdots & [X(t_n)X(t_n)] \end{bmatrix} = \begin{bmatrix} R_{xx}(t_1, t_1) & \cdots & R_{xx}(t_1, t_n) \\ \vdots & & \\ R_{xx}(t_n, t_1) & \cdots & R_{xx}(t_n, t_n) \end{bmatrix}$$

When sampled at $t_k = k\Delta T, \qquad k = 0, 1, \ldots, N$

$$\boldsymbol{R}_{xx} = \begin{bmatrix} R_{xx}(0) & \cdots & R_{xx}(N-1) \\ \vdots & & \\ R_{xx}(N-1) & \cdots & R_{xx}(0) \end{bmatrix} \qquad \text{(a positive definite matrix)}$$

9.6.6 Wiener-Kinchin Theorem on Correlation Functions

Let $X(t)$ and $Y(t)$ be two stationary stochastic processes with Fourier transforms $X(j\omega)$ and $Y(j\omega)$, respectively. There auto- and cross-correlation functions are $R_{xx}(\tau)$ and $R_{xy}(\tau)$ We shall prove the following Wiener-Kinchin theorem.

$$R_{xx}(\tau) = \left[F^{-1}\left(X(-j\omega)X(+j\omega)\right)\right]_{t=\tau}$$
$$R_{xy}(\tau) = \left[F^{-1}\left(X(-j\omega)Y(+j\omega)\right)\right]_{t=\tau}$$

where F^{-1} stands for the Fourier Transform Inverse.

Proof:

$$R_{xx}(\tau) = \int_{-\infty}^{+\infty} X(t)X(t+\tau)\,\mathrm{d}t \tag{9.90}$$

$$X(t) = \frac{1}{2\pi}\int_{-\infty}^{+\infty} X(j\omega)e^{+j\omega t}\,\mathrm{d}\omega \tag{9.91}$$

$$X(t+\tau) = \frac{1}{2\pi}\int_{-\infty}^{+\infty} X(j\omega)e^{j\omega(t+\tau)}\,\mathrm{d}\omega = \frac{1}{2\pi}\int_{-\infty}^{+\infty} X(j\omega_1)e^{j\omega_1(t+\tau)}\,\mathrm{d}\omega_1 \tag{9.92}$$

From Eqs. 9.90, 9.91, and 9.92

$$R_{xx}(\tau) = \left(\frac{1}{2\pi}\right)^2 \int_{-\infty}^{+\infty}\left[\int_{-\infty}^{+\infty} X(j\omega)e^{j\omega t}\,\mathrm{d}\omega\right]\left[\int_{-\infty}^{+\infty} X(j\omega_1)e^{j\omega_1(t+\tau)}\,\mathrm{d}\omega_1\right]\mathrm{d}t$$

or

$$R_{xx}(\tau) = \frac{1}{2\pi}\left[\int_{-\infty}^{+\infty}\int_{-\infty}^{+\infty} X(j\omega)X(j\omega_1)e^{j\omega_1\tau}\left(\frac{1}{2\pi}\int_{-\infty}^{+\infty} e^{j(\omega+\omega_1)t}\,\mathrm{d}t\right)\mathrm{d}\omega\,\mathrm{d}\omega_1\right]$$

But

$$\frac{1}{2\pi}\int_{-\infty}^{+\infty} e^{j(\omega+\omega_1)t}\,\mathrm{d}t = \delta(\omega+\omega_1) \qquad \text{(See Chapter 1)}$$

Hence,

$$\begin{aligned} R_{xx}(\tau) &= \frac{1}{2\pi}\int_{-\infty}^{+\infty}\int_{-\infty}^{+\infty} X(j\omega)X(j\omega_1)e^{j\omega_1\tau}\delta(\omega+\omega_1)\,\mathrm{d}\omega\,\mathrm{d}\omega_1 \\ &= \frac{1}{2\pi}\int_{-\infty}^{+\infty} X(j\omega)X(-j\omega)e^{j\omega\tau}\,\mathrm{d}\omega = \frac{1}{2\pi}\int_{-\infty}^{+\infty} X(j\omega)X(-j\omega)e^{-j\omega\tau}\,\mathrm{d}\omega \end{aligned}$$

Thus,

$$R_{xx}(\tau) = \left[F^{-1}\left[X(j\omega)X(-j\omega)\right]\right]_{t=\tau} = \left[F^{-1}\left[|X(j\omega)|^2\right]\right]_{t=\tau}$$

Following the same reasoning

$$R_{xy}(\tau) = \left[F^{-1}\left[X(j\omega)Y(-j\omega)\right]\right]_{t=\tau}$$

when $\tau = 0$

$$R_{xx}(0) = \frac{1}{2\pi}\int_{-\infty}^{+\infty} X(j\omega)X(-j\omega)\,\mathrm{d}\omega = \int_{-\infty}^{+\infty} X^2(t)\,\mathrm{d}t \quad \text{(Persival's Theorem)}$$

$$R_{xy}(0) = \frac{1}{2\pi}\int_{-\infty}^{+\infty} X(j\omega)Y(-j\omega)\,\mathrm{d}\omega = \int_{-\infty}^{+\infty} X(t)Y(t)\,\mathrm{d}t$$

Note: To fully justify the application of a Fourier Transform, the sample $X(t)$ is truncated for some large value of $t = T$ and the truncated signal will have finite energy and spectral density. This is due to the fact that the condition of Fourier transformability,

$$\int_{-\infty}^{+\infty} |X(t)|\,\mathrm{d}t < \infty$$

is never satisfied by any stationary stochastic process. However, consider a modified function $X_T(t)$, such that

$$\begin{aligned} X_T(t) &= X(t), \quad |X(t)| \le M < \infty, \qquad -T \le t \le T \\ &= 0 \qquad |t| > T \end{aligned}$$

This function $X_T(t)$ is a Fourier transformable function, having finite energy. In what follows, we shall assume that $X(t)$ is the truncated function $X_T(t)$ having a Fourier transform.

9.6.7 Spectral Power Density

Eqs. 9.91 and 9.92 can be written as

$$R_{xx}(\tau) = \int_{-\infty}^{+\infty} \frac{|X(j\omega)|^2}{2\pi} e^{j\omega\tau}\, d\omega = \int_{-\infty}^{+\infty} S_{xx}(\omega) e^{j\omega\tau}\, d\omega$$

$$R_{xy}(\tau) = \int_{-\infty}^{+\infty} \left(\frac{X(j\omega)Y(-j\omega)}{2\pi}\right) e^{j\omega\tau}\, d\omega = \int_{-\infty}^{+\infty} S_{xy}(\omega) e^{j\omega\tau}\, d\omega$$

$$R_{xx}(0) = \int_{-\infty}^{+\infty} \frac{|X(j\omega)|^2}{2\pi}\, d\omega = \int_{-\infty}^{+\infty} S_{xx}(\omega)\, d\omega$$

$$R_{xy}(0) = \int_{-\infty}^{+\infty} \left(\frac{X(j\omega)Y(-j\omega)}{2\pi}\right) d\omega = \int_{-\infty}^{+\infty} S_{xy}(\omega)\, d\omega$$

where $S_{xx}(\omega)$ is called the **Power Spectrum** of $X(t)$ and $S_{xy}(\omega)$ as the cross power spectrum of $X(t)$ and $Y(t)$. Taking Fourier transforms of $R_{xx}(\tau)$ and $R_{xy}(\tau)$

$$S_{xx}(\omega) = \frac{1}{2\pi} \int_{-\infty}^{+\infty} R_{xx}(\tau) e^{-j\omega\tau}\, d\tau = \frac{1}{2\pi} |X(j\omega)|^2$$

$$S_{xy}(\omega) = \frac{1}{2\pi} \int_{-\infty}^{+\infty} R_{xy}(\tau) e^{-j\omega\tau}\, d\tau = \frac{1}{2\pi} (X(j\omega)Y(-j\omega))$$

$$S_{xx}(\omega) = S_{xx}(-\omega) = S_{xx}(\omega^2) \quad \text{even and real functon of } \omega$$

Example 9.17:

$$S_{xx}(\omega^2) = K, \qquad \text{constant spectrum}$$

Such a stochastic process $X(t)$ is called **White Noise** as a reminder that white light spectrum involves all different frequencies of equal magnitude.

The autocorrelation function is

$$R_{xx}(\tau) = K\delta(\tau)$$

Example 9.18:

Determine the spectral density $R_{xx}(\omega)$ of

$$X(t) = A + B\cos(\omega_1 t + \theta)$$

Parameters A and B are constants and the phase θ is a random variable uniformly distributed between 0 to 2π.

Thus,

$$\begin{aligned} \Pr(\theta) &= 1/2\pi \qquad 0 \le \theta \le 2\pi \\ &= 0 \qquad \text{otherwise} \end{aligned}$$

Solution:

$$\begin{aligned} X(t) &= A + B\cos(\omega_1 t + \theta) = A + B/2e^{j\omega t}e^{j\theta} + B/2e^{-j\omega_1 t}e^{-j\theta} \\ X(j\omega) &= 2\pi\left[A\delta(\omega) + B/2e^{j\theta}\delta(\omega - \omega_1) + B/2e^{-j\theta}\delta(\omega + \omega_1)\right] \\ X(-j\omega) &= 2\pi\left[A\delta(\omega) + B/2e^{j\theta}\delta(\omega + \omega_1) + B/2e^{-j\theta}\delta(\omega - \omega_1)\right] \end{aligned}$$

or

$$\begin{aligned} \frac{E\left[X(j\omega)X(-j\omega)\right]}{(2\pi)^2} &= E\left[A^2\delta^2(\omega) + \frac{B^2}{4}\delta^2(\omega + \omega_1) + \frac{B^2}{4}\delta^2(\omega - \omega_1)\right] \\ &+ E\left[\frac{AB}{2}\delta(\omega)\delta(\omega - \omega_1)e^{j\theta} + \frac{AB}{2}\delta(\omega)\delta(\omega + \omega_1)e^{-j\theta}\right] \\ &+ E\left[\frac{AB}{2}\delta(\omega)\delta(\omega + \omega_1)e^{j\theta} + \frac{B^2}{4}\delta(\omega + \omega_1)\delta(\omega - \omega_1)e^{2j\theta}\right] \\ &+ E\left[\frac{AB}{2}\delta(\omega)\delta(\omega - \omega_1)e^{-j\theta} + \frac{B^2}{4}\delta(\omega + \omega_1)\delta(\omega - \omega_1)e^{-j2\theta}\right] \end{aligned}$$

From the properties of delta functions

$$\delta^2(\omega) \equiv \delta(\omega), \quad \delta(\omega)\delta(\omega \pm \omega_1) = 0, \quad \delta(\omega + \omega_1)\delta(\omega - \omega_1) = 0$$

$$E\left[e^{jn\theta}\right] = \int_0^{2\pi} \Pr(\theta)e^{jn\theta}\, d\theta = \frac{1}{2\pi}\int_0^{2\pi} e^{jn\theta}\, d\theta = \frac{1}{2\pi}\left[\frac{e^{jn\theta}}{jn}\right]_0^{2\pi} = 0, \quad n = \pm 1, \pm 2$$

$$E\left[X(j\omega)X(-j\omega)\right] = (2\pi)^2\left(A^2\delta(\omega) + \frac{B^2}{4}\delta(\omega - \omega_1) + \frac{B^2}{4}\delta(\omega + \omega_1)\right)$$

$$S_{xx}(\omega) = 2\pi\left(A^2\delta(\omega) + \frac{B^2}{4}\delta(\omega - \omega_1) + \frac{B^2}{4}\delta(\omega + \omega_1)\right)$$

Note: Spectral Density and Bilateral Laplace Transform

For the systems analysis applications it is more convenient to use bilateral Laplace transforms for spectral functions, allowing us to have the whole complex s-plane for integration purposes. By definition

$$S_{xx}(s) = \int_{-\infty}^{+\infty} R_{xx}(\tau)e^{-s\tau}\, d\tau = S_{xx}(-s) = S_{xx}(s^2)$$

9.6.8 Karhunen-Loeve (K-L) Expansion of a Random Process

Let $X(t)$ be a random process for $0 \leq t \leq T$. We shall choose a set of real orthonormal functions $\{\phi_n(t)\}_{n=1}^{N}$ with a weighting function $w(t) = 1$ and an interval $(0, T)$. Furthermore, $\{X_n\}_{n=1}^{N}$, are uncorrelated random variables with zero mean.

If the random process is **mean square convergent** such that

$$\lim_{N\to\infty} E\left[X(t) - \sum_{n=1}^{N} X_n\phi_n(t)\right]^2 = 0 \tag{9.93}$$

Then, we can represent this random process by

$$X(t) = \lim_{N \to \infty} \left(\sum_{n=1}^{N} X_n \phi_n(t) \right) \tag{9.94}$$

The random variables X_n computed as

$$X_n = (X(t), \phi_n(t)) = \int_0^T X(t)\phi_n(t)\,\mathrm{d}t \tag{9.95}$$

$$1 = \int_0^T \phi_n^2(t)\,\mathrm{d}t$$

The expansion in Eq. 9.94 is called the **Karhunen-Loeve Expansion.**

Random variables $\{X_n\}_1^N$ are uncorrelated with zero mean, implying

$$E[X_n X_m] = \lambda_n \delta_{nm}, \quad \lambda_n \geq 0 \tag{9.96}$$

$$E[X_n] = 0 \qquad n, m = 1, 2, \ldots, N$$

The autocorrelation function of $X(t)$ is $R_{xx}(t, \tau) = E[(X(t), X(\tau))]$

From Eqs. 9.95 and 9.96, $E\left[X_i X_j\right] = E\left[\left(\int_0^T X(t)\phi_i(t)\,\mathrm{d}t, \int_0^T X(\tau)\phi_j(\tau)\,\mathrm{d}\tau\right)\right]$ (9.97)

The operator "E" operates only on $X(t)$ and $X(\tau)$.

$$E[X_n X_m] = \int_0^T \left[\int_0^T E[(X(t), X(\tau))]\,\phi_n(\tau)\,\mathrm{d}\tau\right] \phi_m(t)\,\mathrm{d}t \tag{9.98}$$

$$E[X_n X_m] = \int_0^T \left(\int_0^T R_{xx}(t, \tau)\phi_n(\tau)\,\mathrm{d}\tau\right) \phi_m(t)\,\mathrm{d}t = \lambda_m \delta_{nm} \tag{9.99}$$

Eq. 9.99 implies that

$$\int_0^T R_{xx}(t,\tau)\phi_n(\tau)\,\mathrm{d}\tau = \lambda_n\phi_n(t) \tag{9.100}$$

This is precisely the **Mercer's Theorem**. The kernel $R_{xx}(t,\tau)$ is positive definite. Eq. 9.100 implies that

$$R_{xx}(t,\tau) = \sum_{n=1}^{\infty} \lambda_n\phi_n(t)\phi_n(\tau) \tag{9.101}$$

$$R_{xx}(t,t) = \sum_{n=1}^{\infty} \lambda_n\phi_n^2(t) \quad \text{Uniformly Convergent Series.}$$

$$\int_0^T R_{xx}(t,t)\,\mathrm{d}t = \text{Mean Energy of } X(t) = \sum_{n=1}^{\infty} \lambda_n$$

Summary of Karhunen-Loeve Random Process Representation

$X(t) =$ Random process specified on $(0,T)$ interval

$\{\phi_n(t)\}_{n=1}^{\infty} =$ a complete set of Orthogonal functions,

with weight $w(t) = 1$ and interval $(0,T)$

$R_{xx}(t,\tau) =$ Autocorrelation function of $X(t)$

$\{X_n\}_{n=1}^{\infty} =$ a set of independent RVs with zero mean and variance λ_n

where λ_n is the eigenvalue of $R_{xx}(t,\tau)$

$$X(t) = \sum_{n=1}^{\infty} X_n\phi_n(t)$$

$$E\left[(X_n, X_m)\right] = \lambda_n\delta_{nm}$$

$$R_{xx}(t,\tau) = \sum_{n=1}^{\infty} \lambda_n\phi_n(t)\phi_n(\tau), \qquad 0 \le t,\tau \le T$$

$$\int_0^T R_{xx}(t,\tau)\phi_n(\tau)\,\mathrm{d}\tau = \lambda_n\phi_n(t), \qquad \lambda_n \ge 0$$

$$E\int_0^T X^2(t)\,\mathrm{d}t = \int_0^T R_{xx}(t,t)\,\mathrm{d}t = \sum_{n=1}^{\infty} E\left[X_n^2\right] = \sum_{n=1}^{\infty}\lambda_n = \text{Mean energy of } X(t).$$

Mean energy of the Random Process $X(t)$ is the infinite sum of eigenvalues λ_n of autocorrelation functions $R_{xx}(\tau)$.

9.6.9 Determination of Eigenvalues and Eigenvectors of $S_{xx}(s^2)$

Let $X(t)$ be a truncated WSS process with zero mean and autocorrelation function:

$$S_{xx}(s^2) = \frac{N(s)N(-s)}{D(s)D(-s)} \tag{9.102}$$

$$N(s) = b_0 + b_1 s + \cdots + b_m s^m$$

$$D(s) = a_0 + a_1 s + \cdots + a_n s^n$$

The polynomial $D(s)$ is considered as Hurwitz, with all its roots in the LHS of the s-plane. The degree m of $N(s)$ is less than the degree n of the denominator $D(s)$. From Eq. 9.101, the eigenvalues and eigenvectors are given as:

$$\lambda_k\phi_k(t) = \int_{-\infty}^{+\infty} R_{xx}(t-\tau)\phi_k(\tau)\,\mathrm{d}\tau, \qquad \lambda_k > 0 \quad k = 1, 2, \ldots \tag{9.103}$$

Taking the Bilateral Laplace Transform of Eq. 9.103

$$\lambda_k\Phi_k(s) = S_{xx}(s)\Phi_k(s)$$

$$\left[\lambda_k - \frac{N(s)N(-s)}{D(s)D(-s)}\right]\Phi_k(s) = 0$$

$$\left[\lambda_k D(s)D(-s) - N(s)N(-s)\right]\Phi_k(s) = 0 \tag{9.104}$$

Eq. 9.103 represents the transform of a linear homogenous differential equation of degree $2n$ resulting in $2n$ independent solutions $\phi_k(t)$ and corresponding values λ_k. The only allowable solutions are those that are stable and real functions namely,

$$\lim_{t\to\infty} \phi_k(t) = 0, \qquad \phi_k(t) \text{ are real functions.} \tag{9.105}$$

Let us examine the eigenvalues for which feasible solutions exist.

Eq. 9.104 can be rewritten in the form

$$P_k(s, \lambda_k)P_k(-s, \lambda_k)\Phi_k(s) = 0 \tag{9.106}$$

If $P_k(s, \lambda_k)$ has a root in the LHS representing a complex stable solution, then $P_k(-s, \lambda_k)$ has a root in the RHS, resulting in an unstable complex solution. Hence, only values of allowable λ_k are those that yield **simple roots on the** $j\omega$ **axis**, which results in the required real function stable solutions. We shall illustrate this by the following example.

Example 9.19:

$$R_{xx}(\tau) = P_0 e^{-\alpha|\tau|}, \qquad \alpha > 0$$

$$S_{xx}(s) = \frac{2\alpha P_0}{\alpha^2 - s^2}$$

From Eq. 9.105

$$\left[\lambda_k\left(\alpha^2 - s^2\right) - 2\alpha P_0\right]\Phi_k(s) = 0$$

or

$$P_k(s, \lambda_k)P_k(-s, \lambda_k) = \left[s^2 + \left(\frac{2\alpha P_0}{\lambda_k} - \alpha^2\right)\right]$$

only allowed values of λ_k are:

$$\frac{2\alpha P_0}{\lambda_k} - \alpha^2 = \beta_k^2 > 0$$

or

$$0 < \lambda_k < \frac{2P_0}{\alpha}$$

Orthogonality in the interval $0 \leq t \leq T$ dictates that β_k can take only values:

$$\beta_k = \frac{\pi}{T}(k - 1/2), \qquad k = 1, 2, \ldots$$

This yields the eigenvectors as

$$\left.\begin{aligned} \phi_{k1}(t) &= e^{j\beta_k t} \\ \phi_{k2}(t) &= e^{-j\beta_k t} \end{aligned}\right\} \quad \text{Complex functions}$$

$$\phi_k(t) = c_{k1} e^{j\beta_k t} + c_{k2} e^{-j\beta_k t}$$

Since $\phi_k(t)$ are real functions, the following two cases are feasible:

(1) $c_{k1} = c_{k2} = c_k$

(2) $c_{k1} = -c_{k2} = -jc_k$

Hence,

$$\phi_k(t) = c_k \cos \beta_k t$$

or

$$\phi_k(t) = c_k \sin \beta_k t$$

The phase information is lost in spectral densities, so choose $\phi_k(t) = c_k \sin \beta_k t$. Coefficient c_k is determined from the fact that

$$\int_0^{+T} \phi_k^2(t)\, dt = 1$$

or

$$c_k^2 \int_0^T \sin^2(\beta_k t)\,\mathrm{d}t = c_k^2 T/2 = 1$$

Thus,

$$c_k = \sqrt{2/T}$$

Conclusion: **The Karhunen-Loeve expansion of a random process** $X(t)$ **with auto-correlation** $R_{xx}(\tau) = P_0 e^{-\alpha|\tau|}$ is given by

$$X(t) = \sqrt{2/T} \sum_{k=1}^{\infty} X_k \sin\left(\frac{\pi}{T}(k - 1/2)t\right), \qquad 0 \le t \le \tau$$

where X_k are random variables with zero mean and variance

$$E\left[X_k^2\right] = \lambda_k = \left(\frac{\alpha^2 + \beta_k^2}{2\alpha P_0}\right), \qquad \beta_k = \pi/T(k - 1/2) \qquad k = 1, 2, \ldots$$

Example 9.20:

Let $W(t)$ be a Wiener process with zero mean and covariance $\sigma^2 t$. Let us obtain its autocorrelation function $R_{ww}(t, \tau)$ and its eigenvalues and eigenfunctions.

Solution:

$W(\tau)$ and $(W(t) - W(\tau))$ are statistically independent for $t \ge \tau > 0$ as well as for $\tau \ge t > 0$. Thus,

$$E\left[((W(t) - W(\tau)), W(\tau))\right] = 0$$

or

$$E\left[W(t), W(\tau)\right] - E\left[(W(\tau), W(\tau))\right] = 0$$

$$R_{ww}(t, \tau) = E\left[(W(t), W(\tau))\right] = \sigma^2 \min(t, \tau)$$

yielding

$$R_{ww}(t, \tau) = \sigma^2 \tau \qquad t \ge \tau$$

$$R_{ww}(t,\tau) = \sigma^2 t \qquad \tau \geq t$$

From the Karhunen-Loeve expansion

$$\lambda_k \phi_k(t) = \int_0^T R_{ww}(t,\tau)\phi_k(\tau)\,\mathrm{d}\tau = \int_0^t R_{ww}(t,\tau)\phi_k(\tau)\,\mathrm{d}\tau + \int_t^T R_{ww}(t,\tau)\phi_k(\tau)\,\mathrm{d}\tau$$

or

$$\lambda_k \phi_k(t) = \sigma^2 \int_0^t \tau\phi_k(\tau)\,\mathrm{d}\tau + \sigma^2 t \int_t^T \phi_k(\tau)\,\mathrm{d}\tau \qquad k = 1, 2, \ldots \tag{9.107}$$

Differentiating Eq. 9.107

$$\lambda_k \dot{\phi}_k = \sigma^2 t\phi_k(t) + \sigma^2 \int_t^T \phi_k(\tau)\,\mathrm{d}\tau - \sigma^2 t\phi_k(t)$$

or

$$\lambda_k \dot{\phi}_k(t) = \sigma^2 \int_t^T \phi_k(\tau)\,\mathrm{d}\tau \tag{9.108}$$

Differentiating the above Eq. 9.108 again, we obtain

$$\lambda_k \ddot{\phi}_k(t) + \sigma^2 \phi_k(t) = 0$$

or

$$\ddot{\phi}_k(t) + \beta_k^2 \phi_k(t) = 0, \qquad \beta_k^2 = \sigma^2/\lambda_k$$

The solution is

$$\phi_k(t) = c_{1k}e^{j\beta_k t} + c_{2k}e^{-j\beta_k t}, \qquad \phi_k(0) = 0$$

As discussed earlier, since c_{1k} and c_{2k} are chosen to yield

$$\phi_k(t) = c_k(\sin\beta_k t) \qquad 0 \le t \le T \tag{9.109}$$

$$\int_0^T \phi_k^2(t)\,dt = 1$$

or

$$c_k^2 T/2 = 1$$

or

$$c_k = \sqrt{2/T}$$

From Eq. 9.108, at $t = T$

$$\lambda_k c_k \beta_k \cos\beta_k T = 0 \tag{9.110}$$

Eq. 9.110 yields

$$\beta_k T = \pi\left(k - \frac{1}{2}\right), k = 1, 2, \ldots$$

$$\beta_k = \frac{\pi}{T}\left(k - \frac{1}{2}\right)$$

$$\lambda_k = \frac{\sigma^2}{\beta_k^2} = \frac{\sigma^2 T^2}{\pi^2\left(k - \frac{1}{2}\right)^2}$$

Thus, the Wiener process is expressed as

$$W(t) = \sum_{k=1}^{\infty} W_k \phi_k(t) = \sum_{k=1}^{\infty} W_k\left(\sqrt{\frac{2}{T}}\sin\left[\frac{\pi t}{T}\left(k - \frac{1}{2}\right)\right]\right), \quad 0 \le t \le T \tag{9.111}$$

$$E\left[W_k^2\right] = \lambda_k = \frac{\sigma^2 T^2}{\pi^2\left(k - \frac{1}{2}\right)^2}, \quad E[W_k] = 0, \qquad k = 1, 2, \ldots$$

Note: We can use the truncated expansion to simulate a Wiener process used in driving financial instruments modelling.

Example 9.21:

Derive the Gaussian white noise from the Wiener process. Using the Karhunen-Loeve expansion, determine the eigenfunctions of its autocorrelation function.

Solution:

Consider the white Gaussian noise with an autocorrelation function $R_{nn}(t,\tau)$ as:

$$R_{nn}(t,\tau) = \sigma^2\delta(t-\tau) \tag{9.112}$$

Another way to define white Gaussian noise through Wiener-Levy is:

$$W(t) = \int_0^t N(u)\,\mathrm{d}u \tag{9.113}$$

where $W(t)$ and $N(t)$ are Wiener process and white noise respectively.

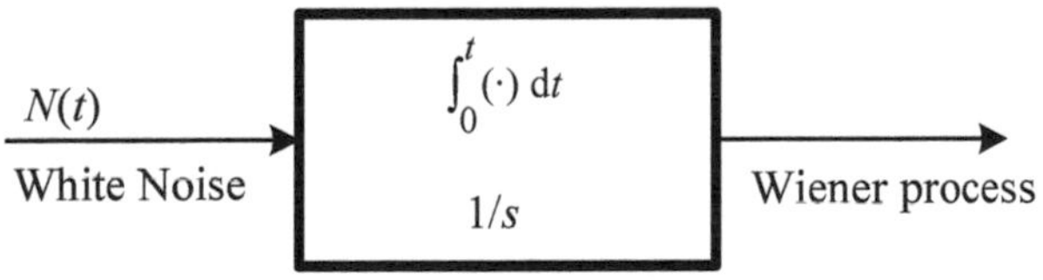

Figure 9.9: Generation of Wiener Process by Integrating White Noise

The Karhunen-Loeve expansion of the Wiener process $W(t)$ is:

$$W(t) = \sum_{k=1}^{\infty} W_k\left(\sqrt{\frac{2}{T}}\sin\left[\frac{\pi t}{T}\left(k-\frac{1}{2}\right)\right]\right), \quad 0 \le t \le T$$

Taking the derivative of $W(t)$

$$N(t) = \frac{\mathrm{d}}{\mathrm{d}t}W(t) = \sum_{k=1}^{\infty} W_k\frac{\pi}{T}\left(k-\frac{1}{2}\right)\left(\sqrt{\frac{2}{T}}\cos\left[\frac{\pi t}{T}\left(k-\frac{1}{2}\right)\right]\right)$$

or

$$N(t) = \sum_{k=1}^{\infty} N_k \left(\sqrt{\frac{2}{T}} \cos\left[\frac{\pi t}{T} \left(k - \frac{1}{2} \right) \right] \right)$$

where

$$N_k = \left(k - \frac{1}{2} \right) \frac{\pi}{T} W_k$$

$$E\left[N_k^2 \right] = \left(\left(k - \frac{1}{2} \right)^2 \frac{\pi^2}{T^2} \right) \left(\frac{\sigma^2 T^2}{\pi^2 \left(k - \frac{1}{2} \right)^2} \right) = \sigma^2$$

The autocorrelation function of the Gaussian white noise is

$$R_{nn}(t, \tau) = E\left[N(t) N(\tau) \right] = \frac{\partial^2}{\partial t \partial \tau} E\left[W(t) W(\tau) \right] = \frac{\partial^2}{\partial t \partial \tau} \left[\sigma^2 \min(t, \tau) \right]$$

$$\min(t, \tau) = (t - \tau) u(t - \tau), \qquad t \geq \tau$$

$$\min(t, \tau) = (\tau - t) u(\tau - t), \qquad \tau \geq t$$

Hence,

$$R_{nn}(t, \tau) = \sigma^2 \delta(t - \tau)$$

Applying the Karhunen-Loeve expansion

$$\lambda_k \phi_k(t) = \int_0^T \sigma^2 \delta(t - \tau) \phi_k(\tau)\, \mathrm{d}\tau, \qquad t < T$$

or

$$\lambda_k \phi_k(t) = \sigma^2 \phi_k(t), \qquad \lambda_k = \sigma^2 \quad \text{for all } k \tag{9.114}$$

Conclusion: **Any complete set of sinusoidal real orthonormal functions $\phi_k(t)$ are eigenfunctions of white noise.**

The total energy of white noise is

$$I = \sum_{k=1}^{\infty} \lambda_k = \sum_{k=1}^{\infty} \sigma^2 \rightarrow \infty$$

White noise is not physically realizable due to the infinite energy associated with it. However, we can approximate white Gaussian noise as:

$$N(t) = \sum_{k=1}^{M} N_k \left(\sqrt{\frac{2}{T}} \cos\left[\frac{\pi t}{T} \left(k - \frac{1}{2} \right) \right] \right), \quad 0 \le t \le T \tag{9.115}$$

N_k are iid RVs with zero mean and variance σ^2.

When the white noise $N(t)$ is passed through an integrator, its output is Wiener or Brownian motion.

9.6.10 LTI System Response to Stochastic Processes

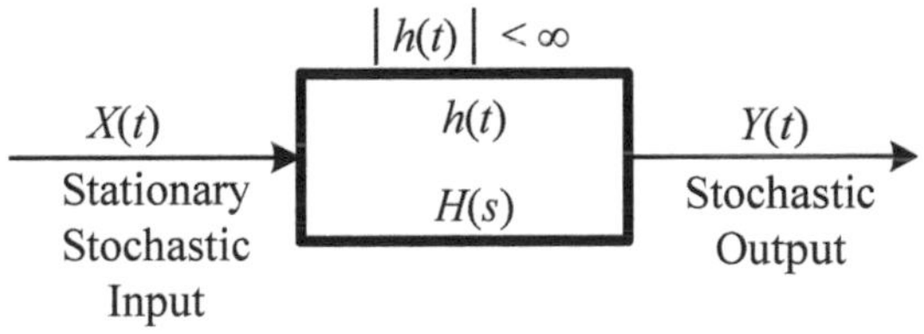

Figure 9.10: Response to Stochastic Input

Let $X(t)$ be a zero mean stochastic input to a LTI System characterized by its impulse response $h(t)$ or transfer function $H(s)$. Initial conditions are zero. The output process $Y(t)$ is obtained as

$$Y(t) = \int_{0}^{\infty} h(\sigma) X(t - \sigma) \, d\sigma = h(t) * X(t) = X(t) * h(t)$$

- **Input autocorrelation function and spectral density**

$$E\left[(X(t), X(t+\tau))\right] = R_{xx}(\tau) = R_{xx}(-\tau)$$

$$S_{xx}(s) = \int_{-\infty}^{+\infty} R_{xx}(\tau)e^{-s\tau}\,\mathrm{d}\tau = S_{xx}(s) = S_{xx}(s^2)$$

- **Input-Output correlation function and cross-spectral density**

$$E\left[(X(t), Y(t+\tau))\right] = R_{xy}(\tau) = E\left[\int_{-\infty}^{+\infty} (X(t), h(\sigma)X(t+\tau-\sigma))\,\mathrm{d}\sigma\right]$$

or

$$R_{xy}(\tau) = \int_{-\infty}^{+\infty} h(\sigma)R_{xx}(\tau-\sigma)\,\mathrm{d}\sigma = h(\tau)\,{}^{*}R_{xx}(\tau) = R_{xx}(\tau)\,{}^{*}h(\tau)$$

Taking the Bilateral Laplace Transform:

$$S_{xy}(s) = H(s)S_{xx}(s^2)$$

Similarly

$$R_{yx}(\tau) = \int_{-\infty}^{+\infty} h(\sigma)R_{xx}(\sigma+\tau)\,\mathrm{d}\sigma = H(-\tau)\,{}^{*}R_{xx}(\tau)$$

$$S_{yx}(s) = H(-s)S_{xx}(s^2)$$

- **Output autocorrelation function and spectral density**

$$E\left[(Y(t), Y(t+\tau))\right] = R_{yy}(\tau) = R_{yy}(-\tau)$$

or

$$R_{yy}(\tau) = E\left[\left(\left(\int_{-\infty}^{+\infty} h(\sigma_1)X(t-\sigma_1)\,\mathrm{d}\sigma_1\right), \left(\int_{-\infty}^{+\infty} h(\sigma_2)X(t+\tau-\sigma_2)\,\mathrm{d}\sigma_2\right)\right)\right]$$

or

$$R_{yy}(\tau) = \int_{-\infty}^{+\infty} h(\sigma_1)\left\{\int_{-\infty}^{+\infty} h(\sigma_2)E\left[(X(t-\sigma_1), X(t+\tau-\sigma_2))\right]\mathrm{d}\sigma_2\right\}\mathrm{d}\sigma_1$$

or

$$R_{yy}(\tau) = \int_{-\infty}^{+\infty} h(\sigma_1)\left\{\int_{-\infty}^{+\infty} h(\sigma_2)R_{xx}(\sigma_2-\sigma_1-\tau)\,\mathrm{d}\sigma_2\right\}\mathrm{d}\sigma_1$$

Taking the Bilateral Laplace transform, we obtain

$$S_{yy}(s^2) = H(s)H(-s)S_{xx}(s^2)$$

Note: The Laplace transform of $X(t)$ and $Y(t)$ are not of any significance, but rather the Laplace Transform of their correlation functions as spectral densities.

Thus

$$R_{yy}(\tau) = \frac{1}{2\pi j}\int_{-j\infty}^{+j\infty} H(s)H(-s)S_{xx}(s^2)e^{+s\tau}\,\mathrm{d}\tau$$

$$R_{yy}(0) = E\left[Y^2(t)\right] = \sigma_y^2 = \frac{1}{2\pi j}\int_{-j\infty}^{+j\infty} H(s)H(-s)S_{xx}(s^2)\,\mathrm{d}s$$

$$= \frac{1}{2\pi}\int_{-\infty}^{+\infty} |H(j\omega)|^2\, S_{xx}(\omega^2)\,\mathrm{d}\omega$$

In Chapter 4, we have discussed how to compute the above integral.

Two commonly used methods are :

(1) Method of residue

(2) Use of Mean Square Integral in Chapter 4

The following example illustrates the computation algorithm for the above integral.

Example 9.22:

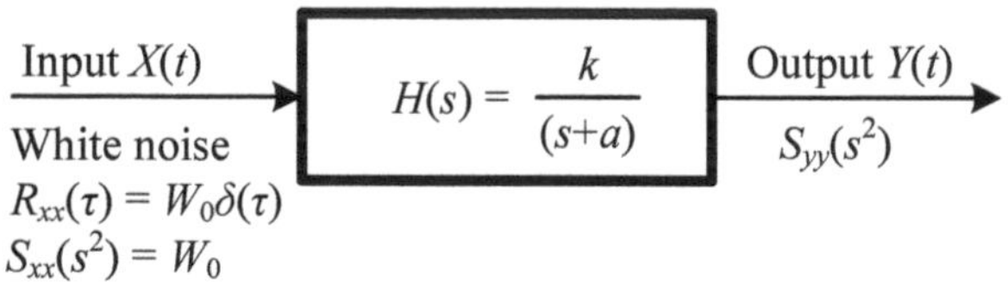

Figure 9.11: White Noise through a Lag Network

Consider the white noise input through a lag filter. The output covariance is given by

$$\sigma_y^2 = I = \frac{1}{2\pi j}\int_{-j\infty}^{+j\infty} H(s)H(-s)S_{xx}(s^2)\,\mathrm{d}s = \frac{1}{2\pi j}\int_{-j\infty}^{+j\infty} \frac{B(s^2)}{D(s)D(-s)}\,\mathrm{d}s$$

$$B(s^2) = k^2 W_0$$

$$D(s) = (s+a) \qquad \text{(Hurwitz Polynomial)}$$

$$I = \frac{(-1)^0}{2a_0}\frac{N}{D} = \frac{k^2}{2a}W_0$$

We can also compute I by taking the residue of the poles of the integrand in the LHS. We leave this as an exercise for the reader.

Example 9.23:

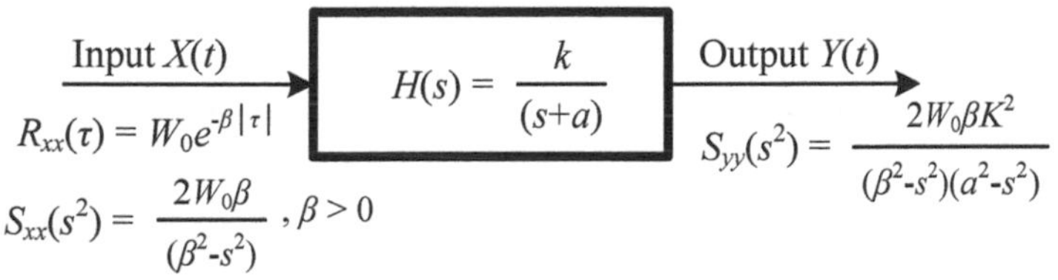

Figure 9.12: Signal with Given Autocorrelation Function through a Lag Network

Using the terminology of Chapter 4,

$$B(s^2) = 2W_0\beta k^2$$

$$D(s) = (s + a)(s + \beta) \qquad \text{(Hurwitz Polynomial)}$$

$$\mu_y^2 = I = \frac{1}{2\pi j}\int_{c-j\infty}^{c+j\infty} S_{yy}(s^2)\,\mathrm{d}s = \frac{1}{2\pi j}\int_{c-j\infty}^{c+j\infty} \frac{B(s^2)}{D(s)D(-s)}\,\mathrm{d}s$$

Once again, the reader should verify the above result using both methods.

$$\mu_y^2 = \frac{W_0\beta k^2}{(a+\beta)}$$

9.7 Wiener Filters

9.7.1 Optimal Estimation with Noise (Memoryless System)

Let us consider the following optimal estimation problem

Given:

$$\boldsymbol{Y} = \boldsymbol{CX} + \boldsymbol{N} \quad \text{Stochastic Vector Equation} \tag{9.116}$$

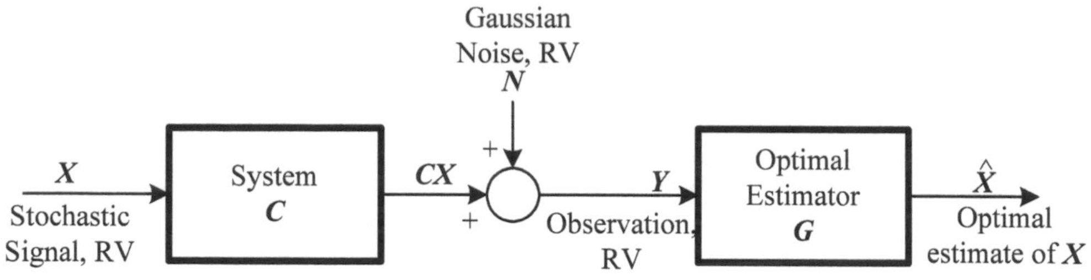

Figure 9.13: Optimal Estimator—Memoryless System

where

$\boldsymbol{X} =$ n-vector stochastic variable with zero mean and covariance matrix $\boldsymbol{P}$

$\boldsymbol{C} = m \times n$ matrix

$\boldsymbol{N} =$ m-vector Gaussian noise with zero mean and covariance matrix $\boldsymbol{Q}$

$\boldsymbol{Y} =$ m-vector Observation variable

$\boldsymbol{X}$ and $\boldsymbol{N}$ are uncorrelated.

Note: $\boldsymbol{X}$, $\boldsymbol{Y}$, and $\boldsymbol{N}$ are vectors and not matrices.

Determine the optimal estimator matrix $\boldsymbol{G}$ such that $\hat{\boldsymbol{X}} = \boldsymbol{G}\boldsymbol{Y}$. The solution depends upon what we consider as an "optimum" criterion. Two optimization criteria considered are:

(1) Minimum mean square estimate, I_1

(2) Minimum variance estimate, I_2

In general, these two criteria will yield different estimates.

(1) **Minimum Mean-Square Error Estimator (Gaussian Estimator)**

The performance index is:

$$I_1 = E\left[\left(\boldsymbol{C}\hat{\boldsymbol{X}} - \boldsymbol{Y}\right)^T \left(\boldsymbol{C}\hat{\boldsymbol{X}} - \boldsymbol{Y}\right)\right]$$

where $\hat{X}$ is the optimal value of X. Taking the gradient of I_1 wrt $\hat{X}$ and equating it to zero. The resulting equation is:

$$0 = E\left[C^T\left(C\hat{X} - Y\right)\right]$$

or

$$C^T C E\left[\hat{X}\right] = C^T E[Y]$$

Let

$$E\left[\hat{X}\right] = GE[Y] \tag{9.117}$$

Then

$$C^T C G E[Y] = C^T E[Y]$$

Thus

$$\left.\begin{aligned} G &= \left(C^T C\right)^{-1} C^T \\ \hat{X} &= \left(C^T C\right)^{-1} C^T Y \end{aligned}\right\} \text{Minimum Mean-Square Estimator} \tag{9.118}$$

Let

$$X_e = X - \hat{X}$$

From Eqs. 9.116 and 9.118

$$X_e = X - \left(C^T C\right)^{-1} C^T C X - \left(C^T C\right)^{-1} C^T N$$

or

$$X_e = -\left(C^T C\right)^{-1} C^T N$$

The error covariance matrix is:

$$E\left[X_e X_e^T\right] = \left(C^T C\right)^{-1} C^T E\left[N N^T\right] C \left(C^T C\right)^{-1} \tag{9.119}$$

Let

$$E\left[N N^T\right] = Q \quad \text{noise covariance matrix}$$
$$E\left[X_e X_e^T\right] = \boldsymbol{\Sigma}_e \quad \text{signal error covariance matrix}$$

Thus,

$$\boldsymbol{\Sigma}_e = \left(C^T C\right)^{-1} C^T Q C \left(C^T C\right)^{-1} \tag{9.120}$$

For $\boldsymbol{Q} = \boldsymbol{I}$, identity matrix

$$\boldsymbol{\Sigma}_e = \left(C^T C\right)^{-1} \tag{9.121}$$

Special Case

Consider the Eq. 9.116 when $\boldsymbol{N} \equiv \boldsymbol{0}$. We can treat this case on a purely deterministic basis. The problem can be reformulated as following:

Given:

$$\boldsymbol{CX} = \boldsymbol{Y} \tag{9.122}$$

where $\boldsymbol{X}, \boldsymbol{Y}$, and $\boldsymbol{C}$ are defined in Eq. 9.116. $\hat{\boldsymbol{X}}$ is the best estimate of $\boldsymbol{X}$ such that $\frac{1}{2}\hat{\boldsymbol{X}}^T \hat{\boldsymbol{X}}$ is maximum. Restating the problem as follows:

$$\text{Maximize,} \quad I = \frac{1}{2}\hat{\boldsymbol{X}}^T \hat{\boldsymbol{X}}$$

subject to the constraint $\boldsymbol{Y} - \boldsymbol{C}\hat{\boldsymbol{X}} = \boldsymbol{0}$

As discussed in the Chapter 7 on the Calculus of Variations, we maximize:

$$I_1(\hat{\boldsymbol{X}}) = \frac{1}{2}\hat{\boldsymbol{X}}^T \hat{\boldsymbol{X}} + \lambda^T(\boldsymbol{Y} - \boldsymbol{C}\hat{\boldsymbol{X}}), \quad \lambda \text{ is the Lagrange's multiplier.}$$

Taking the gradient of I_1 wrt $\hat{\boldsymbol{X}}$

$$\nabla_{\hat{X}}\left(I_1(\hat{\boldsymbol{X}})\right) = \hat{\boldsymbol{X}} - \boldsymbol{C}^T \boldsymbol{\lambda} = \boldsymbol{0}$$

$$\hat{\boldsymbol{X}} = \boldsymbol{C}^T \boldsymbol{\lambda}$$

$$\boldsymbol{C}\hat{\boldsymbol{X}} = \boldsymbol{C}\boldsymbol{C}^T \boldsymbol{\lambda} = \boldsymbol{Y}$$

Therefore,

$$\boldsymbol{\lambda} = \left(\boldsymbol{C}\boldsymbol{C}^T\right)^{-1} \boldsymbol{Y}$$

$$\hat{\boldsymbol{X}} = \boldsymbol{C}^T \left(\boldsymbol{C}\boldsymbol{C}^T\right)^{-1} \boldsymbol{Y}$$

This is the same result as in Eq. 9.118. Lagrange's multiplier device.

(2) **Minimum Error Variance Estimator**

The performance index is

$$I_2 = \text{Trace } E\left[(\boldsymbol{X} - \boldsymbol{G}\boldsymbol{Y})(\boldsymbol{X} - \boldsymbol{G}\boldsymbol{Y})^T\right]$$

Taking the gradient of I_2 wrt the matrix $\boldsymbol{G}$ and setting it equal to zero

$$\boldsymbol{0} = E\left[\boldsymbol{X}\boldsymbol{Y}^T\right] - \boldsymbol{G}E\left[\boldsymbol{Y}\boldsymbol{Y}^T\right] \tag{9.123}$$

$$E\left[\boldsymbol{X}\boldsymbol{Y}^T\right] = E\left[\boldsymbol{X}(\boldsymbol{C}\boldsymbol{X} + \boldsymbol{N})^T\right] = E\left[\boldsymbol{X}\boldsymbol{X}^T\right]\boldsymbol{C} + E\left[\boldsymbol{X}\boldsymbol{N}^T\right]$$

But

$$E\left[\boldsymbol{X}\boldsymbol{N}^T\right] = \boldsymbol{0} \quad \text{(signal and noise are uncorrelated)}$$

Similarly

$$E\left[\boldsymbol{Y}\boldsymbol{Y}^T\right] = E\left[(\boldsymbol{C}\boldsymbol{X} + \boldsymbol{N})(\boldsymbol{C}\boldsymbol{X} + \boldsymbol{N})^T\right] = \boldsymbol{C}E\left[\boldsymbol{X}\boldsymbol{X}^T\right]\boldsymbol{C}^T + E\left[\boldsymbol{N}\boldsymbol{N}^T\right]$$

Let

$$E\left[\boldsymbol{X}\boldsymbol{X}^T\right] = \boldsymbol{P} \quad (n \times n \text{ matrix})$$

$$E\left[\boldsymbol{N}\boldsymbol{N}^T\right] = \boldsymbol{Q} \quad (m \times m \text{ diagonal matrix})$$

Eq. 9.123 can be rewritten as

$$\boldsymbol{G}\left(\boldsymbol{C}\boldsymbol{P}\boldsymbol{C}^T + \boldsymbol{Q}\right) = \boldsymbol{P}\boldsymbol{C}^T \tag{9.124}$$

$$\left.\begin{aligned} \boldsymbol{G} &= \boldsymbol{P}\boldsymbol{C}^T\left(\boldsymbol{C}\boldsymbol{P}\boldsymbol{C}^T + \boldsymbol{Q}\right)^{-1} \\ \hat{\boldsymbol{X}} &= \boldsymbol{G}\boldsymbol{Y} \end{aligned}\right\} \text{Minimum Variance Estimator} \tag{9.125}$$

Comparing Eqs. 9.124 and 9.125, if $\boldsymbol{P} = \boldsymbol{I}$ (Identity) and $\boldsymbol{Q} = 0$ then

$$\boldsymbol{G} = \boldsymbol{C}^T\left(\boldsymbol{C}\boldsymbol{C}^T\right)^{-1} \quad \text{Minimum Variance Estimator}$$

which is different from Eq. 9.118 which is minimum mean squared optimal estimator. $\boldsymbol{G}$ is often referred to as the open loop estimator. From Eq. 9.124, it is easy to see that covariance matrix $\boldsymbol{\Sigma}_e$ is

$$\boldsymbol{\Sigma}_e = \boldsymbol{P} - \boldsymbol{G}\boldsymbol{C}\boldsymbol{P} = \boldsymbol{P} - \boldsymbol{P}\boldsymbol{C}^T\left(\boldsymbol{C}\boldsymbol{P}\boldsymbol{C}^T + \boldsymbol{Q}\right)^{-1}\boldsymbol{C}\boldsymbol{P} \tag{9.126}$$

$$\text{Trace } \boldsymbol{\Sigma}_e = \text{Trace } E\left[\boldsymbol{X}_e\boldsymbol{X}_e^T\right] = \sum_{i=1}^{n} E\left[\boldsymbol{X}_{ei}^2\right] = E\left[\boldsymbol{X}_e^T\boldsymbol{X}_e\right] \tag{9.127}$$

Observation:

We will show that optimal gain matrix $\boldsymbol{G}$ can also be written as:

$$\boldsymbol{G} = \left(\boldsymbol{C}^T\boldsymbol{Q}^{-1}\boldsymbol{C} + \boldsymbol{P}^{-1}\right)^{-1}\boldsymbol{C}^T\boldsymbol{Q}^{-1} \tag{9.128}$$

which implies

$$G = \left(C^T Q^{-1} C + P^{-1}\right)^{-1} C^T Q^{-1} = PC^T \left(CPC^T + Q\right)^{-1} \tag{9.129}$$

This can be seen as the following: Let

$$T_1 = C^T Q^{-1} \left(CPC^T + Q\right) = C^T Q^{-1} CPC^T + C^T$$

$$T_2 = \left(C^T Q^{-1} C + P^{-1}\right) PC^T = C^T Q^{-1} CPC^T + C^T$$

Hence, T_1 and T_2 are equal. Thus,

$$C^T Q^{-1} \left(CPC^T + Q\right) = \left(C^T Q^{-1} C + P^{-1}\right) PC^T$$

$$\left(C^T Q^{-1} C + P\right)^{-1} C^T Q^{-1} = PC^T \left(CPC^T + Q\right)^{-1} = G$$

From Eq. 9.128 an alternate expression for Σ_e can be derived as

$$\Sigma_e = (I - GC)\, P = \left(C^T Q^{-1} C + P^{-1}\right)^{-1}$$

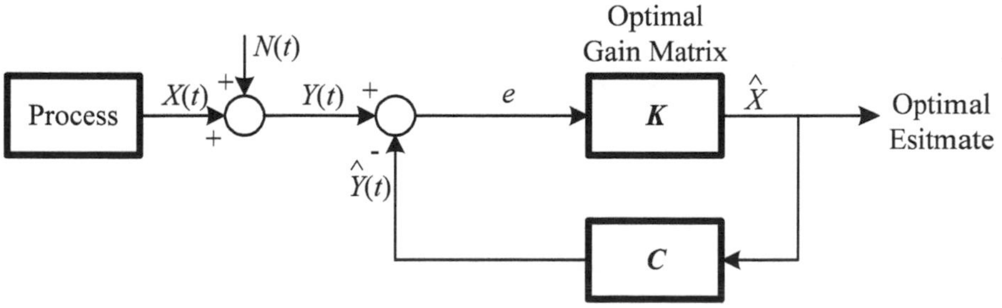

Figure 9.14: Feedback Implementation of Minimum Error Variance Estimation

Note: Often there are advantages to the implementation of an estimator in the feed-

back form. The following is the feedback implementation of the minimum error variance estimation algorithm, which is a forerunner to the "Kalman Filter."
We seek the relationship between $\boldsymbol{G}$ and $\boldsymbol{K}$ matrices.
From Eq. 9.125

$$\hat{\boldsymbol{X}} = \boldsymbol{GY}$$

But from the above feedback formulation

$$\hat{\boldsymbol{X}} = \boldsymbol{K}(\boldsymbol{I} + \boldsymbol{CK})^{-1}\boldsymbol{Y}$$

Thus

$$\boldsymbol{G} = \boldsymbol{K}(\boldsymbol{I} + \boldsymbol{CK})^{-1}$$

or

$$\boldsymbol{K} = (\boldsymbol{I} - \boldsymbol{GC})^{-1}\boldsymbol{G}$$

9.7.2 Wiener Filter Stochastic Signal Estimation without Noise

Consider a system with stochastic input $X(t)$ and a stochastic output $Y(t)$.

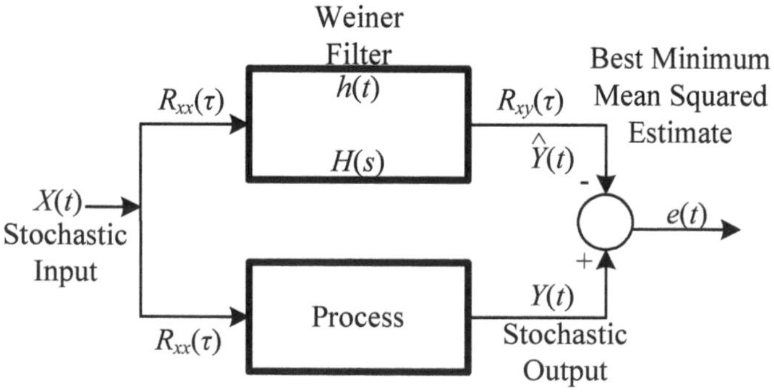

Figure 9.15: Noncausal Wiener Filter

Given:

1. $R_{xx}(\tau)$, the input autocorrelation function of the system input $X(t)$, $-\infty \le \tau < \infty$.

2. $R_{xy}(\tau)$, the input-output cross-correlation function between input $X(t)$ and the output $Y(t)$, $-\infty \le \tau < \infty$.

Required:

A noncausal filter with an impulse response $h(t)$ such that the input $X(t)$ results in the best mean-squared estimate $\hat{Y}(t)$ of the process output $Y(t)$.

Solution:

$$e(t) = Y(t) - \hat{Y}(t)$$

Let us choose a linear estimator:

$$\hat{Y}(t) = \int_{-\infty}^{+\infty} h(\sigma)X(t-\sigma)\,d\sigma$$

From our optimal projection theorem

$$e(t) \perp \hat{Y}(t-\tau) \quad \text{or} \quad e(t) \perp X(t-\tau) \quad \text{for all } \tau$$

Hence,

$$E\left[\left(\left(Y(t) - \int_{-\infty}^{+\infty} h(\sigma)X(t-\sigma)\,d\sigma\right), (X(t-\tau))\right)\right] = 0 \quad \text{for all } \tau \tag{9.130}$$

or

$$E\left[(Y(t), X(t-\tau))\right] = \int_{-\infty}^{+\infty} h(\sigma)E\left[(X(t-\sigma), X(t-\tau))\right]\,d\sigma$$

or

$$\boxed{R_{xy}(\tau) = \int_{-\infty}^{+\infty} h(\sigma)R_{xx}(\tau-\sigma)\,d\sigma \qquad -\infty < \tau < \infty} \tag{9.131}$$

Taking the Bilateral Transform of Eq. 9.131

$$S_{xy}(s) = H(s)S_{xx}(s^2)$$

yielding

$$\boxed{H(s) = \frac{S_{xy}(s)}{S_{xx}(s^2)} \qquad \text{(optimal filter)}} \tag{9.132}$$

This transfer function $H(s)$ represents a noncausal system and is therefore not physically realizable having poles on both sides of the s-plane. Later on, we shall show how to obtain the realizable part of this filter.

9.7.3 Wiener Filter Estimation of the Signal with Additive Noise

Case (a) Noncausal Filter

Let us consider the problem of extracting an optimal estimate of a stochastic signal that is corrupted with additive noise. We assume that the noise and the signal are uncorrelated.

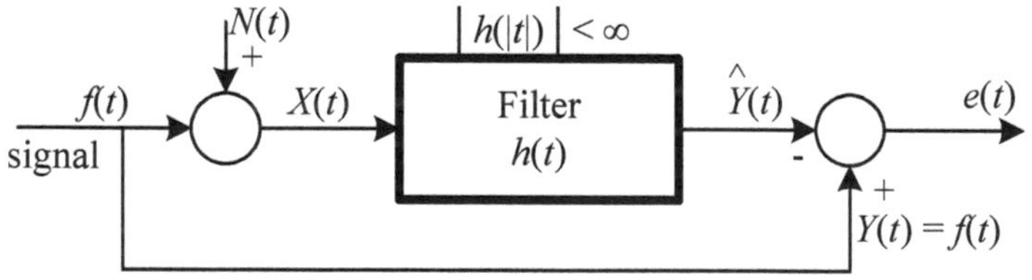

Figure 9.16: Wiener Filter with Uncorrelated Noise Signal and Noise

Let

$$X(t) = f(t) + N(t), \qquad Y(t) = f(t)$$

where $f(t)$ is the signal and $N(t)$ is the uncorrelated noise implying

$$f(t) \perp N(t)$$

$$E\left[(f(t), N(t-\tau))\right] = 0 \quad \text{for all } \tau, \quad \text{thus,}$$

$$R_{xx}(\tau) = E\left[((f(t) + N(t)), (f(t-\tau) + N(t-\tau)))\right] \quad \text{for all } \tau$$

$$R_{xx}(\tau) = R_{ff}(\tau) + R_{nn}(\tau) \tag{9.133}$$

$$R_{xy}(\tau) = R_{ff}(\tau) \tag{9.134}$$

From Eqs. 9.132, 9.133, and 9.134

$$\boxed{H(s) = H(-s) = \frac{S_{ff}(s^2)}{S_{ff}(s^2) + S_{nn}(s^2)}} \tag{9.135}$$

Furthermore,

$$S_{\hat{y}\hat{y}}(s^2) = H(s)H(-s)S_{xx}(s^2) = \left[\frac{S_{ff}(s^2)}{S_{ff}(s^2) + S_{nn}(s^2)}\right]^2 \left[S_{ff}(s^2) + S_{nn}(s^2)\right]$$

or

$$S_{\hat{y}\hat{y}}(s^2) = \frac{S_{ff}^2(s^2)}{S_{ff}(s^2) + S_{nn}(s^2)}$$

But

$$S_{yy}(s^2) = S_{ff}(s^2)$$

Thus

$$S_{ee}(s^2) = S_{yy}(s^2) - S_{\hat{y}\hat{y}}(s^2) = S_{ff}(s^2) - \frac{S_{ff}^2(s^2)}{S_{ff}(s^2) + S_{nn}(s^2)} = \frac{S_{ff}(s^2)S_{nn}(s^2)}{S_{ff}(s^2) + S_{nn}(s^2)}$$

Also

$$R_{ee}(0) = E\left[e^2(t)\right] = \frac{1}{2\pi j}\int_{c-j\infty}^{c+j\infty} S_{ee}(s^2)\,\mathrm{d}s = \frac{1}{2\pi j}\int_{c-j\infty}^{c+j\infty} \left[\frac{S_{ff}(s^2)S_{nn}(s^2)}{S_{ff}(s^2) + S_{nn}(s^2)}\right] \mathrm{d}s$$

Example 9.24:

Let

$$S_{ff}(s^2) = \frac{1}{1 - s^2}$$
$$S_{fn}(s) = S_{nf}(s) = 0$$
$$S_{nn}(s^2) = N$$

From Eq. 9.135

$$H(s) = \frac{1}{1 + N(1 - s^2)} = \frac{1}{N}\left(\frac{1}{((N+1)/N) - s^2}\right)$$
$$h(t) = \frac{1}{2\sqrt{N(N+1)}}\left[e^{-\sqrt{(N+1)/N}\,t}u(t) + e^{\sqrt{(N+1)/N}\,t}u(-t)\right]$$

Note that $h(t)$ represents a noncausal filter.

Case (b) Causal Wiener Filter—Wiener Hopf Factorization

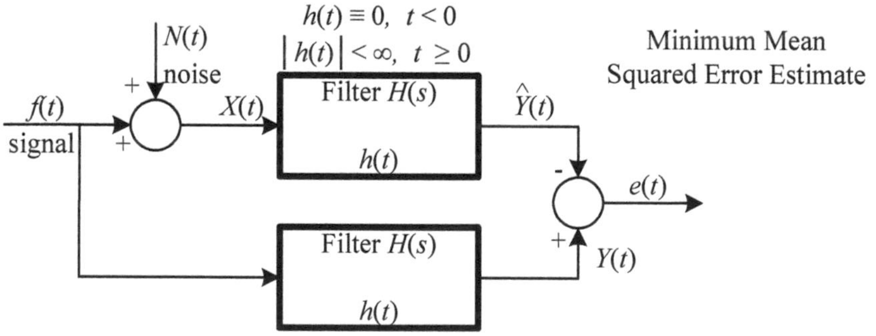

Figure 9.17: Causal Wiener Filter

Given:

(1) Autocorrelation function $R_{ff}(\tau)$ of the stochastic signal $f(t)$, $0 \le \tau < \infty$

(2) Autocorrelation function $R_{nn}(\tau)$ of the additive noise $N(t)$, $0 \le \tau < \infty$

(3) Cross-correlation function $R_{fy}(\tau)$ of the system input $f(t)$ and output $Y(t)$, $0 \leq \tau < \infty$

(4) $f(t) \perp N(t)$ implying signal and noise are uncorrelated and $R_{fy}(\tau) = R_{xy}(\tau)$

Required: A stable causal filter with an impulse response $h(t)$, the mean squared error is minimized.

Solution:

Let the output of the required causal filter be given by:

$$\hat{Y}(t) = \int_0^\infty h(\sigma)X(t-\sigma)\,\mathrm{d}\sigma, \qquad h(\sigma) \equiv 0 \quad \text{for } -\infty < \sigma < 0$$

$$e(t) = Y(t) - \hat{Y}(t)$$

From the optimal projection theorem:

$$e(t) \perp X(t-\tau), \qquad 0 < \tau < \infty$$

Thus,

$$E\left[\left(\left(Y(t) - \int_0^\infty h(\sigma)X(t-\sigma)\,\mathrm{d}\sigma\right), (X(t-\tau))\right)\right] = 0 \qquad 0 < \tau < \infty$$

This leads to the famous Wiener-Hopf Equation (WHE).

$$\left[R_{xy}(\tau) - \int_0^\infty h(\sigma)R_{xx}(\tau-\sigma)\,\mathrm{d}\sigma\right] = 0 \qquad \text{for } 0 \leq \tau < \infty \qquad (9.136)$$

Eq. 9.136 looks similar to Eq. 9.131, but in reality it is quite different. While Eq. 9.131 is valid for all τ, $-\infty < \tau < \infty$, the Eq. 9.136 is valid for only

$0 \leq \tau < \infty$. Furthermore, $h(t) \equiv 0$ for negative values of the time t. We shall solve Eq. 9.136 using the **Spectral Factorization Method**. The key to this method lies in looking for two functions $h(t)$ and $z^-(t)$ such that:

(1) $h(t) \equiv 0$ for $-\infty < t < 0$ and $|h(t)| < \infty$ for $0 < t < \infty$ (Causal function). This implies that all the poles of the double-sided Laplace Transform $H(s)$ of $h(t)$ lie inside LHS of the complex plane-s.
The Denominator of $H(s)$ is **Hurwitz**

(2) $z^-(t) \equiv 0,\ 0 \leq t < \infty, |z^-(-t)| < \infty,\ -\infty < t < 0$ (Anticausal function). All the poles of the double-sided Laplace Transform $Z^-(s)$ of $z^-(t)$ lie inside the RHS of the complex plane-s.
The Denominator of $Z^-(-s)$ is Hurwitz.

Solution of Wiener-Hopf Eq. 9.136 (Spectral Factorization)

Let

$$z^-(\tau) = R_{xy}(\tau) - \int_0^\infty h(\sigma) R_{xx}(\tau - \sigma)\, \mathrm{d}\sigma, \qquad -\infty < \tau < \infty \qquad (9.137)$$

$$z^-(\tau) = 0 \quad 0 < \tau < \infty$$

Eq. 9.137 replaces the equation Eq. 9.136 for $0 \leq \tau < \infty$.

Taking the double-sided Laplace Transform of Eq. 9.137, yields:

$$Z^-(s) = S_{xy}(s) - H(s) S_{xx}(s^2)$$

Let

$$S_{xx}(s^2) = P_x^+(s) P_x^-(-s)$$

$$S_{xy}(s) = S_{xy}^+(s) + S_{xy}^-(s)$$

where

$P_x^+(s)$ has all its poles in the LHS of the s-plane

$S_{xy}^+(s)$ has all its poles in the LHS of the s-plane

$P_x^-(s)$ has all its poles in the RHS of the s-plane

$S_{xy}^-(s)$ has all its poles in the RHS of the s-plane

Furthermore, $H(s)$ has all its poles in the LHS.

Thus,

$$Z^-(s) = S_{xy}^+(s) + S_{xy}^-(s) - H(s)P_x^+(s)P_x^-(s)$$

or

$$\left(\frac{Z^-(s)}{P_x^-(s)} - \frac{S_{xy}^-(s)}{P_x^-(s)}\right) = \left(\frac{S_{xy}^+(s)}{P_x^-(s)} - H(s)P_x^+(s)\right)$$

Let

$$\left(\frac{S_{xy}^+(s)}{P_x^-(s)}\right) = \left[\frac{S_{xy}^+(s)}{P_x^-(s)}\right]^+ + \left[\frac{S_{xy}^+(s)}{P_x^-(s)}\right]^-$$

where $\left[\frac{S_{xy}^+(s)}{P_x^+(s)}\right]^+$ represents only the analytic part of $\frac{S_{xy}^+(s)}{P_x^-(s)}$ for Re$s > 0$ and has all its poles in the LHS while $\left[\frac{S_{xy}^+(s)}{P_x^-(s)}\right]^{-1}$ has all its poles in the RHS of the s-plane.

Thus,

$$\left[\frac{Z^-(s)}{P_x^-(s)} - \frac{S_{xy}^-(s)}{P_x^-(s)} - \left[\frac{S_{xy}^+}{P_x^-(s)}\right]^-\right] = \left[\frac{S_{xy}^+(s)}{P_x^-(s)}\right]^+ - H(s)P_x^+(s) \qquad (9.138)$$

Now LHS of the Eq. 9.138 has all its poles in the LHS of the s-plane while the RHS of the Eq. 9.138 has all its poles in RHS of s-plane. The only possible solution to Eq. 9.138 is

$$\left[\frac{Z^-(s)}{P_x^-(s)} - \frac{S_{xy}^-(s)}{P_x^-(s)} - \left[\frac{S_{xy}^+}{P_x^-(s)}\right]^-\right] = \left[\frac{S_{xy}^+(s)}{P_x^-(s)}\right]^+ - H(s)P_x^+(s) = 0$$

or

$$H(s) = \frac{1}{P_x^+(s)}\left[\frac{S_{xy}^+(s)}{P_x^-(s)}\right]^+ \tag{9.139}$$

Transfer function of the optimal causal Wiener filter.

Example 9.25:

Let us compute the causal filter from the following given data

$$R_{ff}(\tau) = 4e^{-3|\tau|}, \quad S_{ff}(s^2) = \frac{4}{(9 - s^2)}$$

$$R_{nn}(\tau) = \frac{1}{4}\delta(\tau), \quad S_{nn}(s^2) = \frac{1}{4}$$

$$R_{xn}(\tau) = 0, \quad S_{xn}(s) = 0$$

$$R_{xy}(\tau) = R_{ff}(\tau), \quad S_{xy}(s) = S_{ff}(s^2) = \frac{4}{(9 - s^2)}$$

we obtain:

$$S_{xx}(s^2) = \frac{4}{9 - s^2} + \frac{1}{4} = \frac{(25 - s^2)}{4(9 - s^2)} = \frac{1}{4}\left(\frac{5 + s}{3 + s}\right)\left(\frac{5 - s}{3 - s}\right)$$

$$P_x^-(s) = \frac{1}{2}\left(\frac{5 - s}{3 - s}\right), \quad P_x^+(s) = \frac{1}{2}\left(\frac{5 + s}{3 + s}\right)$$

$$S_{xy}^+(s) = \frac{2}{3}\left(\frac{1}{3 + s}\right), \quad S_{xy}^-(s) = \frac{2}{3}\left(\frac{1}{3 - s}\right)$$

$$\left(\frac{S_{xy}^+(s)}{P_x^-(s)}\right) = \frac{4}{3}\left(\frac{1}{5 - s}\right)\left(\frac{3 - s}{3 + s}\right)$$

$$\left[\frac{S^+(s)}{P_x^-(s)}\right]^+ = \frac{1}{(3 + s)}\left(\frac{4}{3}\right)\left(\frac{6}{8}\right) = \frac{1}{(3 + s)}$$

$$H(s) = \frac{2}{(5+s)}, \quad h(t) = 2e^{-5t}u(t)$$

Note: The use of partial fractions and the residue theorem makes the calculations reasonably simple.

9.8 Estimation, Control, Filtering and Prediction

9.8.1 Estimation and Control

In the following section, we shall use the lower case bold letters for random vector. This shall not result in any confusion or contradiction. Let us consider a general estimation and control problem. We follow the approach by Saaty and Bram [Saaty, T.L.] with slight modification for easier understanding.

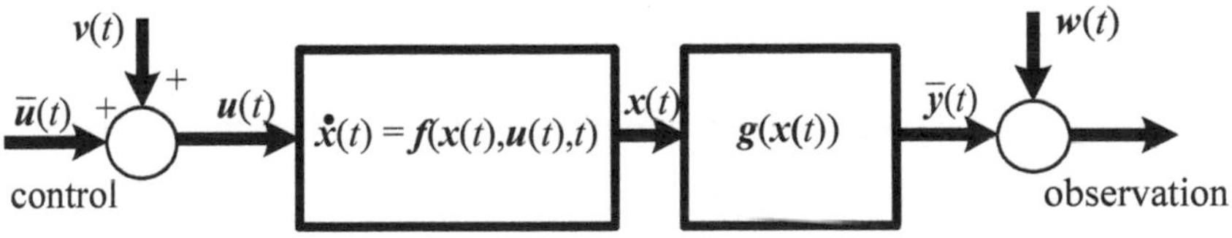

Figure 9.18: Dynamical System—Kalman Filtering Problem

Given:

System equations are:

$$\dot{\boldsymbol{x}}(t) = \boldsymbol{f}(\boldsymbol{x}(t), \boldsymbol{u}(t), t) \tag{9.140}$$

where $\boldsymbol{x}(t), \boldsymbol{f}(t)$ are n-vectors while $\boldsymbol{u}(t)$ is an r-vector ($r \leq n$). The input $\boldsymbol{u}(t)$, represents the control vector with the mean deterministic value $\bar{\boldsymbol{u}}(t)$, contaminated with the Gaussian noise $\boldsymbol{v}(t)$ such that

$$E\left[\boldsymbol{v}(t)\boldsymbol{v}^T(t)\right] = E\left[(\boldsymbol{u}(t) - \bar{\boldsymbol{u}}(t))(\boldsymbol{u}(\tau) - \bar{\boldsymbol{u}}(\tau)^T)\right] = \boldsymbol{Q}(t)\delta(t-\tau), \;\; 0 \leq \tau, t \leq t_f$$

$\boldsymbol{x}(t)$ represents the state vector with Gaussian distributed random initial conditions vector $\boldsymbol{x}(0)$ with $\bar{\boldsymbol{x}}(0)$ mean and covariance matrix:

$$E\left[(\boldsymbol{x}(0)-\bar{\boldsymbol{x}}(0))(\boldsymbol{x}(0)-\bar{\boldsymbol{x}}(0))^T\right]=\boldsymbol{P}(0) \tag{9.141}$$

The state-output relationship is given by:

$$\bar{\boldsymbol{y}}(t)=\boldsymbol{g}(\boldsymbol{x},t) \qquad \text{(without noise)} \tag{9.142}$$

$$\boldsymbol{y}(t)=\bar{\boldsymbol{y}}(t)+\boldsymbol{w}(t) \qquad \text{(observation, } m\text{-vector)} \tag{9.143}$$

$$E\left[\boldsymbol{w}(t)\boldsymbol{w}^T(t)\right]=E\left[(\boldsymbol{y}(t)-\bar{\boldsymbol{y}}(t))(\boldsymbol{y}(\tau)-\bar{\boldsymbol{y}}(\tau))^T\right]=\boldsymbol{R}(t)\delta(t-\tau) \tag{9.144}$$

Required:

Determine the minimum mean-squared value of the initial condition $\hat{\boldsymbol{x}}(t_0)$ and optimal control $\boldsymbol{u}(t)$, $0 \le t \le t_f$. Once $\hat{\boldsymbol{x}}(t_0)$ and $\boldsymbol{u}(t)$ is known, the differential equation, Eq. 9.140, yields the optimum estimate of $\boldsymbol{x}(t)$ defined as $\hat{\boldsymbol{x}}(t)$.

We shall use the tools developed in the Chapter 7 on Calculus of Variations to arrive at the optimal estimate $\hat{\boldsymbol{x}}(t)$ having observed $\boldsymbol{y}(t)$, $t_0 \le t \le t_f$.

Objective Function to be Maximized.

The objective function to be maximized is the joint probability density of Gaussian iid random variables given by:

$$I=e^{-J(0,t_f)}, \quad \text{where} \tag{9.145}$$

$$J(0,t_f)=\frac{1}{2}\int_0^{t_f}\Big[(\boldsymbol{y}(t)-\bar{\boldsymbol{y}}(t))R^{-1}(t)(\boldsymbol{y}(t)-\bar{\boldsymbol{y}}(t))^T$$
$$+(\boldsymbol{u}(t)-\bar{\boldsymbol{u}}(t))Q^{-1}(t)(\boldsymbol{u}(t)-\bar{\boldsymbol{u}}(t))^T\Big]\,\mathrm{d}t+J_0$$

$J_0=(\boldsymbol{x}(t_0)-\bar{\boldsymbol{x}}(t_0))\,\boldsymbol{P}_0^{-1}\,(\boldsymbol{x}(0)-\bar{\boldsymbol{x}}(0))^T$ a positive scalar function.

The function $e^{-J(0,\,t_f)}$, is maximized subject to the constraint

$$\dot{\boldsymbol{x}}(t) - \boldsymbol{f}(\boldsymbol{x}(t), \boldsymbol{u}(t), t) = \boldsymbol{0}$$
$$\bar{\boldsymbol{y}}(t) - \boldsymbol{g}(\boldsymbol{x}(t)) = \boldsymbol{0}$$

This is exactly the type of problem we have dealt with in Chapter 8 on Calculus of Variations. (It is obvious that maximizing $e^{-J(0,\,t_f)}$ is the same as minimizing the function $J(0, t_f)$.)

Solution:

Introducing a new function $J_1(0, t_f)$

$$J_1(0,t_f) = J(0,t_f) + \int_{t_0}^{t_f} \boldsymbol{\lambda}_1^T(t)\,(\dot{\boldsymbol{x}}(t) - \boldsymbol{f}(\boldsymbol{x}(t), \boldsymbol{u}(t), t))\;\mathrm{d}t + \int_0^{t_f} \boldsymbol{\lambda}_2^T(t)\,(\bar{\boldsymbol{y}}(t) - \boldsymbol{g}(\boldsymbol{x}(t)))\;\mathrm{d}t$$

(Lagrange multipliers, $\boldsymbol{\lambda}_1(t)$ and $\boldsymbol{\lambda}_2(t)$ are vectors or dimension n and m respectively.)

We seek the minimum of $J_1(0, t_f)$ with respect to the variables $\boldsymbol{x}(0), \boldsymbol{x}(t), \boldsymbol{u}(t), \bar{\boldsymbol{y}}(t), \boldsymbol{\lambda}_1(t)$ and $\boldsymbol{\lambda}_2(t)$. The deterministic control $\bar{\boldsymbol{u}}(t)$ and initial condition $\bar{\boldsymbol{x}}(t_0)$ are known along with the covariance matrices, $\boldsymbol{P}_0, \boldsymbol{Q}(t), \boldsymbol{R}(t)$ and the functions $\boldsymbol{f}(\cdot)$ and $\boldsymbol{g}(\cdot)$.

The necessary Euler-Lagrange conditions for the minimum are:

$$\left.\begin{aligned}
\dot{\hat{\boldsymbol{x}}}(t) &= \boldsymbol{f}(\hat{\boldsymbol{x}}(t), \hat{\boldsymbol{u}}(t), t)\\
\dot{\hat{\boldsymbol{\lambda}}}_1(t) &= -\left(\boldsymbol{\nabla}_x \boldsymbol{f}^T\right)\hat{\boldsymbol{\lambda}}_1(t) - \left(\boldsymbol{\nabla}_x \boldsymbol{g}^T\right)\hat{\boldsymbol{\lambda}}_2(t)\\
\hat{\boldsymbol{\lambda}}_2(t) &= -\boldsymbol{R}^{-1}(t)\,[\boldsymbol{y}(t) - \bar{\boldsymbol{y}}(t)]\\
\hat{\boldsymbol{y}}(t) &= \boldsymbol{g}(\hat{\boldsymbol{x}}(t), t)\\
\hat{\boldsymbol{x}}(0) &= \bar{\boldsymbol{x}}(0) + \boldsymbol{P}_0\hat{\boldsymbol{\lambda}}(0)\\
\hat{\boldsymbol{\lambda}}_1(t_f) &= \boldsymbol{0}\\
\hat{\boldsymbol{u}}(t) &= \bar{\boldsymbol{u}}(t) + \boldsymbol{Q}(t)\left(\boldsymbol{\nabla}_u \boldsymbol{f}^T\right)\hat{\boldsymbol{\lambda}}_1(t)
\end{aligned}\right\} \qquad (9.146)$$

There are two sets of nonlinear differential equations involving $\hat{\boldsymbol{x}}(t)$ and $\hat{\boldsymbol{\lambda}}_1(t)$ along with two algebraic equations involving $\hat{\boldsymbol{\lambda}}_2(t_f)$ and $\bar{\boldsymbol{y}}(t)$ and three mixed boundary conditions. Initial conditions $\hat{\boldsymbol{x}}(t_0)$ involves $\hat{\boldsymbol{\lambda}}_1(t_0)$, which is unknown. Instead $\hat{\boldsymbol{\lambda}}_1(t_f)$ is known. Hence, we are dealing with a two-point boundary value problem that was discussed in Chapter 8. This problem in its present form is computationally intensive for more than one or two variables. However, when $\boldsymbol{f}(\boldsymbol{x}(t), \boldsymbol{u}(t), t)$ is a linear function of $\boldsymbol{x}(t)$ and $\boldsymbol{u}(t)$, a practical and tractable solution is available, resulting in the Kalman Filter and the Observer theory developed in Chapter 7 (State Space). Considering the linear system:,

$$\left.\begin{aligned} &\boldsymbol{f}(\hat{\boldsymbol{x}}(t), \hat{\boldsymbol{u}}(t), t) = \boldsymbol{A}\hat{\boldsymbol{x}}(t) + \boldsymbol{B}\hat{\boldsymbol{u}}(t) \\ &\boldsymbol{g}(\hat{\boldsymbol{x}}(t)) = \boldsymbol{C}\hat{\boldsymbol{x}}(t), \quad \boldsymbol{R}(t) = \boldsymbol{R}, \quad \boldsymbol{Q}(t) = \boldsymbol{Q} \\ &\hat{\boldsymbol{\lambda}}_1(t) = \hat{\boldsymbol{\lambda}}(t) \end{aligned}\right\} \tag{9.147}$$

Then

$$\begin{aligned} &\hat{\boldsymbol{\lambda}}_2(t) = \boldsymbol{R}^{-1}(\boldsymbol{y}(t) - \boldsymbol{C}\hat{\boldsymbol{x}}(t)) \\ &\hat{\boldsymbol{u}}(t) = \bar{\boldsymbol{u}}(t) + \boldsymbol{Q}\boldsymbol{B}^T\hat{\boldsymbol{\lambda}}_1(t) = \bar{\boldsymbol{u}}(t) + \boldsymbol{Q}\boldsymbol{B}^T\hat{\boldsymbol{\lambda}}(t) \end{aligned}$$

From Eqs. 9.146 and 9.147, the system equations take the form

$$\begin{aligned} &\dot{\hat{\boldsymbol{x}}}(t) = \boldsymbol{A}\hat{\boldsymbol{x}}(t) + \boldsymbol{B}\boldsymbol{Q}\boldsymbol{B}^T\hat{\boldsymbol{\lambda}}(t) + \boldsymbol{B}\bar{\boldsymbol{u}}(t) \\ &\dot{\hat{\boldsymbol{\lambda}}}(t) = \boldsymbol{C}^T\boldsymbol{R}^{-1}\boldsymbol{C}\hat{\boldsymbol{x}}(t) - \boldsymbol{A}^T\hat{\boldsymbol{\lambda}}(t) - \boldsymbol{C}^T\boldsymbol{R}^{-1}\boldsymbol{y}(t) \end{aligned} \tag{9.148}$$

Boundary conditions are

$$\hat{\boldsymbol{x}}(0) = \bar{\boldsymbol{x}}(0) + \boldsymbol{P}_0\boldsymbol{\lambda}(0), \qquad \boldsymbol{\lambda}(t_f) = 0, \qquad (\boldsymbol{\lambda}(0) \text{ is still unknown.})$$

Eq. 9.148 can be written in a $2n \times 2n$ matrix form:

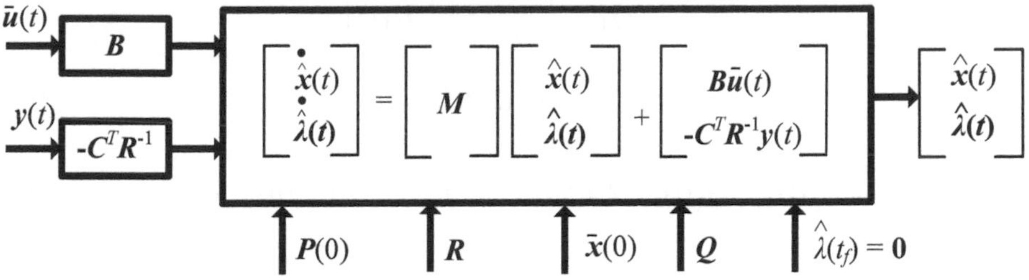

Figure 9.19: Optimal Estimate Equations via Calculus of Variations

$$\frac{\mathrm{d}}{\mathrm{d}t}\begin{bmatrix}\hat{\boldsymbol{x}}(t)\\ \hat{\boldsymbol{\lambda}}(t)\end{bmatrix}=\begin{bmatrix} \boldsymbol{M} \end{bmatrix}\begin{bmatrix}\hat{\boldsymbol{x}}(t)\\ \hat{\boldsymbol{\lambda}}(t)\end{bmatrix}+\begin{bmatrix}\boldsymbol{B}\bar{\boldsymbol{u}}(t)\\ -\boldsymbol{C}^T\boldsymbol{R}^{-1}\boldsymbol{y}(t)\end{bmatrix} \tag{9.149}$$

$$\begin{bmatrix} \boldsymbol{M} \end{bmatrix}=\begin{bmatrix}\boldsymbol{A} & \boldsymbol{BQB}^T\\ \boldsymbol{C}^T\boldsymbol{R}^{-1}\boldsymbol{C} & -\boldsymbol{A}^T\end{bmatrix}$$

$$\hat{\boldsymbol{x}}(0)=\bar{\boldsymbol{x}}(0)+\boldsymbol{P}(0)\hat{\boldsymbol{\lambda}}(0),\qquad \hat{\boldsymbol{\lambda}}(t_f)=\boldsymbol{0}$$

Solution of the Eq. 9.149

Due to the nonhomegenous nature of the Eq. 9.149, we seek a solution made up of particular function and complementary function as follows:

$$\hat{\boldsymbol{x}}(t)=\hat{\boldsymbol{x}}_p(t)+\hat{x}_c(t)$$
$$\hat{\boldsymbol{\lambda}}(t)=\hat{\boldsymbol{\lambda}}_p(t)+\hat{\boldsymbol{\lambda}}_c(t)$$

where

$$\frac{\mathrm{d}}{\mathrm{d}t}\begin{bmatrix}\hat{\boldsymbol{x}}_p(t)\\ \hat{\boldsymbol{\lambda}}_p(t)\end{bmatrix}=\begin{bmatrix} \boldsymbol{M} \end{bmatrix}\begin{bmatrix}\hat{\boldsymbol{x}}_p(t)\\ \hat{\boldsymbol{\lambda}}_p(t)\end{bmatrix}+\begin{bmatrix}\boldsymbol{B}\bar{\boldsymbol{u}}(t)\\ -\boldsymbol{C}^T\boldsymbol{R}^{-1}\boldsymbol{y}(t)\end{bmatrix} \tag{9.150}$$

$$\hat{\boldsymbol{x}}_p(0)=\bar{\boldsymbol{x}}(0),\qquad \hat{\boldsymbol{\lambda}}_p(0)=\boldsymbol{0}$$

and

$$\frac{\mathrm{d}}{\mathrm{d}t}\begin{bmatrix}\hat{\boldsymbol{x}}_c(t)\\ \hat{\boldsymbol{\lambda}}_c(t)\end{bmatrix}=\begin{bmatrix} & \boldsymbol{M} & \end{bmatrix}\begin{bmatrix}\hat{\boldsymbol{x}}_c(t)\\ \hat{\boldsymbol{\lambda}}_c(t)\end{bmatrix} \tag{9.151}$$

$$\hat{\boldsymbol{x}}_c(0)=\boldsymbol{P}(0)\hat{\boldsymbol{\lambda}}(0) \tag{9.152}$$

$$\hat{\boldsymbol{\lambda}}_c(0)=\hat{\boldsymbol{\lambda}}(0)-\hat{\boldsymbol{\lambda}}_p(0)=\hat{\boldsymbol{\lambda}}(0)=\boldsymbol{I}\hat{\boldsymbol{\lambda}}(0),\qquad \boldsymbol{I}\text{ is an identity matrix.} \tag{9.153}$$

The solution to the nonhomogenous equation, Eq. 9.150, is:

$$\begin{bmatrix}\hat{\boldsymbol{x}}_p(t)\\ \hat{\boldsymbol{\lambda}}_p(t)\end{bmatrix}=\begin{bmatrix} & \boldsymbol{e}^{\boldsymbol{M}t} & \end{bmatrix}\begin{bmatrix}\hat{\boldsymbol{x}}_p(0)\\ \hat{\boldsymbol{\lambda}}_p(0)\end{bmatrix}+\int_0^t\left(\begin{bmatrix} & \boldsymbol{e}^{\boldsymbol{M}(t-\sigma)} & \end{bmatrix}\begin{bmatrix}\boldsymbol{B}\hat{\boldsymbol{x}}(\sigma)\\ -\boldsymbol{C}^T\boldsymbol{R}^{-1}\boldsymbol{y}(\sigma)\end{bmatrix}\right)\mathrm{d}\sigma \tag{9.154}$$

On the other hand, the solution to the homogenous Eq. 9.151 takes the form:

$$\hat{\boldsymbol{x}}_c(t)=\boldsymbol{P}_x(t)\hat{\boldsymbol{\lambda}}(0),\qquad \boldsymbol{P}_x(0)=\boldsymbol{P}_0\qquad (n\times n\text{ matrix})$$

$$\hat{\boldsymbol{\lambda}}_c(t)=\boldsymbol{P}_\lambda(t)\hat{\boldsymbol{\lambda}}(0),\qquad \boldsymbol{P}_\lambda(0)=\boldsymbol{I}\qquad (n\times n,\text{ identity matrix})$$

where

$$\frac{\mathrm{d}}{\mathrm{d}t}\begin{bmatrix}\boldsymbol{P}_x(t)\\ \boldsymbol{P}_\lambda(t)\end{bmatrix}=\begin{bmatrix} & \boldsymbol{M} & \end{bmatrix}\begin{bmatrix}\boldsymbol{P}_x(t)\\ \boldsymbol{P}_\lambda(t)\end{bmatrix} \tag{9.155}$$

yielding:

$$\begin{bmatrix}\boldsymbol{P}_x(t)\\ \boldsymbol{P}_\lambda(t)\end{bmatrix}=\begin{bmatrix} & \boldsymbol{e}^{\boldsymbol{M}t} & \end{bmatrix}\begin{bmatrix}\boldsymbol{P}_0\\ \boldsymbol{I}\end{bmatrix} \tag{9.156}$$

Total solution:

$$\left.\begin{aligned}\hat{\boldsymbol{x}}(t)&=\hat{\boldsymbol{x}}_p(t)+\boldsymbol{P}_x(t)\hat{\boldsymbol{\lambda}}(0)\\ \hat{\boldsymbol{\lambda}}(t)&=\hat{\boldsymbol{\lambda}}_p(t)+\boldsymbol{P}_\lambda(t)\hat{\boldsymbol{\lambda}}(0)\end{aligned}\right\} \tag{9.157}$$

At $t=t_f$, $\hat{\boldsymbol{\lambda}}(t_f)=0$. Using Eq. 9.157

$$\hat{\boldsymbol{\lambda}}(t_f)=\hat{\boldsymbol{\lambda}}_p(t_f)+\boldsymbol{P}_\lambda(t_f)\hat{\boldsymbol{\lambda}}(0)=0$$

or

$$\hat{\lambda}(0) = -P_{\lambda}^{-1}(t_f)\hat{\lambda}_p(t_f)$$

Hence, the final expression for $\hat{x}(t)$ and $\hat{\lambda}(t)$ is:

$$\boxed{\begin{aligned} \hat{x}(t) &= \hat{x}_p(t) - P_x(t)P_{\lambda}^{-1}(t_f)\hat{\lambda}_p(t_f) \\ \hat{\lambda}(t) &= \hat{\lambda}_p(t) - P_{\lambda}(t)P_{\lambda}^{-1}(t_f)\hat{\lambda}_p(t_f) \end{aligned}} \tag{9.158}$$

With the above computation algorithm for $\hat{x}(t)$ and $\hat{\lambda}(t)$, the optimal control $\hat{u}(t)$ and the optimal estimate $\hat{y}(t)$ are

$$\boxed{\begin{aligned} &\hat{u}(t) = \bar{u} + QB^T\hat{\lambda}(t) \\ &\hat{y}(t) = C\hat{x}(t) \\ &0 \le t \le t_f \end{aligned}} \tag{9.159}$$

The problem discussed above is computation intensive, but yields the optimal estimate $\hat{x}(t)$ $(0 \le t \le t_f)$ for the given control vector $\bar{u}(t)$ contaminated with the noise $v(t)$ and the observation vector $y(t)$ under the influence of noise $w(t)$. This is called the **Smoothing Problem.** Its main disadvantage is the final boundary conditions that prohibit online computation.

9.8.2 Filtering-Prediction Problem (Kalman-Bucy Filter)

In this case we are only interested in the estimate at time $t = t_f$ namely, $\hat{x}(t_f)$ as well as the predicted value $\hat{x}(t),\ t > t_f$. The solution to the linear differential Eq. 9.148 involving mixed boundary conditions is converted into a nonlinear differential equation involving initial conditions only. Repeating Eq. 9.148

$$\frac{\mathrm{d}}{\mathrm{d}t}\begin{bmatrix}\dot{\hat{x}} \\ \dot{\hat{\lambda}}\end{bmatrix} = \begin{bmatrix} A & BQB^T \\ C^T R^{-1} C & -A^T \end{bmatrix}\begin{bmatrix}\hat{x} \\ \hat{\lambda}\end{bmatrix} + \begin{bmatrix} B\bar{u} \\ -C^T R^{-1} y \end{bmatrix} \tag{9.160}$$

$$\hat{\boldsymbol{x}}(0) = \bar{\boldsymbol{x}}(0) + \boldsymbol{P}(0)\hat{\boldsymbol{\lambda}}(0)$$

$$\hat{\boldsymbol{\lambda}}(t_f) = \boldsymbol{0}$$

This is the filtering problem involving the computation of $\hat{\boldsymbol{x}}(t_f)$. Since the boundary variables are mixed, and differential equations are adjoint, for this filtering problem, we introduce a new variable:

$$z(t) = \hat{\boldsymbol{x}}(t) - \boldsymbol{K}(t)\hat{\boldsymbol{\lambda}}(t), \quad z(0) = \bar{\boldsymbol{x}}(0) \tag{9.161}$$

$$z(t_f) = \hat{\boldsymbol{x}}(t_f) - \boldsymbol{K}(t_f)\hat{\boldsymbol{\lambda}}(t_f) = \hat{\boldsymbol{x}}(t_f)$$

Furthermore,

$$\boldsymbol{K}(0) = \boldsymbol{P}(0), \quad z(0) = \bar{\boldsymbol{x}}(0)$$

The variable $z(t)$ at $t = t_f$ yields the optimal filter estimate $\hat{\boldsymbol{x}}(t_f)$. The filter equations are obtained by differentiating Eq. 9.161 and using Eq. 9.160

$$\dot{z}(t) = \dot{\hat{\boldsymbol{x}}}(t) - \dot{\boldsymbol{K}}(t)\hat{\boldsymbol{\lambda}}(t) - \boldsymbol{K}(t)\dot{\hat{\boldsymbol{\lambda}}}(t)$$

$$\begin{aligned}\dot{z}(t) = {} & \left(\boldsymbol{A} - \boldsymbol{K}(t)\boldsymbol{C}^T\boldsymbol{R}^{-1}\boldsymbol{C}\right)z(t) + \boldsymbol{B}\bar{\boldsymbol{u}}(t) + \boldsymbol{K}(t)\boldsymbol{C}^T\boldsymbol{R}^{-1}\boldsymbol{y}(t) \\ & - \left(\dot{\boldsymbol{K}}(t) - \boldsymbol{A}\boldsymbol{K}(t) - \boldsymbol{K}(t)\boldsymbol{A}^T + \boldsymbol{K}(t)\boldsymbol{C}^T\boldsymbol{R}^{-1}\boldsymbol{C}\boldsymbol{K}(t) - \boldsymbol{B}\boldsymbol{Q}\boldsymbol{B}^T\right)\hat{\boldsymbol{\lambda}}(t)\end{aligned} \tag{9.162}$$

Setting the sum of all the terms involving $\hat{\boldsymbol{\lambda}}(t)$ in the above equation to zero and utilizing the initial condition

$$\dot{z}(t) = \boldsymbol{A}z(t) + \boldsymbol{B}\bar{\boldsymbol{u}}(t) + \boldsymbol{K}(t)\boldsymbol{C}^T\boldsymbol{R}^{-1}(y(t) - \boldsymbol{C}z(t)), \; z(0) = \bar{\boldsymbol{x}}(0), \; \text{(Kalman Filter)}$$

$$\dot{\boldsymbol{K}}(t) = \boldsymbol{A}\boldsymbol{K}(t) + \boldsymbol{K}(t)\boldsymbol{A}^T - \boldsymbol{K}(t)\boldsymbol{C}^T\boldsymbol{R}^{-1}\boldsymbol{C}\boldsymbol{K}(t) + \boldsymbol{B}\boldsymbol{Q}\boldsymbol{B}^T, \; \boldsymbol{K}(0) = \boldsymbol{P}(0), \; \text{(Riccati)}$$

9.8.3 Prediction Problem

For the filtering problem we see that for any time t, $z(t)$ is the minimum variance estimate (maximum likelihood estimate) $\hat{\boldsymbol{x}}(t)$ of $\boldsymbol{x}(t)$, given the observation $\boldsymbol{y}(t)$, $0 \leq t \leq t_f$. Now having observed $\boldsymbol{y}(t)$ for $0 \leq t \leq t_f$, we are interested in determining the optimal estimate of $\boldsymbol{x}(t)$ for $t > t_f$. That means no more data is available for time $t > t_f$. This is known as the prediction problem and is solved as follows:

1. The observation data $\boldsymbol{y}(t)$ is available between $0 \leq t \leq t_f$. We use the Kalman-Bucy filter and obtain the best estimate $\hat{\boldsymbol{x}}(t_f)$ of the state $\boldsymbol{x}(t)$ and then compute the predicted values for $t > t_f$.

2. Define a new variable $\tau \geq 0$ such that $\tau = t - t_f$.

The new system equations are:

$$\frac{\mathrm{d}}{\mathrm{d}\tau}\boldsymbol{x}(\tau + t_f) = \boldsymbol{A}\boldsymbol{x}(\tau + t_f) + \boldsymbol{B}\bar{\boldsymbol{u}}(\tau + t_f) + \boldsymbol{B}\boldsymbol{w}(t_f + \tau)$$

$$\boldsymbol{x}(t_f) = \hat{\boldsymbol{x}}(t_f)$$

$\boldsymbol{w}(t_f + \tau)$ is "noise" with zero mean.

Since no more data is available for $\tau \geq 0$, the optimal estimate $\hat{\boldsymbol{x}}(\tau + t_f)$ is:

$$\hat{\boldsymbol{x}}(\tau + t_f) = \boldsymbol{e}^{\boldsymbol{A}\tau}\boldsymbol{x}(\hat{t}_f) + \int_0^{\tau} \boldsymbol{e}^{\boldsymbol{A}(\tau-\sigma)}\boldsymbol{B}\bar{\boldsymbol{u}}(\tau + t_f - \sigma)\,\mathrm{d}\sigma \quad \text{Prediction estimate}$$

$$\tau \geq 0,\ t_f \geq 0$$

9.8.4 Discrete Time Kalman Filter

Statement of the Problem

$$\boldsymbol{x}(k+1) = \boldsymbol{\Phi}(k)\boldsymbol{x}(k) + \boldsymbol{w}(k) \tag{9.163}$$

$$\boldsymbol{y}(k) = \boldsymbol{C}(k)\boldsymbol{x}(k) + \boldsymbol{v}(k) \tag{9.164}$$

We assume that $\bar{\boldsymbol{u}}(k)$ and $\bar{\boldsymbol{x}}(0)$ are zero since their contribution is well behaved and easily computable. The noise vectors $\boldsymbol{w}(t)$ and $\boldsymbol{v}(t)$ have a zero mean.

Furthermore,

$$\left.\begin{aligned} &E\left[\boldsymbol{w}(k)\boldsymbol{w}^T(k)\right] = \boldsymbol{Q} \\ &E\left[\boldsymbol{v}(k)\boldsymbol{v}^t(k)\right] = \boldsymbol{R} \\ &E\left[\boldsymbol{x}(k)\boldsymbol{w}^T(k)\right] = \boldsymbol{0} \\ &E\left[\boldsymbol{x}(k)\boldsymbol{v}^T(k)\right] = \boldsymbol{0} \\ &E\left[\boldsymbol{v}(k)\boldsymbol{w}^T(k)\right] = E\left[\boldsymbol{w}(k)\boldsymbol{v}^T(k)\right] = 0 \\ &E\left[\boldsymbol{x}(0)\boldsymbol{x}^T(0)\right] = \boldsymbol{P}(0) \end{aligned}\right\} \tag{9.165}$$

The observations $\boldsymbol{y}(k)$, are used for the best estimate $\hat{\boldsymbol{x}}(k)$, $k = 0, 1, \ldots$

Solution:

Step 1. At the instant k, the best estimate $\hat{\boldsymbol{x}}(k)$ is available.

The new observation $\boldsymbol{y}(k+1)$ is not available until instant $(k+1)$. So given $\hat{\boldsymbol{x}}(k)$, we compute the best **extrapolated** or *a priori* value $(\boldsymbol{x}^*(k+1))$ at the instant $(k+1)$ without the knowledge of $\boldsymbol{y}(k+1)$ and $\boldsymbol{w}(k+1)$, yielding

$$\boldsymbol{x}^*(k+1) = \boldsymbol{\Phi}(k)\hat{\boldsymbol{x}}(k) \tag{9.166}$$

The error vector $\boldsymbol{e}(k)$ and its covariance matrix are:

$$\boldsymbol{e}(k) = (\boldsymbol{x}(k) - \hat{\boldsymbol{x}}(k)) \tag{9.167}$$

$$P_e(k) = E\left[e(k)e^T(k)\right] = E\left[(x(k) - \hat{x}(k))\,(x(k) - \hat{x}(k))^T\right] \tag{9.168}$$

Similarly

$$e^*(k+1) = (x^*(k+1) - x(k+1)) \tag{9.169}$$

$$P_e^*(k+1) = E\left[e^*(k+1)e^*(k+1)^T\right] \tag{9.170}$$

From Eqs. 9.163 and 9.166

$$x^*(k+1) = x(k+1) - [\Phi(k)e(k) + w(k)] \tag{9.171}$$

Thus,

$$e^*(k+1) = -\,[\Phi(k)e(k) + w(k)]$$

$$P_e^*(k+1) = E\left[(\Phi(k)e(k) + w(k))\,(\Phi(k)e(k) + w(k))^T\right]$$

Using Eq. 9.165

$$P_e^*(k+1) = \Phi(k)P(k)\Phi^T(k) + Q(k) \tag{9.172}$$

Step 2. Updated new best estimate $\hat{x}(k+1)$ after observation $y(k+1)$.

This is the Kalman Filter algorithm. The updated best estimate $\hat{x}(k+1)$ at the instant $(k+1)$ is:

$$\hat{x}(k+1) = x^*(k+1) + K(k+1)\,(y(k+1) - C(k+1)x^*(k+1)) \tag{9.173}$$

where

$$K(k+1) = \text{Updated Kalman Gain Matrix.}$$

Optimal $K(k+1)$ is derived in terms of $R, Q, P_e(k+1), P_e^*(k+1)$.

Step 3. Derivation of the Kalman-Gain Matrix $\boldsymbol{K}(k+1)$.

From Eqs. 9.164 and 9.173

$$\hat{\boldsymbol{x}}(k+1) = \boldsymbol{x}^*(k+1)+\boldsymbol{K}(k+1)\left(\boldsymbol{C}(k+1)\boldsymbol{x}(k+1)+\boldsymbol{w}(k+1)-\boldsymbol{C}(k+1)\boldsymbol{x}^*(k+1)\right)$$

Subtracting $\boldsymbol{x}(k+1)$ from both sides

$$\boldsymbol{e}(k+1) = (\boldsymbol{I}-\boldsymbol{K}(k+1)\boldsymbol{C}(k+1))\,\boldsymbol{e}^*(k+1)-\boldsymbol{K}(k+1)\boldsymbol{w}(k+1) \quad (9.174)$$

$\boldsymbol{e}^*(k+1)$ and $\boldsymbol{e}(k+1)$ represent the "a priori" and "a posteriori" error vector. Substituting Eq. 9.174 into $(k+1)$th update of Eq. 9.168 and utilizing the uncorrelated properties from Eq. 9.165, we obtain,

$$\boldsymbol{D}(k+1) = \boldsymbol{I}-\boldsymbol{K}(k+1)\boldsymbol{C}(k+1)$$

$$\boldsymbol{P}_e(k+1) = E\left[\boldsymbol{D}(k+1)\boldsymbol{e}^*(k+1)\boldsymbol{e}^*(k+1)^T\boldsymbol{D}(k+1)^T\right]+\boldsymbol{K}(k+1)\boldsymbol{R}\boldsymbol{K}(k+1)$$

or

$$\boldsymbol{P}_e(k+1) = \boldsymbol{D}(k+1)\boldsymbol{P}_e^*(k+1)\boldsymbol{D}(k+1)^T+\boldsymbol{K}(k+1)\boldsymbol{R}\boldsymbol{K}(k+1) \quad (9.175)$$

Our task is to find $\boldsymbol{K}(k+1)$ that minimizes:

$$\boldsymbol{I} = E\left[\boldsymbol{e}^T(k+1)\boldsymbol{e}(k+1)\right] \quad (9.176)$$

This is achieved if we minimize the sum of the diagonal elements of $E\left[\boldsymbol{e}(k+1)\boldsymbol{e}^T(k+1)\right]$. The sum of the diagonal elements of a matrix is known as the trace of a matrix. Thus,

$$\boldsymbol{I} = E\left[\boldsymbol{e}^T(k+1)\boldsymbol{e}(k+1)\right] = \text{Trace}\boldsymbol{P}(k+1) \quad (9.177)$$

From Eq. 9.175

$$\begin{aligned} \boldsymbol{P}_e(k+1) = \; & \boldsymbol{P}_e^*(k) - \boldsymbol{K}(k+1)\boldsymbol{C}(k+1)\boldsymbol{P}_e^*(k+1) \\ & -\boldsymbol{P}_e^*(k+1)\boldsymbol{C}^T(k+1)\boldsymbol{K}(k+1) \\ & +\boldsymbol{K}(k+1)\left(\boldsymbol{C}(k+1)\boldsymbol{P}_e^*(k+1)\boldsymbol{C}^T(k+1)\right)\boldsymbol{K}(k+1) \end{aligned}$$

Taking the Trace of the above equation:

$$\begin{aligned} \text{Trace}\,[\boldsymbol{P}_e(k+1)] &= \text{Trace}\,[\boldsymbol{P}_e^*(k)] - 2\,\text{Trace}\,[\boldsymbol{K}(k+1)\boldsymbol{C}(k+1)\boldsymbol{P}_e^*(k+1)] \\ &+ \text{Trace}\left[\boldsymbol{K}(k+1)\left(\boldsymbol{C}(k+1)\boldsymbol{P}_e^*(k+1)\boldsymbol{C}^T(k+1) + \boldsymbol{R}(k+1)\right)\boldsymbol{K}(k+1)\right] \end{aligned}$$

Differentiating with respect to $\boldsymbol{K}(k+1)$

$$0 = -2\boldsymbol{P}_e^*(k+1)\boldsymbol{C}^T(k+1) + 2\boldsymbol{K}(k+1)\left(\boldsymbol{C}(k+1)\boldsymbol{P}_e^*(k+1)\boldsymbol{C}^T(k+1) + \boldsymbol{R}(k+1)\right)$$

$$\boldsymbol{K}(k+1) = \boldsymbol{P}_e^*(k+1)\boldsymbol{C}^T(k+1)\left(\boldsymbol{C}(k+1)\boldsymbol{P}_e^*(k+1)\boldsymbol{C}^T(k+1) + \boldsymbol{R}(k+1)\right)^{-1}$$

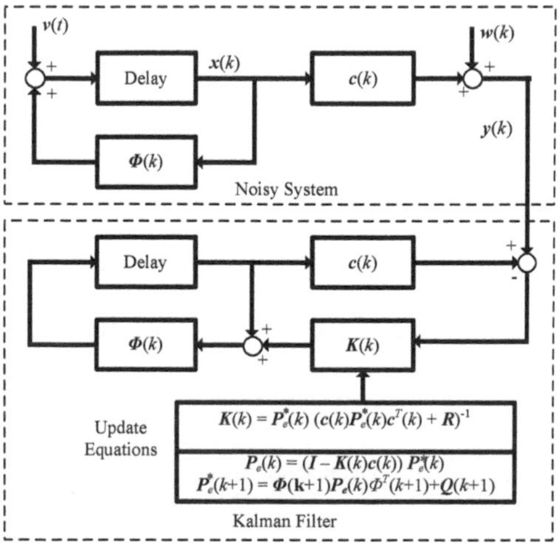

Figure 9.20: Discrete Kalman Filter

9.8.5 Wiener Filter in State Variable—Kalman Form

As discussed earlier the impulse function $\delta(t)$ plays a very important role in the construction of a deterministic process. Any causal deterministic signal $x(t)$ with its Laplace Transform $X(s)$ can be represented as:

$$x(t) = \int_0^\infty x(t-\sigma)\delta(\sigma)\,d\sigma \qquad L[x(t)] = X(s)$$

It is reasonable to look for a function **equivalent** to $\delta(t)$ that can be used to represent a Wide Sense Stationary (WSS) stochastic process $X(t)$. This is accomplished through a **white noise process** $I(t)$, known as the innovation process described as follows:

$$X(t) = \int_0^\infty X(t-\tau)I(\tau)\,d\tau, \quad X(s) = G_x(s)I(s)$$

$$E\,[(I(t), I(t-\tau))] = R_{ii}(\tau) = \delta(\tau)$$

$$E\,[X(t), X(t-\tau)] = R_{xx}(\tau)$$

$$\int_{-\infty}^{+\infty} R_{xx}(\tau)e^{-s\tau}\,d\tau = S_{xx}(s^2)$$

$$S_{xx}(s^2) = G_x(s)G_x(-s)$$

The function $G_x(s)$ represents a **minimum phase** transfer function with all its poles and zeros in the LHS of the s-plane. For the sake of simplicity, we shall deal with only the scalar case. This restriction can be very easily removed.

Consider the following WSS process with noise and its associated Wiener Filter:

$$E\,[W(t)] = 0$$

$$E\,[(W(t), W(t-\tau))] = R\delta(\tau)$$

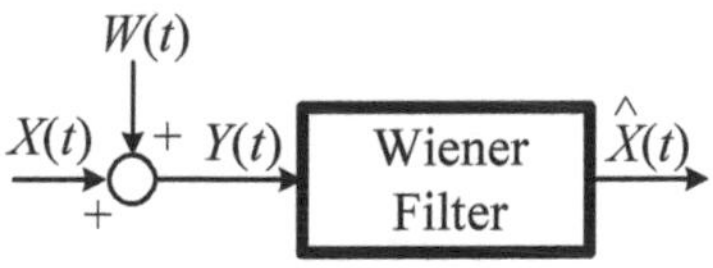

Figure 9.21: WSS Process with Noise and Wiener Filter

Spectral density of $X(t)$,

$$S_{xx}(s^2) = G_x(s)G_x(-s) = \frac{N(s)N(-s)}{D(s)D(-s)}$$

$$\left.\begin{aligned} D(s) &= s^n + a_1 s^{n-1} + \cdots + a_n \\ N(s) &= b_0 s^m + b_1 s^{m-1} + \cdots + b_m, \quad n > m \end{aligned}\right\} \quad \begin{aligned} &D(s) \text{ and } N(s) \\ &\text{are Hurwitz polynomials.} \end{aligned}$$

Figure 9.22 shows the equivalent innovation representation of the process $X(t)$.

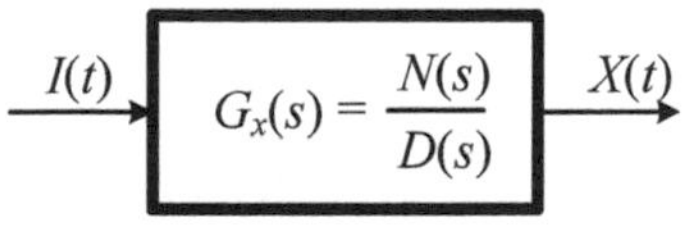

Figure 9.22: Equivalent Innovation Representation of the Random Process

Thus, the representation of stochastic process via innovation is:

$$X(s) = \frac{N(s)}{D(s)} I(s) \tag{9.178}$$

This represents an n-th order differential equation with n initial conditions that are determined by the coefficients $a_i, \quad i = 1, 2, \ldots, n$, $b_k, \quad k = 0, 1, 2, \ldots, m$. Dropping the symbol distinction between random signals $X(t)$, $Y(t)$, and $I(t)$ and the deterministic signals $\boldsymbol{x}(t)$, $y(t)$, and $\delta(t)$, we obtain the state variable innovation formulation of the stochastic process given in Figure 9.21 as:

$$\dot{\boldsymbol{x}} = \boldsymbol{A}\boldsymbol{x} + \boldsymbol{b}\delta(t)$$

$$y(t) = \boldsymbol{c}\boldsymbol{x} + w(t)$$

where

$$\boldsymbol{A} = \begin{bmatrix} 0 & 1 & 0 & \cdots & 0 \\ 0 & 0 & 1 & \cdots & 0 \\ \vdots & & & & \\ 0 & 0 & 0 & \cdots & 1 \\ -a_n & -a_{n-1} & -a_{n-2} & \cdots & -a_1 \end{bmatrix}, \quad \boldsymbol{c} = [\,1\;0\;0 \cdots 0\,]\,, \quad \boldsymbol{b} = \boldsymbol{x}(0)$$

$$\boldsymbol{M}(s) = D(s)(s\boldsymbol{I} - \boldsymbol{A})^{-1}$$

$$\boldsymbol{c}\boldsymbol{M}(s)\boldsymbol{x}(0) = N(s)$$

where $\boldsymbol{x}(0)$ is obtained by equating coefficients of s in the above equation. The corresponding Wiener Filter equations in the Kalman form are:

$$\dot{\hat{\boldsymbol{x}}} = \left(\boldsymbol{A} - \boldsymbol{K}\boldsymbol{c}^T R^{-1}\boldsymbol{c}\right)\hat{\boldsymbol{x}} + \boldsymbol{K}\boldsymbol{c}^T R^{-1}\boldsymbol{y}\,, \quad \hat{\boldsymbol{x}}(0) = \boldsymbol{x}(0)$$

$$\boldsymbol{A}\boldsymbol{K} + \boldsymbol{K}\boldsymbol{A}^T - \boldsymbol{K}\boldsymbol{c}^T \boldsymbol{R}^{-1}\boldsymbol{c}\boldsymbol{K} + \boldsymbol{b}\boldsymbol{b}^T = \boldsymbol{0}$$

Example 9.26:

Let

$$S_{xx}(s^2) = \frac{2}{\alpha^2 - s^2}, \qquad S_{ww}(s^2) = R > 0, \qquad S_{ii}(s^2) = Q = 1$$

Then

$$G_x(s) = \frac{\sqrt{2}}{(\alpha + s)}, \quad x(0) = \sqrt{2}$$

The state variable equation of the random processes are:

$$\dot{x}(t) = -\alpha x(t) + \sqrt{2}\delta(t)$$

$$y(t) = x(t) + w(t)$$

The Wiener Filter equations are:

$$\dot{\hat{x}}(t) = (-\alpha - k/R)\,\hat{x}(t) + (k/R)\,y, \quad k > 0, \quad \hat{x}(0) = \sqrt{2}$$

The Ricati equation for k is:

$$-2\alpha k - \frac{k^2}{R} + 2 = 0$$

yielding

$$k = -\alpha R \pm \sqrt{(2R + \alpha^2 R^2)}$$

Since we are seeking only positive values for k, the allowable value is:

$$k = -\alpha R + \sqrt{(2R + \alpha^2 R^2)}$$

The corresponding Wiener Filter Transfer Function with an initial condition is:

$$\frac{\hat{X}(s)}{Y(s)} = \frac{k/R}{(\alpha + k/R + s)}, \qquad \hat{x}(0) = \sqrt{2}$$

Bibliography

[Brammer] Brammer, Karl and Siffling, G. *Kalman-Bucy Filters-Deterministische Beobachtung und Stochastische Filterung,* 2. Auflage, R. Olderboung Verlag Muenchen Wien, 1985.

[Davenport, W.B.] Davenport, W.B. and Root, W.L. *An Introduction to the Theory of Random Signals and Noise,* New York: McGraw-Hill Inc., 1956.

[Feller, W.] Feller, W. *In Introduction to Probability Theory and Its Applications,* Vol. 1, Vol. 2, New York: John Wiley and Sons, 1968.

[Gauss, C.F.] Gauss, C.F. *Nachlos, Theoria Interpolationis Methods Novo Tracta,* Werke Band 3, 265–327 Gottingen, Konigliche Gesellschaft Der Wissenshaft.

[Heidman, M.T.] Hiedeman, M.T. and Johnson, H.D. Gauss and History of the Fast Fourier Transform. *IEEE ASSP Magazine* 1, 14–21, 1984.

[Kailath, T.] Kailath, T. The Innovation Approach to Detection and Estimation Theory. *Proceedings of IEEE,* 58, 680–695.

[Kailath, T.] Kailath, T. A View of Three decades of Linear Filtering Theory. *IEEE Trans. on Information Theory,* IT–20(2), 146–181.

[Kailath, T] Kailath, T. Some Extensions of Innovations Theorem. *Bell Systems Tech. Journal,* 50, 1487–1494, 1971.

[Kalman, R.E.] Kalman, R.E. and Bucy, R.S. New Results in Lincar Filtering and Prediction Theory. Trans. ASME, Series D, *J. Basic Engg,* 83, 95–108, 1961.

[Kleinrock, L.] Kleinrock, L. *Queueing Systems Theory,* Vol. 1, New York: John Wiley and Sons, 1975.

[Kleinrock, L.] Kleinrock, L. *Queueing Systems, Computer Applications,* Vol. 2, New York: John Wiley and Sons, 1976.

[Laning, J.H.] Laning, J.H. and Battin, R.H. *Random Processes in Automatic Control,* New York: McGraw-Hill, Inc.

[Papoulis, A.] Papoulis, A. *Probability, Random Variables and Stochastic Processes,* New York: McGraw-Hill, Inc., 1991.

[Rozanov, Y.A.] Rozanov, Y.A. *Stationary Random Processes,* San Francisco: Holden-Day, 1967.

[Saaty, T.L.] Saaty, T.L. and Bram, J. *Nonlinear Mathematics,* New York: McGraw-Hill Book Company, 358–368, 1964.

[Welaratna, S.] Welaratna, S. 30 Years of FFT Analyzers. *Sound and Viberation* (January 1997, 30th Anniversary Issue). A historical review of hardware FFT devices.

[Wiener, N.] Wiener, N. *Extrapolation, Interpolaction and Smoothing of Stationary Time Series,* New York: John Wiley and Sons, 1949.

[Yaglon, A.M.] Yaglon, A.M. *Correlation Theory of Stationary and Related Random Functions,* 2 Vols., New York: Springer.

Index

Abcissa of Convergence, 423
Accumulator, 482
Adder Element, 482
Aliasing, 534
Amplification property, 5
Analytic Function, 299
Annihilating Polynomial, 131
AutoCorrelation Properties, 801

Balanced Transformations, 647
Banach Space, 61, 70, 86, 88, 189
Banach Space Norm, 85
Band-Limited, 532
Bang-Bang Control, 707
Basic Vectors Coordinate Representation, 55
Basis, 99
Bellman Principles, 717
Benedixon Limit Cycle Criterion, 255
Bessel's Inequality, 59, 79
Bilateral Laplace Transforms, 82
Bilinear Transformation, 375
Binet-Cauchy Theorem, 173
Birkoffs Ergodic Theorem, 799
Blaschke Product, 193
Bolza's Form, 701
Bordering Matrices, 170
Boundedness, 95
Brachistochrone Problem, 692
Branch Points, 350
Bromwich Integral, 410
Bromwich Line, 410
Butterfly, 566

Calculus of Variations, 677
Canonical Realizations, 598
Cascaded Systems, 37
Cauchy Sequence, 60
Cauchy–Schwartz Inequality, 51, 66
Cauchy-Reimann, 301
Causal, 7
Causal System Response, 421
Causal Systems, 7, 511
Causal Wiener Filter, 836
Cayley-Hamilton, 612, 621, 669, 779
Center Points, 258, 262

Characteristic Equation, 129, 218
Characteristic Polynomial, 218, 511, 598, 622
 Co-efficients, 154
Chebychev Norm, 83
Cholesky Decomposition, 650
Cholesky-Decomposition, 173
Choletsky Decomposition, 163
Chords, 591
Classical Techniques, 242
Clifford Algebra, 108
Closed-Loop Poles Placement, 659
Codomain, 192
Comb Function, 543
Common Matrix Norms, 113
Companion Matrices, 162
Companion Matrix, 607
Complete Vector Space, 61
Complex Analytic Time Function, 457
Complex Integration, 156
Complex Variables, 298
 Analytic Function, 299
 Bilinear Transformation, 375
 Branch Points, 350
 Cauchy's Integral Formula, 310
 Cauchy's Integration Theorem, 307
 Cauchy's Principle Value, 327
 Cauchy-Reimann, 301
 Contour Integration, 298, 314
 Fundamental Theorem of Algebra, 368
 Derivative, 300
 Evaluation of Real Integrals, 327
 Green's Theorem, 306
 Jordan's Lemmas 1 & 2, 328
 Laurent Series, 321
 Maximum Principle, 370
 Meromorphic Functions, 371
 Minimum Principle, 371
 Path of Integration, 302
 Poisson's Integral, 359
 Poisson-Jensen's Integral, 365
 Positive Real Functions, 374
 Residue, 313
 Singular Points, 312
 Singularity at infinity, 316
 Taylor Series Expansion, 310
Complimentary Function, 214
Compound Operator, 95
Conservation of Momentum, 710
Continuity, 95
Contour Integration, 298, 314, 406, 508
Control Theory Concepts, 67
Control Variables, 702

Controllability, 598, 611
Criterion, 612
Definition, 612
Geometric, 618
Geometric Interpretation, 614
Grammian, 636, 637
Properties, 639
LTV System, 656
Multi-Input, 623
Time-Varying Systems, 654
Controllability Realization, 601, 609
Controllable Decomposition, 626
Controller Realization, 600, 609
Convergent Series, 156
Convolution Integral, 31, 35
Convolution Integrals, 421
Convolution Properties, 37
Convolution Theorem, 397
Frequency Domain, 399
Cooley-Tuckey Radix 2, 571
Cramer's Rule, 126
Critical Point, 256
Cross-correlation Functions, 802
Curse of Dimensionality, 717
Curse of Multiplication, 561
Cutset, 591
Cycloid Equation, 694
Decimation Algorithm, 562
Decimation-in-Frequency–Algorithm, 565
Decimation-in-Time–Algorithm, 565
Definite Functions, 165
Delta Function Properties, 16
Delta Functions, 11
Derivative of the Determinant, 117
Derivatives and Gradients, 176
Determinant
Product, 111
Deterministic Signals, 8
Diagonal Matrices, 644
Diagonalization, 141
Difference Equation
First Order, 215
Matrix Formulation, 235
Difference Equations, 478
k-th Order, 225
Differential Equation
Constant Coefficient, 218
Matrix Formulation, 230
Differential Equations
Constant Coefficient, 410
Stability, 252
Time Varying, 238
Variable Parameter, 463

Differential Operator
  Eigenfunctions, 219
Differentiation
  Chain-rule, 692
Digital Transfer Function, 513
Dimension of a Vector Space, 54
Direct Method, 264
Dirichlet Conditions, 537
Dirichlet Kernel, 15, 20
Discrete Delta Function, 21
Discrete Fourier Series (DFS), 547
Discrete Fourier Transform
  Aliasing Error, 555
  Numerical Computation, 563
Discrete Fourier Transform (DFT), 550
Discrete System
  Realization, 483
Discrete Systems, 478
  Transfer Function, 513
Distributed Parameter System, 8
Divide and Conquer, 562
Domain, 4, 94
Dual Space, 57
Duality Theorem, 643
Dynamic Programming, 717, 725

Eigenfunction, 45
Eigenfunctions, 44
Eigenvalue Decomposition, 141, 145
Eigenvalues, 44
  Complex Conjugate Pair, 250
Elementary Transformations, 177
Energy Density, 427
Energy Relation, 26
Equivalent Matrices, 180
Erdmann-Weierstrass Conditions, 701
Estimation and Control, 841
Euclidean Space, 54
Euclidean Space Operator as a Matrix Norm, 113
Euclidian Vector Space, 96
Euler-Abel Transformation, 572
Euler-Lagrange
  Alternate Form, 692
Euler-Lagrange Equations, 691
Euler-Lagrange-Pontryagin Equations, 705
EVD (see Eigenvalue Decomposition), 145
Events
  Independent, 731
  Mutually Exclusive, 731
Evolution Equation, 698
Exterior Algebra, 104
Exterior Product, 107

Extrema
  Euler-Lagrange Variable Endpoints, 687
  Functional, 687
  Multiple Constraints, 698

Fast Fourier Transform (FFT), 547, 561
  Computation Algorithm, 568
  Two Dimensions, 571
Filtering, 841
Final Value Theorem, 503
Focus Points, 258
Fourier Series, 537
  Properties, 540
Fourier Transform, 20, 431
  Derivation, 382
  Frequency Convolution, 443
  Inverse, 451
  Properties, 439
Frequency Convolution, 423
Frequency Sampling Theorem, 560
Frobenius Norm, 113
Function
  Extrema
    Multiple Constraints, 686
    Several Variables, 679
    Single Constraint, 682
    Single Variable, 677
Fundamental Matrix, 239
Fundamental Solution, 213, 215
Fundamental Theorem of Algebra, 368

Gain Element, 482
Gain Matrix, 722
Gaussian Estimator, 826
Gaussian Function, 29
GCD (see Greatest Common Divisor), 129
General Energy-type Function, 264
General Fourier series, 73
General Metric Space, 60
General Signal Classifications, 22
Generalized Coordinates, 696
Generalized Eigenvectors, 139, 249
Generalized Function, 11
Geometric Product, 106
Gerschgorin Circles, 170
Gradient, 680
Gram Determinant, 59
Gram-Schmidt Orthonormalization, 71
Gray Code, 571
Greatest Common Divisor, 129

H-J Equations, 697
Haar Functions, 75
Hadamard Product, 172
Hadamard's Inequality, 114

Hamilton-Jacobi Equations, 697, 704
Hamilton-Jacobi-Bellman (HJB), 719
Hamiltonian, 697, 704
Hamiltonian Principle, 696
Hankel Matrices, 119
Hankel Matrix, 665
Hardy Space, 85, 88, 93
Heaviside Formula, 409
Hermitian Matrices, 165
Hermitian Matrix, 140
Hessian, 680
Hilbert Norm, 204
Hilbert Space, 63, 70, 86, 88, 93, 189
Hilbert Space Basis, 72
Hilbert Space Norm, 84
Hilbert Transform, 455
    Contour Integration, 460
    Derivation, 456
    Pair, 456
    Physical Realization, 459
    Quadrature Filter, 460
    Singular Integrals, 458
Hilbert-Schmidt Norm, 113
Homogeneous Part, 218
Homogenous System of Equations, 122

Impulse, 11
Impulse Response, 44, 419, 513
Impulse Train Function, 29
Indefinite Matrix, 681
Infimum, 62
Infinite Dimensional Space, 187
Inflection, 680
Inhomogenous System of Equations, 126
Initial Value Theorem, 502
Inner Function, 193
Inner Product, 107
Inner Product Space, 60, 63
Input–Output Description, 9
Inversion Formula, 791
Invertible Matrix, 614

Jacobi and Gauss-Seidel Methods, 164
Jordan Canonical Form, 149
Jordan Canonical form, 251
Jordan's Lemmas 1 & 2, 328

Kalman Decomposition, 632, 635
Kalman Filter, 832
Kalman Form, 854
Karhunen-Loeve (K-L) Expansion, 810
Kautz Polynomials, 74, 429
Kautz, Laguerre and Legendre Polynomials, 73
Kernel, 94
Kernel Functions, 465

Euler, 465
Mellin Transform, 465
Sommerfeld, 465
Kirchoff's Laws, 4
Kronecker Product, 172
Krylov Spaces, 154
Krylov Vectors, 139

L-R Elementary Transformation, 179
Lagrangian, 585, 697
Laguerre Functions, 74
Laguerre Polynomials, 430
Lancasters Formula, 173
Laplace Transform, 20
Bilateral, 385
Inverse, 406
Single-Sided, 407
Inverse Bilateral, 412
Single-Sided, 388
Tables, 400
Laurent Series, 321
Laurent series, 89
Law of Large Numbers, 782
Least Action Principle, 704
Least Squares, 165
Lebesque Spaces, 85
Legendre Polynomials, 74
Liapunov, 637
Liapunov Functions, 262
Liapunov Stability Theorem, 166
Liapunov's First Method, 262, 263
Liapunov's Second Method, 264, 268
Limit Cycles, 254
Linear Independence of Vectors, 53
Linear Operators, 93
Linear System, 5
Linear Operator, 5
Linear Time Varying (LTV), 582, 654
Linear Time-Invariant (LTI), 582
Linear Time-Invariant System
Convolution, 420
Linear Vector Spaces, 47
Linearly Independent, 123
Links, 591
Low-Pass Recovery Filter, 535
LTI, 6, 32
LTI Stable, 46
LTV, 6
Lumped Parameter System, 8

Matrix
Addition, 101
Adjoint, 102
Adjugate, 109
Cofactors, 103
Commutator, 103

Companion, 231
Determinant, 103
Eigenvalues, 128
Eigenvectors, 128
Elementary Operation, 177
Function
Computation Algorithm, 137
Jordan Matrices, 157
Functions, 169
Convergence Conditions, 115
Exponential, 116
Geometric Series, 115
Trignometric, 116
Fundamental Concepts, 101
Hankel, 119
Hermitian, 102
Inverse, 102
Kernel, 112
Minors, 111
Multiplication, 101
Nilpotent, 157
Polynomial Function, 155
Positive Definite, 175, 637
Projection, 112
Pseudo-Inverse, 144
Range, 112
Rank, 112
Row Echelon, 181
Semidefinite, 175
Singular Value Decomposition, 144
Stochastic, 121
Toeplitz, 118
Trace, 111
Transition, 234
Transpose, 102
Vandermonde, 117, 142
Matrix Algebra, 96
Matrix Exponential, 156
Matrix Functions, 169
Maxima, 677
Maximum Likelihood Estimate, 849
Maximum Principle, 370
Mayer's Form, 699
McMillan Degree, 579
Mean Squared Quadratic Computation, 636
Memory Systems, 8
Memoryless Systems, 7
Meromorphic Function, 516
Meromorphic Functions, 371
Method of Residues, 520
Method of Variation of Parameters, 214
MIMO, 583, 621, 625
Minimal Realization, 668

State Variable Form, 671
Minima, 677
Minimal Polynomial, 131
Minimal Realization, 635
Minimal Realization Algorithm, 663
Minimization Problem, 687
Minimum Mean Square Estimator, 826
Minimum Principle, 371
Model Matrix, 251
Model Reduction Problem, 187
Multi-Input Multi-Output, 583
Multiple Eigenvalues, 149, 616
Multiplication Operator, 194
Multiplier Method, 521
Multivectors, 108

Negative Definite Matrix, 681
Nehari's Problem, 205
Node, 258
Nodes, 591
Noise, 610
Nonautonomous System, 277
Nonlinear Property, 5
Nonlinear System, 277
Nonsingular Matrices, 121
Normal Tree, 591
Nullity, 112

Observability, 598, 611
Criterion, 613, 658
Definition, 613
Geometric, 618
Geometric Interpretation, 614
Grammian, 636, 641
Proof, 656
Properties, 642
Multi-Output, 624
Time-Varying Systems, 654
Observability Realization, 604, 610
Observable Decomposition, 626, 630
Observer Realization, 603, 609
Operator, 5
Adjoint, 183
Backward Shift, 186, 193
Differential, 185
Eigenvalues, 184
Finite Dimensional, 182
Flip, 186
Forward Shift, 193
Hankel, 196
Infinite Dimensional, 182
Involution, 186
Non Singular Inverse, 183
Projection, 182
Toeplitz, 191
Unitary, 184

Operator Norm, 94
Optimal Control, 702
Order Reduction, 647
Orthogonal Signal, 428
  Generation Algorithm, 428
Orthogonality, 24
Orthonormal Basis Vectors, 56

Parallelogram Law, 51
Parseval Equality, 58
Parseval's Theorem, 426
Partial Fraction, 507
Partial Fraction Expansion, 402
Periodic Solutions, 254
Perturbation Equation, 252
Phase Portrait, 259
Planar Geometric Algebra, 106
Planimetric Problem, 683
Poisson's Bracket, 698
Poisson's Equation, 695
Poisson's Integral, 359
Pontryagin's Extremum Principle, 705
Pontryagin's Maximum Principle, 718
Positive Definite and Semidefinite Matrices, 175
Positive definiteness, 50
Power Series Convergence, 572
Power Spectrum, 808
Prediction, 841
Prediction Problem, 849
Principle of Invariance, 717
Principle of Optimality, 717
Projection Matrices, 112, 159
Projection Theorem, 78
Pseudo Inverse Problem, 165
Pythegoras Theorem, 79

Quadratic Performance, 651
Quasi-Diagonal Form, 99

Ramp Function, 27
Range, 4, 94
Rectangular Pulse Function, 27
Recursive Inverse Algorithm, 110
Residue Theorem, 528
Riccati Differential Equations, 721
Riccati Equation, 721
Row-Reduced Echelon Form, 181

S-N Decomposition, 158
Saddle Point, 680
Saddle Points, 258, 260
Sampling Theorem, 536
Scalar Difference Equation, 244
Schmidt Transformation, 662
Schur-Cohen Criteria, 174
Schur-Cohen Criterion, 268

Second Method of Liapunov, 262, 263
Semidefinite Matrix, 681
Sherman-Morrison Formula, 167
Sifting, 13
Signal Definition, 3
Signal Function Norms, 83
Signal Power, 24
Signal Reconstruction, 531
Signum Function, 28
Similarity Transformations, 595, 644
Simultaneous Diagonalization, 166
Simultaneous Equations
    Geometric Viewpoint, 122
Sinc Function, 28
Sine Integral Function, 30
Single-Input Single-Output, 583
Singular Inhomogenous System, 127
Singular Points, 257
    Classification, 258
Singular Value Decomposition, 144, 174
SISO, 583, 598, 648
    Controller Design, 659
    Geometric Viewpoint, 632
    Kalman Decomposition, 632
    Minimal Realization, 664
Squared Sampled Sequence, 516
Stability
    Definitions, 253
Stable
    Asymptotically Stable, 253
    Liapunov, 253
    Nonlinear Difference Equations, 268
    Third Order, 275
State Estimator, 651
State Observer, 651
State Space, 577
    Concepts, 595
    Definition, 578
    Formulation, 578
    Similarity Transformations, 595
    Transfer Function Matrix, 596
State Space Representation, 583
State Tree, 591
State Variable
    State Observers, 651
State Variable Formulation, 580
State Variables, 578
    Derivation, 584
    Formulation, 581
    Linear Graph Theory, 588
    Mechanical System, 586, 587
    Selection, 583
Stationary Ergodic Process, 801

Stationary Points, 677
Step Function, 27
Sterling's Formula, 782
Stochastic Signals, 8
Strong Law of Large Numbers, 782
Structured Matrices, 117
Sum of Squares, 516
Summation of Finite Series, 116
Superposition
  Integral, 31
  Property, 5
Superposition Integral, 35
Supremum, 62
SVD (see Singular Value Decomposition), 144
Switching Curves, 707
Sylvester Theorem 1, 140
Sylvester's Law of Nullity, 112
Sylvester's Theorem 2, 143
System Classifications, 3
System Impulse Response, 33
System Input–Output Relations, 31
System of Difference Equations, 212
System of Differential Equation
  First Order, 212
System of Differential Equations, 212
Systems of Linear Algebraic Equations, 122

Taylor Series, 116, 262, 504
Taylor series, 89
Taylor Series Expansion, 310
Tchebychev's Inequality Theorem, 781
Time Domain Techniques, 242
Time Function Recovery, 400
Time Signals, 10
Time-Limited Aperiodic Signal, 551
TPBV Problem, 709
TPBVP (see Two-Point Boundary Value Problem), 716, 723
Trace, 240
Transfer Function Matrix, 419
Transient Response, 214
Transition Matrix, 234, 723
Transversality Conditions, 691, 699
Triangular Inequality, 50
Triangular Pulse Function, 30
Tridiagonal Form, 173
Two-Dimensional Space, 107
Two-Point Boundary Value Problem, 716

Van Der Pol equation, 256
Variable Parameter, 463
Variational Principles, 702
Vector Space, 47

Banach Space, 61, 70
Basis Vectors, 55
Orthonormal, 56
Complete, 61
Dimension, 54
Fields, 48
Hilbert Space, 63, 70
Inner Product, 51
Inner Product Space, 63
Linear Independence, 53
Properties, 48
Vector Norm, 50
Vector Operations, 49
Vertices, 591

Walsh Functions, 76
Weak Law of Large Numbers, 782
White Noise, 800
White Noise Stochastic Process, 800
Wide Sense Stationary, 800, 854
Wiener Filters, 825
Wiener Hopf Factorization, 836
Wiener-Hopf Equation, 837
Wiener-Kinchin Theorem, 805
Wronskian, 232
WSS (see Wide Sense Stationary), 800, 854

Z-Transform, 477
Bilateral, 526
Definition, 484
Discrete Systems, 484
Frequency Convolution, 500
Inverse Bilateral, 527
Series Summation, 529
Time Convolution, 498
Z-Transforms
Difference Equations, 510
Properties, 486

Milton Keynes UK
Ingram Content Group UK Ltd.
UKHW051538141024
449569UK00028B/1522

9 781138 374188